FUNDAMENTALS OF CLASSICAL THERMODYNAMICS

English/SI Version

FUNDAMENTALS OF CLASSICAL THERMODYNAMICS

English/SI Version

Third Edition

Gordon J. Van Wylen

President
Hope College, Holland, Michigan

and

Richard E. Sonntag

Professor of Mechanical Engineering
University of Michigan, Ann Arbor, Michigan

John Wiley & Sons

New York Chichester Brisbane Toronto Singapore

Library of Congress Cataloging in Publication Data:

Van Wylen, Gordon John.
 Fundamentals of classical thermodynamics.

 Bibliography: p.
 Includes index.
 1. Thermodynamics. I. Sonntag, Richard Edwin.
II. Title.
TJ265.V23 1986 536'.7 85-26338
ISBN 0-471-86173-1

Printed in the United States of America

10 9 8 7 6 5 4

PREFACE

In this third edition we have retained the basic objective of the first two editions: to present a comprehensive and rigorous treatment of classical thermodynamics while retaining an engineering perspective and, in doing so, to lay the groundwork for subsequent studies in such fields as fluid mechanics, heat transfer, and statistical thermodynamics, and also to prepare the student to effectively use thermodynamics in the practice of engineering.

We have deliberately directed our presentation to students. New concepts and definitions are presented in the context where they are first relevant. The first thermodynamic properties to be defined (Chapter 2) are those that can be readily measured: pressure, specific volume, and temperature. In Chapter 3, tables of thermodynamic properties are introduced, but only in regard to these measurable properties. Internal energy and enthalpy are introduced in connection with the first law, entropy with the second law, and the Helmholtz and Gibbs functions in the chapter on availability. Many examples have been included in the book to assist the student in gaining an understanding of thermodynamics, and the problems at the end of each chapter have been carefully sequenced to correlate with the subject matter and provide some progression in difficulty.

We have attempted to cover fairly comprehensively the basic subject matter of classical thermodynamics, and believe that the book provides adequate preparation for study of the application of thermodynamics to the various professional fields as well as for study of more advanced topics in thermodynamics, such as those related to materials, surface phenomena, plasmas, and cryogenics. We also recognize that a number of colleges offer a single introductory course in thermodynamics for all departments, and we have tried to cover those topics that the various departments might wish to have included in such a course. However, since specific courses vary considerably in prerequisites, specific objectives, duration, and background of the students, we have arranged the material, particularly in the later chapters, so that there is considerable flexibility in the amount of material that may be covered.

Throughout the book we have attempted to maintain an engineering perspective, particularly through the choice of examples and problems. One of the major changes from the previous edition is in the problems. Although a number of the earlier problems have been retained, most are new or have been significantly revised. A number of new problems requiring a computer-based solution have been added throughout the book, beginning in Chapter 2, so that instructors who wish to incorporate these problems into the course may do so at any time throughout the term. Several different classes of applications of thermodynamics are introduced in the first chapter, and these then serve as a basis for many of the problems in later chapters. We have retained a chapter on cycles (Chapter 9) because we find that many students enjoy this subject, and it can also serve to effectively

v

strengthen the student's understanding of the first and second laws of thermodynamics, and introduce him or her to engineering design and practice. Many of the computer-related problems are in this chapter, as are problems concerning a number of modern and relevant applications: heat pumps, cogeneration, topping and bottoming cycles, two-stage systems, combined cycles, and many others. We have also included comments and problems that relate thermodynamics to environmental concerns.

Chapter 14, which includes an introduction to compressible flow and flow through blade passages, has been retained for those students who would not otherwise undertake extensive studies in fluid flow. We recognize that in many cases the material in this chapter will not be included in the thermodynamics courses but may instead be covered elsewhere in the curriculum.

Several significant changes have been made in the text portion of the book. One concerns the inclusion of a formal problem analysis and solution technique within the framework of the example problems in the text. Following its introduction in Chapter 5, this procedure is utilized, where appropriate, throughout the remainder of the book. This technique has been used extensively in undergraduate engineering classes over a period of several years, and has been included in the latest edition of our book *Introduction to Thermodynamics, Classical and Statistical*. It has been well received and proved to be quite successful in helping students to learn to deal with the solution of unfamiliar problems. There has been extensive revision of the treatment of real mixtures, in both the partial-property approach and the pseudo-pure substance approach, to simplify the former and to make it easier to introduce the latter independently, but especially to amplify the concept and use of reference states. This material also affects the coverage of chemical reactions and of thermodynamic equilibrium in subsequent chapters. Other changes in this edition include further simplifications in the control volume developments, an introduction to the concept called second-law efficiency in connection with availability analysis, additional material concerning solids and liquids, several changes concerning the presentation of thermodynamic equilibrium, changes in the Appendix tables, and general updating of the book.

In regard to symbols used in this text, we have tried to maintain consistency throughout the book. In a limited number of cases, we have used a given symbol for more than one purpose. We believe, however, that the context will clarify the meaning of the symbol in these cases.

The units in this version of the Third Edition are mixed, with approximately half of the text material, examples, and problems expressed in English Engineering units and the other half in SI (Système International) units. There is a slight emphasis on the English system in the chapter on cycles, and an emphasis on SI units in the chapters dealing with equations of state, chemical reactions, and equilibrium. The Appendix tables are presented in both sets of units, with the exception of tables for the cryogenic substances nitrogen and methane, and several other later tables in which duplicate listings are unnecessary. In the charts, the psychrometric chart is presented in English units and the generalized enthalpy and entropy

charts in SI units; a conversion factor is necessary when using these charts in different unit systems.

In dealing with English units, there is the problem of distinguishing between force and mass units. To emphasize this distinction, the symbols lbf and lbm have been used for pound force and pound mass in the English system. Such symbols as lbf/in.2 have been used for pressure (rather than psi) and ft^3/lbm for specific volume (rather than cu ft/lb) in order to stress the fundamental units involved in the various parameters. The force-mass conversion constant g_c is treated as being implicit, to enable equations including kinetic or potential energy terms to be handled in a consistent matter with regard to units. In dealing with SI units, we have commonly used the basic units for pressure (pascal) and volume (cubic meter), although we have used the liter as a convenient volume unit much more extensively than in the SI version of the previous edition. Others may wish to use the bar more extensively as the pressure unit, and we trust that such flexibility in these units will present no particular difficulties to the student. Concerning the extensive properties, a lowercase letter (v, u, h, s) designates the property per unit mass (either pound mass or kilogram); an uppercase letter (V, U, H, S), the property for the entire system; a lowercase letter with a bar (\bar{v}, \bar{u}, \bar{h}, \bar{s}), the property per unit mole (either the pound-mole or lb mol, or the kilogram-mole or kmol, in this text); and an uppercase letter with a bar (\bar{V}, \bar{U}, \bar{H}, \bar{S}), the partial molal property in a mixture. Following this pattern, we have found it convenient to designate the total heat transfer as Q, the heat transfer per unit mass of the system as q, the total work as W, and the work per unit mass of the system as w.

Furthermore, we represent the rate of flow across a system boundary or control surface by a dot over the given quantity. Thus, \dot{Q} represents a rate of heat transfer across the system boundary; \dot{W}, the rate at which work crosses the system boundary (that is, the power); and \dot{m}, the mass rate of flow across a control surface (\dot{n} is used when the mass rate of flow is expressed in moles per unit time). We realize that we have departed from the usual mathematical use of a dotted symbol, in which the dotted symbol typically refers to a derivative with respect to time. However, we have used the dotted symbol only to indicate a flow of heat and work across a system boundary, and a flow of heat, work, and mass across a control surface, and not in any other context. We believe that it has contributed to a simple and consistent use of symbols for this book.

We acknowledge with appreciation the suggestions, counsel, and encouragement of many colleagues, both at The University of Michigan and elsewhere. This assistance has been most helpful to us during the writing of this edition, as was the case with the earlier editions of the book. Several secretaries have aided us immeasurably throughout, and we acknowledge this assistance with many thanks. Both undergraduate and graduate students have been of particular assistance, as their perceptive questions have often caused us to rewrite or rethink a given portion of the text, or to try to develop a better way of presenting the material in order to anticipate such questions or difficulties. Finally, for each of us, the encouragement and patience of our wives and families have been indispensable, and

have made this time of writing pleasant and enjoyable in spite of the pressures of the project.

Our hope is that this book will contribute to the effective teaching of thermodynamics to students who face very significant challenges and opportunities during their professional careers. Your comments, criticism, and suggestions will also be appreciated.

GORDON J. VAN WYLEN

RICHARD E. SONNTAG

Ann Arbor, August, 1985

CONTENTS

SYMBOLS

a	acceleration
a	activity
a, A	specific Helmholtz function and total Helmholtz function
AF	air-fuel ratio
c	velocity of sound
C_D	coefficient of discharge
C_p	constant-pressure specific heat
C_v	constant-volume specific heat
C_{po}	zero-pressure constant-pressure specific heat
C_{vo}	zero-pressure constant-volume specific heat
e, E	specific energy and total energy
f	fugacity
\bar{f}_i	fugacity of component i in a mixture
F	force
FA	fuel-air ratio
g	acceleration due to gravity
g, G	specific Gibbs function and total Gibbs function
g_c	a constant that relates force, mass, length, and time
h, H	specific enthalpy and total enthalpy
i	electrical current
I	irreversibility
J	proportionality factor to relate units of work to units of heat
k	specific heat ratio: C_p/C_v
K	equilibrium constant
KE	kinetic energy
L	length
lw, LW	lost work per unit mass and total lost work
m	mass
\dot{m}	mass rate of flow
M	molecular weight
M	Mach number
mf	mass fraction
n	number of moles

n	polytropic exponent
P	pressure
P_i	partial pressure of component i in a mixture
PE	potential energy
P_r	relative pressure as used in gas tables
q, Q	heat transfer per unit mass and total heat transfer
\dot{Q}	rate of heat transfer
Q_H, Q_L	heat transfer with high-temperature body and heat transfer with low-temperature body; sign determined from context
R	gas constant
\bar{R}	universal gas constant
s, S	specific entropy and total entropy
t	time
T	temperature
u, U	specific internal energy and total internal energy
v, V	specific volume and total volume
v_r	relative specific volume as used in gas tables
vf	volume fraction
V	velocity
V_r	relative velocity
w, W	work per unit mass and total work
\dot{W}	rate of work, or power
w_rev	reversible work between two states assuming heat transfer with surroundings
x	quality
x	liquid-phase or solid-phase mole fraction
y	vapor-phase mole fraction
Z	elevation
Z	compressibility factor
Z	electrical charge

SCRIPT LETTERS

\mathscr{A}	area
\mathscr{C}	number of components
\mathscr{E}	electrical potential

\mathscr{H}	magnetic field intensity
\mathscr{M}	magnetization
\mathscr{P}	number of phases
\mathscr{S}	surface tension
\mathscr{T}	tension
\mathscr{V}	variance

GREEK LETTERS

α	residual volume
α_P	volume expansivity
β	coefficient of performance for a refrigerator
β'	coefficient of performance for a heat pump
β_S	adiabatic compressibility
β_T	isothermal compressibility
η	efficiency
μ	chemical potential
μ_J	Joule-Thomson coefficient
ν	stoichiometric coefficient
ρ	density
ϕ	relative humidity
ϕ	availability for a system
ψ	availability associated with a steady-state, steady-flow process
ω	humidity ratio or specific humidity

SUBSCRIPTS

c	property at the critical point
c.v.	control volume
e	state of a substance leaving a control volume
f	formation
f	property of saturated liquid
fg	difference in property for saturated vapor and saturated liquid
g	property of saturated vapor
i	state of a substance entering a control volume
i	property of saturated solid

ig	difference in property for saturated vapor and saturated solid
r	reduced property
s	isentropic process
0	property of the surroundings
0	stagnation property

SUPERSCRIPTS

⁻	bar over symbol denotes property on a molal basis (over *V*, *H*, *S*, *U*, *A*, *G*, the bar denotes partial molal property)
°	property at standard-state condition
*	ideal gas
*	property at the throat of a nozzle
L	liquid-phase
S	solid-phase
V	vapor-phase

SOME INTRODUCTORY COMMENTS

In the course of our study of thermodynamics, a number of the examples and problems presented refer to processes that occur in such equipment as a steam power plant, a fuel cell, a vapor-compression refrigerator, a thermoelectric cooler, a rocket engine, and an air separation plant. In this introductory chapter a brief description of this equipment is given. There are at least two reasons for including such a chapter. First, many students have had limited contact with such equipment, and the solution of problems will be more significant and relevant when they have some familiarity with the actual processes and the equipment involved. Second, this chapter will provide an introduction to thermodynamics, including the use of certain terms (which will be more formally defined in later chapters), some of the problems for which thermodynamics is relevant, and some accomplishments that have resulted, at least in part, from the application of thermodynamics.

It should be emphasized that thermodynamics is relevant to many other processes than those cited in this chapter. It is basic to the study of materials, chemical reactions, and plasmas. The student should bear in mind that this chapter is only a brief and necessarily very incomplete introduction to the subject of thermodynamics.

1.1 THE SIMPLE STEAM POWER PLANT

A schematic diagram of a simple steam power plant is shown in Fig. 1.1. High-pressure superheated steam leaves the boiler, which is also referred to as a steam generator, and enters the turbine. The steam expands in the turbine and, in doing so, does work, which enables the turbine to drive the electric generator. The low-pressure steam leaves the turbine and enters the condenser, where heat is transferred from the steam (causing it to condense) to the cooling water. Since large quantities of cooling water are required, power plants are frequently located near rivers or lakes. It is this transfer of heat to the water in lakes and rivers that leads to the thermal pollution problem, which is now being extensively studied. During our study of thermodynamics we will gain an understanding of why this heat transfer is necessary and ways in which it can be minimized. When the supply of cooling water is limited, a cooling tower may be used. In the cooling tower some of the cooling water evaporates in such a way as to lower the temperature of the water that remains as a liquid.

The pressure of the condensate leaving the condenser is increased in the pump, thus enabling the condensate to flow into the steam generator. In many steam generators an economizer is used. An economizer is simply a heat exchanger in which heat is transferred from the products of combustion (just before they leave

FIGURE 1.1

Schematic diagram of a steam power plant.

the steam generator) to the condensate, with the result that the temperature of the condensate is increased, but no evaporation takes place. In other sections of the steam generator, heat is transferred from the products of combustion to the water, causing it to evaporate. The temperature at which evaporation occurs is called the saturation temperature. The steam then flows through another heat exchanger known as a superheater, where the temperature of the steam is increased well above the saturation temperature.

In many power plants the air that is used for combustion is preheated in the air preheater by transferring heat from the stack gases as they are leaving the furnace. This air is then mixed with fuel—which might be coal, fuel oil, natural gas, or other combustible material—and combustion takes place in the furnace. As the products of combustion pass through the furnace, heat is transferred to the water in the superheater, the boiler, the economizer, and to the air in the air preheater. The products of combustion from power plants are discharged to the atmosphere, and this constitutes one of the facets of the air pollution problem we now face.

A large power plant will have many other pieces of equipment, some of which will be considered in later chapters.

Figure 1.2 shows a steam turbine and the generator that it drives. Steam turbines vary in capacity from less than 10 kilowatts to 1 000 000 kilowatts,

FIGURE 1.2

A large steam turbine (Courtesy General Electric Co.).

FIGURE 1.3
A condenser used in a large power plant (Courtesy Westinghouse Electric Corp.).

Figure 1.3 shows a cutaway view of a condenser. The steam enters at the top and the condensate is collected in the hot well at the bottom while the cooling water flows through the tubes. A large condenser has a tremendous number of tubes, as shown in Fig. 1.3.

Figure 1.4 shows a large steam generator. The flow of air and products of combustion are indicated. The condensate, also called the boiler feedwater, enters at the economizer inlet, and the superheated steam leaves at the superheater outlet.

The number of nuclear power plants in operation has increased substantially in recent years. In these power plants the reactor replaces the steam generator of the conventional power plant, and the radioactive fuel elements replace the coal, oil, or natural gas.

There are several different reactor designs in current use. One of these is the boiling-water reactor, such as the system shown in Fig. 1.5. In other nuclear power plants a secondary fluid circulates from the reactor to the steam generator, where heat is transferred from the secondary fluid to the water, which in turn goes through a conventional steam cycle. Safety considerations and the necessity to

FIGURE 1.4
A large steam generator (Courtesy Babcock and Wilcox Co.).

keep the turbine, condenser, and related equipment from becoming radioactive are always major considerations in the design and operation of a nuclear power plant.

1.2 FUEL CELLS

When a conventional power plant is viewed as a whole, as shown in Fig. 1.6, we see that fuel and air enter the power plant and products of combustion leave the unit. There is also a transfer of heat to the cooling water, and work is done in the form of the electrical energy leaving the power plant. The overall objective of a power plant is to convert the availability (to do work) of the fuel into work (in the

	Water
	Steam–water mixture
	Steam
- - - - -	Electric
	Shielding

FIGURE 1.5

Schematic diagram of the Big Rock Point nuclear plant of Consumers Power Company at Charlevoix, Michigan (Courtesy Consumers Power Company).

form of electrical energy) in the most efficient manner, consistent with such considerations as cost, space, and safety.

We might well ask if all of the equipment in the power plant, such as the steam generator, the turbine, the condenser, and the pump, is necessary. Is it not possible to produce electrical energy from the fuel in a more direct manner?

The fuel cell is a device in which this objective is accomplished. Figure 1.7 shows a schematic arrangement of a fuel cell of the ion-exchange membrane type. In this fuel cell hydrogen and oxygen react to form water. Let us consider the general features of the operation of this type of fuel cell.

FIGURE 1.6

Schematic diagram of a power plant.

The flow of electrons in the external circuit is from anode to cathode. Hydrogen enters at the anode side and oxygen enters at the cathode side. At the surface of the ion-exchange membrane the hydrogen is ionized according to the reaction

$$2H_2 \rightarrow 4H^+ + 4e^-$$

The electrons flow through the external circuit and the hydrogen ions flow through the membrane to the cathode, where the following reaction takes place.

$$4H^+ + 4e^- + O_2 \rightarrow 2H_2O$$

There is a potential difference between the anode and cathode, and thus there is a flow of electricity through a potential difference, which, in thermodynamic terms, is called work. There may also be a transfer of heat between the fuel cell and the surroundings.

At the present time the fuel used in fuel cells is usually either hydrogen or a mixture of gaseous hydrocarbons and hydrogen. The oxidizer is usually oxygen. However, current research is directed toward the development of fuel cells that use hydrocarbon fuels and air. Although the conventional (or nuclear) steam power plant is still used in large-scale power-generating systems, and conventional piston engines and gas turbines are still used in most transportation power systems, the fuel cell may eventually become a serious competitor. The fuel cell is already being used to produce power for certain space applications.

FIGURE 1.7
Schematic arrangement of an ion-exchange membrane type of fuel cell.

Thermodynamics plays a vital role in the analysis, development, and design of all power-producing systems, including reciprocating internal combustion engines and gas turbines. Considerations such as the increase of efficiency, improved design, optimum operating conditions, and alternate methods of power generation involve, among other factors, the careful application of the fundamentals of thermodynamics.

1.3 THE VAPOR-COMPRESSION REFRIGERATION CYCLE

A simple vapor-compression refrigeration cycle is shown schematically in Fig. 1.8. The refrigerant enters the compressor as a slightly superheated vapor at a low pressure. It then leaves the compressor and enters the condenser as a vapor at some elevated pressure, where the refrigerant is condensed as a result of heat transfer to cooling water or to the surroundings. The refrigerant then leaves the condenser as a high-pressure liquid. The pressure of the liquid is decreased as it flows through the expansion valve and, as a result, some of the liquid flashes into vapor. The remaining liquid, now at a low pressure, is vaporized in the evaporator as a

FIGURE 1.8

Schematic diagram of a simple refrigeration cycle.

result of heat transfer from the refrigerated space. This vapor then enters the compressor.

In a typical home refrigerator the compressor is located in the rear near the bottom of the unit. The compressors are usually hermetically sealed, that is, the motor and compressor are mounted in a sealed housing, and the electric leads for the motor pass through this housing. This is done to prevent leakage of the refrigerant. The condenser is also located at the back of the refrigerator and is arranged so that the air in the room flows past the condenser by natural convection. The expansion valve takes the form of a long capillary tube, and the evaporator is located around the outside of the freezing compartment inside the refrigerator.

Figure 1.9 shows a large centrifugal unit that is used to provide refrigeration for an air-conditioning unit. In this unit, water is cooled and then circulated to provide cooling where needed.

1.4 THE THERMOELECTRIC REFRIGERATOR

We may well ask the same question about the vapor-compression refrigerator that we asked about the steam power plant—isn't it possible to accomplish our objective in a more direct manner? Isn't it possible, in the case of a refrigerator, to use the electrical energy (which goes to the electric motor that drives the compressor) to produce cooling in a more direct manner, and avoid the cost of the compressor, condenser, evaporator, and all the related piping?

Centrifugal refrigerant compressor

Compressor inlet

Compressed refrigerant condenses around condenser water tubes

Economizer refrigerant flashes to low pressure

Cold refrigerant liquid boils around water cooler tubes

Cold refrigerant liquid flow control to water cooler

Compressor drive motor

Condenser water tubes

Water cooler tubes

FIGURE 1.9

A refrigeration unit for an air-conditioning system (Courtesy Carrier Air Conditioning Co.).

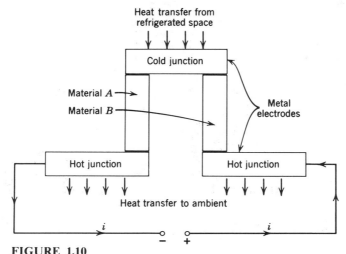

FIGURE 1.10

A thermoelectric refrigerator.

The thermoelectric refrigerator is such a device. This is shown schematically in Fig. 1.10. The thermoelectric device, like the conventional thermocouple, utilizes two dissimilar materials. There are two junctions between these two materials in a thermoelectric refrigerator. One is located in the refrigerated space and the other in ambient surroundings. When a potential difference is applied, as indicated, the temperature of the junction located in the refrigerated space will decrease and the temperature of the other junction will increase. Under steady-state operating conditions, heat will be transferred from the refrigerated space to the cold junction. The other junction will be at a temperature above the ambient, and heat will be transferred from the junction to the surroundings.

It should be emphasized that a thermoelectric device can also be used to generate power by replacing the refrigerated space with a body that is at a temperature above the ambient. Such a system is shown in Fig. 1.11.

The thermoelectric refrigerator cannot yet compete economically with the conventional vapor-compression units. However, in certain special applications, the thermoelectric refrigerator is already in use and, in view of the major research and development efforts under way in this field, it is quite possible that the use of thermoelectric refrigerators will be much more extensive in the future.

1.5 THE AIR SEPARATION PLANT

One process of great industrial significance is the air separation plant, in which air is separated into its various components. The oxygen, nitrogen, argon, and rare gases so produced are used extensively in various industrial, research, space, and

FIGURE 1.11

A thermoelectric power generation device.

consumer goods applications. The air separation plant may be considered as an example from two major fields, the chemical process industry and the field of cryogenics. Cryogenics is a term that refers to technology, processes, and research at very low temperatures (in general, below 150 K). In both chemical processing and cryogenics, thermodynamics is basic to an understanding of many phenomena that occur and to the design and development of processes and equipment.

A number of different designs of air separation plants have been developed. Consider Fig. 1.12, which shows a somewhat simplified sketch of a type of plant that is frequently used. Air from the atmosphere is compressed to a pressure of 20 to 30 atmospheres. It is then purified, particularly to remove carbon dioxide (which would plug the flow passages as it solidifies when the air is cooled to its liquefaction temperature). The air is then compressed to a pressure of 150 to 200 atmospheres, cooled to the ambient temperature in the aftercooler, and dried to remove the water vapor (which would also plug the flow passages as if freezes).

The basic refrigeration in the liquefaction process is provided by two different processes. One involves expansion of the air in the expansion engine. During this process the air does work and as a result the temperature of the air is reduced. The other refrigeration process involves passing the air through a throttle valve that is so designed and so located that there is a substantial drop in the pressure of the air and, associated with this, a substantial drop in the temperature of the air.

As shown in Fig. 1.12, the dry, high-pressure air enters a heat exchanger. The air temperature drops as it flows through the heat exchanger. At some intermediate point in the heat exchanger, part of the air is bled off and flows through the expansion engine. The remaining air flows through the rest of the heat exchanger and through the throttle valve. The two streams join, both being at the pressure of

FIGURE 1.12

A simplified diagram of a liquid oxygen plant.

5 to 10 atmospheres, and enter the bottom of the distillation column, which is re-
ferred to as the high-pressure column. The function of the distillation column is
to separate the air into its various components, principally oxygen and nitrogen.
Two streams of different composition flow from the high-pressure column through
throttle valves to the upper column (also called the low-pressure column). One of
these is an oxygen-rich liquid that flows from the bottom of the lower column and
the other is a nitrogen-rich stream that flows through the subcooler. The separa-
tion is completed in the upper column, with liquid oxygen leaving from the bottom
of the upper column and gaseous nitrogen from the top of the column. The nitro-
gen gas flows through the subcooler and the main heat exchanger. It is the heat
transfer to this cold nitrogen gas that causes the cooling of the high-pressure air
entering the heat exchanger.

Not only is a thermodynamic analysis essential to the design of the sytem as
a whole, but essentially every component of such a system, including the com-
pressors, the expansion engine, the purifiers and driers, and the distillation col-
umn, involves thermodynamics. In this separation process we are also concerned
with the thermodynamic properties of mixtures and the principles and procedures
by which these mixtures can be separated. This is the type of problem encountered
in the refining of petroleum and many other chemical processes. It should also be

noted that cryogenics is particularly relevant to many aspects of the space program, and a thorough knowledge of thermodynamics is essential for creative and effective work in cryogenics.

1.6 THE CHEMICAL ROCKET ENGINE

The advent of missiles and satellites has brought to prominence the use of the rocket engine as a propulsion power plant. Chemical rocket engines may be classified as either liquid propellant or solid propellant, according to the fuel used.

Figure 1.13 shows a simplified schematic diagram of a liquid propellant rocket. The oxidizer and fuel are pumped through the injector plate into the combustion chamber where combustion takes place at high pressure. The high-pressure, high-temperature products of combustion expand as they flow through the nozzle, and as a result they leave the nozzle with a high velocity. The momentum change associated with this increase in velocity gives rise to the forward thrust on the vehicle.

FIGURE 1.13

Simplified schematic diagram of a liquid-propellant rocket engine.

The oxidizer and fuel must be pumped into the combustion chamber, and some auxiliary power plant is necessary to drive the pumps. In a large rocket this auxiliary power plant must be very reliable and have a relatively high power output, yet it must be light in weight. The oxidizer and fuel tanks occupy the largest part of the volume of an actual rocket, and the range of a rocket is determined largely by the amount of oxidizer and fuel that can be carried. Many different fuels and oxidizers have been considered and tested, and much effort has gone into the development of fuels and oxidizers that will give a higher thrust per unit mass rate of flow of reactants. Liquid oxygen is frequently used as the oxidizer in liquid-propellant rockets.

Much work has also been done on solid-propellant rockets. They have been very successfully used for jet-assisted takeoffs of airplanes, military missiles, and space vehicles. They are much simpler in both the basic equipment required for operation and the logistic problems involved in their use.

SOME CONCEPTS AND DEFINITIONS

One very excellent definition of thermodynamics is that it is the science of energy and entropy. However, since we have not yet defined those terms, an alternate definition in terms with which we are already familiar is: thermodynamics is the science that deals with heat and work and those properties of substances that bear a relation to heat and work. Like all sciences, the basis of thermodynamics is experimental observation. In thermodynamics these findings have been formalized into certain basic laws, which are known as the first, second, and third laws of thermodynamics. In addition to these, the zeroth law of thermodynamics, which in the logical development of thermodynamics precedes the first law, has been set forth.

In the chapters that follow, we will present these laws and the thermodynamic properties related to these laws, and apply them to a number of representative examples. The objective of the student should be to gain a thorough understanding of the fundamentals and an ability to apply these fundamentals to thermodynamic problems. The purpose of the examples and problems is to further this twofold objective. It should be emphasized that it is not necessary for the student to memorize numerous equations, for problems are best solved by the application of the definitions and laws of thermodynamics. In this chapter some concepts and definitions basic to thermodynamics are presented.

2.1 THE THERMODYNAMIC SYSTEM AND THE CONTROL VOLUME

A thermodynamic system is defined as a quantity of matter of fixed mass and identity on which attention is focused for study. Everything external to the system is the surroundings, and the system is separated from the surroundings by the system boundaries. These boundaries may be either movable or fixed.

In Fig. 2.1 the gas in the cylinder is considered the system. If a Bunsen burner is placed under the cylinder, the temperature of the gas will increase and the piston will rise. As the piston rises, the boundary of the system moves. As we will see later, heat and work cross the boundary of the system during this process, but the matter that comprises the system can always be identified.

An isolated system is one that is not influenced in any way by the surroundings. This means that no heat or work cross the boundary of the system.

FIGURE 2.1

Example of a system.

FIGURE 2.2

Example of a control volume.

In many cases a thermodynamic analysis must be made of a device, such as an air compressor, that involves a flow of mass into and/or out of the device, as shown schematically in Fig. 2.2. The procedure that is followed in such an analysis is to specify a control volume that surrounds the device under consideration. The surface of this control volume is referred to as a control surface. Mass, as well as heat and work (and momentum), can flow across the control surface.

Thus a system is defined when dealing with a fixed quantity of mass, and a control volume is specified when an analysis is to be made that involves a flow of mass. The difference in these two approaches is considered in detail in Chapter 5. It should be noted that the terms closed system and open system are sometimes used as the equivalent of the terms system (fixed mass) and control volume (involving a flow of mass). The procedure that will be followed in the presentation of the first and second laws of thermodynamics is first to present these laws for a system and then to make the necessary transformations to apply them to a control volume.

2.2 MACROSCOPIC VERSUS MICROSCOPIC POINT OF VIEW

An investigation into the behavior of a system may be undertaken from either a microscopic or macroscopic point of view. Let us briefly consider the problem we

have if we attempt to describe a system from a microscopic point of view. Consider a system consisting of a cube 1 inch on a side and containing a monatomic gas at atmospheric pressure and temperature. This volume contains approximately 10^{20} atoms. To describe the position of each atom, three coordinates must be specified; to describe the velocity of each atom, three velocity components must be specified.

Thus, to completely describe the behavior of this system from a microscopic point of view, it would be necessary to deal with at least 6×10^{20} equations. Even with a large digital computer, this is a quite hopeless computational task. However, there are two approaches to this problem that reduce the number of equations and variables to a few that can be handled relatively easily in performing computations. One of these approaches is the statistical approach in which, on the basis of statistical considerations and probability theory, we deal with "average" values for all particles under consideration. This is usually done in connection with a model of the atom under consideration. This is the approach used in the disciplines known as kinetic theory and statistical mechanics.

The other approach that reduces the number of variables to a few that can be handled is the macroscopic point of view of classical thermodynamics. As the word macroscopic implies, we are concerned with the gross or average effects of many molecules. Furthermore, these effects can be perceived by our senses and measured by instruments. In so doing, however, what we really perceive and measure is the time-averaged influence of many molecules. For example, consider the pressure a gas exerts on the walls of its container. This pressure results from the change in momentum of the molecules as they collide with the wall. However, from a macroscopic point of view, we are not concerned with the action of the individual molecules but with the time-averaged force on a given area, which can be measured by a pressure gage. In fact, these macroscopic observations are completely independent of our assumptions regarding the nature of matter.

Although the theory and development in this book will be presented from a macroscopic point of view, a few supplementary remarks regarding the significance of the microscopic perspective are included as an aid to the understanding of the physical processes involved. Another book in this series on Thermal and Transport Sciences, namely, *Fundamentals of Statistical Thermodynamics*, by R. E. Sonntag and G. J. Van Wylen, deals with thermodynamics from the microscopic and statistical point of view.

A few remarks should be made regarding the continuum. From the macroscopic point of view, we are always concerned with volumes that are very large compared to molecular dimensions and, therefore, with systems that contain many molecules. Since we are not concerned with the behavior of individual molecules, we can treat the substance as being continuous, disregarding the action of individual molecules, and this is called a continuum. The concept of a continuum, of course, is only a convenient assumption that loses validity when the mean free path of the molecules approaches the order of magnitude of the dimensions of the vessel, as, for example, in high-vacuum technology. In much engineering work the assumption of a con-

tinuum is valid and convenient, and goes hand in hand with the macroscopic point of view.

2.3 PROPERTIES AND STATE OF A SUBSTANCE

If we consider a given mass of water, we recognize that this water can exist in various forms. If it is a liquid initially, it may become a vapor when it is heated, or solid when it is cooled. Thus we speak of the different phases of a substance. A phase is defined as a quantity of matter that is homogeneous throughout. When more than one phase is present, the phases are separated from each other by the phase boundaries. In each phase the substance may exist at various pressures and temperatures or, to use the thermodynamic term, in various states. The state may be identified or described by certain observable, macroscopic properties; some familiar ones are temperature, pressure, and density. In later chapters other properties will be introduced. Each of the properties of a substance in a given state has only one definite value, and these properties always have the same value for a given state, regardless of how the substance arrived at that state. In fact, a property can be defined as any quantity that depends on the state of the system and is independent of the path (that is, the prior history) by which the system arrived at the given state. Conversely, the state is specified or described by the properties, and later we will consider the number of independent properties a substance can have, that is, the minimum number of properties that must be specified in order to fix the state of the substance.

Thermodynamic properties can be divided into two general classes, intensive and extensive properties. An intensive property is independent of the mass; the value of an extensive property varies directly with the mass. Thus, if a quantity of matter in a given state is divided into two equal parts, each part will have the same value of intensive properties as the original, and half the value of the extensive properties. Pressure, temperature, and density are examples of intensive properties. Mass and total volume are examples of extensive properties. Extensive properties per unit mass, such as specific volume, are intensive properties.

Frequently we will refer not only to the properties of a substance but to the properties of a system. When we do so we necessarily imply that the value of the property has significance for the entire system, and this implies what is called equilibrium. For example, if the gas that comprises the system in Fig. 2.1 is in thermal equilibrium, the temperature will be the same throughout the entire system, and we may speak of the temperature as a property of the system. We may also consider mechanical equilibrium and this is related to pressure. If a system is in mechanical equilibrium, there is no tendency for the pressure at any point to change with time as long as the system is isolated from the surroundings. There will be a variation in pressure with elevation, because of the influence of gravitational forces, although under equilibrium conditions there will be no tendency for

the pressure at any location to change. However, in many thermodynamic problems, this variation in pressure with elevation is so small that it can be neglected. Chemical equilibrium is also important and will be considered in Chapter 13.

When a system is in equilibrium regarding all possible changes of state, we say that the system is in thermodynamic equilibrium.

2.4 PROCESSES AND CYCLES

Whenever one or more of the properties of a system change, we say that a change in state has occurred. For example, when one of the weights on the piston in Fig. 2.3 is removed, the piston rises and a change in state occurs, for the pressure decreases and the specific volume increases. The path of the succession of states through which the system passes is called the process.

Let us consider the equilibrium of a system as it undergoes a change in state. The moment the weight is removed from the piston in Fig. 2.3, mechanical equilibrium does not exist, and as a result the piston is moved upward until mechanical equilibrium is again restored. The question that arises is this: Since the properties describe the state of a system only when it is in equilibrium, how can we describe the states of a system during a process if the actual process occurs only when equilibrium does not exist? One step in the answer to this question concerns the definition of an ideal process, which we call a quasiequilibrium process. A quasiequilibrium process is one in which the deviation from thermodynamic equilibrium is infinitesimal, and all the states the system passes through during a quasiequilibrium process may be considered as equilibrium states. Many actual processes closely approach a quasiequilibrium process, and may be so treated with essentially no error. If the weights on the piston in Fig. 2.3 are small and are taken off one by one, the process could be considered quasiequilibrium. On the other hand, if all the weights were removed at once, the piston would rise rapidly until it hit the stops. This would be a nonequilibrium process, and the system would not be in equilibrium at any time during this change of state.

FIGURE 2.3

Example of a system that may undergo a quasiequilibrium process.

For nonequilibrium processes, we are limited to a description of the system before the process occurs and after the process is completed and equilibrium is restored. We are not able to specify each state through which the system passes, nor the rate at which the process occurs. However, as we will see later, we are able to describe certain overall effects that occur during the process.

Several processes are described by the fact that one property remains constant. The prefix iso- is used to describe this. An isothermal process is a constant-temperature process, an isobaric (sometimes called isopiestic) process is a constant-pressure process, and an isochoric process is a constant-volume process.

When a system in a given initial state goes through a number of different changes of state or processes and finally returns to its initial state, the system has undergone a cycle. Therefore, at the conclusion of a cycle, all the properties have the same value they had at the beginning. Steam (water) that circulates through a steam power plant undergoes a cycle.

A distinction should be made between a thermodynamic cycle, which has just been described, and a mechanical cycle. A four-stroke cycle internal-combustion engine goes through a mechanical cycle once every two revolutions. However, the working fluid does not go through a thermodynamic cycle in the engine, since air and fuel are burned and changed to products of combustion that are exhausted to the atmosphere. In this text the term cycle will refer to a thermodynamic "cycle" unless otherwise designated.

2.5 UNITS FOR MASS, LENGTH, TIME, AND FORCE

Since we are considering thermodynamic properties from a macroscopic perspective, we are dealing with quantities which can, either directly or indirectly, be measured and counted. Therefore, the matter of units becomes an important consideration. In the remaining sections of this chapter we will define certain thermodynamic properties and the basic units involved. The relation between force and mass is often a difficult matter for students, and is considered in this section in some detail.

Force, mass, length, and time are related by Newton's second law of motion, which states that the force acting on a body is proportional to the product of the mass and the acceleration in the direction of the force.

$$F \propto ma$$

The concept of time is well established. The basic unit of time is the second (s), which in the past has been defined in terms of the solar day, the time interval for one complete revolution of the earth relative to the sun. Since this period will vary with the season of the year, an average value over a one-year period is called the mean solar day, and the mean solar second is 1/86 400 of the mean solar day. (The measurement of the earth's rotation is sometimes made relative to a fixed

star, in which case the period is called a sidereal day.) In 1967, the General Conference of Weights and Measures (CGPM) adopted a definition of the second in terms of a resonator using a beam of cesium-133 atoms. The time required for 9 192 631 770 cycles of the cesium resonator is now accepted as the definition of the second.

For periods of time less than a second, the prefixes milli, micro, nano, or pico, as listed in Table 2.1, are commonly used. For larger periods of time, the units minute (min.), hour (h), or day (day) are frequently used. It should be pointed out that the prefixes in Table 2.1 are used with many other units as well.

TABLE 2.1

Unit Prefixes

Factor	Prefix	Symbol	Factor	Prefix	Symbol
10^{12}	tera	T	10^{-3}	milli	m
10^9	giga	G	10^{-6}	micro	μ
10^6	mega	M	10^{-9}	nano	n
10^3	kilo	k	10^{-12}	pico	p

The concept of length is also well established. The basic unit of length is the meter (m), and for many years the accepted standard was the International Prototype Meter, the distance between two marks on a platinum–iridium bar under certain prescribed conditions. This bar is maintained at the International Bureau of Weights and Measures, Sevres, France. In 1960, the CGPM adopted a definition of the meter as a length equal to 1 650 763.73 wavelengths in vacuo of the orange-red line of krypton-86. Then, in 1983, the CGPM adopted a more precise definition of the meter in terms of the speed of light (which therefore is now a fixed constant) as follows: the meter is the length of the path traveled by light in a vacuum during a time interval of 1/299 792 458 of a second.

The fundamental unit of mass is the kilogram (kg). As adopted by the first CGPM in 1889, and restated in 1901, it is the mass of a certain platinum-iridium cylinder maintained under prescribed conditions at the International Bureau of Weights and Measures. A related unit that is used frequently in thermodynamics is the mole (mol), defined as an amount of substance containing as many elementary entities as there are atoms in 0.012 kg of carbon-12. These elementary entities must be specified, and may be atoms, molecules, electrons, ions, or other particles or specific groups. For example, one mole of diatomic oxygen, having a molecular weight of 32 (compared to 12 for carbon), is a mass of 0.032 kg. The mole is often termed a gram mole, since it is an amount of substance in grams numerically equal to the molecular weight. In this text, when using the metric SI system we will find it more convenient to use the kilomole (kmol); the amount of substance in kilograms numerically equal to the molecular weight, in preference to the mole.

The system of units in use presently throughout most of the world is the metric International System, commonly referred to as SI units (from Le Système Inter-

national d'Unités). In this system, the second, meter, and kilogram are the basic units for time, length, and mass, respectively, as defined above, and the unit of force is defined directly from Newton's second law.

Therefore, a proportionality constant is unnecessary, and we may write that law as an equality:

$$F = ma$$

The unit of force is the newton (N), which by definition is the force required to accelerate a mass of one kilogram at the rate of one meter per second per second.

$$1 \text{ N} = 1 \text{ kg m/s}^2$$

(It is worth noting that SI units derived from proper nouns utilize capital letters for symbols; others utilize the lowercase letters. An exception is the liter, with the symbol L.)

The traditional system of units used in the United States is termed the **English Engineering System**. In this system, the unit of time is the second, which has been discussed above. The basic unit of length is the foot (ft), which at present is defined in terms of the meter as

$$1 \text{ ft} = 0.3048 \text{ m}$$

The inch (in.) is defined in terms of the foot

$$12 \text{ in.} = 1 \text{ ft}$$

The unit of mass in this system is the pound mass (lbm), originally the mass of a certain platinum cylinder kept in the Tower of London, but now defined in terms of the kilogram as

$$1 \text{ lbm} = 0.453 \ 592 \ 37 \text{ kg}$$

A related unit is the pound mole (lb mol), which is an amount of substance in pounds mass numerically equal to the molecular weight of that substance. It is important to distinguish between a pound mole and a mole (gram mole).

In the English Engineering System of Units, the concept of force is not defined from Newton's second law, but is instead established as an independent quantity. Thus, for this system, unlike other systems of units, a conversion constant g_c is necessary in Newton's second law. The unit for force is defined in terms of an experimental procedure as follows. Let the standard pound mass be suspended in the earth's gravitational field at a location where the acceleration due to gravity is 32.1740 ft/s². The force with which the standard pound mass is attracted to the earth (the buoyant effects of the atmosphere on the standard pound mass must also be standardized) is defined as the unit for force and is termed a pound force (lbf). Note that we must be careful to distinguish between a lbm and a lbf, and we do not use the term pound alone. Now, since we have independently defined the

units for force, mass, length, and time, we have the relation, from Newton's second law,

$$F = \frac{ma}{g_c}$$

or

$$1 \text{ lbf} = \frac{1 \text{ lbm} \times 32.174 \text{ ft/s}^2}{g_c}$$

or

$$g_c = 32.174 \frac{\text{lbm ft}}{\text{lbf s}^2}$$

Note that the conversion constant g_c has both a numerical value and dimensions. We also note that for gravitational accelerations that are not too different from the standard value (32.174 ft/s^2 or 9.806 65 m/s^2), the lbf is approximately numerically equal to the lbm. This is the reason that people often drop the designation, which can lead to confusion and inconsistencies. To illustrate, let us calculate the force due to gravity on a one-pound mass at a location where the acceleration due to gravity is 32.14 ft/s^2 (about 10 000 ft above sea level).

$$F = \frac{ma}{g_c} = \frac{1 \text{ lbm} \times 32.14 \text{ ft/s}^2}{32.174 \text{ lbm ft/lbf s}^2}$$

$$= 0.999 \text{ lbf}$$

We further note that the pound force could very well have been defined as

$$1 \text{ lbf} = 32.174 \text{ lbm ft/s}^2$$

in a manner analogous to that for the newton in the SI system. It then follows that

$$1 = 32.174 \frac{\text{lbm ft}}{\text{lbf s}^2}$$

Comparing this with the conversion constant g_c,

$$g_c = 32.174 \frac{\text{lbm ft}}{\text{lbf s}^2} = 1$$

Since a pure number can be substituted into an equation at any point, we can therefore substitute this unit conversion constant g_c, as necessary. Throughout the remainder of this text, we will follow this philosophy, and will not include the constant g_c explicitly in our equations, even though the majority of our attention is focused on the English Engineering system of units.

The term "weight" is often used with respect to a body, and is sometimes confused with mass. Weight is really correctly used only as a force. When we say a body weighs so much we mean that this is the force with which it is attracted to the earth (or some other body), that is, the product of its mass and the local

gravitational acceleration. The mass of a substance remains constant with elevation, but its weight varies with elevation.

2.6 SPECIFIC VOLUME

The specific volume of a substance is defined as the volume per unit mass, and is given the symbol v. The density of a substance is defined as the mass per unit volume, and is therefore the reciprocal of the specific volume. Density is designated by the symbol ρ. Specific volume and density are intensive properties.

The specific volume of a system in a gravitational field may vary from point to point. For example, considering the atmosphere as a system, the specific volume increases as the elevation increases. Therefore the definition of specific volume involves the specific volume of a substance at a point in a system.

Consider a small volume δV of a system, and let the mass be designated δm. The specific volume is defined by the relation

$$v = \lim_{\delta V \to \delta V'} \frac{\delta V}{\delta m}$$

where $\delta V'$ is the smallest volume for which the system can be considered a continuum.

Thus in a given system we should speak of the specific volume or density at a point in the system, and recognize that this may vary with elevation. However, most of the systems that we consider are relatively small, and the change in specific volume with elevation is not significant. In this case, we can speak of one value of specific volume or density for the entire system.

In this text, the specific volume and density will be given on either a mass or on a mole basis. A bar over the symbol (lowercase) will be used to designate the property on a mole basis. Thus \bar{v} will designate molal specific volume and $\bar{\rho}$ will designate the molal density. In English units, the units for specific volume are ft^3/lbm and $\text{ft}^3/\text{lb mol}$; for density the corresponding units are lbm/ft^3 and lb mol/ft^3. In SI units, the units for specific volume are m^3/kg and m^3/kmol; for density they are kg/m^3 and kmol/m^3.

While the SI unit for volume is the cubic meter, a commonly used volume unit is the liter (L), which is special name given to a volume of 0.001 cubic meters, that is, $1 \text{ L} = 10^{-3} \text{ m}^3$.

2.7 PRESSURE

When dealing with liquids and gases we ordinarily speak of pressure; in solids we speak of stresses. The pressure in a fluid at rest at a given point is the same in all directions, and we define pressure as the normal component of force per unit area. More specifically, if $\delta \mathscr{A}$ is a small area, and $\delta \mathscr{A}'$ is the smallest area over which we can consider the fluid a continuum, and δF_n is the component of

force normal to $\delta\mathscr{A}$, we define pressure, P, as

$$P = \lim_{\delta\mathscr{A} \to \delta\mathscr{A}'} \frac{\delta F_n}{\delta\mathscr{A}}$$

The pressure P at a point in a fluid in equilibrium is the same in all directions. In a viscous fluid in motion the variation in the state of stress with orientation becomes an important consideration. These considerations are beyond the scope of this book, and we will consider pressure only in terms of a fluid in equilibrium.

In most thermodynamic investigations we are concerned with absolute pressure. Most pressure and vacuum gages, however, read the difference between the absolute pressure and the atmospheric pressure existing at the gage, and this is referred to as gage pressure. This is shown graphically in Fig. 2.4, and the following examples illustrate the principles involved. Pressures below atmospheric and slightly above atmospheric and pressure differences (for example, across an orifice in a pipe) are frequently measured with a manometer, which contains water, mercury, alcohol, oil, or other fluids. From the principles of hydrostatics one concludes that for a difference in level of L units the pressure difference is calculated by the relation

$$\Delta P = \rho L g$$

where ρ is the density of the fluid and g is the local acceleration due to gravity. The accepted standard value for gravitational acceleration is

$$g = 32.174 \text{ ft/s}^2 = 9.806\,65 \text{ m/s}^2$$

but the value varies with location and elevation. The use of a manometer is illustrated in Fig. 2.5.

FIGURE 2.4

Illustration of terms used in pressure measurement.

FIGURE 2.5

Example of pressure measurement using a column of fluid.

To distinguish between absolute and gage pressure in this text, the term pressure will always refer to absolute pressure. Any gage pressure will then be indicated as such.

In English units, the unit for pressure that is generally consistent with the other units used in this text is pounds force per square foot (lbf/ft^2). However, in common parlance and general experimental work, pressures are often measured in pounds force per square inch ($lbf/in.^2$). Therefore, the student should be careful in numerical calculations to introduce the conversion 144 $in.^2$ = 1 ft^2, as necessary.

In SI units, the unit for pressure is the force of one newton acting on a one square-meter area, which is termed the pascal (Pa). That is,

$$1 \text{ Pa} = 1 \text{ N/m}^2$$

A number of other units are used for pressure in connection with the English or SI systems. The height of a column of fluid, such as expressed in Fig. 2.5 may be used as a unit—for example mm or in. of mercury, or ft of water. The standard atmosphere (atm), which is defined as the pressure exerted by a column of mercury (density of 13 595 kg/m^3) 760 mm in height, is therefore equivalent to other units as follows:

$$1 \text{ atm} = 14.696 \text{ lbf/in.}^2 = 101 \, 325 \text{ Pa}$$

Another unit, only slightly different in magnitude is the bar, where

$$1 \text{ bar} = 10^5 \text{ Pa} = 100 \text{ kPa} = 0.1 \text{ MPa}$$

is frequently used in connection with the SI system.

2.8 EQUALITY OF TEMPERATURE

Although temperature is a property with which we are all familiar, an exact definition of it is difficult. We are aware of "temperature" first of all as a sense of hotness or coldness when we touch an object. We also learn early in our experience

that when a hot body and a cold body are brought into contact, the hot body becomes cooler and the cold body becomes warmer. If these bodies remain in contact for some time, they usually appear to have the same hotness or coldness. However, we also realize that our sense of hotness or coldness is very unreliable. Sometimes very cold bodies may seem hot, and bodies of different materials that are at the same temperature appear to be at different temperatures.

Because of these difficulties in defining temperature, we define equality of temperature. Consider two blocks of copper, one hot and the other cold, each of which is in contact with a mercury-in-glass thermometer. If these two blocks of copper are brought into thermal communication, we then observe that the electrical resistance of the hot block decreases with time and for the cold block it increases with time. After a period of time has elapsed, however, no further changes in resistance are observed. Similarly, when the blocks are first brought in thermal communication, the length of a side of the hot block decreases with time, whereas for the cold block it increases with time. After a period of time, no further change in length of either of the blocks is perceived. Also, the mercury column of the thermometer in the hot block drops at first and in the cold block it rises, but after a period of time no further changes in height are observed. We may say, therefore, that two bodies have equality of temperature when no change in any observable property occurs when they are in thermal communication.

2.9 THE ZEROTH LAW OF THERMODYNAMICS

Now consider the same two blocks of copper, and also another thermometer. Let one block of copper be brought into contact with the thermometer until equality of temperature is established, and then removed. Then let the second block of copper be brought into contact with the thermometer, and suppose that no change in the mercury level of the thermometer occurs during this operation with the second block. Then we can say that both blocks are in thermal equilibrium with the given thermometer.

The zeroth law of thermodynamics states that when two bodies have equality of temperature with a third body, they in turn have equality of temperature with each other. This seems very obvious to us because we are so familiar with this experiment. However, since this fact is not derivable from other laws, and since in the logical presentation of thermodynamics it precedes the first and second laws of thermodynamics, it has been called the zeroth law of thermodynamics. This law is really the basis of temperature measurement, for numbers can be placed on the mercury thermometer, and every time a body has equality of temperature with the thermometer, we can say that the body has the temperature we read on the thermometer. The problem remains, however, of relating temperatures that we might read on different mercury thermometers, or that we obtain when using different temperature-measuring devices, such as thermocouples and resistance thermometers. This suggests the need for a standard scale for temperature measurements.

2.10 TEMPERATURE SCALES

There are two commonly used scales for measuring temperature, namely the Fahrenheit (after Gabriel Fahrenheit, 1686–1736) and Celsius scales. The Celsius scale was formerly called the Centrigrade scale, but is now designated the Celsius scale, after Anders Celsius (1701–1744) the Swedish astronomer who devised this scale.

Until 1954 each of these scales was based on two fixed, easily duplicated points, the ice point and the steam point. The temperature of the ice point is defined as the temperature of a mixture of ice and water which is in equilibrium with saturated air at a pressure of 1 atm. The temperature of the steam point is the temperature of water and steam which are in equilibrium at a pressure of 1 atm. On the Fahrenheit scale these two points are assigned the numbers 32 and 212, respectively, and on the Celsius scale the respective points are numbered 0 and 100. The basis for numbers on the Fahrenheit scale has an interesting background. In searching for an easily reproducible point, Fahrenheit selected the temperature of the human body and assigned it the number 96. He assigned the number 0 to the temperature of a certain mixture of salt, ice and salt solution. On this scale the ice point was approximately 32. When this scale was slightly revised and fixed in terms of the ice point and steam point, the normal temperature of the human body was found to be 98.6 F.

In this text the symbols F and °C will denote the Fahrenheit and Celsius scales, respectively. The symbol T will refer to temperature on all temperature scales.

At the tenth CGPM in 1954, the Celsius scale was redefined in terms of a single fixed point and the ideal-gas temperature scale. The single fixed point is the triple point of water (the state in which the solid, liquid, and vapor phases of water exist together in equilibrium). The magnitude of the degree is defined in terms of the ideal-gas temperature scale, which is discussed in Chapter 6. The essential features of this new scale are a single fixed point and a definition of the magnitude of the degree. The triple point of water is assigned the value 0.01°C. On this scale the steam point is experimentally found to be 100.00°C. Thus, there is essential agreement between the old and new temperature scales.

It should be noted that we have not yet considered an absolute scale of temperature. The possibility of such a scale arises from the second law of thermodynamics and is discussed in Chapter 6. On the basis of the second law of thermodynamics a temperature scale that is independent of any thermometric substance can be defined. This absolute scale is usually referred to as the thermodynamic scale of temperature. However, it is very complicated to use this scale directly, and therefore a more practical scale, the International Practical Temperature Scale, which closely represents the thermodynamic scale, has been adopted.

The absolute scale related to the Celsius scale is referred to as the Kelvin scale (after William Thomson, 1824–1907, who is also known as Lord Kelvin), and is designated K (without the degree symbol). The relation between these scales is

$$K = °C + 273.15$$

In 1967, the CGPM defined the kelvin as 1/273.16 of the temperature at the triple point of water. The Celsius scale is then defined by the equation above, instead of by its earlier definition.

The absolute scale related to the Fahrenheit scale is referred to as the Rankine scale and is designated R. The relation between these scales is

$$R = F + 459.67$$

2.11 THE INTERNATIONAL PRACTICAL TEMPERATURE SCALE

In 1968 the International Committee on Weights and Measures adopted a revised International Practical Temperature Scale, IPTS-68, which is described below. This scale, similar to earlier ones of 1927 and 1948, has been extended in range and made to conform more closely to the thermodynamic temperature scale. It is based on a number of fixed and easily reproducible points that are assigned definite numerical values of temperature, and on specified formulas relating temperature to the readings on certain temperature-measuring instruments for the purpose of interpolation between the fixed points. The primary fixed points and a summary of the interpolation techniques are listed here for the sake of completeness, although the student will have limited need for them at this time.

The primary fixed-point temperatures in degrees Celsius are as follows:

1.	Triple point (equilibrium between solid, liquid, and vapor phases) of equilibrium-hydrogen.	-259.34
2.	Boiling point (equilibrium between liquid and vapor phases) of equilibrium-hydrogen at 25/76 atm (33.33 kPa) pressure.	-256.108
3.	Normal boiling point (1 atm pressure) of equilibrium-hydrogen.	-252.87
4.	Normal boiling point of neon.	-246.048
5.	Triple point of oxygen.	-218.789
6.	Normal boiling point of oxygen.	-182.962
7.	Triple point of water.	0.01
8.	Normal boiling point of water.	100
9.	Normal freezing point (equilibrium between solid and liquid phases at 1 atm pressure) of zinc.	419.58
10.	Normal freezing point of silver.	961.93
11.	Normal freezing point of gold.	1064.43

The means available for measurement and interpolation lead to a division of the temperature scale into four general ranges.

1. The range from $-259.34°C$ to $0°C$ is based on measurements on a platinum resistance thermometer, with temperature expressed in terms of a 20th degree

reference function equation. This range is subdivided into four parts. In each, the difference between measured resistance ratios of a particular thermometer and the reference function at the fixed points is used to determine the constants in a specified polynomial interpolation equation.

2. The range from 0 to 630.74°C (the normal freezing point of antimony, a secondary fixed point) is also based on a platinum resistance thermometer, with constants in a polynomial interpolating equation determined by calibration at the three fixed points in this range.

3. The range from 630.74 to 1064.43°C is based on measurements on a standard platinum versus rhodium–platinum thermocouple, and a three-term equation expressing electrical potential (EMF) as a function of temperature. The constants are determined by a platinum resistance thermometer measurement at the antimony point and by calibration at the two primary fixed points in this range.

4. The range above 1064.43°C is based on measurements of the intensity of visible-spectrum radiation compared with that of the same wavelength at the gold point, and on Planck's equation for blackbody radiation.

PROBLEMS

2.1 A steady force of 5 lbf acts on a mass of 20 lbm. What is the acceleration of this mass?

2.2 Two pound moles of diatomic nitrogen gas are enclosed in a container. A force of 1 lbf now accelerates this system. Neglecting the mass of the container, what is the acceleration?

2.3 The "standard" acceleration (at sea level and 45 degrees latitude) due to gravity is 9.806 65 m/s². Calculate the force due to "standard" gravity acting on a mass of 50 kg.

2.4 With what force is a mass of 100 kg attracted to the earth at a location where the gravitational acceleration is 9.65 m/s²? What is its weight in lbf? What is its mass in lbm?

2.5 A one pound mass is "weighed" with a beam balance at a point where $g = 31.0$ ft/s². What reading is expected? If this mass is weighed with a spring scale that reads correctly for standard gravity (see Problem 2.3), what reading would be obtained?

2.6 A certain body is "weighed" at an elevation of 30 000 ft (where $g = 32.08$ ft/s²) by a spring balance that was calibrated at sea level. The reading on the spring balance is 10 pounds. What is the mass of this body?

2.7 A 25 kg steel tank having a volume capacity of 250 L is filled with liquid water. What force is needed to accelerate this combined system at a rate of 2 m/s²? Assume that the density of the water is 1000 kg/m³.

2.8 A gas is contained in a vertical cylinder fitted with a piston, as shown in Fig. 2.6. Atmospheric pressure outside is 1 bar, and the piston area is 10 in.². What is the mass of the piston, if the gas pressure inside is 17 lbf/in.²? Assume standard gravitational acceleration.

FIGURE 2.6
Sketch for Problem 2.8.

2.9 Consider an orbiting space station in which an artificial gravity of 5 ft/s² is induced by rotating the station. How much would a 175-lbm man weigh inside?

2.10 The reading on a pressure gage is 65 lbf/in.², and the local barometer reading is 29.1 in. Hg. Calculate the absolute pressure that is being measured.

2.11 The height of a mercury manometer column that is used to measure a vacuum is 700 mm, and the local barometer reading is 95 kPa. Calculate the absolute pressure that is being measured, in lbf/in.², assuming that the density of mercury is 13 600 kg/m³.

2.12 Consider a manometer containing a fluid with a density of 800 kg/m³. What pressure difference in lbf/in.² is indicated if the difference in height of the two columns is 20 in.? What would the column height difference be if this same pressure difference were measured by a manometer containing mercury, having a density of 13 600 kg/m³?

2.13 A mercury column is used to measure a pressure difference of 25 lbf/in.² in a piece of experimental apparatus that is located outdoors. The minimum temperature at this location in winter is 0 F and the maximum temperature in the summer is 100 F. What will be the difference in height of this mercury column in the summer as compared to that in the winter when measuring the given pressure difference? Assume standard gravitational acceleration. The following data are given for the density of mercury:

T°C	Density, kg/m³
−10	13 620
0	13 595
10	13 570
20	13 546
30	13 521

2.14 A piece of experimental apparatus is located in an airplane flying at 20 000 m, where $g = 9.75$ m/s². The air flow in this apparatus is measured

by determining the pressure drop across an orifice with a mercury manometer, as shown in Fig. 2.7. The manometer level difference reading is 250 mm, and the local temperature is $-10°C$ (see Problem 2.13). What is the pressure drop in kPa?

FIGURE 2.7
Sketch for Problem 2.14.

2.15 The level of the water in an enclosed water tank is 120 ft above ground level. The pressure in the air space above the water is 17.5 $lbf/in.^2$, and the density of the water is 62.4 lbm/ft^3. What is the water pressure at ground level?

2.16 Two cylinders A and B are connected by a piston of two different diameters, as shown in Fig. 2.8. Cylinder B contains a gas, while cylinder A contains oil that is pumped up to a pressure of 300 kPa by a hydraulic pump. The mass of the piston is 10 kg. Calculate the gas pressure in cylinder B.

FIGURE 2.8
Sketch for Problem 2.16.

2.17 A particular platinum vs. rhodium-platinum thermocouple is to be calibrated and used in the temperature range described as range 3 in connection with the discussion of the IPTS-68 in Section 2.11. During calibration, the thermocouple EMF readings in microvolts are 5552, 9147 and 10 333 at the freezing points of antimony, silver and gold, respectively. Assuming a

polynomial relation of the form

$$EMF = C_0 + C_1 * T + C_2 * T^2$$

determine the temperature at a location where this thermocouple is used for a measurement and gives a reading of 7500 microvolts.

2.18 Write a computer program to solve the thermocouple calibration problem discussed in Problem 2.17, in which any three sets of calibration readings can be given, and the result is a polynomial equation with specified coefficients that can be used at any temperature in the calibrated range.

3
PROPERTIES OF A PURE SUBSTANCE

In the previous chapter we considered three familiar properties of a substance—specific volume, pressure, and temperature. We now turn our attention to pure substances and consider some of the phases in which a pure substance may exist, the number of independent properties a pure substance may have, and methods of presenting thermodynamic properties.

3.1 THE PURE SUBSTANCE

A pure substance is one that has a homogeneous and invariable chemical composition. It may exist in more than one phase, but the chemical composition is the same in all phases. Thus liquid water, a mixture of liquid water and water vapor (steam), or a mixture of ice and liquid water are all pure substances, for every phase has the same chemical composition. On the other hand, a mixture of liquid air and gaseous air is not a pure substance, since the composition of the liquid phase is different from that of the vapor phase.

Sometimes a mixture of gases, such as air, is considered a pure substance as long as there is no change of phase. Strictly speaking, this is not true, but instead, as we will see later, we should say that a mixture of gases such as air exhibits some of the characteristics of a pure substance as long as there is no change of phase.

In this text the emphasis will be on those substances that may be called simple compressible substances. By this we understand that surface effects, magnetic effects, and electrical effects are not significant when dealing with these substances. On the other hand changes in volume, such as those associated with the expansion of a gas in a cylinder, are most important. However, reference will be made to other substances in which surface, magnetic, or electrical effects are important. We will refer to a system consisting of a simple compressible substance as a simple compressible system.

3.2 VAPOR–LIQUID–SOLID PHASE EQUILIBRIUM IN A PURE SUBSTANCE

Consider as a system 1 lbm of water contained in the piston–cylinder arrangement of Fig. 3.1a. Suppose that the piston and weight maintain a pressure 14.7 lbf/in.2 in the cylinder, and that the initial temperature is 60 F. As heat is transferred to

a *b* *c*

FIGURE 3.1

Constant-pressure change from liquid to vapor phase for a pure substance.

the water the temperature increases appreciably, the specific volume increases slightly, and the pressure remains constant. When the temperature reaches 212 F, additional heat transfer results in a change of phase, as indicated in Fig. 3.1*b*. That is, some of the liquid becomes vapor, and during this process both the temperature and pressure remain constant, but the specific volume increases considerably. When the last drop of liquid has vaporized, further transfer of heat results in an increase in both temperature and specific volume of the vapor, Fig. 3.1*c*.

The term saturation temperature designates the temperature at which vaporization takes place at a given pressure, and this pressure is called the saturation pressure for the given temperature. Thus for water at 212 F the saturation pressure is 14.7 lbf/in.2, and for water at 14.7 lbf/in.2 the saturation temperature is 212 F. For a pure substance there is a definite relation between saturation pressure and saturation temperature, a typical curve being shown in Fig. 3.2. This is called the vapor-pressure curve.

If a substance exists as liquid at the saturation temperature and pressure, it is called saturated liquid. If the temperature of the liquid is lower than the saturation temperature for the existing pressure, it is called either a subcooled liquid

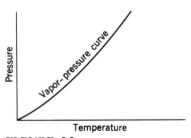

FIGURE 3.2

Vapor-pressure curve of a pure substance.

(implying that the temperature is lower than the saturation temperature for the given pressure) or a compressed liquid (implying that the pressure is greater than the saturation pressure for the given temperature). Either term may be used, but the latter term will be used in this text.

When a substance exists as part liquid and part vapor at the saturation temperature, its quality is defined as the ratio of the mass of vapor to the total mass. Thus, in Fig. 3.1b, if the mass of the vapor is 0.2 lbm and the mass of the liquid is 0.8 lbm, the quality is 0.2 or 20%. The quality may be considered as an intensive property, and it has the symbol x. Quality has meaning only when the substance is in a saturated state, that is, at saturation pressure and temperature.

If a substance exists as vapor at the saturation temperature, it is called saturated vapor. (Sometimes the term dry saturated vapor is used to emphasize that the quality is 100 percent.) When the vapor is at a temperature greater than the saturation temperature, it is said to exist as superheated vapor. The pressure and temperature of superheated vapor are independent properties, since the temperature may increase while the pressure remains constant. Actually, the substances we call gases are highly superheated vapors.

Consider Fig. 3.1 again, and let us plot on the temperature–volume diagram of Fig. 3.3 the constant-pressure line that represents the states through which the water passes as it is heated from the initial state of 14.7 lbf/in.² and 60 F. Let state A represent the initial state, B the saturated-liquid state (212 F), and line AB the process in which the liquid is heated from the initial temperature to the saturation temperature. Point C is the saturated vapor state, and line BC is the constant-temperature process in which the change of phase from liquid to vapor occurs.

FIGURE 3.3

Temperature–volume diagram for water showing liquid and vapor phases (not to scale).

Line CD represents the process in which the steam is superheated at constant pressure. Temperature and volume both increase during this process.

Now let the process take place at a constant pressure of 100 lbf/in.2, beginning from an initial temperature of 60 F. Point E represents the initial state, the specific volume being slightly less than that at 14.7 lbf/in.2 and 60 F. Vaporization now begins at point F, where the temperature is 327.9 F. Point G is the saturated-vapor state, and line GH the constant-pressure process in which the steam is superheated.

In a similar manner, a constant pressure of 1000 lbf/in.2 is represented by line $IJKL$, the saturation temperature being 544.7 F.

At a pressure of 3203.6 lbf/in.2, represented by line MNO, we find, however, that there is no constant-temperature vaporization process. Instead, point N is a point of inflection with a zero slope. This point is called the critical point, and at the critical point the saturated-liquid and saturated-vapor states are identical. The temperature, pressure, and specific volume at the critical point are called the critical temperature, critical pressure, and critical volume. The critical-point data for some substances are given in Table 3.1, and more extensive data are given in Table A.6 in the Appendix.

TABLE 3.1

Some Critical-Point Data

	Critical Temperature F	Critical Pressure lbf/in.2	Critical Volume ft^3/lbm
Water	705	3203.6	0.0505
Carbon dioxide	88	1071	0.0348
Oxygen	-203	735	0.0364
Hydrogen	-400	188	0.534

A constant-pressure process at a pressure greater than the critical pressure is represent by line PQ. If water at 5500 lbf/in.2, 60 F is heated in a constant-pressure process in a cylinder as shown in Fig. 3.1, there will never be two phases present, and the state shown in Fig. 3.1b will never exist. Rather, there will be a continuous change in density and at all times there will be only one phase present. The question then arises as to when do we have a liquid and when do we have a vapor? The answer is that this is not a valid question at supercritical pressures. Instead, we simply term the substance a fluid. However, rather arbitrarily, at temperatures below the critical temperature we usually refer to it as a compressed liquid and at temperatures above the critical temperature as a superheated vapor. It should be emphasized, however, that at pressures above the critical pressure we never have a liquid and vapor phase of a pure substance existing in equilibrium.

In Fig. 3.3 line $NJFB$ represents the saturated-liquid line and line $NKGC$ represents the saturated-vapor line.

Let us consider another experiment with the piston–cylinder arrangement. Suppose that the cylinder contains 1 lbm of ice at 0 F, 14.7 lbf/in.2. When heat is

TABLE 3.2

Some Solid-Liquid-Vapor Triple Point Data

	Temperature F	Pressure atm
Hydrogen (normal)	−435	0.071
Nitrogen	−346	0.1237
Oxygen	−362	0.00150
Mercury	−38	0.0000000013
Water	32	0.00603
Zinc	786	0.05
Silver	1760	0.0001
Copper	1981	0.00000078

transferred to the ice, the pressure remains constant, the specific volume increases slightly, and the temperature increases until it reaches 32 F, at which point the ice melts while the temperature remains constant. In this state the ice is called saturated solid. For most substances the specific volume increases during this melting process but for water the specific volume of the liquid is less than the specific volume of the solid. When all of the ice has melted, a further heat transfer causes an increase in temperature of the liquid.

If the initial pressure of the ice at 0 F is 0.0505 lbf/in.2, heat transfer to the ice first results in an increase in temperature to 20 F. At this point, however, the ice would pass directly from the solid phase to the vapor phase in the process known as sublimation. Further heat transfer would result in superheating of the vapor.

Finally, consider an initial pressure of the ice of 0.0887 lbf/in.2 and a temperature of 0 F. As a result of heat transfer let the temperature increase until it reaches 32.02 F. At this point, however, further heat transfer may result in some of the ice becoming vapor and some becoming liquid, for at this point it is possible to have the three phases in equilibrium. This is called the triple point, which is defined as the state in which three phases may all be present in equilibrium. The pressure and temperature at the triple point for a number of substances are given in Table 3.2.

This whole matter is best summarized by the diagram of Fig. 3.4, which shows how the solid, liquid, and vapor phases may exist together in equilibrium. Along the sublimation line the solid and vapor phases are in equilibrium, along the fusion line the solid and liquid phases are in equilibrium, and along the vaporization line the liquid and vapor phases are in equilibrium. The only point at which all three phases may exist in equilibrium is the triple point. The vaporization line ends at the critical point because there is no distinct change from the liquid phase to the vapor phase above the critical point.

Consider a solid in state A, Fig. 3.4. When the temperature is increased while the pressure (which is less than the triple-point pressure) is constant, the substance passes directly from the solid to the vapor phase. Along the constant-pressure line EF, the substance first passes from the solid to the liquid phase at one temperature,

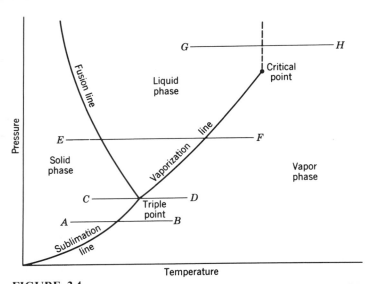

FIGURE 3.4

Pressure–temperature diagram for a substance such as water.

and then from the liquid to the vapor phase at a higher temperature. Constant-pressure line *CD* passes through the triple point, and it is only at the triple point that the three phases may exist together in equilibrium. At a pressure above the critical pressure, such as *GH*, there is no sharp distinction between the liquid and vapor phases.

Although we have made these comments with rather specific reference to water (only because of our familiarity with water), all pure substances exhibit the same general behavior. However, the triple-point temperature and critical temperature vary greatly from one substance to another. For example, the critical temperature of helium, as given in Table A.6, is 9.5 R. Therefore, the absolute temperature of helium at ambient conditions is over 50 times greater than the critical temperature. On the other hand, water has a critical temperature of 705.4 F (1165 R) and at ambient conditions the temperature of water is less than half the critical temperature. Most metals have a much higher critical temperature than water. In considering the behavior of a substance in a given state, it is often helpful to think of this state in relation to the critical state or triple point. For example, if the pressure is greater than the critical pressure, it is impossible to have a liquid and a vapor phase in equilibrium. Or, to consider another example, the states at which vacuum melting a given metal is possible can be ascertained by a consideration of the properties at the triple point. In the case of iron at a pressure just above 0.000 05 atm (the triple-point pressure), iron would melt at a temperature of about 1535°C (the triple-point temperature).

It should also be pointed out that a pure substance can exist in a number of different solid phases. A transition from one solid phase to another is called an

5-6 $\boxed{.87 = \dfrac{W_{aT}}{W_{sT}}} = \dfrac{c_p(T_4 - T_6)}{c_p(T_5 - T_{6s})}$

$\dfrac{T_{6s}}{T_5} = \left(\dfrac{P_6}{P_5}\right)^{.286}$ $T_{6s} = \left(\dfrac{102}{231.3}\right)^{.286} \cdot 1086$

$T_{6s} = 859.5$

$W_{aT} = .87 \cdot 1.0035 (1086 - 859.5)$

$W_{aT} = 197.7 \; kJ/kg$

$\dot{m} = \dfrac{100 \, kW}{197.7} = .506 \; kg/s$

c) consider the generator

$a\eta = \dfrac{T_3 - T_2}{T_{3'} - T_2}$ $T_{3'} = T_6$

$.7 = \dfrac{T_3 - 514.3}{888.95 - 514.3}$ $.87 = \dfrac{T_5 - T_6}{T_5 - T_{6s}}$

$.87 = \dfrac{1086 - T_6}{1086 - 859.5}$

$T_3 = 776.56$

$q_H = c_p(T_4 - T_3)$ $T_6 = 888.95$

$= 1.0035(1300 - 776.56)$

$= 525.27$

$\eta = \dfrac{197.7}{525.27} = \dfrac{W_{aT}}{q}$

FIGURE 3.5

Estimated pressure–temperature diagram for iron (From *Phase Diagrams in Metallurgy* by F. N. Rhines, copyright 1956, McGraw-Hill Book Company; used by permission).

allotropic transformation. Figure 3.5 is a pressure–temperature diagram for iron that shows three solid phases, the liquid phase, and the vapor phase. Figure 3.6 shows a number of solid phases for water. It is evident that a pure substance can have a number of triple points, but only one triple point involves solid, liquid, and vapor equilibrium. Other triple points for a pure substance can involve two solid phases and a liquid phase, two solid phases and a vapor phase, or three solid phases.

3.3 INDEPENDENT PROPERTIES OF A PURE SUBSTANCE

One important reason for introducing the concept of a pure substance is that the state of a simple compressible pure substance (that is, a pure substance in the absence of motion, gravity, and surface, magnetic, or electrical effects) is defined by two independent properties. This means, for example, that if the specific volume and temperature of superheated steam are specified, the state of the steam is determined.

To understand the significance of the term independent property, consider the saturated-liquid and saturated-vapor states of a pure substance. These two states

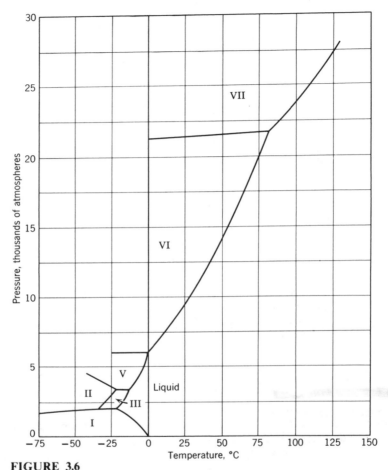

FIGURE 3.6

Phase diagram of water (Adapted from the *American Institute of Physics Handbook*, 2nd ed., 1963, McGraw-Hill).

have the same pressure and the same temperature, but are definitely not the same state. In a saturation state, therefore, pressure and temperature are not independent properties. Two independent properties such as pressure and specific volume, or pressure and quality, are required to specify a saturation state of a pure substance.

The reason for mentioning previously that a mixture of gases, such as air, has the same characteristics as a pure substance as long as only one phase is present, concerns precisely this point. The state of air, which is a mixture of gases of definite composition, is determined by specifying two properties as long as it remains in the gaseous phase, and in this regard air can be treated as a pure substance.

3.4 EQUATIONS OF STATE FOR THE VAPOR PHASE OF A SIMPLE COMPRESSIBLE SUBSTANCE

From experimental observations it has been established that the P-v-T behavior of gases at low density is closely given by the following equation of state.

$$P\bar{v} = \bar{R}T \tag{3.1}$$

where \bar{R} is the universal gas constant, the value of which is

$$\bar{R} = 8.3144 \frac{\text{kN m}}{\text{kmol K}} = 1545 \frac{\text{ft lbf}}{\text{lb mol R}}$$

Dividing Eq. 3.1 by M, the molecular weight, we have the equation of state on a unit mass basis.

$$\frac{P\bar{v}}{M} = \frac{\bar{R}T}{M}$$

or

$$Pv = RT \tag{3.2}$$

where

$$R = \frac{\bar{R}}{M} \tag{3.3}$$

R is a constant for a particular gas. The value of R for a number of substances is given in Table A.8 of the Appendix. It follows from Eqs. 3.1 and 3.2 that this equation of state can be written in terms of the total volume.

$$PV = n\bar{R}T$$
$$PV = mRT \tag{3.4}$$

It should also be noted that Eq. 3.4 can alternately be written in the form

$$\frac{P_1 V_1}{T_1} = \frac{P_2 V_2}{T_2} \tag{3.5}$$

That is, gases at low density closely follow the well-known Boyle's and Charles' laws. Boyle and Charles, of course, based their statements on experimental observations. (Strictly speaking, neither of these statements should be called a law, since they are only approximately true and even then only under conditions of low density.)

The equation of state given by Eq. 3.1 (or Eq. 3.2) is referred to as the ideal gas equation of state. At very low density all gases and vapors approach ideal gas behavior, with the P-v-T relationship being given by the ideal gas equation of

state. At higher densities the behavior may deviate substantially from the ideal gas equation of state.

Because of its simplicity, the ideal gas equation of state is very convenient to use in thermodynamic calculations. However, two questions can appropriately be raised. The ideal gas equation of state is a good approximation at low density. But what constitutes low density? Or, expressed in other words, over what range of density will the ideal gas equation of state hold with accuracy? The second question is, how much does an actual gas at a given pressure and temperature deviate from ideal gas behavior?

To answer both questions we introduce the concept of the compressibility factor, Z, which is defined by the relation

$$Z = \frac{P\bar{v}}{\bar{R}T}$$

or

$$P\bar{v} = Z\bar{R}T \tag{3.6}$$

Note that for an ideal gas $Z = 1$, and the deviation of Z from unity is a measure of the deviation of the actual relation from the ideal gas equation of state.

Figure 3.7 shows a skeleton compressibility chart for nitrogen. From this chart we make three observations. The first is that at all temperatures $Z \to 1$ as $P \to 0$. That is, as the pressure approaches zero, the P-v-T behavior closely approaches

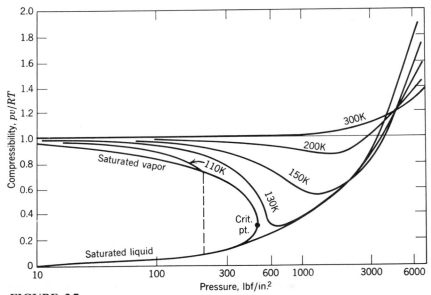

FIGURE 3.7

Compressibility of nitrogen.

Problem 9.57

$$\frac{P_2}{P_1} = 8 = \frac{P_4}{P_3} = \frac{P_6}{P_7} = \frac{P_8}{P_9}$$

$P_1 = 14.7 \, psi \, , \quad T_1 = T_3 = 70°F$

$T_6 = T_8 = 2000 \, F$

$\eta_{regenerator} = 100\% \Rightarrow T_5 = T_4 \text{ and } T_{10} = T_4$

assume constant specific heats

consider compressor, processes ① → ② and ③ → ④

$\frac{T_2}{T_1} = \left(\frac{P_2}{P_1}\right)^{\frac{k-1}{k}}$ and $\frac{T_4}{T_3} = \left(\frac{P_4}{P_3}\right)^{\frac{k-1}{k}}$

since $T_1 = T_3 \Rightarrow T_2 = T_4 = (70 + 460)(8)^{\frac{1.40 - 1.0}{1.40}} = 960.6 \, R$

$W_{compressor})_{total} = W_{comp})_1 + W_{comp})_2 = 2(W_{comp}) = 2\, C_p(T_1 - T_2)$

$= (2)(0.240 \frac{BTU}{\ell_{bm} R})(530 - 960.6)R = -206.7 \frac{BTU}{\ell_{bm}}$

consider turbine, processes ⑥ → ⑦ and ⑧ → ⑨

$\frac{T_7}{T_6} = \left(\frac{P_7}{P_6}\right)^{\frac{k-1}{k}}$ and $\frac{T_9}{T_8} = \left(\frac{P_9}{P_8}\right)^{\frac{k-1}{k}}$

since $T_6 = T_8 \Rightarrow T_7 = T_9 = (2000 + 460)\left(\frac{1}{8}\right)^{\frac{1.4 - 1.0}{1.4}} = 1357.2 \, R$

$W_{turbine})_{total} = W_{turbine})_1 + W_{turbine})_2 = 2 W_{turbine} = (2) \, C_p(T_6 - T_7)$

$= (2)(0.240 \frac{BTU}{\ell_{bm}})(2460 - 1357.2)R = 529.3 \frac{BTU}{\ell_{bm}}$

$W_{net} = (529.3 - 206.7) = 322.6 \frac{BTU}{\ell_{bm}}$

$\eta = \frac{W_{net}}{q_h}$ and $q_h = (h_6 - h_5) + (h_8 - h_7) = C_p(T_6 - T_5) + C_p(T_8 - T_7$

since $T_5 = T_4 = T_7 \Rightarrow q_h = 2 C_p(T_6 - T_5) = 2(0.24)(2460 - 1357.2)$

$T_8 = T_6 \qquad = 529.3 \, BTU/\ell_{bm}$

$\eta = \frac{322.6}{529.3} = 60.9\%$

that predicted by the ideal gas equation of state. Note also that at temperatures of 300 K and above (that is, room temperature and above) the compressibility factor is near unity up to pressures above 1000 lbf/in.2. This means that the ideal gas equation of state can be used for nitrogen (and, as it happens, also air) over this range with considerable accuracy.

Now suppose we reduce the temperature from 300 K while keeping the pressure constant at 600 lbf/in.2. The density will increase and we note a sharp decrease below unity in the value of the compressibility factor. Values of $Z < 1$ mean that the actual density is greater than would be predicted by ideal gas behavior. The physical explanation of this is as follows: As the temperature is reduced from 300 K while pressure remains constant at 600 lbf/in.2, the molecules are brought closer together. In this range of intermolecular distances, and at this pressure and temperature, there is an attractive force between the molecules. The lower the temperature the greater is this intermolecular attractive force. This attractive force between the molecules means that the density is greater than would be predicted by the ideal gas behavior, which assumes no intermolecular forces. Note also from the compressibility chart that at very high densities, for pressures above 4000 lbf/in.2, the compressibility factor is always greater than unity. In this range the intermolecular distances are very small, and there is a repulsive force between the molecules. This tends to make the density less than would otherwise be expected.

The precise nature of intermolecular forces is a rather complex matter. These forces are a function of the temperature as well as the density. The preceding discussion should be considered as a qualitative analysis to assist in gaining some understanding of the ideal gas equation of state and how the P-v-T behavior of actual gases deviates from this equation.

For other gases, the behavior of Z with respect to temperature and pressure is very similar to that of nitrogen, at least in a qualitative sense. To quantify this relation, we divide the temperature by the critical temperature of the substance, calling the result the reduced temperature, T_r. In the same manner, the pressure divided by the critical pressure is termed reduced pressure, P_r. Now, a plot of Z vs. P_r at various T_r is found to be in reasonably close agreement quantitatively for many different gases. This plot is called a generalized compressibility chart, and is included in the Appendix, Fig. A.5. Critical constants for various gases are listed in Table A.6, as mentioned previously.

The generalized compressibility chart will be discussed and used in greater detail in Chapter 10. For the present, the chart is useful to us as an aid in deciding whether, in a given circumstance, it is reasonable to assume ideal gas behavior as a model. For example, we note from the chart that if the pressure is very low (that is, $\ll P_c$), the ideal gas model can be assumed with good accuracy, regardless of the temperature. Furthermore, at temperatures that are high (that is, greater than about twice T_c), the ideal gas model can be assumed with good accuracy to pressures as high as four or five times P_c. When the temperature is less than about twice the critical temperature and the pressure is not extremely low, then we are in a region, commonly termed superheated vapor, in which the deviation from ideal gas behavior may be considerable. In this region, it is preferable to use

tables of thermodynamic properties or charts for a particular substance. These tables are considered in the following section.

Instead of the ideal gas model to represent gas behavior, or even the generalized compressibility chart, which is approximate, it is desirable to have an equation of state that accurately represents the P-v-T behavior for a particular gas over the entire superheated vapor region. Such an equation is necessarily more complicated and consequently more difficult to use, and many such equations have been proposed and used to correlate the observed behavior of gases. To illustrate the nature and complexity of these equations, we refer to one of the best known—the Beattie–Bridgeman equation of state. This equation is

$$P = \frac{\bar{R}T(1 - \epsilon)}{\bar{v}^2}(\bar{v} + B) - \frac{A}{\bar{v}^2} \tag{3.7}$$

where $A = A_0(1 - a/\bar{v})$, $B = B_0(1 - b/\bar{v})$, $\epsilon = c/\bar{v}T^3$, and A_0, a, B_0, b, and c are constants for different gases. The values of these constants for various substances are given in Table 3.3.

The matter of equations of state will be discussed further in Chapter 10. The observation to be made here in particular is that an equation of state that accurately describes the relation between pressure, temperature, and specific volume is rather cumbersome and the solution requires considerable time. When using a digital computer, it is often most convenient to determine the thermodynamic properties in a given state from such equations. However, in hand calculations, it is much more convenient to tabulate values of pressure, temperature, specific volume, and other thermodynamic properties for various substances. The Appendix includes summary tables and graphs of the thermodynamic properties of water, ammonia, Freon-12, nitrogen, methane, and oxygen. The tables of the properties of water are usually referred to as the "steam tables" and are extracted from "Steam Tables" by Keenan, Keyes, Hill, and Moore. The method for compiling the P-v-T data for such a table is to find an equation of state that accurately fits the experimental data, and then to solve the equation of state for the values listed in the table.

TABLE 3.3

Constants of the Beattie–Bridgeman Equation of State (Pressure in kilopascals; specific volume in m³/kmol; temperature in Kelvin; $\bar{R} = 8.3144$ N m/mol K.)

Gas	A_0	a	B_0	b	$10^{-4}c$
Helium	2.1886	0.059 84	0.014 00	0.0	0.0040
Argon	130.7802	0.023 28	0.039 31	0.0	5.99
Hydrogen	20.0117	−0.005 06	0.020 96	−0.043 59	0.0504
Nitrogen	136.2315	0.026 17	0.050 46	−0.006 91	4.20
Oxygen	151.0857	0.025 62	0.046 24	0.004 208	4.80
Air	131.8441	0.019 31	0.046 11	−0.001 101	4.34
Carbon dioxide	507.2836	0.071 32	0.104 76	0.072 35	66.00

Example 3.1

What is the mass of air contained in a room 20 ft × 30 ft × 12 ft if the pressure is 14.7 lbf/in.2, and the temperature is 80 F? Assume air to be an ideal gas.

By using Eq. 3.4, and the value of R from Table A.8

$$m = \frac{PV}{RT} = \frac{(14.7 \times 144) \text{ lbf/ft}^2 \times 7200 \text{ ft}^3}{53.34 \text{ ft-lbf/lbm R} \times 540 \text{ R}} = 529 \text{ lbm}$$

Example 3.2

A tank has a volume of 15 ft^3 and contains 20 lbm of an ideal gas having a molecular weight of 24. The temperature is 80 F. What is the pressure?

The gas constant is first determined:

$$R = \frac{\bar{R}}{M} = \frac{1545 \text{ ft-lbf/lb mole R}}{24 \text{ lbm/lb mole}} = 64.4 \text{ ft-lbf/lbm R}$$

We now solve for P.

$$P = \frac{mRT}{V} = \frac{20 \text{ lbm} \times 64.4 \text{ ft-lbf/lbm R} \times 540R}{144 \text{ in.}^2/\text{ft}^2 \times 15 \text{ ft}^3} = 321 \text{ lbf/in.}^2$$

3.5 TABLES OF THERMODYNAMIC PROPERTIES

Tables of thermodynamic properties of many substances are available, and in general all these have the same form. In this section we will refer to the steam tables. The steam tables are selected both as a vehicle for presenting thermodynamic tables and because steam is used extensively in power plants and industrial processes. Once the steam tables are understood, other thermodynamic tables can be readily used.

Several different versions of steam tables have been published over the years. The Appendix of this book includes a summary of "Steam Tables" by Keenan, Keyes, Hill, and Moore, published in 1969. This reference is a revision of a very extensively used volume by Keenan and Keyes, originally published in 1936.

In Table A.1.1, the first column after the temperature gives the corresponding saturation pressure in pounds force per square inch. The next two columns give specific volume in cubic feet per pound mass. The first of these gives the specific volume of the saturated liquid, v_f; the second column gives the specific volume of saturated vapor, v_g. The difference between these two, $v_g - v_f$, represents the increase in specific volume when the state changes from saturated liquid to saturated vapor, and is designated v_{fg}.

The specific volume of a substance having a given quality can be found by utilizing the definition of quality. Quality has already been defined as the ratio of the mass of vapor to total mass of liquid plus vapor when a substance is in a saturation state. Let us consider a mass m having a quality x. The volume is the sum of the volume of the liquid and the volume of the vapor.

$$V = V_{\text{liq}} + V_{\text{vap}} \tag{3.8}$$

In terms of the masses, Eq. 3.8 can be written in the form

$$mv = m_{\text{liq}} v_f + m_{\text{vap}} v_g$$

Dividing by the total mass and introducing the quality x,

$$v = (1 - x)v_f + xv_g \tag{3.9}$$

Using the definition

$$v_{fg} = v_g - v_f$$

Eq. 3.9 can also be written in the form

$$v = v_f + xv_{fg} \tag{3.10}$$

As an example, let us calculate the specific volume of saturated steam at 500 F having a quality of 70 per cent. Using Eq. 3.9,

$$v = 0.3(0.0204) + 0.7(0.6761) = 0.4794 \text{ ft}^3/\text{lbm}$$

In Table A.1.2, the first column after the pressure lists the saturation temperature for each pressure. The next columns list specific volume in a manner similar to Table A.1.1. When necessary, v_{fg} can readily be found by subtracting v_f from v_g.

Table 3 of the steam tables, which is summarized in Table A.1.3 in the Appendix, gives the properties of superheated vapor. In the superheated region, pressure and temperature are independent properties and, therefore, for each pressure a large number of temperatures is given, and for each temperature four thermodynamic properties are listed, the first one being specific volume. Thus the specific volume of steam at a pressure of 100 lbf/in.² and 500 F is 5.587 ft³/lbm.

Table 4 of the steam tables, summarized in Table A.1.4 in the Appendix, gives the properties of the compressed liquid. To demonstrate the use of this table, consider a piston and a cylinder (as shown in Fig. 3.8) that contains 1 lbm of saturated-liquid water at 200 F. Its properties are given in Table A.1.1, and we note that the pressure is 11.53 lbf/in.² and the specific volume is 0.016 63 ft³/lbm. Suppose the pressure is increased to 1000 lbf/in.² while the temperature is held constant at 200 F by the necessary transfer of heat, Q. Since water is slightly compressible, we would expect a slight decrease in specific volume during this process. Table A.1.4 gives this specific volume as 0.016 58 ft³/lbm. Note that this is only a slight decrease, and only a small error would be made if one assumed that the volume of a compressed liquid was equal to the specific volume of the saturated

Heat transfer
(in an amount
that will maintain
constant temperature)

Liquid

FIGURE 3.8
Illustration of compressed liquid state.

liquid at the same temperature. In many cases this is the most convenient procedure, particularly in those cases when compressed-liquid data are not available.

Furthermore, since specific volume does change rapidly with temperature, care should be exercised in interpolation over the wide temperature ranges in Table A.1.4. (In some cases it may be more accurate to use the saturated-liquid data from Table A.1.1, and interpolate differences between Table A.1.1, the saturated-liquid data, and Table A.1.4, the compressed-liquid data.)

Table 6 of the steam tables, which is summarized in Table A.1.5 in the Appendix, gives the properties of saturated solid and saturated vapor that are in equilibrium. The first column gives the temperature and the second column gives the corresponding saturation pressure. As would be expected, all these pressures are less than the triple-point pressure. The next two columns give the specific volume of the saturated solid and saturated vapor (note that the tabulated value is $v_i \times 10^3$).

3.6 THERMODYNAMIC SURFACES

The matter discussed in this chapter can be well summarized by a consideration of a pressure-specific volume–temperature surface. Two such surfaces are shown in Figs. 3.9 and 3.10. Figure 3.9 shows a substance such as water in which the specific volume increases during freezing, and Fig. 3.10 shows a substance in which the specific volume decreases during freezing.

In these diagrams the pressure, specific volume, and temperature are plotted on mutually perpendicular coordinates, and each possible equilibrium state is thus represented by a point on the surface. This follows directly from the fact that a pure substance has only two independent intensive properties. All points along a quasiequilibrium process lie on the P-v-T surface, since such a process always passes through equilibrium states.

The regions of the surface that represent a single phase—the solid, liquid, and vapor phases—are indicated, these surfaces being curved. The two-phase regions—the solid–liquid, solid–vapor, and liquid–vapor regions—are ruled surfaces. By

FIGURE 3.9

Pressure–volume–temperature surface for a substance that expands on freezing.

this we understand that they are made up of straight lines parallel to the specific volume axis. This, of course, follows from the fact that in the two-phase region, lines of constant pressure are also lines of constant temperature, though the specific volume may change. The triple point actually appears as the triple line on the P-v-T surface, since the pressure and temperature of the triple point are fixed, but the specific volume may vary, depending on the proportion of each phase.

It is also of interest to note the pressure–temperature and pressure–volume projections of these surfaces. We have already considered the pressure–temperature diagram for a substance such as water. It is on this diagram that we observe the triple point. Various lines of constant temperature are shown on the pressure–

FIGURE 3.10

Pressure–volume–temperature surface for a substance that contracts on freezing.

volume diagram, and the corresponding constant-temperature sections are lettered identically on the P-v-T surface. The critical isotherm has a point of inflection at the critical point.

One notices that with a substance such as water, which expands on freezing, the freezing temperature decreases with an increase in pressure. With a substance that contracts on freezing, the freezing temperature increases as the pressure increases. Thus, as the pressure of vapor is increased along the constant-temperature line *abcdef* in Fig. 3.9, a substance that expands on freezing first becomes solid and then liquid. For the substance that contracts on freezing, the corresponding constant-temperature line, Fig. 3.10, indicates that as the pressure on the vapor is increased, it first becomes liquid and then solid.

Example 3.3

A vessel having a volume of 10 ft³ contains 3.0 lbm of a liquid water and water vapor mixture in equilibrium at a pressure of 100 lbf/in.². Calculate

1. The volume and mass of liquid.
2. The volume and mass of vapor.

The specific volume is called first:

$$v = \frac{10.0}{3.0} = 3.333 \text{ ft}^3/\text{lbm}$$

From the steam tables (Appendix Table A.1.2)

$$v_{fg} = 4.434 - 0.017\,74 = 4.416$$

The quality can now be calculated, using Eq. 3.10

$$3.333 = 0.177\,74 + x4.416$$

$$x = 0.751$$

Therefore the mass of liquid is

$$3(0.249) = 0.747 \text{ lbm}$$

The mass of vapor is

$$3(0.751) = 2.253 \text{ lbm}$$

The volume of liquid is

$$V_{\text{liq}} = m_{\text{liq}}v_f = 0.747(0.017\,74) = 0.0133 \text{ ft}^3$$

The volume of the vapor is

$$V_{\text{vap}} = m_{\text{vap}}v_g = 2.253(4.434) = 9.99 \text{ ft}^3$$

Example 3.4

A pressure vessel contains saturated Freon-12 vapor at 60 F. Heat is transferred to the system until the temperature reaches 200 F. What is the final pressure?

Since the volume does not change during this process, the specific volume also remains constant. From the Freon-12 Tables, Table A.3

$$v_1 = v_2 = 0.558 \text{ ft}^3/\text{lbm}$$

We know one other property in the final state, namely the temperature, $T_2 = 200$ F, and therefore the final state is determined.

Since v_g at 200 F is less than 0.558 ft²/lbm, it is evident that in the final state the Freon-12 is superheated vapor. By interpolating between the 90 lbf/in.² and 100 lbf/in.² columns of Table A.3, we find that

$$P_2 = 97.8 \text{ lbf/in.}^2$$

PROBLEMS

3.1 Consider a spherical balloon with a diameter of 15 ft. The local ambient conditions are 14.7 lbf/in.2 pressure and 70 F temperature.
- (a) Calculate the mass in lbm and the number of lb mol of ambient air displaced by this balloon.
- (b) The balloon is filled with helium at 14.7 lbf/in.2 and 70 F. Calculate the mass in lbm and the number of lb mol of helium inside the balloon.

3.2 A vertical cylinder fitted with a frictionless piston contains argon gas at 100°C. The piston has a mass of 5 kg and a diameter of 100 mm, and the ambient pressure outside the cylinder is 97 kPa. If the cylinder volume is 2 L, what is the mass of argon inside?

3.3 Air is contained in a vertical cylinder fitted with a frictionless piston and a set of stops, as shown in Fig. 3.11. The piston cross-sectional area is 0.5 ft^2, and the air inside is initially at 30 lbf/in.2, 800 F. The air is then cooled as a result of heat transfer to the surroundings.
- (a) What is the temperature of the air inside the cylinder when the piston reaches the stops?
- (b) The cooling is now continued until the temperature reaches 70 F. What is the pressure at this state?

FIGURE 3.11
Sketch for Problem 3.3.

3.4 A hollow metal sphere of 6 in. inside diameter is weighed on a precision beam balance when evacuated and then again when filled to 110 lbf/in.2 with an unknown gas. The difference in reading is 0.005 lbm. The room temperature is 80 F. What is the gas, assuming it to be a pure substance?

3.5 A vacuum pump is used to pump a vacuum over a bath of liquid helium. The volume rate of flow into the vacuum pump is 4500 cubic feet per minute. The pressure at the vacuum pump inlet is 0.5 mm Hg and the temperature is −30 F. What is the mass flow rate of helium entering the pump?

3.6 A balloon initially deflated and flat is connected by a valve to a tank containing helium gas at 150 lbf/in.2 at ambient temperature, 65 F. The valve is now opened, and the balloon is inflated at constant pressure (equal to

ambient pressure, 14.7 lbf/in.²), until it becomes spherical at $D_1 = 6$ ft. Beyond this size, the elasticity of the balloon material is such that the pressure inside is

$$P = P_0 + C(D - D_1)^2$$

This balloon is to be inflated slowly to a final diameter of 12 ft, at which point the pressure inside is 60 lbf/in.². All temperatures remain at 65 F during the process. Determine the minimum volume of the helium tank to inflate this balloon.

3.7 A rigid vessel A is connected to a spherical elastic balloon B as shown in Fig. 3.12. Both contain air at the ambient temperature 75 F. The volume of vessel A is 3.5 ft³ and the initial pressure is 45 lbf/in.². The initial diameter of the balloon is 1.5 ft and the pressure inside is 15 lbf/in.². The valve connecting A and B is now opened, and remains open. It may be assumed that the pressure inside the balloon is directly proportional to its diameter, and also that the final temperature of the air is uniform throughout at 75 F. Determine the final pressure in the system and the final volume of the balloon.

FIGURE 3.12
Sketch for Problem 3.7.

3.8 Is it reasonable to assume that at the given states the substance behaves as an ideal gas:
(a) Nitrogen at 90 F, 500 lbf/in.².
(b) Carbon dioxide at 90 F, 500 lbf/in.².
(c) Water at 1800 F, 500 lbf/in.².
(d) Water at 120 F, 1 lbf/in.².
(e) Water at 90 F, 1 lbf/in.².

3.9 A 200-L tank contains carbon dioxide gas at 35°C, 2 MPa.
(a) Determine the mass inside the tank.
(b) Estimate the per cent error in (a) if the ideal gas model is assumed.

3.10 Repeat Problem 3.9 for argon gas instead of carbon dioxide.

3.11 Determine whether water at each of the following states is a compressed liquid, a superheated vapor, or a mixture of saturated liquid and vapor: 250 F, 20 lbf/in.²; 50 lbf/in.², 8 ft³/lbm; 300 F, 8 ft³/lbm; 30 lbf/in.², 230 F; 400 F, 0.02 ft³/lbm; 0.5 lbf/in.², 60 F.

3.12 Plot the following vapor-pressure curves (liquid–vapor saturation pressure versus saturation temperature):
(a) Water on Cartesian coordinates, -40 F to 60 F.
(b) Water on Cartesian coordinates, 0 to 3500 lbf/in.2.
(c) Water, Freon-12, and ammonia, on semi-log paper (pressure on log scale), 1.0 to 1000 lbf/in.2, -50 F to 300 F.

3.13 Determine the quality (if saturated) or temperature (if superheated) of the following substances in the given states:
(a) Ammonia, 75 F, 1.25 ft^3/lbm; 80 lbf/in.2, 5.2 ft^3/lbm.
(b) Freon-12, 50 lbf/in.2, 0.6 ft^3/lbm; 50 lbf/in.2, 1.0 ft^3/lbm.
(c) Water, 20°C, 1 m^3/kg; 1000 lbf/in.2, 0.4 ft^3/lbm.
(d) Nitrogen, 80 K, 0.14 m^3/kg; 0.5 MPa, 0.08 m^3/kg.
(e) Methane, 0.5 MPa, 1.0 m^3/kmol, 3 MPa, 1.0 m^3/kmol

3.14 Calculate the following specific volumes:
(a) Ammonia, 80 F, 15% quality.
(b) Freon-12, 50°C, 80% quality.
(c) Water, 1000 lbf/in.2, 92% quality.
(d) Nitrogen, 90 K, 50% quality.

3.15 Plot a pressure-specific volume diagram on log-log paper (2 × 3 cycles) for Freon-12, showing the following lines:
(a) Saturated liquid.
(b) Saturated vapor.
(c) The following constant-temperature lines: 0 F, 100 F, 230 F, 300 F.
(d) The following constant-quality lines: 10%, 50%, 90%.

3.16 Plot a pressure-specific volume diagram on log-log paper (3 × 5 cycles) for water, showing the following lines:
(a) Saturated liquid.
(b) Saturated vapor.
(c) The following constant-temperature lines, including the compressed liquid region: 300 F, 500 F, 700 F, 800 F, 1000 F.
(d) The following constant-quality lines: 10%; 50%; 90%.

3.17 It is desired to calculate the mass of nitrogen gas in a 250 L tank at 6 MPa, 150 K, and four different methods are suggested. Which is the most accurate, and what per cent differences are the other three?
(a) Beattie–Bridgeman equation of state.
(b) Nitrogen tables, Table A.4.
(c) Ideal gas.
(d) Generalized compressibility chart, Fig. A.5.

3.18 A water storage tank contains liquid and vapor in equilibrium at 500 F. The distance from the bottom of the tank to the liquid level is 20 ft. What is the difference in pressure reading between the top of the tank and the bottom of the tank?

3.19 What is the percent error in specific volume if the ideal gas model is used to represent the behavior of superheated vapor ammonia at 200 F, 280 lbf/in.2? What if the generalized compressibility chart, Fig. A.5, is used instead?

3.20 A spacecraft storage vessel of 500-L capacity contains 100 kg of saturated (liquid + vapor) nitrogen at 90 K. Determine the percentages of liquid and vapor in the vessel on a mass basis and also on a volume basis.

3.21 A 200 m^3 storage tank is being constructed to hold LNG, liquified natural gas, which may be assumed to be essentially pure methane. If the tank is to contain 95% liquid and 5% vapor, by volume, at 100 kPa, what mass in kg of LNG will the tank hold? What is the quality in the tank?

3.22 A rigid 2-ft^3 tank contains equal volumes of liquid and vapor Freon-12 at 100 F. Additional Freon-12 is charged into the tank until the final mass is 100 lbm. If the temperature remains at 100 F, what is the final volume of liquid? How much mass enters the tank during this process?

3.23 Consider compressed liquid water at 100 F. What pressure is required to decrease the specific volume by 1% from its saturated liquid value at the same temperature?

3.24 A boiler feed pump delivers 100 lbm/s of water at 300 F, 3000 lbf/in.2. What is the volume rate of flow in ft^3/min? What would be the percent error if the properties of saturated liquid water at 300 F were used in the calculation? What if the properties of saturated liquid water at 3000 lbf/in.2 were used?

3.25 Consider 1 lbm of water at its triple point. The volume of the liquid phase is equal to that of the solid phase, and the volume of the vapor phase is 10^4 times that of the liquid phase. Determine the mass of water in each phase at this state.

3.26 A 100-L rigid tank containing saturated vapor water at 200 kPa is cooled to $-20°$C. What percentage, on a volume basis, of the water is solid at this temperature?

3.27 For a certain experiment, Freon-12 vapor is contained in a sealed glass tube at 80 F. It is desired to know the pressure at this condition, but there is no means for measuring it, since the tube is sealed. However, if the tube is cooled to 20 F, small droplets of liquid are observed on the glass walls. What is the pressure inside at 80 F?

3.28 A rigid vessel contains saturated (liquid + vapor) water at 14.7 lbf/in.2. Find the percentage of liquid, on a volume basis, at this state such that the water will pass through the critical point when this system is heated.

3.29 A rigid tank fitted with a sight glass, as shown in Fig. 3.13, contains Freon-12 at 70 F. Liquid is slowly withdrawn from the bottom, during which the temperature inside remains constant. If the cross-sectional area of the tank is 100 in.2 and the liquid level drops 8 in., determine the mass of Freon-12 withdrawn during the process.

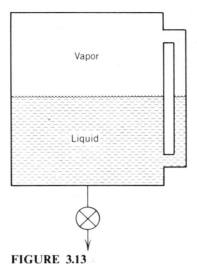

FIGURE 3.13

Sketch for Problem 3.29.

3.30 A rigid vessel having a volume of 0.25 ft³ contains 10 lbm of water (liquid + vapor) at 100 F. The vessel is now slowly heated. Will the liquid level inside eventually rise to the top or drop to the bottom of the vessel? What if the vessel contains 1 lbm instead of 10 lbm?

3.31 A container of liquid methane at 500 kPa pressure has a cross-sectional area of 0.50 m² (Fig. 3.14). As a result of heat transfer to the liquid methane, some of the liquid evaporates, and in one hour the liquid level in the tank drops 30 mm. The vapor that leaves the container passes through a heater and then exits at 500 kPa, 275 K. Calculate the volume rate of flow of methane gas exiting the heater.

FIGURE 3.14

Sketch for Problem 3.31.

3.32 Tank *A*, shown in Fig. 3.15, has a volume of 0.1 ft³ and contains Freon-12 at 80 F, 10% liquid and 90% vapor by volume, while tank B is evacuated. The valve connecting the tanks is now opened, and the tanks eventually come to a uniform state, 80 F and 30 lbf/in.². What is the volume of tank B?

FIGURE 3.15
Sketch for Problems 3.32 and 3.33.

3.33 Consider two tanks, A and B, connected by a valve, as shown in Fig. 3.15. Each has a volume of 200 L. Tank A contains Freon-12 at 25°C, 10% liquid and 90% vapor by volume, while tank B is evacuated. The valve connecting the tanks is now opened, and saturated vapor flows out of A until the pressure in B reaches that in A, at which point the valve is closed. This process occurs slowly, such that all temperatures remain at 25°C throughout the process. How much has the quality changed in tank A as a result of this process?

3.34 A cylinder fitted with a frictionless piston contains water, as shown in Fig. 3.16. The mass of water is 1 lbm and the piston cross-sectional area is 2 ft². At the initial state the water is at 230 F with a quality of 90%, and the spring just touches the piston, but exerts no force on it. Now, heat is transferred to the water, and the piston begins to rise. During this process, the resisting force of the spring is proportional to the distance moved, with a force of 50 lbf/in. Calculate the pressure inside the cylinder when the temperature of the water reaches 320 F.

FIGURE 3.16
Sketch for Problem 3.34.

3.35 Write a computer program to solve the following problem. For any specified substance, it is desired to calculate the pressure according to the Beattie–Bridgeman equation of state for any given set (or sets) of temperature and specific volume and to compare the result with the ideal gas equation of state.

3.36 Write a computer program to solve the following problem. For any speci-
fied substance, it is desired to solve the Beattie-Bridgeman equation of
state for specific volume at any given set (or sets) or pressure and temper-
ature, and to compare the result with the ideal gas equation of state.

3.37 Write a computer program to solve the following problem. For any speci-
fied substance that is one of those in Tables A.1–A.5, it is desired to fit a
polynomial equation of degree n to tabular data for pressure as a function
of density along any given isotherm in the superheated vapor region.

4

WORK AND HEAT

In this chapter we will consider work and heat. It is essential for the student of thermodynamics to understand clearly the definitions of both work and heat, because the correct analysis of many thermodynamic problems depends on distinguishing between them.

4.1 DEFINITION OF WORK

Work is usually defined as a force F acting through a displacement x, the displacement being in the direction of the force. That is,

$$W = \int_1^2 F \cdot dx \tag{4.1}$$

This is a very useful relationship because it enables us to find the work required to raise a weight, to stretch a wire, or to move a charged particle through a magnetic field.

However, when treating thermodynamics from a macroscopic point of view, it is advantageous to tie in the definition of work with the concepts of systems, properties, and processes. We therefore define work as follows: work is done by a system if the sole effect on the surroundings (everything external to the system) could be the raising of a weight. Notice that the raising of a weight is in effect a force acting through a distance. Notice, also, that our definition does not state that a weight was actually raised, or that a force actually acted through a given distance but that the sole effect external to the system could be the raising of a weight. Work done *by* a system is considered positive and work done *on* a system is considered negative. The symbol W designates the work done by a system.

In general, we will speak of work as a form of energy. No attempt will be made to give a rigorous definition of energy. Rather, since the concept is familiar, the term energy will be used as appropriate, and various forms of energy will be identified. Work is the form of energy that fulfills the definition given above.

Let us illustrate this definition of work with a few examples. Consider as a system the battery and motor of Fig. 4.1a and let the motor drive a fan. Does work cross the boundary of the system? To answer this question using the definition of work given above, let the fan be replaced with a pulley and weight arrangement shown in Fig. 4.1b. As the motor turns, the weight is raised, and the

a

FIGURE 4.1

Example of work crossing the boundary of a system.

sole effect external to the system is the raising of a weight. Thus, for our original system of Fig. 4.1*a*, we conclude that work is crossing the boundary of the system since the sole effect external to the system could be the raising of a weight.

Let the boundaries of the system be changed now to include only the battery shown in Fig. 4.2. Again we ask the question, does work cross the boundary of the system? In answering this question, we will be answering a more general question: Does the flow of electrical energy across the boundary of a system constitute work?

The only limiting factor in having the sole external effect be the raising of a weight is the inefficiency of the motor. However, as we design a more efficient motor, with lower bearing and electrical losses, we recognize that we can approach a certain limit, which does meet the requirement of having the only external effect the raising of a weight. Therefore, we can conclude that when there is a flow of electricity across the boundary of a system, as in Fig. 4.2, it is work with which we are concerned.

FIGURE 4.2

Example of work crossing the boundary of a system because of a flow of an electric current across the system boundary.

4.2 UNITS FOR WORK

As already noted, we consider work done *by* a system, such as that done by a gas expanding against a piston, as positive, and work done *on* a system, such as that done by a piston compressing a gas, as negative. Thus positive work means that energy leaves the system and negative work means that energy is added to the system.

Our definition of work involves the raising of a weight. The unit of work should therefore be defined in terms of a unit force acting through a unit distance. In English units, this corresponds to a one-pound force acting through a distance of one foot, such that the unit for work is termed a foot pound force (ft lbf). In SI units, the unit corresponds to a one-newton force acting through a distance of one meter, which is termed a joule (J). That is,

$$1\,J = 1\,N\,m$$

Power is the time rate of doing work, and is designated by the symbol \dot{W}.

$$\dot{W} \equiv \frac{\delta W}{dt}$$

In English units, one familiar unit of power is the horsepower (hp), which is defined as

$$1\ hp = 550\ ft\ lbf/s$$

In SI units, the unit for power is a rate of work of one joule per second, which is termed a watt (W),

$$1\ W = 1\ J/s$$

It is often convenient to speak of the work per unit mass of the system. This quantity is designated w and is defined

$$w \equiv \frac{W}{m}$$

4.3 WORK DONE AT THE MOVING BOUNDARY OF A SIMPLE COMPRESSIBLE SYSTEM IN A QUASIEQUILIBRIUM PROCESS

We have already noted that there are a variety of ways in which work can be done on or by a system. These include work done by a rotating shaft, electrical work, and the work done by the movement of the system boundary, such as the work done in moving the piston in a cylinder. In this section we will consider in some detail the work done at the moving boundary of a simple compressible system during a quasiequilibrium process.

Consider as a system the gas contained in a cylinder and piston, as in Fig. 4.3. Let one of the small weights be removed from the piston, causing the piston to move upward a distance dL. We can consider this quasiequilibrium process and calculate the amount of work W done by the system during this process. The

FIGURE 4.3

Example of work done at the moving boundary of a system in a quasiequilibrium process.

total force on the piston is $P\mathscr{A}$, where P is the pressure of the gas and \mathscr{A} is the area of the piston. Therefore, the work δW is

$$\delta W = P\mathscr{A}\, dL$$

But $\mathscr{A}\, dL = dV$, the change in volume of the gas. Therefore,

$$\delta W = P\, dV \qquad\qquad \textbf{(4.2)}$$

The work done at the moving boundary during a given quasiequilibrium process can be found by integrating Eq. 4.2. However, this integration can be performed only if we know the relationship between P and V during this process. This relationship might be expressed in the form of an equation, or it might be shown in the form of a graph.

Let us consider a graphical solution first, using as an example a compression process such as that which occurs during the compression of air in a cylinder, Fig. 4.4. At the beginning of the process the piston is at position 1, the pressure being relatively low. This state is represented on a pressure–volume diagram (usually referred to as a P-V diagram) as shown. At the conclusion of the process the piston is in position 2, and the corresponding state of the gas is shown at

FIGURE 4.4

Use of pressure–volume diagram to show work done at the moving boundary of a system in a quasiequilibrium process.

point 2 on the *P-V* diagram. Let us assume that this compression was a quasi-equilibrium process, and that during the process the system passed through the states shown by the line connecting states 1 and 2 on the *P-V* diagram. The assumption of a quasiequilibrium process is essential here because each point on line 1–2 represents a definite state, and these states will correspond to the actual state of the system only if the deviation from equilibrium is infinitesimal. The work done on the air during this compression process can be found by integrating Eq. 4.2.

$$_1W_2 = \int_1^2 \delta W = \int_1^2 P\,dV \tag{4.3}$$

The symbol $_1W_2$ is to be interpreted as the work done during the process from state 1 to state 2. It is clear from examining the *P-V* diagram that the work done during this process,

$$\int_1^2 P\,dV$$

is represented by the area under the curve 1–2, area *a*–1–2–*b*–*a*. In this example the volume decreased, and the area *a*–1–2–*b*–*a* represents work done on the system. If the process had proceeded from state 2 to state 1 along the same path, the same area would respresent work done by the system.

Further consideration of a *P-V* diagram, Fig. 4.5, leads to another important conclusion. It is possible to go from state 1 to state 2 along many different quasi-equilibrium paths, such as *A*, *B*, or *C*. Since the area underneath each curve represents the work for each process, it is evident that the amount of work involved in each case is a function not only of the end states of the process, but in addition is dependent on the path that is followed in going from one state to another. For this reason work is called a path function or, in mathematical parlance, δW is an inexact differential.

This leads to a brief consideration of point and path functions or, to use another term, exact and inexact differentials. Thermodynamic properties are point functions, a name that arises from the fact that for a given point on a diagram (such as Fig. 4.5) or surface (such as Fig. 3.9.), the state is fixed, and thus there

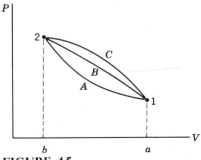

FIGURE 4.5

Various quasiequilibrium processes between two given states, indicating that work is a path function.

is a definite value of each property corresponding to this point. The differentials of point functions are exact differentials, and the integration is simply

$$\int_1^2 dV = V_2 - V_1$$

Thus we can speak of the volume in state 2 and the volume in state 1, and the change in volume depends only on the initial and final states.

Work, on the other hand, is a path function for, as has been indicated, the work done in a quasiequilibrium process between two given states depends on the path followed. The differentials of path functions are inexact differentials, and the symbol δ will be used in this text to designate inexact differentials (in contrast to d for exact differentials). Thus for work we would write

$$\int_1^2 \delta W = {}_1W_2$$

It would be more precise to use the notation ${}_1W_{2,A}$, which would indicate the work done during the change from state 1 to 2 along path A. However, implied in the notation ${}_1W_2$ is that the process between states 1 and 2 has been specified. It should be noted that we never speak about the work in the system in state 1 or state 2, and thus we would never write $W_2 - W_1$.

In evaluating the integral of Eq. 4.3, we should always keep in mind that we are concerned with determining the area under the curve on Fig. 4.5. In connection with this point, we identify the following two classes of problems:

1. The relationship between P and V is given in terms of experimental data or in graphical form (as, for example, the trace on an oscilloscope). In this case, we may evaluate the integral, Eq. 4.3, by graphical or numerical integration.

2. The relationship between P and V is such that it is possible to fit an analytical relationship between them, and we may then integrate directly.

One common example of this second type of functional relationship is a process called a polytropic process, one in which

$$PV^n = \text{constant}$$

throughout the process. The exponent n may possibly be any value from $-\infty$ to $+\infty$ depending on the particular process. For this type of process, we can integrate Eq. 4.3, as follows:

$$PV^n = \text{constant} = P_1V_1{}^n = P_2V_2{}^n$$

$$P = \frac{\text{constant}}{V^n} = \frac{P_1V_1{}^n}{V^n} = \frac{P_2V_2{}^n}{V^n}$$

$$\int_1^2 P\,dV = \text{constant} \int_1^2 \frac{dV}{V^n} = \text{constant} \left(\frac{V^{-n+1}}{-n+1} \right)\Bigg|_1^2$$

$$= \frac{\text{constant}}{1-n} (V_2^{1-n} - V_1^{1-n}) = \frac{P_2V_2{}^n V_2^{1-n} - P_1V_1{}^n V_1^{1-n}}{1-n}$$

$$\int_1^2 P\,dV = \frac{P_2V_2 - P_1V_1}{1-n} \tag{4.4}$$

Note that this result, Eq. 4.4, is valid for any exponent n except $n = 1$. In the case where $n = 1$, then

$$PV = \text{constant} = P_1 V_1 = P_2 V_2$$

and

$$\int_1^2 P\,dV = P_1 V_1 \int_1^2 \frac{dV}{V} = P_1 V_1 \ln \frac{V_2}{V_1} \tag{4.5}$$

Note that in Eqs. 4.4 and 4.5 we did not say that the work is equal to the expressions given in those equations. Those expressions give us the value of a certain integral, that is, a mathematical result. Whether or not that integral equals the work in a particular process depends on the result of a thermodynamic analysis of that process. It is important to keep the mathematical result separate from the thermodynamic analysis, as there are many situations in which work is not given by Eq. 4.3.

The polytropic process as described above demonstrates one special functional relationship between P and V during a process. There are many other possible relations, some of which will be examined in problems at the end of this chapter.

Example 4.1

Consider as a system the gas contained in the cylinder shown in Fig. 4.6, which is fitted with a piston on which a number of small weights are placed. The initial pressure is 20 lbf/in.² and the initial volume of the gas is 1 ft³.

(a) Let a Bunsen burner be placed under the cylinder, and let the volume of the gas increase to 3 ft³ while the pressure remains constant. Calculate the work done by the system during this process.

$$_1W_2 = \int_1^2 P\,dV$$

Since the pressure is constant, we conclude from Eq. 4.3 that

$$_1W_2 = P \int_1^2 dV = P(V_2 - V_1)$$

$$_1W_2 = 20 \text{ lbf/in.}^2 \times 144 \text{ in.}^2/\text{ft}^2 \times (3 - 1)\text{ft}^3 = 5760 \text{ ft lbf}$$

Gas

FIGURE 4.6

Sketch for Example 4.1.

(b) Consider the same system and initial conditions and the same final volume but at the same time that the Bunsen burner is under the cylinder and the piston is rising, let weights be removed from the piston at such a rate that, during the process, the temperature of the gas remains constant.

If we assume that the ideal gas model is valid, then, from Eq. 3.4,

$$PV = mRT$$

and we note that this is a polytropic process with exponent $n = 1$. From our analysis, we conclude that the work is given by Eq. 4.3, and we note that the integral in that equation is given by Eq. 4.5. Therefore,

$$_1W_2 = \int_1^2 P\,dV = P_1 V_1 \ln \frac{V_2}{V_1}$$

$$= 20\ \text{lbf/in.}^2 \times 144\ \text{in.}^2/\text{ft}^2 \times 1\ \text{ft}^3 \times \ln 3 = 3164\ \text{ft lbf}$$

(c) Consider the same system, but during the heat transfer let the weights be removed at such a rate that the expression $PV^{1.3} = $ constant describes the relation between pressure and volume during the process. Again the final volume is 3 ft^3 Calculate the work.

This is a polytropic process in which $n = 1.3$. Analyzing the process, we conclude again that the work is given by Eq. 4.3, the integral being given in this case by Eq. 4.4. Therefore,

$$P_2 = 20\left(\frac{1}{3}\right)^{1.3} = 4.8\ \text{lbf/in.}^2$$

$$_1W_2 = \int_1^2 P\,dV = \frac{P_2 V_2 - P_1 V_1}{1 - 1.3} = \frac{(4.80 \times 144 \times 3) - (20 \times 144 \times 1)}{1 - 1.3}$$

$$= 2688\ \text{ft lbf}$$

(d) Consider the system and initial state given in the first three examples, but let the piston be held by a pin so that the volume remains constant. In addition, let heat be transferred from the system until the pressure drops to 10 lbf/in.2 Calculate the work.

Since $\delta W = P\,dV$ for a quasiequilibrium process, the work is zero, because in this case there is no change in volume.

The process for each of the four examples is shown on the P-V diagram of Fig. 4.7. Process 1–2a is a constant-pressure process, and area 1–2a–f–e–1 represents the work. Similarly, line 1–2b represents the process in which $PV = $ constant, line 1–2c the process in which $PV^{1.3} = $ constant, and line 1–2d represents the constant-volume process. The student should compare the relative areas under each curve with the numerical results obtained above.

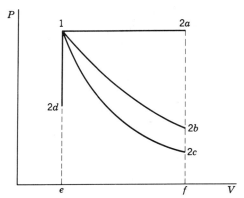

FIGURE 4.7
Pressure–volume diagram showing work done
in the various processes of Example 4.1.

In this section, we have discussed boundary movement work in a quasiequilibrium process. We should also realize that there may very well be boundary movement work in a nonequilibrium process. In such a case, the total force exerted on the piston by the gas inside the cylinder, $P\mathscr{A}$, does not equal the external force F_{ext}, and the work is not given by Eq. 4.2. The work can, however, be evaluated in terms of F_{ext} or, dividing by area, an equivalent external pressure, P_{ext}. The boundary movement work in this case is

$$\delta W = F_{ext} \, dL = P_{ext} \, dV \qquad (4.6)$$

Evaluation of Eq. 4.6 in any particular case requires a knowledge of how the external force or pressure changes during the process.

4.4 SOME OTHER SYSTEMS INVOLVING WORK AT A MOVING BOUNDARY

In the preceding section we considered the work done at the moving boundary of a simple compressible system during a quasiequilibrium process, and also during a nonequilibrium process. There are other types of systems that involve work at a moving boundary, and in this section we briefly consider two such systems, a stretched wire and a surface film.

Consider as a system a stretched wire that is under a given tension \mathscr{T}. When the length of the wire changes by the amount dL, the work done by the system is

$$\delta W = -\mathscr{T} \, dL \qquad (4.7)$$

The minus sign is necessary because work is done by the system when dL is negative. This can be integrated to give

$$_1W_2 = -\int_1^2 \mathscr{T} \, dL \qquad (4.8)$$

The integration can be performed either graphically or analytically if the relation between \mathcal{T} and L is known. The stretched wire is a simple example of the type of problem in solid body mechanics that involves the calculation of work.

Example 4.2

A metallic wire of initial length L_0 is stretched. Assuming elastic behavior, determine the work done in terms of the modulus of elasticity and the strain.

Let σ = stress, e = strain, and E = modulus of elasticity.

$$\sigma = \frac{\mathcal{T}}{\mathcal{A}} = Ee$$

Therefore,

$$\mathcal{T} = \mathcal{A}Ee$$

From the definition of strain,

$$de = \frac{dL}{L_0}$$

Therefore,

$$\delta W = -\mathcal{T}\,dL = -\mathcal{A}EeL_0\,de$$

$$W = -\mathcal{A}EL_0 \int_{e=0}^{e} e\,de = -\frac{\mathcal{A}EL_0}{2}(e)^2$$

Now consider a system that consists of a liquid film having a surface tension \mathcal{S}. A schematic arrangement of such a film is shown in Fig. 4.8, where a film is maintained on a wire frame, one side of which can be moved. When the area of the film is changed, for example, by sliding the movable wire along the frame, work is done on or by the film. When the area changes by an amount $d\mathcal{A}$, the work done by the system is

$$\delta W = -\mathcal{S}\,d\mathcal{A} \tag{4.9}$$

FIGURE 4.8

Schematic arrangement showing work done on a surface film.

For finite changes,

$$_1W_2 = -\int_1^2 \mathscr{S} \, d\mathscr{A} \tag{4.10}$$

4.5 SYSTEMS THAT INVOLVE OTHER MODES OF WORK

There are systems that involve other modes of work, and in this section we will consider two of these—systems involving magnetic and systems involving electrical modes of work. We will consider a quasiequilibrium process for these systems, and present expressions for the work done during such a process.

 To visualize how work can be accomplished by magnetic effects, let us briefly describe magnetic cooling, or adiabatic demagnetization, which is a process used to produce temperatures well below 1 K. A temperature of 1.0 K can be produced by pumping a vacuum over a bath of liquid helium (helium has the lowest normal boiling point of any substance, namely, 4.2 K at one atmosphere pressure). An apparatus in which the magnetic cooling is accomplished is shown schematically in Fig. 4.9. The paramagnetic salt is the magnetic substance in which temperatures well below 1 K are achieved. When the magnetic field is slowly increased, work

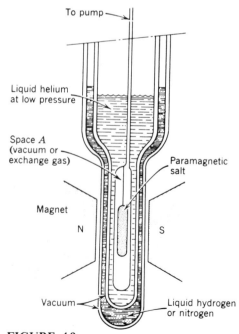

FIGURE 4.9

Schematic arrangement for magnetic cooling.

is done on the paramagnetic salt. From a microscopic point of view, this work is associated with the fact that in the presence of the magnetic field the ions in the salt tend to align themselves with their magnetic axes in the direction of the field. As a result of this work done on the salt, the temperature of the salt tends to increase. However, at this point in the experiment space A is filled with low-pressure helium gas, and heat is transferred from the paramagnetic salt to the liquid helium, which is maintained at about 1 K. When the magnetic field is at full strength, and the paramagnetic salt is at the temperature of the liquid helium, space A is evacuated, thus insulating the paramagnetic salt. The magnetic field is now reduced to zero, and in this process work is done by the paramagnetic salt, and its temperature drops sharply. This entire process may be compared to the compression of a gas that is initially at ambient pressure and temperature. As the result of the compression, the temperature of the gas tends to increase. However, the high-pressure gas can be cooled to the ambient temperature. If this gas is now isolated from the surroundings and allowed to expand and do work (against a piston for example) the temperature of the gas will decrease below the ambient temperature during the expansion process.

The basic parameters in this process are the intensity of the magnetic field and the magnetization. It may be shown that in a reversible quasiequilibrium process, the work done on a simple magnetic substance is

$$\delta W = -\mu_0 \mathscr{H} \, d(V\mathscr{M}) \tag{4.11}$$

where:

$$\mu_0 = \text{permeability of free space}$$
$$V = \text{volume}$$
$$\mathscr{H} = \text{intensity of the magnetic field}$$
$$\mathscr{M} = \text{magnetization}$$

The minus sign indicates that as the magnetization increases, work is done on the simple magnetic substance.

We have already noted that electrical energy flowing across the boundary of a system is work. However, we can gain further insight into such a process by considering a system in which the only work mode is electrical. As an example of such a system we can think of a charged condenser, an electrolytic cell, or the type of fuel cell described in Chapter 1. Consider a quasiequilibrium process for such a system, and during this process let the potential difference be \mathscr{E} and the amount of electrical charge that flows into the system be dZ. For this quasiequilibrium process the work is given by the relation

$$\delta W = -\mathscr{E} \, dZ \tag{4.12}$$

Since the current, i, equals dZ/dt (where $t = $ time) we can also write

$$\delta W = -\mathscr{E} i \, dt$$
$$_1W_2 = -\int_1^2 \mathscr{E} i \, dt \tag{4.13}$$

Equation 4.13 may also be written as a rate equation for work (the power).

$$\frac{\delta W}{dt} = -\mathscr{E}i \qquad (4.14)$$

Since the ampere (electric current) is one of the fundamental units in the International System, and the watt has been defined previously, this relation serves as the definition of the unit for electric potential, the volt (V), which is one watt divided by one ampere.

4.6 SOME CONCLUDING REMARKS REGARDING WORK

The similarity between the expressions for work in the two processes mentioned in Section 4.5 and the three processes involving a moving boundary should be noted. In each of these quasiequilibrium processes, the work is given by the integral of the product of an intensive property and the change of an extensive property. These are summarized below:

Simple compressible system $\qquad {}_1W_2 = \int_1^2 P\,dV$

Stretched wire $\qquad {}_1W_2 = -\int_1^2 \mathscr{T}\,dL$

Surface film $\qquad {}_1W_2 = -\int_1^2 \mathscr{S}\,d\mathscr{A} \qquad (4.15)$

System involving magnetic work only $\qquad {}_1W_2 = -\int_1^2 \mu_0\mathscr{H}\,d(V\mathscr{M})$

System involving electrical work only $\qquad {}_1W_2 = -\int_1^2 \mathscr{E}\,dZ$

Although we will deal primarily with systems involving one mode of work, it is quite possible to have more than one work mode involved in a given process. Thus we could write

$$\delta W = P\,dV - \mathscr{T}\,dL - \mathscr{S}\,d\mathscr{A} - \mu_0\mathscr{H}\,d(V\mathscr{M}) - \mathscr{E}\,dZ + \cdots \qquad (4.16)$$

where the dots represent other products of an intensive property and the derivative of a related extensive property. In each term, the intensive property can be viewed as the driving force that causes a change to occur in the related extensive property, which is often termed the displacement.

It should also be noted that there are many other forms of work that can be identified in processes that are not quasiequilibrium processes. An example of these is the work done by shearing forces in a process involving friction in a viscous fluid or the work done by a rotating shaft that crosses the system boundary.

The identification of work is an important aspect of many thermodynamic problems. We have already noted that work can be identified only at the boundaries of the system. For example, consider Fig. 4.10, which shows a gas separated from the vacuum by a membrane. Let the membrane rupture and the gas fill the

FIGURE 4.10

Example of a process involving a change of volume for which the work is zero.

entire volume. Neglecting any work associated with the rupturing of the membrane, we can ask if there is work involved in the process. If we take as our system the gas and the vacuum space, we readily conclude that there is no work involved, since no work can be identified at the system boundary. If we take the gas as a system we do have a change of volume, and we might be tempted to calculate the work from the integral

$$\int_1^2 P\, dV$$

However, this is not a quasiequilibrium process, and therefore the work cannot be calculated from this relation. Instead, since there is no resistance at the system boundary as the volume increases, we conclude that for this system there is no work involved in this process.

Another example can be cited with the aid of Fig. 4.11. In Fig. 4.11a the system consists of the container plus the gas. Work crosses the boundary of the system at the point where the system boundary intersects the shaft, and can be associated with the shearing forces in the rotating shaft. In Fig. 4.11b the system includes shaft and weight as well as the gas and the container. In this case there is no work crossing the system boundary as the weight moves downward. As we will see in the next chapter, we can identify a change of potential energy within the system, but this should not be confused with work crossing the system boundary.

4.7 DEFINITION OF HEAT

The thermodynamic definition of heat is somewhat different from the everyday understanding of the word. Therefore, it is essential to understand clearly the definition of heat given here, because it is involved in so many thermodynamic problems.

If a block of hot copper is placed in a beaker of cold water, we know from experience that the block of copper cools down and the water warms up until the copper and water reach the same temperature. What causes this decrease in the temperature of the copper and the increase in the temperature of the water? We say that it is the result of the transfer of energy from the copper block to the water. It is out of such a transfer of energy that we arrive at a definition of heat.

FIGURE 4.11
Example showing how selection of the system determines whether work is involved in a process.

Heat is defined as the form of energy that is transferred across the boundary of a system at a given temperature to another system (or the surroundings) at a lower temperature by virtue of the temperature difference between the two systems. That is, heat is transferred from the system at the higher to the system at the lower temperature, and the heat transfer occurs solely because of the temperature difference between the two systems. Another aspect of this definition of heat is that a body never contains heat. Rather, heat can be identified only as it crosses the boundary. Thus heat is a transient phenomenon. If we consider the hot block of copper as one system and the cold water in the beaker as another system, we recognize that originally neither system contains any heat (they do contain energy, of course). When the copper is placed in the water and the two are in thermal communication, heat is transferred from the copper to the water, until equilibrium of temperature is established. At that point we no longer have heat transfer, since there is no temperature difference. Neither of the systems contains heat at the conclusion of the process. It also follows that heat is identified at the boundary

of the system, for heat is defined as energy being transferred across the system boundary.

4.8 UNITS OF HEAT

As discussed above, heat, like work, is a form of energy transfer to or from a system. Therefore, the units for heat, and to be more general for any other form of energy as well, are the same as the units for work, or are at least directly proportional to them. In the International System the unit for heat (energy) is therefore the joule. Similarly, in the English System, the foot pound force is an appropriate unit for heat. However, another unit arose naturally over the years, the result of an association with the process of heating water, such as that used in connection with defining heat in the previous section. Consider as a system 1 lbm of water at 59.5 F, and let a block of hot copper of appropriate mass and temperature be placed in the water, such that when thermal equilibrium is established the temperature of the water is 60.5 F. This unit amount of heat transferred from the copper to the water in this process is called the British thermal unit (Btu). More specifically, it is called the 60-degree Btu, defined as the amount of heat required to raise 1 lbm of water from 59.5 F to 60.5 F. (The Btu as used today is actually defined in terms of the standard SI units.) It is worth noting here, that a unit of heat in metric units, the calorie, arose naturally in a manner analogous to the Btu in the English system, the calorie being defined as the amount of heat required to raise 1 gram of water from 14.5°C to 15.5°C.

Heat transferred *to* a system is considered to be positive, and heat transferred *from* a system is negative. Thus positive heat represents energy transferred to a system, and negative heat represents energy transferred from a system. The symbol Q is used to represent heat.

A process in which there is no heat transfer ($Q = 0$) is called an adiabatic process.

From a mathematical perspective, heat, like work, is a path function and is recognized as an inexact differential. That is, the amount of heat transferred when a system undergoes a change of state from state 1 to state 2 depends on the path that the system follows during the change of state. Since heat is an inexact differential, the differential is written δQ. On integrating, we write

$$\int_1^2 \delta Q = {}_1Q_2$$

In words, ${}_1Q_2$ is the heat transferred during the given process between state 1 and state 2.

The rate at which heat is transferred to a system is designated by the symbol \dot{Q}.

$$\dot{Q} \equiv \frac{\delta Q}{dt}$$

It is also convenient to speak of the heat transfer per unit mass of the system, q, which is defined as

$$q \equiv \frac{Q}{m}$$

4.9 COMPARISON OF HEAT AND WORK

At this point it is evident that there are many similarities between heat and work, and these are summarized here:

1. Heat and work are both transient phenomena. Systems never possess heat or work, but either or both cross the system boundary when a system undergoes a change of state.
2. Both heat and work are boundary phenomena. Both are observed only at the boundaries of the system, and both represent energy crossing the boundary of the system.
3. Both heat and work are path functions and inexact differentials.

It should also be noted that in our sign convention, $+Q$ represents heat transferred *to* the system, and thus is energy added to the system, and $+W$ represents work done *by* the system, and thus represents energy leaving the system.

A final illustration may be helpful to indicate the difference between heat and work. Figure 4.12 shows a gas contained in a rigid vessel. Resistance coils are wound around the outside of the vessel. When current flows through the resistance coils, the temperature of the gas increases. Which crosses the boundary of the system, heat or work?

In Fig. 4.12*a* we consider only the gas as the system. In this case the energy crosses the boundary of the system because the temperature of the walls is higher than the temperature of the gas. Therefore, we recognize that heat crosses the boundary of the system.

In Fig. 4.12*b* the system includes the vessel and the resistance heater. Electricity crosses the boundary of the system, and as indicated earlier, this is work.

FIGURE 4.12

An example showing the difference between heat and work.

PROBLEMS

4.1 A cylinder fitted with a frictionless piston contains 5 lbm of superheated water vapor at 200 lbf/in.2, 500 F. This system is now cooled at constant pressure until the water reaches a quality of 50%. Calculate the work done during this process.

4.2 One-tenth kilogram of oxygen gas at 150 kPa, 20°C is contained in a cylinder fitted with a piston. Weights are now slowly added to the piston, and the gas is compressed isothermally to a final pressure of 600 kPa. Calculate the work done during this process.

4.3 Consider the system shown in Fig. 4.13. The initial volume inside the cylinder is 1 ft^3, at which state the pressure inside is 15 lbf/in.2. The spring is touching the piston at this state but exerts no force on it. Heat is now transferred to the system, causing it to expand until the volume is doubled, at which state the pressure in the cylinder is 50 lbf/in.2. During the process the spring force is proportional to the displacement of the piston from its initial position.
(a) Show the process on a P–V diagram. *Area Under curve*
(b) Considering the air inside the cylinder as the system, calculate the work done by the system. What percentage of this work is done against the spring?

FIGURE 4.13
Sketch for Problem 4.3.

4.4 Methane gas is contained in a cylinder fitted with a piston at 10°C, 200 kPa, at which point the volume is 200 L. The gas is now compressed slowly in an isothermal process to 600 kPa.
(a) Show that it is reasonable to assume that the methane behaves as an ideal gas during this process.
(b) Determine the work done during the process. *$P.V, \; \ln\left(\frac{V_2}{V_1}\right)$*

4.5 A balloon that is initially flat is inflated by filling it with air from a tank of compressed air. The final volume of the balloon is 120 ft^3. The barometer

reads 29.4 in. Hg. Considering the tank, the balloon, and the connecting pipe as a system, calculate the work done during the process.

4.6 Water initially at 50 kPa, 100°C is contained in a piston and cylinder arrangement with an initial volume of 3 m³. The water is then slowly compressed according to the relation $PV = $ constant until a final pressure of 1 MPa is reached. Determine the work for this process.

4.7 The cylinder-piston arrangement shown in Fig. 4.14 contains carbon dioxide at 50 lbf/in.², 350 F, at which point the volume is 5 ft³. Weights are then removed at such a rate that the gas expands according to the relation

$$\frac{PV}{T} \sim$$
$$P, V, ^n$$

$$PV^{1.2} = \text{constant}$$

until the final temperature is 200 F. Determine the work done during this process. $33,840$

$C_V = \left(\frac{\partial u}{\partial t}\right)_V$

$C_p = \left(\frac{\partial h}{\partial t}\right)_p$

$C_p = C_V + R$

FIGURE 4.14
Sketch for Problem 4.7.

4.8 The cylinder shown in Fig. 4.15 contains 1 kg of saturated (liquid + vapor) water at 30°C. The piston has a cross-sectional area of 0.065 m², a mass of 40 kg, and is resting on the stops as shown. The volume at this point is 100 L, atmospheric pressure outside is 94 kPa and the local gravitational

$P = \frac{F}{A}$

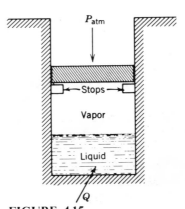

FIGURE 4.15
Sketch for Problem 4.8.

acceleration is 9.75 m/s². Heat is now transferred to the system until the cylinder contains saturated vapor.

(a) What is the temperature of the water when the piston first rises from the stops? *99.6³*

(b) Calculate the work done by the water during the overall process. *159.4*

4.9 A cylinder fitted with a piston contains propane gas at 14.7 lbf/in.², 540 R, at which point the volume is 5 ft³. The gas is now slowly compressed according to the relation

$$PV^{1.1} = \text{constant}$$

$$\frac{P_1 V_1}{T_1} = \frac{P_2 V_2}{T_2}$$

until the final temperature reaches 600 R.

(a) What is the final pressure? *46 psi*

(b) Justify that the ideal gas model is reasonable. $P_1 V_1^n = P_2 V_2^n$

(c) How much work is done during the process? $W = \dfrac{P_2 V_2 - P_1 V_1}{1-n} = -15.13$

4.10 The gas space above the water in a closed storage tank contains nitrogen at 25°C, 100 kPa. The tank has a total volume of 4 m³ and contains 500 kg of water at 25°C. An additional 500 kg of water is now slowly forced into the tank. Assuming that the temperature remains constant, calculate the final pressure of the nitrogen and the work done on the nitrogen during the process.

4.11 Calculate the work done by the helium in the process described in Problem 3.6.

4.12 Ammonia vapor is compressed inside a cylinder by an external force acting on the piston. The ammonia is initially at 100 F, 60 lbf/in.², and the final pressure is 180 lbf/in.². The following data have been measured experimentally for this process:

Pressure, lbf/in.²	Volume, in.³
60	80.0
80	67.2
101	60.1
120	52.5
139	44.8
161	37.6
180	32.5

(a) Determine the work for the process, considering the ammonia as the system.

(b) What is the final temperature of the ammonia?

4.13 A certain balloon is constructed of a material such that the pressure inside is proportional to the balloon diameter squared. This balloon contains 2 kg of ammonia initially at 0°C, 60% quality. The balloon is now heated until the final pressure inside reaches 600 kPa. Considering the ammonia as a system, how much work is done during the process?

4.14 A cylinder having an initial volume of 2 m³ contains 0.1 kg of water at 40°C. The water is compressed in an isothermal quasiequilibrium process until it has a quality of 50%. Calculate the work done during this process

4.15 Saturated vapor water at 400 F is contained in a cylinder fitted with a piston. The initial volume of the cylinder is 0.3 ft³. The steam then expands in a quasiequilibrium isothermal process until the final pressure in the cylinder is 20 lbf/in.², and in so doing does work against the piston.
(a) Determine the work done during the process.
(b) How much error would be made by assuming the steam to behave as an ideal gas?

4.16 A spherical balloon contains 5 lbm of water at 500 F, 60 lbf/in.². The balloon material has an elasticity such that the pressure inside is always proportional to the balloon diameter. The water is now allowed to cool until the volume is one-half the initial volume. $P \alpha D$ $\quad \frac{P_2}{P_3} = \frac{V_2}{V_1}$
(a) Determine the final temperature. $\quad P^3 \alpha V$ $\quad \frac{P_2}{P_3}$
(b) Calculate the work for the process, considering the water as the system.

4.17 In a rocket engine, a bottle of helium gas is used to maintain a constant pressure of 100 kPa on liquid oxygen (1142 kg/m³) as it is fed to the engine pump. The initial conditions are shown in Fig. 4.16. When the tank is drained of liquid, the helium bottle and oyxgen tank are both filled with helium gas at −75°C, 100 kPa.
(a) Calculate the work done by the helium during the process.
(b) Determine the required volume of the helium bottle.

FIGURE 4.16
Sketch for Problem 4.17.

4.18 A spherical elastic balloon initially containing 5 kg of saturated vapor ammonia at 20°C is connected by a valve to a rigid 3-m³ evacuated tank. The balloon material is such that the pressure inside is proportional to the balloon diameter. The valve is now opened, allowing ammonia to flow from the balloon to the tank until the pressure inside the balloon has dropped to 600 kPa, at which point the valve is closed. The final temperature in both the balloon and tank is 20°C. Determine:
(a) The final pressure in the tank.
(b) The work done by the ammonia during the process.

$$\frac{P_1}{P_2} = \left(\frac{T_2}{T_1}\right)^{\frac{n}{1-n}}$$

4.19 Air at 30 lbf/in.², 85 F is contained in a cylinder fitted with a piston. At this point, the cylinder volume is 3.5 ft³ and the pressure balances the outside ambient pressure of 14.7 lbf/in.² plus an externally imposed mechanical force that is proportional to $V^{0.5}$. Now, heat is transferred to the system until the temperature reaches 400 F. Find the final pressure in the cylinder and the work done during the process.

4.20 Tank *A* shown in Fig. 4.17 has a volume of 400 L and contains argon gas at 250 kPa, 30°C. Cylinder *B* contains a frictionless piston of a mass such that a pressure of 150 kPa inside the cylinder is required to raise the piston. The valve connecting the two is now opened, allowing gas to flow into the cylinder. Eventually, the argon reaches a uniform state of 150 kPa, 30°C throughout. Calculate the work done by the argon during the process.

FIGURE 4.17

Sketch for Problem 4.20.

4.21 Consider the cylinder and piston arrangement shown in Fig. 4.18, in which a frictionless piston having a cross-sectional area of 100 in.² rests on stops on the cylinder walls such that the contained volume is 1 ft³. The piston mass is such that 50 lbf/in.² pressure is required to raise the piston against

FIGURE 4.18

Sketch for Problem 4.21.

ambient pressure outside. When the piston has moved to the point where the contained volume is 2.5 ft³, the piston encounters a linear spring that requires 1970 lbf to deflect it 1 in. Initially the cylinder contains 10 lbm of saturated (liquid + vapor) water at 100 F. If this system is heated until the final pressure inside is 1000 lbf/in.², determine the final state of the water and also the work done during the process.

4.22 A cylinder fitted with a piston contains Freon-12 at 10°C, 50% quality, at which point the volume is 50 L. The system is now heated and expands, slowly moving the piston against an external resisting force. It is noted that when the last remaining amount of liquid evaporates the temperature in the cylinder is 25°C. The process continues until the temperature reaches 150°C, at which point the system pressure is 800 kPa. How much work is done during this process?

4.23 A vertical cylinder fitted with a piston contains 0.5 ft³ of Freon-12 at 80 F, 90% quality. The piston has a mass of 200 lbm, cross-sectional area of 10 in.², and is held in place by a pin, as shown in Fig. 4.19. The local ambient pressure is 15 lbf/in.². The pin is now removed, allowing the piston to move. After a period of time, the system comes to equilibrium, with the final temperature being 80 F.
(a) Determine the final pressure and volume of the system. $V = 6.8177$
(b) Calculate the work done by the Freon-12 during this process. $10.45)fe/85$

$$P = \frac{350\,lbm}{10\,in}$$

$$ln\left(\frac{P_1}{P_2}\right) = n\,ln\left(\frac{V_2}{V_1}\right)$$

$$W = -$$

FIGURE 4.19
Sketch for Problem 4.23.

4.24 A cylinder containing 2 lbm of ammonia is fitted with a piston restrained by a variable external force. The initial state of the ammonia is 280 lbf/in.², 360 F. The cylinder is now allowed to cool, and the external force varies in such a manner that when the ammonia reaches the two-phase region its temperature is 105 F. The process continues until the ammonia has cooled to 65 F, at which point the quality is 50%. Calculate the work for the overall process.

4.25 The cylinder shown in Fig. 4.20 is fitted with a piston that is restrained by a spring so arranged that for zero volume in the cylinder the spring is fully extended. The spring force is proportional to the spring displacement

FIGURE 4.20
Sketch for Problem 4.25.

and the weight of the piston is negligible. The enclosed volume in the cylinder is 120 L when the piston encounters the stops. The cylinder contains 4 kg of water initially at 350 kPa, 1% quality, and the water is then heated until it exists as saturated vapor. Show this process on a $P-V$ diagram and determine:

(a) The final pressure in the cylinder.

(b) The work done by the water during the process.

4.26 A steel bar 0.5 in. in diameter and 1 ft long is pulled in tension in a tensile testing machine. How much work is required to produce a strain of 0.1%? The modulus of elasticity of steel is 30×10^6 lbf/in.2.

4.27 At 20°C ethanol has a surface tension of 22.3 mN/m. Suppose that a film of ethanol is maintained on the wire frame shown in Fig. 4.21, one side of which is movable. The initial dimensions of the wire frame are as shown. Considering the film as the system, determine the work done when the wire is moved 10 mm in the direction indicated.

FIGURE 4.21
Sketch for Problem 4.27.

4.28 A sealed vessel having the shape of a rectangular prism with the area of the base A and height L_2, and of negligible mass, is initially floating on a liquid of density rho, as shown in Fig. 4.22. Derive an expression in terms

FIGURE 4.22

Sketch for Problems 4.28 and 4.29.

of the given variables for the work required to move the vessel to the bottom of a very large tank in which depth of the liquid is L_1.

4.29 Repeat Problem 4.28 assuming that the sealed tank has a mass m.

4.30 Consider a paramagnetic salt such as that described in Section 4.5. Assume that the magnetization of this substance is proportional to the magnetic field intensity divided by the temperature. Derive an expression for the work done in changing the magnetization in an isothermal process.

4.31 Repeat Problem 4.30 for a process taking place at constant magnetic field intensity.

4.32 An electric storage battery is well insulated thermally while it is being recharged. The charging voltage is 12.3 volts and the current is 25 amperes. Considering the battery as the system, what are the heat transfer and the work in a 30-minute period?

4.33 Consider a hot-air heating system for a home and examine the following systems for heat transfer:
(a) The combustion chamber and combustion gas side of the heat transfer area.
(b) The furnace as a whole, including the hot and cold air ducts and chimney.

4.34 A room is heated with steam radiators on a winter day. Examine the following systems regarding heat transfer, including sign:
(a) The radiator.
(b) The room.
(c) The radiator and the room together.

4.35 Write a computer program to solve the following problem. Determine the boundary movement work for a specified gas undergoing an isothermal process at temperature T from v_1 to v_2 using the Beattie–Bridgeman equa-

tion of state, and compare the result with that found assuming ideal gas behavior.

4.36 Write a computer program to solve the following problem. Determine the boundary movement work for a specified substance undergoing a process, given a set of data (values of pressure and corresponding volume during the process).

5

THE FIRST LAW
OF THERMODYNAMICS

Having completed our consideration of basic definitions and concepts, we are ready to proceed to a discussion of the first law of thermodynamics. Often this law is called the law of the conservation of energy and, as we will see later, this is essentially true. Our procedure will be to state this law first for a system undergoing a cycle, and then for a change of state of a system. The law of the conservation of matter will also be considered in this chapter, as will the first law for a control volume.

5.1 THE FIRST LAW OF THERMODYNAMICS FOR A SYSTEM UNDERGOING A CYCLE

The first law of thermodynamics states that during any cycle a system undergoes, the cyclic integral of the heat is proportional to the cyclic integral of the work.

To illustrate this law, consider as a system the gas in the container shown in Fig. 5.1. Let this system go through a cycle that is composed of two processes. In the first process work is done on the system by the paddle that turns as the weight is lowered. Let the system then be returned to its initial state by transferring heat from the system until the cycle has been completed.

Historically, work was measured in mechanical units of force times distance, such as, for example, foot pounds-force or joules, while measurements of heat were made in thermal units, such as ·the British thermal unit or the calorie. Measure-

a *b*

FIGURE 5.1

Example of a system undergoing a cycle.

ments of work and heat were made during a cycle for a wide variety of systems and for various amounts of work and heat. When the amounts of work and heat were compared, it was found that these two are always proportional. Observations such as this have led to the formulation of the first law of thermodynamics, which in equation form is written

$$J \oint \delta Q = \oint \delta W \qquad (5.1)$$

The symbol $\oint \delta Q$, which is called the cyclic integral of the heat transfer, represents the net heat transfer during the cycle, and $\oint \delta W$, the cyclic integral of the work, represents the net work during the cycle. J is a proportionality factor, which depends on the units used for work and heat.

The basis of every law of nature is experimental evidence, and this is true also of the first law of thermodynamics. Every experiment that has been conducted thus far has verified the first law either directly or indirectly. It has never been disproved, and has been verified by many different experiments.

As was discussed in Chapter 4, the units for work and for heat, or for any other form of energy, are either the same or are directly proportional. In English units, it was noted that the most common unit for work is the foot pound force, while that for heat is the British thermal unit (Btu). James P. Joule (1818–1889) did the first accurate work in the 1840's on measurement of the proportionality factor J relating these units. Today, the Btu is defined in terms of the basic SI metric units, such that

$$1 \text{ Btu} = 778.17 \text{ ft lbf}$$

and this unit is termed the International British thermal unit. For much engineering work, the accuracy of other data does not warrant more accuracy than the relation 1 Btu = 778 ft lbf, which will be the value used in examples in English units in this text.

In SI units, it was noted in Chapter 4 that the joule is used as the unit for both work and heat, and also for any energy unit. Thus, it is not necessary to use the proportionality factor J in Eq. 5.1. This procedure will be followed in general in this text. That is, the proportionality factor J will not be written explicitly in equations, but the student should realize that, for any system of units, each equation must have consistent units throughout. Therefore, we may write Eq. 5.1 as

$$\oint \delta Q = \oint \delta W \qquad (5.2)$$

which can be considered as the basic statement of the first law of thermodynamics.

Power, the rate of doing work was discussed in Section 4.2. From our discussion above, we realize that units for power apply to any rate of change or transfer of energy, and that units are equivalent. We note that

$$1 \text{ hp} = 550 \text{ ft lbf/s} = 0.7457 \text{ kW} = 42.4 \text{ Btu/min.} = 2544 \text{ Btu/h}$$
$$1 \text{ kW} = 1 \text{ kJ/s} = 1.341 \text{ hp} = 56.9 \text{ Btu/min.} = 3412 \text{ Btu/h}$$

5.2 THE FIRST LAW OF THERMODYNAMICS FOR A CHANGE IN STATE OF A SYSTEM

Equation 5.2 states the first law of thermodynamics for a system during a cycle. Many times, however, we are concerned with a process rather than a cycle, and we now consider the first law of thermodynamics for a system that undergoes a change of state. This can be done by introducing a new property, the energy, which is given the symbol E. Consider a system that undergoes a cycle, changing from state 1 to state 2 by process A, and returning from state 2 to state 1 by process B. This cycle is shown in Fig. 5.2 on a pressure (or other intensive property)– volume (or other extensive property) diagram. From the first law of thermodynamics, Eq. 5.2,

$$\oint \delta Q = \oint \delta W$$

Considering the two separate processes we have

$$\int_1^2 \delta Q_A + \int_2^1 \delta Q_B = \int_1^2 \delta W_A + \int_2^1 \delta W_B$$

Now consider another cycle, the system changing from state 1 to state 2 by process A, as before, and returning to state 1 by process C. For this cycle we can write

$$\int_1^2 \delta Q_A + \int_2^1 \delta Q_C = \int_1^2 \delta W_A + \int_2^1 \delta W_C$$

Subtracting the second of these equations from the first, we have

$$\int_2^1 \delta Q_B - \int_2^1 \delta Q_C = \int_2^1 \delta W_B - \int_2^1 \delta W_C$$

or, by rearranging,

$$\int_2^1 (\delta Q - \delta W)_B = \int_2^1 (\delta Q - \delta W)_C \tag{5.3}$$

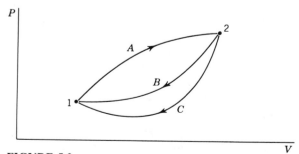

FIGURE 5.2

Demonstration of the existence of thermodynamic property E.

Since B and C represent arbitrary processes between states 1 and 2, we conclude that the quantity $(\delta Q - \delta W)$ is the same for all processes between state 1 and state 2. Therefore, $(\delta Q - \delta W)$ depends only on the initial and final states and not on the path followed between the two states. We conclude that this is a point function, and therefore is the differential of a property of the system. This property is the energy of the system and is given the symbol E. Thus we can write

$$\delta Q - \delta W = dE$$
$$\delta Q = dE + \delta W \tag{5.4}$$

Note that since E is a property, its derivative is written dE. When Eq. 5.4 is integrated from an initial state 1 to final state 2, we have

$$_1Q_2 = E_2 - E_1 + {_1}W_2 \tag{5.5}$$

where $_1Q_2$ is the heat transferred to the system during the process from state 1 to state 2, E_1 and E_2 are the initial and final values of the energy E of the system, and $_1W_2$ is the work done by the system during the process.

The physical significance of the property E is that it represents all the energy of the system in the given state. This energy might be present in a variety of forms, such as the kinetic or potential energy of the system as a whole with respect to the chosen coordinate frame, energy associated with the motion and position of the molecules, energy associated with the structure of the atom, chemical energy such as is present in a storage battery, energy present in a charged condenser, or in any of a number of other forms.

In the study of thermodynamics it is convenient to consider the bulk kinetic and potential energy separately and then to consider all the other energy of the system in a single property that we call the internal energy, and to which we give the symbol U. Thus we would write

$$E = \text{Internal energy} + \text{Kinetic energy} + \text{Potential energy}$$

or

$$E = U + \text{KE} + \text{PE}$$

The reason for doing this is that the kinetic and potential energy of the system are associated with the coordinate frame that we select and can be specified by the macroscopic parameters of mass, velocity, and elevation. The internal energy U includes all other forms of energy of the system and is associated with the thermodynamic state of the system.

Since the terms comprising E are point functions, we can write

$$dE = dU + d(\text{KE}) + d(\text{PE}) \tag{5.6}$$

The first law of thermodynamics for a change of state may therefore be written

$$\delta Q = dU + d(\text{KE}) + d(\text{PE}) + \delta W \tag{5.7}$$

In words this equation states that as a system undergoes a change of state, energy may cross the boundary as either heat or work, and each may be positive or negative. The net change in the energy of the system will be exactly equal to the net energy that crosses the boundary of the system. The energy of the system may change in any of three ways—by a change in internal energy, kinetic energy, or potential energy. Since the mass of a system is fixed, we can also say that a quantity of matter may have energy in three forms, internal energy, kinetic energy, or potential energy.

This section will be concluded by deriving an expression for the kinetic and potential energy of a system. Consider first a system that is initially at rest relative to the earth, which is taken as the coordinate frame, and let this system be acted on by an external horizontal force F that moves the system a distance dx in the direction of the force. Thus there is no change in potential energy. Let there be no heat transfer and no change in internal energy. Then, from the first law, Eq. 5.7,

$$\delta W = -F\,dx = -d\text{KE}$$

But

$$F = ma = m\frac{d\mathbf{V}}{dt} = m\frac{dx}{dt}\frac{d\mathbf{V}}{dx} = m\mathbf{V}\frac{d\mathbf{V}}{dx}$$

Then

$$d\text{KE} = F\,dx = m\mathbf{V}\,d\mathbf{V}$$

Integrating, we obtain

$$\int_{\text{KE}=0}^{\text{KE}} d\text{KE} = \int_{\mathbf{V}=0}^{\mathbf{V}} m\mathbf{V}\,d\mathbf{V}$$

$$\text{KE} = \frac{1}{2}m\mathbf{V}^2 \tag{5.8}$$

An expression for potential energy can be found in a similar manner. Consider a system that is initially at rest and at the elevation of some reference level. Let this system be acted on by a vertical force F that is of such magnitude that it raises (in elevation) the system with constant velocity an amount dZ. Let the acceleration due to gravity at this point be g. From the first law, Eq. 5.7,

$$\delta W = -F\,dZ = -d\text{PE}$$
$$F = ma = mg$$

Then

$$d\text{PE} = F\,dZ = mg\,dZ$$

Integrating,

$$\int_{\text{PE}_1}^{\text{PE}_2} d\text{PE} = m\int_{Z_1}^{Z_2} g\,dZ$$

Assuming that g does not vary with Z (which is a very reasonable assumption for moderate changes in elevation),

$$PE_2 - PE_1 = mg(Z_2 - Z_1) \qquad (5.9)$$

Substituting the expressions for kinetic and potential energy into Eq. 5.6, we have

$$dE = dU + m\mathbf{V}\,d\mathbf{V} + mg\,dZ$$

Integrating for a change of state from state 1 to state 2 with constant g, we have

$$E_2 - E_1 = U_2 - U_1 + \frac{m\mathbf{V}_2{}^2}{2} - \frac{m\mathbf{V}_1{}^2}{2} + mgZ_2 - mgZ_1$$

Similarly, substituting these expressions for kinetic and potential energy into Eq. 5.7, we have

$$\delta Q = dU + \frac{d(m\mathbf{V}^2)}{2} + d(mgZ) + \delta W \qquad (5.10)$$

In the integrated form this equation is, assuming g is a constant,

$$_1Q_2 = U_2 - U_1 + \frac{m(\mathbf{V}_2{}^2 - \mathbf{V}_1{}^2)}{2} + mg(Z_2 - Z_1) + {}_1W_2 \qquad (5.11)$$

Three observations should be made regarding this equation. The first is that the property E, the energy of the system, was found to exist, and we were able to write the first law for a change of state using Eq. 5.5. However, rather than deal with this property E, we find it more convenient to consider the internal energy and the kinetic and potential energies of the system. In general, this will be the procedure followed in the rest of this book.

The second observation is that Eqs. 5.10 and 5.11 are in effect a statement of the conservation of energy. The net change of the energy of the system is always equal to the net transfer of energy across the system boundary as heat and work. This is somewhat analogous to a joint checking account that a husband might have with his wife. There are two ways in which deposits and withdrawals can be made, either by the husband or by the wife, and the balance will always reflect the net amount of the transaction. Similarly, there are two way in which energy can cross the boundary of a system, either as heat or work, and the energy of the system will change by the exact amount of the net energy crossing the system boundary. The concept of energy and the law of the conservation of energy are basic to thermodynamics.

The third observation is that Eqs. 5.10 and 5.11 can give only changes in internal energy, kinetic energy, and potential energy. We can learn nothing about absolute values of these quantities from these equations. If we wish to assign values to internal energy, kinetic energy, and potential energy, we must assume reference states and assign a value to the quantity in this reference state. The kinetic energy of a body with zero velocity relative to the earth is assumed to be zero. Similarly, the value of the potential energy is assumed to be zero when the body is at some

reference elevation. With internal energy, therefore, we must also have a reference state if we wish to assign values of this property. This matter is considered in the following section.

5.3 INTERNAL ENERGY— A THERMODYNAMIC PROPERTY

Internal energy is an extensive property, since it depends on the mass of the system. Similarly, the kinetic and potential energies are extensive properties.

The symbol U designates the internal energy of a given mass of a substance. Following the convention used with other extensive properties the symbol u designates the internal energy per unit mass. We could speak of u as the specific internal energy, as we do in the case of specific volume. However, since the context will usually make it clear whether u or U is referred to, we will simply use the term internal energy to refer to both internal energy per unit mass and the total internal energy.

In Chapter 3 it was noted that in the absence of the effects of motion, gravity, surface effects, electricity, and magnetism, the state of a pure substance is specified by two independent properties. It is very significant that with these restrictions, the internal energy may be one of the independent properties of a pure substance. This means, for example, that if we specify the pressure and internal energy (with reference to an arbitrary base) of superheated steam, the temperature is also specified.

Thus, in a table of thermodynamic properties such as the steam tables, the value of internal energy can be tabulated along with other thermodynamic properties. Tables 1 and 2 of the steam tables by Keenan et al. (Appendix Tables A.1.1 and A.1.2) list the internal energy for saturated states. Included are the internal energy of saturated liquid u_f, the internal energy of saturated vapor u_g, and the difference between the internal energy of saturated liquid and saturated vapor u_{fg}. The values are given in relation to an arbitrarily assumed reference state, which will be discussed later. The internal energy of saturated steam of a given quality is calculated in the same way specific volume is calculated. The relations are

$$U = U_{liq} + U_{vap}$$

or

$$mu = m_{liq}u_f + m_{vap}u_g$$

Dividing by m and introducing the quality x, this becomes

$$u = (1 - x)u_f + xu_g$$

$$u = u_f + xu_{fg}$$

As an example, the specific internal energy of saturated steam having a pressure of 80 lbf/in.2 and a quality of 95% can be calculated as follows:

$$u = u_f + u_{fg} = 281.95 + 0.95 \times 820.6 = 1061.5 \text{ Btu/lbm}$$

Example 5.1

A tank containing a fluid is stirred by a paddle wheel. The work input to the paddle wheel is 5090 Btu. The heat transfer from the tank is 1500 Btu. Considering the tank and the fluid as the system, determine the change in the internal energy of the system.

The first law of thermodynamics is (Eq. 5.11)

$$_1Q_2 = U_2 - U_1 + m\frac{(\mathbf{V}_2{}^2 - \mathbf{V}_1{}^2)}{2} + mg(Z_2 - Z_1) + {}_1W_2$$

Since there is no change in kinetic and potential energy, this reduces to

$$_1Q_2 = U_2 - U_1 + {}_1W_2$$
$$-1500 = U_2 - U_1 - 5090$$
$$U_2 - U_1 = 3590 \text{ Btu}$$

Example 5.2

Consider a system composed of a stone having a mass of 10 kg and a bucket containing 100 kg of water. Initially the stone is 10.2 m above the water and the stone and water are at the same temperature. The stone then falls into the water.

Determine ΔU, ΔKE, ΔPE, Q, and W for the following changes of state, assuming standard gravitational acceleration, 9.806 65 m/s^2.

1. The stone is about to enter the water.
2. The stone has just come to rest in the bucket.
3. Heat has been transferred to the surroundings in such an amount that the stone and water are at the same temperature as they were initially.

The first law of thermodynamics is

$$_1Q_2 = U_2 - U_1 + m\frac{(\mathbf{V}_2{}^2 - \mathbf{V}_1{}^2)}{2} + mg(Z_2 - Z_1) + {}_1W_2$$

1. The stone is about to enter the water: assuming no heat transfer to or from the stone as it falls, we conclude that during the change from the initial state to the state that exists at the moment the stone enters the water,

$$_1Q_2 = 0 \qquad {}_1W_2 = 0 \qquad \Delta U = 0$$

Therefore the first law reduces to

$$-\Delta KE = \Delta PE = mg(Z_2 - Z_1)$$
$$= 10 \text{ kg} \times 9.806\,65\,\frac{m}{s^2} \times (-10.2 \text{ m})$$
$$= -1000 \text{ J} = -1 \text{ kJ}$$

That is,

$$\Delta KE = 1 \text{ kJ} \quad \text{and} \quad \Delta PE = -1 \text{ kJ}$$

2. Just after the stone comes to rest in the bucket,

$$_1Q_2 = 0 \qquad _1W_2 = 0 \qquad \Delta KE = 0$$

Then

$$\Delta PE = -\Delta U = mg(Z_2 - Z_1) = -1 \text{ kJ}$$
$$\Delta U = 1 \text{ kJ} \qquad \Delta PE = -1 \text{ kJ}$$

3. After enough heat has been transferred so that the stone and water are at the same temperature as they were initially, we conclude that $\Delta U = 0$. Therefore in this case

$$\Delta U = 0 \qquad \Delta KE = 0 \qquad _1W_2 = 0$$
$$_1Q_2 = \Delta PE = mg(Z_2 - Z_1) = -1 \text{ kJ}$$

5.4 PROBLEM ANALYSIS AND SOLUTION TECHNIQUE

At this point in our study of thermodynamics we have progressed sufficiently far (that is, we have accumulated sufficient tools to work with) that it is worthwhile to develop a somewhat formal technique or procedure for analyzing and solving thermodynamic problems. For the time being it may not seem entirely necessary to use such a rigorous procedure for many of our problems, but we should keep in mind that as we acquire still more analytical tools the problems that we are capable of dealing with will become much more complicated. Thus it is appropriate that we begin to practice this technique now in anticipation of these future problems.

Our problem analysis and solution technique is contained within the framework of the following set of questions that must be answered in the process of an orderly solution of a thermodynamic problem:

1. What is the system or control volume? It may be helpful to draw a sketch of the system at this point.
2. What do we know about the initial state (which properties)?
3. What do we know about the final state?
4. What do we know about the process that takes place? Is anything constant or zero? Is there some known functional relation between two properties?
5. Is it helpful to draw a diagram of the information in 2 to 4 above (for example, a T-v or P-v diagram)?
6. What is our model for the behavior of the substance (for example, steam tables, ideal gas, et cetera)?

7. What is our analysis of the problem (examine boundaries for various work modes, first law, et cetera)?

8. What is our solution technique (in other words, from what we have done so far in 1 to 7, how do we proceed to find whatever it is that is desired)? Is a trial-and-error solution necessary?

It is not always necessary to write out all of these steps, and in the majority of the examples throughout this text we will not take the space required to do so. However, when faced with a new and unfamiliar problem, the student should always at least think through this set of questions in order to develop the ability to be able to solve increasingly more challenging problems. In solving the following example, we will utilize this technique in detail.

Example 5.3

A vessel having a volume of 100 ft^3 contains 1 ft^3 of saturated liquid water and 99 ft^3 of saturated water vapor at 14.7 lbf/in^2. Heat is transferred until the vessel is filled with saturated vapor. Determine the heat transfer for this process.

System: All the water inside the vessel.

Sketch: Fig. 5.3.

Initial state: Pressure, volume of liquid, volume of vapor; therefore, state 1 is fixed.

Final state: Somewhere along the saturated-vapor curve; the water was heated, so $P_2 > P_1$.

Process: Constant volume and mass; therefore, constant specific volume.

VAP H$_2$O

$_1Q_2$

LIQ — H$_2$O

FIGURE 5.3
Sketch for Example 5.3.

Diagram: Fig. 5.4.

Model: Steam tables.

Analysis: First law:

$$_1Q_2 = U_2 - U_1 + m \frac{(\mathbf{V_2}^2 - \mathbf{V_1}^2)}{2} + mg(Z_2 - Z_1) + {}_1W_2$$

From examining the boundaries for various work modes, we conclude that the work for this process is zero. Furthermore, the system is not moving, so there is no change in kinetic energy. There is a small change in center of mass of the system but we will assume that the corresponding change in potential energy is negligible (compared to other terms). Therefore,

$$_1Q_2 = U_2 - U_1$$

Solution

The heat transfer will be found from the first law above. State 1 is known, so U_1 can be calculated. Also, the specific volume at state 2 is known (from state 1 and the process). Since state 2 is saturated vapor, state 2 is fixed, as is seen from Fig. 5.4. Therefore, U_2 can also be found.

 The solution proceeds as follows:

$$m_{1 \text{ liq}} = \frac{V_{\text{liq}}}{v_f} = \frac{1}{0.01672} = 59.81 \text{ lbm}$$

$$m_{1 \text{ vap}} = \frac{V_{\text{vap}}}{v_g} = \frac{99}{26.80} = 3.69 \text{ lbm}$$

Then,

$$U_1 = m_{1 \text{ liq}} u_{1 \text{ liq}} + m_{1 \text{ vap}} u_{1 \text{ vap}}$$
$$= 59.8(180.1) + 3.69(1077.6) = 14\,740 \text{ Btu}$$

FIGURE 5.4

Sketch for Example 5.3.

To determine u_2 we need to know two thermodynamic properties, since this determines the final state. The properties we know are the quality, $x = 100\%$, and v_2, the final specific volume, which can readily be determined.

$$m = m_{1 \text{ liq}} + m_{1 \text{ vap}} = 59.81 + 3.69 = 63.50 \text{ lbm}$$

$$v_2 = \frac{V}{m} = \frac{100}{63.50} = 1.575 \text{ ft}^3/\text{lbm}$$

In Table 2 of the steam tables we find, by interpolation, that at a pressure of 294 lbf/in.2, $v_g = 1.575$ ft^3/lbm. The final pressure of the steam is therefore 294 lbf/in.2. Then,

$$u_2 = 1117.0 \text{ Btu/lbm}$$
$$U_2 = mu_2 = 63.50(1117.0) = 70\,930 \text{ Btu}$$
$$_1Q_2 = U_2 - U_1 = 70\,930 - 14\,740 = 56\,190 \text{ Btu}$$

Up to this point, we have discussed the manner in which internal energies are listed in the saturation tables of the steam tables. The internal energy of superheated vapor steam is listed as a function of temperature and pressure in Table A.1.3. Similarly, Tables A.1.4 and A.1.5 list values in the compressed liquid and the saturated solid–vapor regions, respectively. In summary, all the values of internal energy are listed in the tables in the same manner as are values of specific volume.

5.5 THE THERMODYNAMIC PROPERTY ENTHALPY

In analyzing specific types of processes, we frequently encounter certain combinations of thermodynamic properties, which are therefore also properties of the substance undergoing the change of state. To demonstrate one such situation in which this occurs, let us consider a system undergoing a quasiequilibrium constant-pressure process, as shown in Fig. 5.5. We further assume that there are

FIGURE 5.5

The constant-pressure quasi-equilibrium process.

no changes in kinetic or potential energy and also that the only work done during the process is that associated with the boundary movement. Taking the gas as our system, and applying the first law, Eq. 5.11,

$$_1Q_2 = U_2 - U_1 + _1W_2$$

The work can be calculated from the relation

$$_1W_2 = \int_1^2 P \, dV$$

Since the pressure is constant,

$$_1W_2 = P \int_1^2 dV = P(V_2 - V_1)$$

Therefore

$$_1Q_2 = U_2 - U_1 + P_2V_2 - P_1V_1$$
$$= (U_2 + P_2V_2) - (U_1 + P_1V_1)$$

We find that in this very restricted case, the heat transfer during the process is given in terms of the change in the quantity $U + PV$ between the initial and final states. Inasmuch as all of these quantities are thermodynamic properties, functions only of the state of the system, their combination must also have these same characteristics. Therefore, we find it convenient to define a new extensive property, called the enthalpy:

$$H = U + PV \tag{5.12}$$

or, per unit mass,

$$h = u + Pv \tag{5.13}$$

As in the case of internal energy, we could speak of specific enthalpy, h, and total enthalpy, H. However, we will simply refer to both as enthalpy, since the context will make it clear which is referred to.

In order to add the internal energy and pressure volume product, the units for both must be the same. In this text the usual unit for enthalpy and internal energy in English units is Btu/lbm. When pressure is in lbf/ft^2 and specific volume in ft^3/lbm, the factor $J = 778$ ft lbf/Btu must be introduced in the denominator (i.e., Pv/J) in order to express the Pv term in Btu/lbm. However, the factor J will not be carried along in the equations because the only essential consideration is that consistent units be used, and many other units than Btu/lbm can be used.

As a result of our definition, the heat transfer in a constant-pressure quasiequilibrium process is equal to the change in enthalpy, which includes both the change in internal energy and the work for this particular process. This is by no means a general result. It is valid for this special case only because the work done during the process is equal to the difference in the PV product for the final and initial states. This would not be true if the pressure had not remained constant during the process.

The significance and use of enthalpy is not restricted to the special process described above. Other cases in which this same combination of properties $u + Pv$ appear will be developed later, notably in Section 5.11, in which we discuss control volume analyses. Our reason for introducing enthalpy at this time is that while the steam tables list values for internal energy, most other tables and charts of thermodynamic properties give values for enthalpy but not for the internal energy. In these cases it is necessary to calculate the internal energy at a state using the tabulated values and Eq. 5.13:

$$u = h - Pv$$

Students often become confused about the validity of this calculation when analyzing system processes that do not occur at constant pressure, for which enthalpy has no physical significance. We must keep in mind that enthalpy, being a property, is a state or point function, and its use in calculating internal energy at the same state is not related to, or dependent on, any processs that may be taking place.

Tabular values of enthalpy, such as those included in Appendix Tables A.1 to A.5, are all relative to some arbitrarily selected base. In the steam tables, the internal energy of saturated liquid at 32.02 F (or 0.01°C) is the reference state and is given a value of zero. For refrigerants, such as ammonia and Freon-12, the reference state is saturated liquid at -40 F $(-40°C)$, the enthalpy in this reference state being assigned the value of zero. Thus it is possible to have negative values of enthalpy, as is the case for saturated solid water in Table A.1.5 of the Appendix. It should be pointed out that when enthalpy and internal energy are given values relative to the same reference state, as is the case in essentially all thermodynamic tables, the difference between internal energy and enthalpy at the reference state is equal to Pv. Since the specific volume of the liquid is very small, this product is negligible as far as the significant figures of the tables are concerned, but the principle should be kept in mind, as in certain cases it is significant.

In the superheat region of most thermodynamic tables, values of the specific internal energy u are not given. As mentioned above, these can be readily calculated from the relation $u = h - Pv$, though it is important to keep the units in mind. As an example, let us calculate the internal energy u of superheated Freon-12 at 100 lbf/in.2, 200 F.

$$u = h - Pv$$
$$= 105.63 - \frac{100 \times 144 \times 0.5441}{778} = 95.55 \text{ Btu/lbm}$$

At 0.4 MPa, 70°C, we find

$$u = h - Pv$$
$$= 232.230 - 400 \times 0.056\,014$$
$$= 209.824 \text{ kJ/kg}$$

The enthalpy of a substance in a saturation state and having a given quality is found in the same way that the specific volume and internal energy were found. The enthalpy of saturated liquid has the symbol h_f, saturated vapor h_g, and the increase in enthalpy during vaporization h_{fg}. For a saturation state, the enthalpy can be calculated by one of the following relations:

$$h = (1 - x)h_f + xh_g$$
$$h = h_f + xh_{fg}$$

The enthalpy of compressed water may be found from Table A.1.4, while for other substances for which compressed-liquid tables are not available the enthalpy is taken as that of saturated liquid at the same temperature.

Example 5.4

A cylinder fitted with a piston has a volume of 0.1 m³ and contains 0.5 kg of steam at 0.4 MPa. Heat is transferred to the steam until the temperature is 300°C, while the pressure remains constant.

Determine the heat transfer and the work for this process.

System: Water inside cylinder.

Initial state: P_1, V_1, m; therefore v_1 is known, state 1 is fixed (at P_1, v_1, check steam tables—two-phase region).

Final state: P_2, T_2; therefore state 2 is fixed (superheated).

Process: Constant pressure.

Diagram: Fig. 5.6.

Model: Steam tables.

Analysis: No change in kinetic energy, no change in potential energy. Work is boundary movement. Assume process to be quasiequilibrium. Then, since the pres-

FIGURE 5.6
Diagram for Example 5.4.

sure is constant,

$$_1W_2 = \int_1^2 P\,dV = P\int_1^2 dV = P(V_2 - V_1) = m(P_2v_2 - P_1v_1)$$

Therefore, the first law is

$$_1Q_2 = m(u_2 - u_1) + {_1W_2}$$
$$= m(u_2 - u_1) + m(P_2v_2 - P_1v_1) = m(h_2 - h_1)$$

Solution

There is a choice of procedures to follow. State 1 is known, so v_1 and h_1 (or u_1) can be found. Also, state 2 is known, so v_2 and h_2 (or u_2) can be found. Using the first law and the work equation, the heat transfer and work can be calculated. Using the enthalpies,

$$v_1 = \frac{V_1}{m} = \frac{0.1}{0.5} = 0.2 = 0.001\,084 + x_1 0.4614$$

$$x_1 = \frac{0.1989}{0.4614} = 0.4311$$

$$h_1 = h_f + x_1 h_{fg}$$
$$= 604.74 + 0.4311 \times 2133.8 = 1524.6$$

$$h_2 = 3066.8$$

$$_1Q_2 = 0.5(3066.8 - 1524.6) = 771.1 \text{ kJ}$$

$$_1W_2 = mP(v_2 - v_1) = 0.5 \times 400(0.6548 - 0.2) = 91.0 \text{ kJ}$$

Therefore

$$U_2 - U_1 = {_1Q_2} - {_1W_2} = 771.1 - 91.0 = 680.1 \text{ kJ}$$

The heat transfer could also have been found from u_1 and u_2.

$$u_1 = u_f + x_1 u_{fg}$$
$$= 604.31 + 0.4311 \times 1949.3 = 1444.6$$

$$u_2 = 2804.8$$

and

$$_1Q_2 = U_2 - U_1 + {_1W_2}$$
$$= 0.5(2804.8 - 1444.6) + 91.0 = 771.1 \text{ kJ}$$

5.6 THE CONSTANT-VOLUME AND CONSTANT-PRESSURE SPECIFIC HEATS

In this section, we will consider a homogeneous phase of a substance of constant composition. This phase may be a solid, a liquid, or a gas, but no change of phase will occur. We will then define a variable termed the specific heat, the amount of

heat required per unit mass to raise the temperature by one degree. Since it would be of interest to examine the relation between the specific heat and other thermodynamic variables, we note first that the heat transfer is given by Eq. 5.10. Neglecting changes in kinetic and potential energies, and also assuming a simple compressible substance and quasiequilibrium process, for which the work in Eq. 5.10 is given by Eq. 4.2, the result is

$$\delta Q = dU + \delta W = dU + P\,dV$$

We find that this expression can be evaluated for two separate special cases:

1. Constant volume, for which the work term $(P\,dV)$ is zero, such that the specific heat (at constant volume) is

$$C_v = \frac{1}{m}\left(\frac{\delta Q}{\delta T}\right)_v = \frac{1}{m}\left(\frac{\partial U}{\partial T}\right)_v = \left(\frac{\partial u}{\partial T}\right)_v \qquad \textbf{(5.14)}$$

2. Constant pressure, for which the work term can be integrated and the resulting PV terms at the initial and final states associated with the internal energy terms, as in Section 5.5, thereby leading to the conclusion that the heat transfer can be expressed in terms of the enthalpy change. The corresponding specific heat (at constant pressure) is then

$$C_p = \frac{1}{m}\left(\frac{\delta Q}{\delta T}\right)_p = \frac{1}{m}\left(\frac{\partial H}{\partial T}\right)_p = \left(\frac{\partial h}{\partial T}\right)_p \qquad \textbf{(5.15)}$$

Note that in each of these special cases, the resulting expression, Eq. 5.14 or 5.15, contains only thermodynamic properties, from which we conclude that the constant-volume and constant-pressure specific heats must themselves be thermodynamic properties. This means that although we began this discussion by considering the amount of heat transfer required to cause a unit temperature change, and then proceeded through a very specific development leading to Eq. 5.14 (or 5.15), that result ultimately expresses a relation among a set of thermodynamic properties and therefore constitutes a definition that is independent of the particular process leading to it (in the same sense that the definition of enthalpy in the previous section is independent of the process used to illustrate one situation in which the property is useful in a thermodynamic analysis). As an example, consider the two identical systems shown in Fig. 5.7. In the first, 100 kJ of heat is transferred to the system, and in the second, 100 kJ of work is done on the system. Thus the change of internal energy is the same in each, and therefore the final state and the final temperature are the same in each. In accordance with Eq. 5.14, therefore, exactly the same value for the average constant-volume specific heat would be found for this substance for the two processes, even though the two processes are very different as far as heat transfer is concerned.

It should also be noted that the above development was carried out for a simple compressible substance, for which the work was given by Eq. 4.2. In cases in which there are different work modes, such as those described in Sections 4.4–4.6, different specific heats become appropriate. For any particular quasiequilibrium work

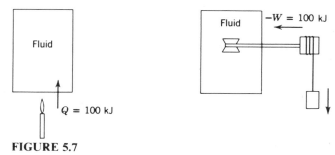

FIGURE 5.7

Sketch showing two ways in which a given ΔU may be achieved.

mode, there is one specific heat at constant displacement and one at constant driving force, as these terms were discussed in connection with Eq. 4.16. For example, in a system involving magnetic effects, it is convenient to utilize a specific heat at constant magnetization and one at constant magnetic field intensity.

Example 5.5

Estimate the constant-pressure specific heat of steam at 100 lbf/in.2, 400 F,

Solution

If we consider a change of state at constant pressure, Eq. 5.15 may be written

$$C_p \approx \left(\frac{\Delta h}{\Delta T}\right)_p$$

From the steam tables,

$$\text{at } 100 \text{ lbf/in.}^2, 360 \text{ F}, h = 1205.9$$
$$\text{at } 100 \text{ lbf/in.}^2, 440 \text{ F}, h = 1248.5$$

Since we are interested in C_p at 100 lbf/in.2, 400F,

$$C_p \simeq \frac{42.6}{80} = 0.533 \text{ Btu/lbm R}$$

As a special case, let us consider either a solid or a liquid. For both of these phases the specific volume is very small, and consequently, for any process,

$$dh = du + d(Pv) \simeq du \qquad \textbf{(5.16)}$$

and we can further say that

$$dh \simeq du \simeq C\,dT \qquad \textbf{(5.17)}$$

where C is either the constant-volume or the constant-pressure specific heat, as the two would be nearly the same. In many processes involving a solid or a liquid,

we might further assume that the specific heat in Eq. 5.17 is constant (unless the process is at low temperature or over a wide range of temperature). In this case, Eq. 5.17 can be integrated to

$$h_2 - h_1 \simeq u_2 - u_1 \simeq C(T_2 - T_1) \tag{5.18}$$

Specific heats for various solids and liquids are listed in Table A.7 of the Appendix.

In other processes for which it is not possible to assume constant specific heat, there may be a known relation for C as a function of temperature, in which case Eq. 5.17 could also be integrated.

5.7 THE INTERNAL ENERGY, ENTHALPY, AND SPECIFIC HEAT OF IDEAL GASES

At this point certain comments about the internal energy, enthalpy, and the constant-pressure and constant-volume specific heats of an ideal gas should be made. An ideal gas has been defined in Chapter 3 as a gas at sufficiently low density so that intermolecular forces and the associated energy are negligibly small. As a result, an ideal gas has the equation of state

$$Pv = RT$$

It can be shown that for an ideal gas, the internal energy is a function of the temperature only. That is, for an ideal gas,

$$u = f(T) \tag{5.19}$$

This means that an ideal gas at a given temperature has a certain definite specific internal energy u, regardless of the pressure. This will be demonstrated mathematically using the methods of classical thermodynamics in Chapter 10.

In 1843 Joule demonstrated this fact when he conducted the following experiment, which is one of the classical experiments in thermodynamics. Two pressure vessels (Fig. 5.8), connected by a pipe and valve, were immersed in a bath of water.

FIGURE 5.8

Apparatus for conducting Joule's experiment.

Initially vessel A contained air at 22 atm pressure and vessel B was highly evacuated. When thermal equilibrium was attained the valve was opened, allowing the pressures in A and B to equalize. No change in the temperature of the bath was detected during or after this process. Because there was no change in the temperature of the bath, Joule concluded that there had been no heat transferred to the air. Since the work was also zero, he concluded from the first law of thermodynamics that there was no change in the internal energy of the gas. Since the pressure and volume changed during this process, one concludes that internal energy is not a function of pressure and volume. Because air does not conform exactly to the definition of an ideal gas, a small change in temperature will be detected when very accurate measurements are made in Joule's experiment.

The relation between the internal energy u and the temperature can be established by using the definition of constant-volume specific heat given by Eq. 5.14.

$$C_v = \left(\frac{\partial u}{\partial T}\right)_v$$

Since the internal energy of an ideal gas is not a function of volume, for an ideal gas we can write

$$C_{vo} = \frac{du}{dT}$$

$$du = C_{vo}\,dT \tag{5.20}$$

where the subscript o denotes the specific heat of an ideal gas. For a given mass m,

$$dU = mC_{vo}\,dT \tag{5.21}$$

From the definition of enthalpy and the equation of state of an ideal gas, it follows that

$$h = u + Pv = u + RT \tag{5.22}$$

Since R is a constant and u is a function of temperature only, it follows that the enthalpy, h, of an ideal gas is also a function of temperature only. That is,

$$h = f(T) \tag{5.23}$$

The relation between enthalpy and temperature is found from the constant-pressure specific heat as defined by Eq. 5.15.

$$C_p = \left(\frac{\partial h}{\partial T}\right)_p$$

Since the enthalpy of an ideal gas is a function of the temperature only, and is independent of the pressure, it follows that

$$C_{po} = \frac{dh}{dT}$$

$$dh = C_{po}\,dT \tag{5.24}$$

For a given mass m,

$$dH = mC_{po}\,dT \qquad \qquad (5.25)$$

The consequences of Eqs. 5.20 and 5.24 are demonstrated by Fig. 5.9, which shows two lines of constant temperature. Since internal energy and enthalpy are functions of temperature only, these lines of constant temperature are also lines of constant internal energy and constant enthalpy. From state 1 the high temperature can be reached by a variety of paths, and in each case the final state is different. However, regardless of the path, the change in internal energy is the same, as is the change in enthalpy, for lines of constant temperature are also lines of constant u and constant h.

Because the internal energy and enthalpy of an ideal gas are functions of temperature only, it also follows that the constant-volume and constant-pressure specific heats are also functions of temperature only. That is,

$$C_{vo} = f(T); \qquad C_{po} = f(T) \qquad \qquad (5.26)$$

Since all gases approach ideal gas behavior as the pressure approaches zero, the ideal-gas specific heat for a given substance is often called the zero-pressure specific heat, and the zero-pressure, constant-pressure specific heat is given the symbol C_{po}. The zero-pressure, constant-volume specific heat is given the symbol C_{vo}. \bar{C}_{po} as a function of temperature for a number of different substances is shown in Fig. 5.10. These values are determined by using the techniques of statistical thermodynamics, and will not be discussed in this text. However, a brief qualitative discussion at this point should provide some insight into this behavior, and is also beneficial in determining under what conditions an assumption of constant specific heat is justified.

From a molecular point of view, the energy possessed by molecules may be stored in several forms. The translational and rotational energies increase linearly with temperature, which means that those contributions to the specific heat are

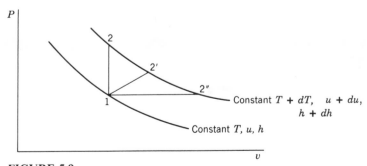

FIGURE 5.9

Pressure–volume diagram for an ideal gas.

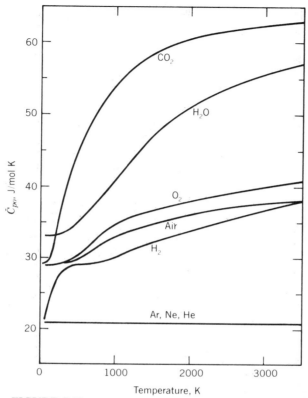

FIGURE 5.10

Constant-pressure specific heats for a number of gases at zero pressure.

not temperature dependent. Contributions from vibrational and electronic modes, on the other hand, are temperature dependent (the electronic usually being very small). From Fig. 5.10, it is evident that the specific heat of a diatomic gas (such as hydrogen or oxygen) increases with an increase in temperature, due primarily to the vibration. A polyatomic gas (such as carbon dioxide or water) shows a much greater increase in specific heat as the temperature increases, and this is due to the additional vibrational modes of a polyatomic molecule. A monatomic gas (such as helium, argon, or neon), possessing only translational and electronic energies, shows little or no variation of specific heat over a wide range of temperatures.

A very important relation between the constant-pressure and constant-volume specific heats of an ideal gas may be developed from the definition of enthalpy.

$$h = u + Pv = u + RT$$

Differentiating, and substituting Eqs. 5.20 and 5.24, we have

$$dh = du + R \, dT$$
$$C_{po} \, dT = C_{vo} \, dT + R \, dT$$

Therefore

$$C_{po} - C_{vo} = R \tag{5.27}$$

On a mole basis this equation would be written

$$\bar{C}_{po} - \bar{C}_{vo} = \bar{R} \tag{5.28}$$

This tells us that difference between the constant-pressure and constant-volume specific heats of an ideal gas is always constant, though both are functions of temperature. Thus we need only examine the temperature dependency of one, and the other is given by Eq. 5.27.

Let us consider the specific heat C_{po}. There are three possibilities to examine. The simplest situation results if we assume constant specific heat, that is, no temperature dependence. In this case it is possible to integrate Eq. 5.24 directly to

$$h_2 - h_1 = C_{po}(T_2 - T_1) \tag{5.29}$$

We note from Fig. 5.10 the circumstances under which this will be an accurate model. It should be added, however, that it may be a reasonable approximation under other conditions, especially if an average specific heat in the particular temperature range is used in Eq. 5.29. Values of specific heat at room temperature and gas constants for various gases are given in Table A.8 of the Appendix.

The second possibility for the specific heat is to utilize an analytical equation for C_{po} as a function of temperature. Since the results of specific-heat calculations from statistical thermodynamics do not lend themselves to convenient mathematical forms, these results have been approximated empirically, and the equations for C_{po} as a function of temperature are listed in Table A.9 for a number of gases.

The third possibility is to integrate the results of the calculations of statistical thermodynamics from an arbitrary reference temperature to any other temperature T, and define a function

$$h_T = \int_{T_0}^{T} C_{po} \, dT$$

This function can then be tabulated in a single-entry (temperature) table. Then, between any two states 1 and 2,

$$h_2 - h_1 = \int_{T_0}^{T_2} C_{po} \, dT - \int_{T_0}^{T_1} C_{po} \, dT = h_{T_2} - h_{T_1} \tag{5.30}$$

and it is seen that the reference temperature cancels out. This function h_T (and a similar function $u_T = h_T - RT$) is listed for air in Table A.10, and the enthalpy function (with respect to a reference temperature of 77 F or 25°C) is listed for various other gases in Table A.11.

In summary of the three possibilities presented above, it is noted that the Ideal Gas Tables, A.10 or A.11, would be the most accurate, with the equations in Table

A.9 giving a close empirical approximation. Constant specific heat would be less accurate, except for monatomic gases and for other gases below room temperature. It should be remembered that all of these are a part of the ideal-gas model, which in many of our problems is not a valid assumption for the behavior of the substance.

Example 5.6

Calculate the change of enthalpy as 1 kg of oxygen is heated from 300 to 1500 K, assuming ideal-gas behavior.

Solution

For an ideal gas, the enthalpy change is given by Eq. 5.24. However, we also need to make an assumption about the dependence of specific heat on temperature. Let us solve this problem in several ways and compare the answers.

Our most accurate answer for the ideal-gas enthalpy change for oxygen between 300 and 1500 K would be from the Ideal Gas Tables, A.11. This result is, using Eq. 5.30,

$$h_2 - h_1 = \frac{\bar{h}_{1500} - \bar{h}_{300}}{M} = \frac{40\,610 - 54}{32} = 1267.4 \text{ kJ/kg}$$

The empirical equation from Table A.9 should give a good approximation to this result. Integrating Eq. 5.24, we have

$$\bar{h}_2 - \bar{h}_1 = \int_{T_1}^{T_2} \bar{C}_{po}\,dT = \int_{\theta_1}^{\theta_2} \bar{C}_{po}(\theta) \times 100\,d\theta$$

$$\bar{h}_{1500} - \bar{h}_{300} = 100\left(37.432\theta + \frac{0.020\,102}{2.5}\theta^{2.5} + \frac{178.57}{0.5}\theta^{-0.5} - 236.88\theta^{-1}\right)\Big|_{\theta_1=3}^{\theta_2=15}$$

$$= 40\,525 \text{ kJ/kmol}$$

$$h_2 - h_1 = \frac{\bar{h}_{1500} - \bar{h}_{300}}{M} = \frac{40\,525}{32}$$

$$= 1266.4 \text{ kJ/kg}$$

which is different than the first result by less than 0.1%.

If we assume constant specific heat, we must be concerned about what value we are going to use. If we use the value at 300 K from Table A.8, we find, from Eq. 5.29,

$$h_2 - h_1 = C_{po}(T_2 - T_1) = 0.9216 \times 1200 = 1105.9 \text{ kJ/kg}$$

which is low by 12.7%. On the other hand, suppose that we assume that the specific heat is constant at its value at 900 K, the average temperature. Substituting 900 K into the equation for specific heat from Table A.9,

$$\bar{C}_{po} = 37.432 + 0.020\,102(9)^{1.5} - 178.57(9)^{-1.5} + 236.88(9)^{-2}$$

$$= 34.2855 \text{ kJ/kmol K}$$

or

$$C_{po} = \frac{34.2855}{32} = 1.0714 \text{ kJ/kg K}$$

Substituting this value into Eq. 5.29 gives the result

$$h_2 - h_1 = 1.0714 \times 1200 = 1285.7 \text{ kJ/kg}$$

which is high by about 1.4%, a much closer result than that when using the room-temperature specific heat. It should be kept in mind that part of the model involving ideal gas with constant specific heat is the choice of what value is to be used.

Example 5.7

A cylinder fitted with a piston has an initial volume of 2 ft³ and contains nitrogen at 20 lbf/in.², 80 F. The piston is moved, compressing the nitrogen until the pressure is 160 lbf/in.² and the temperature is 300 F. During this compression process heat is transferred from the nitrogen, and the work done on the nitrogen is 9.15 Btu. Determine the amount of this heat transfer.

System: Nitrogen.
Initial state: P_1, T_1, V_1; state 1 fixed.
Final state: P_2, T_2; state 2 fixed.
Process: Work input known.
Model: Ideal gas, constant specific heat with value at 540 R, Table A.8.
Analysis: First law:

$$_1Q_2 = m(u_2 - u_1) + {_1W_2}$$

Solution

The mass of nitrogen is found from the equation of state with the value of R from Table A.8.

$$m = \frac{PV}{RT} = \frac{20 \times 144 \times 2}{55.15 \times 540} = 0.1934 \text{ lbm}$$

Assuming constant specific heat as given in Table A.8,

$$\begin{aligned}
_1Q_2 &= mC_{vo}(T_2 - T_1) + {_1W_2} \\
&= 0.1934(0.177)(300 - 80) - 9.15 \\
&= 7.53 - 9.15 = -1.62 \text{ Btu}
\end{aligned}$$

It would of course be somewhat more accurate to use Table A.11 than to assume constant—room temperature—specific heat, but often the slight increase in accuracy does not warrant the added difficulties of interpolation in the tables.

5.8 THE FIRST LAW AS A RATE EQUATION

We frequently find it desirable to use the first law as a rate equation that expresses either the instantaneous or average rate at which energy crosses the system boundary as heat and work and the rate at which the energy of the system changes. In so doing we are departing from a strictly classical point of view, because basically classical thermodynamics deals with systems that are in equilibrium, and time is not a relevant parameter for systems that are in equilibrium. However, since these rate equations are developed from the concepts of classical thermodynamics, and are used in many applications of thermodynamics, they are included in this book. This rate form of the first law will be used in the development of the first law for the control volume in Section 5.11, and in this form the first law finds extensive applications in thermodynamics, fluid mechanics, and heat transfer.

Consider a time interval δt during which an amount of heat δQ crosses the system boundary, an amount of work δW is done by the system, the internal energy change is ΔU, the kinetic energy change is ΔKE, and the potential energy change is ΔPE. From the first law we can write

$$\delta Q = \Delta U + \Delta \text{KE} + \Delta \text{PE} + \delta W$$

Dividing by δt we have the average rate of energy transfer as heat and work and increase of the energy of the system.

$$\frac{\delta Q}{\delta t} = \frac{\Delta U}{\delta t} + \frac{\Delta \text{KE}}{\delta t} + \frac{\Delta \text{PE}}{\delta t} + \frac{\delta W}{\delta t}$$

Taking the limit for each of these quantities as δt approaches zero, we have

$$\lim_{\delta t \to 0} \frac{\delta Q}{\delta t} = \dot{Q}, \quad \text{the heat transfer rate}$$

$$\lim_{\delta t \to 0} \frac{\delta W}{\delta t} = \dot{W}, \quad \text{the power}$$

$$\lim_{\delta t \to 0} \frac{\Delta U}{\delta t} = \frac{dU}{dt} \quad \lim_{\delta t \to 0} \frac{\Delta(\text{KE})}{\delta t} = \frac{d(\text{KE})}{dt} \quad \lim_{\delta t \to 0} \frac{\Delta(\text{PE})}{\delta t} = \frac{d(\text{PE})}{dt}$$

Therefore, the rate equation form of the first law is

$$\dot{Q} = \frac{dU}{dt} + \frac{d(\text{KE})}{dt} + \frac{d(\text{PE})}{dt} + \dot{W} \tag{5.31}$$

We could also write this in the form

$$\dot{Q} = \frac{dE}{dt} + \dot{W} \tag{5.32}$$

Example 5.8

During the charging of a storage battery the current is 20 A and the voltage is 12.8 V. The rate of heat transfer from the battery is 10 W. At what rate is the internal energy increasing?

Solution

Since changes in kinetic and potential energy are not significant, the first law can be written as a rate equation in the form, Eq. 5.31,

$$\dot{Q} = \frac{dU}{dt} + \dot{W}$$

$$\dot{W} = -\mathscr{E}i = -20 \times 12.8 = -256 \text{ W}$$

Therefore,

$$\frac{dU}{dt} = \dot{Q} - \dot{W} = -10 \text{ W} - (-256) = 246 \text{ J/s}$$

5.9 CONSERVATION OF MASS

In the previous section we considered the first law of thermodynamics for a system undergoing a change of state. A system is defined as a fixed quantity of mass. The question now arises, does the mass of the system change when the energy of a system changes? If it does, then our definition of a system as a fixed quantity of mass is no longer valid when the energy of the system changes.

We know from relativistic considerations that mass and energy are related by the well-known equation

$$E = mc^2 \tag{5.33}$$

where c = velocity of light and E = energy. We conclude from this equation that the mass of a system does change when its energy changes. Let us calculate the magnitude of this change of mass for a typical problem and determine whether this change in mass is significant.

Consider as a system a rigid vessel that contains a 1-kg stoichiometric mixture of a hydrocarbon fuel (such as gasoline) and air. From our knowledge of combustion, we know that after combustion takes place it will be necessary to transfer about 2900 kJ from the system in order to restore the system to its initial temperature. From the first law

$$_1Q_2 = U_2 - U_1 + {}_1W_2$$

we conclude, since $_1W_2 = 0$ and $_1Q_2 = -2900$ kJ, that the internal energy of the system decreases by 2900 kJ during the heat transfer process. Let us now calculate the decrease in mass during this process using Eq. 5.33.

The velocity of light, c, is 2.9979×10^8 m/s. Therefore,

$$2900 \text{ kJ} = 2\,900\,000 \text{ J} = m \text{ (kg)} \times (2.9979 \times 10^8 \text{ m/s})^2$$

$$m = 3.23 \times 10^{-11} \text{ kg}$$

Therefore, when the energy of the system decreases by 2900 kJ, the decrease in mass is 3.23×10^{-11} kg.

A change in mass of this magnitude cannot be detected by even our most accurate chemical balance. And, certainly, a fractional change in mass of this mag-

nitude is beyond the accuracy required in essentially all engineering calculations. Therefore, if we use the laws of conservation of mass and conservation of energy as separate laws, we will not introduce significant error into most thermodynamic problems, and our definition of a system as having a fixed mass can be used even though the energy of the system changes.

5.10 CONSERVATION OF MASS AND THE CONTROL VOLUME

A control volume is a volume in space in which one has interest for a particular study or analysis. The surface of this control volume is referred to as a control surface and always consists of a closed surface. The size and shape of the control volume are completely arbitrary, and are so defined as to best suit the analysis that is to be made. The surface may be fixed, or it may move or expand. However, the surface must be defined relative to some coordinate system. In some analyses it may be desirable to consider a rotating or moving coordinate system, and to describe the position of the control surface relative to such a coordinate system.

Mass as well as heat and work can cross the control surface and the mass in the control volume, as well as the properties of this mass, can change with time. Figure 5.11 shows a schematic diagram of a control volume, with heat transfer, shaft work, accumulation of mass within the control volume, and a moving boundary.

FIGURE 5.11

Schematic diagram of a control volume showing mass and energy transfers and accumulation.

Let us first consider the law of the conservation of mass as it relates to the control volume. In doing so we consider the mass flow into and out of the control volume and the net increase of mass within the control volume. During a time interval δt let the mass δm_i, as shown in Fig. 5.12, enter the control volume, and let the mass δm_e leave the control volume. Further, let us designate the mass within the control volume at the beginning of this time interval as m_t and the mass within the control volume after this interval as $m_{t+\delta t}$. Then, from the law of the conservation of mass we can write

$$m_t + \delta m_i = m_{t+\delta t} + \delta m_e$$

We can also look at this from the point of view of the net flow across the control surface and the change of mass within the control volume.

a System and control volume at time t

b System and control volume at time $t + \delta t$

FIGURE 5.12

Schematic diagram of a control volume for analysis of the continuity equation as applied to a control volume. (a) System and control volume at time t. (b) System and control volume at time $t + \delta t$.

Net flow into the control volume during δt = increase of mass within the control volume during δt:

$$(\delta m_i - \delta m_e) = m_{t+\delta t} - m_t$$

or

$$(m_{t+\delta t} - m_t) + (\delta m_e - \delta m_i) = 0 \qquad \textbf{(5.34)}$$

As written, this equation simply states that the change of mass within the control volume during δt—$m_{t+\delta t} - m_t$—and the net mass flow into the control volume during δt—$\delta m_i - \delta m_e$—are equal. However, in many problems requiring a thermodynamic analysis, we find it very convenient to have the law of the conservation of mass (as well as the first and second laws of thermodynamics and the momentum equation) expressed as a rate equation for the control volume. This involves the instantaneous rate of mass flow across the control surface and the instantaneous rate of change of mass within the control volume. Let us therefore write an expression for the average rate of change of mass within the control volume during δt and the average mass rates of flow across the control surface during δt by dividing Eq. 5.34 by δt.

$$\left(\frac{m_{t+\delta t} - m_t}{\delta t}\right) + \frac{\delta m_e}{\delta t} - \frac{\delta m_i}{\delta t} = 0 \qquad \textbf{(5.35)}$$

To obtain the rate equation for the control volume, we find the limit of each term as δt is made to approach zero, at which point the system and control volume coincide.

$$\lim_{\delta t \to 0} \left(\frac{m_{t+\delta t} - m_t}{\delta t}\right) = \frac{dm_{c.v.}}{dt}$$

$$\lim_{\delta t \to 0} \left(\frac{\delta m_e}{\delta t}\right) = \dot{m}_e$$

$$\lim_{\delta t \to 0} \left(\frac{\delta m_i}{\delta t}\right) = \dot{m}_i$$

The symbol $m_{c.v.}$ is used to denote the instantaneous mass inside the control volume; \dot{m}_i is the instantaneous rate of flow entering the control volume across area \mathscr{A}_i; and \dot{m}_e is that leaving across area \mathscr{A}_e. We recognize that, in practice, there may well be several areas on the control surface across which flow occurs (in mixing processes or processes involving chemical reactions, for example). We account for this possibility by including summation signs, implying a summation of flows over the various discrete flow areas. Thus the instantaneous rate form becomes

$$\frac{dm_{c.v.}}{dt} + \sum \dot{m}_e - \sum \dot{m}_i = 0 \qquad \textbf{(5.36)}$$

FIGURE 5.13

Flow across a control surface.

Equation 5.36 for the conservation of mass is commonly termed the continuity equation. While this form of the equation is sufficient for the majority of applications in thermodynamics, it is frequently rewritten in terms of local fluid properties in the study of fluid mechanics and heat transfer. In this text, we consider Eq. 5.36 as the general expression of the continuity equation.

Let us consider one further aspect of mass flow across a control surface. For simplicity, we assume that a fluid is flowing uniformly inside a pipe or duct, as shown in Fig. 5.13.

We wish to examine the flow in terms of the amount of mass crossing the surface \mathscr{A} during the time interval δt. As seen from Fig. 5.13, the fluid moves a distance dx during this time, and therefore the volume of fluid crossing surface \mathscr{A} is $\mathscr{A}\,dx$. Consequently, the mass crossing surface \mathscr{A} is given by

$$\delta m = \frac{\mathscr{A}\,dx}{v}$$

Now, if we divide both sides of this expression by δt and take the limit as $\delta t \to 0$, the result is

$$\dot{m} = \frac{\mathscr{A}\mathbf{V}}{v} = \rho A \bar{V} \qquad v = \frac{1}{\rho} \qquad (5.37)$$

It should be noted that this result, Eq. 5.37, has been developed for a stationary control surface \mathscr{A}, and that we tacitly assumed that the flow was normal to that surface and was uniform across the surface. It should also be pointed out that Eq. 5.37 applies to any of the various flow streams entering or leaving the control volume, subject to the assumptions mentioned.

Example 5.9

Air is flowing in a 6-in.-diameter pipe at a uniform velocity of 30 ft/min. The temperature and pressure are 80 F and 20 lbf/in.². Determine the mass flow rate.

Solution

From Eq. 5.37,

$$\dot{m} = \frac{\mathscr{A}\mathbf{V}}{v}$$

For air, using R from Table A.8,

$$v = \frac{RT}{P} = \frac{53.34 \times 540}{20 \times 144} = 10.0 \text{ ft}^3/\text{lbm}$$

The cross-sectional area is

$$\mathscr{A} = \frac{\pi}{4}\left(\frac{6}{12}\right)^2 = 0.196 \text{ ft}^2$$

Therefore,

$$\dot{m} = \frac{0.196 \times 30}{100} = 0.588 \text{ lbm/min}$$

5.11 THE FIRST LAW OF THERMODYNAMICS FOR A CONTROL VOLUME

We have already considered the first law of thermodynamics for a system, which consists of a fixed quantity of mass, and noted, Eq. 5.5, that it may be written

$$_1Q_2 = E_2 - E_1 + {}_1W_2$$

We have also noted that this may be written as an average rate equation over the time interval δt by dividing by δt:

$$\frac{\delta Q}{\delta t} = \frac{E_2 - E_1}{\delta t} + \frac{\delta W}{\delta t} \qquad \textbf{(5.38)}$$

To write the first law as a rate equation for a control volume, we proceed in a manner analogous to that used in developing a rate equation for the law of the conservation of mass. A system and control volume are shown in Fig. 5.14. The system consists of all the mass initially in the control volume plus the mass δm_i.

Consider the changes that take place in the system and control volume during the time interval δt. During this time δt, the mass δm_i enters the control volume through the discrete area \mathscr{A}_i, and the mass δm_e leaves the control volume through the area \mathscr{A}_e.

In our analysis we assume that the increment of mass, δm_i, has uniform properties, and similarly that δm_e has uniform properties. The total work done by the system during this process, δW, is that associated with the masses δm_i and δm_e crossing the control surface, usually referred to as flow work, and the work $\delta W_{\text{c.v.}}$, which includes all other forms of work, such as work associated with a rotating shaft that crosses the system boundary, shear forces, electrical, magnetic, or surface effects, or expansion or contraction of the control volume. An amount of heat δQ crosses the boundary during δt.

Let us now consider each of the terms of the first law as it is written for the system and transform each term into an equivalent form that applies to the control volume. Consider first the term $E_2 - E_1$.

FIGURE 5.14

Schematic diagram for a first law analysis of control volume, showing heat and work as well as mass crossing the control surface.

Let

$$E_t = \text{the energy in the control volume at time } t.$$

$$E_{t+\delta t} = \text{the energy in the control volume at time } t + \delta t$$

Then

$$E_1 = E_t + e_i \delta m_i = \text{the energy of the system at time } t$$

$$E_2 = E_{t+\delta t} + e_e \delta m_e = \text{the energy of the system at time } t + \delta t$$

Therefore,

$$E_2 - E_1 = E_{t+\delta t} + e_e \delta m_e - E_t - e_i \delta m_i$$

$$= (E_{t+\delta t} - E_t) + (e_e \delta m_e - e_i \delta m_i) \tag{5.39}$$

The term $(e_e \, \delta m_e - e_i \, \delta m_i)$ represents the net flow of energy that crosses the control surface during δt as the result of the masses δm_e and δm_i crossing the control surface.

Let us consider in more detail the work associated with the masses δm_i and δm_e crossing the control surface. Work is done by the normal force (normal to area \mathscr{A}) acting on δm_i and δm_e as these masses cross the control surface. This normal force is equal to the product of the normal tensile stress, $-\sigma_n$, and the area, \mathscr{A}. The work done is

$$-\sigma_n \mathscr{A} \, dl = -\sigma_n \, \delta V = -\sigma_n v \, \delta m \qquad (5.40)$$

A complete analysis of the nature of the normal stress, σ_n, for an actual fluid involves both static pressure and viscous effects, and is beyond the scope of this book. We will assume in this book that the normal stress, σ_n, at a point is equal to the static pressure at this point, P, and simply note that in many applications this is a very reasonable assumption and yields results of good accuracy.

With this assumption the work done on mass δm_i as it enters the control volume is $P_i v_i \delta m_i$, and the work done by mass δm_e as it leaves the control volume is $P_e v_e \delta m_e$. We refer to these terms as the flow work. Various other terms are encountered in the literature such as flow energy and work of introduction and work of expulsion.

Thus the total work done by the system during time δt is

$$\delta W = \delta W_{\text{c.v.}} + (P_e v_e \, \delta m_e - P_i v_i \, \delta m_i) \qquad (5.41)$$

Let us now divide Eqs. 5.39 and 5.41 by δt and substitute into the first law, Eq. 5.38. Combining terms and rearranging,

$$\frac{\delta Q}{\delta t} + \frac{\delta m_i}{\delta t}(e_i + P_i v_i) = \left(\frac{E_{t+\delta t} - E_t}{\delta t}\right) + \frac{\delta m_e}{\delta t}(e_e + P_e v_e) + \frac{\delta W_{\text{c.v.}}}{\delta t} \qquad (5.42)$$

We recognize that each of the flow terms in this expression can be rewritten in the form

$$e + Pv = u + Pv + \frac{\mathbf{V}^2}{2} + gZ = h + \frac{\mathbf{V}^2}{2} + gZ \qquad (5.43)$$

using the definition of the thermodynamic property enthalpy, Eq. 5.13. It should be apparent that the appearance of the combination $u + Pv$ whenever mass flows across a control surface is the principal reason for defining the property enthalpy. Its introduction earlier in conjunction with the constant-pressure process was to facilitate use of the tables of thermodynamic properties at that time.

Utilizing Eq. 5.43 for the masses entering and leaving the control volume, Eq. 5.42 becomes

$$\frac{\delta Q}{\delta t} + \frac{\delta m_i}{\delta t}\left(h_i + \frac{\mathbf{V}_i^2}{2} + gZ_i\right) = \left(\frac{E_{t+\delta t} - E_t}{\delta t}\right) + \frac{\delta m_e}{\delta t}\left(h_e + \frac{\mathbf{V}_e^2}{2} + gZ_e\right) + \frac{\delta W_{\text{c.v.}}}{\delta t}$$

$$(5.44)$$

To reduce this expression to a rate equation, we consider what happens to each of the terms in Eq. 5.44 as δt is made to approach zero. The heat transfer and work terms become rate transfer quantities, as was the case in Section 5.8. Similarly, the two mass quantities become mass flow rates, as in Section 5.10, and the energy term becomes the time rate of change of energy in the control volume, in a manner analogous to the mass term in the conservation of mass. In addition, we originally assumed uniform properties for the incremental mass δm_i entering the control volume across area \mathcal{A}_i, and made a similar assumption for δm_e leaving across area \mathcal{A}_e. Consequently, in taking the limits above, this reduces to the restriction of uniform properties across area \mathcal{A}_i and also across \mathcal{A}_e at an instant of time. These may of course be time dependent.

In utilizing the limiting values to express the rate equation of the first law for a control volume, we again include summation signs on the flow terms to account for the possibility of additional flow streams entering or leaving the control volume. Therefore, the result is

$$\dot{Q}_{c.v.} + \sum \dot{m}_i \left(h_i + \frac{\mathbf{V}_i^2}{2} + gZ_i \right) = \frac{dE_{c.v.}}{dt} + \sum \dot{m}_e \left(h_e + \frac{\mathbf{V}_e^2}{2} + gZ_e \right) + \dot{W}_{c.v.} \quad (5.45)$$

which is, for our purposes, the general expression of the first law of thermodynamics. In words, this equation says that the rate of heat transfer into the control volume plus rate of energy flowing in as a result of mass transfer is equal to the rate of change of energy inside the control volume plus rate of energy flowing out as a result of mass transfer plus the power output associated with shaft, shear, and electrical effects and other factors that have already been mentioned.

Equation 5.45 can be integrated over the total time of a process to give the total energy changes that occur during that period. However, to do so requires knowledge of the time dependency of the various mass flow rates and the states at which mass enters and leaves the control volume. An example of this type of process will be considered in Section 5.14.

Another point to be noted is that if there is no mass flow into or out of the control volume, those terms simply drop out of Eq. 5.45, which then reduces to the rate equation form of the first law for a system:

$$\dot{Q} = \frac{dE}{dt} + \dot{W}$$

discussed in Section 5.8.

Since the control volume approach is more general, and reduces to the usual statement of the first law for a system when there is no mass flow across the control surface, we will, when making a general statement of the first law, utilize Eq. 5.45, the rate form of the first law for a control volume.

The term flux should also be introduced at this point. Flux is defined as a rate of flow of any quantity per unit area across a control surface. Thus the mass flux is the rate of mass flow per unit area and heat flux is the heat flow rate per unit area across the control surface.

Finally, it should be pointed out that in fluid mechanics and heat transfer the control volume expression for the first law is commonly written in terms of local properties, as was the case with the conservation of mass. That form will not be introduced in this text, and for our purposes we will consider Eq. 5.45 as the general expression of the first law for a control volume analysis.

5.12 THE STEADY-STATE, STEADY-FLOW PROCESS

Our first application of the control volume equations will be to develop a suitable analytical model for the long-term steady operation of devices such as turbines, compressors, nozzles, boilers, condensers—a very large class of problems of interest in thermodynamic analysis. This model will not include the short-term transient start-up or shutdown of such devices, but only the steady operating period of time.

Let us consider a certain set of assumptions (beyond those leading to Eqs. 5.36 and 5.45) that lead to a reasonable model for this type of process, which we refer to as the steady-state, steady-flow process. For convenience, we often refer to this process as the SSSF process.

1. The control volume does not move relative to the coordinate frame.

2. As for the mass in the control volume, the state of the mass at each point in the control volume does not vary with time.

3. As for the mass that flows across the control surface, the mass flux and the state of this mass at each discrete area of flow on the control surface do not vary with time. The rates at which heat and work cross the control surface remain constant.

As an example of a steady-state, steady-flow process consider a centrifugal air compressor that operates with constant mass rate of flow into and out of the compressor, constant properties at each point across the inlet and exit ducts, a constant rate of heat transfer to the surroundings, and a constant power input. At each point in the compressor the properties are constant with time, even though the properties of a given elemental mass of air vary as it flows through the compressor. Usually such a process is referred to simply as a steady-flow process, since we are concerned primarily with the properties of the fluid entering and leaving the control volume. On the other hand, in the analysis of certain heat transfer problems in which the same assumptions apply, we are primarily interested in the spatial distribution of properties, particularly temperature, and such a process is often referred to as a steady-state process. Since this is an introductory book we use the term steady-state, steady-flow process to emphasize the basic assumptions involved. The student should realize that the terms steady-state process and steady-flow process are both used extensively in the literature.

Let us now consider the significance of each of these assumptions for the SSSF process.

1. The assumption that the control volume does not move relative to the co-ordinate frame means that all velocities measured relative to the coordinate frame are also velocities relative to the control surface, and there is no work associated with the acceleration of the control volume.

2. The assumption that the state of the mass at each point in the control volume does not vary with time requires that

$$\frac{dm_{c.v.}}{dt} = 0$$

and also

$$\frac{dE_{c.v.}}{dt} = 0$$

Therefore, we conclude that for the SSSF process we can write, from Eqs. 5.36 and 5.45,

Continuity equation:

$$\sum \dot{m}_i = \sum \dot{m}_e \qquad (5.46)$$

First law:

$$\dot{Q}_{c.v.} + \sum \dot{m}_i \left(h_i + \frac{\mathbf{V}_i^2}{2} + gZ_i \right) = \sum \dot{m}_e \left(h_e + \frac{\mathbf{V}_e^2}{2} + gZ_e \right) + \dot{W}_{c.v.} \qquad (5.47)$$

3. The assumption that the various mass flows, states, and rates at which heat and work cross the control surface remain constant requires that every quantity in Eqs. 5.46 and 5.47 is steady with time. This means that application of Eqs. 5.46 and 5.47 to the operation of some device is independent of time.

Many of the applications of the SSSF model are such that there is only one flow stream entering and one leaving the control volume. For this type of process, we can write

Continuity equation:

$$\dot{m}_i = \dot{m}_e = \dot{m} \qquad (5.48)$$

First law:

$$\dot{Q}_{c.v.} + \dot{m} \left(h_i + \frac{\mathbf{V}_i^2}{2} + gZ_i \right) = \dot{m} \left(h_e + \frac{\mathbf{V}_e^2}{2} + gZ_e \right) + \dot{W}_{c.v.} \qquad (5.49)$$

Rearranging this equation we have

$$q + h_i + \frac{\mathbf{V}_i^2}{2} + gZ_i = h_e + \frac{\mathbf{V}_e^2}{2} + gZ_e + w \qquad (5.50)$$

where, by definition,

$$q = \frac{\dot{Q}_{c.v.}}{\dot{m}} \quad \text{and} \quad w = \frac{\dot{W}_{c.v.}}{\dot{m}} \tag{5.51}$$

Note that the units for q and w are kJ/kg. From their definition, q and w can be thought of as the heat transfer and work (other than flow work) per unit mass flowing into and out of the control volume for this particular SSSF process.

The symbols q and w are also used for the heat transfer and work per unit mass of a system. However, since it is always evident from the context whether it is a system (fixed mass) or control volume (involving a flow of mass) with which we are concerned, the significance of the symbols q and w will also be readily evident in each situation.

The SSSF process is often used in the analysis of reciprocating machines, such as reciprocating compressors or engines. In this case the rate of flow. which may actually be pulsating, is considered to be the average rate of flow for an integral number of cycles. A similar assumption is made regarding the properties of the fluid flowing across the control surface, and the heat transfer and work crossing the control surface. It is also assumed that for an integral number of cycles the reciprocating device undergoes, the energy and mass within the control volume do not change.

A number of examples are now given to illustrate the analysis of SSSF processes.

Example 5.10

The mass rate of flow into a steam turbine is 1.5 kg/s, and the heat transfer from the turbine is 8.5 kW. The following data are known for the steam entering and leaving the turbine.

	Inlet Conditions	Exit Conditions
Pressure	2.0 MPa	0.1 MPa
Temperature	350°C	
Quality		100%
Velocity	50 m/s	200 m/s
Elevation above reference plane	6 m	3 m
$g = 9.8066$ m/s		

Determine the power output of the turbine.

Control volume: Turbine (Fig. 5.15).

Inlet state: Fixed (above).

Exit state: Fixed (above).

Process: SSSF.

Model: Steam tables.

$\dot{m}_i = 1.5$ kg/s
$P_i = 2$ MPa
$T_i = 350°C$
$\mathbf{V}_i = 50$ m/s
$Z_i = 6$ m

Control surface

Turbine

$\dot{m}_e = 1.5$ kg/s
$P_e = 0.1$ MPa
$x_e = 100\%$
$\mathbf{V}_e = 200$ m/s
$Z_e = 3$ m

FIGURE 5.15

Illustration for Example 5.10.

Analysis: First law (Eq. 5.49):

$$\dot{Q}_{c.v.} + \dot{m}\left(h_i + \frac{\mathbf{V}_i^2}{2} + gZ_i\right) = \dot{m}\left(h_e + \frac{\mathbf{V}_e^2}{2} + gZ_e\right) + \dot{W}_{c.v.}$$

with

$$\dot{Q}_{c.v.} = -8.5 \text{ kW}$$

Solution

$h_i = 3137.0$ kJ/kg (from the steam tables)

$$\frac{\mathbf{V}_i^2}{2} = \frac{50 \times 50}{2 \times 1000} = 1.25 \text{ kJ/kg}$$

$$gZ_i = \frac{6 \times 9.8066}{1000} = 0.059 \text{ kJ/kg}$$

Similarly, $h_e = 2675.5$ kJ/kg

$$\frac{\mathbf{V}_e^2}{2} = \frac{200 \times 200}{2 \times 1000} = 20.0 \text{ kJ/kg}$$

$$gZ_e = \frac{3 \times 9.8066}{1000} = 0.029 \text{ kJ/kg}$$

Therefore, substituting into Eq. 5.49,

$$-8.5 + 1.5(3137 + 1.25 + 0.059) = 1.5(2675.5 + 20.0 + 0.029) + \dot{W}_{c.v.}$$
$$\dot{W}_{c.v.} = -8.5 + 4707.5 - 4043.3 = 655.7 \text{ kW}$$

If Eq. 5.50 is used, the work per kilogram of fluid flowing is found first.

$$q + h_i + \frac{\mathbf{V}_i^2}{2} + gZ_i = h_e + \frac{\mathbf{V}_e^2}{2} + gZ_e + w$$

$$q = \frac{-8.5}{1.5} = -5.667 \text{ kJ/kg}$$

Therefore, substituting into Eq. 5.50,

$$-5.667 + 3137 + 1.25 + 0.059 = 2675.5 + 20.0 + 0.029 + w$$
$$w = 437.11 \text{ kJ/kg}$$
$$\dot{W}_{c.v.} = 1.5 \text{ kg/s} \times 437.11 \text{ kJ/kg} = 655.7 \text{ kW}$$

Two further observations can be made by referring to this example. First, in many engineering problems, potential energy changes are insignificant when compared with the other energy quantities. In the above example the potential energy change did not affect any of the significant figures. In most problems where the change in elevation is small the potential energy terms may be neglected.

Second, if velocities are small, say, under 100 ft/s, in many cases the kinetic energy is insignificant compared with other energy quantities. Furthermore, when the velocities entering and leaving the system are essentially the same, the change in kinetic energy is small. Since it is the change in kinetic energy that is important in the SSSF energy equation, the kinetic energy terms can usually be neglected when there is no significant difference between the velocity of the fluid entering and leaving the control volume. Thus, in many thermodynamic problems, one must make judgments as to which quantities may be negligible for a given analysis.

Example 5.11

Steam at 100 lbf/in.2, 400 F, enters an insulated nozzle with a velocity of 200 ft/s. It leaves at a pressure of 20 lbf/in.2 and a velocity of 2000 ft/s. Determine the final temperature if the steam is superheated in the final state, and the quality if it is saturated.

Control volume: Nozzle.

Inlet state: Fixed (see Fig. 5.16).

Exit state: P_e known.

Process: SSSF.

Model: Steam tables.

Analysis:

$$\dot{Q}_{c.v.} \approx 0 \text{ (nozzle insulated)}$$
$$\dot{W}_{c.v.} = 0 \qquad PE_i \approx PE_e$$

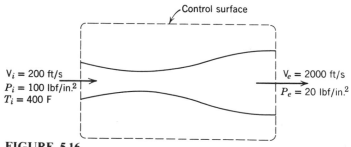

FIGURE 5.16
Illustration for Example 5.11.

First law (Eq. 5.50):

$$h_i + \frac{\mathbf{V}_i^2}{2} = h_e + \frac{\mathbf{V}_e^2}{2}$$

Solution

$$h_e = 1227.5 + \frac{(200)^2}{2 \times 32.17 \times 778} - \frac{(2000)^2}{2 \times 32.17 \times 778} = 1148.3 \text{ Btu/lbm}$$

The two properties of the fluid leaving that we now know are pressure and enthalpy, and therefore the state of this fluid is determined. Since h_e is less than h_g at 20 lbf/in.2, the quality is calculated.

$$h = h_f + xh_{fg}$$
$$1148.3 = 196.26 + x_e 960.1$$
$$x_e = 0.992$$

Example 5.12

In a refrigeration system, in which Freon-12 is the refrigerant, the Freon-12 enters the compressor at 150 kPa, −10°C and leaves at 1.0 MPa, 90°C. The mass rate of flow is 0.016 kg/s, and the power input to the compressor is 1 kW.

On leaving the compressor the Freon-12 enters a water-cooled condenser at 1.0 MPa, 80°C and leaves as a liquid at 0.95 MPa, 35°C. Water enters the condenser at 10°C and leaves at 20°C. Determine:

1. The heat transfer rate from the compressor.
2. The rate at which cooling water flows through the condenser.

First control volume: Compressor.
Inlet state: P_i, T_i; state fixed.
Exit state: P_e, T_e; state fixed.

Process: SSSF.

Model: Freon-12 tables.

Analysis: Since the velocity of the vapor entering a refrigeration compressor is low and not greatly different from the velocity leaving, kinetic energy changes, as well as potential energy changes, can be neglected. Therefore, the first law, Eq. 5.50, is

$$q + h_i = h_e + w$$

Solution

$$w = \frac{-1}{0.016} = -62.5 \text{ kJ/kg}$$

From the Freon-12 tables,

$$h_i = 184.62 \text{ kJ/kg} \qquad h_e = 240.10 \text{ kJ/kg}$$

Therefore,

$$q = 240.10 - 62.5 - 184.62 = -7.02 \text{ kJ/kg}$$
$$\dot{Q}_{c.v.} = 0.016 \text{ kg/s} \times (-7.02 \text{ kJ/kg}) = -0.1123 \text{ kW}$$

Second control volume: Condenser.

Sketch: Fig. 5.17.

Inlet states: Freon-12—fixed; water—fixed.

Exit states: Freon-12—fixed; water—fixed.

Process: SSSF.

Model: Freon-12 tables; steam tables.

Analysis: With this control volume we have two fluid streams, the Freon and the water, entering and leaving the control volume. It is reasonable to assume that the kinetic and potential energy changes are negligible. We note that the

FIGURE 5.17

Schematic diagram of a Freon-12 condenser.

work is zero, and we make the other reasonable assumption that there is no heat transfer across the control surface. Therefore, the first law, Eq. 5.47, reduces to

$$\sum \dot{m}_i h_i = \sum \dot{m}_e h_e$$

Using the subscript r for refrigerant and w for water,

$$\dot{m}_r(h_i)_r + \dot{m}_w(h_i)_w = \dot{m}_r(h_e)_r + \dot{m}_w(h_e)_w$$

Solution

From the Freon-12 and steam tables,

$$(h_i)_r = 232.74 \text{ kJ/kg} \qquad (h_i)_w = 42.01 \text{ kJ/kg}$$
$$(h_e)_r = 69.49 \text{ kJ/kg} \qquad (h_e)_w = 83.96 \text{ kJ/kg}$$

Solving the above equation for \dot{m}_w, the rate of flow of water,

$$\dot{m}_w = \dot{m}_r \frac{(h_i - h_e)_r}{(h_e - h_i)_w} = 0.016 \text{ kg/s} \frac{(232.74 - 69.49) \text{ kJ/kg}}{(83.96 - 42.01) \text{ kJ/kg}} = 0.0623 \text{ kg/s}$$

This problem can also be solved by considering two separate control volumes, one of which has the flow of Freon-12 across its control surface, and the other having the flow of water across its control surface. Further there is heat transfer from one control volume to the other. This is shown schematically in Fig. 5.18.

The heat transfer for the control volume involving Freon is calculated first. In this case the SSSF energy equation, Eq. 5.49, reduces to

$$\dot{Q}_{c.v.} = \dot{m}_r(h_e - h_i)_r$$
$$\dot{Q}_{c.v.} = 0.016 \text{ kg/s} \times (69.49 - 232.74) \text{ kJ/kg} = -2.612 \text{ kW}$$

This is also the heat transfer to the other control volume, for which $\dot{Q}_{c.v.} = +2.612 \text{ kW}$.

$$\dot{Q}_{c.v.} = \dot{m}_w(h_e - h_i)_w$$
$$\dot{m}_w = \frac{2.612 \text{ kW}}{(83.96 - 42.01) \text{ kJ/kg}} = 0.0623 \text{ kg/s}$$

Freon-12 vapor in Cooling water in

Control surfaces

$Q_{c.v.}$

Freon-12 liquid out Cooling water out

FIGURE 5.18

Schematic diagram of Freon-12 condenser considering two control surfaces.

Example 5.13

Consider the simple steam power plant, as shown in Fig. 5.19. The following data are for such a power plant.

Location	Pressure	Temperature or Quality
Leaving boiler	300 lbf/in.2	600 F
Entering turbine	280 lbf/in.2	550 F
Leaving turbine, entering condenser	2 lbf/in.2	93%
Leaving condenser, entering pump	1.9 lbf/in.2	110 F
Pump work = 3 Btu/lbm		

Determine the following quantities per pound mass flowing through the unit.

1. Heat transfer in line between boiler and turbine.
2. Turbine work.
3. Heat transfer in condenser.
4. Heat transfer in boiler.

There is a certain advantage in assigning a number to various points in the cycle. For this reason the subscripts i and e in the steady-state, steady-flow energy equation are often replaced by appropriate numbers.

Since there are several control volumes to be considered in the solution to this problem, let us consolidate our solution procedure somewhat in this example. Using the notation of Fig. 5.19:

All processes: SSSF.
Model: Steam tables.

FIGURE 5.19
Simple steam power plant.

From the steam tables:

$$h_1 = 1314.5 \text{ Btu/lbm}$$
$$h_2 = 1288.6 \text{ Btu/lbm}$$
$$h_3 = 94.0 + 0.93 \times 1022.1 = 1044.6 \text{ Btu/lbm}$$
$$h_4 = 78.0 \text{ Btu/lbm}$$

All analyses: No changes in kinetic or potential energy will be considered in the solution. In each case, the first law is given by Eq. 5.50.

Now, we proceed to answer the specific questions raised in the problem statement.

1. *Control volume* Pipe line between the boiler and the turbine.
 First law and solution:

$$_1q_2 + h_1 = h_2$$
$$_1q_2 = h_2 - h_1 = 1288.6 - 1314.5 = -25.9 \text{ Btu/lbm}$$

2. *Control volume* Turbine.
 First law and solution: A turbine is essentially an adiabatic machine. Therefore, it is reasonable to neglect heat transfer, so that

$$h_2 = h_3 + _2w_3$$
$$_2w_3 = 1288.6 - 1044.6 = 244.0 \text{ Btu/lbm}$$

3. *Control volume* Condenser.
 First law and solution: There is no work for this control volume. Therefore,

$$_3q_4 + h_3 = h_4$$
$$_3q_4 = 78.0 - 1044.6 = -966.6 \text{ Btu/lbm}$$

4. *Control volume* Boiler.
 First law: The work is equal to zero, so that

$$_5q_1 + h_5 = h_1$$

A solution requires a value for h_5, which can be found by taking a control volume around the pump:

$$h_4 = h_5 + _4w_5$$
$$h_5 = 3.0 + 78.0 = 81.0 \text{ Btu/lbm}$$

Therefore, for the boiler,

$$_5q_1 + h_5 = h_1$$
$$_5q_1 = 1314.5 - 81.0 = 1233.5 \text{ Btu/lbm}$$

Example 5.14

The centrifugal air compressor of a gas turbine receives air from the ambient atmosphere where the pressure is 14.5 lbf/in.2 and the temperature is 80 F. At the discharge of the compressor the pressure is 54 lbf/in.2, the temperature is 400 F, and the velocity is 300 ft/s. The mass rate of flow into the compressor is 2000 lbm/ min. Determine the power required to drive the compressor.

Control volume: We consider a control volume around the compressor, and locate the control volume at some distance from the compressor so that the air crossing the control surface has a very low velocity and is essentially at ambient conditions. If we located our control volume directly across the inlet section, it would be necessary to know the temperature and velocity at the compressor inlet.

Sketch: Fig. 5.20.

Inlet and exit states: Both states fixed.

Process: SSSF.

Model: Ideal gas with constant specific heat, value from Table A.8 (540 R).

Analysis: A large, high-flow rate compressor such as this is essentially an adiabatic machine, inasmuch as the air passes through the compressor very quickly and there is no particular effort made to promote any heat transfer. Therefore, we assume the process to be adiabatic. We also neglect any change in potential energy as well as the inlet kinetic energy. The first law, Eq. 5.50, reduces to

$$h_i = h_e + \frac{\mathbf{V}_e^2}{2} + w$$

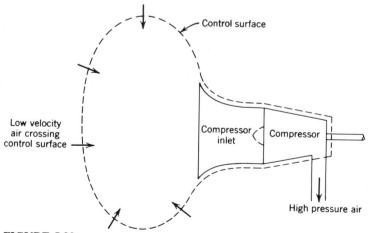

FIGURE 5.20
Sketch for Example 5.14.

Solution

$$-w = h_e - h_i + \frac{\mathbf{V}_e^2}{2} = C_{po}(T_e - T_i) + \frac{\mathbf{V}_e^2}{2}$$

$$= 0.24(400 - 80) + \frac{(300)^2}{2 \times 32.17 \times 778}$$

$$= 76.8 + 1.8 = 78.6 \text{ Btu/lbm}$$

$$-\dot{W}_{c.v.} = \frac{78.6(2000)}{42.4} = 3708 \text{ hp}$$

A more accurate model for the behavior of the air in this process would be the ideal gas and air tables, A.10. For this model, the solution is

$$h_i = 129.06 \text{ Btu/lbm} \qquad h_e = 206.46 \text{ Btu/lbm}$$

$$-w = h_e - h_i + \frac{\mathbf{V}_e^2}{2}$$

$$= 206.46 - 129.06 + \frac{(300)^2}{2 \times 32.17 \times 778}$$

$$-\dot{W}_{c.v.} = \frac{79.2(2000)}{42.4} = 3736 \text{ hp}$$

5.13 THE JOULE–THOMSON COEFFICIENT AND THE THROTTLING PROCESS

The Joule–Thomson coefficient, μ_J, is defined by the relation

$$\mu_J \equiv \left(\frac{\partial T}{\partial P}\right)_h \tag{5.52}$$

As in the definition of specific heats in Section 5.6, this quantity is defined in terms of thermodynamic properties, and therefore is itself a property of a substance.

The significance of the Joule–Thomson coefficient may be demonstrated by considering a throttling process that is a steady-state, steady-flow process across a restriction, with a resulting drop in pressure. A typical example is the flow through a partially opened valve or a restriction in the line. In most cases this occurs so rapidly and in such a small space that there is neither sufficient time nor a large enough area for much heat transfer. Therefore, we usually may assume such processes to be adiabatic.

If we consider the control surface shown in Fig. 5.21 we can write the SSSF energy equation for this process. There is no work, no change in potential energy, and we make the reasonable assumption that there is no heat transfer. The SSSF

FIGURE 5.21

The throttling process.

energy equation, Eq. 5.50, reduces to

$$h_i + \frac{\mathbf{V}_i^2}{2} = h_e + \frac{\mathbf{V}_e^2}{2}$$

If the fluid is a gas, the specific volume always increases in such a process and, therefore, if the pipe is of constant diameter, the kinetic energy of the fluid increases. In many cases, however, this increase in kinetic energy is small (or perhaps the diameter of the exit pipe is larger than that of the inlet pipe) and we can say with a high degree of accuracy that in this process the final and initial enthalpies are equal. Unless otherwise specified, we will assume this to be the case for throttling processes throughout the remainder of this text.

It is for such a process that the Joule–Thomson coefficient, μ_J, is significant. A positive Joule–Thomson coefficient means that the temperature drops during throttling, and when the Joule–Thomson coefficient is negative the temperature rises during throttling.

Example 5.15

Steam at 800 kPa, 300°C is throttled to 200 kPa. Changes in kinetic energy are negligible for this process. Determine the final temperature of the steam, and the average Joule–Thomson coefficient.

Control volume: Throttle valve (or other restriction).

Inlet state: P_i, T_i known; state fixed.

Exit state: P_e known.

Process: SSSF.

Model: Steam tables.

Analysis: First law, Eq. 5.50. Work equals zero. Neglect heat transfer and changes in kinetic and potential energies. Therefore, the first law reduces to

$$h_i = h_e$$

Solution

Since

$$h_e = h_i = 3056.5 \text{ kJ/kg}$$

and

$$P_e = 200 \text{ kPa}$$

these two properties determine the final state. From the superheat table for steam,

$$T_e = 292.4°C$$

$$\mu_{J(av)} = \left(\frac{\Delta T}{\Delta P}\right)_h = \frac{-7.6}{-600} = 0.0127 \text{ K/kPa}$$

Frequently a throttling process involves a change in the phase of the fluid. A typical example is the flow through the expansion valve of a vapor-compression refrigeration system. The following example deals with this problem.

Example 5.16

Consider the throttling process across the expansion valve or through the capillary tube in a vapor-compression refrigeration cycle. In this process the pressure of the refrigerant drops from the high pressure in the condenser to the low pressure in the evaporator, and during this process some of the liquid flashes into vapor. If we consider this process to be adiabatic, the quality of the refrigerant entering the evaporator can be calculated.

Consider the following process, in which ammonia is the refrigerant. The ammonia enters the expansion valve at a pressure of 225 lbf/in.2 and a temperature of 90 F. Its pressure on leaving the expansion valve is 38.5 lbf/in.2. Calculate the quality of the ammonia leaving the expansion valve.

Control volume: Expansion valve or capillary tube.

Inlet state: P_i, T_i known; state fixed.

Exit state: P_e known.

Process: SSSF.

Model: Ammonia tables.

Analysis: Standard throttling process analysis and assumptions, as in Example 5.15. The first law reduces to

$$h_i = h_e$$

Solution

From the ammonia tables

$$h_i = 143.6 \text{ Btu/lbm}$$

(The enthalpy of a slightly compressed liquid is essentially equal to the enthalpy of saturated liquid at the same temperature.)

$$h_e = h_i = 143.6 = 53.8 + x_e(561.1)$$

$$x_e = \frac{89.8}{561.1} = 0.160 = 16.0\%$$

5.14 THE UNIFORM-STATE, UNIFORM-FLOW PROCESS

In Section 5.12 we considered the steady-state, steady-flow process and several examples of its application. Many processes of interest in thermodynamics involve unsteady flow, and do not fit into this category. A certain group of these—for example, filling closed tanks with a gas or liquid, or discharge from closed vessels—can be reasonably represented to a first approximation by another simplified model. We call this process the uniform-state, uniform-flow process, or for convenience, the USUF process. The basic assumptions are as follows:

1. The control volume remains constant relative to the coordinate frame.
2. The state of the mass within the control volume may change with time, but at any instant of time the state is uniform throughout the entire control volume (or over several identifiable regions that make up the entire control volume).
3. The state of the mass crossing each of the areas of flow on the control surface is constant with time although the mass flow rates may be time varying.

Let us examine the consequence of these assumptions and derive an expression for the first law that applies to this process. The assumption that the control volume remains stationary relative to the coordinate frame has already been discussed in Section 5.12. The remaining assumptions lead to the following simplifications for the continuity equation and the first law.

The overall process occurs during time t. At any instant of time during the process, the continuity equation is

$$\frac{dm_{c.v.}}{dt} + \sum \dot{m}_e - \sum \dot{m}_i = 0$$

where the summation is over all areas on the control surface through which flow occurs. Integrating over time t gives the change of mass in the control volume during the overall process.

$$\int_0^t \left(\frac{dm_{c.v.}}{dt} \right) dt = (m_2 - m_1)_{c.v.}$$

The total mass leaving the control volume during time t is

$$\int_0^t \left(\sum \dot{m}_e \right) dt = \sum m_e$$

and the total mass entering the control volume during time t is

$$\int_0^t \left(\sum \dot{m}_i \right) dt = \sum m_i$$

Therefore, for this period of time t, we can write the continuity equation for the USUF process as

$$(m_2 - m_1)_{c.v.} + \sum m_e - \sum m_i = 0 \qquad (5.53)$$

In writing the first law for the USUF process we consider Eq. 5.45, which applies at any instant of time during the process.

$$\dot{Q}_{c.v.} + \sum \dot{m}_i\left(h_i + \frac{\mathbf{V}_i^2}{2} + gZ_i\right) = \frac{dE_{c.v.}}{dt} + \sum \dot{m}_e\left(h_e + \frac{\mathbf{V}_e^2}{2} + gZ_e\right) + \dot{W}_{c.v.}$$

Since at any instant of time the state within the control volume is uniform, the first law for the USUF process becomes

$$\dot{Q}_{c.v.} + \sum \dot{m}_i\left(h_i + \frac{\mathbf{V}_i^2}{2} + gZ_i\right) = \sum \dot{m}_e\left(h_e + \frac{\mathbf{V}_e^2}{2} + gZ_e\right)$$
$$+ \frac{d}{dt}\left[m\left(u + \frac{\mathbf{V}^2}{2} + gZ\right)\right]_{c.v.} + \dot{W}_{c.v.}$$

Let us now integrate this equation over time t, during which time we have

$$\int_0^t \dot{Q}_{c.v.}\, dt = Q_{c.v.}$$

$$\int_0^t \left[\sum \dot{m}_i\left(h_i + \frac{\mathbf{V}_i^2}{2} + gZ_i\right)\right] dt = \sum m_i\left(h_i + \frac{\mathbf{V}_i^2}{2} + gZ_i\right)$$

$$\int_0^t \left[\sum \dot{m}_e\left(h_e + \frac{\mathbf{V}_e^2}{2} + gZ_e\right)\right] dt = \sum m_e\left(h_e + \frac{\mathbf{V}_e^2}{2} + gZ_e\right)$$

$$\int_0^t \dot{W}_{c.v.}\, dt = W_{c.v.}$$

$$\int_0^t \frac{d}{dt}\left[m\left(u + \frac{\mathbf{V}^2}{2} + gZ\right)\right]_{c.v.} dt = \left[m_2\left(u_2 + \frac{\mathbf{V}_2^2}{2} + gZ_2\right)\right.$$
$$\left. - m_1\left(u_1 + \frac{\mathbf{V}_1^2}{2} + gZ_1\right)\right]_{c.v.}$$

Therefore, for this period of time t, we can write the first law for the uniform-state, uniform-flow process as

$$Q_{c.v.} + \sum m_i\left(h_i + \frac{\mathbf{V}_i^2}{2} + gZ_i\right)$$
$$= \sum m_e\left(h_e + \frac{\mathbf{V}_e^2}{2} + gZ_e\right)$$
$$+ \left[m_2\left(u_2 + \frac{\mathbf{V}_2^2}{2} + gZ_2\right) - m_1\left(u_1 + \frac{\mathbf{V}_1^2}{2} + gZ_1\right)\right]_{c.v.} + W_{c.v.} \quad (5.54)$$

As an example of the type of problem for which these assumptions are valid and Eq. 5.54 is appropriate, let us consider the classic problem of flow into an evacuated vessel. This is the subject of Example 5.17.

Example 5.17

Steam at a pressure of 1.4 MPa, 300°C is flowing in a pipe, Fig. 5.22. Connected to this pipe through a valve is an evacuated tank. The valve is opened and the tank fills with steam until the pressure is 1.4 MPa, and then the valve is closed.

FIGURE 5.22

Flow into an evacuated vessel—control volume analysis.

The process takes place adiabatically and kinetic energies and potential energies are negligible. Determine the final temperature of the steam.

Control volume: Tank, as shown in Fig. 5.22.

Initial state (in tank): Evacuated, mass $m_1 = 0$.

Final state: P_2 known.

Inlet state: P_i, T_i (in line) known.

Process: USUF.

Model: Steam tables.

Analysis: First law, Eq. 5.54:

$$Q_{c.v.} + \sum m_i \left(h_i + \frac{\mathbf{V}_i^2}{2} + gZ_i \right)$$

$$= \sum m_e \left(h_e + \frac{\mathbf{V}_e^2}{2} + gZ_e \right)$$

$$+ \left[m_2 \left(u_2 + \frac{\mathbf{V}_2^2}{2} + gZ_2 \right) - m_1 \left(u_1 + \frac{\mathbf{V}_1^2}{2} + gZ_1 \right) \right]_{c.v.} + W_{c.v.}$$

We note that $Q_{c.v.} = 0$, $W_{c.v.} = 0$, $m_e = 0$, and $(m_1)_{c.v.} = 0$. We further assume that changes in kinetic and potential energy are negligible. Therefore, the statement of the first law for this process reduces to

$$m_i h_i = m_2 u_2$$

From the continuity equation for this process, Eq. 5.53, we conclude that

$$m_2 = m_i$$

Therefore, combining the continuity equation with the first law, we have

$$h_i = u_2$$

That is, the final internal energy of the steam in the tank is equal to the enthalpy of the steam entering the tank.

FIGURE 5.23
Flow into an evacuated vessel—system analysis.

Solution

From the steam tables

$$h_i = u_2 = 3040.4 \text{ kJ/kg}$$

Since the final pressure is given as 1.4 MPa, we know two properties at the final state and therefore the final state is determined. The temperature corresponding to a pressure of 1.4 MPa and an internal energy of 3040.4 kJ/kg is found to be 452°C. Had this problem involved a substance for which internal energies are not listed in the thermodynamic tables, it would have been necessary to calculate a few values for u before an interpolation could be made for the final temperature.

This problem can also be solved by considering the steam that enters the tank and the evacuated space as a system, as indicated in Fig. 5.23.

The process is adiabatic, but we must examine the boundaries for work. If we visualize a piston between the steam that is included in the system and the steam that flows behind, we readily recognize that the boundaries of the system move and that the steam in the pipe does work on the steam that comprises the system. The amount of this work is

$$-W = P_1 V_1 = m P_1 v_1$$

Writing the first law for the system, Eq. 5.11, and noting that kinetic and potential energies can be neglected, we have

$$_1 Q_2 = U_2 - U_1 + {}_1 W_2$$
$$0 = U_2 - U_1 - P_1 V_1$$
$$0 = m u_2 - m u_1 - m P_1 v_1 = m u_2 - m h_1$$

Therefore,

$$u_2 = h_1$$

which is the same conclusion that was reached using a control volume analysis.

The two other examples that follow illustrate further the uniform-state, uniform-flow process.

Example 5.18

Let the tank of the previous example have a volume of 0.4 m^3 and initially contain saturated vapor at 350 kPa. The valve is then opened and steam from the line at 1.4 MPa, 300°C flows into the tank until the pressure is 1.4 MPa.

Calculate the mass of steam that flows into the tank.

Control volume: Tank, as in Fig. 5.22.

Initial state: P_1, saturated vapor; state fixed.

Final state: P_2.

Inlet state: P_i, T_i; state fixed.

Process: USUF.

Model: Steam tables.

Analysis: Same as in Example 5.17, except that the tank is not evacuated initially. Again we note that $Q_{c.v.} = 0$, $W_{c.v.} = 0$, $m_e = 0$, and we assume that changes in kinetic energy and potential energy are zero. The statement of the first law for this process, Eq. 5.54, reduces to

$$m_i h_i = m_2 u_2 - m_1 u_1$$

The continuity equation, Eq. 5.53, reduces to

$$m_2 - m_1 = m_i$$

Therefore, combining the continuity equation and the first law, we have

$$(m_2 - m_1)h_i = m_2 u_2 - m_1 u_1$$
$$m_2(h_i - u_2) = m_1(h_i - u_1) \tag{a}$$

There are two unknowns in this equation—m_2 and u_2. However, we have one additional equation:

$$m_2 v_2 = V = 0.4 \text{ m}^3 \tag{b}$$

Substituting (b) into (a) and rearranging, we have

$$\frac{V}{v_2}(h_i - u_2) - m_1(h_i - u_1) = 0 \tag{c}$$

in which the only unknowns are v_2 and u_2, both functions of T_2 and P_2. Since T_2 is unknown, it means that there is only one value of T_2 for which Eq. (c) will be satisfied, and we must find it by trial and error.

Solution

$$v_1 = 0.5243 \text{ m}^3/\text{kg} \qquad m_1 = \frac{0.4}{0.5243} = 0.763 \text{ kg}$$

$$h_i = 3040.4 \text{ kJ/kg} \qquad u_1 = 2548.9 \text{ kJ/kg}$$

Assume

$$T_2 = 342°C$$

For this temperature and the known value of P_2,

$$v_2 = 0.1974 \text{ m}^3/\text{kg} \qquad u_2 = 2855.8 \text{ kJ/kg}$$

Substituting into (c),

$$\frac{0.4}{0.1974}(3040.4 - 2855.8) - 0.763(3040.4 - 2548.9) \simeq 0$$

and we conclude that the assumed value for $T_2 = 342°C$ is correct. The final mass inside the tank is

$$m_2 = \frac{0.4}{0.1974} = 2.026 \text{ kg}$$

and the mass of steam that flows into the tank is

$$m_i = m_2 - m_1 = 2.026 - 0.763 = 1.263 \text{ kg}$$

Example 5.19

A tank of 50 ft³ volume contains saturated ammonia at a pressure of 200 lbf/in.². Initially the tank contains 50 per cent liquid and 50 per cent vapor by volume. Vapor is withdrawn from the top of the tank until the pressure is 100 lbf/in.². Assuming that only vapor (i.e., no liquid) leaves and that the process is adiabatic, calculate the mass of ammonia that is withdrawn.

Control volume: Tank.

Initial state: T_1, V_{liq}, V_{vap}; state fixed.

Final state: T_2.

Exit state: Saturated vapor (temperature changing).

Process: USUF.

Model: Ammonia tables.

Analysis: In the first law, Eq. 5.54, we note that $Q_{\text{c.v.}} = 0$, $W_{\text{c.v.}} = 0$, $m_i = 0$, and we assume that changes in kinetic and potential energy are negligible. However, the enthalpy of saturated vapor varies with temperature, and therefore we cannot simply assume that the enthalpy of the vapor leaving the tank remains constant. We note that at 200 lbf/in.², $h_g = 632.7$ Btu/lbm and at 100 lbf/in.², $h_g = 626.5$ Btu/lbm. Since the change in h_g during this process is small, we may accurately assume that h_e is the average of the two values given above. Therefore

$$(h_e)_{\text{av}} = 629.6 \text{ Btu/lbm}$$

and the first law reduces to

$$m_e h_e + m_2 u_2 - m_1 u_1 = 0$$

and the continuity equation (from Eq. 5.53)

$$(m_2 - m_1)_{\text{c.v.}} + m_e = 0$$

Combining these two equations we have

$$m_2(h_e - u_2) = m_1 h_e - m_1 u_1$$

The following values are from the ammonia tables:

$$v_{f1} = 0.02732 \text{ ft}^3/\text{lbm} \qquad v_{g1} = 1.502 \text{ ft}^3/\text{lbm}$$

$$v_{f2} = 0.02584 \qquad\qquad v_{g2} = 2.952$$

$$u_{f1} = 150.9 - \frac{200 \times 144 \times 0.0273}{778} = 149.9 \text{ Btu/lbm}$$

$$u_{f2} = 104.7 - \frac{100 \times 144 \times 0.0258}{778} = 104.2$$

$$u_{g1} = 632.7 - \frac{200 \times 144 \times 1.502}{778} = 577.1$$

$$u_{g2} = 626.5 - \frac{100 \times 144 \times 2.952}{778} = 571.9$$

$$u_{fg2} = 571.9 - 104.2 = 467.7$$

Calculating first the initial mass, m_1, in the tank, the mass of the liquid initially present, m_{f1}, is

$$m_{f1} = \frac{25}{0.02732} = 915 \text{ lbm}$$

Similarly, the initial mass of vapor, m_{g1}, is

$$m_{g1} = \frac{25}{1.502} = 16.6 \text{ lbm}$$

$$m_1 = m_{f1} + m_{g1} = 915 + 17 = 932 \text{ lbm}$$

$$m_1 h_e = 932 \times 629.6 = 586\,800 \text{ Btu}$$

$$m_1 u_1 = (mu)_{f1} + (mu)_{g1} = 915 \times 149.9 + 16.6 \times 577.0 = 146\,700 \text{ Btu}$$

Substituting these into the first law,

$$m_2(h_e - u_2) = m_1 h_e - m_1 u_1 = 586\,800 - 146\,700 = 440\,100 \text{ Btu}$$

There are two unknowns, m_2 and u_2, in this equation. However,

$$m_2 = \frac{V}{v_2} = \frac{50}{0.0258 + x_2(2.952 - 0.0258)}$$

and

$$u_2 = 104.2 + x_2(467.7)$$

both functions only of x_2, the quality at the final state. Consequently,

$$\frac{50(629.6 - 104.2 - 467.7x_2)}{0.0258 + 2.926x_2} = 440\,100$$

Solving, $x_2 = 0.01137$

Therefore,

$$v_2 = 0.0258 + 0.01137(2.926) = 0.0590 \text{ ft}^3/\text{lbm}$$

$$m_2 = \frac{50}{0.0590} = 847 \text{ lbm}$$

and the mass of ammonia withdrawn, m_e, is

$$m_e = m_1 - m_2 = 932 - 847 = 85 \text{ lbm}$$

PROBLEMS

5.1 The rate of heat transfer to the surroundings from a person at rest is about 400 Btu/h. Suppose that the ventilation system fails in an auditorium containing 100 people.

(a) How much will the internal energy of the air in the auditorium increase during the first ten minutes after the ventilation system fails?

(b) Considering the auditorium and all the people in it as a system, how much does the internal energy of this system change? How do you explain the fact that the temperature of the air increases?

5.2 A "bomb calorimeter", a closed rigid vessel, is to be used to measure the energy released by a certain chemical reaction. This calorimeter initially contains the appropriate chemicals and is located in a large tank of water. When the chemicals react, heat is transferred from the bomb to the water, causing its temperature to rise. The electric power input to a stirrer used to circulate the water is 0.05 kW. In a 25-minute period the heat transfer from the bomb to the water is 1400 kJ and the heat transfer from the water tank to the surrounding air is 70 kJ. Assuming that no water evaporates, what is the increase in the internal energy of the water during this period of time?

5.3 A closed, rigid 50-L tank contains water at 120°C, 60% quality. The tank is then cooled to −10°C. Calculate the heat transfer during the process.

5.4 A radiator of a steam heating system has a volume of 0.5 ft³. At a time when this radiator is filled with saturated vapor steam at 30 lbf/in.², both valves to the radiator are closed. How much heat will have been transferred to the room when the steam pressure in the radiator has dropped to 14.7 lbf/in.²?

5.5 Calculate the heat transfer for the process described in Problem 4.6.

5.6 A rigid 15-ft³ tank contains ammonia at 180 lbf/in.², 200 F. The tank is now cooled to 40 F. Determine the heat transfer for this process.

5.7 Consider the 100 L dewar (a rigid, double-walled vessel for storing cryogenic liquids) shown in Fig. 5.24. The dewar contains liquid nitrogen at 77.35 K,

$$\Delta h = C_p \Delta T$$
$$\Delta u = C_v \Delta T$$

FIGURE 5.24

Sketch for Problem 5.7.

90% liquid and 10% vapor by volume. The insulation holds heat transfer into the dewar from the ambient to a very low rate, 5 J/s. The vent valve is now accidentally closed, so that the pressure inside slowly rises. It is estimated that the dewar will rupture when the pressure reaches 500 kPa. How long would it take to reach this pressure?

5.8 A closed tube has a volume of 5 in.3 and contains the appropriate fraction of liquid and vapor water in equilibrium at 100 F such that when the tube is heated the water will pass through the critical point. Calculate the heat transfer to the tube to change the water from 100 F to the critical point state.

5.9 Consider the insulated vessel shown in Fig. 5.25. This vessel has an evacuated compartment separated by a partition from a second compartment that contains 1 lbm of water at 180 F, 100 lbf/in.2. The partition now ruptures and the water expands to fill the entire volume, with a resulting pressure of 2 lbf/in.2. Determine the final temperature of the water and the volume of the vessel.

FIGURE 5.25

Sketch for Problem 5.9.

5.10 Calculate the heat transfer for the process described in Problem 4.13.

5.11 An insulated cylinder fitted with a piston contains Freon-12 at 25°C with a quality of 90%. The volume at this state is 30 L. The piston is then allowed to move, and the Freon-12 expands until it exists as saturated vapor. During this process, the Freon-12 does 4.0 kJ of work against the piston. Determine the final temperature, assuming the process to be adiabatic.

5.12 A vertical cylinder fitted with a piston contains 10 lbm of Freon-12 at 60 F. Heat is now transferred to the system, causing the piston to rise until it reaches a set of stops, at which point the volume has doubled. Additional heat is transferred until the temperature inside reaches 200 F, at which point the pressure inside the cylinder is 175 lbf/in.2.
(a) What is the quality at the initial state?
(b) Calculate the heat transfer for the overall process.

5.13 A mass of 5 lbm of Freon-12 at 50 F, 30% quality, is contained inside a spherical balloon that supports an internal pressure directly proportional to its diameter. This system is now heated until the pressure in the balloon reaches 90 lbf/in.2. What is the heat transfer for the process?

5.14 A cylinder fitted with a piston restrained by a linear spring (force proportional to distance) has a diameter of 0.25 m and initial volume of 20 L, at which point the cylinder contains ammonia at 1 MPa, 60°C. The spring constant is 150 kN/m. Heat is now rejected from the system, and the piston moves until 5.0 kJ of work has been done on the ammonia.
(a) Find the final temperature of the ammonia.
(b) Calculate the heat transfer for the process.

5.15 Consider the system shown in Fig. 5.26. Tank A has a volume of 100 L and contains saturated vapor Freon-12 at 25°C. When the valve is cracked open, Freon-12 flows slowly into cylinder B. The piston mass is such that a pressure of 150 kPa in cylinder B is required to raise the piston. The process ends when the pressure in tank A has fallen to 150 kPa. During this process, heat is transferred with the surroundings such that the temperature of all the

$u_1 = h_1 - P \cdot v_1$

$W = m P(v_5 - v_1)$

$\sim 3.7249 50(.9324 - .026834)$

FIGURE 5.26
Sketch for Problem 5.15.

Freon-12 always remains at 25°C. Calculate the heat transfer for this process.

5.16 Two heavily insulated tanks are connected by a valve. Tank A has a volume of 25 ft³ and contains water at 20 lbf/in.², 400 F. Tank B has a volume of 15 ft³ and contains water at 80 lbf/in.², 85% quality. The valve is now opened, and the two tanks eventually come to a uniform state. Assuming the process to be adiabatic, what is the final pressure? $m_A u_A + m_B u_B = m_T u_T$

5.17 A rigid 1-m³ tank *A* contains Freon-12 at room temperature, 20°C, at a quality of 70%. This tank is connected by a valve to vertical cylinder *B*, which is initially evacuated and has a volume of 0.5 m³. A pressure of 150 kPa is required to raise the piston from the stops in *B*. The valve is now opened and when half the original mass in *A* has been transferred to *B*, the valve is closed. Eventually, both *A* and *B* return to room temperature. Find:
(a) The final pressure in tank *A*.
(b) The work done by the Freon-12 in cylinder *B*.
(c) The heat transfer to the Freon-12.

5.18 An insulated cylinder containing water has a piston held by a pin, as shown in Fig. 5.27. The water is initially saturated vapor at 65°C, and the volume is 5 L. The piston and weights have a total mass of 10 kg, the piston cross-sectional area is 3000 mm², and the ambient pressure is 100 kPa. The pin is now released, allowing the piston to move. Determine the final state of the water, assuming the process to be adiabatic.

FIGURE 5.27
Sketch for Problem 5.18.

5.19 A cylinder fitted with a piston contains 4 lbm of Freon-12 at 50 F, 90% quality. The system now undergoes a quasiequilibrium, polytropic expansion to 15 lbf/in.², during which the system receives a heat transfer of 50 Btu. What is the final temperature of the Freon-12?

5.20 A cylinder contains 0.5 lbm of saturated vapor water at 230 F, as shown in Fig. 5.28. At this state the spring is touching the piston but exerts no force on it. Heat is now transferred to the water causing the piston to rise, and during this process the spring resisting force is proportional to the distance

$$\ell_{sp} = P_2 - P_1 = 1822$$

FIGURE 5.28

Sketch for Problem 5.20.

moved, with a spring constant of 300 lbf/in. The piston cross-sectional area is 80 in.2.

(a) What is the temperature in the cylinder when the pressure inside reaches 40 lbf/in.2? 887

(b) Calculate the heat transfer for this process. $122.3 Btu$

5.21 Calculate the heat transfer for the process described in Problem 4.22.

5.22 Consider the cylinder-piston arrangement shown in Fig. 5.29, in which a frictionless piston is free to move between two sets of stops. When the piston is resting on the lower stops the enclosed volume is 10 ft^3, and when the piston reaches the upper stops the volume is 15 ft^3. The cylinder initially contains water at 14.7 lbf/in.2, 20% quality. This system is now heated until the water eventually exists as a saturated vapor. If the mass of the piston is

FIGURE 5.29

Sketch for Problem 5.22.

such that 50 lbf/in.² pressure is required to move it against the outside ambient pressure, determine:

(a) The final pressure in the cylinder. 53.1

(b) The heat transfer and work for the overall process. 1420

5.23 A cylinder fitted with a piston restrained by a linear spring contains 2 kg of water at 400 kPa, 700°C. The piston cross-sectional area is 0.1 m² and the spring constant is 5 kN/m. Heat is now transferred from the water to the surroundings, and the process continues until the pressure inside the cylinder has dropped to 125 kPa.

(a) Determine the final state of the water.

(b) Calculate the heat transfer for the process.

5.24 Calculate the heat transfer for the process described in Problem 4.24.

5.25 A 20 lbm bar of a certain alloy having essentially the same specific heat and density as iron is quenched from 1300 F in a 3.5 ft³ oil bath initially at 70 F. Assuming no heat transfer with the surroundings, what is the final temperature? $m = \rho V \quad 0 = U_2 - U_1 \quad ;0 = U_4 - U_1 + U_2 - U_3 \quad U = C_V m T$
$C_{V/H2}(T_{f3} - T_2) = C_{V/H2}(T_f - T_1)$

5.26 A hollow spherical aluminum vessel has an inside diameter of 0.5 m and a 10-mm-thick wall. This vessel contains water at 25°C, with a quality of 1%. The vessel is now heated until the water inside is saturated vapor. Considering the vessel and water together as a system, calculate the heat transfer for the process.

5.27 An ideal gas is heated from 500 F to 1500 F. The enthalpy change per pound mass is to be calculated assuming constant specific heat, value from Table A.8. Calculate this value, and discuss the accuracy of the result, if the gas is:

(a) Helium.

(b) Nitrogen.

(c) Carbon dioxide.

5.28 A mass of 1 kg of water at 50 kPa, 50% quality is heated to 200 kPa, 1600°C. Calculate the change in internal energy for this process, using whatever model (or models) that is appropriate.

5.29 One kilogram of nitrogen gas is heated from 30 to 1500°C. Calculate the change in enthalpy by the following methods:

(a) Constant specific heat, value from Table A.8.

(b) Constant specific heat, value at the average temperature from the equation, Table A.9.

(c) Variable specific heat, integrating the equation, Table A.9.

(d) Variable specific heat, values from the ideal gas tables, Table A.11.

5.30 A rigid 2-ft³ tank contains methane gas at 1000 F, 100 lbf/in.². The tank is then cooled to −100 F.

(a) Calculate the final pressure and the heat transfer for the process.

(b) What is the percent error in the heat transfer if the specific heat is assumed constant at the room-temperature value?

5.31 A cylinder with a piston restrained by a linear spring contains 1 lbm of carbon dioxide at 60 lbf/in.2, 800 F. This system is now cooled to 100 F, at which point the pressure is 50 lbf/in.2. Calculate the heat transfer for the process.

5.32 Calculate the heat transfer for the process described in Problem 4.9.

5.33 A cylinder fitted with a piston contains 0.15 kg of air initially at 100 kPa, 30°C. The air is now compressed in a quasiequilibrium, polytropic process, with polytropic exponent n of 1.25, until the final volume is one-tenth of the initial volume. Determine the work and the heat transfer for this process.

5.34 A closed cylinder is divided into two parts by a frictionless piston held in place by a pin. Side A is 0.3 ft^3 and contains air at 14.7 lbf/in.2, 90 F and side B is 10 ft^3 and contains saturated vapor water at 90 F. The pin is now pulled, releasing the piston. Eventually, both sides come to equilibrium at 90 F. Considering the air and water together as a system, determine
(a) the work done by the system.
(b) the heat transfer to the system.

5.35 A cylinder fitted with a piston contains 1 kg of methane gas at 700 kPa, 40°C. The piston cross-sectional area is 0.5 m^2 and the total external force restraining the piston is directly proportional to the cylinder volume squared. Heat is now transferred to the methane until its temperature reaches 1100°C. Determine the final pressure inside the cylinder, the work done by the methane, and the heat transfer during the process.

5.36 A certain balloon will support an internal pressure equal to $P_o = 100$ kPa until the volume reaches $V_o = 20$ m^3, above which

$$P = P_o + C(V - V_o)^2$$

This balloon contains helium gas at 20°C, 100 kPa, with a 15 m^3 volume. The balloon is then heated until the helium reaches 400°C, at which point the pressure is 150 kPa. Calculate the work done by the helium and the heat transfer.

5.37 A cylinder fitted with a piston contains argon gas at 20 lbf/in.2, 50 F, at which point the volume is 5 ft^3. The gas is now compressed in a polytropic process to 100 lbf/in.2, 350 F. Calculate the heat transfer during the process.

5.38 Heat is transferred at a given rate to a mixture of liquid and vapor in equilibrium in a closed container. Determine the rate of change of temperature as a function of the thermodynamic properties of the liquid and vapor and the masses of liquid and vapor.

5.39 Argon gas at 280 K, 125 kPa, is flowing in a 100-mm-diameter circular duct at a uniform velocity of 3 m/s. What is the mass rate of flow?

5.40 Air is heated electrically in a constant-diameter tube in a steady-flow process. At the tube entrance the air is at 75 lbf/in.2, 70 F, with a velocity of 10 ft/s, and at the exit the air is at 70 lbf/in.2, 240 F. Calculate the velocity of the air at the tube exit.

5.41 Saturated vapor Freon-12 exits a refrigeration evaporator at $-10°C$, with a steady mass flow rate of 0.1 kg/s. What is the smallest diameter tubing that can be used at this location if the velocity of the refrigerant is not to exceed 10 m/s?

5.42 Nitrogen gas enters a steady-state, steady-flow heater at 90 lbf/in.2, 60 F, and exits at 85 lbf/in.2, 1800 F. Changes in kinetic and potential energies are negligible. Calculate the required heat transfer per lbm of nitrogen flowing through the heater.

5.43 Superheated vapor steam enters a turbine nozzle at 3 MPa, 350°C, with a low velocity, and exits the nozzle at 1.6 MPa with a velocity of 550 m/s. The steam mass flow rate is 0.5 kg/s. Calculate the quality or temperature of the steam exiting the nozzle and also the exit area of the nozzle.

5.44 A small, high-speed turbine operating on compressed air produces a power output of 0.1 hp. The inlet state is 60 lbf/in.2, 70 F, and the exit state is 14.7 lbf/in.2, -60 F. Assuming the velocities to be low, find the required mass flow rate of air through the turbine.

5.45 The compressor of a large gas turbine receives air from the ambient at 14.0 lbf/in.2, 70 F, with a low velocity. At the compressor discharge, air exits at 170 lbf/in.2, 720 F, with a velocity of 300 ft/s. The power input to the compressor is 5000 hp. Determine the mass flow rate of air through the unit.

5.46 A number of years ago, a well-known architect designed a 1600-meter-tall building. Suppose that in such a building steam for the heating system enters a pipe at ground level as saturated vapor at 200 kPa. On the top floor of the building, the pressure in the pipe is 100 kPa, and the heat transfer from the steam as it flows up the pipe is 100 kJ/kg. What is the quality of the steam at the top of the pipe?

5.47 Two steady steams of air enter a control volume: one enters at point 1 at 350 kPa, 150°C, with low velocity at the rate of 0.025 kg/s; the other enters at point 2 at 350 kPa, 15°C, with low velocity. A single stream of air exits the control volume at point 3 at 100 kPa, $-40°C$, through a 25-mm-diameter pipe. The control volume rejects heat to the surroundings at the rate of 1.2 kW and also produces 4.5 kW of power output. Determine the flow rate of air entering the control volume at point 2.

5.48 Consider a powerplant steam generator in which compressed liquid water at 1000 lbf/in.2, 100 F enters a 1-in.-diameter tube at the rate of 5 ft^3/min.. Saturated vapor exits this tube at 900 lbf/in.2. Calculate the rate of heat transfer to the water.

5.49 In certain situations, when only superheated vapor steam at a given state is available, a need for saturated steam may arise for a specific purpose. This can be accomplished in a steady-flow desuperheater, in which liquid water is sprayed into the superheated steam in the appropriate amount such that the steam exiting the unit is saturated vapor. Consider such a desuperheater, in which superheated steam at 3.5 MPa, 400°C enters at the rate of 0.5 kg/s,

and liquid water enters the unit at 3.5 MPa, 40°C. If saturated vapor exits at 3 MPa, determine the required rate of liquid water input to the desuperheater unit.

5.50 Consider a steady-state, steady-flow mixing process in which liquid ammonia at 180 lbf/in.², 60 F is mixed with saturated vapor ammonia at 180 lbf/in.². The two inlet stream mass flow rates are equal, and exiting the mixer is ammonia at 140 lbf/in.² with a quality of 85%. Calculate the heat transfer per lbm of ammonia exiting the mixer.

5.51 The following data are for a simple steam power plant as shown in Fig. 5.30.

$$P_1 = 900 \text{ lbf/in.}^2$$
$$P_2 = 890 \text{ lbf/in.}^2, \quad T_2 = 115 \text{ F}$$
$$P_3 = 860 \text{ lbf/in.}^2, \quad T_3 = 350 \text{ F}$$
$$P_4 = 830 \text{ lbf/in.}^2, \quad T_4 = 920 \text{ F}$$
$$P_5 = 800 \text{ lbf/in.}^2, \quad T_5 = 900 \text{ F}$$
$$P_6 = 1.5 \text{ lbf/in.}^2, \quad x_6 = 0.92, \quad V_6 = 600 \text{ ft/s}$$
$$P_7 = 1.4 \text{ lbf/in.}^2, \quad T_7 = 110 \text{ F}$$

Rate of steam flow \doteq 200 000 lbm/h

Power input to pump = 400 hp

Pipe diameters:
 Steam generator to turbine: 8 in.
 Condenser to steam generator: 3 in.

Calculate:
(a) Power output of the turbine.
(b) Heat transfer rates in condenser, economizer, and steam generator.
(c) Flow rate of cooling water through the condenser, if the temperature of the cooling water increases from 55 F to 75 F in the condenser.

$$\dot{m} = \frac{A\bar{V}}{\nu}$$

FIGURE 5.30

Sketch for Problem 5.51.

5.52 A somewhat simplified flow diagram for the nuclear power plant shown in Fig. 1.5 is given in Fig. 5.31, and mass flow rates and state of the water at various points in the cycle are shown in the table below.

Point	\dot{m} kg/s	P kPa	T°C	h kJ/kg
1	75.6	7240	sat vap	
2	75.6	6900		2765
3	62.874	345		2517
4		310		
5		7		2279
6	75.6	7	33	
7		415		140
8	2.772	35		2459
9	4.662	310		558
10		35	54	
11	75.6	380	68	
12	8.064	345		2517
13	75.6	330		
14				349
15	4.662	965	139	584
16	75.6	7930		565
17	4.662	965		2593
18	75.6	7580		688
19	1386	7240	277	
20	1386	7410		1221
21	1386	7310		

FIGURE 5.31
Sketch for Problem 5.52.

This cycle involves a number of "heaters", in which heat is transferred from steam that leaves the turbine at some intermediate pressure to liquid water that is being pumped from the condenser to the steam drum. The rate of heat transfer to the water in the reactor is 157 MW.

(a) Assuming no heat transfer from the moisture separator between the high-pressure and low-pressure turbines, determine the enthalpy per kilogram of steam and the quality of the steam entering the low-pressure turbine.

(b) Determine the power output of the high-pressure turbine, assuming no heat transfer from the steam as it flows through the turbine.

(c) Determine the power output of the low-pressure turbine, assuming no heat transfer from the steam as it flows through the turbine.

(d) Determine the quality of the steam leaving the reactor.

(e) Determine the temperature of the water leaving the intermediate-pressure heater, assuming no heat transfer from the heater to the surroundings.

(f) Determine the rate of heat transfer to the condenser cooling water.

(g) What is the ratio of the total power output of the two turbines to the heat transferred to the water in the reactor?

5.53 A schematic arrangement for a proposed procedure for producing fresh water from saltwater that would operate in conjunction with a large steam power plant and utilize a flash evaporator is shown in Fig. 5.32. Cooling water at the rate of 2 5000 000 lbm/h enters the condenser, where its temperature is increased from 63 F to 85 F. It then enters a flash evaporator, where the pressure is reduced to that corresponding to a saturation temperature of 76.5 F. During this process some of the liquid flashes into vapor, and the remainder is pumped back to the sea. The vapor is then condensed

FIGURE 5.32
Sketch for Problem 5.53.

at 74 F to form the desired fresh water. This takes place by utilizing the sea water that enters at 63 F and exits at 69.5 F. Calculate:

(a) The amount of fresh water that is produced per hour. Assume that the mixture that is formed when the liquid at 85 F is throttled (as it enters the flash evaporator) is in equilibrium at 76.5 F, and that it is perfectly separated.

(b) The amount of cooling sea water that enters at point 4.

5.54 A steam turbine is used to drive a nitrogen compressor, as shown in Fig. 5.33. The states are as given in the diagram, and the mass flow rates are 0.25 kg/s and 0.04 kg/s through the turbine and compressor, respectively. Heat transfer from the turbine is small and can be neglected. The turbine delivers 42 kW to the compressor and the balance of its output to an electric power generator. Determine:

(a) The power available to drive the generator.

(b) The rate of heat transfer from the nitrogen as it flows through the compressor and aftercooler.

FIGURE 5.33

Sketch for Problem 5.54.

5.55 Methane at 3 MPa, 300 K, is throttled to 100 kPa. Calculate the exit temperature, assuming:

(a) Ideal gas behavior.

(b) Real gas behavior.

5.56 Helium is throttled from 175 lbf//in.2, 75 F, to a pressure of 15 lbf/in.2. The diameter of the exit pipe is sufficiently larger than that of the inlet pipe so that the exit and inlet velocities are equal. Determine the exit temperature of the helium and also the ratio of pipe diameters.

5.57 A small steam turbine operating at part load produces a 110-kW output with a steam flow rate of 0.25 kg/s. Steam at 1.4 MPa, 250°C, is throttled to 1.1 MPa before entering the turbine, and the turbine exhaust pressure is 10 kPa. Find the steam quality (or temperature, if superheated) at the turbine outlet.

5.58 It is proposed to use a geothermal supply of hot water to operate a steam turbine, as shown in Fig. 5.34. The high-pressure water at 200 lbf/in.2, 350 F, is throttled into an insulated flash evaporator forming liquid and vapor at a lower pressure, 60 lbf/in.2. The liquid exits at the bottom, while the vapor is drawn off and fed to the turbine, as shown. Steam exits from the turbine at 1 lbf/in.2, 90% quality. If the turbine produces a power output of 1000 hp, how much hot water is required from the geothermal source per hour?

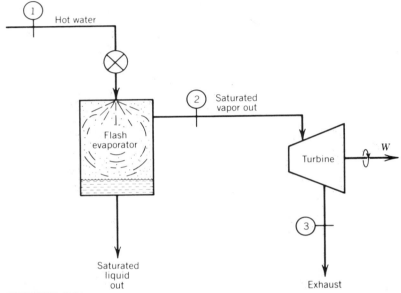

FIGURE 5.34
Sketch for Problem 5.58.

5.59 Water at 200 lbf/in.2, 300 F is throttled adiabatically through a valve to 30 lbf/in.2. The inlet velocity is 5 ft/s and the inlet and exit cross-sectional areas are the same. Determine the state and velocity of the water at the exit.

5.60 The following data are for the Freon-12 heat pump cycle shown in Fig. 5.35:

$$P_1 = 1250 \text{ kPa}, \quad T_1 = 120°C$$
$$P_2 = 1230 \text{ kPa}, \quad T_2 = 110°C$$
$$P_3 = 1200 \text{ kPa}, \quad T_3 = 45°C$$
$$P_4 = 320 \text{ kPa}$$
$$P_5 = 300 \text{ kPa}, \quad T_5 = 0°C$$
$$P_6 = 290 \text{ kPa}, \quad T_6 = 5°C$$

Rate of flow of Freon-12 = 0.05 kg/s
Power input to compressor = 4 kW

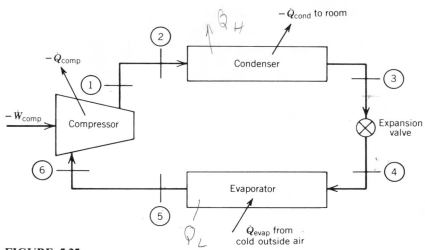

FIGURE 5.35
Sketch for Problem 5.60.

Calculate:
(a) The heat transfer from the compressor.
(b) The heat transfer from the Freon-12 in the condenser.
(c) The heat transfer to the Freon-12 in the evaporator.

5.61 Air flows in a pipeline at 80 lbf/in.2, 80 F. The line is connected by a valve to an evacuated vessel. The valve is now opened, and air flows into the vessel until the pressure inside reaches 80 lbf/in.2.
(a) If this process occurs adiabatically, what is the final temperature of the air inside the vessel?
(b) Develop an expression for the relation between the line temperature and the final temperature in the vessel.

5.62 An insulated 2 m^3 tank containing ammonia at $-20°C$, 80% quality, is connected by a valve to a line flowing ammonia at 2 MPa, 60°C. The valve is now opened, allowing ammonia to flow into the tank. At what pressure should the valve be closed if it is desired to have 15 kg of ammonia inside at the final state?

5.63 A tank having a volume of 25 ft^3 contains saturated vapor steam at a pressure of 25 lbf/in.2. Attached to this tank is a line flowing steam at 120 lbf/in.2, 400 F. The valve is now opened, and steam flows into the tank until the pressure inside reaches 120 lbf/in.2, at which point the valve is closed. If there is no heat transfer from the tank during this process, find the mass of steam that enters the tank.

5.64 A steam power plant for an automotive vehicle is proposed as shown in Fig. 5.36. The boiler has a volume of 50 L and initially contains 90% liquid

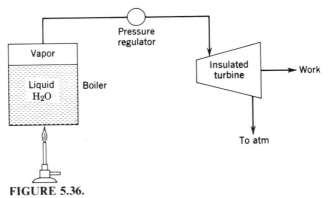

FIGURE 5.36.
Sketch for Problem 5.64.

and 10% vapor by volume at 100 kPa. The burner is now turned on, and when the pressure in the boiler reaches 700 kPa, the pressure-regulating valve holds the boiler pressure constant at this value, and saturated vapor at this pressure passes to the turbine. Saturated vapor steam at 100 kPa exits the turbine and is discharged to the atmosphere. When the liquid in the boiler is depleted, the burner shuts off automatically. Find the total work output of the turbine and the total heat transfer for one charge of the boiler.

5.65 A 20-ft^3 tank contains ammonia at 20 lbf/in.2, 80 F. The tank is attached to a line flowing ammonia at 180 lbf/in.2, 140 F. The valve is now opened, and mass flows in until the tank is half full of liquid, by volume, at 80 F. Calculate the heat transfer from the tank during this process.

5.66 An evacuated, insulated 15-m^3 tank is connected to a line flowing steam at 2 MPa, 400°C. The valve is now opened, and steam flows into the tank until the tank contains 20 kg of steam, at which point the valve is closed. What is the final pressure inside the tank?

5.67 An evacuated 150-L tank is connected to a line flowing air at room temperatuure, 25°C, and 8 MPa pressure. The valve is now opened, allowing air to flow into the tank until the pressure inside reaches 6 MPa, at which time the valve is closed. This filling process occurs rapidly and is essentially adiabatic. The tank is then allowed to sit for a long time with the valve closed, and eventually returns to room temperature. What is the final pressure inside?

5.68 Air is contained in the insulated cylinder shown in Fig. 5.37. At this point the air is at 20 lbf/in.2, 80 F, and the cylinder volume is 0.5 ft^3. The piston cross-sectional area is 0.5 ft^2, and the spring force varies linearly with distance moved, with a spring constant of 200 lbf/in. The valve is now opened, and air from the line at 100 lbf/in.2, 80 F flows into the cylinder until the

FIGURE 5.37
Sketch for Problem 5.68.

pressure inside reaches 100 lbf/in.², at which point the valve is closed. What is the temperature inside the cylinder at the final state?

5.69 A 2-m³ insulated vessel contains saturated vapor steam at 4 MPa pressure. A valve on the top of the tank is now opened and steam is allowed to escape. During the process, any liquid formed collects at the bottom of the vessel, such that only saturated vapor exits. Calculate the total mass that has escaped when the pressure inside reaches 1.0 MPa.

5.70 An insulated 500-L tank contains air at 40°C, 2 MPa. A valve on the tank is now opened, and air escapes until half the original mass is gone, at which point the valve is closed. What is the pressure inside at this point?

5.71 An insulated, 35-ft³ rigid steel tank having a mass of 90 lbm contains air at 75 lbf/in.², and both are at 70 F. The tank is connected by a valve to a line flowing air at 300 lbf/in.², 70 F. The valve is now opened, allowing air to flow into the tank until the pressure is 250 lbf/in.², at which point the valve is closed. Assuming that the tank and the air inside are always at the same temperature, find the final temperature (at the time that the valve is closed).

5.72 Freon-12 is contained in a 1 ft³ tank as indicated in Fig. 5.38. The tank initially contains 90% liquid, 10% vapor, by volume, at 80 F. The valve is then opened and Freon-12 vapor flows out, leaving the nozzle at 15 lbf/in.², 0 F, with a velocity of 700 ft/s. During this process heat is transferred to the tank such that the Freon-12 inside remains at 80 F. Determine:
(a) The total mass exhausted when the tank contains saturated vapor.
(b) The heat transferred to the tank during this process.

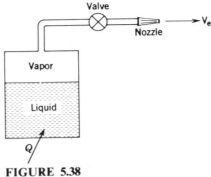

FIGURE 5.38
Sketch for Problem 5.72.

5.73 A spherical elastic balloon that supports an internal pressure proportional to the balloon diameter initially contains water at 200°C, 50% quality, and the diameter is 1 m. The balloon is attached by a valve to a line flowing steam at 5 MPa, 300°C. The valve is now opened, and steam flows into the balloon until the pressure inside reaches 2.5 MPa, at which point the valve is closed. The final mass of water inside the balloon is twice the initial mass. How much heat is transferred to the balloon during the process?

5.74 A cylinder fitted with a piston restrained by a linear spring contains 2 lbm of Freon-12 at 220 F, 125 lbf/in.². The spring constant is 285 lbf/in. and the piston cross-sectional area is 75 in.². A valve on the cylinder is now opened and Freon-12 flows out. During this process, heat is transferred such that when half the initial mass has escaped, the temperature inside the cylinder is 50 F. Find
(a) The final state inside the cylinder.
(b) The heat transfer to or from the cylinder.

5.75 A rigid 750-L tank initially contains water at 250°C, 50% liquid and 50% vapor, by volume. A valve on the bottom of the tank is now opened, and liquid is withdrawn slowly. During the process, heat is transferred such that the temperature inside remains constant. Calculate the amount of heat transferred by the time that half the initial mass has been withdrawn.

5.76 A cylinder fitted with a piston restrained by a linear spring contains 2 lbm of ammonia at 180 lbf/in.², 140 F. The piston cross-sectional area is 100 in.² and the spring constant is 200 lbf/in. A valve on the cylinder is now opened, allowing ammonia to escape. During this process heat is transferred such that when the pressure inside has dropped to 90 lbf/in.², only half of the initial mass remains. Determine:
(a) The final temperature in the cylinder.
(b) The work and heat transfer during the process.

SUPPLEMENTARY PROBLEMS

5.77 An insulated and otherwise evacuated 20-L vessel contains a 1-L capsule of water at 700 kPa, 150°C. The capsule now breaks and its contents fill the entire volume. What is the final pressure?

5.78 A spherical balloon initially 6 in. in diameter and containing Freon-12 at 15 lbf/in.2 is connected to an uninsulated 1-ft^3 rigid tank containing Freon-12 at 80 lbf/in.2. Both are at ambient temperature, 80 F. The valve connecting the two is now opened slightly and remains opened until the pressures equalize. During this process, heat is transferred such that the temperature of all the Freon-12 remains at 80 F, and it may also be assumed that the pressure inside the balloon is proportional to its diameter at any time. Calculate:
(a) The final pressure.
(b) Work done by the Freon-12 during the process.
(c) Heat transfer to the Freon-12 during the process.

5.79 A frictionless, thermally conducting piston separates the air and water in the cylinder shown in Fig. 5.39. The initial volumes of A and B are each 500 L, and the initial pressure on each side is 700 kPa. The volume of the liquid in B is 2% of the volume of B at this state. Heat is now transferred to both A and B until all the liquid in B evaporates. Determine:
(a) The total heat transfer during the process.
(b) The work done by the piston on the air and the heat transfer to the air during the process.

FIGURE 5.39
Sketch for Problem 5.79.

5.80 Liquids are often transferred from storage by pressurization with a gas. As an example of this procedure, consider the process shown in Fig. 5.40, in which tanks A and B both contain Freon-12. Tank A has a volume of 3.2 ft^3 and B has a volume of 2.5 ft^3. The Freon-12 in A is initially at 80 F, saturated vapor, and that in B is at −40 F with a quality of 0.01. The valve is now opened slightly, allowing gas to flow from A to B. When the pressure inside B has built up to 23.85 lbf/in.2, liquid begins to flow out. The liquid transfer

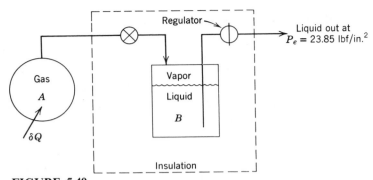

FIGURE 5.40
Sketch for Problem 5.80.

continues until the pressure in A has dropped to 23.85 lbf/in.2. During the process, heat is transferred to the gas inside tank A such that its temperature always remains at 80 F, while tank B is insulated. Determine:

(a) The quality in tank B at the end of the process.
(b) The mass of liquid transferred from tank B.

5.81 A 2-m^3 storage tank contains 90% liquid and 10% vapor, by volume, of LNG (liquified natural gas) at 160 K, as shown in Fig. 5.41. It may be assumed that LNG has the same properties as pure methane. Heat is now transferred to the tank, and saturated vapor at 160 K flows into a steady-flow heater in which the vapor is heated at constant pressure to 300 K. The process continues until all the liquid in the storage tank is gone. Calculate the total amount of heat transferred to the tank and also the total amount of heat transferred to the heater.

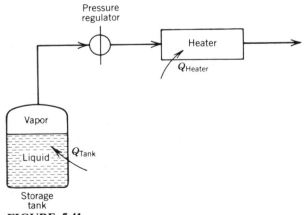

FIGURE 5.41
Sketch for Problem 5.81.

5.82 Consider the system shown in Fig. 5.42. Tank B, which is heavily insulated, has a volume of 10 ft^3 and is initially evacuated. Cylinder A, which is un-insulated, has an initial volume of 5 ft^3 and contains air at 70 F, 500 lbf/in.2. The spring constant is 250 lbf/in. and the piston cross-sectional area is 50 in.2. The valve is now opened, and air flows from A to B until the pressure in B reaches 200 lbf/in.2, at which point the valve is closed. It may be assumed that the spring force varies linearly with distance, and also that heat is transferred to A during the process such that the temperature inside A remains constant. Determine:
(a) The final temperature and mass in B.
(b) The final pressure and volume in cylinder A.
(c) The heat transfer to A during the process.

FIGURE 5.42
Sketch for Problem 5.82.

5.83 Write a computer program to solve the following problem. For one of the substances listed in Table A.9, it is desired to compare the enthalpy change between any two temperatures T_1 and T_2 as calculated by integrating the specific heat equation, by assuming constant specific heat at the average temperature, and by assuming constant specific heat at temperature T_1.

5.84 Write a computer program to solve the following problem. An insulated tank of volume V contains a specified ideal gas (with constant specific heat) at P_1, T_1. A valve is now opened, allowing the gas to flow out until the pressure inside drops to P_2. Determine T_2 and m_2 using a stepwise solution in incre-ments of pressure between P_1 and P_2, where the number of increments is variable.

5.85 Write a computer program to solve the following problem. It is desired to fit a polynomial expression of degree n in the temperature for ideal gas spe-cific heat, using as data the ideal gas enthalpy values for one of the substances listed in Table A.11. Accuracy of the correlation should be studied as a func-tion of the temperature range of the fit, as well as of the polynomial degree.

6

THE SECOND LAW
OF THERMODYNAMICS

The first law of thermodynamics states that during any cycle that a system undergoes, the cyclic integral of the heat is equal to the cyclic integral of the work. The first law, however, places no restrictions on the direction of flow of heat and work. A cycle in which a given amount of heat is transferred from the system and an equal amount of work is done on the system satisfies the first law just as well as a cycle in which the flows of heat and work are reversed. However, we know from our experience that the fact that a proposed cycle does not violate the first law does not ensure that the cycle will actually occur. It is this kind of experimental evidence that has led to the formulation of the second law of thermodynamics. Thus a cycle will occur only if both the first and second laws of thermodynamics are satisfied.

In its broader significance the second law involves the fact that processes proceed in a certain direction but not in the opposite direction. A hot cup of coffee cools by virtue of heat transfer to the surroundings, but heat will not flow from the cooler surroundings to the hotter cup of coffee. Gasoline is used as a car drives up a hill, but on coasting down the hill, the fuel level in the gasoline tank cannot be restored to its original level. Such familiar observations as these, and a host of others, are evidence of the validity of the second law of thermodynamics.

We will first consider this second law for a system undergoing a cycle and in the next chapter will extend the principles to a system undergoing a change of state and then to a control volume.

6.1 HEAT ENGINES AND REFRIGERATORS

Consider the system and the surroundings previously cited in the development of the first law, as shown in Fig. 6.1. Let the gas constitute the system and, as in our discussion of the first law, let this system undergo a cycle in which work is first done on the system by the paddle wheel as the weight is lowered. Then let the cycle be completed by transferring heat to the surroundings.

We know from our experience, however, that we cannot reverse this cycle. That is, if we transfer heat to the gas, as shown by the dotted arrow, the temperature of the gas will increase, but the paddle wheel will not turn and raise the weight. With the given surroundings (the container, the paddle wheel, and the weight) this

162

FIGURE 6.1

A system that undergoes a cycle involving work and heat.

system can operate in a cycle in which the heat transfer and work are both negative, but it cannot operate in a cycle in which both the heat transfer and work are positive, even though this would not violate the first law.

Consider another cycle, which we know from our experience is impossible to actually accomplish. Let two systems, one at a high temperature and the other at a low temperature, undergo a process in which a quantity of heat is transferred from the high-temperature system to the low-temperature system. We know that this process can take place. We also know that the reverse process, in which heat is transferred from the low-temperature system to the high-temperature system, does not occur, and that it is impossible to complete the cycle by heat transfer only. This is illustrated in Fig. 6.2.

These two illustrations lead us to the consideration of the heat engine and refrigerator, which is also referred to as a heat pump. With the heat engine we can have a system that operates in a cycle and has a net positive work and a net positive heat transfer. With the heat pump we can have a system that operates in a cycle and has heat transferred to it from a low-temperature body and heat transferred from it to a high-temperature body, though work is required to do this. Three simple heat engines and two simple refrigerators will be considered.

FIGURE 6.2

An example showing the impossibility of completing a cycle by transferring heat from a low-temperature body to a high-temperature body.

The first heat engine is shown in Fig. 6.3, and consists of a cylinder fitted with appropriate stops and a piston. Let the gas in the cylinder constitute the system. Initially the piston rests on the lower stops, with a weight on the platform. Let the system now undergo a process in which heat is transferred from some high-temperature body to the gas, causing it to expand and raise the piston to the upper stops. At this point the weight is removed. Now let the system be restored to its initial state by transferring heat from the gas to a low-temperature body, thus completing the cycle. Since the weight was raised during the cycle, it is evident that work was done by the gas during the cycle. From the first law we conclude that the net heat transfer was positive and equal to the work done during the cycle.

Such a device is called a heat engine, and the substance to which and from which heat is transferred is called the working substance or working fluid. A heat engine may be defined as a device that operates in a thermodynamic cycle and does a certain amount of net positive work as a result of heat transfer from a high-temperature body and to a low-temperature body. Often the term heat engine is used in a broader sense to include all devices that produce work, either through heat transfer or combustion, even though the device does not operate in a thermodynamic cycle. The internal-combustion engine and the gas turbine are examples of such devices, and calling these heat engines is an acceptable use of the term. In this chapter, however, we are concerned with the more restricted form of heat engine, as defined above, which operates on a thermodynamic cycle.

A simple steam power plant is an example of a heat engine in this restricted sense. Each component in this plant may be analyzed by a steady-state, steady-flow process, but considered as a whole it may be considered a heat engine (Fig. 6.4) in which water (steam) is the working fluid. An amount of heat, Q_H, is transferred from a high-temperature body, which may be the products of combustion

FIGURE 6.3

A simple heat engine.

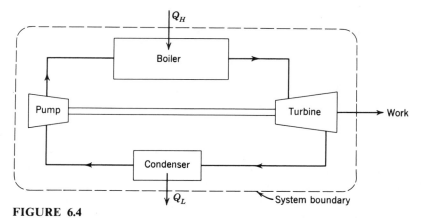

FIGURE 6.4

A heat engine involving steady-state, steady-flow processes.

in a furnace, a reactor, or a secondary fluid that in turn has been heated in a reactor. In Fig. 6.4 the turbine is shown schematically as driving the pump, indicating that what is significant is the net work that is delivered during the cycle. The quantity of heat Q_L is rejected to a low-temperature body, which is usually the cooling water in a condenser. Thus the simple steam power plant is a heat engine in the restricted sense, for it has a working fluid, to which and from which heat is transferred, and which does a certain amount of work as it undergoes a cycle.

Another example of a heat engine is the thermoelectric power-generation device that was discussed in Chapter 1 and shown schematically in Fig. 1.11. Heat is transferred from a high-temperature body to the hot junction (Q_H) and heat is transferred from the cold junction to the surroundings (Q_L). Work is done in the form of electrical energy. Since there is no working fluid we usually do not think of this as a device that operates in a cycle. However, if one adopted a microscopic point of view one could think of a cycle as regards the flow of electrons. Furthermore, as in the case of the steam power plant, the state at each point in the thermoelectric power generator does not change with time under steady-state conditions.

Thus, by means of a heat engine, we are able to have a system operate in a cycle and have the net work and net heat transfer both positive, which we were not able to do with the system and surroundings of Fig. 6.1.

One should note that in using the symbols Q_H and Q_L we have departed from our sign connotation for heat, because for a heat engine Q_L is negative when the working fluid is considered as the system. In this chapter it will be advantageous to use the symbol Q_H to represent the heat transfer to or from the high-temperature body, and Q_L the heat transfer to or from the low-temperature body. The direction of the heat transfer will be evident in each case from the context.

At this point it is appropriate to introduce the concept of thermal efficiency of a heat engine. In general we say that efficiency is the ratio of output (the energy

sought) to input (the energy that costs), but these must be clearly defined. At the risk of oversimplification we may say that in a heat engine the energy sought is the work, and the energy that costs money is the heat from the high-temperature source (indirectly, the cost of the fuel). Thermal efficiency is defined as

$$\eta_{\text{thermal}} = \frac{W \text{ (energy sought)}}{Q_H \text{ (energy that costs)}} = \frac{Q_H - Q_L}{Q_H} = 1 - \frac{Q_L}{Q_H} \tag{6.1}$$

The second cycle we were not able to complete was the one that involved the impossibility of transferring heat directly from a low-temperature body to a high-temperature body. This can of course be done with a refrigerator or heat pump. A vapor-compression refrigerator cycle, which was introduced in Chapter 1 and shown in Fig. 1.8, is also shown in Fig. 6.5. The working fluid is the refrigerant, such as a Freon or ammonia, which goes through a thermodynamic cycle. Heat is transferred to the refrigerant in the evaporator, where its pressure and temperature are low. Work is done on the refrigerant in the compressor and heat is transferred from it in the condenser, where its pressure and temperature are high. The pressure drop occurs as the refrigerant flows through the throttle valve or capillary tube.

Thus, in a refrigerator or heat pump, we have a device that operates in a cycle, that requires work, and that accomplishes the objective of transferring heat from a low-temperature body to a high-temperature body.

The thermoelectric refrigerator, which was discussed in Chapter 1 and is shown schematically in Fig. 1.10, is another example of a device that meets our definition of a refrigerator. The work input to the thermoelectric refrigerator is in the form of electrical energy, and heat is transferred from the refrigerated space to the cold junction (Q_L) and from the hot junction to the surroundings (Q_H).

The "efficiency" of a refrigerator is expressed in terms of the coefficient of performance, which we designate with the symbol β. In the case of a refrigerator the objective (that is, energy sought) is Q_L, the heat transferred from the refrigerated

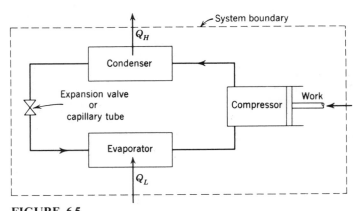

FIGURE 6.5

A simple refrigeration cycle.

space, and the energy that costs is the work W. Thus the coefficient of performance, β,[1] is

$$\beta = \frac{Q_L \text{ (energy sought)}}{W \text{ (energy that costs)}} = \frac{Q_L}{Q_H - Q_L} = \frac{1}{Q_H/Q_L - 1} \tag{6.2}$$

Before stating the second law, the concept of a thermal reservoir should be introduced. A thermal reservoir is a body to which and from which heat can be transferred indefinitely without change in the temperature of the reservoir. Thus a thermal reservoir always remains at constant temperature. The ocean and the atmosphere approach this definition very closely. Frequently it will be useful to designate a high-temperature reservoir and a low-temperature reservoir. Sometimes a reservoir from which heat is transferred is called a source, and a reservoir to which heat is transferred is called a sink.

6.2 SECOND LAW OF THERMODYNAMICS

On the basis of the matter considered in the previous section, we are now ready to state the second law of thermodynamics. There are two classical statements of the second law, known as the Kelvin–Planck statement and the Clausius statement.

The Kelvin–Planck statement: It is impossible to construct a device that will operate in a cycle and produce no effect other than the raising of a weight and the exchange of heat with a single reservoir.

This statement ties in with our discussion of the heat engine, and, in effect, it states that it is impossible to construct a heat engine that operates in a cycle and receives a given amount of heat from a high-temperature body and does an equal amount of work. The only alternative is that some heat must be transferred from the working fluid at a lower temperature to a low-temperature body. Thus work can be done by the transfer of heat only if there are two temperature levels involved, and heat is transferred from the high-temperature body to the heat engine

[1] It should be noted that a refrigeration or heat pump cycle can be used with either of two objectives in mind. It can be used as a refrigerator, in which case the primary objective is Q_L, the heat transferred to the refrigerant from the refrigerated space. It also can be used as a heating system (in which case it usually is referred to as a heat pump), the objective being Q_H, the heat transferred from the refrigerant to the high-temperature body, which is the space to be heated. Q_L is transferred to the refrigerant from the ground, the atmospheric air, or well water. The coefficient of performance is this case, β', is

$$\beta' = \frac{Q_H \text{ (energy sought)}}{W \text{ (energy that costs)}} = \frac{Q_H}{Q_H - Q_L} = \frac{1}{1 - Q_L/Q_H}$$

It also follows that for a given cycle,

$$\beta' - \beta = 1$$

Unless otherwise specified, the term coefficient of performance will always refer to a refrigerator as defined by Eq. 6.2.

and also from the heat engine to the low-temperature body. This implies that it is impossible to build a heat engine that has a thermal efficiency of 100%.

✳ The Clausius statement: It is impossible to construct a device that operates in a cycle and produces no effect other than the transfer of heat from a cooler body to a hotter body.

This statement is related to the refrigerator or heat pump, and in effect states that it is impossible to construct a refrigerator that operates without an input of work. This also implies that the coefficient of performance is always less than infinity.

In regard to these two statements, three observations should be made. The first is that both are negative statements. It is of course impossible to "prove" a negative statement. However, we can say that the second law of thermodynamics (like every other law of nature) rests on experimental evidence. Every relevant experiment that has been conducted has either directly or indirectly verified the second law, and no experiment has ever been conducted that contradicts the second law. The basis of the second law is therefore experimental evidence.

A second observation is that these two statements of the second law are equivalent. Two statements are equivalent if the truth of each statement implies the truth of the other, or if the violation of each statement implies the violation of the other. That a violation of the Clausius statement implies a violation of the Kelvin–Planck statement may be shown as follows. The device at the left in Fig. 6.6 is a refrigerator that requires no work and thus violates the Clausius statement. Let an amount of heat Q_L be transferred from the low-temperature reservoir to this refrigerator, and let the same amount of heat Q_L be transferred to the high-temperature reservoir. Let an amount of heat Q_H, which is greater than Q_L, be transferred from the high-temperature reservoir to the heat engine, and let the engine reject the amount of heat Q_L as it does an amount of work W (which equals $Q_H - Q_L$). Since there is no net heat transfer to the low-temperature reservoir, the low-temperature reservoir, the heat engine, and the refrigerator can be considered

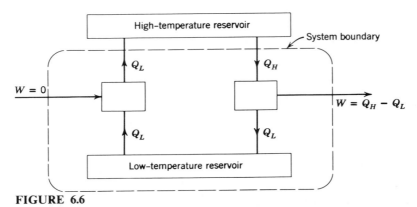

FIGURE 6.6

Demonstration of the equivalence of the two statements of the second law.

together as a device that operates in a cycle and produces no effect other than the raising of a weight (work) and the exchange of heat with a single reservoir. Thus a violation of the Clausius statement implies a violation of the Kelvin–Planck statement. The complete equivalence of those two statements is established when it is also shown that a violation of the Kelvin–Planck statement implies a violation of the Clausius statement. This is left as an exercise for the student.

The third observation is that frequently the second law of thermodynamics has been stated as the impossibility of constructing a perpetual-motion machine of the second kind. A perpetual-motion machine of the first kind would create work from nothing or create mass-energy, thus violating the first law. A perpetual-motion machine of the second kind would violate the second law, and a perpetual-motion machine of the third kind would have no friction, and thus run indefinitely but would produce no work.

A heat engine that violated the second law could be made into a perpetual-motion machine of the second kind as follows. Consider Fig. 6.7, which might be the power plant of a ship. An amount of heat Q_L is transferred from the ocean to a high-temperature body by means of a heat pump. The work required is W', and the heat transferred to the high-temperature body is Q_H; let the same amount of heat be transferred to a heat engine that violates the Kelvin–Planck statement of the second law, and does an amount of work $W = Q_H$. Of this work an amount of work $Q_H - Q_L$ is required to drive the heat pump, leaving the net work $(W_{net} = Q_L)$ available for driving the ship. Thus we have a perpetual motion machine in the sense that work is done by utilizing freely available sources of energy such as the ocean or atmosphere.

6.3 THE REVERSIBLE PROCESS

The question that now logically arises is this. If it is impossible to have a heat engine of 100% efficiency, what is the maximum efficiency one can have? The first

FIGURE 6.7

A perpetual-motion machine of the second kind.

step in the answer to this question is to define an ideal process, which is called a reversible process.

A reversible process for a system is defined as a process that, once having taken place, can be reversed and in so doing leaves no change in either system or surroundings.

Let us illustrate the significance of this definition for a gas contained in a cylinder that is fitted with a piston. Consider first Fig. 6.8, in which a gas (which we define as the system) at high pressure is restrained by a piston that is secured by a pin. When the pin is removed, the piston is raised and forced abruptly against the stops. Some work is done by the system, since the piston has been raised a certain amount. Suppose we wish to restore the system to its initial state. One way of doing this would be to exert a force on the piston, thus compressing the gas until the pin could again be inserted in the piston. Since the pressure on the face of the piston is greater on the return stroke than on the initial stroke, the work done on the gas in this reverse process is greater than the work done by the gas in the initial process. An amount of heat must be transferred from the gas during the reverse stroke in order that the system have the same internal energy it had originally. Thus the system is restored to its initial state, but the surroundings have changed by virtue of the fact that work was required to force the piston down and heat was transferred to the surroundings. Thus the initial process is an irreversible one because it could not be reversed without leaving a change in the surroundings.

In Fig. 6.9 let the gas in the cylinder comprise the system and let the piston be loaded with a number of weights. Let the weights be slid off horizontally one at a time, allowing the gas to expand and do work in raising the weights that remain on the piston. As the size of the weights is made smaller and their number is increased, we approach a process that can be reversed, for at each level of the piston during the reverse process there will be a small weight that is exactly at the level of the platform and thus can be placed on the platform without requiring work. In the limit, therefore, as the weights become very small, the reverse process can be accomplished in such a manner that both the system and surroundings are in exactly the same state they were initially. Such a process is a reversible process.

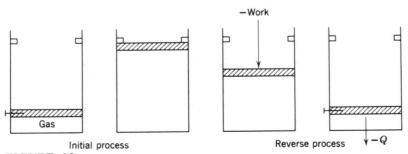

Initial process Reverse process

FIGURE 6.8

An example of an irreversible process.

FIGURE 6.9

An example of a process that approaches being reversible.

6.4 FACTORS THAT RENDER PROCESSES IRREVERSIBLE

There are many factors that make processes irreversible, four of which are considered in this section.

Friction

It is readily evident that friction makes a process irreversible, but a brief illustration may amplify the point. Let a block and an inclined plane comprise a system, Fig. 6.10, and let the block be pulled up the inclined plane by weights that are lowered. A certain amount of work is required to do this. Some of this work is required to overcome the friction between the block and the plane, and some is required to increase the potential energy of the block. The block can be restored to its initial position by removing some of the weights, thus allowing the block to slide down the plane. Some heat transfer from the system to the surroundings will no doubt be required to restore the block to its initial temperature. Since the surroundings are not restored to their initial state at the conclusion of the reverse process, we conclude that friction has rendered the process irreversible. Another

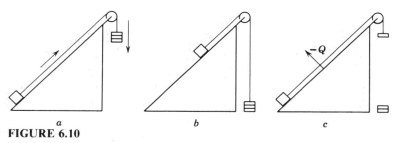

a *b* *c*

FIGURE 6.10

Demonstration of the fact that friction makes processes irreversible.

type of frictional effect is that associated with the flow of viscous fluids in pipes and passages and in the movement of bodies through viscous fluids.

Unrestrained Expansion

The classic example of an unrestrained expansion is shown in Fig. 6.11, in which a gas is separated from a vacuum by a membrane. Consider the process that occurs when the membrane breaks and the gas fills the entire vessel. It can be shown that this is an irreversible process by considering the process that would be necessary to restore the system to its original state. This would involve compressing the gas and transferring heat from the gas until its initial state was reached. Since the work and heat transfer involve a change in the surroundings, the surroundings are not restored to their initial state, indicating that the unrestrained expansion was an irreversible process. The process described in Fig. 6.8 is also an example of an unrestrained expansion.

In the reversible expansion of a gas there must be only an infinitesimal difference between the force exerted by the gas and the restraining force, so the rate at which the boundary moves will be infinitesimal. In accordance with our previous definition, this is a quasiequilibrium process. However, actual cases involve a finite difference in forces, which gives rise to a finite rate of movement of the boundary, and thus are irreversible in some degree.

FIGURE 6.11

Demonstration of the fact that unrestrained expansion makes processes irreversible.

Heat Transfer Through a Finite Temperature Difference

Consider as a system a high-temperature body and a low-temperature body, and let heat be transferred from the high-temperature body to the low-temperature body. The only way in which the system can be restored to its initial state is to provide refrigeration, which requires work from the surroundings, and some heat transfer to the surroundings will also be necessary. Because of the heat transfer and the work, the surroundings are not restored to their original state, indicating that the process was irreversible.

An interesting question now arises. Heat is defined as energy that is transferred due to a temperature difference. We have just shown that heat transfer through

a temperature difference is an irreversible process. Therefore, how can we have a reversible heat-transfer process? A heat-transfer process approaches a reversible process as the temperature difference between the two bodies approaches zero. Therefore, we define a reversible heat-transfer process as one in which the heat is transferred through an infinitesimal temperature difference. We realize of course that to transfer a finite amount of heat through an infinitesimal temperature difference would require an infinite amount of time, or infinite area. Therefore, all actual heat-transfer processes are through a finite temperature difference and hence are irreversible, and the greater the temperature difference the greater the irreversibility. We will find, however, that the concept of reversible heat transfer is very useful in describing ideal processes.

Mixing of Two Different Substances

This process is illustrated in Fig. 6.12 in which two different gases are separated by a membrane. Let the membrane break and a homogeneous mixture of oxygen and nitrogen fill the entire volume. This process will be considered in some detail in Chapter 11. We can say here that this may be considered as a special case of an unrestrained expansion, for each gas undergoes an unrestrained expansion as it fills the entire volume. A certain amount of work is necessary to separate these gases. Thus an air-separation plant such as described in Chapter 1 requires an input of work in order that the separation may be accomplished.

Other Factors

There are a number of other factors that make processes irreversible but they will not be considered in detail here. Hysteresis effects and the $i^2 R$ loss encountered in electrical circuits are both factors that make processes irreversible. A combustion process as it ordinarily takes place is also an irreversible process.

It is frequently advantageous to distinguish between internal and external irreversibility. Figure 6.13 shows two identical systems to which heat is transferred. Assuming each system to be a pure substance, the temperature remains constant during the heat-transfer process. In one the heat is transferred from a reservoir at

FIGURE 6.12

Demonstration of the fact that the mixing of two different substances is an irreversible process.

FIGURE 6.13

Illustration of the difference between an internally and externally reversible process.

a temperature $T + dT$, and in the other the reservoir is at a much higher temperature, $T + \Delta T$, than the system. The first is a reversible heat-transfer process and the second is an irreversible heat-transfer process. However, as far as the system itself is concerned, it passes through exactly the same states in both processes, which we assume are reversible. Thus we can say in the second case that the process is internally reversible but externally irreversible because the irreversibility occurs outside the system.

One should also note the general interrelation of reversibility, equilibrium, and time. In a reversible process, the deviation from equilibrium is infinitesimal, and therefore it occurs at an infinitesimal rate. Since it is desirable that actual processes proceed at a finite rate, the deviation from equilibrium must be finite, and therefore the actual process is irreversible in some degree. The greater the deviation from equilibrium, the greater the irreversibility, and the more rapidly the process will occur. It should also be noted that the quasiequilibrium process, which was described in Chapter 2, is a reversible process, and hereafter the term reversible process will be used.

6.5 THE CARNOT CYCLE

Having defined the reversible process and considered some factors that make processes irreversible, let us again pose the question raised in Section 6.3. If the efficiency of all heat engines is less than 100% what is the most efficient cycle we can have? Let us answer this question for a heat engine that receives heat from a high-temperature reservoir and rejects heat to a low-temperature reservoir. Since we are dealing with reservoirs we recognize that both the high temperature and the low temperature are constant and remain constant regardless of the amount of heat transferred.

Let us assume that this heat engine, which operates between the given high-temperature and low-temperature reservoirs, operates in a cycle in which every process is reversible. If every process is reversible, the cycle is also reversible, and if the cycle is reversed, the heat engine becomes a refrigerator. In the next section

we will show that this is the most efficient cycle that can operate between two constant-temperature reservoirs. It is called the Carnot cycle, and is named after a French engineer, Nicolas Leonard Sadi Carnot (1796–1832), who expressed the foundations of the second law of thermodynamics in 1824.

We now turn our attention to a consideration of the Carnot cycle. Figure 6.14 shows a power plant that is similar in many respects to a simple steam power plant and that we assume operates on the Carnot cycle. Assume the working fluid to be a pure substance, such as steam. Heat is transferred from the high-temperature reservoir to the water (steam) in the boiler. For this to be a reversible heat transfer, the temperature of the water (steam) must be only infinitesimally lower than the temperature of the reservoir. This also implies, since the temperature of the reservoir remains constant, that the temperature of the water must remain constant. Therefore, the first process in the Carnot cycle is a reversible isothermal process in which heat is transferred from the high-temperature reservoir to the working fluid. A change of phase from liquid to vapor at constant pressure is of course an isothermal process for a pure substance.

The next process occurs in the turbine. It occurs without heat transfer and is therefore adiabatic. Since all processes in the Carnot cycle are reversible, this must be a reversible adiabatic process, during which the temperature of the working fluid decreases from the temperature of the high-temperature reservoir to the temperature of the low-temperature reservoir.

In the next process heat is rejected from the working fluid to the low-temperature reservoir. This must be a reversible isothermal process in which the temperature of the working fluid is infinitesimally higher than that of the low-temperature reservoir. During this isothermal process some of the steam is condensed.

FIGURE 6.14

Example of a heat engine that operates on a Carnot cycle.

The final process, which completes the cycle, is a reversible adiabatic process in which the temperature of the working fluid increases from the low temperature to the high temperature. If this were to be done with water (steam) as the working fluid, it would involve taking a mixture of liquid and vapor from the condenser and compressing it. (This would be very inconvenient in practice and therefore in all power plants the working fluid is completely condensed in the condenser, and the pump handles only the liquid phase.)

Since the Carnot heat engine cycle is reversible, every process could be reversed, in which case it would become a refrigerator. The refrigerator is shown by the dotted lines and parentheses in Fig. 6.14. The temperature of the working fluid in the evaporator would be infinitesimally less that the temperature of the low-temperature reservoir, and in the condenser it is infinitesimally higher than that of the high-temperature reservoir.

It should be emphasized that the Carnot cycle can be executed in many different ways. Many different working substances can be used, such as a gas, a thermoelectric device, or a paramagnetic substance in a magnetic field such as was described in Chapter 4. There are also various possible arrangements of machinery. For example, a Carnot cycle can be devised that takes place entirely within a cylinder, using a gas as a working substance, as shown in Fig. 6.15.

The important point to be made here is that the Carnot cycle, regardless of what the working substance may be, always has the same four basic processes. These are:

1. A reversible isothermal process in which heat is transferred to or from the high-temperature reservoir.

2. A reversible adiabatic process in which the temperature of the working fluid decreases from the high temperature to the low temperature.

3. A reversible isothermal process in which heat is transferred to or from the low-temperature reservoir.

4. A reversible adiabatic process in which the temperature of the working fluid increases from the low temperature to the high temperature.

FIGURE 6.15

Example of a gaseous system operating on a Carnot cycle.

6.6 TWO PROPOSITIONS REGARDING THE EFFICIENCY OF A CARNOT CYCLE

There are two important propositions regarding the efficiency of a Carnot cycle.

First Proposition

It is impossible to construct an engine that operates between two given reservoirs and is more efficient than a reversible engine operating between the same two reservoirs.

The proof of this statement involves a "thought experiment." An initial assumption is made, and it is then shown that this assumption leads to impossible conclusions. The only possible conclusion is that the initial assumption was incorrect.

Let us assume that there is an irreversible engine operating between two given reservoirs that has a greater efficiency than a reversible engine operating between the same two reservoirs. Let the heat transfer to the irreversible engine be Q_H, the heat rejected be Q'_L, and the work be W_{IE} (which equals $Q_H - Q'_L$) as shown in Fig. 6.16. Let the reversible engine operate as a refrigerator (since it is reversible this is possible) and let the heat transfer with the low-temperature reservoir be Q_L, the heat transfer with the high-temperature reservoir be Q_H, and the work required be W_{RE} (which equals $Q_H - Q_L$).

Since the initial assumption was that the irreversible engine is more efficient, it follows (because Q_H is the same for both engines) that $Q'_L < Q_L$ and $W_{IE} > W_{RE}$. Now the irreversible engine can drive the reversible engine and still deliver the net work W_{net} (which equals $W_{IE} - W_{RE} = Q_L - Q'_L$). However, if we consider the two

FIGURE 6.16

Demonstration of the fact that the Carnot cycle is the most efficient cycle operating between two fixed temperature reservoirs.

engines and the high-temperature reservoir as a system as indicated in Fig. 6.16, we have a system that operates in a cycle, exchanges heat with a single reservoir, and does a certain amount of work. However, this would constitute a violation of the second law and we conclude that our initial assumption (that the irreversible engine is more efficient than the reversible engine) is incorrect, and therefore we cannot have an irreversible engine that is more efficient than a reversible engine operating between the same two reservoirs.

Second Proposition

All engines that operate on the Carnot cycle between two given constant-temperature reservoirs have the same efficiency. The proof of this proposition is similar to the proof outlined above, and involves the assumption that there is one Carnot cycle that is more efficient than another Carnot cycle operating between the same temperature reservoirs. Let the Carnot cycle with the higher efficiency replace the irreversible cycle of the previous argument, and the Carnot cycle with the lower efficiency operate as the refrigerator. The proof proceeds with the same line of reasoning as in the first proposition. The details are left as an exercise for the student.

6.7 THE THERMODYNAMIC TEMPERATURE SCALE

In discussing the matter of temperature in Chapter 2, it was pointed out that the zeroth law of thermodynamics provides a basis for temperature measurement, but that a temperature scale must be defined in terms of a particular thermometer substance and device. A temperature scale that is independent of any particular substance, which might be called an absolute temperature scale, would be most desirable. In the last paragraph we noted that the efficiency of a Carnot cycle is independent of the working substance and depends only on the temperature. This fact provides the basis for such an absolute temperature scale, which we will call the thermodynamic scale.

The concept of this temperature scale may be developed with the aid of Fig. 6.17, which shows three reservoirs and three engines that operate on the Carnot cycle. T_1 is the highest temperature, T_3 is the lowest temperature, and T_2 is an intermediate temperature, and the engines operate between the various reservoirs as indicated. Q_1 is the same for both A and C and, since we are dealing with reversible cycles, Q_3 is the same for B and C.

Since the efficiency of a Carnot cycle is a function of only the temperature we can write

$$\eta_{\text{thermal}} = 1 - \frac{Q_L}{Q_H} = \psi(T_L, T_H) \tag{6.3}$$

where ψ designates a functional relation.

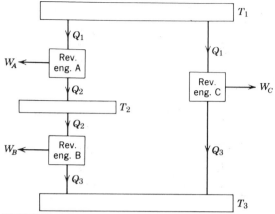

FIGURE 6.17

Arrangement of heat engines to demonstrate the thermodynamic temperature scale.

Let us apply this to the three Carnot cycles of Fig. 6.17.

$$\frac{Q_1}{Q_2} = \psi(T_1, T_2)$$

$$\frac{Q_2}{Q_3} = \psi(T_2, T_3)$$

$$\frac{Q_1}{Q_3} = \psi(T_1, T_3)$$

Since

$$\frac{Q_1}{Q_3} = \frac{Q_1 Q_2}{Q_2 Q_3}$$

it follows that

$$\psi(T_1, T_3) = \psi(T_1, T_2) \times \psi(T_2, T_3) \tag{6.4}$$

Note that the left side is a function of T_1 and T_3 (and not T_2), and therefore the right side of this equation must also be a function of T_1 and T_3 (and not T_2). From this we can conclude that the form of the function ψ must be such that

$$\psi(T_1, T_2) = \frac{f(T_1)}{f(T_2)}$$

$$\psi(T_2, T_3) = \frac{f(T_2)}{f(T_3)}$$

for in this way $f(T_2)$ will cancel from the product of $\psi(T_1, T_2) \times \psi(T_2, T_3)$. Therefore, we conclude that

$$\frac{Q_1}{Q_3} = \psi(T_1, T_3) = \frac{f(T_1)}{f(T_3)} \tag{6.5}$$

In general terms,

$$\frac{Q_H}{Q_L} = \frac{f(T_H)}{f(T_L)} \tag{6.6}$$

Now there are several functional relations that will satisfy this equation. The one that has been selected, which was originally proposed by Lord Kelvin, for the thermodynamic scale of temperature, is the relation

$$\frac{Q_H}{Q_L} = \frac{T_H}{T_L} \tag{6.7}$$

With absolute temperatures so defined, the efficiency of a Carnot cycle may be expressed in terms of the absolute temperatures.[2]

$$\eta_{\text{thermal}} = 1 - \frac{Q_L}{Q_H} = 1 - \frac{T_L}{T_H} \tag{6.8}$$

This means that if the thermal efficiency of a Carnot cycle operating between two given constant-temperature reservoirs is known, the ratio of the two absolute temperatures is also known.

[2] Lord Kelvin also proposed a logarithmic scale of the form

$$\frac{Q_H}{Q_L} = \frac{e^{``T_H"}}{e^{``T_L"}}$$

where "T_H" and "T_L" designate the absolute temperatures on this proposed logarithmic scale. This relation can also be written

$$\ln \frac{Q_H}{Q_L} = ``T_H" - ``T_L"$$

The form Kelvin actually proposed was

$$\log_{10} \frac{Q_H}{Q_L} = ``T_H" - ``T_L"$$

Thus the relation between the scale in use and the proposed logarithmic scale is

$$``T" = \log_{10} T + L$$

where L is a constant that determines the level of temperature that corresponds to zero on the logarithmic scale. On this logarithmic scale temperatures range from $-\infty$ to $+\infty$, whereas on the thermodynamic scale in use they vary from 0 to $+\infty$ for ordinary systems.

It should be noted that Eq. 6.7 gives us a ratio of absolute temperatures, but it does not give us information about the magnitude of the degree. Let us first consider a qualitative approach to this matter and then a more rigorous statement.

Suppose we had a heat engine operating on the Carnot cycle that received heat at the temperature of the steam point and rejected heat at the temperature of the ice point. (Since a Carnot cycle involves only reversible processes, it is impossible to construct such a heat engine and perform the proposed experiment. However, we can follow the reasoning as a "thought experiment" and gain a further understanding of the thermodynamic temperature scale.) If the efficiency of such an engine could be measured, it would be found to be 26.80%. Therefore, from Eq. 6.8,

$$\eta_{th} = 1 - \frac{T_L}{T_H} = 1 - \frac{T_{ice\ point}}{T_{steam\ point}} = 0.2680$$

$$\frac{T_{ice\ point}}{T_{steam\ point}} = 0.7320$$

This gives us one equation involving the two unknowns T_H and T_L. The second equation comes from an arbitrary decision regarding the magnitude of the degree on the thermodynamic temperature scale. If we wish to have the magnitude of the degree on the absolute scale correspond to the magnitude of the degree on the Fahrenheit scale, we can write

$$T_{steam\ point} - T_{ice\ point} = 180$$

This scale, the absolute Fahrenheit scale, is referred to as the Rankine scale, and temperatures on this scale are designated by R.

Solving these two equations simultaneously we find

$$T_{steam\ point} = 671.67\ R \qquad T_{ice\ point} = 491.67\ R$$

It follows that temperatures on the Fahrenheit and Rankine scales are related as follows:

$$T(F) + 459.67 = T(R)$$

The absolute scale related to the Celsius scale is the Kelvin scale, designated by K. On both these scales there are 100 degrees between the ice point and the steam point. Therefore, if we wished to use our engine operating on the Carnot cycle between the steam point and the ice point we would have the relations:

$$T_{steam\ point} - T_{ice\ point} = 100$$

$$\frac{T_{ice\ point}}{T_{steam\ point}} = 0.7320$$

Solving these two equations simultaneously, we find

$$T_{\text{steam point}} = 373.15 \text{ K} \qquad T_{\text{ice point}} = 273.15 \text{ K}$$

It follows that

$$T(^{\circ}\text{C}) + 273.15 = T(\text{K})$$

As already noted, the measurement of efficiencies of Carnot cycles is, however, not a practical way to approach the problem of temperature measurement on the thermodynamic scale of temperature. The actual approach used is based on the ideal gas thermometer and an assigned value for the triple point of water. At the Tenth Conference on Weights and Measures, which was held in 1954, the temperature of the triple point of water was assigned the value 273.16 K. [The triple point of water is approximately 0.01°C above the ice point. The ice point is defined as the temperature of a mixture of ice and water at a pressure of 1 atm (101.3 kPa) of air which is saturated with water vapor.]

Let us now briefly consider the ideal gas scale of temperature. This scale is based on the fact that as the pressure of a gas approaches zero, its equation of state approaches the ideal gas equation of state:

$$Pv = RT$$

Consider how an ideal gas might be used to measure temperature in a constant-volume gas thermometer, which is shown schematically in Fig. 6.18. Let the gas bulb be placed in the location where the temperature is to be measured, and let the mercury column be so adjusted that the level of mercury stands at the reference mark A. Thus the volume of the gas remains constant. Assume that the gas in the capillary tube is at the same temperature as the gas in the bulb. Then the

FIGURE 6.18

Schematic diagram of a constant-volume gas thermometer.

pressure of the gas, which is indicated by the height L of the mercury column, is an indication of the temperature.

Let the pressure that is associated with the temperature of the triple point of water (273.16 K) also be measured and let us designate this pressure $P_{t.p.}$. Then, from the definition of an ideal gas, any other temperature T could be determined from a pressure measurement P by the relation

$$T = 273.16\left(\frac{P}{P_{t.p.}}\right)$$

The temperature so measured is referred to as the ideal-gas temperature, and it can be shown that the temperature so measured is exactly equal to the thermodynamic temperature.

From a practical point of view we have the problem that no gas behaves exactly like an ideal gas. However, we do know that as the pressure approaches zero, the behavior of all gases approaches that of an ideal gas. Suppose then, that a series of measurements is made with varying amounts of gas in the gas bulb. This means that the pressure measured at the triple point, and also the pressure at any other temperature, will vary. If the indicated temperature T_i (obtained by assuming that the gas is ideal) is plotted against the pressure of gas with the bulb at the triple point of water, a curve like the one shown in Fig. 6.19 is obtained. When this curve is extrapolated to zero pressure, the correct ideal-gas temperature is obtained. Different curves might result from different gases, but they would all indicate the same temperature at zero pressure.

We have outlined only the general features and principles for measuring temperature on the ideal gas scale of temperatures. Precision work in this field is difficult and laborious, and there are only a few laboratories in the world where this precision work is carried on. The International Practical Scale of Temperature, which was presented in Chapter 2, closely approximates the thermodynamic temperature scale and is much easier to work with in actual temperature measurement.

FIGURE 6.19

Sketch showing how the ideal-gas temperature is determined.

The significance of absolute zero can be indicated by considering a Carnot cycle heat engine that receives a given amount of heat from a given high-temperature reservoir. As the temperature at which heat is rejected from the cycle is lowered, the work increases and the amount of heat rejected decreases. In the limit, the heat rejected is zero, and the temperature of the reservoir corresponding to this limit is absolute zero.

Similarly, in the case of a Carnot cycle refrigerator, the amount of work required to produce a given amount of refrigeration increases as the temperature of the refrigerated space decreases. Absolute zero represents the limiting temperature that can be achieved, and the amount of work required to produce a finite amount of refrigeration approaches infinity as the temperature at which refrigeration is provided approaches zero.

PROBLEMS

6.1 Calculate the thermal efficiency of the steam power plant cycle described in Problem 5.51.

6.2 Calculate the coefficient of performance of the Freon-12 heat pump cycle described in Problem 5.60. $\beta_2 = \dfrac{Q_L}{Q_H - Q_L}$

6.3 Prove that a cyclic device that violates the Kelvin–Planck statement of the second law also violates the Clausius statement of the second law.

6.4 Discuss the factors that would make the power plant cycle described in Problem 5.51 an irreversible cycle.

6.5 Discuss the factors that would make the heat pump cycle described in Problem 5.60 an irreversible cycle.

6.6 Calculate the thermal efficiency of a Carnot cycle heat engine operating between 920 F and 110 F, and compare the result with that of Problem 6.1.

6.7 Calculate the coefficient of performance of a Carnot cycle heat pump operating between 0 and 45°C, and compare the result with that of Problem 6.2.

6.8 The field of geothermal power, utilizing an underground source of hot water or steam as the energy source for a power plant, has received increased attention in recent years. Consider a supply of saturated liquid water at 300 F. What is the maximum possible thermal efficiency of a cyclic heat engine using this source of energy and operating in a 70 F environment? Would it be more desirable to locate a source of saturated vapor at 300 F than one of saturated liquid?

6.9 It is proposed to heat a house during the winter using a heat pump. The house is to be maintained at 68 F at all times. It is estimated than when the ambient temperature outside drops to 15 F, the rate of heat loss from

the house will be 50 000 Btu/h. What is the minimum electrical power required to drive the heat pump unit?

6.10 A Carnot cycle refrigeration cycle operates in a room in which the temperature is 20°C. It is required to transfer heat from the cold space at the rate of 5 kW in order to maintain its temperature at −30°C. What power motor is required to operate this refrigerator?

6.11 It is proposed to construct a cyclic heat engine to operate in the ocean at a location where the water temperature is 60 F near the surface and 40 F at some greater depth. What is the maximum possible thermal efficiency of such an engine?

6.12 An inventor claims to have developed a refrigeration unit that maintains the refrigerated space at 20 F while operating in a room where the temperature is 80 F, and that has a coefficient of 8.5. How do you evaluate this claim? How would you evaluate the claim of a coefficient of performance of 7.5?

6.13 A certain solar energy collector produces a maximum temperature of 100°C. The energy collected is to be used as the heat source in a cyclic heat engine. What is the maximum thermal efficiency of the engine if it operates in a 10°C environment? What if the collector is designed to focus and concentrate the energy to produce a maximum temperature of 300°C?

6.14 Liquid sodium leaves a nuclear reactor at 1500 F and is to be used as the energy source in a steam power plant. The condenser cooling water is recirculated using a cooling tower and leaves the tower at 60 F. Determine the maximum possible thermal efficiency of this power plant. Is it misleading to use the 1500 F and 60 F temperatures to calculate this value?

6.15 A cyclic machine is used to transfer heat from a higher- to a lower-temperature reservoir, as shown in Fig. 6.20. Determine whether this machine, with energy transfer values as shown, is reversible, irreversible, or impossible.

FIGURE 6.20
Sketch for Problem 6.15.

6.16 Helium has the lowest normal boiling point of any of the elements, namely 4.2 K. At this temperature helium has an enthalpy of evaporation of 83.3 kJ/kmol. A Carnot refrigeration cycle is to be analyzed for the production of 1 kmol of liquid helium at 4.2 K from saturated vapor at the same temperature. What is the work input to the refrigerator and the coefficient of performance of this refrigeration cycle, assuming that the ambient temperature is 300 K?

6.17 A temperature of about 0.01 K can be achieved by the technique of magnetic cooling, which was described in Section 4.5. In this process a strong magnetic field is imposed on a paramagnetic salt that is maintained at 1 K by transferring heat to liquid helium, which is boiling at very low pressure. The salt is then thermally isolated from the helium, the magnetic field is removed, and the temperature of the salt drops. Assume that 1 mJ of energy is to be removed from the paramagnetic salt at an average temperature of 0.1 K, and, for purposes of establishing the theoretical limit, the necessary refrigeration is produced by a Carnot cycle refrigerator. Calculate the work input to the refrigerator and the coefficient of performance for this refrigeration cycle, assuming that the ambient temperature is 300 K.

6.18 The lowest temperature that has been achieved at the present time is about 1×10^{-6} K. Achieving this temperature involved an additional stage to that described in Problem 6.17, namely, nuclear cooling. This is similar to magnetic cooling, but involves the magnetic moment associated with the nucleus rather than that associated with certain ions in the paramagnetic salt. Suppose that 10 μJ were to be removed from a specimen at an average temperature of 10^{-5} K (10 μJ is about the amount of energy associated with the dropping of a pin through a distance of 3 mm, and is about equal to the energy transferred as heat at this temperature in some experiments). If this amount of refrigeration at an average temperature of 10^{-5} K were produced, for the purpose of determining the theoretical limit, by a Carnot cycle refrigerator, determine the work input and the coefficient of performance for the refrigeration cycle, assuming that the ambient temperature is 300 K.

6.19 It is desired to produce refrigeration at -20 F. A reservoir is available at a temperature of 400 F and the ambient temperature is 90 F. Thus work can be done by a cyclic heat engine operating between the 400 F reservoir and the ambient, and this work can be used to drive the refrigerator. Determine the ratio of the heat transferred from the high-temperature reservoir to the heat transferred from the refrigerated space, assuming that all processes are reversible.

6.20 Consider a Carnot cycle heat engine operating in outer space. The only way in which heat can be rejected from this engine is by thermal radiation. The rate at which heat is transferred by thermal radiation is proportional to the fourth power of the absolute temperature and to the area of the

radiating surface. Show that for a given engine power output and a given T_H, the area of the radiator will be a minimum when the ratio $T_L/T_H = 3/4$.

6.21 A heat pump is to be used to heat a house in the winter and then reversed to cool the house in the summer. The interior temperature is to be maintained at 20°C in the winter and 25°C in the summer. Heat transfer through the walls and roof is estimated to be 2400 kJ per hour per degree temperature difference between the inside and outside.
(a) If the outside temperature in the winter is 0°C, what is the minimum power required to drive the heat pump?
(b) If the power input is the same as that in part (a), what is the maximum outside summer temperature for which the inside of the house can be maintained at 25°C?

6.22 It is proposed to build a 1000 MW electric power plant with steam as the working fluid. The condensers are to be cooled with river water (see Fig. 6.21). The maximum steam temperature will be 1000 F and the pressure in the condensers will be 1 lbf/in.² As an engineering consultant, you are asked to estimate the resulting temperature rise of the river downstream of the power plant. What is your estimate?

$$\eta_c = 1 = \frac{560}{1460} \frac{T_L}{T_H}$$

$$Q_H = \frac{W}{\eta_c}$$

$$1 - \frac{T_L}{T_H} = 1 - \frac{Q_L}{Q_H}$$

Q_L

$Q_L = \dot{m} \, C_p \, \Delta T$

$\dot{m} = A \cdot \text{Rate} \cdot \rho$

FIGURE 6.21

Sketch for Problem 6.22.

(Labels within figure: Power plant; Intake; Discharge; 200 ft; 25 ft deep; River mean speed 30 ft/min)

ENTROPY

Up to this point in our consideration of the second law of thermodynamics we have dealt only with thermodynamic cycles. Although this is a very important and useful approach, we are in many cases concerned with processes rather than cycles. Thus we might be interested in the second-law analysis of processes we encounter daily, such as the combustion process in an automobile engine, the cooling of a cup of coffee, or the chemical processes that take place in our bodies. It would also be most desirable to be able to deal with the second law quantitatively as well as qualitatively.

In our consideration of the first law, we initially stated the law in terms of a cycle, but then defined a property, the internal energy, which enabled us to use the first law quantitatively for processes. Similarly we have stated the second law for a cycle, and we will now find that the second law leads to another property, entropy, which enables us to treat the second law quantitatively for processes. Energy and entropy are both abstract concepts that have been devised to aid in describing certain observations. As we noted in Chapter 2, thermodynamics can be described as the science of energy and entropy. The significance of this statement will now become increasingly evident.

7.1 INEQUALITY OF CLAUSIUS

The first step in our consideration of the property we call entropy is to establish the inequality of Clausius, which is

$$\oint \frac{\delta Q}{T} \leqslant 0$$

The inequality of Clausius is a corollary or consequence of the second law of thermodynamics, and will be demonstrated to be valid for all possible cycles. This includes both reversible and irreversible heat engines and refrigerators. Since any reversible cycle can be represented by a series of Carnot cycles, in this analysis we need only consider a Carnot cycle that leads to the inequality of Clausius.

Consider first a reversible (Carnot) heat engine cycle, operating between reservoirs at temperatures T_H and T_L, as shown in Fig. 7.1.

For this cycle, the cyclic integral of the heat transfer, $\oint \delta Q$, is greater than zero.

$$\oint \delta Q = Q_H - Q_L > 0$$

FIGURE 7.1

Reversible heat engine cycle for demonstration of the inequality of Clausius.

Since T_H and T_L are constant, it follows from the definition of the absolute temperature scale and the fact that this is a reversible cycle that

$$\oint \frac{\delta Q}{T} = \frac{Q_H}{T_H} - \frac{Q_L}{T_L} = 0$$

If $\oint \delta Q$, the cyclic integral of δQ, is made to approach zero (by making T_H approach T_L), while the cycle remains reversible, the cyclic integral of $\delta Q/T$ remains zero. Thus we conclude that for all reversible heat engine cycles

$$\oint \delta Q \geqslant 0$$

and

$$\oint \frac{\delta Q}{T} = 0$$

Now consider an irreversible cyclic heat engine operating between the same T_H and T_L as the reversible engine of Fig. 7.1, and receiving the same quantity of heat Q_H. Comparing the irreversible cycle with the reversible one, we conclude from the second law that

$$W_{\text{irr}} < W_{\text{rev}}$$

Since $Q_H - Q_L = W$ for both the reversible and irreversible cycles, we conclude that

$$Q_H - Q_{L_{\text{irr}}} < Q_H - Q_{L_{\text{rev}}}$$

and therefore

$$Q_{L_{\text{irr}}} > Q_{L_{\text{rev}}}$$

Consequently, for the irreversible cyclic engine,

$$\oint \delta Q = Q_H - Q_{L_{\text{irr}}} > 0$$

$$\oint \frac{\delta Q}{T} = \frac{Q_H}{T_H} - \frac{Q_{L_{\text{irr}}}}{T_L} < 0$$

Suppose that we cause the engine to become more and more irreversible while keeping Q_H, T_H, and T_L fixed. The cyclic integral of δQ then approaches zero, while that for $\delta Q/T$ becomes a progressively larger negative value. In the limit, as the work output goes to zero,

$$\oint \delta Q = 0$$

$$\oint \frac{\delta Q}{T} < 0$$

Thus we conclude that for all irreversible heat engine cycles

$$\oint \delta Q \geqslant 0$$

$$\oint \frac{\delta Q}{T} < 0$$

To complete the demonstration of the inequality of Clausius we must perform similar analyses for both reversible and irreversible refrigeration cycles. For the reversible refrigeration cycle shown in Fig. 7.2,

$$\oint \delta Q = -Q_H + Q_L < 0$$

and

$$\oint \frac{\delta Q}{T} = -\frac{Q_H}{T_H} + \frac{Q_L}{T_L} = 0$$

As the cyclic integral of δQ is made to approach zero reversibly (T_H approaching T_L), the cyclic integral of $\delta Q/T$ remains at zero. In the limit,

$$\oint \delta Q = 0$$

$$\oint \frac{\delta Q}{T} = 0$$

FIGURE 7.2

Reversible refrigeration cycle for demonstration of the inequality of Clausius.

Thus, for all reversible refrigeration cycles,

$$\oint \delta Q \leqslant 0$$

$$\oint \frac{\delta Q}{T} = 0$$

Finally, let an irreversible cyclic refrigerator operate between temperatures T_H and T_L and receive the same amount of heat Q_L as the reversible refrigerator of Fig. 7.2. From the second law, we conclude that the work input required will be greater for the irreversible refrigerator, or

$$W_{irr} > W_{rev}$$

Since $Q_H - Q_L = W$ for each cycle, it follows that

$$Q_{H_{irr}} - Q_L > Q_{H_{rev}} - Q_L$$

and therefore

$$Q_{H_{irr}} > Q_{H_{rev}}$$

That is, the heat rejected by the irreversible refrigerator to the high-temperature reservoir is greater than the heat rejected by the reversible refrigerator. Therefore, for the irreversible refrigerator,

$$\oint \delta Q = -Q_{H_{irr}} + Q_L < 0$$

$$\oint \frac{\delta Q}{T} = -\frac{Q_{H_{irr}}}{T_H} + \frac{Q_L}{T_L} < 0$$

By making this machine progressively more irreversible while keeping Q_L, T_H, and T_L constant, the cyclic integrals of δQ and $\delta Q/T$ both become larger in the negative direction. Consequently, a limiting case as the cyclic integral of δQ approaches zero does not exist for the irreversible refrigerator.

Thus, for all irreversible refrigeration cycles,

$$\oint \delta Q < 0$$

$$\oint \frac{\delta Q}{T} < 0$$

Summarizing, we note that, as regards the sign of $\oint \delta Q$, we have considered all possible reversible cycles (that is, $\oint \delta Q \lessgtr 0$), and for each of these reversible cycles

$$\oint \frac{\delta Q}{T} = 0$$

We have also considered all possible irreversible cycles for the sign of $\oint \delta Q$ (that is, $\oint \delta Q \lessgtr 0$), and for all these irreversible cycles

$$\oint \frac{\delta Q}{T} < 0$$

Thus for all cycles we can write

$$\oint \frac{\delta Q}{T} \leqslant 0 \tag{7.1}$$

where the equality holds for reversible cycles and the inequality for irreversible cycles. This relation, Eq. 7.1, is known as the inequality of Clausius.

The significance of the inequality of Clausius may be illustrated by considering the simple steam power plant cycle shown in Fig. 7.3. This cycle is slightly different from the usual cycle for steam power plants in that the pump handles a mixture of liquid and vapor in such proportions that saturated liquid leaves the pump and enters the boiler. Suppose that someone reports that the pressure and quality at various points in the cycle are as given in Fig. 7.3. Does this cycle satisfy the inequality of Clausius?

Heat is transferred in two places, the boiler and the condenser. Therefore

$$\oint \frac{\delta Q}{T} = \int \left(\frac{\delta Q}{T}\right)_{boiler} + \int \left(\frac{\delta Q}{T}\right)_{condenser}$$

Since the temperature remains constant in both the boiler and condenser, this may be integrated as follows:

$$\oint \frac{\delta Q}{T} = \frac{1}{T_1} \int_1^2 \delta Q + \frac{1}{T_3} \int_3^4 \delta Q = \frac{{}_1Q_2}{T_1} + \frac{{}_3Q_4}{T_3}$$

Let us consider a 1-lb mass as the working fluid.

$_1q_2 = h_2 - h_1 = 889.2 \text{ Btu/lbm};\qquad T_1 = 327.9 \text{ F}$

$_3q_4 = h_4 - h_3 = 196.2 - 1013.9 = -817.7 \text{ Btu/lbm};\quad \cdot\ T_3 = 126.0 \text{ F}$

FIGURE 7.3

A simple steam power plant that demonstrates the inequality of Clausius.

Therefore

$$\oint \frac{\partial Q}{T} = \frac{889.2}{327.9 + 459.7} - \frac{817.7}{126.0 + 459.7} = -0.267 \text{ Btu/lbm-R}$$

Thus this cycle satisfies the inequality of Clausius, which is equivalent to saying that it does not violate the second law of thermodynamics.

7.2 ENTROPY—A PROPERTY OF A SYSTEM

By the use of Eq. 7.1 and Fig. 7.4 it may be shown that the second law of thermodynamics leads to a property of a system that we call entropy. Let a system undergo a reversible process from state 1 to state 2 along path A, and let the cycle be completed along path B, which is also reversible.

Since this is a reversible cycle we can write

$$\oint \frac{\delta Q}{T} = 0 = \int_1^2 \left(\frac{\delta Q}{T}\right)_A + \int_2^1 \left(\frac{\delta Q}{T}\right)_B$$

Now consider another reversible cycle, which has the same initial process, but let the cycle be completed along path C. For this cycle we can write

$$\oint \frac{\delta Q}{T} = 0 = \int_1^2 \left(\frac{\delta Q}{T}\right)_A + \int_2^1 \left(\frac{\delta Q}{T}\right)_C$$

Subtracting the second equation from the first, we have

$$\int_2^1 \left(\frac{\delta Q}{T}\right)_B = \int_2^1 \left(\frac{\delta Q}{T}\right)_C$$

Since the $\int \delta Q/T$ is the same for all reversible paths between states 2 and 1, we conclude that this quantity is independent of the path and is a function of the end states only, and is therefore a property. This property is called entropy, and is designated S. It follows that entropy may be defined as a property of a substance

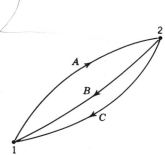

1

FIGURE 7.4

Two reversible cycles demonstrating the fact that entropy is a property of a substance.

in accordance with the relation

$$dS \equiv \left(\frac{\delta Q}{T}\right)_{rev} \tag{7.2}$$

Entropy is an extensive property, and the entropy per unit mass is designated s. It is important to note that entropy is defined here in terms of a reversible process.

The change in the entropy of a system as it undergoes a change of state may be found by integrating Eq. 7.2. Thus

$$S_2 - S_1 = \int_1^2 \left(\frac{\delta Q}{T}\right)_{rev} \tag{7.3}$$

To perform this integration, the relation between T and Q must be known, and illustrations will be given subsequently. The important point to note here is that since entropy is a property, the change in the entropy of a substance in going from one state to another is the same for all processes, both reversible and irreversible, between these two states. Equation 7.3 enables us to find the change in entropy only along a reversible path. However, once it has been evaluated, this is the magnitude of the entropy change for all processes between these two states.

Equation 7.3 enables us to determine changes of entropy, but tells us nothing about absolute values of entropy. However, from the third law of thermodynamics, which is discussed in Chapter 12, it follows that the entropy of all pure substances can be assigned the value of zero at the absolute zero of temperature. This gives rise to absolute values of entropy and is particularly important when chemical reactions are involved.

However, when no change of composition is involved, it is quite adequate to give values of entropy relative to some arbitrarily selected reference state. This is the procedure followed in most tables of thermodynamic properties, such as the steam tables and ammonia tables. Therefore, until absolute entropy is introduced in Chapter 12, values of entropy will always be given relative to some arbitrary reference state.

A word should be added here regarding the role of T as an integrating factor. We noted in Chapter 4 that Q is a path function, and therefore δQ is an inexact differential. However, since $(\delta Q/T)_{rev}$ is a thermodynamic property, it is an exact differential. From a mathematical perspective we note that an inexact differential may be converted to an exact differential by the introduction of an integrating factor. Therefore, $1/T$ serves as the integrating factor in converting the inexact differential δQ to the exact differential $\delta Q/T$ for a reversible process.

7.3 THE ENTROPY OF A PURE SUBSTANCE

Entropy is an extensive property of a system. Values of specific entropy (entropy per unit mass) are tabulated in tables of thermodynamic properties in the same manner as specific volume and specific enthalpy. The units of specific entropy in

the steam tables, Freon-12 tables, and ammonia tables are kJ/kg K, and the values are given relative to an arbitrary reference state. In the steam tables the entropy of saturated liquid at 32.02 F or 0.01°C is given the value of zero. For many refrigerants, such as Freon-12 and ammonia, the entropy of saturated liquid at −40 F or −40°C is assigned the value of zero.

In general we use the term "entropy" to refer to both total entropy and entropy per unit mass, since the context or appropriate symbol will clearly indicate the precise meaning of the term.

In the saturation region the entropy may be calculated using the quality, the relations being similar to those for specific volume and enthalpy.

$$s = (1 - x)s_f + xs_g$$
$$s = s_f + xs_{fg} \tag{7.4}$$

The entropy of a compressed liquid is tabulated in the same manner as the other properties. These properties are primarily a function of the temperature, and are not greatly different from those for saturated liquid at the same temperature. Table 4 of the steam tables, which is summarized in Table A.1.4 of the Appendix, gives the entropy of compressed liquid water in the same manner as for other properties, as has been discussed previously.

The thermodynamic properties of a substance are often shown on a temperature–entropy diagram, and an enthalpy–entropy diagram, which is also called a Mollier diagram, after Richard Mollier (1863–1935) of Germany. Figures 7.5 and 7.6 show the essential elements of temperature–entropy and enthalpy–entropy diagrams for steam. The general features of such diagrams are the same for all

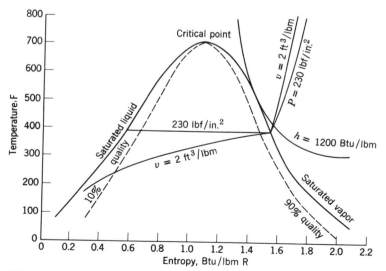

FIGURE 7.5

Temperature–entropy diagram for steam.

pure substances. A more complete temperature–entropy diagram for steam is shown in Fig. A.1.

These diagrams are valuable both as a means of presenting thermodynamic data and also because they enable one to visualize the changes of state that occur in various processes, and as our study progresses the student should acquire facility in visualizing thermodynamic processes on these diagrams. The temperature–entropy diagram is particularly useful for this purpose.

One more observation should be made here concerning the compressed-liquid lines on the temperature–entropy diagram for water. Reference to Table 4 of the steam tables (Appendix, Table A.1.4) indicates that the entropy of the compressed liquid is less than that of saturated liquid at the same temperature for all the temperatures listed except 0°C. It can be shown that each constant-pressure line crosses the saturated-liquid line at the point of maximum density, which is about 4°C for water (the exact temperature at which maximum density occurs varies with pressure). For temperatures less than the temperature at which the density is a maximum, the entropy of compressed liquid is greater than that of the saturated liquid. Thus lines of constant pressure in the compressed-liquid region appear (on a magnified scale) as shown in Fig. 7.7a. It is important to understand the general shape of these lines when showing the pumping process for liquids.

Having made this observation for water, it should be stated that for most substances the difference in the entropy between compressed liquid and saturated liquid at the same temperature is so small that usually a process in which liquid is heated at constant pressure is shown as coinciding with the saturated liquid line until the saturation temperature is reached (Fig. 7.7b). Thus, if water at 10 MPa is heated from 0°C to the saturation temperature, it would be shown by line *ABD*, which coincides with the saturated-liquid line.

FIGURE 7.6

Enthalpy–entropy diagram for steam.

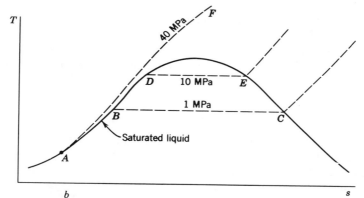

FIGURE 7.7

Temperature–entropy diagram to show properties of a compressed liquid, water.

7.4 ENTROPY CHANGE IN REVERSIBLE PROCESSES

Having established the fact that entropy is a thermodynamic property of a system, its significance in various processes will now be considered. In this section we will limit ourselves to systems that undergo reversible processes and consider the Carnot cycle, reversible heat-transfer processes, and reversible adiabatic processes.

Let the working fluid of a heat engine operating on the Carnot cycle comprise the system. The first process is the isothermal transfer of heat to the working fluid from the high-temperature reservoir. For this we can write

$$S_2 - S_1 = \int_1^2 \left(\frac{\delta Q}{T}\right)_{rev}$$

Since this is a reversible process in which the temperature of the working fluid remains constant, it can be integrated to give

$$S_2 - S_1 = \frac{1}{T_H} \int_1^2 \delta Q = \frac{{}_1Q_2}{T_H}$$

This process is shown in Fig. 7.8a, and the area under line 1–2, area 1–2–b–a–1, represents the heat transferred to the working fluid during the process.

The second process of a Carnot cycle is a reversible adiabatic one. From the definition of entropy:

$$dS = \left(\frac{\delta Q}{T}\right)_{\text{rev}}$$

it is evident that the entropy remains constant in a reversible adiabatic process. A constant-entropy process is called an isentropic process. Line 2–3 represents this process, and this process is concluded at state 3 when the temperature of the working fluid reaches T_L.

The third process is the reversible isothermal process in which heat is transferred from the working fluid to the low-temperature reservoir. For this we can write

$$S_4 - S_3 = \int_3^4 \left(\frac{\delta Q}{T}\right)_{\text{rev}} = \frac{{}_3Q_4}{T_L}$$

Since during this process the heat transfer is negative (as regards the working fluid), the entropy of the working fluid decreases. Also, since the final process 4–1, which completes the cycle, is a reversible adiabatic process (and therefore isentropic), it is evident that the entropy decrease in process 3–4 must exactly equal the entropy increase in process 1–2. The area under line 3–4 area 3–4–a–b–3, represents the heat transferred from the working fluid to the low-temperature reservoir.

Since the net work of the cycle is equal to the net heat transfer, it is evident that area 1–2–3–4–1 represents the net work of the cycle. The efficiency of the cycle may also be expressed in terms of areas.

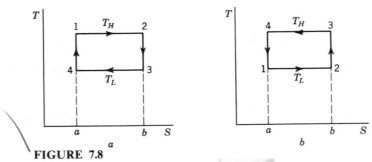

FIGURE 7.8

The Carnot cycle on the temperature–entropy diagram.

$$\eta_{th} = \frac{W_{net}}{Q_H} = \frac{\text{area } 1-2-3-4-1}{\text{area } 1-2-b-a-1}$$

Some statements made earlier about efficiencies may now be understood graphically. For example, increasing T_H while T_L remains constant increases the efficiency. Decreasing T_L as T_H remains constant increases the efficiency. It is also evident that the efficiency approaches 100% as the absolute temperature at which heat is rejected approaches zero.

If the cycle is reversed, we have a refrigerator or heat pump, and the Carnot cycle for a refrigerator is shown in Fig. 7.8b. Notice in this case that the entropy of the working fluid increases at T_L, since heat is transferred to the working fluid at T_L. The entropy decreases at T_H because of heat transfer from the working fluid.

Let us next consider reversible heat-transfer processes. Actually we are concerned here with processes that are internally reversible, that is, processes that involve no irreversibilities within the boundary of the system. For such processes the heat transfer to or from a system can be shown as an area on a temperature–entropy diagram. For example, consider the change of state from saturated liquid to saturated vapor at constant pressure. This would correspond to the process 1–2 on the T-s diagram of Fig. 7.9 (note that absolute temperature is required here), and the area 1–2–b–a–1 represents the heat transfer. Since this is a constant-pressure process, the heat transfer per unit mass is equal to h_{fg}. Thus,

$$s_2 - s_1 = s_{fg} = \frac{1}{m} \int_1^2 \left(\frac{\delta Q}{T} \right)_{rev} = \frac{1}{mT} \int_1^2 \delta Q = \frac{_1q_2}{T} = \frac{h_{fg}}{T}$$

This relation gives a clue as to how s_{fg} is calculated for tabulation in tables of thermodynamic properties. For example, consider steam at 10 MPa. From the

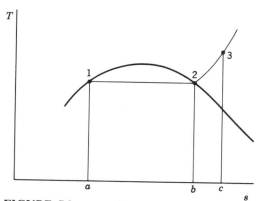

FIGURE 7.9

A temperature–entropy diagram to show areas that represent heat transfer for an internally reversible process.

steam tables we have

$$h_{fg} = 1317.1 \text{ kJ/kg}$$
$$T = 311.96 + 273.15 = 584.21 \text{ K}$$

Therefore,

$$s_{fg} = \frac{h_{fg}}{T} = \frac{1317.1}{584.21} = 2.2544 \text{ kJ/kg K}$$

This is the value listed for s_{fg} in the steam tables.

If heat is transferred to the saturated vapor at constant pressure, the steam is superheated along line 2–3. For this process we can write

$$_2q_3 = \frac{1}{m} \int_2^3 \delta Q = \int_2^3 T \, ds$$

Since T is not constant this cannot be integrated unless we know a relation between temperature and entropy. However, we do realize that the area under line 2–3, area 2–3–c–b–2, represents $\int_2^3 T \, ds$, and therefore represents the heat transferred during this reversible process.

The important conclusion to draw here is that for processes that are internally reversible, the area underneath the process line on a temperature–entropy diagram represents the quantity of heat transferred. This is not true for irreversible processes, as will be demonstrated later.

There are many situations in which essentially adiabatic processes take place. We have already noted that in such cases the ideal process, which is a reversible adiabatic process, is isentropic. We will consider here an example of a reversible adiabatic process for a system and consider later in the chapter the reversible adiabatic process for a control volume. We will also note in a later section that by comparing an actual process with the ideal or isentropic process, we have a basis for defining the efficiency of certain classes of machines.

Example 7.1

Consider a cylinder fitted with a piston which contains saturated Freon-12 vapor at 20 F. Let this vapor be compressed in a reversible adiabatic process until the pressure is 150 lbf/in.2 Determine the work per pound of Freon-12 for this process.

System: Freon-12.
Initial state: T_1, saturated vapor; state fixed.
Final state: P_2 known.
Process: Reversible and adiabatic.
Model: Freon-12 tables.

Analysis: First law—adiabatic.

$$_1q_2 = u_2 - u_1 + {}_1w_2 = 0$$
$$_1w_2 = u_1 - u_2$$

Second law—reversible and adiabatic.

$$s_1 = s_2$$

Therefore we know entropy and pressure in the final state, which is sufficient to specify the final state, since we are dealing with a pure substance.

Solution

From the Freon-12 tables

$$u_1 = h_1 - P_1v_1 = 79.385 - \frac{35.736 \times 144 \times 1.0988}{778} = 72.11 \text{ Btu/lbm}$$

$$s_1 = s_2 = 0.16719 \text{ Btu/lbm R}$$
$$P_2 = 150 \text{ lbf/in.}^2$$

Therefore, from the superheat tables for Freon 12

$$T_2 = 122.9 \text{ F}; \quad h_2 = 90.330; \quad v_2 = 0.28270$$

$$u_2 = 90.330 - \frac{150 \times 144 \times 0.28270}{778} = 82.48 \text{ Btu/lbm}$$

$$_1w_2 = u_1 - u_2 = 72.11 - 82.48 = -10.37 \text{ Btu/lbm}$$

7.5 TWO IMPORTANT THERMODYNAMIC RELATIONS

At this point we derive two important thermodynamic relations for a simple compressible substance. These relations are

$$T\,dS = dU + P\,dV$$
$$T\,dS = dH - V\,dP$$

The first of these relations can be derived by considering a simple compressible substance in the absence of motion or gravitational effects. The first law for a change of state under these conditions can be written

$$\delta Q = dU + \delta W$$

The equations we are deriving here deal first with those changes of state in which the state of the substance can be identified at all times. Thus we must consider a quasiequilibrium process or, to use the term introduced in the last chapter, a reversible process. For a reversible process of a simple compressible substance, we can write

$$\delta Q = T\,dS \quad \text{and} \quad \delta W = P\,dV$$

Substituting these relations into the first law equation we have

$$T\,dS = dU + P\,dV \tag{7.5}$$

which is the equation we set out to derive. Note that this equation was derived by assuming a reversible process, and thus this equation can be integrated for any reversible process, for during such a process the state of the substance can be identified at any point during the process. We also note that Eq. 7.5 deals only with properties. Suppose we have an irreversible process taking place between given initial and final states. The properties of a substance depend only on the state, and therefore the change in the properties during a given change of state are the same for an irreversible process as for a reversible process. Therefore Eq. 7.5 is often applied to an irreversible process between two given states, but the integration of Eq. 7.5 is performed along a reversible path between the same two states.

Since enthalpy is defined as

$$H = U + PV$$

it follows that

$$dH = dU + P\,dV + V\,dP$$

Substituting this relation into Eq. 7.5 we have

$$T\,dS = dH - V\,dP \tag{7.6}$$

which is the second equation that we set out to derive.

These equations can also be written for a unit mass.

$$T\,ds = du + P\,dv$$
$$T\,ds = dh - v\,dP \tag{7.7}$$

or on a mole basis:

$$T\,d\bar{s} = d\bar{u} + P\,d\bar{v}$$
$$T\,d\bar{s} = d\bar{h} - \bar{v}\,dP \tag{7.8}$$

These equations will be used extensively in certain subsequent sections of this book.

If we consider substances of fixed composition other than a simple compressible substance, we can write other "$T\,dS$" equations than those given above for a simple compressible substance. In Chapter 4, Eq. 4.16, we noted that for a reversible process we can write the following expression for work:

$$\delta W = P\,dV - \mathcal{T}\,dL - \mathcal{S}\,d\mathcal{A} - \mu_0 \mathcal{H}\,d(V\mathcal{M}) - \mathcal{E}\,dZ + \cdots$$

It follows that a more general expression for the "$T\,dS$" equation would be

$$T\,dS = dU + P\,dV - \mathcal{T}\,dL - \mathcal{S}\,d\mathcal{A} - \mu_0 \mathcal{H}\,d(V\mathcal{M}) - \mathcal{E}\,dZ + \cdots \tag{7.9}$$

7.6 ENTROPY CHANGE OF A SYSTEM DURING AN IRREVERSIBLE PROCESS

Consider a system that undergoes the cycles shown in Fig. 7.10. The cycle made up of the reversible processes A and B is a reversible cycle. Therefore we can write

$$\oint \frac{\delta Q}{T} = \int_1^2 \left(\frac{\delta Q}{T}\right)_A + \int_2^1 \left(\frac{\delta Q}{T}\right)_B = 0$$

The cycle made up of the reversible process A and the irreversible process C is an irreversible cycle. Therefore, for this cycle the inequality of Clausius may be applied, giving the result

$$\oint \frac{\delta Q}{T} = \int_1^2 \left(\frac{\delta Q}{T}\right)_A + \int_2^1 \left(\frac{\delta Q}{T}\right)_C < 0$$

Subtracting the second equation from the first and rearranging, we have

$$\int_2^1 \left(\frac{\delta Q}{T}\right)_B > \int_2^1 \left(\frac{\delta Q}{T}\right)_C$$

Since path B is reversible, and since entropy is a property,

$$\int_2^1 \left(\frac{\delta Q}{T}\right)_B = \int_2^1 dS_B = \int_2^1 dS_C$$

Therefore,

$$\int_2^1 dS_C > \int_2^1 \left(\frac{\delta Q}{T}\right)_C$$

For the general case we can write

$$dS \geq \frac{\delta Q}{T}$$

$$S_2 - S_1 \geq \int_1^2 \frac{\delta Q}{T} \tag{7.10}$$

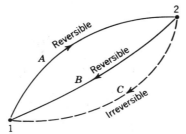

FIGURE 7.10

Entropy change of a system during an irreversible process.

In these equations the equality holds for a reversible process and the inequality for an irreversible process.

This is one of the most important equations of thermodynamics and is used to develop a number of concepts and definitions. In essence this equation states the influence of irreversibility on the entropy of a system. Thus, if an amount of heat δQ is transferred to a system at temperature T in a reversible process, the change of entropy is given by the relation

$$dS = \frac{\delta Q}{T}$$

However, if while the amount of heat δQ is transferred to the system at temperature T there are any irreversible effects occurring, the change of entropy will be greater than for the reversible process, and we would write

$$dS > \frac{\delta Q}{T}$$

Equation 7.10 holds when $\delta Q = 0$ or when $\delta Q < 0$ as well as when $\delta Q > 0$. If δQ is negative, the entropy will tend to decrease as the result of the heat transfer. However, the influence of irreversibilities is still to increase the entropy of the system, and from the absolute numerical perspective we can still write for $\delta Q < 0$:

$$dS \geqslant \frac{\delta Q}{T}$$

7.7 LOST WORK

The significance of the change of entropy for an irreversible process can be amplified by introducing the concept of lost work, to which we give the symbol LW. This concept can be described with the aid of Fig. 7.11, in which a gas is separated from a vacuum by a membrane. Let the membrane have a small rupture so that the gas fills the entire volume, and let the necessary heat transfer take place so that the final temperature is the same as the initial temperature. (How much heat transfer would be necessary if this were an ideal gas?) Since the work is zero for

(a) (b)

FIGURE 7.11

A process that demonstrates lost work.

this process, we conclude from the first law that

$$\delta Q = dU$$

We would like to compare this irreversible process with a reversible process between the same two states. This could be achieved by having the gas expand against a frictionless piston while heat is transferred to the gas in the amount necessary to maintain constant temperature. For this process we conclude from the first law that

$$\delta Q = dU + \delta W$$

and, since this is a reversible process, we can write

$$\delta Q = T\,dS \qquad \text{and} \qquad \delta W = P\,dV$$

The major difference between these two processes is that in the irreversible process the work is zero, whereas in the reversible process the greatest possible work is done. Thus we might speak of the lost work for an irreversible process. Every irreversible process has associated with it a certain amount of lost work. In Fig. 7.11 we have shown the two extreme cases, a reversible process and one in which the work is zero. Between these two extremes are processes that have various degrees of irreversibility, that is, the lost work may vary from zero to some maximum value. For a simple compressible substance we can therefore write

$$P\,dV = \delta W + \delta LW \qquad\qquad (7.11)$$

where δLW is the lost work and δW the actual work.

Substituting the relation, Eq. 7.5,

$$T\,dS = dU + P\,dV$$

into Eq. 7.11 we have

$$T\,dS = dU + \delta W + \delta LW$$

But, from the first law,

$$\delta Q = dU + \delta W$$

Therefore,

$$T\,dS = \delta Q + \delta LW$$

and

$$dS = \frac{\delta Q}{T} + \frac{\delta LW}{T} \qquad\qquad (7.12)$$

Thus we have an expression for the change of entropy for an irreversible process as an equality, whereas in the last section we had the inequality

$$dS \geqslant \frac{\delta Q}{T} \qquad\qquad (7.13)$$

In a reversible process the lost work is zero and therefore both Eqs. 7.12 and 7.13 reduce to

$$dS = \left(\frac{\delta Q}{T}\right)_{rev}$$

for a reversible process.

Some important conclusions can now be drawn from Eqs. 7.12 and 7.13. First, there are two ways in which the entropy of a system can be increased—by transferring heat to it and by having it undergo an irreversible process. Since the lost work cannot be less than zero, there is only one way in which the entropy of a system can be decreased, and that is to transfer heat from the system.

Second, the change in entropy of a system can be separated into the change due to heat transfer and the change due to internal irreversibilities. Frequently the increase in entropy due to irreversibilities is called the irreversible production of entropy.

Finally, as we have already noted for an adiabatic process, $\delta Q = 0$, and in this case the increase in entropy is always associated with the irreversibilities.

One other point, which involves the representation of irreversible processes on P-V and T-S diagrams, should be made. The work for an irreversible process is not equal to $\int P\, dV$ and the heat transfer is not equal to $\int T\, dS$. Therefore, the area underneath the path does not represent work and heat on the P-V and T-S diagrams, respectively. In fact, in many cases we are not certain of the exact state through which a system passes when it undergoes an irreversible process. For this reason it is advantageous to show irreversible processes as dotted lines and reversible processes as solid lines. Thus the area underneath the dotted line will never represent work or heat. For example, the processes of Fig. 7.11 would be shown on P-V and T-S diagrams as in Fig. 7.12.

Figure 7.12a shows an irreversible process and, since the heat transfer and work for this process are zero, the area underneath the dashed line has no significance. Figure 7.12b shows the reversible process, and area 1–2–b–a–1 represents the work on the P-V diagram and the heat transfer on the T-S diagram.

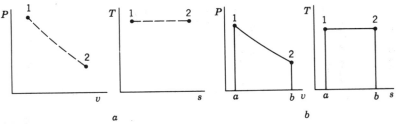

FIGURE 7.12

Reversible and irreversible processes on pressure–volume and temperature–entropy diagrams.

7.8 PRINCIPLE OF THE INCREASE OF ENTROPY

In this section we consider the total change in the entropy of a system and its surroundings when the system undergoes a change of state. This consideration leads to the principle of the increase of entropy.

Consider the process shown in Fig. 7.13 in which a quantity of heat δQ is transferred from the surroundings at temperature T_0 to the system at temperature T, and let the work done by the system during this process be δW. For this process we can apply Eq. 7.10 to the system and write

$$dS_{\text{system}} \geqslant \frac{\delta Q}{T}$$

For the surroundings, δQ is negative and we can write

$$dS_{\text{surr}} = \frac{-\delta Q}{T_0}$$

The total net change of entropy is therefore

$$dS_{\text{net}} = dS_{\text{system}} + dS_{\text{surr}} \geqslant \frac{\delta Q}{T} - \frac{\delta Q}{T_0}$$

$$\geqslant \delta Q\left(\frac{1}{T} - \frac{1}{T_0}\right) \tag{7.14}$$

Since $T_0 > T$, the quantity $(1/T - 1/T_0)$ is positive and we conclude that

$$dS_{\text{net}} = dS_{\text{system}} + dS_{\text{surr}} \geqslant 0$$

If $T > T_0$, the heat transfer is from the system to the surroundings, and both δQ and the quantity $(1/T) - (1/T_0)$ are negative, thus yielding the same result.

Thus we conclude that for all possible processes that a system in a given surroundings can undergo,

$$dS_{\text{net}} = dS_{\text{system}} + dS_{\text{surr}} \geqslant 0 \tag{7.15}$$

FIGURE 7.13

Entropy change for the system plus surroundings.

where the equality holds for reversible processes and the inequality for irreversible processes. This is a very important equation, not only for thermodynamics but also for philosophical thought, and is referred to as the principle of the increase of entropy. The great significance is that the only processes that can take place are those in which the net change in entropy of the system plus its surroundings increases (or in the limit, remains constant). The reverse process, in which both the system and surroundings are returned to their original state, can never be made to occur. In other words, Eq. 7.15 dictates the single direction in which any process can proceed. Thus the principle of the increase of entropy can be considered as a quantitative general statement of the second law from the macroscopic point of view and applies to the combustion of fuel in our automobile engines, the cooling of our coffee, and the processes that take place in our body.

Sometimes this principle of the increase of entropy is stated in terms of an isolated system, one in which there is no interaction between the system and its surroundings. In this case, there is no change in entropy of the surroundings, and one then concludes that

$$dS_{\text{isolated system}} \geqslant 0 \qquad (7.16)$$

That is, in an isolated system, the only processes that can occur are those that have an increase in entropy associated with them.

Example 7.2

Suppose 1 lbm of saturated water vapor at 212 F is condensed to saturated liquid at 212 F in a constant-pressure process by heat transfer to the surrounding air, which is at 80 F. What is the net increase in entropy of the system plus surroundings?

Solution

For the system, from the steam tables,

$$\Delta S_{\text{system}} = -s_{fg} = -1.4446 \text{ Btu/lbm R}$$

Considering the surroundings

$$Q_{\text{to surroundings}} = h_{fg} = 970.3 \text{ Btu/lbm}$$

$$\Delta S_{\text{surr}} = \frac{Q}{T_0} = \frac{970.3}{540} = 1.7980 \text{ Btu/lbm R}$$

$$\Delta S_{\text{system}} + \Delta S_{\text{surr}} = -1.4446 + 1.7980 = 0.3534 \text{ Btu/lbm R}$$

This increase in entropy is in accordance with the principle of the increase of entropy, and tells, as does our experience, that this process can take place.

It is interesting to note how this heat transfer from the water to the surroundings might have taken place reversibly. Suppose that an engine operating on the Carnot cycle received heat from the water and rejected heat to the surroundings,

FIGURE 7.14

Reversible heat transfer with the surroundings.

as shown in Fig. 7.14. The decrease in the entropy of the water is equal to the increase in the entropy of the surroundings.

$$\Delta S_{\text{system}} = -1.4446 \text{ Btu/lbm R}$$
$$\Delta S_{\text{surr}} = 1.4446 \text{ Btu/lbm R}$$
$$Q_{\text{to surroundings}} = T_0 \Delta S = 540(1.4446) = 780.1 \text{ Btu/lbm}$$
$$W = Q_H - Q_L = 970.3 - 780.1 = 190.2 \text{ Btu/lbm}$$

Since this is a reversible cycle, the engine could be reversed and operated as a heat pump. For this cycle the work input to the heat pump would be 190.2 Btu/lbm.

7.9 ENTROPY CHANGE OF A SOLID OR LIQUID

In Section 5.6 we considered the calculation of the internal energy and enthalpy changes with temperature for solids and liquids and found that, in general, it is possible to express both in terms of the specific heat in a simple manner, as Eq. 5.17, and in most cases in the integrated form of Eq. 5.18. We can now use this result and the thermodynamic property relation, Eq. 7.7, to calculate the entropy change for a solid or liquid. Note that for such a phase the specific volume term in Eq. 7.7 is very small, such that substitution of Eq. 5.17 yields

$$ds \simeq \frac{du}{T} \simeq \frac{C}{T} dT \tag{7.17}$$

Now, as was mentioned in Section 5.6, for many processes involving a solid or liquid, we may assume that the specific heat remains constant, in which case Eq. 7.17 can be integrated, with the result being

$$s_2 - s_1 \simeq C \ln\left(\frac{T_2}{T_1}\right) \tag{7.18}$$

If the specific heat is not constant, then commonly C is known as a function of T, in which case Eq. 7.17 can also be integrated to find the entropy change.

Example 7.3

One kilogram of liquid water is heated from 20 to 90°C. Calculate the entropy change assuming constant specific heat and compare with the result found using the steam tables.

System: Water.

Initial and final states: Known.

Model: Constant specific heat, value at room temperature.

Solution

For constant specific heat, from Eq. 7.18,

$$s_2 - s_1 = 4.184 \ln \left(\frac{363.2}{293.2} \right) = 0.8958 \text{ kJ/kg K}$$

Comparing this result with that using the steam tables,

$$s_2 - s_1 = s_{f\,90°C} - s_{f\,20°C} = 1.1925 - 0.2966$$
$$= 0.8959 \text{ kJ/kg K}$$

7.10 ENTROPY CHANGE OF AN IDEAL GAS

Two very useful equations for computing the entropy change of an ideal gas can be developed from Eq. 7.7 by substituting Eqs. 5.20 and 5.24, as follows:

$$T\,ds = du + P\,dv$$

For an ideal gas

$$du = C_{vo}\,dT \quad \text{and} \quad \frac{P}{T} = \frac{R}{v}$$

Therefore,

$$ds = C_{vo}\frac{dT}{T} + \frac{R\,dv}{v} \tag{7.19}$$

$$s_2 - s_1 = \int_1^2 C_{vo}\frac{dT}{T} + R \ln \frac{v_2}{v_1} \tag{7.20}$$

Similarly

$$T\,ds = dh - v\,dP$$

For an ideal gas

$$dh = C_{po}\, dT \qquad \text{and} \qquad \frac{v}{T} = \frac{R}{P}$$

Therefore,

$$ds = C_{po}\frac{dT}{T} - R\frac{dP}{P} \tag{7.21}$$

$$s_2 - s_1 = \int_1^2 C_{po}\frac{dT}{T} - R\ln\frac{P_2}{P_1} \tag{7.22}$$

To integrate Eqs. 7.20 and 7.22, the temperature dependence of the specific heats must be known. However, recalling that their difference is always constant as expressed by Eq. 5.27, we realize that we need to examine the temperature dependence of only one of the specific heats.

As in Section 5.7, let us consider the specific heat C_{po}. Again, there are three possibilities to examine, the simplest of which is the assumption of constant specific heat. In this case it is possible to integrate Eq. 7.22 directly, to

$$s_2 - s_1 = C_{po}\ln\left(\frac{T_2}{T_1}\right) - R\ln\left(\frac{P_2}{P_1}\right) \tag{7.23}$$

Similarly, integrating Eq. 7.20 for constant specific heat,

$$s_2 - s_1 = C_{vo}\ln\left(\frac{T_2}{T_1}\right) + R\ln\left(\frac{v_2}{v_1}\right) \tag{7.24}$$

The second possibility for the specific heat is to utilize an analytical equation for C_{po} as a function of temperature, such as one of those listed in Table A.9. The third possibility is to integrate the results of the calculations of statistical thermodynamics from reference temperature T_0 to any other temperature T, and define a function

$$s_T^{\,0} = \int_{T_0}^T \frac{C_{po}}{T}\, dT \tag{7.25}$$

This function can then be tabulated in the single-entry (temperature) ideal-gas table, as for air in Table A.10 or for various other gases in Table A.11. The entropy change between any two states 1 and 2 is then given by

$$s_2 - s_1 = (s_{T_2}^{\,0} - s_{T_1}^{\,0}) - R\ln\left(\frac{P_2}{P_1}\right) \tag{7.26}$$

As was the case with the energy functions discussed in Section 5.7, the Ideal Gas Tables, A.10 or A.11, would give the most accurate results, with the equations listed in Table A.9 giving a close empirical approximation. Constant specific heat would be less accurate, except for monatomic gases and for other gases below room temperature. Again, it should be remembered that all of these results are

part of the ideal-gas model, which may or may not be appropriate in any particular problem.

Example 7.4

Consider Example 5.6, in which oxygen is heated from 300 to 1500 K. Assume that during this process the pressure dropped from 200 to 150 kPa. Calculate the change in entropy per kilogram.

Solution

The most accurate answer for the entropy change, assuming ideal-gas behavior, would be from the Ideal Gas Tables, A.11. This result is, using Eq. 7.26,

$$\bar{s}_2 - \bar{s}_1 = (258.078 - 205.322) - 8.3144 \ln \left(\frac{150}{200} \right)$$

$$= 52.756 + 2.392 = 55.148 \text{ kJ/kmol K}$$

$$s_2 - s_1 = \frac{55.148}{32} = 1.7234 \text{ kJ/kg K}$$

The empirical equation from Table A.9 should give a good approximation to this result. Integrating Eq. 7.22,

$$\bar{s}_2 - \bar{s}_1 = \int_{T_1}^{T_2} \bar{C}_{po} \frac{dT}{T} - \bar{R} \ln \frac{P_2}{P_1}$$

$$= \left(37.432 \ln \theta + \frac{0.020\ 102}{1.5} \theta^{1.5} \right.$$

$$\left. + \frac{178.57}{1.5} \theta^{-1.5} - \frac{236.88}{2} \theta^{-2} \right) \Bigg|_{\theta_1 = 3}^{\theta_2 = 15} - 8.3144 \ln \left(\frac{150}{200} \right)$$

$$= 52.726 + 2.392 = 55.118 \text{ kJ/kmol K}$$

$$s_2 - s_1 = \frac{\bar{s}_2 - \bar{s}_1}{M} = \frac{55.118}{32} = 1.7224 \text{ kJ/kg K}$$

which is within 0.1% of the previous value. For constant specific heat, using the value at 300 K from Table A.8,

$$s_2 - s_1 = 0.9216 \ln \left(\frac{1500}{300} \right) - 0.259\ 83 \ln \left(\frac{150}{200} \right)$$

$$= 1.4833 + 0.0747 = 1.558 \text{ kJ/kg K}$$

which is too low by 9.6%. If, on the other hand, we assume that the specific heat is constant at its value at 900 K, the average temperature, as in Example 5.6, then

$$s_2 - s_1 = 1.0714 \ln \left(\frac{1500}{300} \right) + 0.0747 = 1.7991 \text{ kJ/kg K}$$

which is high by 4.4%.

Example 7.5

Calculate the change in entropy per pound as air is heated from 540-R to 1200-R while pressure drops from 50 lbf/in.2 to 40 lbf/in.2, assuming

1. Constant specific heat.
2. Variable specific heat.

Solution

1. From Table A.8 for air at 80 F,

$$C_{po} = 0.24 \text{ Btu/lbm R}$$

Therefore, using Eq. 7.23,

$$s_2 - s_1 = 0.24 \ln \left(\frac{1200}{540}\right) - \frac{53.34}{778} \ln \left(\frac{40}{50}\right) = 0.2068 \text{ Btu/lbm R}$$

2. From Table A.10,

$$s_{T_1}^0 = 0.6008 \text{ Btu/lbm R} \qquad s_{T_2}^0 = 0.7963 \text{ Btu/lbm R}$$

Using Eq. 7.26,

$$s_2 - s_1 = 0.7963 - 0.6008 - \frac{53.34}{778} \ln \frac{40}{50} = 0.2108 \text{ Btu/lbm R}$$

The *Air Tables* can be used for reversible adiabatic processes by employing the relative pressure P_r and relative specific volume v_r. The definition of these terms and the derivation follow.

For the reversible adiabatic process

$$T \, ds = dh - v \, dP = 0$$

Therefore

$$dh = C_{po} \, dT = v \, dP = RT \frac{dP}{P}$$

$$\frac{dP}{P} = \frac{C_{po}}{R} \frac{dT}{T}$$

Let this equation be integrated between a reference state having a temperature T_0 and a pressure P_0, and a given arbitrary state having a temperature T and a pressure P. Then

$$\ln \frac{P}{P_0} = \frac{1}{R} \int_{T_0}^{T} C_{po} \frac{dT}{T}$$

The right side of this equation is a function of temperature only. The relative

pressure P_r is defined as

$$\ln P_r \equiv \ln \frac{P}{P_0} = \frac{1}{R} \int_{T_0}^{T} C_{po} \frac{dT}{T} = \frac{s_T^0}{R} \tag{7.27}$$

Thus a value of P_r can be tabulated as a function of temperature.

If we consider two states, 1 and 2, along a constant-entropy line it follows from Eq. 7.27 that

$$\frac{P_1}{P_2} = \left(\frac{P_{r1}}{P_{r2}}\right)_{s=\text{constant}} \tag{7.28}$$

This equation states that the ratio of the relative pressures for two states having the same entropy is equal to the ratio of the absolute pressures.

The development of the relative specific volume is similar, and the ratio of the relative specific volumes v_r in an isentropic process is equal to the ratio of the specific volumes. That is,

$$\frac{v_1}{v_2} = \left(\frac{v_{r1}}{v_{r2}}\right)_{s=\text{constant}} \tag{7.29}$$

The isentropic functions P_r and v_r are tabulated for air, Table A.10, but are not listed in the other gas tables in the Appendix, A.11. In analyzing an isentropic process for these gases, it is necessary to use Eq. 7.26 with the left side of the equation equal to zero and the standard state entropies from Table A.11.

Example 7.6

One kilogram of air is contained in a cylinder fitted with a piston, at a pressure of 400 kPa and a temperature of 600 K. The air is now expanded to 150 kPa in a reversible, adiabatic process. Calculate the work done by the air.

System: Air.

Initial state: P_1, T_1; state 1 fixed.

Final state: P_2.

Process: Reversible and adiabatic.

Model: Ideal gas and Air Tables, A.10.

Analysis:

First law: $\qquad\qquad 0 = u_2 - u_1 + w$

Second law: $\qquad\qquad s_2 = s_1$

Solution

From Table A.10,

$$u_1 = 434.80 \text{ kJ/kg} \qquad P_{r1} = 16.278$$

From Eq. 7.28,

$$P_{r2} = P_{r1} \times \frac{P_2}{P_1} = 16.278 \times \frac{150}{400} = 6.104$$

From Table A.10,

$$T_2 = 457 \text{ K} \qquad u_2 = 327.79$$

Therefore

$$w = 434.80 - 327.79 = 107.01 \text{ kJ/kg}$$

At this point it is advantageous to introduce the specific heat ratio k, which is defined as the ratio of the constant-pressure to the constant-volume specific heat at zero pressure.

$$k = \frac{C_{po}}{C_{vo}} \tag{7.30}$$

Since the difference between C_{po} and C_{vo} is a constant, Eq. 5.27, and since C_{po} and C_{vo} are functions of temperature, it follows that k is also a function of temperature. However, when we consider the specific heat to be constant, k is also constant.

From the definition of k and Eq. 5.27, it follows that

$$C_{vo} = \frac{R}{k-1} \qquad C_{po} = \frac{kR}{k-1} \tag{7.31}$$

Some very useful and simple relations for the reversible adiabatic process can be developed when the specific heats are assumed to be constant.

For the reversible adiabatic process, $ds = 0$. Therefore,

$$T\,ds = du + P\,dv = C_{vo}\,dT + P\,dv = 0$$

From the equation of state for an ideal gas,

$$dT = \frac{1}{R}(P\,dv + v\,dP)$$

Therefore,

$$\frac{C_{vo}}{R}(P\,dv + v\,dP) + P\,dv = 0$$

Substituting Eq. 7.31 into the expression and rearranging,

$$\frac{1}{k-1}(P\,dv + v\,dP) + P\,dv = 0$$

$$v\,dP + kP\,dv = 0$$

$$\frac{dP}{P} + k\frac{dv}{v} = 0$$

Since k is constant when the specific heat is constant, this equation can be integrated under these conditions to give

$$Pv^k = \text{constant} \tag{7.32}$$

Equation 7.32 holds for all reversible adiabatic processes involving an ideal gas with constant specific heat. It is usually advantageous to express this constant in terms of the initial and final states.

$$Pv^k = P_1 v_1{}^k = P_2 v_2{}^k = \text{constant} \tag{7.33}$$

From this equation and the ideal gas equation of state the following expressions relating the initial and final states of an isentropic process can be derived.

$$\frac{P_2}{P_1} = \left(\frac{v_1}{v_2}\right)^k = \left(\frac{V_1}{V_2}\right)^k \tag{7.34}$$

$$\frac{T_2}{T_1} = \left(\frac{P_2}{P_1}\right)^{(k-1)/k} = \left(\frac{v_1}{v_2}\right)^{k-1} \tag{7.35}$$

With the assumption of constant specific heat, some convenient equations can be derived for the work done by an ideal gas during an adiabatic process. Consider first a system consisting of an ideal gas that undergoes a process in which work is done only at the moving boundary.

$$_1Q_2 = m(u_2 - u_1) + {}_1W_2 = 0$$
$$_1W_2 = -m(u_2 - u_1) = -mC_{vo}(T_2 - T_1)$$
$$= \frac{mR}{1-k}(T_2 - T_1) = \frac{P_2 V_2 - P_1 V_1}{1-k} \tag{7.36}$$

It should be noted that Eq. 7.36 applies to adiabatic processes only. Since no assumption was made regarding reversibility, it applies to both reversible and irreversible processes. Frequently, Eq. 7.36 is derived for reversible processes by starting with the relation

$$_1W_2 = \int_1^2 P\,dV$$

for the system.

7.11 THE REVERSIBLE POLYTROPIC PROCESS FOR AN IDEAL GAS

When a gas undergoes a reversible process in which there is heat transfer, the process frequently takes place in such a manner that a plot of log P vs. log V is a straight line as shown in Fig. 7.15. For such a process $PV^n = \text{constant}$.

This is called a polytropic process. An example is the expansion of the combustion gases in the cylinder of a water-cooled reciprocating engine. If the pres-

FIGURE 7.15

Example of a polytropic process.

sure and volume during a polytropic process are measured during the expansion stroke, as might be done with an engine indicator, and the logarithms of the pressure and volume are plotted, the result would be similar to Fig. 7.15. From this figure it follows that

$$\frac{d \ln P}{d \ln V} = -n$$

$$d \ln P + n d \ln V = 0$$

If n is a constant (which implies a straight line on the log P vs. log V plot), this can be integrated to give the following relation:

$$PV^n = \text{constant} = P_1 V_1{}^n = P_2 V_2{}^n \tag{7.37}$$

From this it is evident that the following relations can be written for a polytropic process.

$$\frac{P_2}{P_1} = \left(\frac{V_1}{V_2}\right)^n$$

$$\frac{T_2}{T_1} = \left(\frac{P_2}{P_1}\right)^{(n-1)/n} = \left(\frac{V_1}{V_2}\right)^{n-1} \tag{7.38}$$

For a system consisting of an ideal gas, the work done at the moving boundary during a reversible polytropic process can be derived from the relations

$$_1W_2 = \int_1^2 P\,dV \quad \text{and} \quad PV^n = \text{constant.}$$

$$_1W_2 = \int_1^2 P\,dV = \text{constant} \int_1^2 \frac{dV}{V^n}$$

$$= \frac{P_2 V_2 - P_1 V_1}{1 - n} = \frac{mR(T_2 - T_1)}{1 - n} \tag{7.39}$$

for any value of n except $n = 1$.

FIGURE 7.16

Polytropic processes on P-v and T-s diagrams.

The polytropic processes for various values of n are shown in Fig. 7.16 on P-v and T-s diagrams. The values of n for some familiar processes are given below.

Isobaric process	$n = 0$
Isothermal process	$n = 1$
Isentropic process	$n = k$
Isochoric process	$n = \infty$

Example 7.7

Nitrogen is compressed in a reversible process in a cylinder from 14.7 lbf/in.2, 60 F to 60 lbf/in.2 During the compression process the relation between pressure and volume is $PV^{1.3}$ = constant. Calculate the work and heat transfer per pound, and show this process on P-v and T-s diagrams.

System: Nitrogen.

Initial state: P_1, T_1; state 1 known.

Final state: P_2.

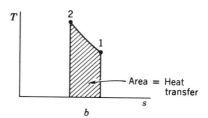

FIGURE 7.17

Diagram for Example 7.7.

Process: Reversible; polytropic with exponent $n < k$.

Diagram: Fig. 7.17.

Model: Ideal gas, constant specific heat—value at 80 F.

Analysis: Boundary movement work. From Eq. 7.39,

$$_1W_2 = \int_1^2 P\,dV = \frac{P_2V_2 - P_1V_1}{1 - n} = \frac{mR(T_2 - T_1)}{1 - n}$$

First law:
$$_1q_2 = u_2 - u_1 + {}_1w_2 = C_{vo}(T_2 - T_1) + {}_1w_2$$

Solution

From Eq. 7.38,

$$\frac{T_2}{T_1} = \left(\frac{P_2}{P_1}\right)^{(n-1)/n} = \left(\frac{60}{14.7}\right)^{(1.3-1)/1.3} = 1.383$$

$$T_2 = 520 \times 1.383 = 719 \text{ R}$$

Then,

$$_1w_2 = \frac{R(T_2 - T_1)}{1 - n} = \frac{55.15(719 - 520)}{(1 - 1.3)778} = -47.0 \text{ Btu/lbm}$$

and from the first law,

$$_1q_2 = C_{vo}(T_2 - T_1) + {}_1w_2$$
$$= 0.177(719 - 520) - 47.0 = -11.8 \text{ Btu/lbm}$$

The reversible isothermal process for an ideal gas is of particular interest. In this case

$$PV = \text{constant} = P_1V_1 = P_2V_2 \qquad (7.40)$$

The work done at the boundary of a simple compressible system during a reversible isothermal process can be found by integrating the equation

$$_1W_2 = \int_1^2 P\,dV$$

The integration is as follows:

$$_1W_2 = \int_1^2 P\,dV = \text{constant} \int_1^2 \frac{dV}{V} = P_1 V_1 \ln \frac{V_2}{V_1} = P_1 V_1 \ln \frac{P_1}{P_2} \qquad (7.41)$$

or

$$_1W_2 = mRT \ln \frac{V_2}{V_1} = mRT \ln \frac{P_1}{P_2} \qquad (7.42)$$

Since there is no change in internal energy or enthalpy in an isothermal process, the heat transfer is equal to the work (neglecting changes in kinetic and potential energy). Therefore, we could have derived Eq. 7.41 by calculating the heat transfer. For example, using Eq. 7.7,

$$\int_1^2 T\,ds = {}_1q_2 = \int_1^2 du + \int_1^2 P\,dv$$

But $du = 0$ and $Pv = \text{constant} = P_1 v_1 = P_2 v_2$

$$_1q_2 = \int_1^2 P\,dv = P_1 v_1 \ln \frac{v_2}{v_1}$$

which yields the same result as Eq. 7.41.

7.12 THE SECOND LAW OF THERMODYNAMICS FOR A CONTROL VOLUME

The second law of thermodynamics can be applied to a control volume by a similar procedure to that used in writing the first law for a control volume in Section 5.11. The second law for a system analysis has been stated, Eq. 7.12, in the form

$$dS = \frac{\delta Q}{T} + \frac{\delta LW}{T}$$

For a change in entropy, $S_2 - S_1$, that occurs during a time interval δt, we can write

$$\frac{S_2 - S_1}{\delta t} = \frac{1}{\delta t}\left(\frac{\delta Q}{T}\right) + \frac{1}{\delta t}\left(\frac{\delta LW}{T}\right) \qquad (7.43)$$

Consider the system and control volume shown in Fig. 7.18. During time δt the mass δm_i enters the control volume across the discrete area \mathscr{A}_i, the mass δm_e leaves across the discrete area \mathscr{A}_e, an amount of heat δQ is transferred to the system across an element of area where the surface temperature is T, and the work δW is done by the system. As in the analysis of Section 5.11 we assume that the increment of mass δm_i has uniform properties, and similarly that δm_e has uniform properties.

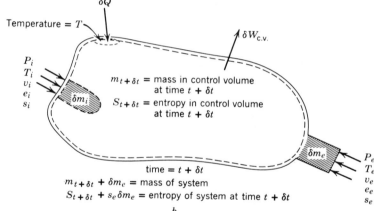

FIGURE 7.18
Schematic diagram for a second law analysis of a control volume.

Let

$$S_t = \text{the entropy in the control volume at time } t$$
$$S_{t+\delta t} = \text{the entropy in the control volume at time } t + \delta t$$

Then

$$S_1 = S_t + s_i\,\delta m_i = \text{entropy of the system at time } t$$
$$S_2 = S_{t+\delta t} + s_e\,\delta m_e = \text{entropy of the system at time } t + \delta t$$

Therefore

$$S_2 - S_1 = (S_{t+\delta t} - S_t) + (s_e\,\delta m_e - s_i\,\delta m_i) \tag{7.44}$$

The term $s_e \delta m_e - s_i \delta m_i$ represents the net flow of entropy out of the control volume during δt as the result of the masses δm_e and δm_i crossing the control surface.

The significance of the remaining terms of Eq. 7.43 must be carefully considered when applied to a control volume. When we write the terms $\delta Q/T$ and $\delta LW/T$ for a system, we normally consider a quantity of mass at uniform temperature at any instant of time. As noted before, in the case of a control volume analysis, we deviate from a strictly classical concept of thermodynamics and are prepared to consider a variation in temperature throughout the control volume, the assumption of "local" equilibrium.

As a consequence, for the heat transfer term, we must consider each element of area on the surface across which heat flows and the surface temperature of this local area. Therefore, this term in Eq. 7.43 must be written as the summation of all local $\delta Q/T$ terms over the control surface, or

$$\frac{1}{\delta t}\left(\frac{\delta Q}{T}\right) = \frac{1}{\delta t}\sum_{\text{c.v.}}\left(\frac{\delta Q}{T}\right)_{\text{c.v.}} \tag{7.45}$$

Similarly, for the lost work term, which represents irreversibilities occurring inside the control volume, the result must be written as the summation of all local $\delta LW/T$ terms throughout the interior of the control volume, or

$$\frac{1}{\delta t}\left(\frac{\delta LW}{T}\right) = \frac{1}{\delta t}\sum_{\text{c.v.}}\left(\frac{\delta LW}{T}\right)_{\text{c.v.}} \tag{7.46}$$

Substituting Eqs. 7.44, 7.45, and 7.46 into Eq. 7.43, we have

$$\left(\frac{S_{t+\delta t} - S_t}{\delta t}\right) + \frac{s_e \delta m_e}{\delta t} - \frac{s_i \delta m_i}{\delta t} = \frac{1}{\delta t}\sum_{\text{c.v.}}\left(\frac{\delta Q}{T}\right)_{\text{c.v.}} + \frac{1}{\delta t}\sum_{\text{c.v.}}\left(\frac{\delta LW}{T}\right)_{\text{c.v.}} \tag{7.47}$$

To reduce this expression to a rate equation, we consider what happens to each of the terms in Eq. 7.47 as δt is made to approach zero. As was the case in Section 5.11, the two mass quantities become mass flow rates, and the entropy term becomes the time rate of change of entropy in the control volume, in a manner analogous to the energy term in the first-law equation. Similarly, the heat transfer and lost work terms become rate quantities, with each being divided by the appropriate local temperature.

In utilizing these limiting values to express the rate form of the entropy equation for a control volume, we again include summation signs on the flow terms, to account for the possibility of additional flow streams entering or leaving the control volume. The resulting expression is

$$\frac{dS_{\text{c.v.}}}{dt} + \sum \dot{m}_e s_e - \sum \dot{m}_i s_i = \sum_{\text{c.v.}}\left(\frac{\dot{Q}_{\text{c.v.}}}{T}\right) + \sum_{\text{c.v.}}\left(\frac{L\dot{W}_{\text{c.v.}}}{T}\right) \tag{7.48}$$

which, for our purposes, is the general expression of the second law entropy equation. In words, this expression states that the rate of change of entropy inside the control volume plus the net rate of entropy flow out is equal to the sum of

two terms, the overall heat transfer surface term and the positive internal irreversibility production term. Since this latter term is necessarily positive (or zero) and is difficult if not impossible to evaluate quantitatively, Eq. 7.48 is frequently written in the form

$$\frac{dS_{c.v.}}{dt} + \sum \dot{m}_e s_e - \sum \dot{m}_i s_i \geqslant \sum_{c.v.} \left(\frac{\dot{Q}_{c.v.}}{T} \right) \tag{7.49}$$

where the equality applies to internally reversible processes and the inequality to internally irreversible processes.

Note that if there is no mass flow into or out of the control volume and if temperature is considered uniform at any instant of time, then Eq. 7.48 reduces to a rate form of the system entropy equation, Eq. 7.12. In this sense, Eq. 7.48 (or 7.49) can be considered a general expression, as was also pointed out for the control volume form of the first law.

7.13 THE STEADY-STATE, STEADY-FLOW PROCESS AND THE UNIFORM-STATE, UNIFORM-FLOW PROCESS

We now consider the application of the second-law control volume equation, Eq. 7.48 or 7.49, to the two control volume model processes developed in Chapter 5.

For the steady-state, steady-flow process, which has been defined in Section 5.12, we conclude that there is no change with time of the entropy per unit mass at any point within the control volume, and therefore the first term of Eq. 7.49 equals zero. That is,

$$\frac{dS_{c.v.}}{dt} = 0 \tag{7.50}$$

so that, for the SSSF process,

$$\sum \dot{m}_e s_e - \sum \dot{m}_i s_i \geqslant \sum_{c.v.} \left(\frac{\dot{Q}_{c.v.}}{T} \right) \tag{7.51}$$

in which the various mass flows, heat transfer rate, and states are all constant with time.

If in a steady-state, steady-flow process there is only one area over which mass enters the control volume at a uniform rate and only one area over which mass leaves the control volume at a uniform rate, we can write

$$\dot{m}(s_e - s_i) \geqslant \sum_{c.v.} \left(\frac{\dot{Q}_{c.v.}}{T} \right) \tag{7.52}$$

For an adiabatic process with these assumptions it follows that

$$s_e \geqslant s_i \tag{7.53}$$

where the equality holds for a reversible adiabatic process.

Example 7.8

Steam enters a steam turbine at a pressure of 1 MPa, a temperature of 300°C, and a velocity of 50 m/s. The steam leaves the turbine at a pressure of 150 kPa and a velocity of 200 m/s. Determine the work per kilogram of steam flowing through the turbine, assuming the process to be reversible and adiabatic.

Control volume: Turbine.

Sketch: Fig. 7.19.

Inlet state: Fixed (Fig. 7.19).

Exit state: P_e, \mathbf{V}_e known.

Process: SSSF.

Model: Steam tables.

Analysis:

Continuity eq.: $$\dot{m}_e = \dot{m}_i = \dot{m}$$

First law: $$h_i + \frac{\mathbf{V}_i^2}{2} = h_e + \frac{\mathbf{V}_e^2}{2} + w$$

Second law: $$s_e = s_i$$

Solution

From the steam tables,

$$h_i = 3051.2 \text{ kJ/kg} \qquad s_i = 7.1229 \text{ kJ/kg K}$$

The two properties known in the final state are pressure and entropy.

$$P_e = 0.15 \text{ MPa} \qquad s_e = s_i = 7.1229 \text{ kJ/kg K}$$

Therefore, the quality and enthalpy of the steam leaving the turbine can be determined.

$P_i = 1$ MPa
$T_i = 300$°C
$V_i = 50$ m/s

W

$P_e = 150$ kPa
$V_e = 200$ m/s

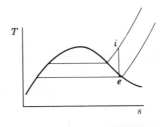

FIGURE 7.19
Sketch for Example 7.8.

$$s_e = 7.1229 = s_f + x_e s_{fg} = 1.4336 + x_e 5.7897$$
$$x_e = 0.9827$$
$$h_e = h_f + x_e h_{fg} = 467.1 + 0.9827(2226.5)$$
$$= 2655.0 \text{ kJ/kg}$$

Therefore, the work per kilogram of steam for this isentropic process may be found using the equation for the first law as given above.

$$w = 3051.2 + \frac{50 \times 50}{2 \times 1000} - 2655.0 - \frac{200 \times 200}{2 \times 1000} = 377.5 \text{ kJ/kg}$$

Example 7.9

Consider the reversible adiabatic flow of steam through a nozzle. Steam enters the nozzle at 100 lbf/in.2, 500 F, with a velocity of 100 ft/s. The pressure of the steam at the nozzle exit is 40 lbf/in.2 Determine the exit velocity of the steam from the nozzle, assuming a reversible adiabatic, steady-state, steady-flow process.

Control volume: Nozzle.

Sketch: Fig. 7.20.

Inlet state: Fixed (Fig. 7.20).

Exit state: P_e known.

Process: SSSF.

Model: Steam tables.

Analysis: Since this is a steady-state, steady-flow process in which the work, the heat transfer, and the changes in potential energy are zero, we can write

Continuity eq.: $$\dot{m}_e = \dot{m}_i = \dot{m}$$

First law: $$h_i + \frac{V_i^2}{2} = h_e + \frac{V_e^2}{2}$$

Second law: $$s_e = s_i$$

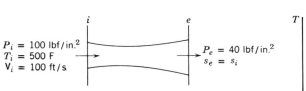

$P_i = 100$ lbf/in.2
$T_i = 500$ F
$V_i = 100$ ft/s
$P_e = 40$ lbf/in.2
$s_e = s_i$

Solution

From the steam tables,

$$h_i = 1279.1 \text{ Btu/lbm}; \quad s_i = 1.7085 \text{ Btu/lbm R}$$

The two properties that we know in the final state are

$$s_e = s_i = 1.7085 \text{ Btu/lbm R}; \quad P_e = 40 \text{ lbf/in.}^2$$

Therefore,

$$T_e = 314.2 \text{ F}; \quad h_e = 1193.9 \text{ Btu/lbm}$$

Substituting into the equation for the first law as given above, we have

$$\frac{\mathbf{V}_e^{\,2}}{2} = h_i - h_e + \frac{\mathbf{V}_i^{\,2}}{2}$$

$$= 1279.1 - 1193.9 + \frac{100 \times 100}{2 \times 32.17 \times 778} = 85.4 \text{ Btu/lbm}$$

$$\mathbf{V}_e = \sqrt{2 \times 32.17 \times 778 \times 85.4} = 2070 \text{ ft/s}$$

Example 7.10

An inventor reports that he has a refrigeration compressor that receives saturated Freon-12 vapor at 0 F and delivers the vapor at 150 lbf/in.2, 120 F. The compression process is adiabatic. Does the process described violate the second law?

Control volume: Compressor.
Inlet state: Fixed (saturated vapor at T_i).
Exit state: Fixed (P_e, T_e known).
Process: SSSF, adiabatic.
Model: Freon-12 tables.
Analysis: Since this is a steady-state, steady-flow, adiabatic process we can write
Second law: $s_e \geqslant s_i$

Solution

From the Freon-12 tables,

$$s_e = 0.16629 \text{ Btu/lbm R}; \quad s_i = 0.16888 \text{ Btu/lbm R}$$

Therefore, $s_e < s_i$, whereas, for this process, the second law requires that $s_e \geqslant s_i$. The process described would involve a violation of the second law, and would not be possible.

Example 7.11

Air expands in an air turbine from-a pressure of 450 kPa and a temperature of 550 K to an exhaust pressure of 150 kPa. Assume the process to be reversible and adiabatic, with negligible changes in kinetic and potential energy. Calculate the work per kilogram of air flowing through the turbine.

Control volume: Turbine.

Inlet state: P_i, T_i known; state fixed.

Exit state: P_e known.

Process: SSSF.

Model: Ideal gas, Air Tables, A.10.

Analysis: Since this is a steady-state, steady-flow reversible process we can write

Continuity eq.: $\qquad\qquad\qquad\qquad \dot{m}_e = \dot{m}_i = \dot{m}$

First law: $\qquad\qquad\qquad\qquad h_i = h_e + w$

Second law: $\qquad\qquad\qquad\qquad s_e = s_i$

Solution

From Table A.10,

$$h_i = 554.8 \text{ kJ/kg} \qquad P_{ri} = 11.858$$

From Eq. 7.28,

$$P_{re} = P_{ri}\left(\frac{P_e}{P_i}\right) = 11.858 \, \frac{150}{450} = 3.953$$

Therefore,

$$T_e = 404.2 \text{ K} \qquad h_e = 405.3$$

and from the first law,

$$w = h_i - h_e = 554.8 - 405.3 = 149.5 \text{ kJ/kg}$$

For the uniform-state, uniform-flow process, which was described in Section 5.14, the second law for a control volume, Eq. 7.49, can be written in the following form.

$$\frac{d}{dt}(ms)_{c.v.} + \sum m_e s_e - \sum \dot{m}_i s_i \geqslant \sum_{c.v.} \left(\frac{\dot{Q}_{c.v.}}{T}\right) \qquad (7.54)$$

If this is integrated over the time interval t, we have

$$\int_0^t \frac{d}{dt}(ms)_{c.v.} \, dt = (m_2 s_2 - m_1 s_1)_{c.v.}$$

$$\int_0^t \left(\sum \dot{m}_e s_e\right) dt = \sum m_e s_e \qquad \int_0^t \left(\sum \dot{m}_i s_i\right) dt = \sum m_i s_i$$

Therefore, for this period of time t, we can write the second law for the uniform-state, uniform-flow process as

$$(m_2 s_2 - m_1 s_1)_{\text{c.v.}} + \sum m_e s_e - \sum m_i s_i \geq \int_0^t \sum_{\text{c.v.}} \left(\frac{\dot{Q}_{\text{c.v.}}}{T} \right) dt \qquad \textbf{(7.55)}$$

However, since in this process the temperature is uniform throughout the control volume at any instant of time, the integral on the right reduces to

$$\int_0^t \sum_{\text{c.v.}} \left(\frac{\dot{Q}_{\text{c.v.}}}{T} \right) dt = \int_0^t \frac{1}{T} \sum_{\text{c.v.}} \dot{Q}_{\text{c.v.}} \, dt = \int_0^t \left(\frac{\dot{Q}_{\text{c.v.}}}{T} \right) dt$$

and therefore the second law for the uniform-state, uniform-flow process can be written

$$(m_2 s_2 - m_1 s_1)_{\text{c.v.}} + \sum m_e s_e - \sum m_i s_i \geq \int_0^t \left(\frac{\dot{Q}_{\text{c.v.}}}{T} \right) dt \qquad \textbf{(7.56)}$$

By introducing the lost work we can write this as an equality. In doing so, we note that since the temperature is uniform throughout the control volume at any instant of time,

$$\int_0^t \sum_{\text{c.v.}} \left(\frac{L\dot{W}_{\text{c.v.}}}{T} \right) dt = \int_0^t \frac{1}{T} \sum_{\text{c.v.}} L\dot{W}_{\text{c.v.}} \, dt = \int_0^t \left(\frac{L\dot{W}_{\text{c.v.}}}{T} \right) dt$$

Therefore,

$$(m_2 s_2 - m_1 s_1)_{\text{c.v.}} + \sum m_e s_e - \sum m_i s_i = \int_0^t \left(\frac{\dot{Q}_{\text{c.v.}} + L\dot{W}_{\text{c.v.}}}{T} \right) dt \qquad \textbf{(7.57)}$$

7.14 THE REVERSIBLE STEADY-STATE, STEADY-FLOW PROCESS

An expression can be derived for the work in a reversible adiabatic, steady-state, steady-flow process which is of great help in understanding the significant variables in such a process. We have noted that when a steady-state, steady-flow process involves a single flow of fluid into and out of the control volume, the first law can be written, Eq. 5.50:

$$q + h_i + \frac{\mathbf{V}_i^2}{2} + gZ_i = h_e + \frac{\mathbf{V}_e^2}{2} + gZ_e + w$$

and the second law is, Eq. 7.52,

$$\dot{m}(s_e - s_i) \geq \sum_{\text{c.v.}} \left(\frac{\dot{Q}_{\text{c.v.}}}{T} \right)$$

Let us now consider two types of flow, a reversible adiabatic process and a reversible isothermal process.

Print Page

If the process is reversible and adiabatic, the second law equation above reduces to

$$s_e = s_i$$

and it follows from the property relation

$$T \, ds = dh - v \, dP$$

that

$$h_e - h_i = \int_i^e v \, dP \tag{7.58}$$

Substituting these relations into Eq. 5.50 and noting that $q = 0$ we have

$$w = (h_i - h_e) + \frac{(\mathbf{V}_i{}^2 - \mathbf{V}_e{}^2)}{2} + g(Z_i - Z_e)$$

$$= -\int_i^e v \, dP + \frac{(\mathbf{V}_i{}^2 - \mathbf{V}_e{}^2)}{2} + g(Z_i - Z_e) \tag{7.59}$$

If, instead, the process is reversible and isothermal, the second law reduces to

$$\dot{m}(s_e - s_i) = \frac{1}{T} \sum_{\text{c.v.}} \dot{Q}_{\text{c.v.}} = \frac{\dot{Q}_{\text{c.v.}}}{T}$$

or

$$T(s_e - s_i) = \frac{\dot{Q}_{\text{c.v.}}}{\dot{m}} = q \tag{7.60}$$

and the property relation can be integrated to give

$$T(s_e - s_i) = (h_e - h_i) - \int_i^e v \, dP \tag{7.61}$$

Substituting Eqs. 7.60 and 7.61 into the first law, Eq. 5.50, then results in the same expression as for the reversible adiabatic process, Eq. 7.59. We further note that any other reversible process can be constructed, in the limit, from a series of alternate adiabatic and isothermal processes. Thus we may conclude that Eq. 7.59 is valid for any reversible, steady-state, steady-flow process, without the restriction that it be either adiabatic or isothermal.

This expression has a wide range of application. If we consider a reversible steady-state, steady-flow process in which the work is zero (such as flow through a nozzle) and the fluid is incompressible ($v =$ constant), Eq. 7.59 can be integrated to give

$$v(P_e - P_i) + \frac{(\mathbf{V}_e{}^2 - \mathbf{V}_i{}^2)}{2} + g(Z_e - Z_i) = 0 \tag{7.62}$$

This is known as the Bernoulli equation (after Daniel Bernoulli) and is a very important equation in fluid mechanics.

Equation 7.59 is also frequently applied to the large class of flow processes involving work (such as turbines and compressors) in which changes in kinetic and potential energies of the working fluid are small. The model process for these

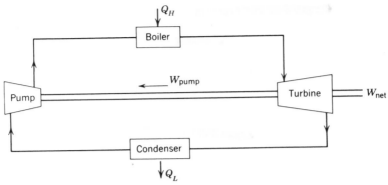

FIGURE 7.21

Simple steam power plant.

machines is then a reversible, SSSF process with no change in kinetic or potential energy (and commonly, although not necessarily, adiabatic as well). For this process, Eq. 7.59 reduces to the form

$$w = -\int_i^e v\, dP \qquad (7.63)$$

From this result, we conclude that the shaft work associated with this type of process is closely related to the specific volume of the fluid during the process. To further amplify this point, consider the simple steam power plant shown in Fig. 7.21. Suppose this is an ideal power plant, with no pressure drop in the piping, the boiler, or the condenser. Thus the pressure increase in the pump is equal to the pressure decrease in the turbine. Neglecting kinetic and potential energy changes, the work done in each of these processes is given by Eq. 7.63. Since the pump handles liquid, which has a very small specific volume as compared to the vapor that flows through the turbine, the power input to the pump is much less than the power output of the turbine, the difference being the net power output of the power plant.

This same line of reasoning can be qualitatively applied to actual devices that involve steady-state, steady-flow processes, even though the processes are not exactly reversible and adiabatic.

Example 7.12

Calculate the work per pound to pump water isentropically from 100 lbf/in.2, 80 F to 1000 lbf/in.2

Control volume: Pump.

Inlet state: P_i, T_i known; state fixed.

Exit state: P_e known.

Process: SSSF.

Model: Steam tables.

Analysis: Since the process is SSSF, reversible, and adiabatic, and changes in kinetic and potential energies can be neglected,

First law: $h_i = h_e + w$

Second law: $s_e - s_i = 0$

Solution

Since P_e and s_e are known, state e is fixed and, therefore, h_e is known and w could be found from the first law. However, the process is reversible, SSSF, with negligible changes in kinetic and potential energies, so that Eq. 7.63 is also valid. Furthermore, since a liquid is being pumped, the specific volume will change very little during the process.

From the steam tables, $v_1 = 0.016\,07$ ft^3/lbm. Assuming the specific volume to remain constant and using Eq. 7.63 we have,

$$-w = \int_1^2 v\,dP = v(P_2 - P_1) = 0.016\,07(1000 - 100) \times \frac{144}{778}$$

$$= 2.68 \text{ Btu/lbm}$$

As a final application of Eq. 7.59, we recall the reversible polytropic process for an ideal gas, discussed in Section 7.11 for a system process. For the SSSF process with no change in kinetic and potential energies, from the relations,

$$w = -\int_i^e v\,dP \qquad \text{and} \qquad Pv^n = \text{constant} = C^n$$

$$w = -\int_i^e v\,dP = -C\int_i^e \frac{dP}{P^{1/n}}$$

$$= -\frac{n}{n-1}(P_e v_e - P_i v_i) = -\frac{nR}{n-1}(T_e - T_i) \tag{7.64}$$

If the process is isothermal, then $n = 1$ and the integral becomes

$$w = -\int_i^e v\,dP = -\text{constant}\int_i^e \frac{dP}{P} = -P_i v_i \ln \frac{P_e}{P_i} \tag{7.65}$$

These evaluations of the integral

$$\int_i^e v\,dP$$

may also be used in conjunction with Eq. 7.59 in cases for which kinetic and potential energy changes are not negligibly small.

7.15 PRINCIPLE OF THE INCREASE OF ENTROPY FOR A CONTROL VOLUME

The principle of the increase of entropy for a system analysis was discussed in Section 7.8. The same general conclusion is reached in the case of a control volume analysis. To demonstrate this, consider a control volume, Fig. 7.22, which

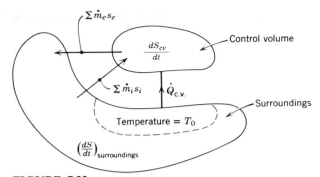

FIGURE 7.22

Entropy change for a control volume plus surroundings.

exchanges both mass and heat with its surroundings. At the point in the sur-
roundings where the heat transfer occurs, the temperature is T_0. From Eq. 7.49
the second law for this process is

$$\frac{dS_{c.v.}}{dt} + \sum \dot{m}_e s_e - \sum \dot{m}_i s_i \geq \sum_{c.v.} \left(\frac{\dot{Q}_{c.v.}}{T} \right)$$

We recall that the first term represents the rate of change of entropy within
the control volume, and the next terms the net entropy flow out of the control
volume as the result of the mass flow. Therefore, for the surroundings, we can
write

$$\frac{dS_{surr}}{dt} = \sum \dot{m}_e s_e - \sum \dot{m}_i s_i - \frac{\dot{Q}_{c.v.}}{T_0} \qquad (7.66)$$

Adding Eqs. 7.49 and 7.66, we have

$$\frac{dS_{net}}{dt} = \frac{dS_{c.v.}}{dt} + \frac{dS_{surr}}{dt} \geq \sum_{c.v.} \left(\frac{\dot{Q}_{c.v.}}{T} \right) - \frac{\dot{Q}_{c.v.}}{T_0} \qquad (7.67)$$

Since $\dot{Q}_{c.v.} > 0$ when $T_0 > T$, and $\dot{Q}_{c.v.} < 0$ when $T_0 < T$, it follows that

$$\frac{dS_{net}}{dt} = \frac{dS_{c.v.}}{dt} + \frac{dS_{surr}}{dt} \geq 0 \qquad (7.68)$$

which can be termed the general statement of the principle of the increase of
entropy.

When we utilize Eq. 7.68 to check any particular process for a possible vio-
lation of the second law, it will be in connection with one of our model processes.
For example, in a steady-state, steady-flow process, as we consider the two terms
in Eq. 7.68 we realize that, in accordance with Eq. 7.50, the first term is zero, the
result being that all of the entropy change due to irreversibilities in this type of
process is observed in the surroundings. That term may then be evaluated using
Eq. 7.66. On the other hand, for the uniform-state, uniform-flow process, there are
both control volume and surroundings terms to evaluate, with each being inte-

grated over the time t of the process, as was done in Section 7.13. Thus Eq. 7.68 is integrated to

$$\Delta S_{net} = \Delta S_{c.v.} + \Delta S_{surr} \tag{7.69}$$

in which the control volume term is

$$\Delta S_{c.v.} = (m_2 s_2 - m_1 s_1)_{c.v.} \tag{7.70}$$

while the result for the surroundings is, after applying Eq. 7.57 to the surroundings and integrating,

$$\Delta S_{surr} = \frac{-Q_{c.v.}}{T_0} + \sum m_e s_e - \sum m_i s_i \tag{7.71}$$

7.16 EFFICIENCY

In Chapter 6 we noted that the second law of thermodynamics led to the concept of thermal efficiency for a heat engine cycle, namely,

$$\eta_{th} = \frac{W_{net}}{Q_H}$$

where W_{net} is the net work of the cycle and Q_H is the heat transfer from the high-temperature body.

In this chapter we have extended our consideration of the second law to control volume processes, and this leads us now to consider the efficiency of a process. For example, we might be interested in the efficiency of a turbine in a steam power plant, or the compressor in a gas turbine engine.

In general we can say that the efficiency of a machine in which a process takes place involves a comparison between the actual performance of the machine under given conditions and the performance that would have been achieved in an ideal process. It is in the definition of this ideal process that the second law becomes a major consideration. For example, a steam turbine is intended to be an adiabatic machine. The only heat transfer is the unavoidable heat transfer that takes place between the given turbine and the surroundings. Also, we note that for a given steam turbine operating in a steady-state, steady-flow manner, the state of the steam entering the turbine and the exhaust pressure are fixed. Therefore the ideal process would be a reversible adiabatic process, which is an isentropic process, between the inlet state and the turbine exhaust pressure. If we denote the actual work done per unit mass of steam flow through the turbine as w_a and the work that would have been done in a reversible adiabatic process between the inlet state and the turbine exhaust pressure as w_s, the isentropic efficiency of the turbine is defined as

$$\eta_{turbine} = \frac{w_a}{w_s} \tag{7.72}$$

The same relation would hold for a gas turbine.

A few other examples may help to clarify this point. In a nozzle the objective is to have the maximum kinetic energy leaving the nozzle for the given inlet conditions and exhaust pressure. The nozzle is also an adiabatic device and therefore the ideal process is a reversible adiabatic or isentropic process. The isentropic efficiency of a nozzle is the ratio of the actual kinetic energy leaving the nozzle, $\mathbf{V}_a^2/2$, to the kinetic energy for an isentropic process between the same inlet conditions and exhaust pressure—$\mathbf{V}_s^2/2$.

$$\eta_{\text{nozzle}} = \frac{\mathbf{V}_a^2/2}{\mathbf{V}_s^2/2} \tag{7.73}$$

In compressors for air or other gases, there are two ideal processes to which the actual performance can be compared. If no effort is made to cool the gas during compression (that is, when the process is adiabatic), the ideal process is a reversible adiabatic or isentropic process between the given inlet state and exhaust pressure. If we denote the work per unit mass of gas flow through the compressor for this isentropic process as w_s, and the actual work as w_a (the actual work input will be greater than the work input for an isentropic process), the isentropic efficiency is defined by the relation

$$\eta_{\text{adiabatic compressor}} = \frac{w_s}{w_a} \tag{7.74}$$

If an attempt is made to cool the air during compression by use of a water jacket or fins, the ideal process is considered a reversible isothermal process. If w_t is the work for the reversible isothermal process between the given inlet state and exhaust pressure, and w_a the actual work, the isothermal efficiency is defined by the relation

$$\eta_{\text{cooled compressor}} = \frac{w_t}{w_a} \tag{7.75}$$

Thus we note that the efficiency of a device that involves a process (rather than a cycle) involves a comparison of the actual performance to that which would be achieved in a related, but well-defined ideal process.

Example 7.13

A steam turbine receives steam at a pressure of 1 MPa, 300°C. The steam leaves the turbine at a pressure of 15 kPa. The work output of the turbine is measured and is found to be 600 kJ/kg of steam flowing through the turbine. Determine the isentropic efficiency of the turbine.

Control volume: Turbine.
Inlet state: P_i, T_i known; state fixed.
Exit state: P_e known.
Process: SSSF.

Model: Steam tables.

Analysis: The isentropic efficiency of the turbine is given by Eq. 7.72.

$$\eta_{\text{turbine}} = \frac{w_a}{w_s}$$

Thus a determination of the turbine efficiency involves a calculation of the work that would be done in an isentropic process between the given inlet state and final pressure. For this isentropic process,

Continuity eq.: $\qquad\qquad \dot{m}_i = \dot{m}_e = \dot{m}$

First law: $\qquad\qquad\quad h_i = h_{es} + w_s$

Second law: $\qquad\qquad s_i = s_{es}$

Solution

From the steam tables,

$$h_i = 3051.2 \text{ kJ/kg} \qquad s_i = 7.1229 \text{ kJ/kg K}$$

Therefore, at $P_e = 15 \text{ kPa}$,

$$s_{es} = s_i = 7.1229 = 0.7549 + x_{es} 7.2536$$
$$x_{es} = 0.8779$$
$$h_{es} = 225.9 + 0.8779(2373.1) = 2309.3 \text{ kJ/kg}$$

From the first law for the isentropic process,

$$w_s = h_i - h_{es} = 3051.2 - 2309.3 = 741.9 \text{ kJ/kg}$$

But, since

$$w_a = 600 \text{ kJ/kg}$$

we find that

$$\eta_{\text{turbine}} = \frac{w_a}{w_s} = \frac{600}{741.9} = 0.809 = 80.9\%$$

In connection with this example, it should be noted that to find the actual state e of the steam exiting the turbine, it is necessary to analyze the real process taking place. For the real process

$$\dot{m}_i = \dot{m}_e = \dot{m}$$
$$h_i = h_e + w_a$$
$$s_e > s_i$$

Therefore, from the first law for the real process,

$$h_e = 3051.2 - 600 = 2451.2 \text{ kJ/kg}$$
$$2451.2 = 225.9 + x_e 2373.1$$
$$x_e = 0.9377$$

7.17 SOME GENERAL COMMENTS REGARDING ENTROPY

It is quite possible at this point that a student may have a good grasp of the material that has been covered, and yet may have only a vague understanding of the significance of entropy. In fact, the question "What is entropy?" is frequently raised by students with the implication that no one really knows! This section has been included in an attempt to give insight into the qualitative and philosophical aspects of the concept of entropy, and to illustrate the broad application of entropy to many different disciplines.

First, we recall that the concept of energy rises from the first law of thermodynamics and the concept of entropy from the second law of thermodynamics. Actually it is just as difficult to answer the question "What is energy?" as it is to answer the question "What is entropy?" However, since we regularly use the term energy and are able to relate this term to phenomena that we observe every day, the word energy has a definite meaning to us and thus serves as an effective vehicle for thought and communication. The word entropy could serve in the same capacity. If, when we observed a highly irreversible process (such as cooling coffee by placing an ice cube in it), we said, "That surely increases the entropy," we would soon be as familiar with the word *entropy* as we are with the word *energy*. In many cases when we speak about a higher efficiency we are actually speaking about accomplishing a given objective with a smaller total increase in entropy.

A second point to be made regarding entropy is that in statistical thermodynamics, the property entropy is defined in terms of probability. Although this topic will not be examined in detail in this text, a few brief remarks regarding entropy and probability may prove helpful. From this point of view the net increase in entropy that occurs during an irreversible process can be associated with a change of state from a less probable state to a more probable state. For instance, to use a previous example, one is more likely to find gas on both sides of the ruptured membrane of Fig. 6.11 than to find a gas on one side and a vacuum on the other. Thus, when the membrane ruptures, the direction of the process is from a less probable state to a more probable state and associated with this process is an increase in entropy. Similarly, the more probable state is that a cup of coffee will be at the same temperature as its surroundings than at a higher (or lower) temperature. Therefore, as the coffee cools as the result of a transferring of heat to the surroundings, there is a change from a less probable to a more probable state, and associated with this is an increase in entropy.

The final point to be made is that the second law of thermodynamics and the principle of the increase of entropy have philosophical implications. Does the second law of thermodynamics apply to the universe as a whole? Are there processes unknown to us that occur somewhere in the universe, such as "continual creation," that have a decrease in entropy associated with them, and thus offset the continual increase in entropy that is associated with the natural processes that are known to us? If the second law is valid for the universe (we of course do not know if the universe can be considered as an isolated system), how did it get in the state

of low entropy? On the other end of the scale, if all processes known to us have an increase in entropy associated with them, what is the future of the natural world as we know it?

Quite obviously it is impossible to give conclusive answers to these questions on the basis of the second law of thermodynamics alone. However, we see the second law of thermodynamics as a description of the prior and continuing work of a creator, who also holds the answer to our future destiny and that of the universe.

PROBLEMS

7.1 Consider a Carnot cycle heat engine that operates between reservoirs at temperatures of 1000 F and 100 F and receives 1000 Btu of heat from the high-temperature reservoir.
 (a) Taking the working fluid as the system, show the cycle on T–S coordinates.
 (b) Calculate the net work and the thermal efficiency of this cycle.
 (c) Calculate the change of entropy of the high-temperature and low-temperature reservoirs.

7.2 A Carnot cycle heat pump utilizes Freon-12 as the working fluid. Heat is transferred from the fluid at 100 F, and during this process the Freon-12 changes from saturated vapor to saturated liquid. Heat transfer to the Freon-12 takes place at 20 F.

 (a) Show this cycle on a T–s diagram.
 (b) Calculate the quality at the beginning and at the end of the isothermal process at 20 F.
 (c) Determine the coefficient of performance for the cycle.

7.3 Consider a particular Carnot cycle with water as the working fluid and having a thermal efficiency of 25%. The heat transfer to the working fluid takes place at 300°C, during which the water changes from saturated liquid to saturated vapor.
 (a) Show this cycle on a T–s diagram.
 (b) Calculate the quality of the working fluid at the beginning and at the end of the isothermal heat-rejection process.
 (c) Determine the net work output per kilogram of water.

7.4 A cylinder fitted with a frictionless piston contains steam at 600 F, 300 lbf/in.², at which point the cylinder volume is 5 ft³. The steam now expands, doing work against the piston, until the final pressure is 30 lbf/in.² How much work is done by the steam during this process, if the process is assumed to be adiabatic?

7.5 A cylinder and piston arrangement contains 2 kg of Freon-12 at 150 kPa, 60°C. The piston is now moved slowly, compressing the Freon-12 until it

exists as saturated vapor. During this process, the necessary heat transfer takes place such that the temperature of the Freon-12 remains constant.
(a) Determine the final pressure and specific volume of the Freon-12.
(b) Calculate the work and heat transfer for the process.

7.6 A 100-in.3 cylinder fitted with a piston contains saturated vapor ammonia at -20 F. The external force acting on the piston is now increased, compressing the ammonia in a reversible adiabatic process until the pressure reaches 100 lbf/in.2 Determine the work done during this process.

7.7 A rigid, insulated vessel contains superheated vapor steam at 200 lbf/in.2, 700 F. A valve on the vessel is now opened, allowing steam to escape. It may be assumed that at any instant of time the steam remaining inside the vessel has undergone a reversible, adiabatic expansion. Determine the fraction of steam that has escaped when the steam remaining inside the vessel reaches the saturated vapor line.

7.8 A heavily insulated cylinder fitted with a piston contains ammonia at 10°C. The piston is now moved, compressing the ammonia in a reversible process until the pressure reaches 2 MPa, at which point the temperature is 70°C. During the process, 400 kJ of work is done on the system. What was the initial volume of the cylinder?

7.9 A 40-kg mass is held in position in a length of 125-mm-diameter tube by a pin, as shown in Fig. 7.23. Below this mass is a volume of 15 L of steam at 5 MPa, 400°C. The pin is now released and the mass accelerates upward, leaving the top of the tube with a certain velocity. If the steam undergoes a reversible adiabatic expansion, until a pressure of 600 kPa is reached, what is the exit velocity of the mass?

FIGURE 7.23

Sketch for Problem 7.9.

7.10 Consider Problem 4.15, involving steam expanding in a reversible isothermal process. Solve the problem without using the relation.

$$_1W_2 = \int_1^2 P\,dV$$

7.11 A cylinder fitted with a piston contains 5 kg of ammonia at 200 kPa, 20°C. The piston is now moved, compressing the ammonia in a reversible isothermal process until the volume is one-tenth of the initial value. Determine the work and the heat transfer for this process.

7.12 The rigid tank A shown in Fig. 7.24 contains 10 lbm of steam at 100 lbf/in.², 600 F, and is connected by a valve to cylinder B, which is fitted with a frictionless piston. A pressure of 20 lbf/in.² is required inside B to raise the piston. The valve is now opened, and steam flows from A to B until the pressure in A has dropped to 20 lbf/in.², at which point the valve is closed. Assume the entire process to be adiabatic, and also that the steam that finally remains in A has undergone a reversible adiabatic expansion. Determine the work done against the piston/surroundings and the final temperature of the steam in cylinder B.

$$0 = m_{2A}u_{2A} + m_{2B}u_{2B} - m_{1A}u_{1A} + P_{2B}m_{2B}\dot{V}_{2B}$$

$$0 = m_{2A}u_{2u} + m_{2B}h_{2B} - m_{1A}u_{1A}$$

FIGURE 7.24

Sketch for Problem 7.12.

7.13 Repeat Problem 7.12, but let the piston initially rest on stops so that the initial volume of the cylinder B is 10 ft³, as shown in Fig. 7.25. It may be assumed that this volume is initially evacuated.

FIGURE 7.25

Sketch for Problem 7.13.

7.14 It is desired to cool a given quantity of material rapidly to a temperature of 10°C, and the process requires a heat rejection of 2000 kJ. One possibility is to immerse the material in a mixture of ice and water, in which

case the heat transfer from the material results in the melting of ice. Another possibility is to cool the material by evaporating Freon-12 at $-20°C$, in which case the heat transfer results in changing Freon-12 from saturated liquid to saturated vapor. A third possibility is to do the same with liquid nitrogen at 101.3 kPa pressure.

(a) Calculate the change of entropy of the cooling medium in each of the three cases. $\quad \Delta S = S_2 - S_1 = m(s_2 - s_1)$

(b) Discuss the significance of these results.

7.15 A cylinder fitted with a piston contains 5 lbm of water at 80 lbf/in.², 1000 F. The piston has a cross-sectional area of 1.0 ft² and is restrained by a linear spring (force proportional to distance moved) with a spring constant of 60 lbf/in. The system is now allowed to cool until the pressure inside drops to 20 lbf/in.², as a result of transferring heat to the room at 70 F. Calculate the net change of entropy, system plus surroundings, for this process.

7.16 A spherical elastic balloon that supports an internal pressure proportional to its diameter contains 10 lbm of Freon-12 at 30 F, 30% quality. This system is now allowed to warm to room temperature, 80 F. Calculate:

(a) The heat transfer during this process. $\quad P v_2 = v_1 \left(\frac{P_2}{P_1}\right)^3$

(b) The net entropy change, system plus surroundings.

7.17 Consider the process shown in Fig. 7.26. Tank A, which is insulated, has a volume of 600 L and contains steam at 1.4 MPa, 300°C. Tank B, which is uninsulated, has a volume of 300 L and contains steam at 200 kPa, 200°C. A valve connecting the two tanks is now opened and steam flows from A to B until the temperature in A reaches 250°C, at which time the valve is closed. During this process, heat is transferred from B to the surroundings at 25°C such that the temperature inside B remains at 200°C. It may be assumed that the steam remaining inside A has undergone a reversible adiabatic process. Determine:

(a) The final pressure in each tank.

(b) The final mass in tank B.

(c) The net entropy change, system plus surroundings, for the process.

FIGURE 7.26

Sketch for Problem 7.17.

7.18 An insulated cylinder fitted with a piston has an initial volume of 0.15 m³ and contains steam at 400 kPa, 200°C. The steam is now expanded adia-

batically, and during this process the work output is carefully measured and found to be 30 kJ. It is claimed that at the final state the steam is in the two-phase (liquid plus vapor) region. What is your evaluation of this claim?

7.19 An insulated cylinder fitted with a piston contains Freon-12 at 90 lbf/in.², 100 F, at which point the volume is 3 ft³. The Freon-12 now expands, moving the piston until the pressure in the cylinder has dropped to 15 lbf/in.² It is claimed that the Freon-12 does 60 Btu of work against the piston during the process. Is this possible?

7.20 A cylinder fitted with a frictionless piston contains water at 20 lbf/in.², 240 F, at which point the volume is 5 ft³. The piston is now moved slowly, compressing the water to a pressure of 100 lbf/in.² During this process

$$\Delta S_{sys} = m(s_2 - s_1) \qquad PV^{0.9} = \text{constant} \qquad \Delta S_{surr} = \frac{Q}{T_{env}}$$

Assuming that the room temperature is 70 F, show that this process does not violate the second law.

7.21 A cylinder fitted with a piston contains ammonia at 200 kPa, 20°C, at which point the volume is 1.5 m³. The ammonia is now compressed to a final temperature of 40°C with a work input of 750 kJ, and during the process 1500 kJ of heat is rejected to the surroundings at 20°C. Is this process impossible?

7.22 A system consisting of 20 lbm of Freon-12 is initially at 210 F, 70% quality. This system then expands to a pressure of 30 lbf/in.² as a result of receiving a heat transfer of 220 Btu from a large heat source at 220 F. It is claimed that the system does 80 Btu of work on its surroundings during this process. Is this possible? $\quad _1Q_2 = m(u_2 - u_1) + {}_1W_2 \qquad \Delta S_{sys} = m(s_2 - s_1) \qquad \Delta S_{surr} = \frac{Q}{T}$

7.23 A vertical cylinder with a frictionless piston contains Freon-12 at $-20°C$, 70% quality, and the volume is 0.1 m³. This cylinder is now brought into a 25°C room, and an electric current of 5 A is passed through a resistor inside the cylinder. The voltage drop across the resistor is 20 V. It is claimed that after 30 minutes the temperature inside the cylinder is 40°C. Is this possible?

7.24 A 10-lbm bar of cast iron initially at 700 F is quenched in a 5-ft³ water tank initially at 60 F. Assuming no heat transfer with the surroundings and no boiling away of liquid water, calculate the net entropy change for the process.

7.25 A 10-lbm bar of lead initially at 70 F is heated at constant pressure to 1000 F, by means of heat transfer from a large constant-temperature heat source at 1000 F. Calculate the net entropy change for this process. The melting point of lead is 620 F, at which the enthalpy change of melting is 10.6 Btu/lbm. The specific heat of the solid is 0.033 Btu/lbm R and that of the liquid is 0.037 Btu/lbm R. $\quad \Delta S_{sys} = \int_1^2 c\frac{dT}{T}$

7.26 A hollow steel sphere having an inside diameter of 0.5 m and a 2-mm wall thickness contains water at 2 MPa, 250°C. This system (steel plus water)

then cools to the ambient temperature, 30°C. Calculate the net entropy change of the system plus surroundings for this process.

7.27 A mass of 5 lbm of air contained in a cylinder at 100 lbf/in.2, 500 F, expands in a reversible isothermal process to 15 lbf/in.2 Calculate:
(a) The heat transfer during the process.
(b) The change of entropy of the air.

7.28 A mass of 5 lbm of air contained in a cylinder at 100 lbf/in.2, 500 F, expands in a reversible adiabatic process to 15 lbf/in.2 Calculate the final temperature and the work done during the process, using: $\frac{T_2}{T_1} = \left(\frac{P_2}{P_1}\right)^{\frac{K-1}{K}}$
(a) Constant specific heat, value from Table A.8.
(b) The ideal gas air tables, Table A.10. $\frac{P_2}{P_1} = \frac{P_{r2}}{P_{r1}}$

7.29 Consider a Carnot cycle heat engine having 1 kg of air in a cylinder and piston arrangement as the working fluid. This engine operates between reservoirs at 1000 K and 300 K. At the beginning of the high-temperature heat addition process, the pressure is 1 MPa, and during this process the volume triples. Analyze each of the four processes in the cycle, and determine:
(a) The pressure, volume, and temperature at each point.
(b) The work and heat transfer for each process.

7.30 An engineer has a problem to solve that involves the entropy of super-heated vapor steam at pressures below 10 kPa, but he finds that the steam table that he has does not give the properties of superheated vapor below this pressure.
(a) How would you advise him to proceed?
(b) Set up, in accordance with whatever assumptions or models are used, a line similar to the 10-kPa line in Table A.1.3. that gives v, u, h, and s at 5 kPa at temperatures of 100°C, 400°C, and 1000°C.

7.31 It is desired to obtain a supply of cold helium gas by the following technique: helium contained in a cylinder at ambient conditions, 14.7 lbf/in.2 and 70 F, is compressed in a reversible isothermal process to 90 lbf/in.2, after which the gas is expanded back to 14.7 lbf/in.2 in a reversible adiabatic process.
(a) Show the processes on a T-s diagram.
(b) Calculate the final temperature and the net work per unit mass of helium.
(c) If a diatomic gas, such as nitrogen or oxygen, were used instead of helium, would the final temperature be higher, lower or the same?

7.32 An insulated, rigid 25-ft^3 tank contains air at 110 lbf/in.2, 75 F. A valve on the tank is now opened and the pressure inside quickly drops to 15 lbf/in.2, at which point the valve is closed.
(a) Assume that the air remaining inside the tank has undergone a revers-ible adiabatic expansion, calculate the mass withdrawn during the process.
(b) Calculate the mass withdrawn by a first law, control volume analysis, and compare the result with part (a).

(c) Repeat part (b) utilizing a stepwise solution procedure, dividing the solution into several parts. Then try to develop this approach into a general first law solution.

7.33 An insulated cylinder having an initial volume of 100 L contains carbon dioxide gas at 300 kPa, 400 K. The gas is then compressed to 1.8 MPa in a reversible adiabatic process. Calculate the final temperature and the work, assuming:

(a) Variable specific heat, Table A.11.
(b) Constant specific heat, value from Table A.8.
(c) Constant specific heat, value at an intermediate temperature from Table A.9.

7.34 A rigid container with a volume of 6 ft^3 is divided into two equal volumes by a partition. Both sides contain nitrogen, one at 300 lbf/in.2, 400 F and the other at 30 lbf/in.2, 200 F. The partition now ruptures and the nitrogen comes to equilibrium at 160 F. Assuming that the temperature of the surroundings is 70 F, determine the work done and the net entropy change for the process.

7.35 Water at 20°C is pumped from a lake to an elevated storage tank. The average elevation of the water inside the tank is 40 m above the surface of the lake, and the volume of the tank is 50 m^3. Initially the tank contains air at 100 kPa, 20°C, and the tank is closed so that the air is compressed as the water enters the bottom of the tank. The pump is operated until the tank is three-quarters full of water, and during the process the temperature of the air and water remain constant. Determine the work input to the pump.

7.36 A cylinder fitted with a piston contains carbon dioxide at 1 MPa, 300°C, at which point the volume is 200 L. The total external force acting on the piston is proportional to V^3. This system is now allowed to cool to room temperature, 20°C. What is the net entropy change, system plus surroundings, for this process?

7.37 Consider the following procedure for measuring the specific heat ratio k of a gas. The gas is initially contained in a rigid vessel at pressure P_1 (slightly higher than ambient pressure P_0), and at the ambient temperature. A valve on the vessel is then opened and the gas pressure quickly drops to P_0, at which point the valve is closed. After a long time, the gas remaining inside the vessel returns to thermal equilibrium with the ambient, at which point the pressure inside is P_2. Develop an expression for k in terms of P_0, P_1, and P_2. It may be assumed that there is no heat transfer to the gas inside during the time that gas is escaping from the vessel.

7.38 An aluminum cylinder fitted with a piston (total mass of 8 lbm) has an initial volume of 2 ft^3 and contains carbon dioxide at 200 lbf/in.2 The system (aluminum and gas) at this state is at a temperature of 400 F. By transferring heat with the surroundings, this system is now allowed to cool at constant

pressure to the ambient temperature, 80 F. What is the net entropy change associated with this process?

7.39 A cylinder fitted with a piston contains 1 lbm of argon gas at 15 lbf/in.2, 540 R. The gas is now compressed reversibly to a pressure of 100 lbf/in.2 Calculate the work required if the process is:

$T_2 = T_1 \left(\frac{P_2}{P_1}\right)^{\frac{k-1}{k}}$ $W = mR(t_2 - t_1)$

(a) Adiabatic.
(b) Isothermal. $W = mRT \ln\left(\frac{P_1}{P_2}\right)$
(c) Polytropic with exponent $n = 1.30$. $T_2 = T_1 \left(\frac{P_2}{P_1}\right)^{\frac{n-1}{n}}$

7.40 The power-stroke expansion of exhaust gases in an internal combustion engine can reasonably be represented as a polytropic expansion. Consider air contained in a cylinder volume of 0.1 L at 7 MPa, 1650 K. The air then expands in a reversible polytropic process, with exponent $n = 1.30$, through a volume ratio of 8:1. Show this process on P-v and T-s coordinates, and calculate the work and heat transfer for the process.

$W = \frac{mR(T_2 - T_1)}{1 - n}$

$\frac{T_2}{T_1} = \left(\frac{V_1}{V_2}\right)^{n-1}$

7.41 An ideal gas having a constant specific heat of

$$\bar{C}_{po} = 10 \text{ Btu/lb mol R}$$

undergoes a reversible polytropic compression in which the polytropic exponent $n = 1.40$. Will the heat transfer for this process be positive, negative, or zero?

7.42 A cylinder fitted with a piston contains air at ambient conditions, 100 kPa and 20°C, at which point the volume is 1 m^3. The air is now compressed to 800 kPa in a reversible polytropic process with exponent $n = 1.20$, after which it is expanded back to 100 kPa in a reversible adiabatic process.

(a) Show the two processes on P-v and T-s diagrams.
(b) Determine the final temperature and the net work.
(c) What is the potential refrigeration capacity (in kJ) of the air at the final state?

7.43 Ten lbm of oxygen gas at 120 lbf/in.2, 400 F, undergoes a reversible polytropic process with $n = 1.20$ until the final temperature equals that of the ambient, 70 F. Calculate the net entropy change of the system plus surroundings for this process. $\frac{T_2}{T_1} = \frac{P_2}{P_1}^{\frac{n-1}{n}}$

7.44 Air is contained in a cylinder fitted with a frictionless piston. Initially the cylinder contains 500 L of air at 150 kPa, 20°C. The air is then compressed reversibly in a polytropic process until the final pressure is 600 kPa, at which point the temperature is 120°C. Determine:

(a) The polytropic exponent n.
(b) The final volume of the air.
(c) The work done on the air and the heat transfer for the process.
(d) The net change in entropy for the process.

7.45 Steam enters a turbine at 200 lbf/in.2, 800 F, expands in a reversible adiabatic process, and exhausts at 2 lbf/in.2 Changes in kinetic and potential energies are small between the inlet and exit conditions of the turbine. The power

output of this turbine is 1000 hp. What is the mass flow rate of steam through the turbine?

7.46 Consider a reversible adiabatic nozzle flowing nitrogen gas at the rate of 0.2 kg/s. The state at the nozzle inlet is 500 kPa, 200°C, with a velocity at this point of 10 m/s, and the pressure at the nozzle exit is 100 kPa. Determine the nozzle exit velocity and cross-sectional area.

7.47 The mass flow rate of Freon-12 in a refrigeration cycle is 0.05 kg/s. The state at the compressor inlet is 250 kPa, 0°C, and the exit pressure is 1.2 MPa. Assuming that the compression process is reversible and adiabatic, what power motor is required to drive the compressor?

7.48 An ideal gas with constant specific heat enters a nozzle at a known pressure, temperature and velocity, and expands in a reversible adiabatic process to a known exit pressure. Develop an expression for the nozzle exit velocity in terms of these variables (and the appropriate fixed constants).

7.49 As a first approximation, consider the turbine process in a gas turbine power plant to be a reversible adiabatic expansion, with the gas having the same properties as air. The turbine inlet conditions are 1800 F, 210 lbf/in.², and the exit pressure is 15 lbf/in.² Calculate the turbine exit temperature and the work output per lbm, assuming:
(a) Variable specific heat, Table A.10.
(b) Constant specific heat, value from Table A.8.

7.50 A diffuser is a steady-state, steady-flow device in which a fluid flowing at high velocity is decelerated in such a way that the pressure increases during the process. Consider a diffuser in which steam enters at 200 kPa, 200°C, with a velocity of 700 m/s, and exits with a velocity of 70 m/s. Assuming that the process is reversible and adiabatic, what are the exit pressure and temperature of the steam?

7.51 Consider the two-stage steam turbine with interstage heating shown in Fig. 7.27. The data are as shown on the diagram. Assuming a constant pipe diameter in the interstage heater, and also that both turbine expansions are

FIGURE 7.27
Sketch for Problem 7.51.

reversible and adiabatic, determine:

(a) The power output of the turbine.

(b) The rate of heat transfer in the interstage heater.

7.52 Air enters a turbine at 90 lbf/in.2, 2000 R, and expands in a reversible adiabatic process to 15 lbf/in.2 Calculate the exit temperature and the work output per lbm of air, using:

(a) The ideal gas air tables, Table A.10.

(b) Constant specific heat, value at 80 F from Table A.8.

(c) Constant specific heat, value at an intermediate temperature from Fig. 5.10. Discuss why the method of part (b) results in a poor value for the exit temperature and yet a relatively good value for the work output.

7.53 Consider a geothermal supply of hot water available as saturated liquid at $P_1 = 1.5$ MPa. This liquid is to be flashed (throttled) to some lower pressure P_2. The saturated liquid and saturated vapor at this pressure are then separated, with the vapor being expanded through a reversible adiabatic turbine to the exhaust pressure $P_3 = 10$ kPa. Study the turbine power output per unit initial mass m_1 as a function of the value of pressure P_2.

7.54 One technique for operating a steam turbine at part-load power output is to throttle the steam to a lower pressure before it enters the turbine, as shown in Fig. 7.28. The steam line conditions are 200 lbf/in.2, 600 F, and the turbine exhaust pressure is fixed at 1 lbf/in.2 Assuming the expansion inside the turbine to be reversible and adiabatic, determine:

(a) The full-load work output of the turbine per lbm of steam.

(b) The pressure to which the steam must be throttled in order to produce 80% of the full-load output. Show both processes on a T-s diagram.

FIGURE 7.28

Sketch for Problem 7.54.

7.55 Saturated liquid methane from a large storage tank at 150 K is pumped to 3 MPa in a steady-state, steady-flow process at the rate of 0.5 kg/s. If the pump process can be assumed to be reversible, what is the required power input?

7.56 Consider the liquid pump in a steam power plant cycle. Water leaves the condenser and enters the pump at 2 lbf/in.², 110 F, and exits from the pump at a pressure of 2000 lbf/in.² Assuming the pumping process to be reversible and adiabatic, calculate:
(a) The work required per lbm of water.
(b) The water temperature at the pump exit.

7.57 Consider a reversible adiabatic steady-state, steady-flow process involving shaft work (either input or output), and no changes in kinetic or potential energies. The fluid may be assumed to be an ideal gas with constant specific heat. Starting with Eq. 7.59, develop an expression for the work per unit mass in terms of the inlet pressure and temperature and the exit pressure as variables (and the appropriate constants).

7.58 Consider the steam turbine power plant operating at supercritical pressure, as shown in Fig. 7.29. As a first approximation, it may be assumed that both the turbine and the pump processes are reversible and adiabatic. Neglecting any changes in kinetic and potential energies, calculate:
(a) The turbine work output, and the state at the exit from the turbine.
(b) The pump work input, and the enthalpy of the liquid at the exit from the pump.
(c) The thermal efficiency of the cycle.

$P_2 = P_3 = 1$ lbf/in.², $T_3 = 90$ F
$P_4 = P_1 = 3500$ lbf/in.², $T_1 = 1200$ F

FIGURE 7.29
Sketch for Problem 7.58.

7.59 Helium gas enters a steady-flow expander at 800 kPa, 300°C and exits at 120 kPa. The mass flow rate is 0.2 kg/s, and the expansion process can be considered reversible and polytropic, with a polytropic exponent $n = 1.30$. Calculate the power output of the expander.

7.60 It is desired to compress air from ambient conditions, 15 lbf/in.² and 80 F, to a pressure of 120 lbf/in.² in a reversible steady-state, steady-flow process.

Calculate the work of compression, the associated heat transfer, and the entropy change of the air, all per lbm for each of the following processes.
(a) Isothermal.
(b) Polytropic with $n = 1.25$.
(c) Adiabatic.
(d) Show all three processes on P-v and T-s diagrams.

7.61 Argon gas expands reversibly through a nozzle from 75 lbf/in.2, 400 F, to a pressure of 15 lbf/in.2 at the mass flow rate of 0.1 lbm/s. During this process, $Pv^{1.4} =$ constant. Find the exit velocity and the rate of heat transfer to the nozzle.

7.62 In a refrigeration plant, liquid ammonia leaves a water-cooled condenser and enters the expansion valve at 150 lbf/in.2, 70 F, and is throttled to a pressure of 20 lbf/in.2 Changes in kinetic and potential energies are negligible. What is the net increase of entropy, per lbm of ammonia, as a result of this process? Show the process on a T-s diagram.

7.63 One type of feedwater heater for preheating the water before entering a boiler operates on the principle of mixing the water with steam. For the states as shown in Fig. 7.30, calculate the rate of net entropy increase for this process, assuming the process to be steady-flow and adiabatic.

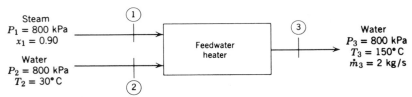

FIGURE 7.30
Sketch for Problem 7.63.

7.64 Air from a line at 1800 lbf/in.2, 60 F, flows into a 20-ft^3 rigid tank that initially contained air at ambient conditions, 14.7 lbf/in.2 and 60 F. This process occurs rapidly and is essentially adiabatic. The valve is closed when the pressure inside reaches some value P_2. The tank eventually cools to room temperature, at which time the pressure inside is 750 lbf/in.2 What is the pressure P_2? What is the net entropy change for the overall process?

7.65 A spherical steel tank having an inside diameter of 3 ft and wall thickness of 0.04 in. contains water at 300 lbf/in.2, 550 F. A valve on the tank is now opened, allowing steam to flow out until the tank and steam remaining inside both reach 300 F, at which point the valve is closed. The steam inside at this point has a quality of 90%.
(a) How much steam was lost from the tank?
(b) Taking a control volume that includes the tank and its contents, calculate the entropy change of the control volume and also that of its surroundings for this process.

7.66 A stream of ammonia enters a steady-flow device at 100 kPa, 50°C at the rate of 1 kg/s. Two streams exit the device at equal mass flow rates, one at 200 kPa, 50°C, and the other as saturated liquid at 10°C. It is claimed that this device operates in a 25°C room on an electrical power input of 250 kW. Is this possible?

7.67 A cylinder with a piston restrained by an external force contains Freon-12 at 50°C, 90% quality, at which point the volume is 100 L. The cylinder is attached to a line flowing Freon-12 at 3 MPa, 150°C. The valve is now opened, and mass flows into the cylinder until the pressure inside reaches 3 MPa, at which point the temperature inside is 100°C. It is claimed that during the process the Freon-12 does 150 kJ of boundary movement work against the external restraining force, and that any heat transfer is with the ambient at 20°C. Does this process violate the second law?

7.68 A spherical steel tank of 5-ft inside diameter and 0.2-in. wall thickness contains ammonia at 30 lbf/in.² and 70 F, the temperature of the surroundings. The tank is connected to a line flowing ammonia at 180 lbf/in.², 100 F. The valve is now opened and ammonia flows into the tank. When the valve is closed, the temperature (tank and contents) is 80 F and the tank contains 5% liquid, 95% vapor by volume.
(a) Considering a control volume that includes the tank and its contents, determine the entropy change of the control volume during the process.
(b) Show that this process does or does not violate the second law of thermodynamics.

7.69 A rigid 1-m³ tank contains 100 kg of Freon-12 at ambient temperature, 25°C. A valve on top of the tank is now opened, and saturated vapor is throttled to ambient pressure, 100 kPa, and discharged. During this process the temperature inside the tank remains at 25°C. The valve is closed when there is no liquid remaining inside. Calculate:
(a) The heat transfer to the tank.
(b) The net entropy change for the process.

7.70 Repeat Problem 7.58 assuming that the turbine and pump each have an isentropic efficiency of 85%.

7.71 Air enters the compressor of a large gas turbine at 95 kPa, 15°C at the rate of 50 m³/s, and exits at 1.2 MPa. The compression process is essentially adiabatic, and changes in kinetic and potential energies are negligible. Calculate the power required to drive this compressor and the exit temperature, assuming that:
(a) The process is reversible.
(b) The compressor has an isentropic efficiency of 87%.

7.72 Repeat Problem 7.51, assuming that each stage of the two-stage turbine has an isentropic efficiency of 86%.

7.73 Liquid water enters a pump at 70 F, 14.7 lbf/in.², and exits at a pressure of 1000 lbf/in.² If the isentropic efficiency of the pump is 75%, determine the enthalpy (steam tables reference) of the water at the pump exit.

7.74 Air enters an insulated compressor at ambient conditions, 14.7 $lbf/in.^2$ and 70 F, at the rate of 0.1 lbm/s and exits at 400 F. The isentropic efficiency of the compressor is 70%. What is the exit pressure? How much power input is required to drive the compressor?

7.75 Steam enters a turbine at 300°C and exhausts at 20 kPa. It is estimated that the isentropic efficiency of the turbine is 70%. What is the maximum turbine inlet pressure if the exhaust is not to be in the two-phase region?

7.76 In one type of heat-powered refrigeration system, part of the working fluid is expanded through a turbine to drive the compressor of the refrigeration cycle. Consider the turbine–compressor unit of such a cycle, as shown in Fig. 7.31. The turbine produces just enough power to drive the compressor, and both exit streams then mix together. Specifying any assumptions, find the ratio of mass flow rates \dot{m}_3/\dot{m}_1 and T_5 (x_5 if in the two-phase region) if:

(a) The turbine and compressor are reversible and adiabatic.

(b) The turbine and compressor each have an isentropic efficiency of 70%.

Freon-12 from boiler

$T_3 = 220$ F
Saturated vapor ③

Freon-12 from evaporator

$T_1 = 0$ F
Saturated vapor ①

Turbine Compressor

④ ②

$P_2 = P_4 = P_5 = 150$ lbf/in^2

⑤

To condenser

FIGURE 7.31
Sketch for Problem 7.76.

7.77 A certain process requires a steady 1 kg/s of air at 300 kPa pressure. What is your best estimate of the power required to drive a compressor to supply this air?

7.78 A small, high-speed air turbine having a 70% isentropic efficiency is required to produce 40 Btu/lbm of work output. The inlet temperature is 80 F and the turbine exhausts to the room. What is the required inlet pressure and what is the exhaust air temperature?

$$\eta_{tur} = \frac{W_a}{W_s}$$

$$.7 = \frac{40}{W_s}$$

$V_L = 0$ $.92 = \dfrac{\frac{200^2}{2}}{\frac{V_5^2}{2}}$

7.79 A nozzle is required to produce a stream of air at 200 m/s at a condition of 20°C, 100 kPa. It is estimated that the nozzle has an isentropic efficiency of 92%. What nozzle inlet pressure and temperature are required?

7.80 A centrifugal compressor takes in ambient air at 14.7 lbf/in.², 60 F, and discharges it at 70 lbf/in.² What is your best estimate of the discharge temperature?

7.81 A jet-ejector pump, shown schematically in Fig. 7.32, is a device in which a low-pressure (secondary) fluid is compressed by entrainment in a high-velocity (primary) fluid stream, the compression resulting from deceleration in a diffuser. For purposes of analysis, an ideal jet pump can be considered as equivalent to the ideal (reversible and adiabatic) turbine-compressor unit shown in Fig. 7.31, with states 1, 3, and 5 corresponding to those in Fig. 7.32. Now consider a steam jet pump in which state 1 is saturated vapor at 5 lbf/in.², state 3 is 40 lbf/in.², 300 F, and the discharge pressure P_5 is 14.7 lbf/in.²

(a) Calculate the ideal mass flow ratio (\dot{m}_1/\dot{m}_3).

(b) The efficiency of a jet pump is defined as

$$\eta_{\text{jet pump}} = \frac{(\dot{m}_1/\dot{m}_3)_{\text{actual}}}{(\dot{m}_1/\dot{m}_3)_{\text{ideal}}}$$

for the same inlet conditions and discharge pressure. For the pump considered above, determine the discharge temperature if the jet pump efficiency is 10%.

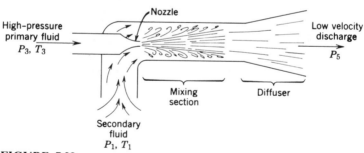

FIGURE 7.32

Sketch for Problem 7.81.

7.82 A geothermal supply of hot water at 80 lbf/in.², 300 F, is fed to an insulated flash evaporator at the rate of 10 000 lbm/h. A stream of saturated liquid at 30 lbf/in.² is drained from the bottom of the flash evaporator and a stream of saturated vapor at 30 lbf/in.² is drawn from the top and fed to a turbine. The turbine has an isentropic efficiency of 70% and an exit pressure of 2 lbf/in.² Considering a control volume that includes the flash evaporator and turbine, evaluate the second law for this process.

SUPPLEMENTARY PROBLEMS

7.83 A cylinder divided by a frictionless, thermally insulated piston contains air on one side and water on the other, as shown in Fig. 7.33. The cylinder is well insulated, except for the end of the side that contains the water. Each volume is initially 100 L at which point the air is at 40°C and the water is at 90°C with a quality of 10%. Heat is now slowly transferred to the water, until at the final state it exists as saturated vapor. Calculate the final pressure and the amount of heat transferred.

FIGURE 7.33
Sketch for Problem 7.83.

7.84 Consider the system shown in Fig. 7.34. Tank A has a volume of 10 ft^3 and initially contains air at 100 lbf/in.2, 100 F. Cylinder B is fitted with a frictionless piston resting on the bottom, at which position the spring is fully extended. The piston has a cross-sectional area of 100 in.2 and mass of 100 lbm, and the spring constant K_S is 100 lbf/in. The atmospheric pressure is 14.7 lbf/in.2 The valve is now opened and air flows into the cylinder until the pressure in A and B become equal, after which the valve is closed. During this process, the air finally remaining in A may be assumed to have undergone a reversible adiabatic process, and the entire process is adiabatic. It may also be assumed that the spring force is proportional to the displacement. Determine the final pressure in the system and the final temperature in cylinder B.

FIGURE 7.34
Sketch for Problem 7.84.

7.85 A nozzle is required to produce a steady stream of Freon-12 at 280 m/s, at which point the Freon-12 is to be at ambient conditions, 100 kPa and 20°C. The isentropic nozzle efficiency may be assumed to be 90%. What pressure and temperature are required in the line upstream of the nozzle?

7.86 An insulated cylinder is divided into two parts A and B by a frictionless, insulated piston, as shown in Fig. 7.35. Part A initially contains air at 14.7 lbf/in.2, 80 F, and part B contains saturated vapor water at 14.7 lbf/in.2 Each part has an initial volume of 1 ft^3. Part A is connected by a valve to a line flowing air at 100 lbf/in.2, 80 F. The valve is now opened and air flows into A until the pressure reaches 100 lbf/in.2, at which point the valve is closed. Calculate the final temperature in both A and B and the amount of mass that enters A during the process.

FIGURE 7.35
Sketch for Problem 7.86.

7.87 Consider the two rigid 1-ft^3 tanks A and B, as shown in Fig. 7.36. Tank A contains saturated vapor Freon-12 at 100 F, and B contains Freon-12 at 0 F, 15 lbf/in.2 Tank B, the line connecting the tanks, and the valve are heavily insulated. The valve is now opened slightly and is then closed when the two tanks have come to the same pressure. During the process, heat is exchanged between the surroundings and tank A such that the Freon-12 inside A remains at 100 F. Calculate the final pressure, the final temperature in tank B, the heat transferred during the process, and the net entropy change.

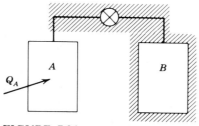

FIGURE 7.36
Sketch for Problem 7.87.

7.88 Consider the scheme shown in Fig. 7.37 for producing fresh water from salt-water. The conditions are as shown in the figure. Assume that the properties of saltwater are the same as for pure water, and also that the pump is reversible and adiabatic.
(a) Determine the ratio (\dot{m}_7/\dot{m}_1), that is, the fraction of saltwater purified by the process.
(b) Determine the input quantities, w_P and q_H.
(c) Make a second law analysis of the overall system.

FIGURE 7.37
Sketch for Problem 7.88.

7.89 A certain industrial process requires a steady stream of saturated vapor steam at 200 kPa at the rate of 2 kg/s. There are two alternatives to be considered for supplying this steam from ambient liquid water at 20°C, 100 kPa. 1. Pump the water to 200 kPa and feed to a steam generator. 2. Pump the water to 5 MPa, feed to a steam generator, and then expand through a turbine, from which the steam exhausts at the desired state.
(a) Compare these two alternatives, making reasonable assumptions about the various processes involved.
(b) What is the rate of net entropy increase for each of the alternatives?

7.90 A 5-ft³ tank containing steam at 14.7 lbf/in.², 1% quality, is fitted with a relief valve. Heat is transferred to the tank from a large heat source at 500 F. When the pressure in the tank reaches 300 lbf/in.², the relief valve opens, and saturated vapor at 300 lbf/in.² is throttled across the valve and discharged at 14.7 lbf/in.² pressure. This process continues until the quality in the tank reaches 90%.
(a) Calculate the mass discharged from the tank.
(b) Determine the heat transfer to the tank during the process.
(c) Considering a control volume that includes the tank and relief valve, calculate the entropy change of the control volume and also that of the surroundings. Show that the process does not violate the second law.

7.91 A turbo-supercharger is to be utilized to boost the inlet air pressure to an automobile engine. This device consists of an exhaust gas-driven turbine directly connected to an air compressor, as shown in Fig. 7.38. At a certain engine load, the conditions are as shown in the figure. If it is assumed that the turbine and compressor are both reversible and adiabatic, calculate:
(a) The turbine exit temperature and power output.
(b) The compressor exit pressure and temperature.
(c) Repeat parts (a) and (b), assuming that the turbine has an isentropic efficiency of 85% and the compressor has an isentropic efficiency of 80%.

FIGURE 7.38
Sketch for Problem 7.91.

7.92 A cylinder fitted with a piston and containing 2 kg of ammonia at −10°C, 90% quality is brought into a 20°C room and attached to a line flowing ammonia at 800 kPa, 40°C. The total restraining force on the piston is proportional to the cylinder volume squared. The valve is now opened and

ammonia flows into the cylinder until the mass inside is twice the initial mass, at which point the valve is closed. An electrical current of 15 A is now passed through a 2-ohm resistor inside the cylinder for 20 minutes. It is claimed that the final pressure in the cylinder is 600 kPa. Is this possible?

7.93 Steam is contained in cylinder A and in rigid tank B at the conditions given in Fig. 7.39. Everything is thermally insulated. The valve is now opened and the pin is removed from the piston at the same time. When the pressures in A and B become equal, the valve is closed. Find the final state in A and in B in terms of the variables given.

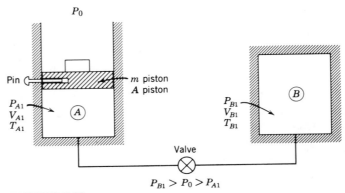

FIGURE 7.39
Sketch for Problem 7.93.

7.94 A tank of volume V_T containing air at the initial state P_1 and T_1 is to be evacuated with a vacuum pump. The local ambient pressure is P_a. Assume that both the tank and the pump operate under isothermal conditions, and also that the pumping process is reversible.
 (a) Calculate the work of evacuation as a function of the pressure inside the tank, and plot a curve of work of evacuation vs. pressure.
 (b) Considering the pump to handle a constant volume per unit time, determine the power as a function of pressure in the tank and also power as a function of time.

7.95 The cylinder shown in Fig. 7.40 is so constructed that compartments A and B are separated by a metal partition, and part A is fitted with a frictionless piston. It may be assumed that the temperatures in A and B differ by only an infinitesimal. Part A initially contains 10 kg of air at 1.4 MPa, 150°C, while part B contains 4 kg of liquid water and 0.5 kg of vapor water at 150°C. The piston is now moved to the right until the pressure of the air in A reaches 2 MPa. There is no heat transfer except across the metal partition. Determine the total change of entropy and of internal energy for this process.

Fixed metal
partition

A Air	B H₂O
10 kg 1.4 MPa 150°C	4 kg liquid 0.5 kg vapor 150°C

FIGURE 7.40

Sketch for Problem 7.95.

7.96 An insulated closed cylinder fitted with a frictionless piston contains nitrogen at 14.7 lbf/in.2, 60 F, as shown in Fig. 7.41. The piston is thermally insulated and rests at the bottom of the cylinder, and the total volume of the cylinder is 2 ft^3. The valve is now opened to a 2-ft^3 rigid tank containing steam at 60 lbf/in.2, 400 F, and the valve is closed when the pressures equalize. During this process, heat is transferred to the steam inside the tank such that its temperature remains at 400 F. It may be assumed that the nitrogen is compressed reversibly in the cylinder. Find:

(a) The final pressure.

(b) The final temperatures in both parts of the cylinder.

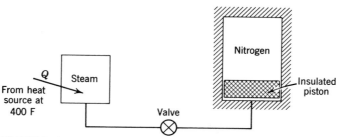

FIGURE 7.41

Sketch for Problem 7.96.

7.97 Write a computer program to solve the following problem. One of the gases listed in Table A.9 undergoes a reversible adiabatic process in a cylinder from P_1, T_1, to P_2. It is desired to calculate the final temperature and the work for the process by integrating the specific heat equation, by assuming constant specific heat at temperature T_1, and finally by assuming constant specific heat at an average temperature (by iteration).

7.98 Write a computer program to solve the following problem. It is desired to extend and generalize Problem 7.48, which involves the expansion of an ideal gas with constant specific heat from a known inlet state P_1, T_1, with

velocity \mathbf{V}_1 and area \mathscr{A}_1, through a reversible adiabatic nozzle. In particular it is desired to map out the nozzle contour, in terms of cross-sectional area \mathscr{A}, as a function of the local pressure P, where P is any value less than the inlet pressure. This should be studied over a range of conditions, for given values of the constants R and k, and the results discussed.

7.99 Write a computer program to solve the following problem. In Problem 5.85 the enthalpy values in the ideal gas tables, Table A.11, were used to curve fit polynomial equations for ideal gas specific heat as functions of temperature. It is desired to use these equations to calculate ideal gas entropy functions, and to compare the results with values listed in Table A.11, and also with polynomial equations that are curve fitted directly to the entropy functions as listed in Table A.11. Why are the results different?

8

IRREVERSIBILITY AND AVAILABILITY

We now turn our attention to irreversibility and availability, two additional concepts that have found increasing use in recent years. These concepts are particularly applicable in the analysis of complex thermodynamic systems, for with the aid of a digital computer, irreversibility and availability are very powerful tools in design and optimization studies of such systems.

8.1 REVERSIBLE WORK

To introduce the concepts of reversible work and irreversibility, let us consider Fig. 8.1, which shows a control volume undergoing a uniform-state, uniform-flow process. There are irreversibilities present as the process takes place, as in every real process. The work crossing the control surface during the process is $W_{c.v.}$ and the heat transfer is $Q_{c.v.}$. All of the heat transfer is with the surroundings at temperature T_0.

We now ask the following questions: What if this working fluid had undergone exactly the same change in state, except in a completely reversible process? How much work would have been done in such a process? We realize, of course, that such a process does not really exist. Any process that actually occurs can do so only with some irreversibilities, of the types discussed in Chapter 6. Thus, we are imagining an ideal process, to which we can compare the real process shown in Fig. 8.1. This situation is similar to our discussion of cycles in Chapter 6. That led us to develop the ideal, completely reversible Carnot cycle, to which we can compare the performance of real cycles. The situation is also similar to that in Section 7.16, in which we compared the performance of real machines (turbines,

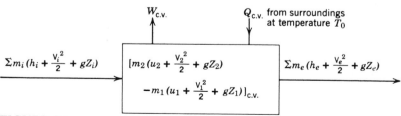

FIGURE 8.1

A uniform-state, uniform-flow process.

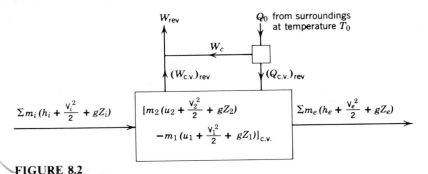

FIGURE 8.2

A reversible process for the same change of state as in Fig. 8.1.

compressors, nozzles) to idealized model processes. In that case, however, the ideal and real processes did not occur between the same end states, which was the question raised here.

We proceed then to Fig. 8.2, which shows a control volume identical to that of the real process, Fig. 8.1. All quantities and states of fluid entering and leaving the control volume are the same as in Fig. 8.1, and exactly the same change of state occurs within the control volume. In contrast to Fig. 8.1, however, in this case all processes are completely reversible. Thus, in Fig. 8.2 we have the (hypothetical) ideal process corresponding to exactly the same change of state that occurred in the real process of Fig. 8.1. The work and heat crossing the control surface for this ideal process will be different than those in Fig. 8.1 (which involved irreversibilities) and these quantities are designated $(W_{c.v.})_{rev}$ and $(Q_{c.v.})_{rev}$. In order that the heat transfer between the control volume and the surroundings may occur reversibly when there is a difference in the temperature in the control volume and the surroundings, it is necessary that this heat transfer take place through a reversible heat engine. The work output of this reversible heat engine is designated W_c. The sum of the work crossing the control surface for the reversible case and the work output of the reversible heat engine is called the reversible work and is designated W_{rev}. That is,

$$W_{rev} = (W_{c.v.})_{rev} + W_c \tag{8.1}$$

The difference between this reversible work W_{rev} and that done in the first case, $W_{c.v.}$, when irreversible processes occur is called the irreversibility, which is designated I.

$$I = W_{rev} - W_{c.v.} \tag{8.2}$$

Quite obviously, if real processes could take place in a completely reversible manner, the irreversibility would be zero.

While we could calculate the reversible work for each problem that we might consider, it is advantageous at this point to develop a general expression for the reversible work. Let us therefore analyze the reversible process shown in Fig. 8.2. We have assumed a uniform-state, uniform-flow process for this con-

trol volume. This process was described in Section 5.14, and for this process the first law can be written, Eq. 5.54, as

$$(Q_{c.v.})_{rev} + \sum m_i \left(h_i + \frac{\mathbf{V}_i^2}{2} + gZ_i \right)$$

$$= (W_{c.v.})_{rev} + \sum m_e \left(h_e + \frac{\mathbf{V}_e^2}{2} + gZ_e \right)$$

$$+ \left[m_2 \left(u_2 + \frac{\mathbf{V}_2^2}{2} + gZ_2 \right) - m_1 \left(u_1 + \frac{\mathbf{V}_1^2}{2} + gZ_1 \right) \right]_{c.v.} \tag{8.3}$$

The significance of each of these terms is shown in Fig. 8.2.

Since all processes are to be reversible, the heat transfer with the surroundings must also be reversible. Therefore, if the temperature within the control volume is different from the surroundings, this heat transfer must take place through a reversible heat engine. This is also shown schematically in Fig. 8.2, and the work done by the reversible engine is designated W_c.

For the reversible heat engine work, W_c, we can write:
First law:

$$W_c = Q_0 - (Q_{c.v.})_{rev}$$

Second law: Since

$$\frac{Q_0}{T_0} = \int_0^t \left(\frac{\dot{Q}_{c.v.}}{T} \right)_{rev} dt$$

it follows that

$$W_c = T_0 \int_0^t \left(\frac{\dot{Q}_{c.v.}}{T} \right)_{rev} dt - (Q_{c.v.})_{rev} \tag{8.4}$$

The second law for the uniform-state, uniform-flow process, Eq. 7.57, for this reversible process is

$$m_2 s_2 - m_1 s_1 + \sum m_e s_e - \sum m_i s_i = \int_0^t \left(\frac{\dot{Q}_{c.v.}}{T} \right)_{rev} dt \tag{8.5}$$

Substituting this into Eq. 8.4 we have

$$W_c = T_0[m_2 s_2 - m_1 s_1 + \sum m_e s_e - \sum m_i s_i] - (Q_{c.v.})_{rev} \tag{8.6}$$

We can now substitute the expression for $(W_{c.v.})_{rev}$ from Eq. 8.3 and for W_c from Eq. 8.6 into Eq. 8.1.

$$W_{rev} = (Q_{c.v.})_{rev} + \sum m_i \left(h_i + \frac{\mathbf{V}_i^2}{2} + gZ_i \right) - \sum m_e \left(h_e + \frac{\mathbf{V}_e^2}{2} + gZ_e \right)$$

$$- \left[m_2 \left(u_2 + \frac{\mathbf{V}_2^2}{2} + gZ_2 \right) - m_1 \left(u_1 + \frac{\mathbf{V}_1^2}{2} + gZ_1 \right) \right]_{c.v.}$$

$$+ T_0[m_2 s_2 - m_1 s_1 + \sum m_e s_e - \sum m_i s_i] - (Q_{c.v.})_{rev}$$

On rearranging this equation, we have

$$W_{rev} = \sum m_i \left(h_i - T_0 s_i + \frac{\mathbf{V}_i^2}{2} + gZ_i \right) - \sum m_e \left(h_e - T_0 s_e + \frac{\mathbf{V}_e^2}{2} + gZ_e \right)$$
$$- \left[m_2 \left(u_2 - T_0 s_2 + \frac{\mathbf{V}_2^2}{2} + gZ_2 \right) - m_1 \left(u_1 - T_0 s_1 + \frac{\mathbf{V}_1^2}{2} + gZ_1 \right) \right]_{c.v.}$$

$$(8.7)$$

Note that $(Q_{c.v.})_{rev}$ cancels and does not appear in Eq. 8.7.

Thus we have an expression for the reversible work of a control volume that exchanges heat with the surroundings at temperature T_0. Note that the reversible work is a function of T_0, the temperature of the surroundings.

Let us now consider two special cases, namely, a system (fixed mass) and a steady-state, steady-flow process for a control volume.

For a system there is no flow across the control surface, and therefore, considering Eq. 8.7,

$$\sum m_i \left(h_i - T_0 s_i + \frac{\mathbf{V}_i^2}{2} + gZ_i \right) = 0;$$

$$\sum m_e \left(h_e - T_0 s_e + \frac{\mathbf{V}_e^2}{2} + gZ_e \right) = 0$$

Thus, for a system the reversible work is

$$\left(\frac{W_{rev}}{m} \right)_1^2 = {}_1W_{rev2} = \left[\left(u_1 - T_0 s_1 + \frac{\mathbf{V}_1^2}{2} + gZ_1 \right) \right.$$
$$\left. - \left(u_2 - T_0 s_2 + \frac{\mathbf{V}_2^2}{2} + gZ_2 \right) \right]$$

$$(8.8)$$

For a steady-state, steady-flow process we note on consideration of Eq. 8.7 that

$$\left[m_2 \left(u_2 - T_0 s_2 + \frac{\mathbf{V}_2^2}{2} + gZ_2 \right) - m_1 \left(u_1 - T_0 s_1 + \frac{\mathbf{V}_1^2}{2} + gZ_1 \right) \right]_{c.v.} = 0$$

Therefore, for a steady-state, steady-flow process, in rate form,

$$\dot{W}_{rev} = \sum \dot{m}_i \left(h_i - T_0 s_i + \frac{\mathbf{V}_i^2}{2} + gZ_i \right) - \sum \dot{m}_e \left(h_e - T_0 s_e + \frac{\mathbf{V}_e^2}{2} + gZ_e \right) \quad (8.9)$$

When there is a single flow of fluid into and out of the control volume in a steady-state, steady-flow process we can write

$$\frac{\dot{W}_{rev}}{\dot{m}} = w_{rev} = \left(h_i - T_0 s_i + \frac{\mathbf{V}_i^2}{2} + gZ_i \right) - \left(h_e - T_0 s_e + \frac{\mathbf{V}_e^2}{2} + gZ_e \right) \quad (8.10)$$

Each expression for reversible work involves the temperature of the surroundings, T_0. Therefore in solving problems this information must be available. Unless otherwise stated we will use 77 F or 25°C for the temperature of the surroundings. This temperature has been selected because thermochemical data are fre-

quently given relative to this base and it is also a reasonable temperature to assume for the surroundings. One must keep in mind that one should use the actual temperature of the surroundings when this information is available.

8.2 IRREVERSIBILITY

In the last section we considered the two control volumes shown in Figs. 8.1 and 8.2, and noted that the irreversibility is defined as

$$I = W_{rev} - W_{c.v.}$$

We also derived Eq. 8.7, which was a general expression for the reversible work for a control volume undergoing a uniform-state, uniform-flow process. We can also derive a general expression for irreversibility. We do so by noting that an expression for $W_{c.v.}$ can be found by applying the first law to the control volume that involves the irreversible process in Fig. 8.1. From Eq. 5.54,

$$W_{c.v.} = \sum m_i \left(h_i + \frac{\mathbf{V}_i^2}{2} + gZ_i \right) - \sum m_e \left(h_e + \frac{\mathbf{V}_e^2}{2} + gZ_e \right)$$
$$- \left[m_2 \left(u_2 + \frac{\mathbf{V}_2^2}{2} + gZ_2 \right) - m_1 \left(u_1 + \frac{\mathbf{V}_1^2}{2} + gZ_1 \right) \right]_{c.v.} + Q_{c.v.} \quad \textbf{(8.11)}$$

Substituting the expression for W_{rev} from Eq. 8.7 and for $W_{c.v.}$ from Eq. 8.11 into Eq. 8.2 we have

$$I = \sum m_e T_0 s_e - \sum m_i T_0 s_i + m_2 T_0 s_2 - m_1 T_0 s_1 - Q_{c.v.} \quad \textbf{(8.12)}$$

This is a general expression for the irreversibility in a uniform-state, uniform-flow process. Once again we consider two particular cases, namely, the system and the steady-state, steady-flow process. For the system,

$$\sum m_e T_0 s_e = 0 \qquad \sum m_i T_0 s_i = 0 \qquad m_2 = m_1 = m$$

and Eq. 8.12 reduces to

$$_1 I_2 = m T_0 (s_2 - s_1) - {}_1 Q_2 \quad \textbf{(8.13)}$$

For the steady-state, steady-flow process,

$$m_2 T_0 s_2 = m_1 T_0 s_1$$

and Eq. 8.12 reduces to

$$I = \sum m_e T_0 s_e - \sum m_i T_0 s_i - Q_{c.v.} \quad \textbf{(8.14)}$$

Returning to the general expression, Eq. 8.12, we can make an interesting comparison between the irreversibility I and the principle of the increase of entropy discussed in Sections 7.8 and 7.15. Equation 8.12 in rate form becomes

$$\dot{I} = T_0 \left[\frac{dS_{c.v.}}{dt} + \sum \dot{m}_e s_e - \sum \dot{m}_i s_i - \frac{\dot{Q}_{c.v.}}{T_0} \right]$$

But substituting Eq. 7.66 for the rate of entropy change of the surroundings, we have

$$\dot{I} = T_0 \left[\frac{dS_{c.v.}}{dt} + \frac{dS_{surr}}{dt} \right] \tag{8.15}$$

The quantity inside the brackets in Eq. 8.15 is the net rate of entropy change for the overall process, and as discussed in Chapter 7, is always greater than zero, approaching zero as the process becomes completely reversible. Thus, the irreversibility I is equal to the product of the temperature of the surroundings and the net entropy change for the process, and is therefore directly related to the second law of thermodynamics. Equation 8.15 is a general result, and reduces to Eq. 8.13 or 8.14 in those special cases.

Example 8.1

1. Considering Fig. 8.3, tank A has a volume of 100 ft^3 and initially contains Freon-12 at a pressure of 10 lbf/in.2 and a temperature of 80 F. The compressor evacuates tank A and charges tank B. Tank B is initially evacuated and is of such a volume that the final pressure of the Freon-12 in tank B is 80 lbf/in.2 when the temperature reaches its final value of 80 F. The temperature of the surroundings is 80 F. Determine the minimum work input to the compressor.

FIGURE 8.3
Sketch for Example 8.1.

2. After tank B is charged and tank A is evacuated, a by-pass valve around the compressor is left open, and the two tanks come to a uniform pressure at a temperature of 80 F. Determine the irreversibility for this process.

For part 1:

System: Tank A plus tank B.
Initial state: P_1, T_1 (in A) known; state fixed.
Final state: P_2, T_2 (in B) known; state fixed.
Process: System, no change in kinetic or potential energy.
Model: Freon-12 tables.
Analysis: From Eq. 8.8,

$$_1(w_{rev})_2 = (u_1 - u_2) - T_0(s_1 - s_2)$$

Solution

$$u_1 = h_1 - P_1 v_1 = 89.596 - \frac{10 \times 144 \times 4.7248}{778} = 80.85 \text{ Btu/lbm}$$

$$s_1 = 0.207\,46 \text{ Btu/lbm R}$$

$$u_2 = 86.316 - \frac{80 \times 144 \times 0.52795}{778} = 78.51 \text{ Btu/lbm}$$

$$s_2 = 0.168\,85 \text{ Btu/lbm R}$$

$$_1(w_{rev})_2 = (80.85 - 78.51) - 540(0.207\,46 - 0.168\,85)$$
$$= 2.34 - 20.85 = -18.51 \text{ Btu/lbm}$$

$$m = \frac{V_1}{v_1} = \frac{100}{4.7248} = 21.15 \text{ lbm}$$

$$_1(W_{rev})_2 = 21.15(-18.51) = -392 \text{ Btu}$$

Note that the reversible work is a negative number. This means that the minimum work input to the compressor to accomplish this change of state is 392 Btu. If there are any irreversibilities this work input will increase, say to 500 Btu. Thus the reversible work can still be thought of as the maximum work, inasmuch as −392 is a larger number than −500.

For part 2:

System: Tank A plus tank B.

Initial state: P_2, T_2 (in B) known; state fixed.

Final state: T_3 (in A and B) known.

Process: System, change of volume.

Model: Freon-12 tables.

Analysis: Since the total mass of the system is known and the total volume at the final state can be found, this fixes the specific volume and therefore the final state. We could then calculate the irreversibility either by calculating the reversible work from Eq. 8.8 and the irreversibility from Eq. 8.2 or by calculating the irreversibility directly from Eq. 8.13. We will carry out both calculations to gain further insight into these concepts.

Solution

We must first find the final pressure in the system. The two properties known are specific volume and temperature. The volume of tank B is found from the results of part 1.

$$V_B = m v_{2B} = 21.15(0.527\,95) = 11.17 \text{ ft}^3$$

$$V_3 = V_A + V_B = 100 + 11.17 = 111.17 \text{ ft}^3$$

$$v_3 = \frac{V_3}{m} = \frac{111.17}{21.15} = 5.26 \text{ ft}^3/\text{lbm}$$

By interpolation from the Freon-12 table, $P_3 = 9.44$ lbf/in.2

$$h_3 = 89.620 \text{ Btu/lbm} \quad s_3 = 0.208\ 71 \text{ Btu/lbm R.}$$

$$u_3 = 89.620 - \frac{9.44 \times 144 \times 5.26}{778} = 80.44 \text{ Btu/lbm}$$

From Eq. 8.8

$$_2(w_{rev})_3 = u_2 - u_3 - T_0(s_2 - s_3)$$
$$= 78.51 - 80.44 - 540(0.168\ 85 - 0.208\ 71)$$
$$= 19.60 \text{ Btu/lbm}$$
$$_2(W_{rev})_3 = 21.15(19.60) = 415 \text{ Btu}$$

The irreversibility can now be calculated from Eq. 8.2.

$$_2I_3 = {_2}(W_{rev})_3 - {_2}W_3 = 415 - 0 = 415 \text{ Btu}$$

The irreversibility can also be calculated from Eq. 8.13.

$$_2I_3 = mT_0(s_3 - s_2) - {_2}Q_3$$

In order to determine $_2Q_3$ we apply the equation for the first law to this process.

$$_2Q_3 = m(u_3 - u_2) + {_2}W_3$$
$$= 21.15(80.44 - 78.51) + 0 = +40.9 \text{ Btu.}$$
$$_2I_3 = 21.15 \times 540(0.208\ 71 - 0.168\ 85) - 40.9 = 415 \text{ Btu.}$$

Example 8.2

Referring to Example 7.13, determine the reversible work and the irreversibility for the actual change of state that takes place as the steam flows through the turbine (using a temperature of the surroundings of 25°C).

Solution

In this solution it will be helpful to refer to the temperature–entropy diagram of Fig. 8.4.

In Example 7.13, we determined the actual state exiting the turbine to be a quality of 0.9377 at a pressure of 15 kPa. Therefore,

$$s_e = 0.7549 + 0.9377(7.2536) = 7.5566 \text{ kJ/kg K}$$

We calculate the reversible work using Eq. 8.10 with no changes in kinetic or potential energy:

$$\frac{\dot{W}_{rev}}{\dot{m}} = (h_i - T_0 s_i) - (h_e - T_0 s_e)$$

$$= (3051.2 - 2451.2) - 298.15(7.1229 - 7.5566)$$
$$= 729.3 \text{ kJ/kg}$$

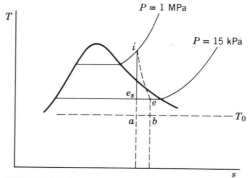

FIGURE 8.4
Sketch for Example 8.2.

The irreversibility can be found using Equation 8.14.

$$\frac{\dot{I}}{\dot{m}} = T_0(s_e - s_i) - \frac{\dot{Q}_{c.v.}}{\dot{m}}$$

$$= 298.15(7.5566 - 7.1229) = 129.3 \text{ kJ/kg}$$

Note also that we could use our basic definition of the irreversibility.

$$\frac{\dot{I}}{\dot{m}} = \frac{\dot{W}_{rev}}{\dot{m}} - \frac{\dot{W}_{c.v.}}{\dot{m}}$$

$$= 729.3 - 600 = 129.3 \text{ kJ/kg}$$

It may be of help in understanding these concepts to conceive of a device by which we could achieve this amount of reversible work. Consider the device shown in Fig. 8.5 in which the steam expands in a reversible adiabatic process until the pressure reaches 15 kPa, followed by a heat transfer process in which heat is transferred to the steam in a constant-pressure process. The heat is transferred from the surroundings, which requires a heat pump, with a work input, W_c.

We already found from Example 7.13 that the work for the reversible adiabatic process is 741.9 kJ/kg. The work required for the heat pump is area $e_s abee_s$ on Fig. 8.5.

$$W_c = (T_e - T_0)(s_e - s_i)$$
$$= (327.12 - 298.15)(7.5566 - 7.1229) = 12.6 \text{ kJ/kg}$$
$$W_{rev} = 741.9 - 12.6 = 729.3 \text{ kJ/kg}$$

Thus, in deriving the expressions for reversible work and irreversibility we have only generalized the calculations, and one can always determine these quantities by applying the fundamentals directly to a given problem.

FIGURE 8.5
Sketch for Example 8.2.

8.3 AVAILABILITY AND SECOND LAW EFFICIENCY

What is the maximum reversible work that can be done by a system in a given state? In Section 8.1 we developed an expression for the reversible work for a given change of state of a system. But the question that arises is, what final state will make this reversible work the maximum?

The answer to this question is that when a system is in equilibrium with the environment, no spontaneous change of state will occur, and the system will not be capable of doing any work. Therefore, if a system in a given state undergoes a completely reversible process until it reaches a state in which it is in equilibrium with the environment, the maximum reversible work will have been done by the system.

If a system is in equilibrium with the surroundings, it must certainly be in pressure and temperature equilibrium with the surroundings—that is, at pressure P_0 and temperature T_0. It must also be in chemical equilibrium with the surroundings, which implies that no further chemical reaction will take place. Equilibrium with the surroundings also requires that the system have zero velocity and minimum potential energy. Similar requirements could be set forth regarding magnetic, electrical, and surface effects if these are relevant to a given problem.

The same general remarks can be made in regard to a quantity of mass that undergoes a steady-state, steady-flow process. With a given state for the mass entering the control volume, the reversible work will be a maximum when this mass leaves the control volume in equilibrium with the surroundings. This means that as the mass leaves the control volume it must be at the pressure and temperature of the surroundings, in chemical equilibrium with the surroundings, and have minimum potential energy, and zero velocity. (The mass leaving the control volume must of necessity have some velocity but it can be made to approach zero.)

Let us first consider the availability associated with a steady-state, steady-flow process. In Eq. 8.10 we noted that when we consider a single flow,

$$w_{rev} = \left(h_i - T_0 s_i + \frac{\mathbf{V}_i^2}{2} + gZ_i \right) - \left(h_e - T_0 s_e + \frac{\mathbf{V}_e^2}{2} + gZ_e \right)$$

This reversible work will be a maximum when the mass leaving the control volume is in equilibrium with the surroundings. If we designate this state in which the fluid is in equilibrium with the surroundings with subscript 0, the reversible work will be a maximum when $h_e = h_0$, $s_e = s_0$, $\mathbf{V}_e = 0$, and $Z_e = 0$. Designating this maximum reversible work per unit mass flow as the availability per unit mass flow, and assigning this the symbol ψ, we have

$$\psi = \left(h - T_0 s + \frac{\mathbf{V}^2}{2} + gZ \right) - \left(h_0 - T_0 s_0 + gZ_0 \right) \tag{8.16}$$

The initial state is designated without a subscript to indicate that this is the availability associated with a substance in any state as it enters a control volume in a steady-state, steady-flow process. It also follows that the reversible work per unit mass flow between any two states is equal to the decrease in availability between these two states.

$$w_{rev} = \psi_i - \psi_e \tag{8.17}$$

If we have more than one flow into and out of the control volume in a steady-state, steady-flow process we can write

$$\dot{W}_{rev} = \sum \dot{m}_i \psi_i - \sum \dot{m}_e \psi_e \tag{8.18}$$

The availability associated with a system is developed in a similar way, except for one factor. When the volume of a system increases, some work is done by the system against the surroundings, and this is not available for doing useful work.

To simplify this analysis let us also assume that the change in kinetic and potential energy of the system is negligible. In this case w_{rev}, as given by Eq. 8.8, is

$$_1(w_{rev})_2 = (u_1 - T_0 s_1) - (u_2 - T_0 s_2)$$

If the final state is in equilibrium with the surroundings $u_2 = u_0$ and $s_2 = s_0$. In this case the reversible work is a maximum. If we designate this maximum reversible work as $(w_{rev})_{max}$, and write it without subscript to indicate a general state, we have

$$(w_{rev})_{max} = (u - T_0 s) - (u_0 - T_0 s_0) \tag{8.19}$$

The availability per unit mass for a system is equal to this maximum reversible work minus the work done against the surroundings. This work done against the surroundings, W_{surr}, is

$$W_{surr} = P_0(V_0 - V) = -m P_0(v - v_0) \tag{8.20}$$

The availability per unit mass for a system in the absence of kinetic and potential energy changes is designated ϕ:

$$
\begin{aligned}
\phi &= (w_{rev})_{max} - w_{surr} \\
&= (u - T_0 s) - (u_0 - T_0 s_0) + P_0(v - v_0) \\
&= (u + P_0 v - T_0 s) - (u_0 + P_0 v_0 - T_0 s_0) \\
&= (u - u_0) + P_0(v - v_0) - T_0(s - s_0)
\end{aligned}
\tag{8.21}
$$

It follows that

$$
1(w{rev})_2 = \phi_1 - \phi_2 - P_0(v_1 - v_2) + \frac{\mathbf{V}_1{}^2 - \mathbf{V}_2{}^2}{2} + g(Z_1 - Z_2)
\tag{8.22}
$$

The use of availability and irreversibility in an actual thermodynamic problem is shown in Fig. 8.6. A theoretical analysis was made of a reciprocating internal-combustion automotive engine to see what happened to the availability of the air–fuel mixture that entered the engine and where the irreversibility occurred during the process. The abscissa on Fig. 8.6 is crank angle, the left side representing bottom dead center when the cylinder is assumed to be filled with an air–fuel mixture having the availability indicated on the ordinate. As the compression process takes place, the availability of this mixture increases as the result of the work done in compressing the mixture. When the piston passes top dead center

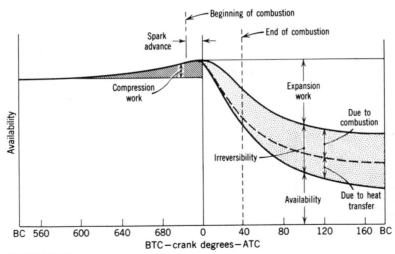

FIGURE 8.6

Availability vs. crank angle of the charge in a spark-ignited internal combustion engine. From D. J. Patterson and G. J. Van Wylen, "A Digital Computer Simulation for Spark Ignited Engine Cycles," *SAE Progress in Technology Series*, **7**, p. 88. Published by SAE Inc., New York, 1964.

the expansion process takes place. The beginning and end of combustion are also indicated on the diagram. During the combustion and expansion process irreversibilities occur. Those that are associated with the combustion process itself and those associated with heat transfer to the cooling water or surroundings are both indicated. The work done during the expansion process and availability at the end of the expansion are indicated on the right ordinate. Note that this availability that remains in the cylinder at the end of the expansion stroke is exhausted to the atmosphere.

The less the irreversibility associated with a given change of state, the greater the work that will be done (or the less work that will be required). This is significant in at least two regards. The first is that availability is one of our natural resources. This availability is found in such forms as oil reserves, coal reserves, and uranium reserves. Suppose we wish to accomplish a given objective that requires a certain amount of work. If this work is produced reversibly while drawing on one of the availability reserves, the decrease in availability would be exactly equal to the reversible work. However, since there are irreversibilities involved in producing this required amount of work, the actual work will be less than the reversible work, and the decrease in availability will be greater (by the amount of the irreversibility) than if this work had been produced reversibly. Thus the more irreversibilities we have in all of our processes, the greater will be the decrease in our availability reserves.[1] The conservation and effective use of these availability reserves is an important responsibility for all of us.

The second reason why it is desirable to accomplish a given objective with the smallest irreversibility is an economic one. Work costs money, and in many cases a given objective can be accomplished at less cost when the irreversibility is less. It should be noted, however, that many factors enter into the total cost of accomplishing a given objective, and an optimization process that involves consideration of many factors is often necessary in arriving at the most economical design. For example, in a heat transfer process, the smaller the temperature difference across which the heat is transferred, the less the irreversibility. However, for a given rate of heat transfer, a smaller temperature difference will require a larger (and therefore more expensive) heat exchanger, and these various factors must all be considered in the development of the optimum and most economical design.

In many engineering decisions other factors, such as the impact on the environment (for example, air pollution and water pollution) and the impact on society, must be considered in developing the optimum design.

In connection with the increased use of availability analysis in recent years, a term called the "second law efficiency" has come into more common use. This involves a comparison of the desired output of a process with the cost, or input, in terms of the thermodynamic availability. Thus, the isentropic turbine efficiency defined in Chapter 7 by Eq. 7.72 as the actual work output divided by the work

[1] In many popular talks reference is made to our energy reserves. From a thermodynamic point of view availability reserves would be a much more acceptable term. There is much energy in the atmosphere and the ocean, but relatively little availability.

for a hypothetical isentropic expansion from the same inlet state to the same exit pressure might well be called a "first law efficiency" in that it is a comparison of two energy quantities. The "second law efficiency," as described above, would be the actual work output of the turbine divided by the decrease in availability from the same inlet state to the same exit state, in this case the real state at the turbine exhaust. In this sense, this concept provides a rating or measure of the real process in terms of the actual change of state, and is simply another convenient way of utilizing the concept of thermodynamic availability. This will be illustrated in the following example.

Example 8.3

A steam turbine, Fig. 8.7, receives 200 000 lbm of steam per hour at 400 lbf/in.2, 600 F. At the point in the turbine where the pressure is 60 lbf/in.2, steam is bled off for use in processing equipment at the rate of 50 000 lbm/h. The temperature of this bled steam is 300 F. The balance of the steam leaves the turbine at 2 lbf/in.2, 90 per cent quality. The heat transfer to the surrounding is 500 000 Btu/h. Determine the availability per pound of the steam entering and at both points at which steam leaves the turbine, the reversible work for the given change of state and the second law efficiency for this process.

FIGURE 8.7

Sketch for Example 8.3.

Control volume: Turbine.

Inlet state: P_1, T_1 known; state fixed.

Exit states: P_2, T_2 known; P_3, x_3 known; both states fixed.

Process: SSSF.

Model: Steam tables.

Analysis: The availability at any point for the steam entering or leaving the turbine is given by Eq. 8.16.

$$\psi = (h - h_0) - T_0(s - s_0) + \frac{\mathbf{V}^2}{2} + g(Z - Z_0)$$

Since changes in kinetic and potential energy are not involved in this problem, this equation reduces to

$$\psi = (h - h_0) - T_0(s - s_0)$$

The reversible work can then be found from Eq. 8.18.

$$\dot{W}_{rev} = \dot{m}_1\psi_1 - \dot{m}_2\psi_2 - \dot{m}_3\psi_3$$

Solution

At the pressure and temperature of the surroundings, 14.7 lbf/in.2, 77 F, the water is a slightly compressed liquid, and the properties of the water are essentially equal to those for saturated liquid at 77 F.

$$h_0 = 45.1 \text{ Btu/lbm}; \quad s_0 = 0.0877 \text{ Btu/lbm R}$$

From Eq. 8.16,

$$\psi_1 = (1306.6 - 45.1) - 537(1.5892 - 0.0877) = 455.5 \text{ Btu/lbm}$$
$$\psi_2 = (1181.9 - 45.1) - 537(1.6495 - 0.0877) = 296.8 \text{ Btu/lbm}$$
$$\psi_3 = (1013.9 - 45.1) - 537(1.7453 - 0.0877) = 79.8 \text{ Btu/lbm}$$

From Eq. 8.18,

$$\dot{W}_{rev} = 200\,000(455.5) - 50\,000(296.8) - 150\,000(79.8)$$
$$= 64\,290\,000 \text{ Btu/h} = 25\,271 \text{ hp}$$

The actual turbine output is, from Eq. 5.49,

$$\dot{W} = -500\,000 + 200\,000(1306.6) - 50\,000(1181.9) - 150\,000(1013.9)$$
$$= 49\,640\,000 \text{ Btu/h} = 19\,513 \text{ hp}$$

such that the second law efficiency is

$$\eta_{2nd\,law} = \frac{19\,513}{25\,271} = 0.772$$

Example 8.4

A lead storage battery of the type used in an automobile is able to deliver 1440 watt-hours of electrical energy. This energy is available for starting the car.

Suppose we wish to use compressed air for doing an equivalent amount of work in starting the car. The compressed air is to be stored at 1000 lbf/in.2, 77 F. What volume of tank would be required to have the compressed air have an availability of 1440 watt-hours?

System: Compressed air tank.

State: P, T known; state fixed.

Model: Ideal gas.

Solution

$$1440 \text{ watt hours} = 1.440 \times 3412 = 4910 \text{ Btu.}$$

Let us first find the availability of air at 1000 lbf/in.², 77 F. From Eq. 8.21,

$$\phi = (u - u_0) - T_0(s - s_0) + P_0(v - v_0)$$

$$v = \frac{RT}{P} = \frac{53.34 \times 537}{1000 \times 144} = 0.1975 \text{ ft}^3/\text{lbm}$$

$$v_0 = \frac{RT_0}{P_0} = \frac{53.34 \times 537}{14.7 \times 144} = 13.38 \text{ ft}^3/\text{lbm}$$

$$\phi = 0 - 537\left(\frac{-53.34}{778} \ln \frac{1000}{14.7}\right) + \frac{14.7 \times 144}{778}(0.1975 - 13.38)$$

$$= 0 + 155.3 - 35.9 = 119.4 \text{ Btu/lbm}$$

In order to have an availability of 4910 Btu, the mass of air is

$$m = \frac{4910}{119.4} = 41.1 \text{ lbm}$$

$$V = \frac{mRT}{P} = \frac{41.1 \times 53.34 \times 537}{1000 \times 144} = 8.21 \text{ ft}^3$$

8.4 CONSIDERATION OF PROCESSES THAT INVOLVE HEAT TRANSFER WITH A BODY OTHER THAN THE ATMOSPHERE

Up to this point we have considered process in which the only heat transfer is with the surroundings. Since many processes do involve heat transfer with bodies at a temperature above or below the temperature of the surroundings, a few remarks should be made regarding them.

The simplest way of handling such processes is to consider two (or more) subsystems, one comprising the body from which heat is transferred, and the other the body to which heat is transferred. These two sub-systems comprise the system under consideration and the reversible work, availability, and irreversibility for for the system will be the sum of these quantities for the sub-systems. The same procedure could be used in the case of a control volume. The two examples that follow illustrate this approach. The first involves the irreversible transfer of heat and the second involves reversible heat transfer.

Example 8.5

In a boiler, heat is transferred from the products of combustion to the steam. The temperature of the products of combustion decreases from 1100°C to 550°C while the pressure remains constant at 0.1 MPa. The average constant-pressure specific heat of the products of combustion is 1.09 kJ/kg K. The water enters at 0.8 MPa, 150°C and leaves at 0.8 MPa, 250°C. Determine the reversible work and the irreversibility for this process per kilogram of water evaporated.

FIGURE 8.8
Sketch for Example 8.5.

Control volume: Overall heat exchanger.

Sketch: Fig. 8.8.

Inlet states: Both known, given in Fig. 8.8.

Exit states: Both known, given in Fig. 8.8.

Process: Overall, adiabatic.

Diagram: Fig. 8.9.

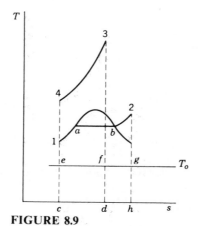

FIGURE 8.9

Temperature-entropy diagram for Example 8.5.

Model: Products—ideal gas, constant specific heat. Water—steam tables.

Analysis: For the products, the entropy change for this constant-pressure process is

$$(s_e - s_i)_{prod} = C_{po} \ln \frac{T_e}{T_i}$$

For this control volume we can write the following governing equations:

Continuity eq.:
$$(\dot{m}_i)_{H_2O} = (\dot{m}_e)_{H_2O} \tag{a}$$
$$(\dot{m}_i)_{prod} = (\dot{m}_e)_{prod} \tag{b}$$

First law (a steady-state, steady-flow process):
$$(\dot{m}_i h_i)_{H_2O} + (\dot{m}_i h_i)_{prod} = (\dot{m}_e h_e)_{H_2O} + (\dot{m}_e h_e)_{prod} \tag{c}$$

Second law (the process is adiabatic for the control volume shown):
$$(\dot{m}_e s_e)_{H_2O} + (\dot{m}_e s_e)_{prod} \geq (\dot{m}_i s_i)_{H_2O} + (\dot{m}_i s_i)_{prod}$$

Solution

From Eqs. a, b, and c, we can calculate the ratio of the mass flow of products to the mass flow of water.

$$\dot{m}_{prod}(h_i - h_e)_{prod} = \dot{m}_{H_2O}(h_e - h_i)_{H_2O}$$

$$\frac{\dot{m}_{prod}}{\dot{m}_{H_2O}} = \frac{(h_e - h_i)_{H_2O}}{(h_i - h_e)_{prod}} = \frac{2950 - 632.2}{1.09(1100 - 550)} = 3.866$$

To find the reversible work we find w_{rev} for the water and for the products and add these together.

We first calculate w_{rev} for the change of state of the water.

$$(w_{rev})_{H_2O} = (h_1 - h_2) - T_0(s_1 - s_2)$$
$$= (632.2 - 2950) - 298.15(1.8418 - 7.0384)$$
$$= -768.4 \text{ kJ/kg}$$

Next we consider the products.

$$(w_{rev})_{prod} = \frac{\dot{m}_{prod}}{\dot{m}_{H_2O}} [(h_3 - h_4) - T_0(s_3 - s_4)]$$

(States 3 and 4 designate the initial and final states of the products.)

$$(w_{rev})_{prod} = 3.866 \left[1.09(1100 - 550) - 298.15 \left(1.09 \ln \frac{1373.15}{823.15} \right) \right]$$
$$= 1674.7 \text{ kJ/kg } H_2O$$

$$\frac{i}{\dot{m}_{H_2O}} = (-768.4 + 1674.7) - 0 = 906.3 \text{ kJ/kg } H_2O$$

It is also of interest to determine the net change of entropy. The change in the entropy of the water is

$$(s_2 - s_1)_{H_2O} = 7.0384 - 1.8418 = 5.1966 \text{ kJ/kg } H_2O \text{ K}$$

The change in the entropy of the products is

$$\frac{\dot{m}_{prod}}{\dot{m}_{H_2O}}(s_4 - s_3)_{prod} = -3.866\left(1.09 \ln \frac{1373.15}{823.15}\right) = -2.1564 \text{ kJ/kg } H_2O \text{ K}$$

Thus there is a net increase in entropy during the process. The irreversibility could also have been calculated from Eq. 8.14:

$$\dot{I} = \sum \dot{m}_e T_0 s_e - \sum \dot{m}_i T_0 s_i - \dot{Q}_{c.v.}$$

For the control volume selected, $\dot{Q}_{c.v.} = 0$, and therefore

$$\frac{\dot{I}}{\dot{m}_{H_2O}} = T_0(s_2 - s_1)_{H_2O} + \frac{\dot{m}_{prod}}{\dot{m}_{H_2O}} T_0(s_4 - s_3)_{prod}$$

$$= 298.15(5.1966) + 298.15(-2.1564)$$

$$= 906.3 \text{ kJ/kg } H_2O$$

These two processes are shown on the T-s diagram of Fig. 8.9. Line 3–4 represents the process for the 3.866 kg of products. Area 3–4–c–d–3 represents the heat transferred from the 3.866 kg of products of combustion, and area 3–4–e–f–3 represents the reversible work for the given change in state of these products. Area 1–a–b–2–h–c–1 represents the heat transferred to the water, and this is equal to are 3–4–c–d–3, which represents the heat transferred from the products of combustion. Area 1–a–b–2–g–e–1 represents the reversible work for the given change in state of the water. The difference between area 3–4–e–f–3 (the reversible work for the products) and area 1–a–b–2–g–e–1 (the reversible work for the water) represents the net reversible work. It is readily shown that this net reversible work is equal to area f–g–h–d–f, or $T_0(\Delta s)_{net}$. Since the actual work is zero, this area also represents the irreversibility, which agrees with our calculation above.

It is essential to note that when the change of state involving heat transfer takes place reversibly, the net change in entropy is zero, and therefore the decrease in the entropy of the body from which heat is transferred must be equal to the increase in entropy of the body to which heat is transferred. This is best demonstrated by considering an example similar to Example 8.5.

Example 8.6

Repeat Example 8.5, but assume that the heat transfer could be made to be entirely reversible, that is, that the heat transfer takes place through a reversible engine.

FIGURE 8.10

Diagram for Example 8.6.

Sketch: Fig. 8.10.

Analysis: Schematically this would involve heat transfer from the products of combustion to reversible engines that reject heat to the water, as shown in Fig. 8.10.

We again write the governing equations:

Continuity eq.:
$$(\dot{m}_i)_{H_2O} = (\dot{m}_e)_{H_2O}$$
$$(\dot{m}_i)_{prod} = (\dot{m}_e)_{prod}$$

First law:

$$(\dot{m}_i h_i)_{H_2O} + (\dot{m}_i h_i)_{prod} = (\dot{m}_e h_e)_{H_2O} + (\dot{m}_e h_e)_{prod} + \dot{W}_{c.v.}$$

Second law: Since this is a reversible adiabatic process,

$$(\dot{m}_e s_e)_{H_2O} + (\dot{m}_e s_e)_{prod} = (\dot{m}_i s_i)_{H_2O} + (\dot{m}_i s_i)_{prod}$$

Solution

From the continuity equation and second law we can determine flow of products per unit flow of water.

$$(s_2 - s_1)_{H_2O} + \frac{\dot{m}_{prod}}{\dot{m}_{H_2O}}(s_4 - s_3)_{prod} = 0$$

$$\frac{\dot{m}_{prod}}{\dot{m}_{H_2O}}(s_3 - s_4)_{prod} = (7.0384 - 1.8418)$$

$$\frac{\dot{m}_{prod}}{\dot{m}_{H_2O}}\left(1.09 \ln \frac{1373.15}{823.15}\right) = 5.1966$$

$$\frac{\dot{m}_{prod}}{\dot{m}_{H_2O}} = 9.317$$

We now calculate the reversible work for the change of state of the water and the products:

$$(w_{rev})_{H_2O} = (h_1 - h_2) - T_0(s_1 - s_2)$$

$$(w_{rev})_{prod} = [(h_3 - h_4) - T_0(s_3 - s_4)] \frac{\dot{m}_{prod}}{\dot{m}_{H_2O}}$$

But, since

$$(s_2 - s_1)_{H_2O} = \frac{\dot{m}_{prod}}{\dot{m}_{H_2O}} (s_3 - s_4)_{prod}$$

it follows that

$$w_{rev} = (h_1 - h_2) + \frac{\dot{m}_{prod}}{\dot{m}_{H_2O}} (h_3 - h_4)$$

We note that this net reversible work is exactly equal to the work that would be determined from the first law. We would expect this, of course, since the process is completely reversible.

$$w_{rev} = (632.2 - 2950) + 9.317 \times 1.09(1100 - 550)$$
$$= 3267.7 \text{ kJ/kg } H_2O$$

Since the net change in entropy is zero, the T-s diagram is as shown in Fig. 8.10. Area 3–4–c–d–3 represents the heat transferred from the products to the engines, and area 1–a–b–2–d–c–1 represents the heat received by the water. Area 3–4–1–a–b–2–3 represents the work done by the heat engines, which is equal to the reversible work for this process.

8.5 PROCESSES INVOLVING CHEMICAL REACTIONS

Although chemical reaction will not be considered in detail until Chapter 12 some preliminary remarks regarding availability in such processes can be made here. In the first place, we observe that the reactants are often in pressure and temperature equilibrium with the surroundings before the reaction takes place, and the same is true of the products after the reaction. An automobile engine would be an example of such a process if we visualized the products being cooled to atmospheric temperature before being discharged from the engine.

Let us first consider for a system the implications of temperature equilibrium with the surroundings during a chemical reaction in a system. The temperature of the system T is equal to T_0, the temperature of the surroundings. Therefore, from Eq. 8.8 we can write (noting that in this case $T_0 = T$)

$$_1(w_{rev})_2 = \left(u_1 - T_1 s_1 + \frac{\mathbf{V}_1^2}{2} + gZ_1 \right) - \left(u_2 - T_2 s_2 + \frac{\mathbf{V}_2^2}{2} + gZ_2 \right)$$

The quantity $(U - TS)$ is a thermodynamic property of a substance and is called the Helmholtz function. It is an extensive property and we designate it by the symbol A. Thus

$$A = U - TS$$
$$a = u - Ts$$

(8.23)

Therefore, when a system undergoes a change of state while in temperature equilibrium with the surroundings, the reversible work is given by the relation

$$_1(w_{rev})_2 = \left(a_1 + \frac{\mathbf{V}_1^2}{2} + gZ_1\right) - \left(a_2 + \frac{\mathbf{V}_2^2}{2} + gZ_2\right)$$

In those cases where the kinetic and potential energy changes are not significant, this reduces to

$$_1(W_{rev})_2 = A_1 - A_2 = m(a_1 - a_2)$$

(8.24)

Let us now consider a system that undergoes a chemical reaction while in both pressure and temperature equilibrium with the surroundings. The availability function ϕ for a system has been defined as

$$\phi = (u + P_0 v - T_0 s) - (u_0 + P_0 v_0 - T_0 s_0)$$

If $P = P_0$ and $T = T_0$, then

$$\phi = (u + Pv - Ts) - (u_0 + P_0 v_0 - T_0 s_0)$$
$$= (h - Ts) - (h_0 - T_0 s_0)$$

(8.25)

The quantity $h - Ts$ is a thermodynamic property and is termed the Gibbs function, designated by the symbol G.

$$G = H - TS$$
$$g = h - Ts$$

(8.26)

Introducing the Gibbs function into Eq. 8.25 we have, when a system is in pressure and temperature equilibrium with the surroundings,

$$\phi = g - g_0$$

From Eq. 8.22 it follows that under these same conditions

$$_1(w_{rev})_2 = (g_1 - g_2) - P_0(v_1 - v_2) + \frac{\mathbf{V}_1^2 - \mathbf{V}_2^2}{2} + g(Z_1 - Z_2)$$

(8.27)

The Gibbs function is also of significance in a steady-state, steady-flow process that takes place in temperature equilibrium with the surroundings. Since in this case $T_i = T_e = T_0$, Eq. 8.9 reduces to

$$\dot{W}_{rev} = \dot{m}_i \left(h_i - T_i s_i + \frac{\mathbf{V}_i^2}{2} + gZ_i\right) - \dot{m}_e\left(h_e - T_e s_e + \frac{\mathbf{V}_e^2}{2} + gZ_e\right)$$

Introducing the Gibbs function we have

$$\dot{W}_{rev} = \dot{m}_i \left(g_i + \frac{\mathbf{V}_i^2}{2} + gZ_i \right) - \dot{m}_e \left(g_e + \frac{\mathbf{V}_e^2}{2} + gZ_e \right) \qquad (8.28)$$

Thus we have introduced two new properties, the Helmholtz function, A, and the Gibbs function, G. Both these functions are very important in the thermodynamics of chemical reactions and will be used extensively in later chapters of this book.

PROBLEMS

With no other information available, assume that the surroundings are at 77 F, 1 atm pressure (English) or 25°C, 100 kPa (SI).

8.1 Liquid Freon-12 enters the expansion valve of a refrigerator at 150 lbf/in.², 80 F, and exits the expansion valve at a temperature of 10 F. Calculate the reversible work and irreversibility for this process. Do you think that it would be worthwhile to attempt to replace the irreversible throttling process with a reversible process in order to reduce the work required for operating the compressor?

8.2 Calculate the irreversibility for the process described in Problem 5.15.

8.3 One means for controlling the power output of a steam turbine is a throttling governor, which throttles the supply steam to a lower pressure before it enters the turbine. Consider a case in which the supply is at 1000 lbf/in.², 1200 F, and the turbine load is such that the steam is throttled to 800 lbf/in.² in the governor. Determine the reversible work and irreversibility for this throttling process.

8.4 Consider the compressor of a gas turbine engine in which air enters at ambient conditions, 100 kPa and 10°C, and exits at 1.2 MPa. The compressor has an insentropic efficiency of 87%. Determine the work of compression, the reversible work for the actual change of state, and the irreversibility of this process per kilogram of air.

8.5 A commercial laundry requires a hot water supply of 10^4 lbm/h at 180 F. This supply is to be produced in a steady-flow process by mixing the 50 F water supply with steam supplied from a boiler as saturated vapor at 1 atm pressure. Determine the rate of irreversibility of the mixing process.

8.6 Superheated vapor Freon-12 enters a compressor at 100 kPa, 25°C and exits at 2.75 MPa, 100°C. It may be assumed that the compression process is reversible and polytropic. Determine the compressor work input and the process irreversibility per kilogram of Freon-12.

8.7 What is the minimum work required to change one kilogram of methane

gas from 500 kPa, 300 K to saturated liquid at 100 kPa? Sketch an apparatus that would be used in such a process and show the process on a temperature–entropy diagram.

8.8 A 10-ft³ tank containing saturated vapor Freon-12 at 20 F is connected to a line flowing liquid Freon-12 at 140 lbf/in.², 80 F. The valve is now opened and Freon-12 flows into the tank. During the process, heat is transferred with the surroundings at 80 F such that when the valve is closed the tank contains 50% liquid, 50% vapor, by volume, at 80 F. Calculate the heat transfer and the irreversibility for this process.

8.9 Consider a closed feedwater heater used in a steam power plant, as shown in Fig. 8.11. Determine the irreversibility per kilogram of feedwater leaving the heater at 2.7 MPa, 90°C.

FIGURE 8.11

Sketch for Problem 8.9.

8.10 Air enters the turbocharger compressor of an automotive engine at 14.7 lbf/in.², 85 F, and exits at 25 lbf/in.² The air is then cooled by 90° in an intercooler before entering the engine. If the isentropic compressor efficiency is 75%, determine the temperature of the air entering the engine, and the irrreversibility of the compression-cooling process.

8.11 A 30-lbm copper vessel contains 3 ft³ of Freon-12 at −10 F, 40% quality. This vessel is now brought indoors and eventually comes to room temperature, 70 F. What is the irreversibility of this process?

8.12 A 400-L evacuated tank is connected by a valve to a line flowing steam at 1.4 MPa, 300°C. The valve is now opened and steam flows into the tank until the pressure inside reaches 1.4 MPa, at which point the valve is closed. Assuming that this process is adiabatic, determine the irreversibility.

8.13 A solid aluminum sphere 0.1 m in diameter and initially at 200°C is allowed to cool to ambient temperature, 25°C. What is the irreversibility of this process?

8.14 Consider the system shown in Fig. 8.12 in which a refrigeration unit transfers heat from the water in cylinder *A* and rejects heat to the water in vessel

FIGURE 8.12

Sketch for Problem 8.14.

B until *A* contains saturated liquid. Cylinder *A* is fitted with a frictionless piston of such a mass that 25 lbf/in.2 is required to balance the force exerted by the piston and atmospheric pressure. Cylinder *A* initially contains 1 lbm of vapor and 1 lbm of liquid, while vessel *B* with a volume of 40 ft^3 contains water at 30 lbf/in.2, 50% quality.

(a) If the refrigerator has a coefficient of performance of 0.5, determine the final temperature in *B*.

(b) If the refrigerator is reversible, determine the final temperature in *B*, and the coefficient of performance for the refrigerator.

8.15 A steady stream of liquid Freon-12 at room temperature, 70 F, and at 100 lbf/in.2 enters a solar collector, and the stream exits at 200 F, 90 lbf/in.2 Calculate the change in availability of the Freon-12 between these two states.

8.16 An ideal gas with constant specific heat expands through an adiabatic turbine. Show that the irreversibility of the process can be expressed in terms of the following: specific heat, ambient temperature, turbine efficiency, and pressure ratio across the turbine.

8.17 A rigid 50-ft^3 tank contains air at 300 lbf/in.2, 300 F. The air is now allowed to cool to ambient temperature, 77 F, by heat transfer with the surroundings. Calculate the availability of the air at the initial and final states and the irreversibility for this process.

8.18 Liquid nitrogen at 101.3 kPa pressure is to be vaporized and delivered to a pipeline at 4 MPa, 275 K. Three possible schemes for doing this are shown in Fig. 8.13. The first scheme involves pumping the liquid and

FIGURE 8.13

Sketch for Problem 8.18.

then vaporizing it. The second involves vaporizing the liquid and super-
heating it just enough so that it would leave a reversible adiabatic
compressor at 4 MPa, 275 K. The third involves vaporizing it and
superheating to 275 K, followed by a reversible isothermal compression
process. Assume an ambient temperature of 275 K.

(a) What is the availability of liquid nitrogen at 101.3 kPa?

(b) What is the minimum work necessary to deliver 1 kg of nitrogen at
4 MPa, 275 K?

(c) Determine the work and irreversibility per kilogram of nitrogen
delivered for each of the three suggested ways of accomplishing this.

8.19 A rigid 10-ft^3 vessel contains air initially at 500 lbf/in.2 and room tem-
perature, 77 F. A valve is now opened and the pressure quickly drops to
200 lbf/in.2 before the valve is closed. It may be assumed that the air that
remains inside the vessel has undergone a reversible adiabatic expansion.
Determine:

(a) The initial availability of the air in the vessel.

(b) The availability of the air in the vessel immediately after the valve is
closed.

(c) The availability of the air in the vessel a long time later, when the air has returned to room temperature.

8.20 A nuclear reactor power plant utilizing water as the working fluid is shown schematically in Fig. 8.14. The high-temperature compressed liquid from the reactor is throttled into the flash evaporator. That fraction of the water that flashes into vapor flows to the turbine, while that which remains liquid flows to the mixing chamber. Determine:

(a) The availability of the water leaving the reactor.
(b) The turbine output per lbm of water leaving the reactor, assuming a reversible, adiabatic process in the turbine.
(c) The irreversibility in the flash evaporator per lbm of water entering.
(d) The irreversibility in the mixing chamber per lbm of water leaving.

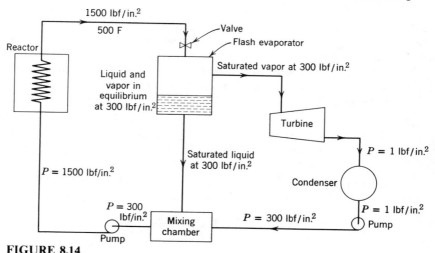

FIGURE 8.14
Sketch for Problem 8.20.

8.21 Helium at ambient conditions, 100 kPa and 20°C, is compressed in a reversible isothermal process to 500 kPa and then expanded back to 100 kPa in a reversible adiabatic process. Both are steady-state, steady-flow processes. What is the availability at the exit?

8.22 Consider a geothermal supply of hot water at 300 F, 500 lbf/in.² What is the availability of this energy resource, with respect to an ambient temperature of 77 F?

8.23 At a certain location in the ocean, the local water temperature is 25°C, compared with an ambient air temperature of 20°C. What is the thermodynamic availability of the water?

8.24 Air enters a compressor at ambient conditions, 14.7 lbf/in.² and 540 R, and exits at 80 lbf/in.² If the isentropic compressor efficiency is 80%, what is the second-law efficiency of the compression process?

8.25 Two iron blocks, each having a mass of 10 kg, are initially at a tempera-
ture of 40°C. A reversible refrigerator receives heat from one block and
rejects heat to the other. Calculate the work required to produce a temper-
ature difference of 100°C between the two blocks.

8.26 Consider two metal blocks, block A being aluminum with a mass of 40 lbm
and initially at 80 F, and block B being copper with a mass of 20 lbm and
initially at 1200 F. The two blocks are now brought to a common final
temperature.
 (a) If the process is reversible, find the final temperature and the work
 done during the process.
 (b) If the process is accomplished by simply bringing the blocks into
 thermal communication, find the final temperature and the irreversibi-
 lity of the process.

8.27 A steady combustion of natural gas results in 0.15 kg/s of products
(having approximately the same properties as air) at 1100°C, 100 kPa. The
products are passed through a heat exchanger and exit at 550°C. What is
the maximum theoretical power output from a cyclic heat engine oper-
ating on the heat rejected from the combustion products, assuming that
the ambient temperature is 20°C?

8.28 At a certain location the temperature of the water supply is 15°C. Ice is
to be made from this water supply by the process shown in Fig. 8.15.
The final temperature of the ice is −10°C, and the final temperature of
the water that is used as cooling water in the condenser is 30°C. What is
the minimum work input required to produce 1000 kg of ice?

FIGURE 8.15
Sketch for Problem 8.28.

8.29 An air preheater is used to cool the products of combustion from a furnace
while heating the air to be used for combustion. The flow rate of products
is 100 000 lbm/h, and the products are cooled from 600 F to 400 F, and
for the products at this temperature $C_{po} = 0.26$ Btu/lbm R. The flow rate
of air is 93 000 lbm/h, the initial air temperature is 100 F, and for the air
$C_{po} = 0.24$ Btu/lbm R.
 (a) What is the initial and final availability of the products?
 (b) What is the irreversibility for this process?

(c) Suppose that this heat transfer from the products took place reversibly through heat engines. What would be the final temperature of the air? What power would be produced by the heat engines?

8.30 One kmol of carbon dust is burned with 1 kmol of oxygen to form 1 kmol of carbon dioxide in a steady-flow process. The temperature of the carbon and oxygen before combustion is 25°C and the temperature of the carbon dioxide after combustion is also 25°C. The heat transferred during the process is $-393\ 522$ kJ, and the entropy of the carbon dioxide after combustion is 2.967 kJ/kmol K higher than the entropy of carbon and oxygen before combustion. Calculate the reversible work and the irreversibility for this process.

9
SOME POWER AND
REFRIGERATION CYCLES

Some power plants, such as the simple steam power plant, which we have considered several times, operate in a cycle. That is, the working fluid undergoes a series of processes and finally returns to the initial state. In other power plants, such as the internal combustion engine and gas turbine, the working fluid does not go through a thermodynamic cycle, even though the engine itself may operate in a mechanical cycle. In this case the working fluid has a different composition or is in a different state at the conclusion of the process than at the beginning. Such equipment is sometimes said to operate on the open cycle (the word cycle is really a misnomer), whereas the steam power plant operates on a closed cycle. The same distinction between open and closed cycles can be made regarding refrigeration devices. For both the open- and closed-cycle type of apparatus, however, it is advantageous to analyze the performance of an idealized closed cycle similar to the actual cycle. Such a procedure is particularly advantageous in determining the influence of certain variables on performance. For example, the spark-ignition internal combustion engine is usually approximated by the Otto cycle. From an analysis of the Otto cycle one concludes that increasing the compression ratio increases the efficiency. This is also true for the actual engine, even though the Otto cycle efficiencies may deviate significantly from the actual efficiencies.

This chapter is concerned with these idealized cycles, both for power and refrigeration apparatuses. The working fluids considered are both vapors and ideal gases. An attempt will be made to point out how the processes in actual apparatuses deviate from the ideal. Consideration is also given to certain modifications of the basic cycles that are intended to improve performance. These involve the use of devices such as regenerators, multistage compressors and expanders, and intercoolers. The order in which the cycles will be considered is: (1) vapor power cycles, (2) vapor refrigeration cycles, (3) air-standard power cycles, and (4) air-standard refrigeration cycles. Various combinations of these types of cycles and also special applications, such as topping cycles, bottoming cycles, and cogeneration of electrical power and energy, can then be studied; these special topics are examined in the chapter problems.

VAPOR POWER CYCLES

9.1 THE RANKINE CYCLE

The ideal cycle for a simple steam power plant is the Rankine cycle, shown in Fig. 9.1. The processes that comprise the cycle are:

1–2: Reversible adiabatic pumping process in the pump.
2–3: Constant-pressure transfer of heat in the boiler.
3–4: Reversible adiabatic expansion in the turbine (or other prime mover such as a steam engine).
4–1: Constant-pressure transfer of heat in the condenser.

The Rankine cycle also includes the possibility of superheating the vapor, as in cycle 1–2–3′–4′–1.

If changes of kinetic and potential energy are neglected, heat transfer and work may be represented by various areas on the T-s diagram. The heat transferred to the working fluid is represented by area a–2–2′–3–b–a, and the heat transferred from the working fluid by area a–1–4–b–a. From the first law we conclude that the area representing the work is the difference between these two areas—area 1–2–2′–3–4–1. The thermal efficiency is defined by the relation

$$\eta_{th} = \frac{w_{net}}{q_H} = \frac{\text{area } 1-2-2'-3-4-1}{\text{area } a-2-2'-3-b-a} \tag{9.1}$$

In analyzing the Rankine cycle it is helpful to think of efficiency as depending on the average temperature at which heat is supplied and the average temperature at which heat is rejected. Any changes that increase the average temperature at which heat is supplied or decrease the average temperature at which heat is rejected will increase the Rankine-cycle efficiency.

FIGURE 9.1

Simple steam power plant that operates on the Rankine cycle.

It should be stated that in analyzing the ideal cycles in this chapter, the changes in kinetic and potential energies from one point in the cycle to another are neglected. In general this is a reasonable assumption for the actual cycles.

It is readily evident that the Rankine cycle has a lower efficiency than a Carnot cycle with the same maximum and minimum temperatures as a Rankine cycle, because the average temperature between 2 and 2′ is less than the temperature during evaporation. We might well ask, why choose the Rankine cycle as the ideal cycle? Why not select the Carnot cycle 1′–2′–3–4–1′? At least two reasons can be given. The first involves the pumping process. State 1′ is a mixture of liquid and vapor, and great difficulties are encountered in building a pump that will handle the mixture of liquid and vapor at 1′ and deliver saturated liquid at 2′. It is much easier to completely condense the vapor and handle only liquid in the pump, and the Rankine cycle is based on this fact. The second reason involves superheating the vapor. In the Rankine cycle the vapor is superheated at constant pressure, process 3–3′. In the Carnot cycle all the heat transfer is at constant temperature, and therefore the vapor is superheated in process 3–3″. Note, however, that during this process the pressure is dropping, which means that the heat must be transferred to the vapor as it undergoes an expansion process in which work is done. This also is very difficult to achieve in practice. Thus the Rankine cycle is the ideal cycle that can be approximated in practice. In the sections that follow we will consider some variations on the Rankine cycle that enable one to more closely approach the Carnot-cycle efficiency.

Before discussing the influence of certain variables on the performance of the Rankine cycle, an example is given.

Example 9.1

Determine the efficiency of a Rankine cycle utilizing steam as the working fluid in which the condenser pressure is 10 kPa. The boiler pressure is 2 MPa. The steam leaves the boiler as saturated vapor.

In solving Rankine-cycle problems we will let w_p denote the work into the pump per unit mass of fluid flowing and q_L the heat rejected from the working fluid per unit mass of fluid flowing.

In solving this problem we consider, in succession, a control surface around the pump, the boiler, the turbine, and the condenser. In each case the thermodynamic model is the steam tables, and the process is SSSF with negligible changes in kinetic and potential energies. Now, in turn,

Control volume: Pump.

Inlet state: P_1 known, saturated liquid; state fixed.

Exit state: P_2 known.

Analysis:

First law: $\qquad\qquad\qquad w_p = h_2 - h_1$

Second law: $\qquad\qquad\qquad s_2 = s_1$

Since

$$s_2 = s_1, \qquad h_2 - h_1 = \int_1^2 v\, dP$$

Solution

Assuming the liquid to be incompressible,

$$w_p = v(P_2 - P_1) = 0.001\,01(2000 - 10) = 2.0 \text{ kJ/kg}$$
$$h_2 = h_1 + w_p = 191.8 + 2.0 = 193.8$$

Control volume: Boiler.
Inlet state: P_2, h_2 known; state fixed.
Exit state: P_3 known, saturated vapor; state fixed.
Analysis:
 First law: $\qquad\qquad\qquad\qquad q_H = h_3 - h_2$

Solution

$$q_H = h_3 - h_2 = 2799.5 - 193.8 = 2605.7 \text{ kJ/kg}$$

Control volume: Turbine.
Inlet state: State 3 known (above).
Exit state: P_4 known.
Analysis:
 First law: $\qquad\qquad\qquad\qquad w_t = h_3 - h_4$
 Second law: $\qquad\qquad\qquad\qquad s_3 = s_4$

Solution

We can determine the quality at state 4 as follow:

$$s_3 = s_4 = 6.3409 = 0.6493 + x_4 7.5009, \qquad x_4 = 0.7588$$
$$h_4 = 191.8 + 0.7588(2392.8) = 2007.5$$
$$w_t = 2799.5 - 2007.5 = 792.0 \text{ kJ/kg}$$

Control volume: Condenser.
Inlet state: State 4 known (above).
Exit state: State 1 known (above).
Analysis:
 First law: $\qquad\qquad\qquad\qquad q_L = h_4 - h_1$

Solution

$$q_L = h_4 - h_1 = 2007.5 - 191.8 = 1815.7 \text{ kJ/kg}$$

We can now calculate the thermal efficiency

$$\eta_{th} = \frac{w_{net}}{q_H} = \frac{q_H - q_L}{q_H} = \frac{w_t - w_p}{q_H} = \frac{792.0 - 2.0}{2605.7} = 30.3\%$$

We could also write an expression for thermal efficiency in terms of properties at various points in the cycle.

$$\eta_{th} = \frac{(h_3 - h_2) - (h_4 - h_1)}{h_3 - h_2} = \frac{(h_3 - h_4) - (h_2 - h_1)}{h_3 - h_2}$$

$$= \frac{2605.7 - 1815.7}{2605.7} = \frac{792.0 - 2.0}{2605.7} = 30.3\%$$

9.2 EFFECT OF PRESSURE AND TEMPERATURE ON THE RANKINE CYCLE

Let us first consider the effect of exhaust pressure and temperature on the Rankine cycle. This effect is shown on the T-s diagram of Fig. 9.2.

Let the exhaust pressure drop from P_4 to P_4', with the corresponding decrease in temperature at which heat is rejected. The net work is increased by area $1-4-4'-1'-2'-2-1$ (shown by the crosshatching). The heat transferred to the steam is increased by area $a'-2'-2-a-a'$. Since these two areas are approximately equal, the net result is an increase in cycle efficiency. This is also evident from the fact that the average temperature at which heat is rejected is decreased. Note, however, that lowering the back pressure causes an increase in the moisture content in the steam leaving the turbine. This is a significant factor because if the moisture in the low-pressure stages of the turbine exceeds about 10%, not only is there a decrease in turbine efficiency but also erosion of the turbine blades may be a very serious problem.

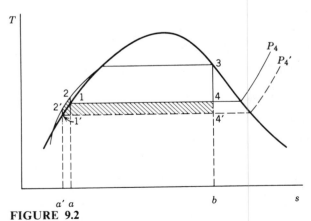

FIGURE 9.2

Effect of exhaust pressure on Rankine-cycle efficiency.

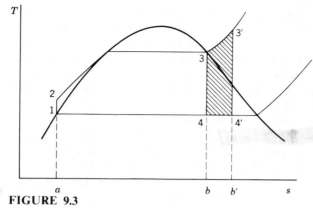

FIGURE 9.3

Effect of superheating on Rankine-cycle efficiency.

Next consider the effect of superheating the steam in the boiler, as shown in Fig. 9.3. It is readily evident that the work is increased by area $3-3'-4'-4-3$, and the heat transferred in the boiler is increased by area $3-3'-b'-b-3$. Since this ratio of these two areas is greater than the ratio of net work to heat supplied for the rest of the cycle, it is evident that for given pressures, superheating the steam increases the Rankine-cycle efficiency. This would also follow from the fact that the average temperature at which heat is transferred to the steam is increased. Note also that when the steam is superheated the quality of the steam leaving the turbine increases.

Finally, the influence of the maximum pressure of the steam must be considered, and this is shown in Fig. 9.4. In this analysis the maximum temperature of the

FIGURE 9.4

Effect of boiler pressure on Rankine-cycle efficiency.

steam, as well as the exhaust pressure, is held constant. The heat rejected decreases by area $b'-4'-4-b-b'$. The net work increases by the amount of the single cross-hatching and decreases by the amount of the double crosshatching. Therefore, the net work tends to remain the same, but the heat rejected decreases, and hence the Rankine-cycle efficiency increases with an increase in maximum pressure. Note that in this case also the average temperature at which heat is supplied increases with an increase in pressure. The quality of the steam leaving the turbine decreases as the maximum pressure increases.

To summarize this section, we can say that the Rankine-cycle efficiency can be increased by lowering the exhaust pressure, increasing the pressure during heat addition, and by superheating the steam. The quality of the steam leaving the turbine is increased by superheating the steam, and decreased by lowering the exhaust pressure and by increasing the pressure during heat addition.

Example 9.2

In a Rankine cycle steam leaves the boiler and enters the turbine at 600 lbf/in.2, 800 F. The condenser pressure is 1 lbf/in.2 Determine the cycle efficiency.

To determine the cycle efficiency we must calculate the turbine work, the pump work, and the heat transfer to the steam in the boiler. We do this by considering a control surface around each of these components in turn. In each case, the thermodynamic model is the steam tables, and the process is SSSF with negligible changes in kinetic and potential energies.

Control volume: Pump.

Inlet state: P_1 known, saturated liquid; state fixed.

Exit state: P_2 known.

Analysis:

First law: $\qquad\qquad\qquad\qquad w_p = h_2 - h_1$

Second law: $\qquad\qquad\qquad\qquad s_2 = s_1$

Since $s_2 = s_1$,

$$h_2 - h_1 = \int_1^2 v\, dP = v(P_2 - P_1)$$

Solution

$$w_p = v(P_2 - P_1) = 0.01614(600 - 1) \times \tfrac{144}{778} = 1.8 \text{ Btu/lbm}$$
$$h_1 = 69.70$$
$$h_2 = 69.7 + 1.8 = 71.5 \text{ Btu/lbm.}$$

Control volume: Turbine.

Inlet state: P_3, T_3 known; state fixed.

Exit state: P_4 known.

Analysis:

First law: $\qquad\qquad\qquad\qquad w_t = h_3 - h_4$

Second law: $\qquad\qquad\qquad\quad s_4 = s_3$

Solution

$$h_3 = 1407.6 \; s_3 = 1.6343$$
$$s_3 = s_4 = 1.6343 = 1.9779 - 1\,(-x)_4\,1.8453$$
$$(1-x)_4 = 0.1861$$
$$h_4 = 1105.8 - 0.1861(1036.0) = 913.0$$
$$w_t = h_3 - h_4 = 1407.6 - 913.0 = 494.6 \text{ Btu/lbm}$$
$$w_\text{net} = w_t - w_p = 494.6 - 1.8 = 492.8 \text{ Btu/lbm}$$

Control volume: Boiler.

Inlet state: P_2, h_2 known; state fixed.

Exit state: State 3 fixed (above).

Analysis:

First law: $\qquad\qquad\qquad\qquad q_H = h_3 - h_2$

Solution

$$q_H = h_3 - h_2 = 1407.6 - 71.5 = 1336.1 \text{ Btu/lbm}$$

$$\eta_\text{th} = \frac{w_\text{net}}{q_H} = \frac{492.8}{1336.1} = 36.9\%$$

The net work could also be determined by calculating the heat rejected in the condenser, q_L, and noting, from the first law, that the net work for the cycle is equal to the net heat transfer. Considering a control surface around the condenser,

$$q_L = h_4 - h_1 = 913.0 - 69.7 = 843.3 \text{ Btu/lbm}$$

Therefore,

$$w_\text{net} = q_H - q_L = 1336.1 - 843.3 = 492.8 \text{ Btu/lbm}$$

9.3 THE REHEAT CYCLE

In the last paragraph we noted that the efficiency of the Rankine cycle could be increased by increasing the pressure during the addition of heat. However, this also increases the moisture content of the steam in the low-pressure end of the turbine. The reheat cycle has been developed to take advantage of the increased efficiency with higher pressures, and yet avoid excessive moisture in the low-pressure stages of the turbine. This cycle is shown schematically and on a *T-s* diagram in Fig. 9.5. The unique feature of this cycle is that the steam is expanded to some intermediate pressure in the turbine and is then reheated in the boiler, after which it expands in the turbine to the exhaust pressure. It is evident from the *T-s*

FIGURE 9.5

The ideal reheat cycle.

diagram that there is very little gain in efficiency from reheating the steam, because the average temperature at which heat is supplied is not greatly changed. The chief advantage is in decreasing the moisture content in the low-pressure stages of the turbine to a safe value. Note also that if metals could be found that would enable one to superheat the steam to 3', the simple Rankine cycle would be more efficient than the reheat cycle, and there would be no need for the reheat cycle.

Example 9.3

Consider a reheat cycle utilizing steam. Steam leaves the boiler and enters the turbine at 600 lbf/in.2, 800 F. After expansion in the turbine to 60 lbf/in.2, the steam is reheated to 800 F and then expanded in the low-pressure turbine to 1 lbf/in.2 Determine the cycle efficiency.

For each control volume analyzed, the thermodynamic model is the steam tables, the process is SSSF, and changes in kinetic and potential energies are negligible.

Control volume: High-pressure turbine.

Inlet state: P_3, T_3 known; state fixed.

Exit state: P_4 known.

Analysis:

First law: $$w_{h-p} = h_3 - h_4$$

Second law: $$s_3 = s_4$$

Solution

$$h_3 = 1407.6 \quad s_3 = 1.6343$$
$$s_4 = s_3 = 1.6343 = 1.6444 - (1 - x)_4 1.2170$$
$$(1 - x)_4 = 0.0083$$
$$h_4 = 1178.0 - 0.0083(915.8) = 1170.4$$

Control volume: Low-pressure turbine.

Inlet state: P_5, T_5 known; state fixed.

Exit state: P_6 known.

Analysis:

First law:
$$w_{l\text{-}p} = h_5 - h_6$$

Second law:
$$s_5 = s_6$$

Solution

$$h_5 = 1431.2 \quad s_5 = 1.9022$$
$$s_5 = s_6 = 1.9022 = 1.9779 - (1 - x)_6 1.8453$$
$$(1 - x)_6 = 0.0411$$
$$h_6 = 1105.8 - 0.0411(1036.0) = 1063.3$$

For the overall turbine, the total work output w_t is the sum of $w_{h\text{-}p}$ and $w_{l\text{-}p}$, so that

$$w_t = (h_3 - h_4) + (h_5 - h_6)$$
$$w_t = (1407.6 - 1170.4) + (1431.2 - 1063.3) = 605.1 \text{ Btu/lbm}$$

Control volume: Pump.

Inlet state: P_1 known, saturated liquid; state fixed.

Exit state: P_2 known.

Analysis:

First law:
$$w_p = h_2 - h_1$$

Second law:
$$s_2 = s_1$$

Since $s_2 = s_1$,

$$h_2 - h_1 = \int_1^2 v \, dP = v(P_2 - P_1)$$

Solution

$$w_p = v(P_2 - P_1) = 0.01614(600 - 1)\tfrac{144}{778} = 1.8 \text{ Btu/lbm}$$
$$h_2 = 69.7 + 1.8 = 71.5 \text{ Btu/lbm}$$

Control volume: Boiler.

Inlet states: States 3 and 5 both known (above).

Exit states: States 2 and 4 both known (above).

Analysis:

First law:
$$q_H = (h_3 - h_2) + (h_5 - h_4)$$

Solution

$$q_H = (h_3 - h_2) + (h_5 - h_4)$$
$$q_H = (1407.6 - 71.5) + (1431.2 - 1170.4) = 1596.9 \text{ Btu/lbm}$$

Therefore,

$$w_{net} = w_t - w_p = 605.1 - 1.8 = 603.3 \text{ Btu/lbm}$$

$$\eta_{th} = \frac{w_{net}}{q_H} = \frac{603.3}{1596.9} = 37.8\%$$

Note by comparison with Example 9.2 that the gain in efficiency from reheating is relatively small, but that the moisture content leaving the turbine is decreased as a result of reheating from 18.6 per cent to 4.1 per cent.

9.4 THE REGENERATIVE CYCLE

Another important variation from the Rankine cycle is the regenerative cycle, which involves the use of feedwater heaters. The basic concepts of this cycle can be demonstrated by considering the Rankine cycle without superheat as shown in Fig. 9.6. During the process between states 2 and 2′ the working fluid is heated while in the liquid phase, and the average temperature of the working fluid is much lower during this process than during the vaporization process 2′–3. This causes the average temperature at which heat is supplied in the Rankine cycle to be lower than in the Carnot cycle 1′–2′–3–4–1′, and consequently the efficiency of the Rankine cycle is less than that of the corresponding Carnot cycle. In the regenerative cycle the working fluid enters the boiler at some state between 2 and 2′, and consequently the average temperature at which heat is supplied is increased.

Consider first an idealized regenerative cycle, shown in Fig. 9.7. The unique feature of this cycle compared to the Rankine cycle is that after leaving the pump, the liquid circulates around the turbine casing, counterflow to the direction of

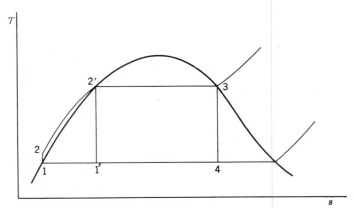

FIGURE 9.6

Temperature–entropy diagram showing the relationship between Carnot-cycle efficiency and Rankine-cycle efficiency.

FIGURE 9.7

The ideal regenerative cycle.

vapor flow in the turbine. Thus it is possible to transfer heat from the vapor as it flows through the turbine to the liquid flowing around the turbine. Let us assume for the moment that this is a reversible heat transfer; that is, at each point the temperature of the vapor is only infinitesimally higher than the temperature of the liquid. In this case line 4–5 on the T-s diagram of Fig. 9.7, which represents the states of the vapor flowing through the turbine, is exactly parallel to line 1–2–3, which represents the pumping process (1–2) and the states of the liquid flowing around the turbine. Consequently areas 2–3–b–a–2 and 5–4–d–c–5 are not only equal but congruous, and these areas, respectively, represent the heat transferred to the liquid and from the vapor. Note also that heat is transferred to the working fluid at constant temperature in process 3–4, and area 3–4–d–b–3 represents this heat transfer. Heat is transferred from the working fluid in process 5–1, and area 1–5–c–a–1 represents this heat transfer. Note that this area is exactly equal to area 1′–5′–d–b–1′, which is the heat rejected in the related Carnot cycle 1′–3–4–5′–1′. Thus this idealized regenerative cycle has an efficiency exactly equal to the efficiency of the Carnot cycle with the same heat-supply and heat-rejection temperatures.

Quite obviously this idealized regenerative cycle is not practical. First, it would not be possible to effect the necessary heat transfer from the vapor in the turbine to the liquid feedwater. Furthermore, the moisture content of the vapor leaving the turbine is considerably increased as a result of the heat transfer, and the disadvantage of this has been noted previously. The practical regenerative cycle involves the extraction of some of the vapor after it has partially expanded in the turbine and the use of feedwater heaters, as shown in Fig. 9.8.

Steam enters the turbine at state 5. After expansion to state 6, some of the steam is extracted and enters the feedwater heater. The steam that is not extracted is expanded in the turbine to state 7 and is then condensed in the condenser. This condensate is pumped into the feedwater heater where it mixes with the steam extracted from the turbine. The proportion of steam extracted is just sufficient to cause the liquid leaving the feedwater heater to be saturated at state 3. Note that

FIGURE 9.8
Regenerative cycle with open feedwater heater.

the liquid has not been pumped to the boiler pressure, but only to the intermediate pressure corresponding to state 6. Another pump is required to pump the liquid leaving the feedwater heater to boiler pressure. The significant point is that the average temperature at which heat is supplied has been increased.

This cycle is somewhat difficult to show on a T-s diagram because the mass of steam flowing through the various components is not the same. The T-s diagram of Fig. 9.8 simply shows the state of the fluid at the various points.

Area 4–5–c–b–4 in Fig. 9.8 represents the heat transferred per unit mass of working fluid. Process 7–1 is the heat-rejection process, but since not all the steam passes through the condenser, area 1–7–c–a–1 represents the heat transfer per unit mass flowing through the condenser, which does not represent the heat transfer per unit mass of working fluid entering the turbine. Note also between states 6 and 7 only part of the steam is flowing through the turbine. The example that follows illustrates the calculations involved in the regenerative cycle.

Example 9.4

Consider a regenerative cycle utilizing steam as the working fluid. Steam leaves the boiler and enters the turbine at 600 lbf/in.^2, 800 F. After expansion to 60 lbf/in.^2 some of the steam is extracted from the turbine for the purpose of heating the feedwater in an open feedwater heater. The pressure in the feedwater heater is 60 lbf/in.^2 and the water leaving it is saturated liquid at 60 lbf/in.^2 The steam not extracted expands to 1 lbf/in.^2 Determine cycle efficiency.

The line diagram and T-s diagram for this cycle are shown in Fig. 9.8.

As in previous examples, for each control volume the model used is the steam tables, the process is SSSF, and kinetic and potential energy changes are negligible.

From Examples 9.2 and 9.3 we have the following properties:

$$h_5 = 1407.6 \quad h_6 = 1170.4 \quad h_7 = 913.0 \quad h_1 = 69.70$$

Control volume: Low-pressure pump.
Inlet state: P_1 known, saturated liquid; state fixed.
Exit state: P_2 known.
Analysis:

First law:
$$w_{p1} = (h_2 - h_1)$$

Second law:
$$s_2 = s_1$$

Therefore,

$$h_2 - h_1 = \int_1^2 v\,dP = v(P_2 - P_1)$$

Solution

$$w_{p1} = v(P_2 - P_1) = 0.01614(60 - 1)\tfrac{144}{778} = 0.2 \text{ Btu/lbm}$$
$$h_2 = h_1 + w_p = 69.7 + 0.2 = 69.9$$

Control volume: Turbine.
Inlet state: P_5, T_5 known; state fixed.
Exit state: P_6 known; P_7 known.
Analysis:

First law:
$$w_t = (h_5 - h_6) + (1 - m_1)(h_6 - h_7)$$

Second law:
$$s_5 = s_6 = s_7$$

Solution

From the second law, the values for h_6 and h_7 given above have been calculated in Examples 9.2 and 9.3.

Control volume: Feedwater heater.
Inlet states: States 2 and 6 both known (above).
Exit state: P_3 known, saturated liquid; state fixed.
Analysis:

First law:
$$m_1(h_6) + (1 - m_1)h_2 = h_3$$

Solution

$$m_1(1170.4) + (1 - m_1)69.9 = 262.2$$
$$m_1 = 0.1747$$

We can now calculate the turbine work.

$$\begin{aligned} w_t &= (h_5 - h_6) + (1 - m_1)(h_6 - h_7) \\ &= (1407.6 - 1170.4) + (1 - 0.1747)(1170.4 - 913.0) \\ &= 449.7 \end{aligned}$$

Control volume: High-pressure pump.

Inlet state: State 3 known (above).

Exit state: P_4 known.

Analysis:

First law: $$w_{p2} = (h_4 - h_3)$$

Second law: $$s_4 = s_3$$

Solution

$$w_{p2} = v(P_4 - P_3) = 0.017\ 68(600 - 60)\tfrac{144}{778} = 1.7\ \text{Btu/lbm}$$
$$h_4 = h_3 + w_{p2} = 262.2 + 1.7 = 263.9$$

Therefore,

$$w_{net} = w_t - (1 - m_1)w_{p1} - w_{p2}$$
$$= 449.7 - 0.825(0.2) - 1.7 = 447.8$$

Control volume: Boiler.

Inlet state: P_4, h_4 known (above); state fixed.

Exit state: State 5 known (above).

Analysis:

First law: $$q_H = h_5 - h_4$$

Solution

$$q_H = h_5 - h_4 = 1407.6 - 263.9 = 1143.7$$
$$\eta_{th} = \frac{w_{net}}{q_H} = \frac{447.8}{1143.7} = 39.1\%$$

Note the increase in efficiency over the Rankine cycle of Example 9.2.

Up to this point the discussion and example have tacitly assumed that the extraction steam and feedwater are mixed in the feedwater heater. Another often-used, type of feedwater heater, known as a closed heater, is one in which the steam and feedwater do not mix, but rather heat is transferred from the extracted steam as it condenses on the outside of tubes as the feedwater flows through the tubes. In a closed heater, a schematic sketch of which is shown in Fig. 9.9, the steam and feedwater may be at considerably different pressures. The condensate may be pumped into the feedwater line, or it may be removed through a trap (a device that permits liquid but not vapor to flow to a region of lower pressure) to a lower-pressure heater or to the main condenser.

Open feedwater heaters have the advantage of being less expensive and having better heat transfer characteristics compared to closed feedwater heaters. They have the disadvantage of requiring a pump to handle the feedwater between each heater.

FIGURE 9.9

Schematic arrangement for a closed feedwater heater.

In many power plants a number of stages of extraction are used, though only rarely more than five. The number is, of course, determined by economic considerations. It is evident that by using a very large number of extraction stages and feedwater heaters, the cycle efficiency would approach that of the idealized regenerative cycle of Fig. 9.7, where the feedwater enters the boiler as saturated liquid at the maximum pressure. However, in practice this could not be economically justified because the savings effected by the increase in efficiency would be more than offset by the cost of additional equipment (feedwater heaters, piping, etc).

A typical arrangement of the main components in an actual power plant is shown in Fig. 9.10. Note that one open feedwater heater is a deaerating feedwater

FIGURE 9.10

Arrangement of heaters in an actual power plant utilizing regenerative feedwater heaters.

heater, and this has the dual purpose of heating and removing the air from the feedwater. Unless the air is removed, excessive corrosion occurs in the boiler. Note also that the condensate from the high-pressure heater drains (through a trap) to the intermediate heater, and the intermediate heater drains to the deaerating feedwater heater. The low-pressure heater drains to the condenser.

In many cases an actual power plant combines one reheat stage with a number of extraction stages. The principles already considered are readily applied to such a cycle.

9.5 DEVIATION OF ACTUAL CYCLES FROM IDEAL CYCLES

Before leaving the matter of vapor power cycles, a few comments are in order regarding the ways in which an actual cycle deviates from an ideal cycle. (The losses associated with the combustion process are considered in a later chapter.) The most important of these are as follows.

Piping Losses

Pressure drop due to frictional effects and heat transfer to the surroundings are the most important piping losses. Consider, for example, the pipe connecting the turbine to the boiler. If only frictional effects occurred, the states a and b in Fig. 9.11 would represent the states of the steam leaving the boiler and entering the turbine, respectively. Note that this causes an increase in entropy. Heat transferred to the surroundings at constant pressure can be represented by process bc. This effect causes a decrease in entropy. Both the pressure drop and heat transfer cause a decrease in the availability of the steam entering the turbine, and the irreversibility of this process can be calculated by the methods outlined in Chapter 8.

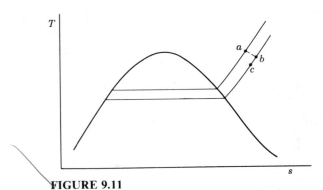

FIGURE 9.11

Temperature–entropy diagram showing effect of losses between boiler and turbine.

A similar loss is the pressure drop in the boiler. Because of this pressure drop, the water entering the boiler must be pumped to a much higher pressure than the desired steam pressure leaving the boiler, and this requires additional pump work.

Turbine Losses

The losses in the turbine are primarily those associated with the flow of the working fluid through the turbine. Heat transfer to the surroundings also represents a loss, but this is usually of secondary importance. The effects of these two losses are the same as those outlined for piping losses, and the process might be as represented in Fig. 9.12, where 4_s represents the state after an isentropic expansion and state 4 represents the actual state leaving the turbine. The governing procedures may also cause a loss in the turbine, particularly if a throttling process is used to govern the turbine.

The isentropic efficiency of the turbine has been defined (in Chapter 7) as

$$\eta_t = \frac{w_t}{h_3 - h_{4s}}$$

where the states are as designated in Fig. 9.12.

Pump Losses

The losses in the pump are similar to those of the turbine and are primarily due to the irreversibilities associated with the fluid flow. Heat transfer is usually a minor loss.

The pump efficiency is defined as

$$\eta_p = \frac{h_{2s} - h_1}{w_p}$$

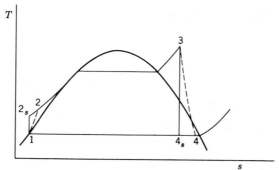

FIGURE 9.12

Temperature–entropy diagram showing effect of turbine and pump inefficiencies on cycle performance.

where the states are as shown in Fig. 9.12, and w_p is the actual work input per unit mass of fluid.

Condenser Losses

The losses in the condenser are relatively small. One of these minor losses is the cooling below the saturation temperature of the liquid leaving the condenser. This represents a loss because additional heat transfer is necessary to bring the water to its saturation temperature.

The influence of these losses on the cycle is illustrated in the following example, which should be compared to Example 9.2.

Example 9.5

A steam power plant operates on a cycle with pressure and temperatures as designated in Fig. 9.13. The efficiency of the turbine is 86 per cent and the efficiency of the pump is 80 per cent. Determine the thermal efficiency of this cycle.

As in previous examples, for each control volume the model used is the steam tables, and each process is SSSF with no changes in kinetic or potential energy. This cycle is shown on the T-s diagram of Fig. 9.14.

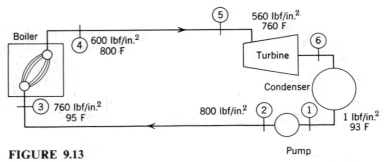

FIGURE 9.13
Schematic diagram for Example 9.5.

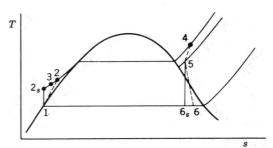

FIGURE 9.14
Temperature–entropy diagram for Example 9.5.

Control volume: Turbine.
Inlet state: P_5, T_5 known; state fixed.
Exit state: P_6 known.
Analysis:

First law: $\qquad\qquad w_t = h_5 - h_6$

Second law: $\qquad\qquad s_{6s} = s_5$

$$\eta_t = \frac{w_t}{h_5 - h_{6s}} = \frac{h_5 - h_6}{h_5 - h_{6s}}$$

Solution

From the steam tables,

$$h_5 = 1386.8 \quad s_5 = 1.6248$$
$$s_{6s} = s_5 = 1.6248 = 1.9779 - (1 - x)_{6s}1.8453$$
$$(1 - x)_{6s} = \frac{0.3531}{1.8453} = 0.1912$$
$$h_{6s} = 1105.8 - 0.1912(1036.0) = 907.6$$
$$w_t = \eta_t(h_5 - h_{6s}) = 0.86(1386.8 - 907.6)$$
$$= 0.86(479.2) = 412.1 \text{ Btu/lbm}$$

Control volume: Pump.
Inlet state: P_1, T_1 known; state fixed.
Exit state: P_2 known.
Analysis:

First law: $\qquad\qquad w_p = h_2 - h_1$

Second law: $\qquad\qquad s_{2s} = s_1$

$$\eta_p = \frac{h_{2s} - h_1}{w_p} = \frac{h_{2s} - h_1}{h_2 - h_1}$$

Since $s_{2s} = s_1$,

$$h_{2s} - h_1 = v(P_2 - P_1)$$

Therefore,

$$w_p = \frac{h_{2s} - h_1}{\eta_p} = \frac{v(P_2 - P_1)}{\eta_p}$$

Solution

$$w_p = \frac{v(P_2 - P_1)}{\eta_p} = \frac{0.016\ 15(800 - 1)144}{0.8 \times 778} = 3.0 \text{ Btu/lbm}$$

Therefore,

$$w_{\text{net}} = w_t - w_p = 412.1 - 3.0 = 409.1 \text{ Btu/lbm}$$

Control volume: Boiler.
Inlet state: P_3, T_3 known; state fixed.
Exit state: P_4, T_4 known; state fixed.
Analysis:
First law: $$q_H = h_4 - h_3$$

Solution

$$q_H = h_4 - h_3 = 1407.6 - 65.1 = 1342.5 \text{ Btu/lbm}$$

$$\eta_{th} = \frac{409.1}{1342.5} = 30.4\%$$

This compares to an efficiency of 36.9% for the Rankine efficiency of the similar cycle of Example 9.2.

VAPOR REFRIGERATION CYCLES

9.6 VAPOR-COMPRESSION REFRIGERATION CYCLES

The ideal cycle for vapor-compression refrigeration is shown in Fig. 9.15 as cycle 1–2–3–4–1. Saturated vapor at low pressure enters the compressor and undergoes a reversible adiabatic compression, 1–2. Heat is then rejected at constant pressure in process 2–3, and the working fluid leaves the condenser as saturated liquid. An adiabatic throttling process follows, process 3–4, and the working fluid is then evaporated at constant pressure, process 4–1, to complete the cycle.

The similarity between this cycle and the Rankine cycle is evident, for it is essentially the same cycle in reverse, except that an expansion valve replaces the pump. This throttling process is irreversible, whereas the pumping process of the Rankine cycle is reversible. The deviation of this ideal cycle from the Carnot cycle 1'–2'–3–4'–1' is evident from the *T-s* diagram. The reason for the deviation is that it is much more expedient to have a compressor handle only vapor than a mixture of liquid and vapor as would be required in process 1'–2' of the Carnot

FIGURE 9.15
The ideal vapor-compression refrigeration cycle.

cycle. It is virtually impossible to compress (at a reasonable rate) a mixture such as that represented by state 1′ and maintain equilibrium between the liquid and vapor, because there must be a heat and mass transfer across the phase boundary. It is also much simpler to have the expansion process take place irreversibly through an expansion valve than to have an expansion device that receives saturated liquid and discharges a mixture of liquid and vapor, as would be required in process 3–4′. For these reasons the ideal cycle for vapor-compression refrigeration is as shown in Fig. 9.15 by cycle 1–2–3–4–1.

As we saw in Chapter 6, the performance of a refrigeration cycle is given in terms of the coefficient of performance, β, which is defined for a refrigeration cycle as

$$\beta = \frac{q_L}{w_c} \tag{9.2}$$

If the cycle is to be used as a heat pump, then the desired energy quantity is q_H instead of q_L, and that quantity is used in the numerator of the expression for coefficient of performance. It is important to note whether any given coefficient of performance refers to the use of the machine as a refrigerator or as a heat pump.

Example 9.6

Consider an ideal refrigeration cycle that utilizes Freon-12 as the working fluid. The temperature of the refrigerant in the evaporator is $-20°C$ and in the condenser it is $40°C$. The refrigerant is circulated at the rate of 0.03 kg/s. Determine the coefficient of performance and the capacity of the plant in rate of refrigeration.

The diagram for this example is as shown in Fig. 9.15. For each control volume analyzed, the thermodynamic model is the Freon-12 tables, and each process is SSSF with no changes in kinetic or potential energy.

Control volume: Compressor.

Inlet state: T_1 known, saturated vapor; state fixed.

Exit state: P_2 known (saturation pressure at T_3).

Analysis:

First law: $w_c = h_2 - h_1$

Second law: $s_2 = s_1$

Solution

At $T_3 = 40°C$,

$$P_g = P_2 = 0.9607 \, \text{MPa}$$

From the Freon-12 tables,

$$h_1 = 178.61, \qquad s_1 = 0.7082$$

Therefore,

$$s_2 = s_1 = 0.7082$$

so that

$$T_2 = 50.8°C \quad \text{and} \quad h_2 = 211.38$$
$$w_c = h_2 - h_1 = 211.38 - 178.61 = 32.77 \text{ kJ/kg}$$

Control volume: Expansion valve.

Inlet state: T_3 known, saturated liquid; state fixed.

Exit state: T_4 known.

Analysis:

First law: $$h_3 = h_4$$

Solution

$$h_4 = h_3 = 74.53$$

Control volume: Evaporator.

Inlet state: State 4 known (above).

Exit state: State 1 known (above).

Analysis:

First law: $$q_L = h_1 - h_4$$

Solution

$$q_L = h_1 - h_4 = 178.61 - 74.53 = 104.08 \text{ kJ/kg}$$

Therefore,

$$\beta = \frac{q_L}{w_c} = \frac{104.08}{32.77} = 3.18$$
$$\text{Capacity} = 104.08 \times 0.03 = 3.12 \text{ kW}$$

9.7 WORKING FLUIDS FOR VAPOR-COMPRESSION REFRIGERATION SYSTEMS

A much larger number of different working fluids (refrigerants) are utilized in vapor-compression refrigeration systems than in vapor power cycles. Ammonia and sulfur dioxide were important in the early days of vapor-compression refrigeration. Today, however, the main refrigerants are the halogenated hydrocarbons, which are marketed under the trade names of Freon and Genatron. For example, dichlorodifluoromethane (CCl_2F_2) is known as Freon-12 and Genatron-12. Two important considerations in selecting a refrigerant are the temperature at which refrigeration is desired (or heating in the case of a heat pump) and the type of equipment to be used.

Since the refrigerant undergoes a change of phase during the heat transfer process, the pressure of the refrigerant will be the saturation pressure during the

heat-supply and heat-rejection processes. Low pressures mean large specific volumes and correspondingly large equipment. High pressures mean smaller equipment but it must be designed to withstand higher pressure. In particular, the pressures should be well below the critical pressure. For extremely low temperature applications a binary fluid system may be used by cascading two separate systems, or a single-fluid system may be operated using a two-stage compressor, in which only part of the working fluid is expanded to the lowest-temperature portion of the refrigeration cycle.

The type of compressor used has a particular bearing on the refrigerant. Reciprocating compressors are best adapted to low specific volumes, which means higher pressures, whereas centrifugal compressors are most suitable for low pressures and high specific volumes.

It is also important that the refrigerants used in domestic appliances be nontoxic. Other important characteristics are tendency to cause corrosion, miscibility with compressor oil, dielectric strength, stability, and cost. Also, for given temperatures during evaporation and condensation, not all refrigerants have the same coefficient of performance for the ideal cycle. It is, of course, desirable to utilize the refrigerant with the highest coefficient of performance, other factors permitting.

9.8 DEVIATION OF THE ACTUAL VAPOR-COMPRESSION REFRIGERATION CYCLE FROM THE IDEAL CYCLE

The actual refrigeration cycle deviates from the ideal cycle primarily because of pressure drops associated with fluid flow and heat transfer to or from the surroundings. The actual cycle might approach the one shown in Fig. 9.16.

The vapor entering the compressor will probably be superheated. During the compression process there are irreversibilities and heat transfer either to or from the surroundings, depending on the temperature of the refrigerant and the surroundings. Therefore, the entropy might increase or decrease during this process, for the irreversibility and heat transfer to the refrigerant cause an increase in entropy, and heat transfer from the refrigerant causes a decrease in entropy. These

FIGURE 9.16

The actual vapor-compression refrigeration cycle.

possibilities are represented by the two dotted lines 1–2 and 1–2′. The pressure of the liquid leaving the condenser will be less than the pressure of the vapor entering, and the temperature of the refrigerant in the condenser will be somewhat above that of the surroundings to which heat is being transferred. Usually the temperature of the liquid leaving the condenser is lower than the saturation temperature, and it might drop somewhat more in the piping between the condenser and expansion valve. This represents a gain, however, because as a result of this heat transfer the refrigerant enters the evaporator with a lower enthalpy, thus permitting more heat transfer to the refrigerant in the evaporator.

There is some drop in pressure as the refrigerant flows through the evaporator. It may be slightly superheated as it leaves the evaporator, and due to heat transfer from the surroundings the temperature will increase in the piping between the evaporator and compressor. This heat transfer represents a loss, because it increases the work of the compressor as a result of the increased specific volume of the fluid entering it.

Example 9.7

A refrigeration cycle utilizes Freon-12 as the working fluid. Following are the properties at various points of the cycle designated in Fig. 9.16.

$$
\begin{aligned}
P_1 &= 125 \text{ kPa} & T_1 &= -10°C \\
P_2 &= 1.2 \text{ MPa} & T_2 &= 100°C \\
P_3 &= 1.19 \text{ MPa} & T_3 &= 80°C \\
P_4 &= 1.16 \text{ MPa} & T_4 &= 45°C \\
P_5 &= 1.15 \text{ MPa} & T_5 &= 40°C \\
P_6 &= P_7 = 140 \text{ kPa} & x_6 &= x_7 \\
P_8 &= 130 \text{ kPa} & T_8 &= -20°C
\end{aligned}
$$

The heat transfer from the Freon-12 during the compression process is 4 kJ/kg. Determine the coefficient of performance of this cycle.

For each control volume, the model is the Freon-12 tables, and each process is SSSF with no changes in kinetic or potential energy.

Control volume: Compressor.

Inlet state: P_1, T_1 known; state fixed.

Exit state: P_2, T_2 known; state fixed.

Analysis:

First law:
$$
q + h_1 = h_2 + w
$$
$$
w_c = -w = h_2 - h_1 - q
$$

Solution

From the Freon-12 tables,

$$
h_1 = 185.16, \qquad h_2 = 245.52
$$

Therefore,

$$w_c = 245.52 - 185.16 - (-4) = 64.36 \text{ kJ/kg}$$

Control volume: Throttling valve plus line.
Inlet state: P_5, T_5 known; state fixed.
Exit state: $P_7 = P_6$ known, $x_7 = x_6$.
Analysis:
 First law: $\qquad\qquad\qquad\qquad h_5 = h_6$
Since $x_7 = x_6$, it follows that $h_7 = h_6$.

Solution

$$h_5 = h_6 = h_7 = 74.53$$

Control volume: Evaporator.
Inlet state: P_7, h_7 known (above).
Exit state: P_8, T_8 known; state fixed.
Analysis:
 First law: $\qquad\qquad\qquad\qquad q_L = h_8 - h_7$

Solution

$$q_L = h_8 - h_7 = 179.12 - 74.53 = 104.59 \text{ kJ/kg}$$

Therefore,

$$\beta = \frac{q_L}{w_c} = \frac{104.59}{64.36} = 1.625$$

9.9 THE AMMONIA-ABSORPTION REFRIGERATION CYCLE

The ammonia-absorption refrigeration cycle differs from the vapor-compression cycle in the manner in which compression is achieved. In the absorption cycle the low-pressure ammonia vapor is absorbed in water and the liquid solution is pumped to a high pressure by a liquid pump. Figure 9.17 shows a schematic arrangement of the essential elements of such a system.

The low-pressure ammonia vapor leaving the evaporator enters the absorber where it is absorbed in the weak ammonia solution. This process takes place at a temperature slightly above that of the surroundings and heat must be transferred to the surroundings during this process. The strong ammonia solution is then pumped through a heat exchanger to the generator where a higher pressure

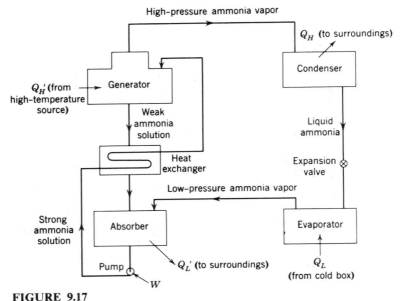

FIGURE 9.17

The ammonia-absorption refrigeration cycle.

and temperature are maintained. Under these conditions ammonia vapor is driven from the solution as a result of heat transfer from a high-temperature source. The ammonia vapor goes to the condenser where it is condensed, as in a vapor-compression system, and then to the expansion valve and evaporator. The weak ammonia solution is returned to the absorber through the heat exchanger.

The distinctive feature of the absorption system is that very little work input is required because the pumping process involves a liquid. This follows from the fact that for a reversible steady-flow process with negligible changes in kinetic and potential energy, the work is equal to $-\int v\,dP$, and the specific volume of the liquid is much less than the specific volume of the vapor. On the other hand, a relatively high temperature source of heat must be available (100 to 200°C). The equipment involved in an absorption system is somewhat greater than in a vapor-compression system and it can usually be economically justified only in those cases where a suitable source of heat is available that would otherwise be wasted. In recent years, the absorption cycle has been given increased attention in connection with alternate energy sources, such as, for example, solar or geothermal energy supplies.

This cycle brings out the important principle that since the work in a reversible steady-flow process with negligible changes in kinetic and potential energy is $-\int v\,dP$, a compression process should take place with the smallest possible specific volume.

AIR-STANDARD POWER CYCLES

9.10 AIR-STANDARD CYCLES

Many work-producing devices (engines) utilize a working fluid that is always a gas. The spark-ignition automotive engine is a familiar example, and the same is true of the Diesel engine and conventional gas turbine. In all of these engines there is a change in the composition of the working fluid, because during combustion it changes from air and fuel to combustion products. For this reason these engines are called internal-combustion engines. In contrast to this the steam power plant may be called an external-combustion engine, because heat is transferred from the products of combustion to the working fluid. External-combustion engines using a gaseous working fluid (usually air) have been built. To date they have had very limited application, but the use of the gas turbine cycle in conjunction with a nuclear reactor has been investigated extensively. Other external-combustion engines are currently receiving serious attention in an effort to combat the air pollution problem.

Because the working fluid does not go through a complete thermodynamic cycle in the engine (even though the engine operates in a mechanical cycle) the internal-combustion engine operates on the so-called open cycle. However, in order to analyze internal-combustion engines, it is advantageous to devise closed cycles that closely approximate the open cycles. One such approach is the air-standard cycle, which is based on the following assumptions:

1. A fixed mass of air is the working fluid throughout the entire cycle, and the air is always an ideal gas. Thus there is no inlet process or exhaust process.

2. The combustion process is replaced by a heat transfer process from an external source.

3. The cycle is completed by heat transfer to the surroundings (in contrast to the exhaust and intake process of an actual engine).

4. All processes are internally reversible.

5. The additional assumption is usually made that air has a constant specific heat.

The main value of the air-standard cycle is to enable us to examine qualitatively the influence of a number of variables on performance. The results obtained from the air-standard cycle, such as efficiency and mean effective pressure, will differ a great deal from those of the actual engine. The emphasis, therefore, in our consideration of the air-standard cycle will be primarily on the qualitative aspects.

The term "mean effective pressure" (mep), which is used in conjunction with reciprocating engines, is defined as the pressure that, if it acted on the piston during the entire power stroke, would do an amount of work equal to that actually done on the piston. The work for one cycle is found by multiplying this mean effective pressure by the area of the piston (minus the area of the rod on the crank end of a double-acting engine) and by the stroke.

9.11 THE AIR-STANDARD CARNOT CYCLE

The air-standard Carnot cycle is shown on the P-v and T-s diagrams of Fig. 9.18. Such a cycle could be achieved in either a reciprocating or steady-flow device, as shown in the same figure.

In Chapter 6 it was noted that the efficiency of a Carnot cycle depends only on the temperatures at which heat is supplied and rejected, and is given by the relation

$$\eta_{th} = 1 - \frac{T_L}{T_H} = 1 - \frac{T_4}{T_1} = 1 - \frac{T_3}{T_2}$$

where the subscripts refer to Fig. 9.18. The efficiency may also be expressed by the pressure ratio or compression ratio during the isentropic processes. This follows from the fact that

$$\text{Isentropic pressure ratio} = r_{ps} = \frac{P_1}{P_4} = \frac{P_2}{P_3} = \left(\frac{T_3}{T_2}\right)^{k/(1-k)}$$

$$\text{Isentropic compression ratio} = r_{vs} = \frac{V_4}{V_1} = \frac{V_3}{V_2} = \left(\frac{T_3}{T_2}\right)^{1/(1-k)}$$

Therefore

$$\eta_{th} = 1 - r_{ps}^{(1-k)/k} = 1 - r_{vs}^{1-k} \tag{9.3}$$

One other important variable in the air-standard Carnot cycle is the amount of heat transferred to the working fluid per cycle. It is evident from Fig. 9.18 that

FIGURE 9.18
The air-standard Carnot cycle.

increasing the heat transfer per cycle at a given T_H causes a larger change in volume during the cycle, and this causes a lower mean effective pressure in a reciprocating engine. In fact, for air-standard Carnot cycles having a minimum pressure between 0.1 and 1 MPa and a reasonable heat transfer per cycle, the mean effective pressure is so low that it would scarcely overcome friction forces.

Another practical difficulty of the Carnot cycle, which applies to both the reciprocating and steady-flow types of cycle, is the difficulty of transferring heat during the isothermal expansion and compression processes. It is virtually impossible to even approach this in an actual machine operating at a reasonable rate of speed. Thus the air-standard Carnot cycle is not practical. Nonetheless it is of value as a standard for comparison with other cycles, and in a later paragraph we will consider the ideal gas turbine cycle with intercooling and reheating and note how its efficiency approaches that of the corresponding Carnot cycle.

Example 9.8

In an air-standard Carnot cycle heat is transferred to the working fluid at 2000 R, and heat is rejected at 500 R. The heat transfer to the working fluid at 2000 R is 50 Btu/lbm. The minimum pressure in the cycle is 1 atm. Assuming constant specific heat of air, determine the cycle efficiency and the mean effective pressure.

System: Carnot heat engine.

Diagram: Fig. 9.18.

State information:

$$T_H = T_1 = T_2 = 2000 \text{ R}$$
$$T_L = T_3 = T_4 = 500 \text{ R}$$
$$P_3 = 14.7 \text{ lbf/in.}^2$$

Process information: Four processes known (Fig. 9.18). Also,

$$q_H = {}_1q_2 = 50 \text{ Btu/lbm}$$

Model: Ideal gas, constant specific heat, value at 80 F.

Analysis: For a Carnot cycle,

$$\eta_{th} = 1 - \frac{T_L}{T_H}$$

Second law and property relation for process 1–2:

$${}_1q_2 = T_1(s_2 - s_1) = RT_1 \ln \frac{V_2}{V_1} = RT_1 \ln \frac{P_1}{P_2}$$

Second law for process 2–3:

$$s_2 = s_3$$

so that

$$\frac{T_2}{T_3} = \left(\frac{P_2}{P_3}\right)^{(k-1)/k}$$

Also,

$$\text{mep}(v_3 - v_1) = w_{\text{net}} = \eta_{\text{th}} \times {}_1q_2$$

Solution

$$\eta_{\text{th}} = 1 - \frac{T_L}{T_H} = 1 - \frac{500}{2000} = 0.75$$

$$\frac{T_2}{T_3} = \left(\frac{P_2}{P_3}\right)^{(k-1)/k} = 4$$

$$\frac{P_2}{P_3} = 128$$

$$P_2 = 14.7(128) = 1882 \text{ lbf/in.}^2$$

$${}_1q_2 = RT_1 \ln \frac{V_2}{V_1} = RT_1 \ln \frac{P_1}{P_2} = \frac{53.34 \times 2000}{778} \ln \frac{P_1}{P_2} = 50$$

Therefore,

$$\frac{P_1}{P_2} = 1.44 \qquad P_1 = 1882(1.44) = 2710 \text{ lbf/in.}^2$$

$$\text{mep}(v_3 - v_1) = w_{\text{net}} = \eta_{\text{th}} \times {}_1q_2 = 0.75(50) = 37.5 \text{ Btu/lbm}$$

$$v_3 = \frac{RT_3}{P_3} = \frac{53.34 \times 500}{14.7 \times 144} = 12.6 \text{ ft}^3/\text{lbm}$$

$$v_1 = \frac{RT_1}{P_1} = \frac{53.34 \times 2000}{2710 \times 144} = 0.273 \text{ ft}^3/\text{lbm}$$

$$\text{mep} = \frac{37.5 \times 778}{(12.6 - 0.273)} = 2360 \text{ lbf/ft}^2 = 16.4 \text{ lbf/in.}^2$$

This is a much lower mean effective pressure than can be effectively utilized in a reciprocating engine.

9.12 THE AIR-STANDARD OTTO CYCLE

The air-standard Otto cycle is an ideal cycle that approximates a spark-ignition internal-combustion engine. This cycle is shown on the *P-v* and *T-s* diagrams of Fig. 9.19. Process 1–2 is an isentropic compression of the air as the piston moves from crank-end dead center to head-end dead center. Heat is then added at constant volume while the piston is momentarily at rest at head-end dead center. (This process corresponds to the ignition of the fuel–air mixture by the spark and the subsequent burning in the actual engine.) Process 3–4 is an isentropic expansion, and process 4–1 is the rejection of heat from the air while the piston is at crank-end dead center.

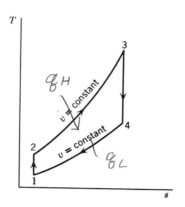

FIGURE 9.19
The air-standard Otto cycle.

The thermal efficiency of this cycle is found as follows, assuming constant specific heat of air.

$$\eta_{th} = \frac{Q_H - Q_L}{Q_H} = \left(1 - \frac{Q_L}{Q_H}\right) = 1 - \frac{mC_v(T_4 - T_1)}{mC_v(T_3 - T_2)}$$

$$= 1 - \frac{T_1}{T_2} \frac{(T_4/T_1 - 1)}{(T_3/T_2 - 1)}$$

We note further that

$$\frac{T_2}{T_1} = \left(\frac{V_1}{V_2}\right)^{k-1} = \left(\frac{V_4}{V_3}\right)^{k-1} = \frac{T_3}{T_4}$$

Therefore,

$$\frac{T_3}{T_2} = \frac{T_4}{T_1}$$

and

$$\eta_{th} = 1 - \frac{T_1}{T_2} = 1 - (r_v)^{1-k} = 1 - \frac{1}{r_v^{k-1}} \tag{9.4}$$

where

$$r_v = \text{compression ratio} = \frac{V_1}{V_2} = \frac{V_4}{V_3}$$

The important thing to note is that the efficiency of the air-standard Otto cycle is a function only of the compression ratio, and that the efficiency is increased

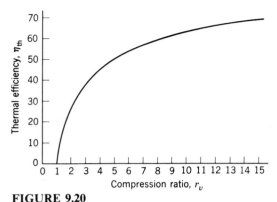

FIGURE 9.20

Thermal efficiency of the Otto cycle as a function of compression ratio.

by increasing the compression ratio. Figure 9.20 is a plot of the air-standard cycle thermal efficiency versus compression ratio. It is also true of an actual spark-ignition engine that the efficiency can be increased by increasing the compression ratio. The trend toward higher compression ratios is prompted by the effort to obtain higher thermal efficiency. In the actual engine there is an increased tendency toward detonation of the fuel as compression ratio is increased. Detonation is characterized by an extremely rapid burning of the fuel and strong pressure waves present in the engine cylinder, which give rise to the so-called spark knock. Therefore, the maximum compression ratio that can be used is fixed by the fact that detonation must be avoided. The advance in compression ratios over the years in the actual engine has been made possible by developing fuels with better antiknock characteristics, primarily through the addition of tetraethyl lead. More recently, however, nonleaded gasolines with fairly good antiknock characteristics have been developed in an effort to reduce atmospheric contamination.

Some of the most important ways in which the actual open-cycle spark-ignition engine deviates from the air-standard cycle are as follows:

1. The specific heats of the actual gases increase with an increase in temperature.
2. The combustion process replaces the heat transfer process at high temperature, and combustion may be incomplete.
3. Each mechanical cycle of the engine involves an inlet and an exhaust process and, because of the pressure drop through the valves, a certain amount of work is required to charge the cylinder with air and exhaust the products of combustion.
4. There will be considerable heat transfer between the gases in the cylinder and the cylinder walls.

9.55

$P_1 = 100$ kPa, $T_1 = 300$K

$P_2/P_1 = 5.0$

$P_7 = 100$ kPa $= P_1$

$T_4 = 1300$ K

assume $c_p = \overline{c}_{po}$
constant specific heats

a) $W_{turbine\ \#1} = -W_{compressor}$

1st law around turbine #1: $W_T = h_4 - h_5 = \overline{c}_{po}(T_4 - T_5)$

around compressor: $W_C = (h_1 - h_2) = \overline{c}_p(T_1 - T_2)$

① → ② isentropic compression

$$T_2 = T_1\left(\frac{P_2}{P_1}\right)^{\frac{k-1}{k}} = (300K)(5)^{\frac{1.4+0}{1.4}} = 475.4K$$

$\therefore \overline{c}_p(T_4 - T_5) = -\overline{c}_{po}(T_1 - T_2)$ or $(1300 - T_5) = -(300 - 475.4)$

$T_5 = 1124.6$ K

④ → ⑤ isentropic ~~compression~~ expansion

$$(T_5/T_4) = (P_5/P_4)^{k-1/k} \implies P_5 = P_4\left(\frac{T_5}{T_4}\right)^{\frac{k}{k-1}}, \text{ with } P_4 = P$$

$$P_5 = (500\ kPa)\left(\frac{1124.6}{1300}\right)^{\frac{1.4}{1.4-1}} = 301.1\ kPa$$

b) net work output as work of turbine #2

1st law $W_T = h_5 - h_6 = \overline{c}_{po}(T_5 - T_6)$

Process ⑤ → ⑥ isentropic compression

$$T_6 = T_5\left(\frac{P_6}{P_5}\right)^{\frac{k-1}{k}}, \text{ with } P_6 = P_7 = P_1 = 100\ kPa$$

$$= (1124.6)\left(\frac{100}{301.1}\right)^{\frac{1.4-1.0}{1.4}} = 820.5\ K$$

$$W_T = (1.0035\ \tfrac{kJ}{kgK})(1124.6 - 820.5)K = 305.1\ \tfrac{kJ}{kg}$$

$$\dot{m} = \frac{\dot{W}}{W} = \frac{100\ kW}{305.1\ kJ/kg} = 0.328\ kg/sec$$

c) Ideal regenerator $T_6 = T_3$, $T_7 = T_2$

$$\eta = \frac{W_{net}}{q_H}, \text{ and } q_H = h_4 - h_3 = \overline{c}_p(T_4 - T_3) = (1.0035)(1300 - 820.5)$$
$$= 481.2$$

$$\eta = \frac{305.1}{481.2} = 63.4\%$$

5. There will be irreversibilities associated with pressure and temperature gradients.

Example 9.9

The compression ratio in an air-standard Otto cycle is 8. At the beginning of the compression stroke the pressure is 14.7 lbf/in.2 and the temperature is 60 F. The heat transfer to the air per cycle is 800 Btu/lbm air. Determine:

1. The pressure and temperature at the end of each process of the cycle.
2. The thermal efficiency.
3. The mean effective pressure.

System: Air inside cylinder.

Diagram: Fig. 9.19.

State information: $P_1 = 14.7$ lbf/in.2 $T_1 = 520$ R

Process information: Four processes known (Fig. 9.19). Also, $r_v = 8$ and $q_H = 800$ Btu/lbm.

Model: Ideal gas, constant specific heat, value at 80 F.

Analysis:

Second law for compression process 1–2:

$$s_2 = s_1$$

so that

$$\frac{T_2}{T_1} = \left(\frac{V_1}{V_2}\right)^{k-1}$$

$$\frac{P_2}{P_1} = \left(\frac{V_1}{V_2}\right)^{k}$$

First law for heat addition process 2–3:

$$q_H = {}_2q_3 = u_3 - u_2 = C_v(T_3 - T_2)$$

Second law for expansion process 3–4:

$$s_4 = s_3$$

so that

$$\frac{T_3}{T_4} = \left(\frac{V_4}{V_3}\right)^{k-1}$$

$$\frac{P_3}{P_4} = \left(\frac{V_4}{V_3}\right)^{k}$$

Also,

$$\eta_{th} = 1 - \frac{1}{r_v^{k-1}}, \qquad \text{mep} = \frac{w_{net}}{(v_1 - v_2)}$$

Solution

$$v_1 = \frac{53.34 \times 520}{14.7 \times 144} = 13.08 \text{ ft}^3/\text{lbm} \qquad Pv = RT$$

$$\frac{T_2}{T_1} = \left(\frac{V_1}{V_2}\right)^{k-1} = 8^{0.4} = 2.3 \qquad T_2 = 2.3(520) = 1197 \text{ R}$$

$$\frac{P_2}{P_1} = \left(\frac{V_1}{V_2}\right)^k = 8^{1.4} = 18.4 \qquad P_2 = 18.4(14.7) = 270.3 \text{ lbf/in.}^2$$

$$v_2 = \frac{13.08}{8} = 1.637 \text{ ft}^3/\text{lbm}$$

$$_2q_3 = C_v(T_3 - T_2) = 800 \text{ Btu/lbm}$$

$$T_3 - T_2 = \frac{800}{0.171} = 4690 \text{ R} \qquad T_3 = 1197 + 4690 = 5887 \text{ R}$$

$$\frac{T_3}{T_2} = \frac{P_3}{P_2} = \frac{5887}{1197} = 4.92 \qquad P_3 = 4.92(270.3) = 1331 \text{ lbf/in.}^2$$

$$\frac{T_3}{T_4} = \left(\frac{V_4}{V_3}\right)^{k-1} = 8^{0.4} = 2.3 \qquad T_4 = \frac{5887}{2.3} = 2558 \text{ R}$$

$$\frac{P_3}{P_4} = \left(\frac{V_4}{V_3}\right)^k = 8^{1.4} = 18.4 \qquad P_4 = \frac{1331}{18.4} = 73.0 \text{ lbf/in.}^2$$

$$\eta_{th} = 1 - \frac{1}{r_v^{k-1}} = 1 - \frac{1}{8^{0.4}} = 1 - \frac{1}{2.3} = 1 - 0.435 = 0.565$$

This can be checked by finding the heat rejected.

$$_4q_1 = C_v(T_1 - T_4) = 0.171(520 - 2558) = -348 \text{ Btu/lbm}$$

$$\eta_{th} = 1 - \frac{348}{800} = 1 - 0.435 = 0.565$$

$$w_{net} = 800 - 348 = 452 \text{ Btu/lbm} = (v_1 - v_2) \text{ mep}$$

$$\text{mep} = \frac{452 \times 778}{(13.08 - 1.637)144} = 213.5 \text{ lbf/in.}^2$$

Note how much higher this mean effective pressure is than for the Carnot cycle of Example 9.8. A low mean effective pressure means a large piston displacement for a given power output, and a large piston displacement means high frictional losses in an actual engine. In fact, the mean effective pressure of the Carnot cycle of Example 9.8 would scarcely overcome the friction in an actual engine.

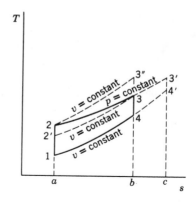

FIGURE 9.21

The air-standard Diesel cycle.

9.13 THE AIR-STANDARD DIESEL CYCLE

The air-standard Diesel cycle is shown in Fig. 9.21. This is the ideal cycle for the Diesel engine, which is also called the compression-ignition engine.

In this cycle the heat is transferred to the working fluid at constant pressure. This process corresponds to the injection and burning of the fuel in the actual engine. Since the gas is expanding during the heat addition in the air-standard cycle, the heat transfer must be just sufficient to maintain constant pressure. When state 3 is reached the heat addition ceases and the gas undergoes an isentropic expansion, process 3–4, until the piston reaches crank-end dead center. As in the air-standard Otto cycle, a constant-volume rejection of heat at crank-end dead center replaces the exhaust and intake processes of the actual engine.

The efficiency of the Diesel cycle is given by the relation

$$\eta_{th} = 1 - \frac{Q_L}{Q_H} = 1 - \frac{C_v(T_4 - T_1)}{C_p(T_3 - T_2)} = 1 - \frac{T_1(T_4/T_1 - 1)}{kT_2(T_3/T_2 - 1)} \tag{9.5}$$

It is important to note that the isentropic compression ratio is greater than the isentropic expansion ratio in the Diesel cycle. Also, for a given state before compression and a given compression ratio (that is, given states 1 and 2), the cycle efficiency decreases as the maximum temperature increases. This is evident from the *T-s* diagram, because the constant-pressure and constant-volume lines converge, and increasing the temperature from 3 to 3′ requires a large addition of heat (area 3–3′–c–b–3) and results in a relatively small increase in work (area 3–3′–4′–4–3).

There are a number of comparisons between the Otto cycle and the Diesel cycle, but here we will note only two. Consider Otto cycle 1–2–3″–4–1 and Diesel cycle 1–2–3–4–1, which have the same state at the beginning of the compression stroke and the same piston displacement and compression ratio. It is evident from the *T-s* diagram that the Otto cycle has the higher efficiency. In practice, however, the Diesel engine can operate on a higher compression ratio than the spark-ignition engine. The reason is that in the spark-ignition engine an air–fuel mixture is compressed, and detonation (spark knock) becomes a serious problem if too high a compression ratio is used. This problem does not exist in the Diesel engine because only air is compressed during the compression stroke.

Therefore, we might compare an Otto cycle with a Diesel cycle and in each case select a compression ratio that might be achieved in practice. Such a comparison can be made by considering Otto cycle 1–2′–3–4–1 and Diesel cycle 1–2–3–4–1. The maximum pressure and temperature is the same for both cycles, which means that the Otto cycle has a lower compression ratio than the Diesel cycle. It is evident from the *T-s* diagram that in this case the Diesel cycle has the higher efficiency. Thus the conclusions drawn from a comparison of these two cycles must always be related to the basis on which the comparison has been made.

The actual compression-ignition open cycle differs from the air-standard Diesel cycle in much the same way that the spark-ignition open cycle differs from the air-standard Otto cycle.

Example 9.10

An air-standard Diesel cycle has a compression ratio of 18, and the heat transferred to the working fluid per cycle is 1800 kJ/kg. At the beginning of the compression process the pressure is 0.1 MPa and the temperature is 15°C. Determine:

1. The pressure and temperature at each point in the cycle.
2. The thermal efficiency.
3. The mean effective pressure.

System: Air inside cylinder.

Diagram: Fig. 9.21.

State information: $P_1 = 0.1$ MPa, $T_1 = 288.2$ K.

Process information: Four processes known (Fig. 9.21). Also, $r_v = 18$ and $q_H = 1800$ kJ/kg.

Model: Ideal gas, constant specific heat, value at 300 K.

Analysis:

Second law for compression process 1–2:

$$s_2 = s_1$$

so that

$$\frac{T_2}{T_1} = \left(\frac{V_1}{V_2}\right)^{k-1}$$

$$\frac{P_2}{P_1} = \left(\frac{V_1}{V_2}\right)^{k}$$

First law for heat addition process 2–3:

$$q_H = {}_2q_3 = C_p(T_3 - T_2)$$

Second law for expansion process 3–4:

$$s_4 = s_3$$

so that

$$\frac{T_3}{T_4} = \left(\frac{V_4}{V_3}\right)^{k-1}$$

Also,

$$\eta_{th} = \frac{w_{net}}{q_H}, \qquad mep = \frac{w_{net}}{v_1 - v_2}$$

Solution

$$v_1 = \frac{0.287 \times 288.2}{100} = 0.827 \ m^3/kg$$

$$v_2 = \frac{v_1}{18} = \frac{0.827}{18} = 0.045\,95 \ m^3/kg$$

$$\frac{T_2}{T_1} = \left(\frac{V_1}{V_2}\right)^{k-1} = 18^{0.4} = 3.1777, \qquad T_2 = 915.8 \ K$$

$$\frac{P_2}{P_1} = \left(\frac{V_1}{V_2}\right)^{k} = 18^{1.4} = 57.2, \qquad P_2 = 5.72 \ MPa$$

$$q_H = {}_2q_3 = C_p(T_3 - T_2) = 1800 \ kJ/kg$$

$$T_3 - T_2 = \frac{1800}{1.0035} = 1794, \qquad T_3 = 2710 \ K$$

$$\frac{V_3}{V_2} = \frac{T_3}{T_2} = \frac{2710}{915.8} = 2.959, \qquad v_3 = 0.135\,98 \ m^3/kg$$

$$\frac{T_3}{T_4} = \left(\frac{V_4}{V_3}\right)^{k-1} = \left(\frac{0.827}{0.135\,98}\right)^{0.4} = 2.0588, \qquad T_4 = 1316 \ K$$

$$q_L = {}_4q_1 = C_v(T_1 - T_4) = 0.7165(288.2 - 1316) = -736.6 \ kJ/kg$$

$$w_{net} = 1800 - 736.6 = 1063.4 \ kJ/kg$$

$$\eta_{th} = \frac{w_{net}}{q_H} = \frac{1063.4}{1800} = 59.1\%$$

$$mep = \frac{w_{net}}{v_1 - v_2} = \frac{1063.4}{(0.827 - 0.045\,95)} = 1362 \ kPa$$

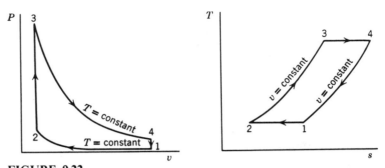

FIGURE 9.22
The air-standard Stirling cycle.

9.14 THE ERICSSON AND STIRLING CYCLES

We briefly consider these two cycles, not so much because of their extensive use, but rather because they serve to demonstrate how a regenerator can often be incorporated in a cycle to give a significant increase in efficiency. This principle finds extensive application in turbines as well as in certain reciprocating devices. The Stirling cycle is shown on the P-v and T-s diagrams of Fig. 9.22. Heat is transferred to the working fluid during the constant-volume process 2–3 and during the isothermal expansion process 3–4. Heat is rejected during the constant-volume process 4–1 and during the isothermal compression process 1–2. The significance of this cycle in conjunction with a regenerator is discussed in the next paragraph.

The Ericsson cycle is shown on the P-v and T-s diagrams of Fig. 9.23. This cycle differs from the Stirling cycle in that the constant-volume processes of the Stirling cycle are replaced by constant-pressure processes. In both cycles there is an isothermal compression and expansion.

The importance of both cycles is the possibility of including a regenerator; by so doing the air-standard Stirling and Ericsson cycles may have an efficiency equal to that of a Carnot cycle operating between the same temperatures. This may be

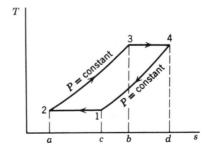

FIGURE 9.23
The air-standard Ericsson cycle.

FIGURE 9.24

Schematic arrangement of an engine operating on the Ericsson cycle and utilizing a regenerator.

demonstrated by considering Fig. 9.24, in which the Ericsson cycle is accomplished in a device that is essentially a gas turbine. If we assume an ideal heat-transfer process in the regenerator, that is, no pressure drop and an infinitesimal temperature difference between the two streams, and reversible compression and expansion processes then this device operates on the Ericsson cycle.

Note that the heat transfer to the gas between states 2 and 3, area 2–3–b–a–2, is exactly equal to the heat transfer from the gas between states 4 and 1, area 1–4–d–c–1. Thus Q_H all takes place in the isothermal turbine between states 3 and 4, and Q_L takes place in the isothermal compressor between states 1 and 2. Since all the heat is supplied and rejected isothermally, the efficiency of this cycle will equal the efficiency of a Carnot cycle operating between the same temperatures. A similar cycle could be developed that would approximate the Stirling cycle.

The difficulties in achieving such a cycle are primarily those associated with heat transfer. It is difficult to achieve an isothermal compression or expansion in a machine operating at a reasonable speed, and there will be pressure drops in the regenerator and a temperature difference between the two streams flowing through the regenerator. However, the gas turbine with intercooling and regenerators, which is described in Section 9.17, is a practical attempt to approach the Ericsson cycle. There have also been attempts to approach the Stirling cycle with the use of regenerators.

9.15 THE BRAYTON CYCLE

The air-standard Brayton cycle is the ideal cycle for the simple gas turbine. The simple open-cycle gas turbine utilizing an internal-combustion process and the simple closed-cycle gas turbine, which utilizes heat-transfer processes, are both shown schematically in Fig. 9.25. The air-standard Brayton cycle is shown on the P-v and T-s diagrams of Fig. 9.26.

The efficiency of the air-standard Brayton cycle is found as follows:

$$\eta_{th} = 1 - \frac{Q_L}{Q_H} = 1 - \frac{C_p(T_4 - T_1)}{C_p(T_3 - T_2)} = 1 - \frac{T_1(T_4/T_1 - 1)}{T_2(T_3/T_2 - 1)}$$

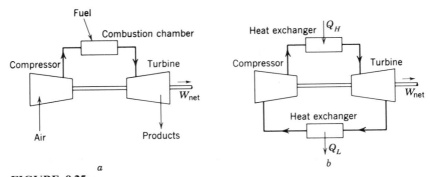

FIGURE 9.25

A gas turbine operating on the Brayton cycle. (a) Open cycle. (b) Closed cycle.

We note, however, that

$$\frac{P_3}{P_4} = \frac{P_2}{P_1}$$

$$\frac{P_2}{P_1} = \left(\frac{T_2}{T_1}\right)^{k/(k-1)} = \frac{P_3}{P_4} = \left(\frac{T_3}{T_4}\right)^{k/(k-1)}$$

$$\frac{T_3}{T_4} = \frac{T_2}{T_1} \quad \therefore \quad \frac{T_3}{T_2} = \frac{T_4}{T_1} \quad \text{and} \quad \frac{T_3}{T_2} - 1 = \frac{T_4}{T_1} - 1$$

$$\eta_{\text{th}} = 1 - \frac{T_1}{T_2} = 1 - \frac{1}{(P_2/P_1)^{(k-1)/k}} \tag{9.6}$$

The efficiency of the air-standard Brayton cycle is therefore a function of isentropic pressure ratio; Fig. 9.27 shows a plot of efficiency versus pressure ratio. The fact that efficiency increases with pressure ratio is evident from the T-s diagram of Fig. 9.26, because increasing the pressure ratio will change the cycle from 1–2–3–4–1 to 1–2′–3′–4–1. The latter cycle has a greater heat supply and the

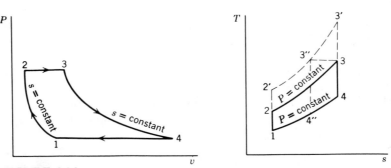

FIGURE 9.26

The air-standard Brayton cycle.

FIGURE 9.27

**Cycle efficiency as a function of pressure ratio
for the Brayton and regenerative cycles.**

same heat rejected as the orginal cycle, and therefore it has a greater efficiency. Note further that the latter cycle has a higher maximum temperature, T'_3, than the original cycle, T_3. In the actual gas turbine the maximum temperature of the gas entering the turbine is fixed by metallurgical considerations. Therefore, if we fix the temperature T_3 and increase the pressure ratio, the resulting cycle is $1-2'-3''-4''-1$. This cycle would have a higher efficiency than the orginal cycle, but the work per unit mass of working fluid is thereby changed.

With the advent of nuclear reactors the closed-cycle gas turbine has become more important. Heat is transferred, either directly or via a second fluid, from the fuel in the nuclear reactor to the working fluid in the gas turbine. Heat is rejected from the working fluid to the surroundings.

The actual gas-turbine engine differs from the ideal cycle primarily because of irreversibilities in the compressor and turbine, and because of pressure drop in the flow passages and combustion chamber (or in the heat exchanger of a closed-cycle turbine). Thus the state points in a simple open-cycle gas turbine might be as shown in Fig. 9.28.

The efficiencies of the compressor and turbine are defined in relation to isentropic processes. Designating the states as in Fig. 9.28, the definitions of isentropic compressor and turbine efficiencies are as follows:

$$\eta_{comp} = \frac{h_{2s} - h_1}{h_2 - h_1} \tag{9.7}$$

$$\eta_{turb} = \frac{h_3 - h_4}{h_3 - h_{4s}} \tag{9.8}$$

One other important feature of the Brayton cycle is the large amount of compressor work (also called back work) compared to the turbine work. Thus the compressor might require from 40 to 80% of the output of the turbine. This is particularly important when the actual cycle is considered, because the effect of the losses is to require a large amount of compression work from a smaller amount

FIGURE 9.28

Effect of inefficiencies on the gas-turbine cycle.

of turbine work, and thus the overall efficiency drops very rapidly with a decrease in the efficiencies of the compressor and turbine. In fact, if these efficiencies drop below about 60%, all the work of the turbine will be required to drive the compressor, and the overall efficiency will be zero. This is in sharp contrast to the Rankine cycle, where only 1 or 2% of the turbine work is required to drive the pump. The reason for this is that for a reversible steady-state, steady-flow process with negligible change in kinetic and potential energy, the work is equal to $-\int v\,dP$. Because we are pumping a liquid in the Rankine cycle, the specific volume is very low compared to the specific volume of the gas in a gas turbine. This matter is illustrated in the following examples.

Example 9.11

In an air-standard Brayton cycle the air enters the compressor at 0.1 MPa, 15°C. The pressure leaving the compressor is 1.0 MPa and the maximum temperature in the cycle is 1100°C. Determine:

1. The pressure and temperature at each point in the cycle.
2. The compressor work, turbine work, and cycle efficiency.

For each of the control volumes analyzed, the model is ideal gas with constant specific heat, value at 300 K, and each process is SSSF with no kinetic or potential energy changes. The diagram for this example is Fig. 9.26.

Control volume: Compressor.
Inlet state: P_1, T_1 known; state fixed.
Exit state: P_2 known.

Analysis:

First law: $w_c = h_2 - h_1$. (Note that the compressor work w_c is here defined as work input to the compressor.)

Second law:

$$s_2 = s_1$$

Therefore

$$\frac{T_2}{T_1} = \left(\frac{P_2}{P_1}\right)^{(k-1)/k}$$

Solution

$$\left(\frac{P_2}{P_1}\right)^{(k-1)/k} = 10^{0.286} = 1.932, \qquad T_2 = 556.8 \text{ K}$$

$$w_c = h_2 - h_1 = C_p(T_2 - T_1)$$
$$= 1.0035(556.8 - 288.2) = 269.5 \text{ kJ/kg}$$

Control volume: Turbine.

Inlet state: $P_3 \; (= P_2)$ known, T_3 known; state fixed.

Exit state: $P_4 \; (= P_1)$ known.

Analysis:

First law: $\qquad\qquad\qquad\quad w_t = h_3 - h_4$

Second law: $\qquad\qquad\qquad\quad s_3 = s_4$

Therefore,

$$\frac{T_3}{T_4} = \left(\frac{P_3}{P_4}\right)^{(k-1)/k}$$

Solution

$$\left(\frac{P_3}{P_4}\right)^{(k-1)/k} = 10^{0.286} = 1.932, \qquad T_4 = 710.8 \text{ K}$$

$$w_t = h_3 - h_4 = C_p(T_3 - T_4)$$
$$= 1.0035(1373.2 - 710.8) = 664.7 \text{ kJ/kg}$$
$$w_{net} = w_t - w_c = 664.7 - 269.5 = 395.2 \text{ kJ/kg}$$

Control volume: High-temperature heat exchanger.

Inlet state: State 2 fixed (above).

Exit state: State 3 fixed (above).

Analysis:

First law: $\qquad\qquad q_H = h_3 - h_2 = C_p(T_3 - T_2)$

Solution

$$q_H = h_3 - h_2 = C_p(T_3 - T_2) = 1.0035(1373.2 - 556.8) = 819.3 \text{ kJ/kg}$$

Control volume: Low-temperature heat exchanger.

Inlet state: State 4 fixed (above).

Exit state: State 1 fixed (above).

Analysis:

First law: $$q_L = h_4 - h_1 = C_p(T_4 - T_1)$$

Solution

$$q_L = h_4 - h_1 = C_p(T_4 - T_1) = 1.0035(710.8 - 288.2) = 424.1 \text{ kJ/kg}$$

Therefore,

$$\eta_{th} = \frac{w_{net}}{q_H} = \frac{395.2}{819.3} = 48.2\%$$

This may be checked by using Eq. 9.6.

$$\eta_{th} = 1 - \frac{1}{(P_2/P_1)^{(k-1)/k}} = 1 - \frac{1}{10^{0.286}} = 48.2\%$$

Example 9.12

Consider a gas turbine with air entering the compressor under the same conditions as in Example 9.11 and leaving at a pressure of 1.0 MPa. The maximum temperature is 1100°C. Assume a compressor efficiency of 80%, a turbine efficiency of 85%, and a pressure drop between the compressor and turbine of 15 kPa. Determine the compressor work, turbine work, and cycle efficiency.

As in the previous example, for each control volume the model is ideal gas, constant specific heat, value at 300 K, and each process is SSSF with no kinetic or potential energy changes. In this example, the diagram is Fig. 9.28.

Control volume: Compressor.

Inlet state: P_1, T_1 known; state fixed.

Exit state: P_2 known.

Analysis:

First law, real process: $w_c = h_2 - h_1$

Second law, ideal process: $s_{2s} = s_1$

Therefore,

$$\frac{T_{2s}}{T_1} = \left(\frac{P_2}{P_1}\right)^{(k-1)/k}$$

Also,

$$\eta_c = \frac{h_{2s} - h_1}{h_2 - h_1} = \frac{T_{2s} - T_1}{T_2 - T_1}$$

Solution

$$\left(\frac{P_2}{P_1}\right)^{(k-1)/k} = \frac{T_{2s}}{T_1} = 10^{0.286} = 1.932, \qquad T_{2s} = 556.8 \text{ K}$$

$$\eta_c = \frac{h_{2s} - h_1}{h_2 - h_1} = \frac{T_{2s} - T_1}{T_2 - T_1} = \frac{556.8 - 288.2}{T_2 - T_1} = 0.80$$

$$T_2 - T_1 = \frac{556.8 - 288.2}{0.80} = 335.8, \qquad T_2 = 624.0 \text{ K}$$

$$w_c = h_2 - h_1 = C_p(T_2 - T_1)$$
$$= 1.0035(624.0 - 288.2) = 337.0 \text{ kJ/kg}$$

Control volume: Turbine.
Inlet state: P_3 (P_2 − drop) known, T_3 known; state fixed.
Exit state: P_4 known.
Analysis:

First law, real process: $w_t = h_3 - h_4$
Second law, ideal process: $s_{4s} = s_3$

Therefore,

$$\frac{T_3}{T_{4s}} = \left(\frac{P_3}{P_4}\right)^{(k-1)/k}$$

Also,

$$\eta_t = \frac{h_3 - h_4}{h_3 - h_{4s}} = \frac{T_3 - T_4}{T_3 - T_{4s}}$$

Solution

$$P_3 = P_2 - \text{pressure drop} = 1.0 - 0.015 = 0.985 \text{ MPa}$$

$$\left(\frac{P_3}{P_4}\right)^{(k-1)/k} = \frac{T_3}{T_{4s}} = 9.85^{0.286} = 1.9236, \qquad T_{4s} = 713.9 \text{ K}$$

$$\eta_t = \frac{h_3 - h_4}{h_3 - h_{4s}} = \frac{T_3 - T_4}{T_3 - T_{4s}} = 0.85$$

$$T_3 - T_4 = 0.85(1373.2 - 713.9) = 560.4$$

$$T_4 = 812.8 \text{ K}$$

$$w_t = h_3 - h_4 = C_p(T_3 - T_4)$$
$$= 1.0035(1373.2 - 812.8) = 562.4 \text{ kJ/kg}$$

$$w_{net} = w_t - w_c = 562.4 - 337.0 = 225.4 \text{ kJ/kg}$$

Control volume: High-temperature heat exchanger.
Inlet state: State 2 fixed (above).
Exit state: State 3 fixed (above).

Analysis:

First law: $$q_H = h_3 - h_2$$

Solution

$$q_H = h_3 - h_2 = C_p(T_3 - T_2)$$
$$= 1.0035(1373.2 - 624.0) = 751.8 \text{ kJ/kg}$$

so that

$$\eta_{th} = \frac{w_{net}}{q_H} = \frac{225.4}{751.8} = 30.0\%$$

The following comparisons can be made between Examples 9.11 and 9.12.

	w_c	w_t	w_{net}	q_H	η_{th}
Example 9.11 (Ideal)	269.5	664.7	395.2	819.3	48.2
Example 9.12 (Actual)	337.0	562.4	225.4	751.8	30.0

As stated previously the result of the irreversibilities is to decrease the turbine work and increase the compressor work. Since the net work is the difference between these two, it decreases very rapidly as compressor and turbine efficiencies decrease. The development of compressors and turbines of high efficiency is therefore an important aspect of the development of gas turbines.

Note also that in the ideal cycle (Example 9.11) about 41% of the turbine work is required to drive the compressor and 59% is delivered as net work. In the actual turbine (Example 9.12) 60% of the turbine work is required to drive the compressor and 40% is delivered as net work. Thus, if the net power of this unit is to be 10 000 kW, a 25 000-kW turbine and a 15 000-kW compressor are required. This demonstrates the statement that a gas turbine has a high back-work ratio.

9.16 THE SIMPLE GAS-TURBINE CYCLE WITH REGENERATOR

The efficiency of the gas-turbine cycle may be improved by introducing a regenerator. The simple open-cycle gas-turbine cycle with regenerator is shown in Fig. 9.29, and the corresponding ideal air-standard cycle with regenerator is shown on the *P-v* and *T-s* diagrams. Note that in cycle 1–2–*x*–3–4–*y*–1, the temperature of the exhaust gas leaving the turbine in state 4 is higher than the temperature of the gas leaving the compressor. Therefore, heat can be transferred from the exhaust gases to the high-pressure gases leaving the compressor. If this is done in a counterflow heat exchanger, which is known as a regenerator, the temperature of the high-pressure gas leaving the regenerator, T_x, may, in the ideal case, have a temperature equal to T_4, the temperature of the gas leaving the turbine. In this case heat transfer from the external source is necessary only to increase the tem-

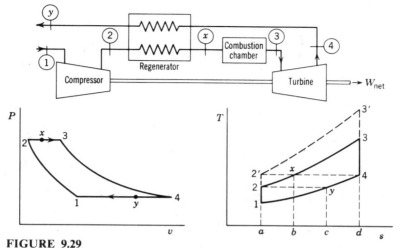

FIGURE 9.29
The ideal regenerative cycle.

perature from T_x to T_3, and this heat transfer is represented by area x–3–d–b–x. Area y–1–a–c–y represents the heat rejected.

The influence of pressure ratio on the simple gas-turbine cycle with regenerator is shown by considering cycle 1–2′–3′–4–1. In this cycle the temperature of the exhaust gas leaving the turbine is just equal to the temperature of the gas leaving the compressor; therefore there is no possibility of utilizing a regenerator. This may be shown more exactly by determining the efficiency of the ideal gas-turbine cycle with regenerator.

The efficiency of this cycle with regeneration is found as follows, where the states are as given in Fig. 9.29.

$$\eta_{th} = \frac{w_{net}}{q_H} = \frac{w_t - w_c}{q_H}$$

$$q_H = C_p(T_3 - T_x)$$

$$w_t = C_p(T_3 - T_4)$$

But for an ideal regenerator $T_4 = T_x$, and therefore $q_H = w_t$. Consequently,

$$\eta_{th} = 1 - \frac{w_c}{w_t} = 1 - \frac{C_p(T_2 - T_1)}{C_p(T_3 - T_4)}$$

$$= 1 - \frac{T_1(T_2/T_1 - 1)}{T_3(1 - T_4/T_3)} = 1 - \frac{T_1}{T_3} \frac{[(P_2/P_1)^{(k-1)/k} - 1]}{[1 - (P_1/P_2)^{(k-1)/k}]}$$

$$\eta_{th} = 1 - \frac{T_1}{T_3}\left(\frac{P_2}{P_1}\right)^{(k-1)/k}$$

FIGURE 9.30

Temperature–entropy diagram to illustrate the definition of regenerator efficiency.

Thus we see that for the ideal cycle with regeneration the thermal efficiency depends not only on the pressure ratio but also on the ratio of the minimum to maximum temperature. We also note, in contrast to the Brayton cycle, that the efficiency decreases with an increase in pressure ratio. The thermal efficiency versus pressure ratio for this cycle is plotted in Fig. 9.27 for a value of

$$\frac{T_1}{T_3} = 0.25$$

The effectiveness or efficiency of a regenerator is given by the term regenerator efficiency. This may best be defined by reference to Fig. 9.30. State x represents the high-pressure gas leaving the regenerator. In the ideal regenerator there would be only an infinitesimal temperature difference between the two streams, and the high-pressure gas would leave the regenerator at temperature T'_x, and $T'_x = T_4$. In an actual regenerator, which must operate with a finite temperature difference, T_x, the actual temperature leaving the regenerator is therefore less than T'_x. The regenerator efficiency is defined by

$$\eta_{reg} = \frac{h_x - h_2}{h'_x - h_2}$$

If the specific heat is assumed to be constant, the regenerator efficiency is also given by the relation

$$\eta_{reg} = \frac{T_x - T_2}{T'_x - T_2}$$

It should also be pointed out that a higher efficiency can be achieved by using a regenerator with greater heat-transfer area. However, this also increases the pressure drop, which represents a loss, and both the pressure drop and the regenerator efficiency must be considered in determining which regenerator gives maximum thermal efficiency for the cycle. From an economic point of view, the cost of the regenerator must be weighed against the saving that can be effected by its use.

Example 9.13

If an ideal regenerator is incorporated into the cycle of Example 9.11, determine the thermal efficiency of the cycle.

The diagram for this example is Fig. 9.30. Values are those from Example 9.11. In this case, for the analysis of the high-temperature heat exchanger (combustion chamber), from the first law:

$$q_H = h_3 - h_x$$

so that the solution is as follows:

$$T_x = T_4 = 710.8 \text{ K}$$

$$q_H = h_3 - h_x = C_p(T_3 - T_x) = 1.0035(1373.2 - 710.8) = 664.7 \text{ kJ/kg}$$

$$w_{net} = 395.2 \text{ kJ/kg (from Example 9.11)}$$

$$\eta_{th} = \frac{395.2}{664.7} = 59.5\%$$

9.17 THE IDEAL GAS-TURBINE CYCLE USING MULTISTAGE COMPRESSION WITH INTERCOOLING, MULTISTAGE EXPANSION WITH REHEATING, AND REGENERATOR

In Section 9.14 it was pointed out that when an ideal regenerator was incorporated into the Ericsson cycle, an efficiency equal to the efficiency of the corresponding Carnot cycle could be attained. It was also pointed out that it is impossible to even closely approach in practice the reversible isothermal compression and expansion required in the Ericsson cycle.

The practical approach to this cycle is to use multistage compression with intercooling between stages, multistage expansion with reheat between stages, and a regenerator. Figure 9.31 shows a cycle with two stages of compression and two stages of expansion. The air-standard cycle is shown on the corresponding T-s diagram. It may be shown that for this cycle the maximum efficiency is obtained if equal pressure ratios are maintained across the two compressors and the two turbines. In this ideal cycle it is assumed that the temperature of the air leaving the intercooler, T_3, is equal to the temperature of the air entering the first stage of compression, T_1, and that the temperature after reheating, T_8, is equal to the temperature entering the first turbine, T_6. Furthermore, in the ideal cycle it is assumed that the temperature of the high-pressure air leaving the regenerator, T_5, is equal to the temperature of the low-pressure air leaving the turbine, T_9.

If a large number of stages of compression and expansion are used, it is evident that the Ericsson cycle is approached. This is shown in Fig. 9.32. In practice the economical limit to the number of stages is usually two or three. The turbine and compressor losses and pressure drops that have already been discussed would be involved in any actual unit employing this cycle.

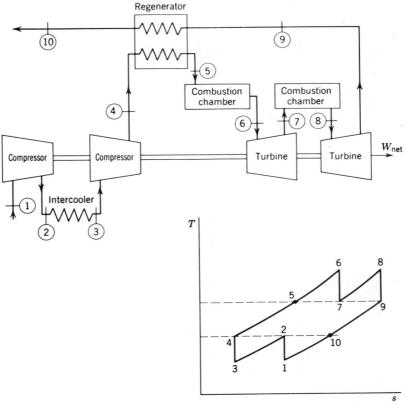

FIGURE 9.31

The ideal gas-turbine cycle utilizing intercooling, reheat, and a regenerator.

FIGURE 9.32

Temperature–entropy diagram that shows how the gas-turbine cycle with many stages approaches the Ericsson cycle.

FIGURE 9.33

Some arrangements of components that may be utilized in stationary gas-turbine power plants.

There are a variety of ways in which the turbines and compressors using this cycle can be utilized. Two possible arrangements for closed cycles are shown in Fig. 9.33. One advantage frequently sought in a given arrangement is ease of control of the unit under various loads. Detailed discussion of this point, however, is beyond the scope of this text.

9.18 THE AIR-STANDARD CYCLE FOR JET PROPULSION

The last air-standard power cycle we will consider is utilized in jet propulsion. In this cycle the work done by the turbine is just sufficient to drive the compressor. The gases are expanded in the turbine to a pressure such that the turbine work is just equal to the compressor work. The exhaust pressure of the turbine will then be above that of the surroundings, and the gas can be expanded in a nozzle to the pressure of the surroundings. Since the gases leave at a high velocity, the change in momentum the gases undergo results in a thrust on the aircraft in which the engine is installed. The air-standard cycle for this is shown in Fig. 9.34.

FIGURE 9.34

The ideal gas-turbine cycle for a jet engine.

Since all the principles involved in this cycle have already been covered, the example that follows will conclude our consideration of air-standard power cycles.

Example 9.14

Consider an ideal cycle in which air enters the compressor at 0.1 MPa, 15°C. The pressure leaving the compressor is 1.0 MPa and the maximum temperature is 1100°C. The air expands in the turbine to a pressure such that the turbine work is just equal to the compressor work. On leaving the turbine the air expands in a reversible adiabatic process in a nozzle to 0.1 MPa. Determine the velocity of the air leaving the nozzle.

The model used is ideal gas, constant specific heat, value at 300 K, and each process is SSF with no potential energy change. The only kinetic energy change occurs in the nozzle. The diagram is shown in Fig. 9.34.

The compressor analysis is the same as in Example 9.11. From the results of that solution,

$$P_1 = 0.1 \text{ MPa}, \qquad T_1 = 288.2 \text{ K}$$
$$P_2 = 1.0 \text{ MPa}, \qquad T_2 = 556.8 \text{ K}$$
$$w_c = 269.5 \text{ kJ/kg}$$

The turbine analysis is also the same as in Example 9.11. However, in this case,

$$P_3 = 1.0 \text{ MPa}, \qquad T_3 = 1373.2 \text{ K}$$
$$w_c = w_t = C_p(T_3 - T_4) = 269.5 \text{ kJ/kg}$$
$$T_3 - T_4 = \frac{269.5}{1.0035} = 268.6, \qquad T_4 = 1104.6 \text{ K}$$

so that

$$\frac{T_3}{T_4} = \left(\frac{P_3}{P_4}\right)^{(k-1)/k} = \frac{1373.2}{1104.6} = 1.2432$$

$$\frac{P_3}{P_4} = 2.142 \qquad P_4 = 0.4668 \text{ MPa}$$

Control volume: Nozzle.
Inlet state: State 4 fixed (above).
Exit state: P_5 known.
Analysis:

First law:
$$h_4 = h_5 + \frac{\mathbf{V}_5^2}{2}$$

Second law:
$$s_4 = s_5$$

Solution

Since P_5 is 0.1 MPa, from the second law we find that $T_5 = 710.8$ K.
$$\mathbf{V}_5^2 = 2C_{po}(T_4 - T_5)$$
$$\mathbf{V}_5^2 = 2 \times 1000 \times 1.0035(1104.6 - 710.8)$$
$$\mathbf{V}_5 = 889 \text{ m/s}$$

AIR-STANDARD REFRIGERATION CYCLE

9.19 THE AIR-STANDARD REFRIGERATION CYCLE

The final section of this chapter concerns the air-standard refrigeration cycle. Its main use in practice is in the liquefaction of air and other gases and in certain special situations that require refrigeration.

The simplest form of the air-standard refrigeration cycle, which is essentially the reverse of the Brayton cycle, is shown in Fig. 9.35. The compressor and expander might be either reciprocating or rotary. After compression from 1 to 2, the air is cooled as a result of heat transfer to the surroundings at temperature T_0. The air is then expanded in process 3–4 to the pressure entering the compressor, and the temperature drops to T_4 in the expander. Heat may then be transferred to the air until temperature T_L is reached. The work for this cycle is represented

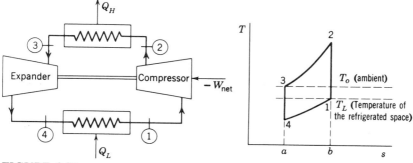

FIGURE 9.35
The air-standard refrigeration cycle.

FIGURE 9.36

An air refrigeration cycle that might be utilized for aircraft cooling.

by area $1-2-3-4-1$, and the refrigeration effect is represented by area $4-1-b-a-4$. The coefficient of performance is the ratio of these two areas.

In practice this cycle has been utilized for the cooling of aircraft in an open cycle, a simplified form of which is shown in Fig. 9.36. Upon leaving the expander the cool air is blown directly into the cabin thus providing the cooling effect where needed.

When counterflow heat exchangers are incorporated, very low temperatures can be obtained. This is essentially the cycle used in low-pressure air liquefaction plants and in other liquefaction devices such as the Collins helium liquefier. The ideal cycle in this case is as shown in Fig. 9.37. It is evident that the expander operates at very low temperature, which presents unique problems to the designer in regard to lubrication and materials.

Example 9.15

Consider the simple air-standard refrigeration cycle of Fig. 9.35. Air enters the compressor at 14.7 lbf/in.^2, 0 F, and leaves at 80 lbf/in.^2 Air enters the expander

FIGURE 9.37

The air refrigeration cycle utilizing a heat exchanger.

at 60 F. Determine:

1. The coefficient of performance for this cycle.
2. The rate at which air must enter the compressor in order to provide 1 kW of refrigeration.

For each control volume in this example, the model is ideal gas with constant specific heat, value at 80 F, and each process is SSSF with no kinetic or potential energy changes. The diagram for this example is Fig. 9.35.

Control volume: Compressor.

Inlet state: P_1, T_1 known; state fixed.

Exit state: P_2 known.

Analysis:

First law: $w_c = h_2 - h_1$ (w_c designates work into the compressor)

Second law: $s_1 = s_2$

Therefore,

$$\frac{T_2}{T_1} = \left(\frac{P_2}{P_1}\right)^{(k-1)/k}$$

Solution

$$\frac{T_2}{T_1} = \left(\frac{P_2}{P_1}\right)^{(k-1)/k} = \left(\frac{80}{14.7}\right)^{0.286} = 1.624$$

$$T_2 = 460(1.624) = 747 \text{ R}$$

$$w_c = h_2 - h_1 = C_p(T_2 - T_1) = 0.24(747 - 460) = 68.9 \text{ Btu/lbm}$$

Control volume: Expander.

Inlet state: P_3 ($=P_2$) known, T_3 known; state fixed.

Exit state: P_4 ($=P_1$) known.

Analysis:

First law: $w_t = h_3 - h_4$

Second law: $s_3 = s_4$

so that

$$\frac{T_3}{T_4} = \left(\frac{P_3}{P_4}\right)^{(k-1)/k}$$

Solution

$$\frac{T_3}{T_4} = \left(\frac{P_3}{P_4}\right)^{(k-1)/k} = \left(\frac{80}{14.7}\right)^{0.286} = 1.624$$

$$T_4 = \frac{520}{1.624} = 320 \text{ R}$$

$$w_t = h_3 - h_4 = 0.24(520 - 320) = 48.0 \text{ Btu/lbm}$$

Control volume: High-temperature heat exchanger.
Inlet state: State 2 known (above).
Exit state: State 3 known (above).
Analysis:

 First law: $\qquad\qquad q_H = h_2 - h_3 \qquad$ (heat rejected)

Solution

$$q_H = h_2 - h_3 = C_p(T_2 - T_3) = 0.24(747 - 520) = 54.5 \text{ Btu/lbm}$$

Control volume: Low-temperature heat exchanger.
Inlet state: State 4 known (above).
Exit state: State 1 known (above).
Analysis:

 First law: $\qquad\qquad\qquad\qquad q_L = h_1 - h_4$

Solution

$$q_L = h_1 - h_4 = C_p(T_1 - T_4) = 0.24(460 - 320) = 33.6 \text{ Btu/lbm}$$

Therefore,

$$w_{net} = w_c - w_t = 68.9 - 48.0 = 20.9 \text{ Btu/lbm}$$

$$\beta = \frac{q_L}{w_{net}} = \frac{33.6}{20.9} = 1.61$$

In order to provide 1 kW of refrigeration capacity. (3412 Btu/h)

$$\dot{m} = \frac{\dot{Q}_L}{q_L} = \frac{3412}{33.6} = 101.5 \text{ lbm/h}$$

PROBLEMS

9.1 Consider a solar energy-powered ideal Rankine cycle using water as the working fluid. Saturated vapor leaves the solar collector at 300 F and the condenser pressure is 1 lbf/in.2 Determine the thermal efficiency of this cycle.

9.2 A supply of geothermal hot water is to be used as the energy source in an ideal Rankine cycle, with Freon-12 as the cycle working fluid. Saturated vapor Freon-12 leaves the boiler at a temperature of 85°C, and the condenser temperature is 40°C.
 (a) Calculate the thermal efficiency of this cycle.
 (b) If 2 kg/s of geothermal water is available at 95°C, what is the maximum power output of the cycle?

9.3 It is desired to study the effect of turbine exhaust pressure on the performance of an ideal steam Rankine cycle. Steam enters the turbine at

500 lbf/in.², 700 F. Calculate the cycle thermal efficiency and the moisture content of the steam leaving the turbine for turbine exhaust pressures of 1 lbf/in.², 2 lbf/in.², 5 lbf/in.², and 14.7 lbf/in.² Plot thermal efficiency versus turbine exhaust pressure for the specified turbine inlet pressure and temperature.

9.4 It is desired to study the effect of turbine inlet pressure on the performance of an ideal steam Rankine cycle. Steam enters the turbine at 700 F and exhausts at 2 lbf/in.² Calculate the cycle thermal efficiency and the moisture content of the steam leaving the turbine for turbine inlet pressures of 100 lbf/in.², 500 lbf/in.², 1000 lbf/in.², and saturated vapor. Plot thermal efficiency versus turbine inlet pressure for the specified turbine inlet temperature and exhaust pressure.

9.5 It is desired to study the effect of turbine inlet temperature on the performance of an ideal steam Rankine cycle. Steam enters the turbine at 500 lbf/in.² and exhausts at 2 lbf/in.² Calculate the cycle thermal efficiency and the moisture content of the steam leaving the turbine for turbine inlet temperatures of saturated vapor, 700 F, 1000 F, and 1400 F. Plot thermal efficiency versus turbine inlet temperature for the specified turbine inlet pressure and exhaust pressure.

9.6 Consider the ammonia Rankine cycle power plant shown in Fig. 9.38, proposed to operate on a difference in ocean water temperature of 25°C near the surface and 5°C at some greater depth.

 (a) Determine the turbine power output and the pump power input for the ammonia cycle.

FIGURE 9.38
Sketch for Problem 9.6.

(b) Determine the mass flow rate of water through each heat exchanger.

(c) What is the thermal efficiency of this power plant?

9.7 Consider an ideal steam reheat cycle in which steam enters the high-pressure turbine at 500 lbf/in.2, 700 F, and expands to 120 lbf/in.2 It is then reheated to 700 F and then expanded to 2 lbf/in.2 in the low-pressure turbine. Calculate the cycle thermal efficiency and the moisture content of the steam leaving the low-pressure turbine.

9.8 It is desired to study the effect of reheat pressure on the ideal steam reheat cycle. Repeat Problem 9.7, using various values for the reheat pressure other than the value of 120 lbf/in.2

9.9 It is desired to study the effect of number of reheat stages on the ideal steam reheat cycle. Repeat Problem 9.7, using two reheat stages, one at 180 lbf/in.2 and the second at 30 lbf/in.2, instead of the single reheat at 120 lbf/in.2 $m_1 = m_2 + m_3 + m_4$

9.10 Consider an ideal steam regenerative cycle in which steam enters the turbine at 500 lbf/in.2, 700 F, and exhausts to the condenser at 2 lbf/in.2 Steam is extracted from the turbine at 120 lbf/in.2 and also at 20 lbf/in.2 for heating the boiler feedwater in two open feedwater heaters. The feedwater leaves each heater at the temperature of the condensing steam. The appropriate pumps are used for the water leaving the condenser and the two feedwater heaters. Calculate the cycle thermal efficiency and the net work per lbm of steam. $w_{turb} = (h_1 - h_2) + (1 - m_2)(h_2 - h_3) + (1 - m_2 - m_3)(h_3 - h$

9.11 Repeat Problem 9.10 assuming closed instead of open feedwater heaters. A single pump is used to pump the water leaving the condenser to the boiler pressure of 500 lbf/in.2 Condensate from the high-pressure heater is drained through a trap to the low-pressure heater, and that from the low-pressure heater is drained through a trap to the condenser.

9.12 It is desired to study the effect of the number of open feedwater heaters on the thermal efficiency of an ideal cycle in which steam leaves the steam generator at 20 MPa, 600°C, and that has a condenser pressure of 10 kPa. Determine the thermal efficiency for each of the following cases.

(a) No feedwater heater.

(b) One feedwater heater, operating at 1.0 MPa.

(c) Two feedwater heaters, one operating at 3.0 MPa and the other at 0.2 MPa.

9.13 Consider an ideal steam combined reheat and regenerative cycle in which steam enters the high-pressure turbine at 500 lbf/in.2, 700 F, and is extracted for feedwater heating at 120 lbf/in.2 The remainder of the steam is reheated at this pressure to 700 F and fed to the low-pressure turbine. Steam is extracted from the low-pressure turbine at 20 lbf/in.2 for feedwater heating. The condenser pressure is 2 lbf/in.2 and both feedwater heaters are open heaters. Calculate the cycle thermal efficiency and the net work per lbm of steam.

9.14 An ideal steam power plant is designed to operate on the combined reheat

$$\dot{m} = \frac{\dot{W}}{W_{net}}$$

and regenerative cycle and produce a net power output of 10 MW. Steam enters the high-pressure turbine at 8 MPa, 550°C, and is expanded to 0.6 MPa, at which pressure some of the steam is fed to an open feedwater heater and the remainder is reheated to 550°C. The reheated steam is then expanded in the low-pressure turbine to 10 kPa.

(a) Determine the steam flow rate to the high-pressure turbine.

(b) Determine the size of motor required to drive each of the pumps.

(c) If the condenser cooling water temperature increase is restricted to a maximum of 10°C, what is the cooling water flow rate?

(d) If the steam velocity in the turbine–condenser connecting pipe is restricted to a maximum of 100 m/s, what is the connecting pipe diameter?

9.15 In a particular nuclear power plant the nuclear reactor is so designed that the maximum temperature in the steam cycle is 450°C. The steam-condensing temperature is to be fixed at 40°C. It is planned to build a steam power plant that has one open feedwater heater. Select what you consider to be a reasonable ideal steam cycle within these specifications, determine the cycle thermal efficiency, and explain why the particular pressures involved were selected.

9.16 Steam leaves a power plant steam generator at 500 lbf/in.2, 650 F, and enters the turbine at 490 lbf/in.2, 625 F. The isentropic turbine efficiency is 88% and the turbine exhaust pressure is 1.7 lbf/in.2 Condensate leaves the condenser and enters the pump at 110 F, 1.7 lbf/in.2 The isentropic pump efficiency is 80% and the discharge pressure is 520 lbf/in.2 The feedwater enters the steam generator at 510 lbf/in.2, 100 F. Calculate the following:

(a) The cycle thermal efficiency.

(b) The irreversibility of the process in the line between the steam generator exit and the turbine inlet, assuming an ambient temperature of 77 F.

9.17 In a particular reheat-cycle power plant, steam enters the high-pressure turbine at 5 MPa, 450°C, and expands to 0.5 MPa, after which it is reheated to 450°C. The steam is then expanded through the low-pressure turbine to 7.5 kPa. Liquid water leaves the condenser at 30°C, is pumped to 5 MPa, and returned to the steam generator. Each turbine is adiabatic, with an isentropic efficiency of 87%, and the pump efficiency is 82%. If the total power output of the turbines is 10 000 kW, determine the following:

(a) Mass flow rate of steam.

(b) Pump power input.

(c) Power plant thermal efficiency.

9.18 If one type of nuclear power plant, heat is transferred in the nuclear reactor to liquid sodium. The liquid sodium is then pumped through a heat exchanger where heat is transferred to boiling water. Saturated vapor steam at 800 lbf/in.2 exits this heat exchanger and is then superheated to 1200 F in an external gas-fired superheater. The steam then enters the turbine, which has one (open type) feedwater extraction at 60 lbf/in.2 The isentropic

turbine efficiency is 87% and the condenser pressure is 1 lbf/in.2 Determine the heat transfer in the reactor and in the superheater to produce a net power output of 1 MW.

9.19 Repeat Problem 9.12 assuming an isentropic turbine efficiency of 88%.

9.20 A certain industrial application has the following steam requirement: one 5-kg/s stream at a pressure of 0.5 MPa and one 5-kg/s stream at 1.4 MPa (both saturated or slightly superheated vapor). Instead of installing a low-pressure boiler to generate this steam, it is planned to obtain it by cogeneration, in which the steam required will be the exhaust from a steam turbine. A high-pressure boiler generating steam at 10 MPa, 500°C, is therefore installed to supply steam to the turbine. The required amount is withdrawn at 1.4 MPa and the remainder expanded in the low-pressure end of the turbine to 0.5 MPa. Assuming that the high-pressure and low-pressure turbines each have an isentropic efficiency of 85%, determine the following:
(a) The power output of the turbine.
(b) The heat transfer rate required in the boiler, compared with that required if the steam were generated in a low-pressure boiler without cogeneration. Assume that in each case 20°C liquid water is pumped to the required pressure and fed to the boiler.

9.21 Consider the following preliminary design specifications for a supercritical steam power plant cycle. The maximum pressure will be 4000 lbf/in.2 and the maximum temperature will be 1100 F. Cooling water temperature is such that 1 lbf/in.2 pressure can be maintained in the condensers. Turbine isentropic efficiencies of at least 87% can be expected.
(a) Do you recommend any reheat for this cycle? If so, how many stages of reheat and at what pressures? Give reasons for your decisions.
(b) Do you recommend feedwater heaters? If so, how many and at what pressures would they operate? Would they be open or closed feedwater heaters?
(c) Estimate the thermal efficiency of the cycle that you recommend.

9.22 In both power and refrigeration cycles that operate over a wide temperature range it is frequently advantageous to use more than one working fluid. In power plants two fluids that have been used in combination are mercury and water, and this is referred to as a binary system. The advantage of this system is that mercury has a much lower vapor pressure than water, and therefore it is possible to have an isothermal boiling process in the mercury take place at a high temperature (much higher than the critical temperature of water) at a moderate pressure. Such a binary system is shown in Fig. 9.39, with the temperatures and pressures for an ideal cycle shown on the corresponding T-s diagram. The maximum temperature in the steam cycle, that leaving the superheater at point 4, is 500°C. Determine:
(a) The kilograms of mercury condensed in the mercury condenser per kilogram of water boiled.
(b) The thermal efficiency of this ideal cycle.

FIGURE 9.39

Sketch for Problem 9.22.

From tables of thermodynamic properties for mercury, the following satu-
ration data are known:

P (MPa)	T_g (°C)	h_f (kJ/kg)	h_g (kJ/kg)	s_f (kJ/kg K)	s_g (kJ/kg K)
0.04	309	42.21	335.64	0.1034	0.6073
1.6	562	75.37	364.04	0.1498	0.4954

9.23 Consider an ideal refrigeration cycle having a condenser temperature of
45°C and an evaporator temperature of −15°C. Determine the coefficient
of performance of this refrigerator for the working fluids Freon-12 and
ammonia.

9.24 Consider an ideal heat pump having a condenser temperature of 120 F and
an evaporator temperature of 30 F. Determine the coefficient of perfor-
mance of this heat pump for the working fluids Freon-12 and ammonia.

9.25 It is desired to study the effect of evaporator temperature on the coefficient
of performance of a heat pump. Consider an ideal cycle with Freon-12 as
the working fluid and a condenser temperature of 40°C. Plot a curve of
coefficient of performance versus evaporator temperature for temperature
values from 15°C to −25°C.

9.26 It is desired to study the effect on cycle power input requirements of the
difference between the temperature of the refrigerant in the condenser and
the temperature of the surroundings. For this study, consider an ideal refri-
geration cycle with Freon-12 as the working fluid, a temperature of the
surroundings of 80 F and an evaporator temperature of 0 F. Plot a curve
of power input per kilowatt of refrigeration for temperature differences of
0 to 60 F between the refrigerant temperature in the condenser and the
temperature of the surroundings.

9.27 It is desired to study the effect on cycle power input requirements of the
difference between the temperature of the refrigerated cold space and the

temperature of the refrigerant in the evaporator. For this study, consider an ideal refrigeration cycle with Freon-12 as the working fluid, a condenser temperature of 120 F and a cold space temperature of 0 F. Plot a curve of power input per kilowatt of refrigeration for temperature differences of 0 to 60 F between the cold space and the refrigerant in the evaporator.

9.28 A small heat pump unit is used to heat water for a hot-water supply. Assume that the unit uses Freon-12 and operates on the ideal refrigeration cycle. The evaporator temperature is 15°C and the condenser temperature is 55°C. If the hot water requirement is 0.1 kg/s, determine the amount of energy saved by using the heat pump instead of direct heating of the water from 15 to 55°C.

9.29 In an ideal refrigeration cycle using Freon-12 as the working fluid, the evaporator temperature is − 10 F, and saturated vapor exits at that temperature and enters the compressor. Vapor is compressed to the condenser pressure, 200 lbf/in.², and liquid exits the condenser and enters the expansion value at 100 F. It is proposed to modify this cycle as shown in Fig. 9.40 by adding a heat exchanger in which the cold vapor cools the liquid before it enters the expansion valve. Compare the coefficient of performance of these two cycles, and discuss the result.

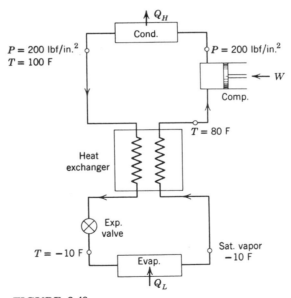

FIGURE 9.40
Sketch for Problem 9.29.

9.30 In both power and refrigeration cycles that operate over a wide temperature range it is frequently advantageous to use more than one working fluid. In refrigeration systems this is commonly referred to as a cascade system. Con-

sider such a system composed of two ideal refrigeration cycles, as shown in Fig. 9.41. The high-temperature cycle uses Freon-12, with saturated liquid leaving the condenser at 40°C, and saturated vapor leaving the heat exchanger at −20°C. The low-temperature cycle uses a different refrigerant, Freon-13, with saturated vapor leaving the evaporator at −70°C, and saturated liquid leaving the heat exchanger at −10°C. Calculate the following:
(a) The ratio of mass flow rates through the two cycles.
(b) The coefficient of performance of the system.
From tables of thermodynamic properties of Freon-13, the following data are known:

T (°C)	P (MPa)	h (kJ/kg)	s (kJ/kg K)	
−70	0.182	188.8	0.9626	(saturated vapor)
−10	1.525	106.5		(saturated liquid)
13.8	1.525	224.7	0.9626	(superheated vapor)

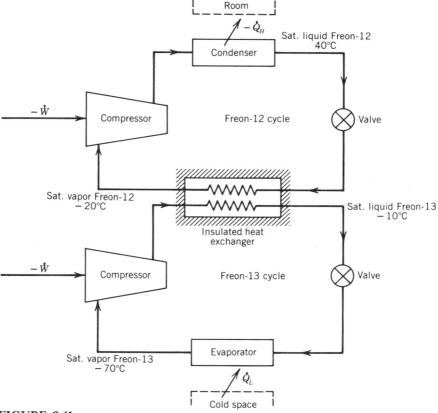

FIGURE 9.41
Sketch for Problem 9.30.

9.31 One means of improving the performance of a refrigeration system that operates over a wide temperature range is to use a two-stage compressor. Consider an ideal refrigeration system of this type using Freon-12 as the working fluid, as shown in Fig. 9.42. Saturated liquid leaves the condenser at 40°C and is throttled to −20°C. The liquid and vapor at this temperature are separated, with the liquid being throttled to the evaporator temperature, −70°C. Vapor leaving the evaporator is compressed to the saturation pressure corresponding to −20°C, after which it is mixed with the vapor leaving the flash chamber. It may be assumed that both the flash chamber and the mixing chamber are well insulated to prevent heat transfer from the ambient. Vapor leaving the mixing chamber is compressed in the second stage of the compressor to the saturation pressure corresponding to the condenser temperature, 40°C. Determine:

(a) The coefficient of performance of the system.

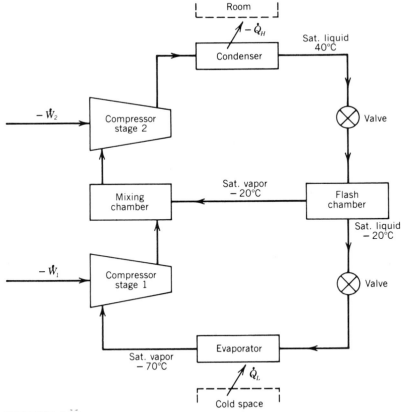

FIGURE 9.42
Sketch for Problem 9.31.

(b) The coefficient of performance of a simple ideal refrigeration cycle operating over the same condenser and evaporator temperature range as the two-stage compressor unit studied in this problem.

9.32 Freon-12 is used as the working fluid in a conventional heat pump cycle. Saturated vapor enters the compressor of this unit at 50 F, and the compressor exit temperature is measured and found to be 185 F. If the isentropic efficiency of the compressor is estimated to be 70%, what is the coefficient of performance of the heat pump?

9.33 In an actual refrigeration cycle using Freon-12 as the working fluid, the refrigerant flow rate is 0.05 kg/s. Vapor enters the compressor at 150 kPa, −10°C, and leaves at 1.2 MPa, 75°C. The power input to the compressor is measured and found to be 2.4 kW. The refrigerant enters the expansion valve at 1.15 MPa, 40°C, and leaves the evaporator at 175 kPa, −15°C. Determine the following:
(a) The irreversibility during the compression process.
(b) The refrigeration capacity.
(c) The coefficient of performance for this cycle.

9.34 Consider an ideal dual-loop heat-powered refrigeration cycle using Freon-12 as the working fluid, as shown in Fig. 9.43. Saturated vapor at 220 F leaves the boiler and expands in the turbine to the condenser pressure. Saturated vapor at 0 F leaves the evaporator and is compressed to the condenser pressure. The ratio of flows through the two loops is such that the turbine produces just enough power to drive the compressor. The two exiting streams mix together and enter the condenser. Saturated liquid at

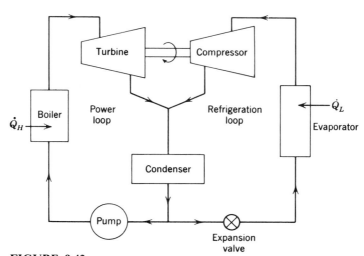

FIGURE 9.43
Sketch for Problem 9.34.

110 F leaving the condenser is then separated into two streams in the necessary proportions. Determine the following:

(a) The ratio of mass flow rate through the power loop to that through the refrigeration loop.

(b) The performance of the cycle, in terms of the ratio \dot{Q}_L/\dot{Q}_H.

9.35 A jet ejector is a device with no moving parts that functions as the equivalent of a coupled turbine–compressor unit (see Problems 7.76 and 7.81). Thus the turbine–compressor in the dual-loop cycle of Fig. 9.43 could be replaced by a jet ejector, in which the primary stream enters from the boiler, the secondary stream enters from the evaporator, and the discharge flows to the condenser. Alternatively, a jet ejector may be used with water as the working fluid, in which case the purpose of the device is to chill water, usually for an air-conditioning system. In this application, the physical setup may be as shown in Fig. 9.44. Using the data given on the diagram, evaluate the performance of this cycle in terms of the ratio \dot{Q}_L/\dot{Q}_H.

(a) Assume an ideal cycle.

(b) Assume an ejector efficiency of 20% (see Problem 7.81).

FIGURE 9.44
Sketch for Problem 9.35.

9.36 It is desired to compare the performance of an ammonia absorption cycle refrigerator with that of a similar vapor compression system. Consider an

absorption system having an evaporator temperature of 10 F and a condenser temperature of 120 F. The generator temperature in this system is 300 F. In this cycle, 0.42 Btu is transferred to the ammonia in the evaporator for each Btu transferred to the ammonia solution in the generator from the high-temperature source. To make the desired comparison, assume that a reservoir is available at 300 F and that heat is transferred from this reservoir to a reversible engine that rejects heat to the surroundings at 77 F. This work is then used to drive an ideal vapor-compression system with ammonia as the refrigerant. Compare the amount of refrigeration that can be achieved per Btu from the high-temperature source in this case with the 0.42 Btu that can be achieved in the absorption system.

9.37 Consider a small, solar-energy-powered ammonia-absorption refrigeration cycle to be used in an air-conditioning application. Saturated vapor ammonia leaves the generator at 50°C, and saturated vapor leaves the evaporator at 10°C. If 7000 kJ of heat is required in the generator (solar collector) per kilogram of ammonia vapor generated, determine the overall performance of this system.

9.38 A stoichiometric mixture of fuel and air has an energy release upon combustion of approximately 2800 kJ/kg of mixture. To approximate an actual spark-ignition engine using such a mixture, consider an air-standard Otto cycle that has a heat addition of 2800 kJ/kg of air, a compression ratio of 9, and a pressure and temperature at the beginning of the compression process of 100 kPa, 20°C. Assuming constant specific heat, value from Table A.8, determine the following:
(a) The maximum cycle pressure and temperature.
(b) The cycle thermal efficiency.
(c) The mean effective pressure.

9.39 Repeat Problem 9.38, assuming variable specific heat. The Ideal Gas Air Tables, Table A.10, are recommended for this calculation (and specific heat from Fig. 5.10 at high temperature).

9.40 It has been determined experimentally that the power stroke expansion in an internal-combustion engine can be approximated reasonably as a polytropic process with a value of the polytropic exponent n somewhat smaller than the specific heat ratio k. Repeat Problem 9.38 assuming that the expansion process is reversible and polytropic (instead of the isentropic expansion of the Otto cycle) with a value of n equal to 1.25.

9.41 In an air-standard Otto cycle, all the heat transfer q_H occurs at constant volume. It would be more realistic to assume that part of the q_H occurs after the piston has started to move downward in the expansion stroke. Therefore, consider a cycle identical to the Otto cycle except that the first two-thirds of the total q_H occurs at constant volume and the final one-third occurs at constant pressure. Assume that the total q_H in this cycle is 1100 Btu/lbm, the pressure and temperature at the beginning of the compression process are 14.7 lbf/in.2, 70 F, and the compression ratio is 10. Calculate

the maximum pressure and temperature and the thermal efficiency of this cycle, and compare the results with a conventional Otto cycle having the same given variables.

9.42 Consider an ideal air-standard Diesel cycle is which the state before the compression process is 14.7 lbf/in.2, 80 F, and the compression ratio is 18. What maximum cycle temperature is required for this cycle to have a thermal efficiency of 60%?

9.43 A particular Diesel engine has a compression ratio of 20. The pressure at the beginning of the compression stroke is 95 kPa, and the temperature at this point is 25°C. If the maximum temperature in this engine is 2800 K, what are the thermal efficiency and the cycle mean effective pressure?

9.44 A simple Rankine cycle with Freon-12 as the working fluid is to be used as a bottoming cycle for a Diesel engine-driven electrical generating facility (the exhaust gas from the Diesel cycle is used as the high-temperature energy source in the Freon-12 boiler). Diesel inlet conditions are 14.7 lbf/in.2, 70 F, the compression ratio is 16, and the maximum cycle temperature is 4500 R. Saturated vapor Freon-12 leaves the bottoming cycle boiler at 230 F, and the condenser temperature is 90 F. The power output of the Diesel engine is 1.0 MW. Assuming ideal cycles throughout, determine the following:
(a) The flow rate required in the Diesel cycle.
(b) The power output of the bottoming cycle, assuming that the Diesel exhaust is cooled to 400 F in the Freon-12 boiler.

9.45 Consider an ideal air-standard Ericsson cycle having an ideal regenerator. The high-temperature heat supplied is 1000 kJ/kg, which occurs at a temperature of 1500 K. Heat rejection in the cycle occurs at a temperature of 300 K, and the cycle pressure at the beginning of the isothermal compression process is 100 kPa. Determine the compressor work and the turbine work per kilogram of air and the thermal efficiency of the cycle.

9.46 Consider an ideal Stirling cycle engine in which the pressure and temperature at the beginning of the isothermal compression process are 14.7 lbf/in.2, 80 F, the compression ratio is 6, and the maximum temperature in the cycle is 2000 F. Calculate the following:
(a) The maximum pressure in the cycle.
(b) The cycle thermal efficiency with and without regenerators.

9.47 Consider an ideal air-standard Stirling cycle with an ideal regenerator. The minimum pressure and temperature in the cycle are 100 kPa and 25°C, the compression ratio is 10, and the maximum temperature in the cycle is 1000°C. Analyze each of the four processes in this cycle for work and heat transfer, and determine the overall performance of the engine.

9.48 Consider an ideal air-standard Brayton cycle in which the pressure and temperature of the air entering the compressor are 14.7 lbf/in.2, 60 F, and the pressure ratio across the compressor is 12 to 1. The maximum temperature in the cycle is 2000 F and the air flow rate is 20 lbm/s. Assume constant specific heat for the air, value from Table A.8.

(a) Determine the compressor work, the turbine work, and the thermal efficiency of the cycle.

(b) If this cycle were to be used for a reciprocating machine, what would be the mean effective pressure? Would you recommend the Brayton cycle for a reciprocating machine?

9.49 Repeat Problem 9.48, assuming variable specific heat for the air.

9.50 An ideal regenerator is incorporated into the ideal air-standard Brayton cycle of Problem 9.48. Calculate the thermal efficiency of the cycle with this modification.

9.51 Consider a simple Brayton cycle gas-turbine engine with air as the working fluid. The pressure ratio across the compressor is 8 to 1, the compressor inlet temperature is 25°C, the isentropic compressor efficiency is 75%, and the isentropic turbine efficiency is 85%.

(a) What is the lowest maximum cycle temperature at which the engine will run (no net power output)?

(b) If the maximum cycle temperature is 1150°C, what is the cycle thermal efficiency?

9.52 Consider a stationary gas-turbine power plant that operates on the ideal Brayton cycle and delivers a power output of 20 MW to an electric generator. The minimum cycle temperature is 520 R and the maximum temperature is 2700 R. The minimum cycle pressure is 14.0 lbf/in.² and the maximum pressure is 140 lbf/in.²

(a) Calculate the power output of the turbine. What fraction of the turbine output is required to drive the compressor?

(b) What is the thermal efficiency of the cycle?

(c) Calculate the irreversibility for each process in the cycle.

9.53 Repeat Problem 9.52 assuming that a regenerator of 75% efficiency is added to the cycle, and also that the compressor has an isentropic efficiency of 85% and the turbine an isentropic efficiency of 87%.

9.54 A gas turbine engine is to be used for pumping natural gas through a cross-country pipeline. The required power input from this engine to the natural gas compressor is 1000 hp. A simple open cycle will be used with a regenerator for this application. Since a supply of fuel is readily available, and since some of these units may be located in relatively isolated places, low maintenance costs are more important than a high efficiency. Select a cycle that you would recommend, making appropriate assumptions for isentropic compressor and turbine efficiencies and also for the regenerator efficiency, and determine, for the gas-turbine engine unit, the power output of the turbine, the power input to the compressor, and the engine thermal efficiency.

9.55 A gas turbine cycle for use as an automotive engine is shown in Fig. 9.45. In the first turbine, the gas expands to just a low enough pressure P_5 for that turbine to drive the compressor. The gas is then expanded through a second turbine connected to the drive wheels. The data for this engine are as shown in the figure. Consider the working fluid to be air

throughout the entire cycle, and assume that all processes are ideal. Determine the following:
(a) The intermediate pressure P_5.
(b) The net work output of the engine, per kilogram of air, and the mass flow rate through the engine.
(c) The air temperature entering the burner T_3, and the thermal efficiency of the engine.

$P_1 = 100$ kPa
$T_1 = 300$ K

$P_2/P_1 = 5.0$
$P_7 = 100$ kPa

$T_4 = 1300$ K

FIGURE 9.45
Sketch for Problem 9.55.

9.56 Repeat Problem 9.55 assuming that the compressor has an efficiency of 82%, that both turbines have efficiencies of 87%, and that the regenerator efficiency is 70%. Also assume that there are pressure drops due to friction in the burner and on both sides of the regenerator. In each case, the pressure drop can be estimated to be 2% of the inlet pressure to that component of the system.

9.57 Consider an ideal gas-turbine cycle with two stages of compression and two stages of expansion. The pressure ratio across each compressor stage and each turbine stage is 8 to 1. The pressure at the entrance to the first compressor is 14.7 lbf/in.², the temperature entering each compressor is 70 F, and the temperature entering each trubine is 2000 F. An ideal regenerator is also incorporated into the cycle. Determine the compressor work, the turbine work, and the cycle thermal efficiency.

9.58 Repeat Problem 9.57 assuming that each compressor stage and each turbine state has an isentropic efficiency of 85%, and also that the regenerator has an efficiency of 70%.

9.59 Consider the combined gas turbine cycle-steam turbine cycle power plant shown in Fig. 9.46. The following data are known for the gas turbine cycle: air enters the compressor at 14.7 lbf/in.², 80 F, the compressor pressure ratio is 14, and the isentropic compressor efficiency is 87%; the heater input rate is 2.0×10^8 Btu/h; the turbine inlet temperature is 2300 F, exhaust pressure is 14.7 lbf/in.², and the isentropic turbine efficiency is 87%; the cycle exhaust temperature from the heat exchanger is 400 F. The following data

FIGURE 9.46

Sketch for Problem 9.59.

are known for the steam turbine cycle: pump inlet state is saturated liquid at 100 F, pump exit pressure is 2000 lbf/in.2 and the isentropic pump efficiency is 85%; turbine inlet temperature is 900 F and the isentropic turbine efficiency is 87%. Determine:

(a) The mass flow rate of air in the gas turbine cycle.
(b) The mass flow rate of water in the steam cycle.
(c) The overall thermal efficiency of the combined cycle.

9.60 Consider an ideal air-standard cycle for a gas-turbine and jet-propulsion unit, such as that shown in Fig. 9.34. The pressure and temperature entering the compressor are 100 kPa, 290 K. The pressure ratio across the compressor is 14 to 1 and the turbine inlet temperature is 1500 K. On leaving the turbine the air enters the nozzle and expands to 100 kPa. Determine the pressure at the nozzle inlet and the velocity of the air leaving the nozzle.

9.61 Repeat Problem 9.60 assuming that the isentropic compressor efficiency is 87%, the isentropic turbine efficiency is 89%, and the isentropic nozzle efficiency is 96%.

9.62 A jet aircraft is flying at an altitude of 16 000 ft, where the ambient pressure is approximately 8 lbf/in.² and the ambient temperature is 0 F. The velocity of the aircraft is 900 ft/s, and the pressure ratio across the compressor is 14. Devise an air-standard cycle that approximates this engine, and determine the velocity (relative to the aircraft) of the air leaving the engine, assuming that it has been expanded to the ambient pressure. Assume a maximum gas temperature in the engine of 2600 R.

9.63 A heat exchanger is incorporated into an ideal air-standard refrigeration cycle, as shown in Fig. 9.47. It may be assumed that both the compression and expansion are reversible adiabatic processes in this ideal case. Determine the coefficient of performance for the cycle.

FIGURE 9.47
Schematic arrangement of Problem 9.63.

9.64 Repeat Problem 9.63 assuming an isentropic efficiency for both the compressor and the expander of 75%.

9.65 Repeat Problems 9.63 and 9.64 assuming that helium is used as the cycle working fluid instead of air. Discuss the significance of the results.

9.66 Write a computer program to solve the following problem. It is desired to determine the effects of varying parameters on the performance of an air-standard Brayton cycle. Consider a compressor inlet condition of 100 kPa, 20°C, and assume constant specific heat. The cycle thermal efficiency and net work output per kilogram are to be determined for all combinations of the following variables:
(a) Compressor pressure ratio of 6, 8, 10, 12, and 14.

(b) Maximum cycle temperature of 900°C, 1100°C, 1300°C, and 1500°C.

(c) Compressor and turbine isentropic efficiencies each 100%, 90%, 80%, and 70%.

9.67 Write a computer program to solve the following problem. It is desired to study the effect of variable specific heat on the results of Problem 9.66, using ideal gas "enthalpy" and "relative pressure" functions that are curve-fitted polynomial functions of temperature from Table A.10 values, using the same procedure as for specific heat values in Problem 5.85.

9.68 Write a computer program to solve the following problem. Consider an ideal Otto cycle operating with nitrogen gas as the working fluid. The condition at the beginning of the compression process is 14.7 lbf/in.², 70 F. Assume variable specific heat according to the equation in Table A.9. Determine the net work output per lbm and the cycle thermal efficiency for various combinations of compression ratio and maximum cycle temperature, and compare the results with those found assuming constant specific heat.

9.69 Write a computer program to solve the following problem. It is desired to compare the performance of the ideal Otto and Diesel cycles. Repeat Problem 9.68 for the Diesel cycle operating with nitrogen gas as the working fluid, and using the specific heat equation given in Table A.9.

9.70 Write a computer program to solve the following problem. It is desired to study the effect of the number of stages on the performance of an ideal gas turbine cycle having intercooling, reheat, and a regenerator, as shown in Fig. 9.31 (for two stages). Assume that the inlet condition, point 1, is 14.7 lbf/in.², 70 F (or 100 kPa, 20°C), and that the total pressure ratio, the maximum cycle temperature, and the number of stages (equal number of compressor and turbine stages) are all variable. Each stage is to have an equal pressure ratio. Determine the net work output per unit mass and the cycle thermal efficiency for various combinations of these variables.

9.71 Write a computer program to solve the following problem. It is desired to study the effect of variable specific heat on the results of Problem 9.70. Repeat that problem, using the ideal gas polynomial functions for air developed in the program of Problem 9.67.

9.72 Write a computer program to solve the following problem. It is desired to study the effect of irreversible component processes on the results of Problem 9.71. Repeat that problem for various values of isentropic compressor and turbine efficiencies and of regenerator efficiency.

9.73 Write a computer program to solve the following problem. It is desired to study the performance of an air-standard refrigeration cycle incorporating a heat exchanger, as shown in Fig. 9.47. The pressure and temperature at the inlet to the compressor are 100 kPa, 15°C. The pressure ratio across the compressor and expander and also the isentropic efficiencies of the compressor and expander are to be taken as variables in this study.

10

THERMODYNAMIC RELATIONS

We have already defined and used several thermodynamic properties. Among these are pressure, specific volume, density, temperature, mass, internal energy, enthalpy, entropy, constant-pressure and constant-volume specific heats, and the Joule–Thomson coefficient. Two other properties, the Helmholtz function and the Gibbs function, have been introduced and will be used more extensively in the following chapters. We have also had occasion to use tables of thermodynamic properties for a number of different substances.

One important question is now raised, namely, which of the thermodynamic properties can be experimentally measured? We can answer this question by considering the measurements we can make in the laboratory. Some of the properties such as internal energy and entropy cannot be measured directly and must be calculated from other experimental data. If we carefully consider all these thermodynamic properties, we conclude that there are only four that can be directly measured: pressure, temperature, volume, and mass.

This leads to a second question, namely, how can values of the thermodynamic properties that cannot be measured be determined from experimental data on those properties that can be measured? In answering this question we will develop certain general thermodynamic relations. In view of the fact that there are millions of such equations that can be written, our study will be limited to certain basic considerations, with particular reference to the determination of thermodynamic properties from experimental data. We will also consider such related matters as generalized charts and equations of state.

10.1 TWO IMPORTANT RELATIONS

This chapter involves partial derivatives, and two important relations are reviewed here. Consider a variable z that is a continuous function of x and y.

$$z = f(x, y)$$
$$dz = \left(\frac{\partial z}{\partial x}\right)_y dx + \left(\frac{\partial z}{\partial y}\right)_x dy$$

It is convenient to write this in the form

$$dz = M\,dx + N\,dy \tag{10.1}$$

$$M = \left(\frac{\partial z}{\partial x}\right)_y$$

= partial derivative of z with respect to x (the variable y being held constant)

$$N = \left(\frac{\partial z}{\partial y}\right)_x$$

= partial derivative of z with respect to y (the variable x being held constant)

The physical significance of partial derivatives as they relate to the properties of a pure substance can be explained by referring to Fig. 10.1, which shows a P-v-T surface of the superheated vapor region of a pure substance. It shows constant-temperature, constant-pressure, and constant-specific volume planes that intersect at point b on the surface. Thus the partial derivative $(\partial P/\partial v)_T$ is the slope of curve

FIGURE 10.1

Schematic representation of partial derivatives.

abc at point *b*. Line *de* represents the tangent to curve *abc* at point *b*. A similar interpretation can be made of the partial derivatives $(\partial P/\partial T)_v$ and $(\partial v/\partial T)_p$.

It should also be noted that if we wish to evaluate the partial derivative along a constant-temperature line the rules for ordinary derivatives can be applied. Thus we can write for a constant-temperature process:

$$\left(\frac{\partial P}{\partial v}\right)_T = \frac{dP_T}{dv_T}$$

and the integration can be performed as usual. This point will be demonstrated later in a number of examples.

Let us return to the consideration of the relation

$$dz = M\,dx + N\,dy$$

If *x*, *y*, and *z* are all point functions (that is, quantities that depend only on the state and are independent of the path), the differentials are exact differentials. If this is the case, the following important relation holds:

$$\left(\frac{\partial M}{\partial y}\right)_x = \left(\frac{\partial N}{\partial x}\right)_y$$

The proof of this is as follows:

$$\left(\frac{\partial M}{\partial y}\right)_x = \frac{\partial^2 z}{\partial x\,\partial y}$$

$$\left(\frac{\partial N}{\partial x}\right)_y = \frac{\partial^2 z}{\partial y\,\partial x}$$

Since the order of differentiation makes no difference when point functions are involved, it follows that

$$\frac{\partial^2 z}{\partial x\,\partial y} = \frac{\partial^2 z}{\partial y\,\partial x}$$

$$\left(\frac{\partial M}{\partial y}\right)_x = \left(\frac{\partial N}{\partial x}\right)_y$$

The second important mathematical relation is

$$\left(\frac{\partial x}{\partial y}\right)_z \left(\frac{\partial y}{\partial z}\right)_x \left(\frac{\partial z}{\partial x}\right)_y = -1 \tag{10.2}$$

The proof of this relation is as follows. Consider three variables *x*, *y*, and *z*. Suppose there exists a relation between the variables of the form

$$x = f(y, z)$$

then

$$dx = \left(\frac{\partial x}{\partial y}\right)_z dy + \left(\frac{\partial x}{\partial z}\right)_y dz \tag{10.3}$$

If this relationship between the three variables is written in the form

$$y = f(x, z)$$

it follows that

$$dy = \left(\frac{\partial y}{\partial x}\right)_z dx + \left(\frac{\partial y}{\partial z}\right)_x dz \qquad (10.4)$$

Substituting Eq. 10.4 into Eq. 10.3, we have

$$dx = \left(\frac{\partial x}{\partial y}\right)_z \left[\left(\frac{\partial y}{\partial x}\right)_z dx + \left(\frac{\partial y}{\partial z}\right)_x dz\right] + \left(\frac{\partial x}{\partial z}\right)_y dz$$

$$= \left(\frac{\partial x}{\partial y}\right)_z \left(\frac{\partial y}{\partial x}\right)_z dx + \left[\left(\frac{\partial x}{\partial y}\right)_z \left(\frac{\partial y}{\partial z}\right)_x + \left(\frac{\partial x}{\partial z}\right)_y\right] dz$$

There are two independent variables, and we select x and z as these variables. Suppose that $dz = 0$ and $dx \neq 0$. It then follows that

$$\left(\frac{\partial x}{\partial y}\right)_z \left(\frac{\partial y}{\partial x}\right)_z = 1 \qquad (10.5)$$

Similarly, suppose that $dx = 0$ and $dz \neq 0$. It then follows that

$$\left(\frac{\partial x}{\partial y}\right)_z \left(\frac{\partial y}{\partial z}\right)_x + \left(\frac{\partial x}{\partial z}\right)_y = 0$$

$$\left(\frac{\partial x}{\partial y}\right)_z \left(\frac{\partial y}{\partial z}\right)_x = -\left(\frac{\partial x}{\partial z}\right)_y$$

$$\left(\frac{\partial x}{\partial y}\right)_z \left(\frac{\partial y}{\partial z}\right)_x \left(\frac{\partial z}{\partial x}\right)_y = -1$$

This is Eq. 10.2, which we set out to derive.

10.2 THE MAXWELL RELATIONS

Consider a simple compressible system of fixed chemical composition. The Maxwell relations, which can be written for such a system, are four equations relating the properties P, v, T, and s.

The Maxwell relations are most easily derived by considering four relations involving thermodynamic properties. Two of these relations have already been derived and are

$$du = T\,ds - P\,dv \qquad (10.6)$$

$$dh = T\,ds + v\,dP \qquad (10.7)$$

The other two are derived from the definition of the Helmholtz function, a, and the Gibbs function, g.

$$a = u - Ts$$

$$da = du - T\,ds - s\,dT$$

Substituting Eq. 10.6 into this relation gives the third relation.

$$da = -P\,dv - s\,dT \qquad (10.8)$$

Similarly,

$$g = h - Ts$$
$$dg = dh - T\,ds - s\,dT$$

Substituting Eq. 10.7 yields the fourth relation.

$$dg = v\,dP - s\,dT \qquad (10.9)$$

Since Eqs. 10.6, 10.7, 10.8, and 10.9 are relations involving properties, we conclude that these are exact differentials and, therefore, are of the general form

$$dz = M\,dx + N\,dy$$

Since

$$\left(\frac{\partial M}{\partial y}\right)_x = \left(\frac{\partial N}{\partial x}\right)_y \qquad (10.10)$$

it follows from Eq. 10.6 that

$$\left(\frac{\partial T}{\partial v}\right)_s = -\left(\frac{\partial P}{\partial s}\right)_v \qquad (10.11)$$

Similarly, from Eqs. 10.7, 10.8, and 10.9 we can write

$$\left(\frac{\partial T}{\partial P}\right)_s = \left(\frac{\partial v}{\partial s}\right)_P \qquad (10.12)$$

$$\left(\frac{\partial P}{\partial T}\right)_v = \left(\frac{\partial s}{\partial v}\right)_T \qquad (10.13)$$

$$\left(\frac{\partial v}{\partial T}\right)_P = -\left(\frac{\partial s}{\partial P}\right)_T \qquad (10.14)$$

These four equations are known as the Maxwell relations for a simple compressible system, and the great utility of these equations will be demonstrated in later sections of this chapter. In particular, it should be noted that pressure, temperature, and specific volume can be measured by experimental methods, whereas entropy cannot be determined experimentally. By using the Maxwell relations changes in entropy can be determined from quantities that can be measured, that is, pressure, temperature, and specific volume.

There are a number of other very useful relations that can be derived from Eqs. 10.6 through 10.9. For example, from Eq. 10.6, we can write the relations

$$\left(\frac{\partial u}{\partial s}\right)_v = T, \qquad \left(\frac{\partial u}{\partial v}\right)_s = -P \qquad (10.15)$$

Similarly, from the other equations, we have the following:

$$\left(\frac{\partial h}{\partial s}\right)_P = T, \qquad \left(\frac{\partial h}{\partial P}\right)_s = v$$

$$\left(\frac{\partial a}{\partial v}\right)_T = -P, \qquad \left(\frac{\partial a}{\partial T}\right)_v = -s$$

$$\left(\frac{\partial g}{\partial P}\right)_T = v, \qquad \left(\frac{\partial g}{\partial T}\right)_P = -s \qquad \textbf{(10.16)}$$

As already noted, the Maxwell relations just presented are written for a simple compressible substance. It is readily evident, however, that similar Maxwell relations can be written for substances involving other effects, such as electrical and magnetic effects. For example, Eq. 7.9 can be written in the form

$$dU = T\,dS - P\,dV + \mathcal{T}\,dL + \mathcal{S}\,d\mathcal{A} + \mu_0 \mathcal{H}\,d(V\mathcal{M}) + \mathcal{E}\,dZ + \cdots \qquad \textbf{(10.17)}$$

Thus, at constant volume for a substance involving only magnetic effects, we can write

$$dU = T\,dS + \mu_0 V\mathcal{H}\,d\mathcal{M}$$

and it follows that for such a substance

$$\left(\frac{\partial T}{\partial \mathcal{M}}\right)_{S,V} = \mu_0 V\left(\frac{\partial \mathcal{H}}{\partial S}\right)_{\mathcal{M},V}$$

Other Maxwell relations similar to Eqs. 10.12 through 10.14 could be written for such a substance. The extension of this approach to other systems, as well as to the interrelation between the various effects that may occur in a given system, is readily evident. For example, suppose a system involved both magnetic and surface effects. For such a system we could consider a constant entropy process and write

$$\left(\frac{\partial \mathcal{S}}{\partial \mathcal{M}}\right)_{S,\mathcal{A},V} = \mu_0 V\left(\frac{\partial \mathcal{H}}{\partial \mathcal{A}}\right)_{S,\mathcal{M},V}$$

This matter becomes much more complex when we consider applying the property relation to a system of variable composition. This subject will be taken up in Chapter 11.

Example 10.1

From an examination of the properties of compressed liquid water, as given in Table A.1.4 of the Appendix, we find that the entropy of compressed liquid is greater than the entropy of saturated liquid for a temperature of 32 F, and is less than that of saturated liquid for all the other temperatures listed. Explain why this follows from other thermodynamic data.

System: Water.

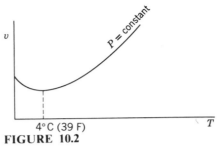

FIGURE 10.2

Sketch for Example 10.1.

Solution

Suppose we increase the pressure of liquid water that is initially saturated, while keeping the temperature constant. The change of entropy for the water during this process can be found by integrating the following Maxwell relation, Eq. 10.14:

$$\left(\frac{\partial s}{\partial P}\right)_T = -\left(\frac{\partial v}{\partial T}\right)_P$$

Therefore, the sign of the entropy change depends on the sign of the term $(\partial v/\partial T)_P$. The physical significance of this term is that it involves the change in specific volume of water as the temperature changes while the pressure remains constant. Now, as water at moderate pressures and 32 F is heated in a constant-pressure process, the specific volume decreases until the point of maximum density is reached at approximately 39 F, after which it increases. This is shown on a v-T diagram in Fig. 10.2. Thus the quantity $(\partial v/\partial T)_P$ is the slope of the curve in Fig. 10.2. Since this slope is negative at 32 F, the quantity $(\partial s/\partial P)_T$ is positive at 32 F. At the point of maximum density the slope is zero and, therefore, the constant-pressure line shown in Fig. 7.7 crosses the saturated-liquid line at the point of maximum density.

10.3 CLAPEYRON EQUATION

The Clapeyron equation is an important relation involving the saturation pressure and temperature, the change of enthalpy associated with a change of phase, and the specific volumes of the two phases. In particular it is an example of how a change in a property that cannot be measured directly, the enthalpy in this case, can be determined from measurements of pressure, temperature, and specific volume. It can be derived in a number of ways. Here we proceed by considering one of the Maxwell relations, Eq. 10.13.

$$\left(\frac{\partial P}{\partial T}\right)_v = \left(\frac{\partial s}{\partial v}\right)_T$$

Consider, for example, the change of state from saturated liquid to saturated vapor of a pure substance. This is a constant-temperature process, and therefore

Eq. 10.13 can be integrated between the saturated-liquid and saturated-vapor state. We also note that when saturated states are involved, pressure and temperature are independent of volume. Therefore,

$$\left(\frac{dP}{dT}\right)_{\text{sat}} = \frac{s_g - s_f}{v_g - v_f} = \frac{s_{fg}}{v_{fg}} = \frac{h_{fg}}{Tv_{fg}} \tag{10.18}$$

The significance of this equation is that $(dP/dT)_{\text{sat}}$ is the slope of the vapor-pressure curve. Thus h_{fg} at a given temperature can be determined from the slope of the vapor-pressure curve and the specific volume of saturated liquid and saturated vapor at the given temperature.

There are several different changes of phase that can occur at constant temperature and constant pressure. If we designate the two phases with superscripts $''$ and $'$, we can write the Clapeyron equation for the general case.

$$\left(\frac{dP}{dT}\right)_{\text{sat}} = \frac{s'' - s'}{v'' - v'}$$

We also note that $T(s'' - s') = h'' - h'$. Therefore

$$\left(\frac{dP}{dT}\right)_{\text{sat}} = \frac{h'' - h'}{T(v'' - v')} \tag{10.19}$$

If the phase designated $''$ is vapor, then at low pressure the equation is usually simplified by assuming that $v'' \gg v'$ and assuming also that $v'' = RT/P$. The relation then becomes

$$\left(\frac{dP}{dT}\right)_{\text{sat}} = \frac{h_g - h'}{T(RT/P)}$$

$$\left(\frac{dP}{P}\right)_{\text{sat}} = \frac{(h_g - h')}{R}\left(\frac{dT}{T^2}\right)_{\text{sat}} \tag{10.20}$$

Example 10.2

Determine the saturation pressure of water vapor at $-60°C$ using data available in the steam tables.

System: Water.

Solution

Table 6 of the steam tables (Appendix Table A.1.5) does not give saturation pressures for temperatures less than $-40°C$. However, we do notice that h_{ig} is relatively constant in this range and, therefore, we proceed to use Eq. 10.20 and integrate between the limits $-40°C$ and $-60°C$.

$$\int_1^2 \frac{dP}{P} = \int_1^2 \frac{h_{ig}}{R}\frac{dT}{T^2} = \frac{h_{ig}}{R}\int_1^2 \frac{dT}{T^2}$$

$$\ln \frac{P_2}{P_1} = \frac{h_{ig}}{R}\left(\frac{T_2 - T_1}{T_1 T_2}\right)$$

Let

$$P_2 = 0.0129 \text{ kPa}, \qquad T_2 = 233.2 \text{ K}$$
$$P_1 = ?, \qquad\qquad T_1 = 213.2 \text{ K}$$

Then

$$\ln \frac{P_2}{P_1} = \frac{2838.9}{0.461\ 52} \left(\frac{233.2 - 213.2}{233.2 \times 213.2} \right) = 2.4744$$
$$P_1 = 0.001\ 09 \text{ kPa}$$

10.4 SOME THERMODYNAMIC RELATIONS INVOLVING ENTHALPY, INTERNAL ENERGY, AND ENTROPY

Let us first derive two equations, one involving C_p and the other involving C_v. We have defined C_p as

$$C_p \equiv \left(\frac{\partial h}{\partial T} \right)_P$$

We have also noted that for a pure substance

$$T\,ds = dh - v\,dP$$

Therefore,

$$C_p = \left(\frac{\partial h}{\partial T} \right)_P = T \left(\frac{\partial s}{\partial T} \right)_P \tag{10.21}$$

Similarly, from the definition of C_v,

$$C_v \equiv \left(\frac{\partial u}{\partial T} \right)_v$$

and the relation

$$T\,ds = du + P\,dv$$

it follows that

$$C_v = \left(\frac{\partial u}{\partial T} \right)_v = T \left(\frac{\partial s}{\partial T} \right)_v \tag{10.22}$$

We will now derive a general relation for the change of enthalpy of a pure substance. We first note that for a pure substance

$$h = h(T, P)$$

Therefore,

$$dh = \left(\frac{\partial h}{\partial T} \right)_P dT + \left(\frac{\partial h}{\partial P} \right)_T dP$$

From the relation

$$T\,ds = dh - v\,dP$$

it follows that

$$\left(\frac{\partial h}{\partial P}\right)_T = v + T\left(\frac{\partial s}{\partial P}\right)_T$$

Substituting the Maxwell relation, Eq. 10.14, we have

$$\left(\frac{\partial h}{\partial P}\right)_T = v - T\left(\frac{\partial v}{\partial T}\right)_P \qquad (10.23)$$

On substituting this equation and Eq. 10.21, we have

$$dh = C_p\,dT + \left[v - T\left(\frac{\partial v}{\partial T}\right)_P\right]dP \qquad (10.24)$$

Along an isobar we have

$$dh_P = C_p\,dT_P$$

and along an isotherm,

$$dh_T = \left[v - T\left(\frac{\partial v}{\partial T}\right)_P\right]dP_T \qquad (10.25)$$

The significance of Eq. 10.24 is that this equation can be integrated to give the change in enthalpy associated with a change of state

$$h_2 - h_1 = \int_1^2 C_p\,dT + \int_1^2 \left[v - T\left(\frac{\partial v}{\partial T}\right)_P\right]dP \qquad (10.26)$$

The information needed to integrate the first term is a constant-pressure specific heat along one (and only one) isobar. The integration of the second integral requires that an equation of state giving the relation between P, v, and T be known. Furthermore, it is advantageous to have this equation of state explicit in v, for then the derivative $(\partial v/\partial T)_P$ is readily evaluated.

This can be further illustrated by reference to Fig. 10.3. Suppose we wished to know the change of enthalpy between states 1 and 2. We might do so along path 1–x–2, which consists of one isotherm, 1–x, and one isobar, x–2. Thus we could integrate Eq. 10.26:

$$h_2 - h_1 = \int_{T_1}^{T_2} C_p\,dT + \int_{P_1}^{P_2} \left[v - T\left(\frac{\partial v}{\partial T}\right)_P\right]dP$$

Since $T_1 = T_x$ and $P_2 = P_x$, this can be written

$$h_2 - h_1 = \int_{T_x}^{T_2} C_p\,dT + \int_{P_1}^{P_x} \left[v - T\left(\frac{\partial v}{\partial T}\right)_P\right]dP$$

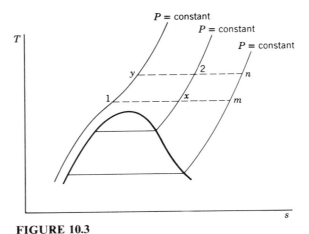

FIGURE 10.3

Sketch showing various paths by which a given change of state can take place.

The second term in this equation gives the change in enthalpy along the isotherm 1–x and the first term the change in enthalpy along the isobar x–2. When these are added together, the result is the net change in enthalpy between 1 and 2. It should be noted that in this case the constant-pressure specific heat must be known along the isobar passing through 2 and x. One could also find the change in enthalpy by following path 1–y–2, in which case the constant-pressure specific heat must be known along the 1–y isobar. If the constant-pressure specific heat is known at another pressure, say, the isobar passing through m–n, the change in enthalpy could be found by following path 1–m–n–2. This would involve calculating the change of enthalpy along two isotherms—1–m and n–2.

Let us now derive a similar relation for the change of internal energy. All the steps in this derivation are given, but without detailed comment. Note that the starting point is to write $u = u(T, v)$, whereas in the case of enthalpy the starting point was $h = h(T, P)$.

$$u = f(T, v)$$

$$du = \left(\frac{\partial u}{\partial T}\right)_v dT + \left(\frac{\partial u}{\partial v}\right)_T dv$$

$$T\, ds = du + P\, dv$$

Therefore,

$$\left(\frac{\partial u}{\partial v}\right)_T = T\left(\frac{\partial s}{\partial v}\right)_T - P \tag{10.27}$$

Substituting the Maxwell relation, Eq. 10.13,

$$\left(\frac{\partial u}{\partial v}\right)_T = T\left(\frac{\partial P}{\partial T}\right)_v - P$$

Therefore,

$$du = C_v \, dT + \left[T \left(\frac{\partial P}{\partial T} \right)_v - P \right] dv \qquad (10.28)$$

Along an isometric this reduces to

$$du_v = C_v \, dT_v$$

and along an isotherm we have

$$du_T = \left[T \left(\frac{\partial P}{\partial T} \right)_v - P \right] dv_T \qquad (10.29)$$

In a manner similar to that outlined above for changes in enthalpy, the change of internal energy for a given change of state for a pure substance can be determined from Eq. 10.28 if the constant-volume specific heat is known along one isometric and an equation of state explicit in P [to obtain the derivative $(\partial P/\partial T)_v$] is available in the region involved. A diagram similar to Fig. 10.3 could be drawn, with the isobars replaced with isometrics, and the same general conclusions would be reached.

To summarize, we have derived Eqs. 10.24 and 10.28:

$$dh = C_p \, dT + \left[v - T \left(\frac{\partial v}{\partial T} \right)_P \right] dP$$

$$du = C_v \, dT + \left[T \left(\frac{\partial P}{\partial T} \right)_v - P \right] dv$$

The first of these involves the change of enthalpy, the constant-pressure specific heat, and is particularly suited to an equation of state explicit in v. The second involves the change of internal energy, the constant-volume specific heat, and is particularly suited to an equation of state explicit in P. If the first of these equations is used to determine the change of enthalpy, the internal energy is readily found by noting that

$$u_2 - u_1 = h_2 - h_1 - (P_2 v_2 - P_1 v_1)$$

If the second equation is used to find changes of internal energy, the change of enthalpy is readily found from this same relation. Which of these two equations is used to determine changes in internal energy and enthalpy will depend on the information available for specific heat and an equation of state (or other P-v-T data).

Two parallel expressions can be found for the change of entropy.

$$s = s(T, P)$$

$$ds = \left(\frac{\partial s}{\partial T} \right)_P dT + \left(\frac{\partial s}{\partial P} \right)_T dP$$

Substituting Eqs. 10.21 and 10.14, we have

$$ds = C_p \frac{dT}{T} - \left(\frac{\partial v}{\partial T}\right)_P dP \qquad (10.30)$$

$$s_2 - s_1 = \int_1^2 C_p \frac{dT}{T} - \int_1^2 \left(\frac{\partial v}{\partial T}\right)_P dP \qquad (10.31)$$

Along an isobar we have

$$(s_2 - s_1)_P = \int_1^2 C_p \frac{dT_P}{T}$$

and along an isotherm,

$$(s_2 - s_1)_T = -\int_1^2 \left(\frac{\partial v}{\partial T}\right)_P dP$$

Note from Eq. 10.31 that if a constant-pressure specific heat is known along one isobar and an equation of state explicit in v is available, the change of entropy can be evaluated. This is analogous to the expression for the change of enthalpy given in Eq. 10.24.

$$s = s(T, v)$$

$$ds = \left(\frac{\partial s}{\partial T}\right)_v dT + \left(\frac{\partial s}{\partial v}\right)_T dv$$

Substituting Eqs. 10.22 and 10.13,

$$ds = C_v \frac{dT}{T} + \left(\frac{\partial P}{\partial T}\right)_v dv \qquad (10.32)$$

$$s_2 - s_1 = \int_1^2 C_v \frac{dT}{T} + \int_1^2 \left(\frac{\partial P}{\partial T}\right)_v dv \qquad (10.33)$$

This expression for change of entropy involves the change of entropy along an isometric where the constant-volume specific heat is known and along an isotherm where an equation of state explicit in P is known, and thus it is analogous to the expression for change of internal energy given in Eq. 10.28.

Example 10.3

Over a certain range of pressures and temperatures the equation of state of a certain substance is given with considerable accuracy by the relation

$$\frac{Pv}{RT} = 1 - C' \frac{P}{T^4}$$

or

$$v = \frac{RT}{P} - \frac{C}{T^3}$$

where C and C' are constants.

Derive an expression for the change of enthalpy and entropy of this substance in an isothermal process.

System: Gas.

Solution

Since the equation of state is explicit in v, Eq. 10.25 would be particularly relevant to this problem. On integrating this equation, we have

$$(h_2 - h_1)_T = \int_1^2 \left[v - T \left(\frac{dv}{dT} \right)_P \right] dP_T$$

From the equation of state,

$$\left(\frac{\partial v}{\partial T} \right)_P = \frac{R}{P} + \frac{3C}{T^4}$$

Therefore,

$$(h_2 - h_1)_T = \int_1^2 \left[v - T \left(\frac{R}{P} + \frac{3C}{T^4} \right) \right] dP_T$$

$$= \int_1^2 \left[\frac{RT}{P} - \frac{C}{T^3} - \frac{RT}{P} - \frac{3C}{T^3} \right] dP_T$$

$$(h_2 - h_1)_T = \int_1^2 -\frac{4C}{T^3} dP_T = -\frac{4C}{T^3} (P_2 - P_1)_T$$

For entropy we use Eq. 10.31, which is particularly relevant for an equation of state explicit in v.

$$(s_2 - s_1)_T = -\int_1^2 \left(\frac{\partial v}{\partial T} \right)_P dP_T = -\int_1^2 \left(\frac{R}{P} + \frac{3C}{T^4} \right) dP_T$$

$$(s_2 - s_1)_T = -R \ln \left(\frac{P_2}{P_1} \right)_T - \frac{3C}{T^4} (P_2 - P_1)_T$$

10.5 SOME THERMODYNAMIC RELATIONS INVOLVING SPECIFIC HEAT

Some important relations involving specific heats can also be developed. We have noted that the specific heat of an ideal gas is a function of the temperature only. For real gases the specific heat varies with pressure as well as temperature, and frequently we are interested in the variation of specific heat with pressure or volume. These relations can be derived as follows. Consider Eq. 10.30:

$$ds = \left(\frac{C_p}{T} \right) dT - \left(\frac{\partial v}{\partial T} \right)_P dP$$

Since this equation is of the general form $dz = Mdx + Ndy$, we can proceed as

follows to find a relation that gives the variation of the constant-pressure specific heat with pressure at constant temperature.

$$\left(\frac{\partial(C_p/T)}{\partial P}\right)_T = -\left[\frac{\partial}{\partial T}\left(\frac{\partial v}{\partial T}\right)_P\right]_P$$

$$\left(\frac{\partial C_p}{\partial P}\right)_T = -T\left(\frac{\partial^2 v}{\partial T^2}\right)_P \qquad\qquad \textbf{(10.34)}$$

The variation of the constant-volume specific heat with volume as the temperature remains constant can be found in a similar manner. Consider Eq. 10.32:

$$ds = \left(\frac{C_v}{T}\right)dT + \left(\frac{\partial P}{\partial T}\right)_v dv$$

Since this is of the form $dz = M\,dx + N\,dy$,

$$\left(\frac{\partial(C_v/T)}{\partial v}\right)_T = \left[\frac{\partial}{\partial T}\left(\frac{\partial P}{\partial T}\right)_v\right]_v$$

$$\left(\frac{\partial C_v}{\partial v}\right)_T = T\left(\frac{\partial^2 P}{\partial T^2}\right)_v \qquad\qquad \textbf{(10.35)}$$

The important thing to note about Eqs. 10.34 and 10.35 is that the variation of the constant-volume and constant-pressure specific heats at constant temperature can be found from the equation of state.

Example 10.4

Determine the variation of C_p with pressure at constant temperature for a substance such as the one in Example 10.3 over the range where the equation of state is given by the relation

$$v = \frac{RT}{P} - \frac{C}{T^3}$$

System: Gas.

Solution

Using Eq. 10.34:

$$\left(\frac{\partial C_p}{\partial P}\right)_T = -T\left(\frac{\partial^2 v}{\partial T^2}\right)_P$$

$$\left(\frac{\partial v}{\partial T}\right)_P = \frac{R}{P} + \frac{3C}{T^4}$$

$$\left(\frac{\partial^2 v}{\partial T^2}\right)_P = -\frac{12C}{T^5}$$

$$\left(\frac{\partial C_p}{\partial P}\right)_T = -T\left(-\frac{12C}{T^5}\right) = \frac{12C}{T^4}$$

A final interesting and useful relation involving the difference between C_p and C_v can be derived by equating Eqs. 10.30 and 10.32.

$$C_p \frac{dT}{T} - \left(\frac{\partial v}{\partial T}\right)_P dP = C_v \frac{dT}{T} + \left(\frac{\partial P}{\partial T}\right)_v dv$$

$$dT = \frac{T(\partial P/\partial T)_v}{C_p - C_v} dv + \frac{T(\partial v/\partial T)_P}{C_p - C_v} dP$$

But

$$T = f(v, P)$$

$$dT = \left(\frac{\partial T}{\partial v}\right)_P dv + \left(\frac{\partial T}{\partial P}\right)_v dP$$

Therefore,

$$\left(\frac{\partial T}{\partial v}\right)_P = \frac{T(\partial P/\partial T)_v}{C_p - C_v}$$

and

$$\left(\frac{\partial T}{\partial P}\right)_v = \frac{T(\partial v/\partial T)_P}{C_p - C_v}$$

When these equations are solved for $C_p - C_v$, they yield the same result.

$$C_p - C_v = T\left(\frac{\partial v}{\partial T}\right)_P \left(\frac{\partial P}{\partial T}\right)_v \qquad (10.36)$$

But from Eq. 10.2,

$$\left(\frac{\partial P}{\partial T}\right)_v = -\left(\frac{\partial v}{\partial T}\right)_P \left(\frac{\partial P}{\partial v}\right)_T$$

Therefore,

$$C_p - C_v = -T\left(\frac{\partial v}{\partial T}\right)_P^2 \left(\frac{\partial P}{\partial v}\right)_T \qquad (10.37)$$

From this equation we draw several conclusions:

1. For liquids and solids, $(\partial v/\partial T)_P$ is usually relatively small and, therefore, for these phases the difference between the constant-pressure and constant-volume specific heats is small. For this reason many tables simply give the specific heat of a solid or a liquid without designating that it is at constant volume or pressure. Further, $C_p = C_v$ exactly when $(\partial v/\partial T)_P = 0$, as is true at the point of maximum density of water.

2. $C_p \rightarrow C_v$ as $T \rightarrow 0$ and, therefore, we conclude that the constant-pressure and constant-volume specific heats are equal at absolute zero.

3. The difference between C_p and C_v is always positive because $(\partial v/\partial T)_P^2$ is always positive and $(\partial P/\partial v)_T$ is negative for all known substances.

10.6 VOLUME EXPANSIVITY AND ISOTHERMAL AND ADIABATIC COMPRESSIBILITY

The student has most likely encountered the coefficient of linear expansion in his or her studies of strength of materials. This coefficient indicates how the length of a solid body is influenced by a change in temperature while the pressure remains constant. In terms of the notation of partial derivatives, the *coefficient of linear expansion*, δ_T, is defined as follows:

$$\delta_T = \frac{1}{L}\left(\frac{\partial L}{\partial T}\right)_P \tag{10.38}$$

A similar coefficient can be defined for changes in volume, and such a coefficient is applicable to liquids and gases as well as to solids. This coefficient of volume expansion, α_P, also called the volume expansivity, is an indication of the change in volume that results from a change in temperature while the pressure remains constant. The definition of *volume expansivity* is

$$\alpha_P \equiv \frac{1}{V}\left(\frac{\partial V}{\partial T}\right)_P = \frac{1}{v}\left(\frac{\partial v}{\partial T}\right)_P \tag{10.39}$$

The isothermal compressibility, β_T, is an indication of the change in volume that results from a change in pressure while the temperature remains constant. The definition of the *isothermal compressibility* is

$$\beta_T \equiv -\frac{1}{V}\left(\frac{\partial V}{\partial P}\right)_T = -\frac{1}{v}\left(\frac{\partial v}{\partial P}\right)_T \tag{10.40}$$

The reciprocal of the isothermal compressibility is called the *isothermal bulk modulus*, B_T.

$$B_T \equiv -v\left(\frac{\partial P}{\partial v}\right)_T \tag{10.41}$$

The *adiabatic compressibility*, β_s, is an indication of the change in volume that results from a change in pressure while the entropy remains constant, and is defined as follows:

$$\beta_s \equiv -\frac{1}{v}\left(\frac{\partial v}{\partial P}\right)_s \tag{10.42}$$

The *adiabatic bulk modulus*, B_s, is the reciprocal of the adiabatic compressibility.

$$B_s \equiv -v\left(\frac{\partial P}{\partial v}\right)_s \tag{10.43}$$

Both the volume expansivity and isothermal compressibility are thermodynamic properties of a substance, and for a simple compressible substance are functions of two independent properties. Values of these properties are found in the standard handbooks of physical properties. The following example gives an indication of the use and significance of the volume expansivity and isothermal compressibility.

Example 10.5

Show that $C_p - C_v$ can be expressed in terms of the volume expansivity α_P, the specific volume v, the temperature T, and the isothermal compressibility β_T, by the relation

$$C_p - C_v = \frac{\alpha_P^2 v T}{\beta_T}$$

System: Pure substance.

Solution

From Eq. 10.37

$$C_p - C_v = -T\left(\frac{\partial v}{\partial T}\right)_P^2 \left(\frac{\partial P}{\partial v}\right)_T$$

$$\alpha_P \equiv \frac{1}{v}\left(\frac{\partial v}{\partial T}\right)_P, \qquad \left(\frac{\partial v}{\partial T}\right)^2 = \alpha_P^2 v^2$$

$$\beta_T \equiv -\frac{1}{v}\left(\frac{\partial v}{\partial P}\right)_T, \qquad \left(\frac{\partial P}{\partial v}\right)_T = -\frac{1}{\beta_T v}$$

Therefore

$$C_p - C_v = -T(\alpha_P^2 v^2)\left(-\frac{1}{\beta_T v}\right) = \frac{\alpha_P^2 v T}{\beta_T}$$

Example 10.6

The pressure on a block of copper having a mass of 1 lbm is increased in a reversible process from 1 atm to 1000 atm while the temperature is held constant at 60 F. Determine the work done on the copper during this process, the change in entropy per pound of copper, the heat transfer, the change of internal energy per pound, and $C_p - C_v$ for this change of state.

Over the range of pressures and temperatures involved in this problem the following data can be used:

> Volume expansivity = $\alpha_P = 2.8 \times 10^{-5} R^{-1}$
> Isothermal compressibility = $\beta_T = 5.9 \times 10^{-8}$ in.2/lbf
> Specific volume = 1.82×10^{-3} ft^3/lbm

System: Copper block.

States: Initial and final states known.

Process: Constant temperature, reversible.

Analysis: The work done during the isothermal expansion is

$$w = \int P \, dv_T$$

The isothermal compressibility has been defined:

$$\beta_T = -\frac{1}{v}\left(\frac{\partial v}{\partial P}\right)_T$$

$$v\beta_T\, dP_T = -dv_T$$

Therefore, for this isothermal process,

$$w = -\int_1^2 v\beta_T P\, dP_T$$

Since v and β_T remain essentially constant, this is readily integrated:

$$w = -\frac{v\beta_T}{2}(P_2{}^2 - P_1{}^2)$$

The change of entropy can be found by considering the Maxwell relation, Eq. 10.14, and the definition of volume expansivity.

$$\left(\frac{\partial s}{\partial P}\right)_T = -\left(\frac{\partial v}{\partial T}\right)_P = -\frac{v}{v}\left(\frac{\partial v}{\partial T}\right)_P = -v\alpha_P$$

$$ds_T = v\alpha_P\, dP_T$$

This can be readily integrated as follows if we assume that v and α_P remain constant:

$$(s_2 - s_1)_T = -v\alpha_P(P_2 - P_1)_T$$

The heat transfer for this reversible isothermal process is

$$q = T(s_2 - s_1)$$

The change in internal energy follows directly from the first law.

$$(u_2 - u_1) = q - w$$

From Example 10.5,

$$C_p - C_v = \frac{\alpha_P{}^2 v T}{\beta_T}$$

Solution

$$w = -\frac{v\beta_T}{2}(P_2{}^2 - P_1{}^2)$$

$$= -1.82 \times 10^{-3}\ \text{ft}^3/\text{lbm} \times 5.9 \times 10^{-8}\ \text{in.}^2/\text{lbf} \times \frac{1000^2 - 1^2}{2}$$

$$\times (14.7)^2\ \text{lbf}^2/\text{in.}^4 \times 144\ \text{in.}^2/\text{ft}^2 = -1.68\ \text{ft lbf}$$

$$(s_2 - s_1)_T = -v\alpha_P(P_2 - P_1)_T$$

$$= -1.82 \times 10^{-3}\ \text{ft}^3/\text{lbm} \times \frac{2.8 \times 10^{-5}}{R} \times (1000 - 1)$$

$$\times 14.7 \times 144\ \text{lbf/ft}^2 = -0.108\ \text{ft lbf/lbm R}$$

$$q = T(s_2 - s_1) = 520(-0.108) = -56.1\ \text{ft lbf}$$

$$(u_2 - u_1) = q - w = -56.1 - (-1.7) = -54.4\ \text{ft lbf}$$

$$C_p - C_v = \left(\frac{2.8}{10^5}\right)^2 \times \frac{1.82}{10^3} \times \frac{10^8}{5.9} \times 520 \times \frac{144}{778} = 0.0023\ \frac{\text{Btu}}{\text{lbm R}}$$

This is consistent with the earlier observation that $C_p - C_v$ is relatively small for solids.

10.7 DEVELOPING TABLES OF THERMODYNAMIC PROPERTIES FROM EXPERIMENTAL DATA

There are many ways in which tables of thermodynamic properties can be developed from experimental data. The purpose of this section is to convey some general principles and concepts by considering only the liquid and vapor phases.

Let us assume that the following data for a pure substance have been obtained in the laboratory.

1. Vapor-pressure data. That is, saturation pressures and temperatures have been measured over a wide range.
2. Pressure, specific volume, temperature data in the vapor region. These data are usually obtained by determining the mass of the substance in a closed vessel (which means a fixed specific volume) and then measuring the pressure as the temperature is varied. This is done for a large number of specific volumes.
3. Density of the saturated liquid and the critical pressure and temperature.
4. Zero-pressure specific heat for the vapor. This might be obtained either calorimetrically or from spectroscopic data.

From these data a complete set of thermodynamic tables for the saturated liquid, saturated vapor, and superheated vapor can be calculated. The first step is to determine an equation for the vapor-pressure curve that accurately fits the data. It may be necessary to use one equation for one portion of the vapor-pressure curve and a different equation for another portion of the curve.

One form of equation that has been used is

$$\ln P_{\text{sat}} = A + \frac{B}{T} + C \ln T + DT$$

Once an equation has been found that accurately represents the data, the saturation pressure for any given temperature can be found by solving this equation. Thus the saturation pressures in Table 1 of the Steam Tables would be determined for the given temperatures. The second step is to determine an equation of state for the vapor region that accurately represents the P-v-T data. There are many possible forms of the equation of state that may be selected. The important considerations are that the equation of state accurately represents the data, and that it be of such a form that the differentiations required can be performed (that is, in some cases it may be desirable to have an equation of state explicit in v, whereas on other occasions an equation of state that is explicit in P may be more desirable).

Once an equation of state has been determined, the specific volume of superheated vapor at given pressures and temperatures can be determined by solving

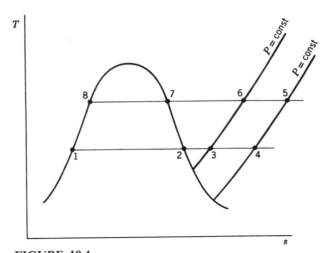

FIGURE 10.4

Sketch showing procedure for developing a table of thermo-dynamic properties from experimental data.

the equation and tabulating the results as in the superheat tables for steam, ammonia, and Freon-12. The specific volume of saturated vapor at a given temperature may be found by finding the saturation pressure from the vapor-pressure curve and substituting this saturation pressure and temperature into the equation of state.

The procedure followed in determining enthalpy and entropy is best explained with the aid of Fig. 10.4. Let us assume that the enthalpy and entropy of saturated liquid in state 1 are zero. The enthalpy of saturated vapor in state 2 can be found from the Clapeyron equation.

$$\left(\frac{dP}{dT}\right)_{\text{sat}} = \frac{h_{fg}}{T(v_g - v_f)}$$

The left side of this equation is found by differentiating the vapor-pressure curve. The specific volume of the saturated vapor is found by the procedure outlined in the last paragraph, and it is assumed the specific volume of the saturated liquid has been measured. Thus the enthalpy of evaporation, h_{fg}, can be found for this particular temperature, and the enthalpy at state 2 is equal to the enthalpy of evaporation (since the enthalpy in state 1 is assumed to be zero). The entropy at state 2 is readily found, since

$$s_{fg} = \frac{h_{fg}}{T}$$

From state 2 we now proceed along this isotherm into the superheated vapor region. The specific volume at 3 is found from the equation of state at this pressure, while the enthalpy and entropy are determined by integrating Eqs. 10.25 and 10.31:

$$h_3 - h_2 = \int_2^3 \left[v - T\left(\frac{\partial v}{\partial T}\right)_P \right] dP_T$$

$$s_3 - s_2 = \int_2^3 -\left(\frac{\partial v}{\partial T}\right)_P dP_T$$

The properties at point 4 are found in exactly the same manner. Pressure P_4 is sufficiently low that the real superheated vapor behaves essentially as an ideal gas (perhaps 10 kPa). Thus we use this constant-pressure line to make all temperature changes for our calculations, as for example to point 5. Since the specific heat C_{po} is known as a function of temperature, the enthalpy and entropy at 5 are found by integrating the ideal-gas relations

$$(h_5 - h_4)_P = \int_4^5 C_{po} \, dT_P$$

$$(s_5 - s_4)_P = \int_4^5 C_{po} \frac{dT_P}{T}$$

The properties at points 6 and 7 are found from those at 5 in the same manner as those at 3 and 4 were found from 2 (the saturation pressure P_7 is calculated from the vapor-pressure equation). Finally, the enthalpy and entropy for saturated liquid at point 8 are found from the properties at point 7 by applying the Clapeyron equation.

Thus values for the pressure, temperature, specific volume, enthalpy, entropy, and internal energy of saturated liquid, saturated vapor, and superheated vapor can be tabulated for the entire region for which experimental data were obtained. It is evident that the accuracy of such a table depends both on the accuracy of the experimental data and the degree to which the equation for the vapor pressure and the equation of state represent the experimental data.

10.8 THE IDEAL GAS

The ideal gas has been discussed at various relevant points in prior chapters, particularly in Sections 3.4, 5.7, and 7.10. Since an understanding of the properties of an ideal gas is essential to a study of the behavior of real substances, equations of state, and tables of thermodynamic properties, a brief summary of the behavior of ideal gases is presented here.

The ideal gas was defined in Section 3.4 as a gas that follows the equation of state, Eq. 3.1,

$$P\bar{v} = \bar{R}T$$

From a microscopicc viewpoint, such an equation results when there are no intermolecular forces, in other words when the molecules are widely separated from one another. Thus, the ideal gas can be a reasonable model of real gas behavior only at very low density.

In Section 5.7 it was stated that the internal energy of an ideal gas is a function only of temperature, as stated by Eq. 5.19. At this point it is appropriate to demonstrate that this is in fact the case. For an ideal gas, from Eq. 3.2,

$$P = \frac{RT}{v}$$

Therefore,

$$\left(\frac{\partial P}{\partial T}\right)_v = \frac{R}{v}$$

Substituting this relation into Eq. 10.29, we have

$$du_T = \left[T\left(\frac{\partial P}{\partial T}\right)_v - P\right]dv_T = \left[T\left(\frac{R}{v}\right) - P\right]dv_T = 0$$

and we find no variation of internal energy along an isotherm. In other words, for the ideal gas Eq. 10.28 reduces to Eq. 5.20,

$$du = C_{vo}\,dT$$

We can also examine various other general thermodynamic relations for the special case of the ideal gas. The procedure is in each case similar to that followed above for the internal energy, and the details are left as an exercise. To summarize the results:

Eqs. 10.34 and 10.35 demonstrate the validity of Eq. 5.26,

$$C_{vo} = f(T); \qquad C_{po} = f(T)$$

Eq. 10.24 reduces to Eq. 5.24,

$$dh = C_{po}\,dT$$

Eq. 10.37 reduces to Eq. 5.27,

$$C_{po} - C_{vo} = R$$

Eq. 10.32 reduces to Eq. 7.19,

$$ds = C_{vo}\frac{dT}{T} + R\frac{dv}{v}$$

and Eq. 10.30 reduces to Eq. 7.21,

$$ds = C_{po}\frac{dT}{T} - R\frac{dP}{P}$$

In discussing the accuracy of the ideal gas equation of state in Section 3.4, we introduced the compressibility factor

$$Z = \frac{P\bar{v}}{\bar{R}T} \tag{10.44}$$

as a useful and important parameter in expressing the nonideality of gases (or other phases). The compressibility factor is always unity (for all pressures and temperatures) for an ideal gas.

Another useful parameter in describing the behavior of a real gas relative to the ideal gas is the residual volume α, which is defined as

$$\alpha = \frac{\bar{R}T}{P} - \bar{v} \qquad (10.45)$$

We note that α is always zero for an ideal gas.

The Joule–Thomson coefficient μ_J was defined in Section 5.13 as

$$\mu_J = \left(\frac{\partial T}{\partial P}\right)_h \qquad (10.46)$$

Since h is a function only of T for an ideal gas, it follows that the Joule–Thomson coefficient is always zero for an ideal gas.

10.9 THE BEHAVIOR OF REAL GASES

In Section 3.4 we discussed briefly the nonideality in P-v-T behavior of a gas, using nitrogen as an example, and in connection with that discussion constructed a skeleton compressibility diagram, Fig. 3.7. A more detailed compressibility diagram for N_2 is shown in Fig. 10.5. In examining this diagram, we note that Z approaches unity for all isotherms as P approaches zero, that the nonideality of the gas phase is especially severe in the vicinity of the critical point (Z at the critical point is about 0.29), and that at high pressures Z increases to values greater than unity for all isotherms.

If we examine compressibility diagrams for other pure substances, we find that the diagrams are all similar in the characteristics described above for nitrogen, at least in a qualitative sense. Quantitatively the diagrams are all different, since the critical temperatures and pressures of different substances vary over wide ranges, as evidenced from the values listed in Table A.7. Is there a way in which we can put all of these substances on a common basis? To do so, we "reduce" the properties with respect to the values at the critical point. The reduced properties are defined as

$$\text{Reduced pressure} = P_r = \frac{P}{P_c}, \qquad P_c = \text{Critical pressure}$$

$$\text{Reduced temperature} = T_r = \frac{T}{T_c}, \qquad T_c = \text{Critical temperature}$$

$$\text{Reduced specific volume} = v_r = \frac{v}{v_c}, \qquad v_c = \text{Critical specific volume}$$

These equations state that the reduced property for a given state is the value of that property in this state divided by the value of that same property at the critical point.

If lines of constant T_r are plotted on a Z vs. P_r diagram, a plot such as Fig. A.5 is obtained. The striking fact is that when such Z vs. P_r diagrams are prepared for a large number of different substances, these diagrams very nearly coincide.

FIGURE 10.5

Compressibility diagram for nitrogen.

This leads to the development of a generalized compressibility chart. Figure A.5 is actually a generalized diagram, which means that it represents the average diagram for a number of different substances. It should be noted that when such a diagram is used for a particular substance, the results might be slightly in error. On the other hand, if P-v-T information is required for a substance in a region where no experimental measurements have been made, the generalized compressibility diagram will in all probability give relatively accurate results. It is only necessary to know the critical pressure and the critical temperature in order to use the generalized chart. It should be emphasized that whenever one has accurate thermodynamic data for a given substance, these should be used rather than the generalized charts.

This behavior of pure substances that leads to the generalized compressibility chart is sometimes referred to as the rule of corresponding states, which may be expressed as

$$v_r = f(P_r, T_r)$$

That is, if the rule of corresponding states held exactly, there would be a single functional relation between v_r, P_r, and T_r that would apply to all substances. The generalized chart is one way of expressing this relationship, and the fact that this is only an approximate relation indicates the approximate nature of the rule of corresponding states.

It follows that the use of only two parameters, P_c and T_c, in developing the generalized chart necessarily results in a common value of compressibility factor at the critical point for all substances. In Fig. A.5, Z_c has been taken as 0.27, which is a good average value. Actually, Z_c varies from about 0.23 to 0.33 for different substances, and therefore the generalized chart is usually not very accurate in the immediate vicinity of the critical point. More accurate three-parameter generalized compressibility charts that utilize Z_c or another parameter as an independent variable have been developed, but will not be discussed further here. It should also be pointed out that the compressibility factor of liquids does not generalize well as a function of only P_r and T_r. As a result, the liquid region is often not included on a generalized chart. Figure A.5 does show a saturated liquid line, and the compressed liquid region is shown as dashed lines. This region is considerably less accurate than the gas phase, and should not be used in preference to tables or data for any particular substance.

The following example illustrates the use of the generalized compressibility chart.

Example 10.7

1. Volume unknown. Calculate the specific volume of propane at a pressure of 7 MPa and a temperature of 150°C, and compare this with the specific volume given by the ideal-gas equation of state.

System: Propane.

State: P, T known.

Model: Generalized chart.

Solution

For propane

$$T_c = 370 \text{ K}, \qquad P_c = 4.26 \text{ MPa}$$
$$R = 0.188 \ 55 \text{ kJ/kg K}$$
$$T_r = \frac{423.2}{370} = 1.144, \qquad P_r = \frac{7}{4.26} = 1.64$$

From the compressibility chart,

$$Z = 0.54$$
$$v = \frac{ZRT}{P} = \frac{0.54 \times 0.188 \ 55 \times 423.2}{7000} = 0.006 \ 16 \text{ m}^3/\text{kg}$$

The ideal-gas equation would give the value

$$v = \frac{0.188 \ 55 \times 423.2}{7000} = 0.0114 \text{ m}^3/\text{kg}$$

2. Pressure unknown. What pressure is required in order that propane have a specific volume of 0.006 16 m³/kg at a temperature of 150°C?

State: T, v known.

Solution

$$T_r = \frac{423.2}{370} = 1.144$$
$$P_r = \frac{P}{P_c} = \frac{ZRT}{vP_c} = Z \times \frac{0.188 \ 55 \times 423.2}{0.006 \ 16 \times 4260} = 3.04Z$$

By a trial-and-error procedure, or by plotting a few points and drawing the curve representing this equation, P_r is found to be 1.64 at the point where $T_r = 1.144$. Therefore,

$$P = P_r P_c = 1.64 \times 4260 = 7000 \text{ kPa} = 7 \text{ MPa}$$

3. Temperature unknown. What will be the temperature of propane when it has a specific volume of 0.006 16 m³/kg and a pressure of 7 MPa?

State: P, v known.

Solution

$$P_r = \frac{7}{4.26} = 1.64$$
$$T_r = \frac{T}{T_c} = \frac{Pv}{ZRT_c} = \frac{7000 \times 0.006 \ 16}{Z \times 0.188 \ 55 \times 370} = \frac{0.6177}{Z}$$

When this line is plotted on the compressibility chart, the state point is given

by the intersection of this line with the known reduced-pressure line ($P_r = 1.64$ in this case). The reduced temperature is thus found to be

$$T_r = 1.144$$
$$T = T_r \times T_c = 1.144 \times 370 = 423.2 \text{ K}$$

In order to gain additional insight into the behavior of gases at low density, let us examine the low-pressure portion of the generalized compressibility chart in greater detail. This behavior is as shown in Fig. 10.6. The isotherms are essentially straight lines in this region, and their slope is of particular importance. Note that the slope increases as T_r increases until a maximum value is reached at a T_r of about 5, and then the slope decreases toward the $Z = 1$ line for higher temperatures. That single temperature, about 2.5 times the critical temperature, for which

$$\lim_{P \to 0} \left(\frac{\partial Z}{\partial P} \right)_T = 0 \qquad \qquad \textbf{(10.47)}$$

is defined as the Boyle temperature of the substance. This is the only temperature at which a gas really behaves exactly as an ideal gas at low, but finite pressures, since all other isotherms go to zero pressure on Fig. 10.6 with a nonzero slope. To amplify this point, let us consider the residual volume α, which was defined according to Eq. 10.45,

$$\alpha = \frac{\bar{R}T}{P} - \bar{v}$$

Multiplying this equation by P we have

$$\alpha P = \bar{R}T - P\bar{v} \qquad \qquad \textbf{(10.48)}$$

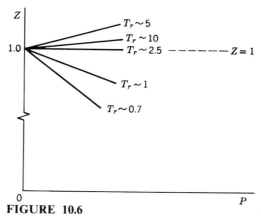

FIGURE 10.6

Low-pressure region of compressibility chart.

Thus the quantity αP is the difference between $\bar{R}T$ and $P\bar{v}$, Now as $P \to 0$, $P\bar{v} \to \bar{R}T$. However, it does not necessarily follow that $\alpha \to 0$ as $P \to 0$. Instead, it is only required that α remain finite. The derivative in Eq. 10.47 can be written as

$$\lim_{P \to 0} \left(\frac{\partial Z}{\partial P} \right)_T = \lim_{P \to 0} \left(\frac{Z - 1}{P - 0} \right)$$

$$= \lim_{P \to 0} \frac{1}{\bar{R}T} \left(\bar{v} - \frac{\bar{R}T}{P} \right)$$

$$= -\frac{1}{\bar{R}T} \lim_{P \to 0} (\alpha) \qquad (10.49)$$

from which we find that α tends to zero as $P \to 0$ only at the Boyle temperature, since that is the only temperature for which the isothermal slope is zero on Fig. 10.6. It is perhaps a somewhat surprising result that in the limit as $P \to 0$, $P\bar{v} \to \bar{R}T$ but in general the quantity $(\bar{R}T/P - \bar{v})$ does not go to zero, but is instead a small difference between two large values. This does have an effect on certain other properties of the gas.

Another aspect of generalized behavior of gases is the behavior of isotherms in the vicinity of the critical point. If we plot experimental data on P-v coordinates, it is found that the critical isotherm is unique in that is goes through a horizontal inflection point at the critical point as shown in Fig. 10.7. Mathematically, this means that the first two derivatives are zero at the critical point,

$$\left(\frac{\partial P}{\partial v} \right)_{T_c} = 0 \qquad \text{at C.P.} \qquad (10.50)$$

$$\left(\frac{\partial^2 P}{\partial v^2} \right)_{T_c} = 0 \qquad \text{at C.P.} \qquad (10.51)$$

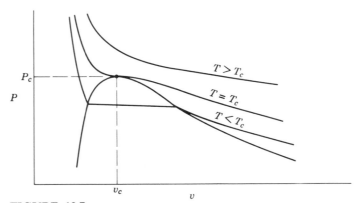

FIGURE 10.7

Plot of isotherms in the region of the critical point on pressure–volume coordinates for a typical pure substance.

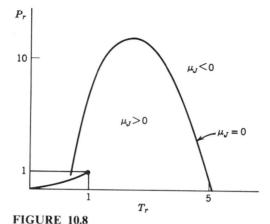

FIGURE 10.8

Joule–Thomson inversion curve.

a feature that is used to constrain many equations of state, as will be discussed in Section 10.10.

The Joule–Thomson coefficient, defined in connection with the throttling process, can be expressed in terms of P-v-T behavior as

$$\mu_J = \left(\frac{\partial T}{\partial P}\right)_h = \frac{T\left(\frac{\partial v}{\partial T}\right)_P - v}{C_p} = \frac{RT^2}{PC_p}\left(\frac{\partial Z}{\partial T}\right)_P \tag{10.52}$$

From Eq. 10.52 and Fig. 10.6 it is evident that at low pressure μ_J is zero only at one temperature, T_r about 5, and is positive at lower temperatures and negative at higher temperatures. By examining higher density P-v-T data, we can determine the locus of points at which μ_J is zero, which is called the Joule–Thomson inversion curve. The result is as shown in Fig. 10.8. Inside the dome μ_J is positive, which means that a substance cools upon being throttled to a lower pressure. Outside the dome μ_J is negative, and the temperature increases during a throttling process.

There are also many other aspects of generalized behavior of gases, some of which are discussed in connection with equations of state in the next section.

10.10 EQUATIONS OF STATE

An accurate equation of state, which is an analytical representation of P-v-T behavior, is often desirable from a computational standpoint. Many different equations of state have been developed. Most of these are accurate only to some density less than the critical density, though a few are reasonably accurate to approximately 2.5 times the critical density. All equations of state fail badly when the density exceeds the maximum density for which the equation was developed.

Three broad classifications of equations of state can be identified, namely, generalized, empirical, and theoretical.

The best known of the generalized equations of state is also the oldest, namely, the van der Waals equation, which was presented in 1873 as a semi-theoretical improvement over the ideal gas equation. The van der Waals equation of state is:

$$P = \frac{RT}{v-b} - \frac{a}{v^2} \qquad (10.53)$$

The constant b is intended to correct for the volume occupied by the molecules, and the term a/v^2 is a correction that accounts for the intermolecular forces of attraction. As might be expected in the case of a generalized equation, the constants a and b are evaluated from the general behavior of gases. In particular these constants are evaluated by noting that the critical isotherm passes through a point of inflection at the critical point, and that the slope is zero at this point. Thus, for the van der Waals equation of state we have

$$\left(\frac{\partial P}{\partial v}\right)_T = -\frac{RT}{(v-b)^2} + \frac{2a}{v^3} \qquad (10.54)$$

$$\left(\frac{\partial^2 P}{\partial v^2}\right)_T = \frac{2RT}{(v-b)^3} - \frac{6a}{v^4} \qquad (10.55)$$

Since both of these derivatives are equal to zero at the critical point we can write

$$-\frac{RT_c}{(v_c-b)^2} + \frac{2a}{v_c^3} = 0$$

$$\frac{2RT_c}{(v_c-b)^3} - \frac{6a}{v_c^4} = 0 \qquad (10.56)$$

$$P_c = \frac{RT_c}{(v_c-b)} - \frac{a}{v_c^2}$$

Solving these three equations we find

$$v_c = 3b$$

$$a = \frac{27}{64}\frac{R^2 T_c^2}{P_c} \qquad (10.57)$$

$$b = \frac{RT_c}{8P_c}$$

The compressibility factor at the critical point for the van der Waals equation is

$$Z_c = \frac{P_c v_c}{RT_c} = \frac{3}{8}$$

which is considerably higher than the actual value for any substance.

Van der Waals' equation can be written in terms of the compressibility factor and the reduced pressure and reduced temperature as follows:

$$Z^3 - \left(\frac{P_r}{8T_r} + 1\right)Z^2 + \left(\frac{27P_r}{64T_r^2}\right)Z - \frac{27P_r^2}{512T_r^3} = 0 \qquad (10.58)$$

It is significant to note that this is of the same form as the generalized compressibility chart, namely, $Z = f(P_r, T_r)$, though the functional relation is quite different from the generalized chart. This concept that different substances will have the same compressibility factor at the same reduced pressure and reduced temperature is another way of expressing the rule of corresponding states.

A simple equation of state that is considerably more accurate than the van der Waals equation is that proposed by Redlich and Kwong in 1949.

$$P = \frac{\bar{R}T}{\bar{v} - b} - \frac{a}{\bar{v}(\bar{v} + b)T^{1/2}} \qquad (10.59)$$

with

$$a = 0.427\,48 \, \frac{\bar{R}^2 T_c^{5/2}}{P_c} \qquad (10.60)$$

$$b = 0.086\,64 \, \frac{\bar{R}T_c}{P_c} \qquad (10.61)$$

The numerical values in the constants have been determined by a procedure similar to that followed in the van der Waals equation. Because of its simplicity this equation could not be expected to be sufficiently accurate to find use in the calculation of precision tables of thermodynamic properties. It has, however, been used frequently for mixture calculations and phase equilibrium correlations with reasonably good success. A number of modified versions of this equation have also been utilized in recent years.

One of the best known empirical equations of state is the Beattie–Bridgeman equation, proposed in 1928. This is a pressure-explicit equation with five constants that are determined by graphical procedure from experimental data for each substance. This equation was introduced in Chapter 3 in the form

$$P = \frac{RT(1 - \epsilon)}{v^2}(v + B) - \frac{A}{v^2} \qquad (10.62)$$

where

$$A = A_0(1 - a/v)$$
$$B = B_0(1 - b/v)$$
$$\epsilon = c/vT^3$$

For use with thermodynamic relations, it is more convenient to rewrite the equation in the form

$$P = \frac{RT}{v} + \frac{\beta}{v^2} + \frac{\gamma}{v^3} + \frac{\delta}{v^4} \qquad (10.63)$$

where

$$\beta = B_0 RT - A_0 - cR/T^2$$
$$\gamma = -B_0 bRT + A_0 a - B_0 cR/T^2$$
$$\delta = B_0 bcR/T^2$$

Values for the five constants have been given for a number of substances in Table 3.3.

The Beattie–Bridgeman equation of state is reasonably accurate for densities less than about 0.8 times the critical density. A more complex equation of state, and one that is suitable for use at higher densities, is that of Benedict, Webb, and Rubin, proposed in 1940. This equation has 8 empirical constants and is essentially an extension of the Beattie–Bridgeman equation through the addition of high-density terms. Strobridge in 1962 further extended this type of equation to 16 constants, and many other empirical equations, some even more complex in form, have been developed and used to represent the P-v-T behavior of various substances.

A different approach to this problem is from the theoretical point of view. The theoretical equation of state, which is derived from kinetic theory or statistical thermodynamics, is written here in the form of a power series in reciprocal volume:

$$Z = \frac{P\bar{v}}{\bar{R}T} = 1 + \frac{B(T)}{\bar{v}} + \frac{C(T)}{\bar{v}^2} + \frac{D(T)}{\bar{v}^3} + \cdots \qquad \textbf{(10.64)}$$

where $B(T)$, $C(T)$, $D(T)$ are temperature dependent and are called virial coefficients. $B(T)$ is termed the second virial coefficient and is due to binary interactions on the molecular level. The general temperature dependence of the second virial coefficient is as shown for nitrogen in Fig. 10.9.

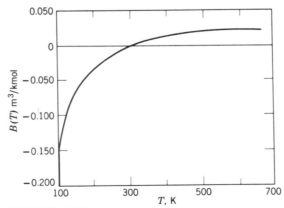

FIGURE 10.9

The second virial coefficient for nitrogen.

If we multiply Eq. 10.64 by $\bar{R}T/P$, the result can be rearranged to the form

$$\frac{\bar{R}T}{P} - \bar{v} = \alpha = -B(T)\frac{\bar{R}T}{P\bar{v}} - C(T)\frac{\bar{R}T}{P\bar{v}^2} - \cdots \qquad (10.65)$$

In the limit, as $P \to 0$,

$$\lim_{P \to 0} \alpha = -B(T) \qquad (10.66)$$

and we conclude from Eqs. 10.47 and 10.49 that the single temperature at which $B(T) = 0$, Fig. 10.9, is the Boyle temperature. The second virial coefficient can be viewed as the first-order correction for nonideality of the gas, and consequently becomes of considerable importance and interest. In fact, the low-density behavior of the isotherms shown in Fig. 10.6 is directly attributable to the second virial coefficient. This is of further interest, since the virial coefficients can be expressed in terms of intermolecular forces from statistical mechanics, and evaluated upon selection of an empirical potential function model. The best known of these is the Lennard–Jones (6–12) potential. This function has two force constants, ϵ/k and b_0, values of which are listed for various substances in Table A.12. The expression for $B(T)$ can be put into dimensionless form in terms of the parameter

$$T^* = \frac{T}{\epsilon/k} \qquad (10.67)$$

and evaluated for the dimensionless second virial coefficient $B^*(T^*)$, defined from the relation

$$B^*(T^*) = \frac{B(T)}{b_0} \qquad (10.68)$$

Values of the dimensionless parameter $B^*(T^*)$, as calculated using the Lennard–Jones (6–12) potential, are also given in Table A.12.

Example 10.8

Use the virial equation of state to determine the error in specific volume if carbon dioxide is assumed to be an ideal gas at 80 F, 10 atm pressure.

System: Carbon dioxide.

State: P, T known.

Model: Virial equation with second virial coefficient; compare with ideal-gas model.

Solution

From Eq. 10.64 at low density,

$$Z = \frac{P\bar{v}}{\bar{R}T} = 1 + \frac{B(T)}{\bar{v}}$$

$$T = 80\ F = 540\ R = 300\ K$$

$$P = 10\ atm = 1013\ kPa$$

For carbon dioxide, from Table A.12,

$$\epsilon/k = 186 \text{ K}, \qquad b_0 = 0.118 \text{ m}^3/\text{kmol}$$

Therefore,

$$T^* = \frac{T}{\epsilon/k} = \frac{300}{186} = 1.612$$

From Table A.12,

$$B^*(T^*) = -1.0365$$

Using Eq. 10.68,

$$B(T) = b_0 B^*(T^*) = 0.118(-1.0365) = -0.122 \text{ m}^3/\text{kmol}$$

Substituting and solving for \bar{v},

$$\frac{1013\bar{v}}{8.3144 \times 300} = 1 - \frac{0.122}{\bar{v}}$$

$$\bar{v} = 2.334 \text{ m}^3/\text{kmol}$$

For an ideal gas,

$$\bar{v} = \frac{\bar{R}T}{P} = \frac{8.3144 \times 300}{1013} = 2.462 \text{ m}^3/\text{kmol}$$

which is in error by 5.5%.

10.11 THE GENERALIZED CHART FOR CHANGES OF ENTHALPY AT CONSTANT TEMPERATURE

In Section 10.5, Eq. 10.25 was derived for the change of enthalpy at constant temperature.

$$(h_2 - h_1)_T = \int_1^2 \left[v - T\left(\frac{\partial v}{\partial T}\right)_P \right] dP_T$$

The essential point about this equation for $(h_2 - h_1)_T$ is that the change of enthalpy at constant temperature is a function of the P-v-T behavior of the gas. This means that if the equation of state is known for a substance, the change in enthalpy at constant temperature can be determined by the appropriate differentiation and integration. It also follows that since the generalized chart is a representation of the P-v-T behavior of the gas, this chart can be used to prepare a generalized chart for the change of enthalpy in an isothermal process.

The procedure for the development of such a chart is as follows: It has been shown that

$$\left(\frac{\partial h}{\partial P}\right)_T = v - T\left(\frac{\partial v}{\partial T}\right)_P$$

But

$$v = \frac{ZRT}{P}$$

and

$$\left(\frac{\partial v}{\partial T}\right)_P = \frac{ZR}{P} + \frac{RT}{P}\left(\frac{\partial Z}{\partial T}\right)_P$$

Therefore,

$$\left(\frac{\partial h}{\partial P}\right)_T = \frac{ZRT}{P} - \frac{ZRT}{P} - \frac{RT^2}{P}\left(\frac{\partial Z}{\partial T}\right)_P = -\frac{RT^2}{P}\left(\frac{\partial Z}{\partial T}\right)_P$$

$$dh_T = -\frac{RT^2}{P}\left(\frac{\partial Z}{\partial T}\right)_P dP_T$$

But $T = T_c T_r$ and $P = P_c P_r$. Therefore,

$$\frac{dT}{T} = \frac{dT_r}{T_r}, \qquad \frac{dP}{P} = \frac{dP_r}{P_r}$$

$$dT = T_c dT_r, \quad dP = P_c dP_r$$

Substituting these values we have

$$dh_T = -\frac{RT_c^2 T_r^2}{P_c P_r}\left(\frac{\partial Z}{T_c \partial T_r}\right)_{P_r} P_c(dP_r)_{T_r}$$

$$= -RT_c T_r^2 \left(\frac{\partial Z}{\partial T_r}\right)_{P_r} (d \ln P_r)_{T_r}$$

Integrating at constant temperature we have

$$\frac{\Delta \bar{h}_T}{T_c} = -\bar{R} \int_1^2 T_r^2 \left(\frac{\partial Z}{\partial T_r}\right)_{P_r} d \ln P_r \qquad \textbf{(10.69)}$$

Let us designate by an asterisk, \bar{h}^*, the enthalpy per mole at a given temperature and a very low pressure, so that $P_r \to 0$, and let us designate without superscript the enthalpy at any state at the same temperature and a given pressure P_r. Then the integration is as follows:

$$\frac{\bar{h}^* - \bar{h}}{T_c} = \bar{R} \int_{P_r=0}^{P_r} T_r^2 \left(\frac{\partial Z}{\partial T_r}\right)_{P_r} d \ln P_r \qquad \textbf{(10.70)}$$

The right side of this equation can be obtained by graphical integration of the compressibility chart and is a function of only the reduced pressure and temperature. The left side represents the change in enthalpy as the pressure is increased from zero to the given pressure while the temperature is held constant. Sometimes this quantity is called the enthalpy departure. Figure A.6 shows a plot of this quantity, $(\bar{h}^* - \bar{h})/T_c$, as a function of P_r for various lines of constant T_r.

In many cases the specific heat of a substance is known at low pressure. This information, along with the generalized residual enthalpy chart, enables one to find the change of enthalpy for any given change of state.

Example 10.9

Nitrogen is throttled from 20 MPa, $-70°C$, to 2 MPa in an adiabatic, steady-state, steady-flow process. Determine the final temperature of the nitrogen.

Control volume: Throttling valve.
Inlet state: P_1, T_1 known; state fixed.
Exit state: P_2 known.
Process: SSSF, throttling process.
Diagram: Figure 10.10.
Model: Generalized charts.
Analysis:
 First law: $$h_1 = h_2$$

Solution

$$P_1 = 20 \text{ MPa}, \qquad P_{r1} = \frac{20}{3.39} = 5.9$$

$$T_1 = 203.2 \text{ K}, \qquad T_{r1} = \frac{203.2}{126.2} = 1.61$$

$$P_2 = 2 \text{ MPa}, \qquad P_{r2} = \frac{2}{3.39} = 0.59$$

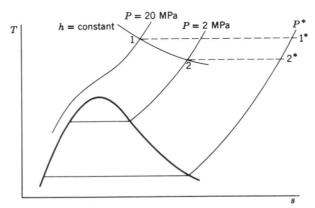

FIGURE 10.10
Sketch for Example 10.9.

From the generalized chart for enthalpy change at constant temperature,

$$\frac{\bar{h}_1^* - \bar{h}_1}{T_c} = 14.2 \text{ kJ/kmol K}$$

$$h_1^* - h_1 = 14.2 \times \frac{126.2}{28} = 64.0 \text{ kJ/kg}$$

It is now necessary to assume a final temperature and check to see if the net change in enthalpy for the process is zero. Let us assume that $T_2 = 156$ K. Then the change in enthalpy between 1* and 2* can be found from the zero-pressure specific-heat data.

$$h_1^* - h_2^* = C_{po}(T_1^* - T_2^*) = 1.0416(203.2 - 156) = +49.2 \text{ kJ/kg}$$

(The variation in C_{po} with temperature can be taken into account when necessary.)

We now find the enthalpy change between 2* and 2.

$$T_{r_2} = \frac{156}{126.2} = 1.236, \qquad P_{r_2} = 0.59$$

Therefore, from the enthalpy departure chart, Fig. A.6,

$$\frac{\bar{h}_2^* - \bar{h}_2}{T_c} = 3.3 \text{ kJ/kmol K}$$

$$h_2^* - h_2 = 3.3 \times \frac{126.2}{28} = 14.9 \text{ kJ/kg}$$

We now check to see if the net change in enthalpy for the process is zero.

$$h_1 - h_2 = 0 = -(h_1^* - h_1) + (h_1^* - h_2^*) + (h_2^* - h_2)$$
$$= -64 + 49.2 + 14.9 = 0.1 \simeq 0$$

This essentially checks, and we conclude that the final temperature is approximately 156 K.

10.12 FUGACITY AND THE GENERALIZED FUGACITY CHART

At this point a new thermodynamic property, the fugacity, f, is introduced. The fugacity is particularly important when considering mixtures and equilibrium, which is discussed in Chapter 13. However, in view of the fact that a generalized fugacity chart can be developed, the fugacity is introduced here.

The fugacity is essentially a pseudopressure. When the fugacity is substituted for pressure, one can, in effect, use the same equations for real gases that one normally uses for ideal gases. The concept of fugacity can be introduced as follows. Consider the relation

$$dg = -s\,dT + v\,dP$$

At constant temperature

$$dg_T = v\,dP_T \qquad\qquad (10.71)$$

For an ideal gas this last equation may be written

$$dg_T = \frac{RT}{P}dP_T = RT\,d(\ln P)_T \tag{10.72}$$

and for a real gas, with the equation of state $Pv = ZRT$, this equation becomes

$$dg_T = ZRT\frac{dP_T}{P} = ZRT\,d(\ln P)_T \tag{10.73}$$

Fugacity, f, is defined as

$$dg_T \equiv RT\,d(\ln f)_T \tag{10.74}$$

with the requirement that

$$\lim_{P \to 0}(f/P) = 1 \tag{10.75}$$

Therefore, as $P \to 0, f \to 0$.

Let us consider the change in the Gibbs function of a real gas during an iso-thermal process at temperature T that involves a change in pressure from a very low pressure, P^* (where ideal gas behavior can be assumed), to a higher pressure P. Let the Gibbs function at this low pressure be designated g^*. The value of the Gibbs function at this temperature T and at the pressure P can be found in terms of the fugacity at P and T and g^*. This is evident if one integrates Eq. 10.74 from the very low pressure P^* to the pressure P.

$$\int_{g^*}^{g} dg_T = \int_{f^*=P^*}^{f} RT(d\ln f)_T \tag{10.76}$$
$$g = g^* + RT\ln(f/P^*)$$

A generalized fugacity coefficient chart based on the generalized compressibility chart, Fig. A.7, can be developed by the following procedure. From Eqs. 10.71 and 10.74,

$$dg_T = v\,dP_T = \frac{ZRT\,dP_T}{P} = RT\,d(\ln f)_T$$

Therefore,

$$Z\,d(\ln P)_T = d(\ln f)_T$$

But

$$\frac{dP}{P} = \frac{dP_r}{P_r}, \qquad d\ln P = d\ln P_r$$

Therefore,

$$Z(d\ln P_r)_T = d(\ln f)_T$$
$$Z(d\ln P_r)_T - d(\ln P_r)_T = d(\ln f)_T - d(\ln P_r)_T$$
$$(Z - 1)(d\ln P_r)_T = d(\ln f/P)_T \tag{10.77}$$

Integrating at constant temperature from $P = 0$ to a finite pressure, we have

$$\int_{f/P=1}^{f/P} d(\ln f/P)_T = \int_0^{P_r} (Z - 1)d \ln P_{rT}$$

$$\ln (f/P) = \int_0^{P_r} (Z - 1)d \ln P_{rT} \tag{10.78}$$

At any temperature the right-hand side of this equation can be integrated graphically using the compressibility chart to find Z at each P_r. The result is the generalized chart shown in Fig. A.7.

The reversible isothermal process is one type of problem in which the fugacity can be used to advantage.

Example 10.10

Calculate the work of compression and the heat transfer per pound when ethane is compressed reversibly and isothermally from 14.7 $lbf/in.^2$ to 1000 $lbf/in.^2$ at a temperature of 110 F in a steady-state, steady-flow process.

Control volume: Compressor.

Inlet state: P_1, T known; state fixed.

Exit state: P_2, T known; state fixed.

Process: SSSF, isothermal and reversible.

Diagram: Figure 10.11.

Model: Generalized charts.

Analysis:

First law: $q + h_1 + KE_1 + PE_1 = h_2 + KE_2 + PE_2 + w$

Since the process is reversible and isothermal,

$$q = T(s_2 - s_1) = T_2 s_2 - T_1 s_1$$

Therefore,

$$-w = h_2 - h_1 - (T_2 s_2 - T_1 s_1) + \Delta KE + \Delta PE$$
$$-w = g_2 - g_1 + \Delta KE + \Delta PE$$

We assume that the changes in kinetic and potential energy are negligible.

$$-w = g_2 - g_1 = RT \ln (f_2/f_1)$$

FIGURE 10.11
Sketch for Example 10.10.

Solution

The fugacity in state 1 and state 2 can be found from the generalized fugacity chart in the Appendix, Fig. A.7.

For ethane,

$$P_c = 708 \ \text{lbf/in.}^2 \qquad T_c = 549.8 \ \text{R}$$

Therefore,

$$P_{r1} = \frac{14.7}{708} = 0.021 \qquad T_{r1} = T_{r2} = \frac{570}{549.8} = 1.037$$

$$P_{r2} = \frac{1000}{708} = 1.412$$

From the generalized enthalpy departure chart, Fig. A.6,

$$\frac{\bar{h}_2^* - \bar{h}_2}{T_c} = 26.0, \qquad \frac{\bar{h}_1^* - \bar{h}_1}{T_c} < 0.1$$

Since $h_1^* = h_2^* = h_1$

$$h_2^* - h_2 = h_1 - h_2 = \frac{26.0}{4.1868} \times \frac{549.8}{30.07} = 113.5 \ \text{Btu/lbm}$$

From the generalized fugacity chart, Fig. A.7,

$$f_2/P_2 = 0.6, \qquad f_2 = 0.6 \times 1000 = 600 \ \text{lbf/in.}^2$$
$$f_1/P_1 = 1.0, \qquad f_1 = 1.0 \times 14.7 = 14.7 \ \text{lbf/in.}^2$$

Therefore,

$$-w = g_2 - g_1 = RT \ln f_2/f_1$$

$$= \frac{51.38}{778} \times 570 \ln \frac{600}{14.7} = 139.6 \ \text{Btu/lbm}$$

$$g_2 - g_1 = (h_2 - h_1) - T(s_2 - s_1) = (h_2 - h_1) - q$$
$$139.6 = -113.5 - q$$
$$q = -253.1 \ \text{Btu/lbm}$$

$$s_2 - s_1 = \frac{-253.1}{570} = -0.444 \ \text{Btu/lbm R}$$

10.13 THE GENERALIZED CHART FOR CHANGES OF ENTROPY AT CONSTANT TEMPERATURE

A generalized chart for entropy can be developed that gives the difference between the entropy of an ideal gas and an actual gas as the pressure is increased from zero pressure along an isotherm. The thermodynamic analysis for the development of such a chart is as follows.

From Eq. 10.30,

$$ds_T = -\left(\frac{\partial v}{\partial T}\right)_P dP_T$$

Integrating at constant temperature, T, from $P = 0$ to P, we have

$$(s_P - s_0^*)_T = -\left[\int_{P=0}^{P} \left(\frac{\partial v}{\partial T}\right)_P dP_T\right] \tag{10.79}$$

The entropy at zero pressure can be designated with an * (which we use to refer to ideal-gas behavior) since ideal-gas behavior is approached as $P \rightarrow 0$. However, $s \rightarrow \infty$ as $P \rightarrow 0$, and therefore this equation is not particularly useful in this form.

If we repeat this integration for an ideal gas, we have

$$(s_P^* - s_0^*)_T = -\left[\int_{P=0}^{P} \left(\frac{\partial v}{\partial T}\right)_P dP_T\right] = -\left[\int_{P=0}^{P} R \frac{dP_T}{P}\right] \tag{10.80}$$

Subtracting Eq. 10.79 from Eq. 10.80,

$$(s_P^* - s_P) = -\int_{P=0}^{P} \left[\frac{R}{P} - \left(\frac{\partial v}{\partial T}\right)_P\right] dP_T$$

Therefore, with the equation of state $Pv = ZRT$,

$$(s_P^* - s_P) = -\int_{P=0}^{P} \left[\frac{R}{P} - \frac{ZR}{P} - \frac{RT}{P}\left(\frac{\partial Z}{\partial T}\right)_P\right] dP_T$$

or

$$(\bar{s}_P^* - \bar{s}_P) = \bar{R} \int_0^{P_r} (Z - 1) \frac{dP_r}{P_r} + \bar{R}T_r \int_0^{P_r} \left(\frac{\partial Z}{\partial T_r}\right)_{P_r} \frac{dP_r}{P_r}$$

But, from Eq. 10.78,

$$\bar{R} \int_0^{P_r} (Z - 1) \frac{dP_r}{P_r} = \bar{R} \ln (f/P)$$

and from Eq. 10.70,

$$\bar{R}T_r \int_0^{P_r} \left(\frac{\partial Z}{\partial T_r}\right)_{P_r} \frac{dP_r}{P_r} = \frac{\bar{h}^* - \bar{h}}{T_c T_r}$$

Therefore,

$$(\bar{s}_P^* - \bar{s}_P) = \bar{R} \ln f/P + \frac{\bar{h}^* - \bar{h}}{T_c T_r}$$

Thus, for the equation of state $Pv = ZRT$, the quantity $\bar{s}_P^* - \bar{s}_P$ can be obtained from the generalized enthalpy departure and the generalized fugacity coefficient charts. Figure A.8 shows such a chart.

Example 10.11

Nitrogen at 8 MPa, 150 K, is throttled to 0.5 MPa. After passing through a short length of pipe the temperature is measured and found to be 125 K. Determine the heat transfer and the change of entropy using the generalized charts and compare these results with those obtained by using the nitrogen tables.

Control volume: Throttle and pipe.

Inlet state: P_1, T_1 known; state fixed.

Exit state: P_2, T_2 known; state fixed.

Process: SSSF.

Diagram: Figure 10.12.

Model: Generalized charts, compare results with nitrogen tables.

Analysis: There is no work done, and we neglect changes in kinetic and potential energies. Therefore, the first law is, per kilogram,

$$q + h_1 = h_2$$
$$q = h_2 - h_1 = -(h_2^* - h_2) + (h_2^* - h_1^*) + (h_1^* - h_1)$$

Solution

$$P_{r_1} = \frac{8}{3.39} = 2.36, \qquad T_{r_1} = \frac{150}{126.2} = 1.189$$

$$P_{r_2} = \frac{0.5}{3.39} = 0.147, \qquad T_{r_2} = \frac{125}{126.2} = 0.99$$

From Fig. A.6 in the Appendix,

$$\frac{\bar{h}_1^* - \bar{h}_1}{T_c} = 20.0 \text{ kJ/kmol K}, \qquad h_1^* - h_1 = 20.0 \times \frac{126.2}{28} = 90.1 \text{ kJ/kg}$$

$$\frac{\bar{h}_2^* - \bar{h}_2}{T_c} = 1.3 \text{ kJ/kmol K}, \qquad h_2^* - h_2 = 1.3 \times \frac{126.2}{28} = 5.9 \text{ kJ/kg}$$

FIGURE 10.12
Sketch for Example 10.11.

Assuming a constant specific heat for the ideal gas,

$$h_2^* - h_1^* = C_{p0}(T_2 - T_1) = 1.0416(125 - 150) = -26.0 \text{ kJ/kg}$$
$$q = -5.9 - 26.0 + 90.1 = 58.2 \text{ kJ/kg}$$

From the nitrogen tables, Table A.4, we can find the change of enthalpy directly.

$$q = h_2 - h_1 = 123.82 - 61.90 = 61.92 \text{ kJ/kg}$$

To calculate the change of entropy using the generalized charts, we proceed as follows:

$$\bar{s}_2 - \bar{s}_1 = -(\bar{s}_{P_2,T_2}^* - \bar{s}_2) + (\bar{s}_{P_2,T_2}^* - \bar{s}_{P_1,T_1}^*) + (\bar{s}_{P_1,T_1}^* - \bar{s}_1)$$

From Fig. A.8 in the Appendix,

$$\bar{s}_{P_1,T_1}^* - \bar{s}_1 = 12.6 \text{ kJ/kmol K}$$
$$\bar{s}_{P_2,T_2}^* - \bar{s}_2 = 1.3 \text{ kJ/kmol K}$$

Assuming a constant specific heat for the ideal gas,

$$\bar{s}_{P_2,T_2}^* - \bar{s}_{P_1,T_1}^* = \bar{C}_{p0} \ln \frac{T_2}{T_1} - \bar{R} \ln \frac{P_2}{P_1}$$

$$= 1.0416 \times 28.013 \ln \frac{125}{150} - 8.3144 \ln \frac{0.5}{8}$$

$$= 17.73 \text{ kJ/kmol K}$$

$$\bar{s}_2 - \bar{s}_1 = -1.3 + 17.73 + 12.6 = 29.03 \text{ kJ/kmol K}$$

$$s_2 - s_1 = \frac{29.03}{28.013} = 1.0363 \text{ kJ/kg K}$$

From Table A.4,

$$s_2 - s_1 = 5.4343 - 4.3518 = 1.0825 \text{ kJ/kg K}$$

PROBLEMS

10.1 Calculate the values h_{fg} and s_{fg} for water at 200 F and also at 400 F from the Clapeyron equation, using the necessary pressure and specific volume values from Table A.1.1.

10.2 The normal (atmospheric pressure) melting-point temperature of methane is 90.67 K. Using this value and any other information from Table A.5, estimate the triple-point pressure of methane.

10.3 Using triple-point data for water from Table A.1.5 and assuming h_{ig} constant, extrapolate the sublimation pressure from the triple point to -40 F. What is the percent error in this calculated value at -40 F?

10.4 At standard atmospheric pressure, 101.3 kPa, helium boils at 4.22 K, and the latent heat of vaporization is 83.3 kJ/kmol. It is desired to estimate

how low a temperature could be obtained by pumping a vacuum over liquid helium, thus causing it to boil at a lower pressure and temperature. Determine what pressure would be necessary to produce a temperature of 1 K and also 0.5 K. What do you think is the lowest temperature that could be obtained by pumping a vacuum over liquid helium?

10.5 A transfer line for liquid oxygen is designed with a double-walled pipe, as shown in Fig. 10.13. At ambient temperature, the annular space in the line is filled with carbon dioxide gas at slightly above ambient pressure. When liquid oxygen at 100 K flows through the transfer line, most of the carbon dioxide in the annular space will freeze on the inner wall, and the pressure in this space will be equal to the sublimation pressure of carbon dioxide at 100 K. There are no thermodynamic tables available at this low a temperature, but it is known that at $-90°C$ the sublimation pressure is 38.1 kPa and the enthalpy of sublimation is 574.5 kJ/kg. Estimate the carbon dioxide pressure in the annular space when liquid oxygen is flowing through the transfer line.

FIGURE 10.13
Sketch for Problem 10.5.

10.6 Consider the situation in which a person weighing 180 lbm is ice skating on "hollow ground" blades that have a total area in contact with the ice of 0.032 in.2, and the local ice temperature is 28 F. Using thermodynamic data for water from Tables A.1.1 and A.1.5, determine whether the ice will melt under the skate blades in this situation.

10.7 A particular pollutant in combustion products appears in the form of very small solid particles. To analyze and evaluate the problem it is necessary to know the sublimation pressure of this substance at various temperatures. Suppose that the only thermodynamic data available are as follows: normal (101.3 kPa) boiling-point temperature and enthalpy of vaporization h_{fg}; and normal melting-point temperature and enthalpy of fusion h_{if}. Develop a procedure for estimating the sublimation pressure as a function of temperature based on this information.

10.8 An experiment is to be conducted at -180 F using Freon-12 as the bath liquid in a rigid closed tank (with a small amount of vapor at the top).
(a) What is the pressure inside the tank?
(b) When the experiment is completed, the tank will be allowed to warm to room temperature, 80 F. If the pressure inside is not to exceed 120 lbf/in.2 at this time, what is the maximum percent liquid (by volume) that can be used in the tank during the experiment at -180 F?

10.9 Consider a Carnot cycle heat engine in which the working fluid changes from saturated liquid to saturated vapor at temperature T (and pressure P) during the heat addition process. The heat rejection occurs at $T - \Delta T$ (and $P - \Delta P$). Using a first and second law analysis, show that the Clapeyron equation, Eq. 10.18, results in the limit as $\Delta T \to dT$, assuming that the four processes in the cycle are system processes with boundary movement work.

10.10 Repeat Problem 10.9, assuming that the four processes are steady-state, steady-flow processes with shaft work.

10.11 Derive expressions for $(\partial u/\partial P)_T$ and for $(\partial u/\partial T)_P$ that do not contain the properties h, u, or s.

10.12 Derive expressions for $(\partial T/\partial v)_u$ and for $(\partial h/\partial s)_v$ that do not contain the properties h, u, or s.

10.13 Develop an expression for the relation between temperature and pressure in an isentropic process that includes only the $P - v - T$ behavior and the specific heat of the substance.

10.14 It is desired to determine the average value (over a range of pressure) of the isothermal compressibility of liquid water at 32 F, and also at 200 F and 500 F, using data from the compressed liquid table, Table A.1.4. Does the isothermal compressibility vary significantly with pressure at a given temperature?

10.15 Show that the following relation is valid, assuming that the volume expansivity α_P is constant:

$$\left(\frac{\partial C_P}{\partial P}\right)_T = -vT\alpha_P{}^2$$

10.16 It can be shown (see Section 14.5) that the velocity of sound c in a fluid is given by the expression

$$c^2 = \left(\frac{\partial P}{\partial \rho}\right)_s$$

(a) Using this expression, show that the velocity of sound can be written as

$$c = \sqrt{\frac{1}{\beta_s \rho}}$$

where β_s is the adiabatic compressibility.

(b) At a temperature of 100 F the velocity of sound in liquid water is 5040 ft/s. Verify this value by determining the adiabatic compressibility of liquid water at 100 F by using Tables A.1.1 and A.1.4.

10.17 The following data are available for ethyl alcohol at 20°C:

$$\rho = 788 \text{ kg/m}^3$$
$$\beta_s = 943 \text{ } \mu\text{m}^2/\text{N}$$
$$\beta_T = 1117 \text{ } \mu\text{m}^2/\text{N}$$

(a) What is the velocity of sound (see Problem 10.16) in ethyl alcohol at 20°C?

(b) The pressure on a cylinder having a volume of 10 L and containing ethyl alcohol at 20°C is increased from 0.1 to 20 MPa while the temperature remains constant. Determine the work done on the ethyl alcohol during this process.

10.18 Suppose that the following data are available for a given pure substance:
A liquid–vapor vapor-pressure curve.
An equation of state (vapor) of the form $P = fn(T, v)$.
Specific volume of saturated liquid.
Critical pressure and temperature.
Constant-volume specific heat (vapor) at one specific volume.
Outline the procedure that you would follow in order to develop a table of thermodynamic properties comparable to Tables 1, 2, and 3 of the steam tables.

10.19 A rigid tank contains 5 lbm of ethylene at 450 lbf/in.², 90 F. This system is now cooled until the ethylene reaches the saturated vapor curve. What is the final temperature?

10.20 A rigid 2-ft³ tank contains propane at 1350 lbf/in.², 1000 R. The propane is then allowed to cool to 600 R due to heat transfer with the surroundings. Determine the quality at the final state and the mass of liquid in the tank.

10.21 A cylinder fitted with a piston contains 5 kg of butane gas at 500 K, 5 MPa. The butane now expands in a reversible polytropic process with polytropic exponent $n = 1.05$ until the final pressure is 3 MPa. Determine:
(a) The final temperature of the butane.
(b) The work done during the process.

10.22 Saturated liquid carbon dioxide at the ambient temperature, 15°C, is contained in a cylinder as shown in Fig. 10.14, at which point the volume is 0.3 L. The piston has a mass of 5 kg and cross-sectional area of 0.015 m²,

FIGURE 10.14
Sketch for Problem 10.22.

and the ambient pressure is 100 kPa. The pin is now released from the piston, allowing it to rise and hit the stops, at which point the volume is 4.3 L. Eventually, the temperature of the carbon dioxide returns to that of the surroundings. Determine the following:

(a) The pressure (if superheated) or quality (if saturated) at the final state. 2.6 mPa

(b) The work done during the process.

10.23 Show that the two expressions for the Joule–Thomson coefficient μ_J given by Eq. 10.52 are valid:

$$\mu_J = \frac{1}{C_p}\left[T\left(\frac{\partial v}{\partial T}\right)_P - v\right] = \frac{RT^2}{PC_p}\left(\frac{\partial Z}{\partial T}\right)_P$$

10.24 Two uninsulated tanks of equal volume are connected by a valve. One tank contains a gas at moderate pressure P_1 and the other tank is evacuated. The valve is now opened, and remains open for a long time. Is the final pressure P_2 greater than, equal to, or less than one-half of P_1?

10.25 One technique for determining the compressibility factor Z of gases that requires measuring only pressure and temperature is shown in Fig. 10.15. Tanks A and B of unknown volumes V_A and V_B are held in a uniform and constant temperature T. Initially, tank A contains a gas at high pressure P_0, and tank B is evacuated. Value I is then opened, and remains open until the pressure has stabilized at a lower value P_1. Valve I is then closed and tank B is re-evacuated through valve II. This expansion process is then repeated by opening valve I, allowing the pressure to come to a new value P_2, then closing valve I and re-evacuating tank B. A continued series of these expansion–re-evacuation processes results in the set of experimental data:

$$P_0, P_1, P_2, \dots, P_j \text{ at temperature } T$$

Develop a graphical-analytical procedure for determining the compressibility at each of these finite pressures (the final value P_j is low but not ideal gas—$Z = 1$ only in the limit at zero pressure). In this solution, it may be assumed that volumes V_A and V_B do not change with pressure.

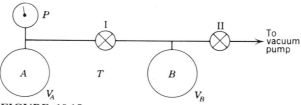

FIGURE 10.15
Sketch for Problem 10.25.

10.26 Using the van der Waals equation of state, develop expressions for the following:
 (a) $(h_2 - h_1)_T$.
 (b) $(s_2 - s_1)_T$.
 (c) $C_p - C_v$.
 (d) $(\partial C_v / \partial v)_T$.

10.27 Repeat Problem 10.26, using the Redlich–Kwong equation of state.

10.28 Repeat Problem 10.26, using the Beattie–Bridgeman equation of state.

10.29 Determine the Boyle temperature as predicted by an equation of state, using:
 (a) The van der Waals equation.
 (b) The Redlich–Kwong equation.

10.30 Consider a straight line connecting the point $P = 0$, $Z = 1$ to the critical point $P = P_c$, $Z = Z_c$ on a Z vs. P compressibility diagram. This straight line will be tangent to one particular isotherm at low pressure. Determine what value of reduced temperature is predicted by an equation of state, using:
 (a) The van der Waals equation.
 (b) The Redlich–Kwong equation.

10.31 Determine the low-pressure Joule–Thomson inversion temperature.

$$\lim_{P \to 0} \mu_J = 0$$

as predicted by an equation of state, using:
 (a) The van der Waals equation.
 (b) The Redlich–Kwong equation.

10.32 A rigid, 50-L vessel contains oxygen gas at 10 MPa, 20°C. The vessel is then cooled to $-100°C$. Use the virial equation of state to find the mass of oxygen and also the final pressure inside the vessel.

10.33 One early attempt to improve on the van der Waals equation of state was an expression of the form

$$P = \frac{RT}{v - b} - \frac{a}{v^2 T}$$

 (a) Solve for the constants a, b, and v_c using the same procedure as for the van der Waals equation.
 (b) Repeat Problems 10.29, 10.30, and 10.31 using this equation of state.

10.34 Calculate the difference in internal energy between 30 atm pressure and the ideal gas value for carbon dioxide at a temperature of 70 F, as determined using the virial equation of state.

10.35 Calculate the heat transfer during the process described in Problem 10.21.

10.36 A 10-ft^3 tank contains propane at 90 F, 90% quality. The tank is now heated to 600 F. Calculate the heat transfer during this process.

10.37 Methane gas at 8 MPa, 40°C, is throttled in a steady-state, steady-flow process to a pressure of 2 MPa. Determine the final temperature of the methane, using:
(a) The generalized charts.
(b) The methane tables, Table A.5.

10.38 A rigid, 200-L tank contains propane at 400 K, 3.5 MPa. A valve is now opened, and propane flows out until half the initial mass has escaped, at which point the valve is closed. During this process, the mass remaining inside the tank expands according to the relation $Pv^{1.4} = $ constant. Calculate the heat transfer to the tank during the process.

10.39 A number of tanks are to be filled with ethylene gas for use in another location. The procedure is as follows: a 2-ft³ evacuated tank is connected to a line flowing C_2H_4 at 2100 lbf/in.² and ambient temperature, 80 F. The valve is then opened, and ethylene flows in rapidly, such that the process is essentially adiabatic. The valve is to be closed when the pressure reaches a certain value P_2, and the tank is then disconnected from the line. After a period of time, the temperature inside the tank will return to ambient temperature because of heat transfer with the surroundings. At this time, the pressure inside the tank must be 1500 lbf/in.² What is the pressure P_2 at which the valve should be closed during the filling process?

10.40 Propane gas flows through a well-insulated, constant-diameter pipeline. The gas enters the pipe at 3 MPa, 400 K, with a velocity of 25 m/s, and exits at a pressure of 0.8 MPa. Determine the temperature and velocity of the propane at the pipe exit.

10.41 An evacuated 10-ft³ rigid tank is connected to a line flowing propane gas at 1200 lbf/in.², 300 F. The valve is now opened, allowing propane to flow into the tank for a period of time, and the valve is then closed. Eventually, the tank cools to ambient temperature, 70 F, at which point it contains 50% liquid, 50% vapor, by volume. Calculate the heat transfer for this process.

10.42 An insulated vertical cylinder is fitted with a piston restrained by a spring and held by a pin, as shown in Fig. 10.16. The cylinder cross-sectional

FIGURE 10.16
Sketch for Problem 10.42.

area is 0.2 m^2, the initial volume is 0.1 m^3, and it contains carbon dioxide at 2.5 MPa, $0°$C. Ambient pressure above the piston plus gravitational force/area on the piston is 250 kPa. The spring is linear (force proportional to distance) with a spring constant of 600 kN/m, and the spring force would be zero at a cylinder volume of 0.05 m^3. The pin is now pulled from the piston. What is the final pressure inside the cylinder?

10.43 An uninsulated, 200-L rigid tank contains ethylene gas at 8 MPa and $20°$C, the temperature of the surroundings. The cylinder valve leaks very slightly, such that after a period of several days the tank pressure has dropped to 5 MPa. Determine the following:
(a) The mass of ethylene that has escaped from the tank.
(b) The heat transfer during the process.

10.44 A cylinder fitted with a piston contains Freon-12 at 300 lbf/in.2, 180 F. The Freon-12 now expands to 35 lbf/in.2 in an irreversible, isothermal process, during which there is a heat transfer of 20 Btu/lbm to the system. Calculate the work done by the Freon-12, per pound, during this process, using the following:
(a) The Freon-12 tables, Table A.3.
(b) The generalized charts.

10.45 Repeat Problem 10.44 using the van der Waals equation of state and also using the Redlich–Kwong equation of state. Which of the four answers is the best?

10.46 A 10-m^3 storage tank contains methane at low temperature. The pressure inside is 700 kPa and the tank contains 25% liquid and 75% vapor, on a volume basis. The tank now warms very slowly because of heat transfer from the ambient.
(a) What is the temperature inside the storage tank when the pressure reaches 10 MPa?
(b) Calculate the heat transferred during this process, using the generalized charts.
(c) Repeat parts (a) and (b), using the methane tables, Table A.5. Discuss the differences in the results.

10.47 Carbon dioxide enters a nozzle at 15 MPa, 400 K, with a velocity of 10 m/s. The cross-sectional area at the nozzle entrance is 2000 mm^2. At the nozzle exit, the area is 200 mm^2 and the pressure is 5 MPa. There is a heat rejection at the rate of 200 kW from the nozzle to the surroundings. What is the exit velocity of the carbon dioxide?

10.48 Consider a reversible steady-state, steady-flow process in which propane is compressed isothermally from 15 atm, 250 F, to a pressure of 55 atm. Calculate the work and the heat transfer per pound of propane for this process.

10.49 Consider a reversible steady-state, steady-flow process in which propane is compressed adiabatically from 15 atm, 250 F, to a pressure of 55 atm. Calculate the work per pound of propane for this process.

10.50 A system consisting of 10 kg of butane gas initially at 80°C, 500 kPa, is compressed in a reversible isothermal process to one-fifth of its initial volume. What is the heat transfer during the process?

10.51 An uninsulated, 4-m³ storage tank that is initially evacuated is connected to a line flowing ethane gas at 10 MPa, 100°C. The valve is now opened, and ethane flows into the tank for a period of time, after which the valve is closed. Eventually the tank cools to ambient temperature, 0°C, at which time it contains one-fourth liquid and three-fourths vapor, by volume. For the overall process, calculate:
(a) The heat transfer from the tank.
(b) The net change of entropy.

10.52 Carbon dioxide gas enters a turbine at 50 atm, 200 F, and exits at 10 atm pressure. If the isentropic efficiency of the turbine is 75%, determine:
(a) The exit temperature.
(b) The second-law efficiency for this process.

10.53 A vertical cylinder with a frictionless piston and weights contains ethylene at 700 kPa with a quality of 50%, at which point the volume is 50 L. The system is now allowed to warm to room temperature, 20°C. Calculate the net entropy change for this process.

10.54 An uninsulated rigid tank contains 20 lbm of propane at 77 F, 80% quality. This tank is connected by a valve to another uninsulated rigid tank, which is initially evacuated. The valve is now opened and the propane eventually comes to equilibrium in the two tanks at 77 F, at which time the pressure is 90% of the initial pressure. Determine the irreversibility for this process.

10.55 An insulated, hollow steel sphere has an inside diameter of 1 m and a wall thickness of 4 mm. The sphere contains saturated vapor ethane at ambient temperature, 300 K. A valve on the top of the sphere is now opened, and ethane flows out rapidly until the pressure reaches 500 kPa, at which point the valve is closed. During this process, the ethane remaining inside the sphere may be assumed to have undergone a reversible, adiabatic expansion. Now, after sitting for a period of time, the sphere and its contents come to a uniform temperature. Assuming no heat transfer with the surroundings, determine the final state of the ethane inside the sphere.

10.56 It is desired to evaluate a recently developed compound for possible use as a working fluid in a portable, closed-cycle power plant, as shown schematically in Fig. 10.17. The only information available for this compound is the following:

$$T_c = 160°C, \qquad P_c = 2.0 \text{ MPa}$$
$$M = 100, \qquad k = 1.10 \text{ (at very low pressure)}$$

The condenser temperature is fixed at 50°C, and the maximum cycle temperature is fixed at 180°C. The isentropic efficiency of the expansion engine is estimated to be 80% and the minimum allowable quality of the fluid exiting the expansion engine is 90%.

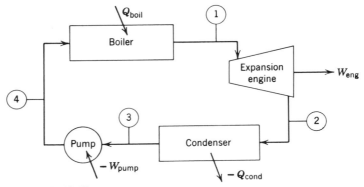

FIGURE 10.17
Sketch for Problem 10.56.

(a) Calculate the heat transfer from the condenser, assuming saturated liquid at the exit.

(b) Determine the maximum cycle pressure, based on the specifications listed above.

10.57 One lbm per second of water enters a solar collector at 100 F and exits at 200 F, as shown in Fig. 10.18. The hot water is sprayed into a direct-contact heat exchanger (no mixing of the two fluids) that is used to boil liquid butane. Pure saturated vapor butane exits at the top at 180 F and is fed to the turbine. If the butane condenser temperature is 90 F and the turbine and pump isentropic efficiencies are each 80%, determine the net power output of the cycle.

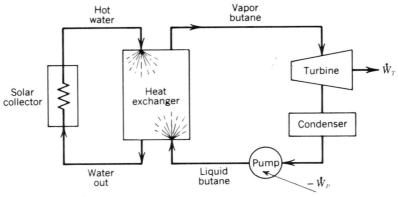

FIGURE 10.18
Sketch for Problem 10.57.

10.58 Write a computer program to solve the following problem. It is desired to obtain a plot of pressure versus specific volume at various temperatures

(all on a generalized reduced basis) as predicted by the van der Waals equation of state. Temperatures less than the critical should be included in the results.

10.59 Write a computer program to solve the following problem. For one of the substances listed in Table 3.3, calculate enthalpy changes along several isotherms at various integral pressures, using the Beattie–Bridgeman equation of state.

10.60 Write a computer program to solve the following problem. It is desired to curve fit a polynomial in reduced temperature T^* to values of the reduced second virial coefficient $B^*(T^*)$ given in Table A.12.1 over a specific range of reduced temperature. For one of the substances listed in Table A.12.2, plot Z versus P at low pressures for various isotherms.

11

MIXTURES AND SOLUTIONS

Up to this point in our development of thermodynamics we have limited our consideration primarily to pure substances. A large number of thermodynamic problems involve mixtures of different pure substances. Sometimes these mixtures are referred to as solutions, particularly in the liquid and solid phases.

In this chapter we shall turn our attention to various thermodynamic considerations of mixtures and solutions. We begin with a consideration of a rather simple problem, mixtures of ideal gases. This leads to a consideration of a simplified but very useful model of certain mixtures, such as air and water vapor, which may involve a condensed (solid or liquid) phase of one of the components. This is followed by certain considerations of mixtures and solutions in general.

An understanding of the material in this chapter is a necessary foundation for the consideration of chemical reactions and chemical and phase equilibrium. These topics are covered in subsequent chapters.

11.1 GENERAL CONSIDERATIONS AND MIXTURES OF IDEAL GASES

In any gaseous mixture, the mole fraction y_i of component i defined as

$$y_i = \frac{n_i}{n} \tag{11.1}$$

where n_i is the number of moles of component i, and n the total number of moles in the mixture.

Similarly, in any gaseous mixture, the mass fraction mf_i is defined as

$$mf_i = \frac{m_i}{m} \tag{11.2}$$

where m_i is the mass of component i and m is the total mass of the mixture.

Consider a mixture of two gases (not necessarily ideal gases) such as shown in Fig. 11.1.

What properties can we experimentally measure for such a mixture? Certainly we can measure the pressure, temperature, volume, and mass of the mixture. We can also experimentally measure the composition of the mixture, and thus determine the mole and mass fractions.

416

FIGURE 11.1

A mixture of two gases.

Suppose that this mixture undergoes a process or a chemical reaction and we wish to perform a thermodynamic analysis of this process or reaction. What type of thermodynamic data would we use in performing such an analysis? One possibility would be to have tables of thermodynamic properties of mixtures. However, the number of different mixtures that is possible, both as regards the substances involved and the relative amounts of each, is such that one would need a library full of tables of thermodynamic properties to handle all possible situations. It would be much simpler if we could determine the thermodynamic properties of a mixture from the properties of the pure components. This is in essence the approach that is used in dealing with ideal gases and certain other simplified models of mixtures.

One exception to this procedure is the case where a particular mixture is encountered very frequently, the most familiar being air. Tables and charts of the thermodynamics properties of air are available. However, even in this case it is necessary to define the composition of the "air" for which the tables are given, since the composition of the atmosphere varies with altitude, with the number of pollutants, and with other variables at a given location. The composition of air on which air tables are usually based is as follows:

Component	% on Mole Basis
Nitrogen	78.10
Oxygen	20.95
Argon	0.92
CO_2 & trace elements	0.03

Consider again the mixture of Fig. 11.1. For the general case the properties of the mixture are defined in terms of the partial molal properties of the individual components. A partial molal property is defined as the value of a property, such as internal energy, for a given component as it exists in the mixture. With this definition, the internal energy of the mixture of Fig. 11.1 would be

$$U_{\text{mix}} = n_A \bar{U}_A + n_B \bar{U}_B \qquad (11.3)$$

where \bar{U} designates the partial molal internal energy. Similar equations can be written for other properties, and this matter will be further developed in Section 11.9.

In this section we focus on mixtures of ideal gases. We assume that each component is uninfluenced by the presence of the other components, and that each component can be treated as an ideal gas. In an actual case of a gaseous mixture at high pressure this assumption would probably not be true because of the nature of the interaction between the molecules of the different components.

There are two models used in conjunction with the mixtures of gases, namely, the Dalton model and the Amagat model.

Dalton Model

In the case of the Dalton model, the properties of each component are considered as though each component existed separately at the volume and temperature of the mixture, as shown in Fig. 11.2.

Consider this model for the special case in which both the mixture and the separated components can be considered an ideal gas.

For the mixture:
$$PV = n\bar{R}T \tag{11.4}$$
$$n = n_A + n_B$$

For the components:
$$P_A V = n_A \bar{R} T$$
$$P_B V = n_B \bar{R} T \tag{11.5}$$

On substituting we have

$$n = n_A + n_B$$
$$\frac{PV}{\bar{R}T} = \frac{P_A V}{\bar{R}T} + \frac{P_B V}{\bar{R}T} \tag{11.6}$$

or

$$P = P_A + P_B$$

where P_A and P_B are referred to as partial pressures.

Thus for a mixture of ideal gases, the pressure is the sum of the partial pressures of the individual components.

It should be stressed that the term partial pressure has relevance only for ideal gases. In effect this concept assumes that the molecules of each component are

FIGURE 11.2

The Dalton model.

uninfluenced by the other components, and that the total pressure is the sum of partial pressures of the individual components. It should also be noted that partial pressure is not a partial molal property in the sense as defined by Eq. 11.3, since partial molal properties relate only to extensive properties.

Amagat Model

In the Amagat model the properties of each component are considered as though each component existed separately at the pressure and temperature of the mixture, as shown in Fig. 11.3. The volumes of A and B under these conditions are V_A and V_B, respectively.

In the general case the sum of the volumes when separated, namely, $V_A + V_B$, need not be equal to the volume of the mixture. However, let us consider the special case in which both the separated components and the mixture are considered to be ideal gases. In this case we can write:

For the mixture:
$$PV = n\bar{R}T$$
$$n = n_A + n_B$$
<div align="right">(11.7)</div>

For the components:
$$PV_A = n_A\bar{R}T$$
$$PV_B = n_B\bar{R}T$$
<div align="right">(11.8)</div>

On substituting we have

$$n = n_A + n_B$$
$$\frac{PV}{\bar{R}T} = \frac{PV_A}{\bar{R}T} + \frac{PV_B}{\bar{R}T}$$

Therefore

$$V_A + V_B = V$$

or

$$\frac{V_A}{V} + \frac{V_B}{V} = 1$$
<div align="right">(11.9)</div>

V_A/V and V_B/V are referred to as the volume fractions.

FIGURE 11.3

The Amagat model.

Thus in the case of ideal gases, Amagat's model leads to the conclusion that the sum of the volume fractions is unity, and that there would be no volume change if the components were mixed while holding the temperature and pressure constant. From Eqs. 11.4, 11.5, 11.7, and 11.8 it is evident that

$$\frac{V_A}{V} = \frac{n_A}{n} = \frac{P_A}{P}$$

$$\frac{V_A}{V} = y_A = \frac{P_A}{P} \tag{11.10}$$

That is, for each component of a mixture of ideal gases, the volume fraction, the mole fraction, and the ratio of the partial pressure to the total pressure are equal.

In determining the internal energy, enthalpy, and entropy of a mixture of ideal gases, the Dalton model proves useful because the assumption is made that each constituent behaves as though it occupied the entire volume by itself. Thus the internal energy, enthalpy, and entropy can be evaluated as the sum of the respective properties of the constituent gases at the condition at which the component exists in the mixture. Since for ideal gases the internal energy and enthalpy are functions only of temperature, it follows that

$$U = n\bar{u} = n_A\bar{u}_A + n_B\bar{u}_B \tag{11.11}$$

$$H = n\bar{h} = n_A\bar{h}_A + n_B\bar{h}_B \tag{11.12}$$

where \bar{u}_A and \bar{h}_A are the internal energy and enthalpy per mole for pure A and \bar{u}_B and \bar{h}_B are the same quantities for pure B, all at the temperature of the mixture.

The entropy of an ideal gas is a function of pressure as well as temperature. Since each component exists in the mixture at its partial pressure,

$$S = n\bar{s} = n_A\bar{s}_A + n_B\bar{s}_B \tag{11.13}$$

where \bar{s}_A is the entropy per mole for pure A at T and P_A (the partial pressure of A), and \bar{s}_B that for pure B at T and P_B.

Example 11.1

A volumetric analysis of a gaseous mixture yields the following results:

CO_2	12.0%
O_2	4.0
N_2	82.0
CO	2.0

Determine the analysis on a mass basis, and the molecular weight and the gas constant on a mass basis for the mixture, using SI units. Assume ideal gas behavior.

System: Gas mixture.

State: Composition known.

Solution

It is convenient to set up and solve the problem as shown below in Table 11.1

TABLE 11.1

Constituent	Percent by Volume	Mole Fraction	Molecular Weight	Mass kg per kmol of Mixture	Analysis on Mass Basis, Percent
CO_2	12	0.12 ×	44.0	= 5.28	$\dfrac{5.28}{30.08} = 17.55$
O_2	4	0.04 ×	32.0	= 1.28	$\dfrac{1.28}{30.08} = 4.26$
N_2	82	0.82 ×	28.0	= 22.96	$\dfrac{22.96}{30.08} = 76.33$
CO	2	0.02 ×	28.0	= 0.56	$\dfrac{0.56}{30.08} = 1.86$
				30.08	100.00

Molecular weight of mixture = 30.08

$$R \text{ for mixture} = \frac{\bar{R}}{M} = \frac{8.3144}{30.08} = 0.2764 \text{ kJ/kg K}$$

If the analysis had been given on a mass basis, and the mole fraction or volumetric analysis was desired, the procedure shown in Table 11.2 could be used.

$$M = \frac{1}{\text{kmol/kg mixture}} = \frac{1}{0.033\ 24} = 30.08$$

$$R = \frac{\bar{R}}{M} = \frac{8.3144}{30.08} = 0.2764 \text{ kJ/kg K}$$

TABLE 11.2

Constituent	Mass Fraction	Molecular Weight	kmol per kg of Mixture	Mole Fraction	Volumetric Analysis, Percent
CO_2	0.1755 ÷	44.0	= 0.003 99	0.120	12.0
O_2	0.0426 ÷	32.0	= 0.001 33	0.040	4.0
N_2	0.7633 ÷	28.0	= 0.027 26	0.820	82.0
CO	0.0186 ÷	28.0	= 0.000 66	0.020	2.0
			0.033 24	1.000	100.0

While this example of conversion between mole and mass basis has been worked in SI units, the procedure would be entirely equivalent in the English system. In either case, the key to the conversion is to keep in mind that the molecular weight of a component or of the mixture has units—kg/kmol in the SI system, and lbm/lb mol in the English system.

Example 11.2

Let n_A moles of gas A at a given pressure and temperature be mixed with n_B moles of gas B at the same pressure and temperature in an adiabatic constant-volume process, shown in Fig. 11.4. Determine the increase in entropy for this process.

System: All gas (A and B).
Initial states: P, T known for A and B.
Final state: P, T of mixture known.
Sketch: Figure 11.4.

Analysis and Solution

The final partial pressure of gas A is P_A and for gas B it is P_B. Since there is no change in temperature, Eq. 8.22 reduces to

$$(S_2 - S_1)_A = -n_A \bar{R} \ln \frac{P_A}{P} = -n_A \bar{R} \ln y_A$$

$$(S_2 - S_1)_B = -n_B \bar{R} \ln \frac{P_B}{P} = -n_B \bar{R} \ln y_B$$

The total change in entropy is the sum of the entropy changes for gases A and B.

$$S_2 - S_1 = -\bar{R}(n_A \ln y_A + n_B \ln y_B) \tag{11.14}$$

This equation can be written for the general case of mixing any number of components at the same pressure and temperature as

$$S_2 - S_1 = -\bar{R} \sum_k n_k \ln y_k \tag{11.15}$$

The interesting thing about this equation is that the increase in entropy depends only on the number of moles of component gases, and is independent of the composition of the gas. For example, when 1 mol of oxygen and 1 mol of nitrogen are mixed, the increase in entropy is the same as when 1 mol of hydrogen and 1 mol of nitrogen are mixed. But we also know that if 1 mol of nitrogen is "mixed" with another mole of nitrogen there is no increase in entropy. The question that arises is how dissimilar must the gases be in order to have an increase in entropy? The answer lies in our ability to distinguish between the two gases. The entropy increases

FIGURE 11.4
Sketch for Example 11.2.

whenever we can distinguish between the gases being mixed. When we cannot distinguish between the gases, there is no increase in entropy.

11.2 A SIMPLIFIED MODEL OF A MIXTURE INVOLVING GASES AND A VAPOR

Let us now consider a simplification, which in many cases is a reasonable one, of the problem involving a mixture of ideal gases that is in contact with a solid or liquid phase of one of the components. The most familiar example is a mixture of air and water vapor in contact with liquid water or ice, such as the problems encountered in air conditioning or drying. We are all familiar with the condensation of water from the atmosphere when it is cooled on a summer day.

This problem, and a number of similar problems, can be analyzed quite simply and with considerable accuracy if the following assumptions are made:

1. The solid or liquid phase contains no dissolved gases.
2. The gaseous phase can be treated as a mixture of ideal gases.
3. When the mixture and the condensed phase are at a given pressure and temperature, the equilibrium between the condensed phase and its vapor is not influenced by the presence of the other component. This means that when equilibrium is achieved the partial pressure of the vapor will be equal to the saturation pressure corresponding to the temperature of the mixture.

Since this approach is used extensively and with considerable accuracy, let us give some attention to the terms that have been defined and the type of problems for which this approach is valid and relevant. In our discussion we will refer to this as a gas–vapor mixture.

The dew point of a gas–vapor mixture is the temperature at which the vapor condenses or solidifies when it is cooled at constant pressure. This is shown on the T-s diagram for the vapor shown in Fig. 11.5. Suppose that the temperature of the

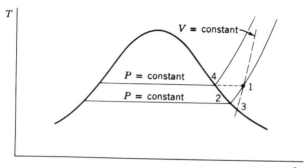

FIGURE 11.5

Temperature–entropy diagram to show definition of the dew point.

gaseous mixture and the partial pressure of the vapor in the mixture are such that the vapor is initially superheated at state 1. If the mixture is cooled at constant pressure, the partial pressure of the vapor remains constant until point 2 is reached, and then condensation begins. The temperature at state 2 is the dew-point temperature. Line 1–3 on the diagram indicates that if the mixture is cooled at constant volume the condensation begins at point 3, which is slightly lower than the dew-point temperature.

If the vapor is at the saturation pressure and temperature, the mixture is referred to as a saturated mixture, and for an air–water vapor mixture, the term "saturated air" is used.

The relative humidity ϕ is defined as the ratio of the mole fraction of the vapor in the mixture to the mole fraction of vapor in a saturated mixture at the same temperature and total pressure. Since the vapor is considered an ideal gas, the definition reduces to the ratio of the partial pressure of the vapor as it exists in the mixture, P_v, to the saturation pressure of the vapor at the same temperature, P_g.

$$\phi = \frac{P_v}{P_g}$$

In terms of the numbers on the T-s diagram of Fig. 11.5, the relative humidity ϕ would be

$$\phi = \frac{P_1}{P_4}$$

Since we are considering the vapor to be an ideal gas, the relative humidity can also be defined in terms of specific volume or density.

$$\phi = \frac{P_v}{P_g} = \frac{\rho_v}{\rho_g} = \frac{v_g}{v_v} \tag{11.16}$$

The humidity ratio ω of an air-water vapor mixture is defined as the ratio of the mass of water vapor m_v to the mass of dry air m_a. The term "dry air" is used to emphasize that this refers only to air and not to the water vapor. The term "specific humidity" is used synonymously with humidity ratio.

$$\omega = \frac{m_v}{m_a} \tag{11.17}$$

This definition is identical for any other gas-vapor mixture, and the subscript a refers to the gas, exclusive of the vapor. Since we are considering both the vapor and the mixture to be ideal gases, a very useful expression for humidity ratio in terms of partial pressures and molecular weights can be developed.

$$m_v = \frac{P_v V}{R_v T} = \frac{P_v V M_v}{\bar{R} T}, \qquad m_a = \frac{P_a V}{R_a T} = \frac{P_a V M_a}{\bar{R} T}$$

Then

$$\omega = \frac{P_v V / R_v T}{P_a V / R_a T} = \frac{R_a P_v}{R_v P_a} = \frac{M_v P_v}{M_a P_a} \tag{11.18}$$

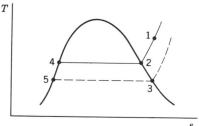

FIGURE 11.6

Temperature–entropy diagram to show the cooling of a gas–vapor mixture at a constant pressure.

For an air-water vapor mixture this reduces to

$$\omega = 0.622 \frac{P_v}{P_a} \qquad\qquad (11.19)$$

The degree of saturation is defined as the ratio of the actual humidity ratio to the humidity ratio of a saturated mixture at the same temperature and total pressure.

An expression for the relation between the relative humidity ϕ and the humidity ratio ω can be found by solving Eqs. 11.16 and 11.19 for P_v and equating them. The resulting relation for an air-water vapor mixture is

$$\phi = \frac{\omega P_a}{0.622 P_g} \qquad\qquad (11.20)$$

A few words should also be said about the nature of the process that occurs when a gas-vapor mixture is cooled at constant pressure. Suppose that the vapor is initially superheated at state 1 in Fig. 11.6. As the mixture is cooled at constant pressure, the partial pressure of the vapor remains constant until the dew point is reached at point 2, at which point the vapor in the mixture is saturated. The initial condensate is at state 4, and is in equilibrium with the vapor at state 2. As the temperature is lowered further, more of the vapor condenses, which lowers the partial pressure of the vapor in the mixture. The vapor that remains in the mixture is always saturated, and the liquid or solid is in equilibrium with it. For example, when the temperature is reduced to T_3, the vapor in the mixture is at state 3, and its partial pressure is the saturation pressure corresponding to T_3. The liquid in equilibrium with it is at state 5.

Example 11.3

Consider 2000 ft³ of an air-water vapor mixture at 14.7 lbf/in.², 90 F, 70 per cent relative humidity. Calculate the humidity ratio, dew point, mass of air, and mass of vapor.

System: Mixture.

State: P, T, ϕ known; state fixed.

Analysis and Solution

From Eq. 11.16 and the steam tables,

$$\phi = 0.70 = \frac{P_v}{P_g}$$

$$P_v = 0.70(0.6988) = 0.4892 \text{ lbf/in.}^2$$

The dew point is the saturation temperature corresponding to this pressure, which is 78.9 F.

The partial pressure of the air is

$$P_a = P - P_v = 14.70 - 0.49 = 14.21 \text{ lbf/in.}^2$$

The humidity ratio can be calculated from Eq. 11.19.

$$\omega = 0.622 \times \frac{P_v}{P_a} = 0.622 \times \frac{0.4892}{14.21} = 0.02135$$

The mass of air is

$$m_a = \frac{P_a V}{R_a T} = \frac{14.21 \times 144 \times 2000}{53.34 \times 550} = 139.6 \text{ lbm}$$

The mass of the vapor can be calculated by using the humidity ratio or by using the ideal gas equation of state.

$$m_v = \omega m_a = 0.02135(139.6) = 2.98 \text{ lbm}$$

$$m_v = \frac{0.4892 \times 144 \times 2000}{85.7 \times 550} = 2.98 \text{ lbm}$$

Example 11.4

Calculate the amount of water vapor condensed if the mixture of Example 11.3 is cooled to 40 F in a constant-pressure process.

System: Mixture.

Initial state: Known (Example 11.3).

Final state: T known.

Process: Constant pressure.

Analysis: At the final temperature, 40 F, the mixture is saturated, since this is below the dew-point temperature. Therefore,

$$P_{v2} = P_{g2}, \qquad P_{a2} = P - P_{v2}$$

and

$$\omega_2 = 0.622 \frac{P_{v2}}{P_{a2}}$$

From the conservation of mass, it follows that the amount of water condensed is equal to the difference between the initial and final mass of water vapor, or

$$\text{Mass of vapor condensed} = m_a(\omega_1 - \omega_2)$$

Solution

$$P_{v2} = P_{g2} = 0.1217 \text{ lbf/in.}^2$$
$$P_{a2} = 14.7 - 0.12 = 14.58 \text{ lbf/in.}^2$$
$$\omega_2 = 0.622 \times \frac{0.1217}{14.58} = 0.00520$$

$$\text{mass of vapor condensed} = m_a(\omega_1 - \omega_2) = 139.6(0.02135 - 0.0052)$$
$$= 2.25 \text{ lbm}$$

11.3 THE FIRST LAW APPLIED TO GAS–VAPOR MIXTURES

In applying the first law of thermodynamics to gas–vapor mixtures, it is helpful to realize that because of our assumption that ideal gases are involved, the various components can be treated separately when calculating changes of internal energy and enthalpy. Therefore, in dealing with air–water vapor mixtures, the changes in enthalpy of the water vapor can be found from the steam tables and the ideal-gas relations can be applied to the air. This is illustrated by the examples that follow.

Example 11.5

An air-conditioning unit is shown in Fig. 11.7, with pressure, temperature, and relative humidity data. Calculate the heat transfer per pound of dry air, assuming that changes in kinetic energy are negligible.

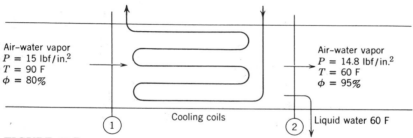

FIGURE 11.7
Sketch for Example 11.5.

Control volume: Duct, excluding cooling coils.

Inlet state: Known (Fig. 11.7).

Exit state: Known (Fig. 11.7).

Process: SSSF, no kinetic or potential energy changes.

Model: Air–ideal gas, constant specific heat, value at 80 F. Water–steam tables. (Since the water vapor at these low pressures is being considered an ideal gas, the enthalpy of the water vapor is a function of the temperature only. Therefore the enthalpy of slightly superheated water vapor is equal to the enthalpy of saturated vapor at the same temperature.)

Analysis: Continuity, air and water:

$$\dot{m}_{a1} = \dot{m}_{a2}$$
$$\dot{m}_{v1} = \dot{m}_{v2} + \dot{m}_{l2}$$

First law:

$$\dot{Q}_{c.v.} + \sum \dot{m}_i h_i = \sum \dot{m}_e h_e$$
$$\dot{Q}_{c.v.} + \dot{m}_a h_{a1} + \dot{m}_{v1} h_{v1} = \dot{m}_a h_{a2} + \dot{m}_{v2} h_{v2} + \dot{m}_{l2} h_{l2}$$

If we divide this equation by \dot{m}_a, introduce the continuity equation for the water, and note that $\dot{m}_v = \omega \dot{m}_a$, we can write the first law in the form

$$\frac{\dot{Q}_{c.v.}}{\dot{m}_a} + h_{a1} + \omega_1 h_{v1} = h_{a2} + \omega_2 h_{v2} + (\omega_1 - \omega_2) h_{l2}$$

Solution

$$P_{v1} = \phi P_{g1} = 0.80(0.6988) = 0.5590 \text{ lbf/in.}^2$$

$$\omega_1 = \frac{R_a}{R_v} \frac{P_{v1}}{P_{a1}} = 0.622 \times \left(\frac{0.5590}{15.00 - 0.56}\right) = 0.0241$$

$$P_{v2} = \phi_2 P_{g2} = 0.95(0.2563) = 0.2435$$

$$\omega_2 = \frac{R_a}{R_v} \times \frac{P_{v2}}{P_{a2}} = 0.622 \times \left(\frac{0.2435}{14.80 - 0.24}\right) = 0.0104$$

$$Q_{c.v.}/m_a + h_{a1} + \omega_1 h_{v1} = h_{a2} + \omega_2 h_{v2} + (\omega_1 - \omega_2) h_{l2}$$

$$Q_{c.v.}/m_a = 0.240(60 - 90) + 0.0104(1087.7) - 0.0241(1100.7)$$

$$+ (0.0241 - 0.0104)28.08$$

$$= -7.20 + 11.32 - 26.53 + 0.38 = -22.03 \text{ Btu/lbm dry air}$$

Example 11.6

A tank has a volume of 0.5 m³ and contains nitrogen and water vapor. The temperature of the mixture is 50°C and the total pressure is 2 MPa. The partial pressure of the water vapor is 5 kPa. Calculate the heat transfer when the contents of the tank are cooled to 10°C.

System: Nitrogen and water.

Initial state: P_1, T_1 known; state fixed.

Final state: T_2 known.

Process: Constant volume.

Model: Ideal gas mixture; constant specific heat for nitrogen; steam tables for water.

Analysis: This is a constant-volume process, and since the work is zero the first law reduces to

$$Q = U_2 - U_1 = m_{N_2}C_{v(N_2)}(T_2 - T_1) + (m_2 u_2)_v + (m_2 u_2)_l - (m_1 u_1)_v$$

This equation assumes that some of the vapor condensed. This must be checked, however, as shown below.

Solution

The mass of nitrogen and water vapor can be calculated using the ideal-gas equation of state.

$$m_{N_2} = \frac{P_{N_2}V}{R_{N_2}T} = \frac{1995 \times 0.5}{0.2968 \times 323.2} = 10.39 \text{ kg}$$

$$m_{v1} = \frac{P_{v1}V}{R_v T} = \frac{5 \times 0.5}{0.461\,52 \times 323.2} = 0.016\,76 \text{ kg}$$

If condensation takes place the final state of the vapor will be saturated vapor at 10°C. In this case

$$m_{v2} = \frac{P_{v2}V}{R_v T} = \frac{1.2276 \times 0.5}{0.461\,52 \times 283.2} = 0.004\,70 \text{ kg}$$

Since this is less than the original mass of vapor there must have been condensation.

The mass of liquid that is condensed, m_{l_2}, is

$$m_{l_2} = m_{v_1} - m_{v_2} = 0.016\,76 - 0.004\,70 = 0.012\,06 \text{ kg}$$

The internal energy of the water vapor is equal to the internal energy of saturated water vapor at the same temperature. Therefore,

$$u_{v1} = 2443.5 \text{ kJ/kg}$$
$$u_{v2} = 2389.2 \text{ kJ/kg}$$
$$u_{l2} = 42.0 \text{ kJ/kg}$$
$$Q_{c.v.} = 10.39 \times 0.7448(10 - 50) + 0.0047(2389.2)$$
$$+ 0.012\,06(42.0) - 0.016\,76(2443.5)$$
$$= -338.8 \text{ kJ}$$

11.4 THE ADIABATIC SATURATION PROCESS

An important process involving an air–water vapor mixture is the adiabatic saturation process, in which an air–vapor mixture comes in contact with a body of water in a well-insulated duct (Fig. 11.8). If the initial relative humidity is less than 100%, some of the water will evaporate and the temperature of the air–vapor mixture will decrease. If the mixture leaving the duct is saturated and if the process is adiabatic, the temperature of the mixture on leaving is known as the adiabatic saturation temperature. In order for this to take place as a steady-flow process, make-up water at the adiabatic saturation temperature is added at the same rate at which water is evaporated. The pressure is assumed to be constant.

Considering the adiabatic saturation process to be a steady-state, steady-flow process, and neglecting changes in kinetic and potential energy, the first law reduces to

$$h_{a1} + \omega_1 h_{v1} + (\omega_2 - \omega_1) h_{l2} = h_{a2} + \omega_2 h_{v2}$$
$$\omega_1(h_{v1} - h_{l2}) = C_{pa}(T_2 - T_1) + \omega_2(h_{v2} - h_{l2})$$
$$\omega_1(h_{v1} - h_{l2}) = C_{pa}(T_2 - T_1) + \omega_2 h_{fg2} \qquad \textbf{(11.21)}$$

The most significant point to be made about the adiabatic saturation process is that the adiabatic saturation temperature, the temperature of the mixture when it leaves the duct, is a function of the pressure, temperature, and relative humidity of the entering air–vapor mixture and of the exit pressure. Thus the relative humidity and the humidity ratio of the entering air–vapor mixture can be determined from the measurements of the pressure and temperature of the air–vapor mixture entering and leaving the adiabatic saturator. Since these measurements are relatively easy to make, this is one means of determining the humidity of an air–vapor mixture.

Example 11.7

The pressure of the mixture entering and leaving the adiabatic saturator is 14.7 lbf/in.2, the entering temperature is 84 F, and the temperature leaving is 70 F, which is the adiabatic saturation temperature. Calculate the humidity ratio and relative humidity of the air-water vapor mixture entering.

Control volume: Adiabatic saturator.

Inlet state: P_1, T_1 known.

Exit state: P_2, T_2 known; $\phi_2 = 100\%$; state fixed.

Process: SSSF, adiabatic saturation (Fig. 11.8).

Model: Ideal-gas mixture; constant specific heat for air; steam tables for water.

Analysis: Continuity and first law, Eq. 11.21.

FIGURE 11.8
The adiabatic saturation process.

Solution

Since the water vapor leaving is saturated, $P_{v2} = P_{g2}$ and ω_2 can be calculated.

$$\omega_2 = 0.622 \times \frac{0.3632}{14.7 - 0.36} = 0.01573$$

ω_1 can be calculated using Eq. 11.21.

$$\omega_1 = \frac{C_{pa}(T_2 - T_1) + \omega_2 h_{fg2}}{(h_{v1} - h_{l2})}$$

$$\omega_1 = \frac{0.24(70 - 84) + 0.01573 \times 1054.0}{1098.1 - 38.1} = \frac{-3.36 + 16.60}{1060.0} = 0.0125$$

$$\omega_1 = 0.622 \times \left(\frac{P_{v1}}{14.7 - P_{v1}}\right) = 0.0125$$

$$P_{v1} = 0.289$$

$$\phi_1 = \frac{P_{v1}}{P_{g1}} = \frac{0.289}{0.584} = 0.495$$

11.5 WET-BULB AND DRY-BULB TEMPERATURES

The humidity of an air–water vapor mixture is usually found from dry-bulb and wet-bulb data. These data are obtained by use of a psychrometer, which involves the flow of air past wet-bulb and dry-bulb thermometers. The bulb of the wet-bulb thermometer is covered with a cotton wick that is saturated with water. The dry-bulb thermometer is used simply to measure the temperature of the air. The flow of air may be maintained by a fan, as in the continuous-flow psychrometer shown in Fig. 11.9, or by moving the thermometer through the air, as in the sling psychrometer, which consists of wet-bulb and dry-bulb thermometers mounted so that they can be whirled.

The processes that take place at the wet-bulb thermometer are somewhat involved. First, if the air–water vapor mixture is not saturated, some of the water in the wick evaporates and diffuses into the surrounding air. A drop in the temperature of the water in the wick will be associated with this evaporation. However, as soon as the temperature of the water drops, heat is transferred to the

FIGURE 11.9

Steady-flow apparatus for measuring wet- and dry-bulb temperatures.

water from both the air and the thermometer. Finally, a steady state, determined by heat and mass transfer rates, will be reached. In general, air velocities exceeding 3 m/s are required in order that the convective heat transfer be large in comparison to the radiant heat transfer.

The psychrometric chart is the most convenient method of determining relative humidity and humidity ratio from wet-bulb and dry-bulb data, although equations that have been developed can also be used. A psychrometric chart is included in the Appendix, Fig. A.4.

The difference between the wet-bulb temperature and adiabatic saturation temperature should be carefully noted. The wet-bulb temperature is influenced by heat and mass transfer rates, whereas the adiabatic saturation temperature simply involves equilibrium between the entering air–vapor mixture and water at the adiabatic saturation temperature. However, it happens that the wet-bulb temperature and the adiabatic saturation temperature are approximately equal for air–water vapor mixtures at atmospheric temperatures and pressure. This is not necessarily true at temperatures and pressures that deviate significantly from ordinary atmospheric conditions, or for other gas–vapor mixtures.

11.6 THE PSYCHROMETRIC CHART

Properties of air-water vapor mixtures are given in graphical form on psychrometric charts. These are available in a number of different forms, and only the main features are considered here.

The basic psychrometric chart consists of a plot of dry-bulb temperature (abscissa) and humidity ratio (ordinate). If we fix the total pressure for which the chart is to be constructed (which is usually one standard atmosphere), lines of constant relative humidity and wet-bulb temperature can be drawn on the chart, because for a given dry-bulb temperature, total pressure, and humidity ratio, the relative humidity and wet-bulb temperature are fixed. The partial pressure of the water vapor is fixed by the humidity ratio and total pressure, and therefore a

second ordinate scale that indicates the partial pressure of the water vapor can be constructed.

Most psychrometric charts give the enthalpy of an air-vapor mixture per pound of dry air. The values given assume that the enthalpy of the dry air is zero at 0 F, and the enthalpy of the vapor is taken from the steam tables (which are based on the assumption that the enthalpy of saturated liquid is zero at 32 F). This procedure is satisfactory because we are usually concerned only with differences in enthalpy. The fact that lines of constant enthalpy are essentially parallel to lines of constant wet-bulb temperature is evident from the fact that the wet-bulb temperature is essentially equal to the adiabatic saturation temperature. Thus, in Fig. 11.8, if we neglect the enthalpy of the liquid entering the adiabatic saturator, the enthalpy of the air-vapor mixture entering is equal to the enthalpy of the saturated air-vapor mixture leaving, and a given adiabatic saturation temperature fixes the enthalpy of the mixture entering.

Some charts are available that give corrections for variation from standard atmospheric pressures. Before using a given chart one should fully understand the assumptions made in constructing it, and that it is applicable to the particular problem at hand.

11.7 INTRODUCTION TO REAL MIXTURES AND SOLUTIONS

In the remainder of this chapter, we will develop certain concepts and relations concerning the behavior of real mixtures and solutions. These considerations are not necessarily restricted to gaseous mixtures alone; they are also valid for solid, liquid, and gaseous solutions. In later chapters we will also include the effects of chemical reactions as well as the equilibrium between different phases, but for the present will concern ourselves with nonreacting mixtures or solutions in a single, homogeneous phase.

In a thermodynamic analysis involving any real mixture or solution, there are two distinct alternatives in the approach to the problem, as far as the expression of extensive thermodynamic properties at a given state (that is, at a given pressure and temperature). The first approach is to treat the mixture as a pseudo-pure substance having its own set of properties, as, for example, has often been used previously for air in cases in which it could be assumed an ideal gas and there was no question about saturation or change of phase. The second approach is to treat the mixture properties as a summation of the contributions of the mixture components, at the conditions at which the components exist in the mixture, that is, using the concept of partial properties. From what we have considered up to this point in the present chapter, it is apparent that ideal gas mixtures can be analyzed using either of these two alternative approaches. Real mixture models, however, follow one of these alternatives or the other. The remainder of this chapter develops these two alternatives and also several related topics concerning the general thermodynamics of real mixtures. In the next section, the matter of pseudo-pure substance models is developed, and it is also necessary to consider the matter

of reference states in connection with this approach to mixture analysis. Subsequent sections then deal with partial properties, including the development of various thermodynamic relations and a number of expressions relating various thermodynamic properties, as well as models for expressing mixture behavior.

11.8 PSEUDO-PURE SUBSTANCE MODELS FOR REAL GAS MIXTURES

A basic prerequisite to the treatment of real gas mixtures in terms of pseudo-pure substance models is the concept and use of appropriate reference states. As an introduction to this topic, let us consider several preliminary reference state questions for a pure substance undergoing a change of state, for which it is desired to calculate the entropy change. We can express the entropy at the initial state 1 and also at the final state 2 in terms of a reference state 0, in a manner similar to that followed when dealing with the generalized-charts corrections in Chapter 10. It follows that

$$s_1 = s_0 + (s_{P_0 T_0}^* - s_0) + (s_{P_1 T_1}^* - s_{P_0 T_0}^*) + (s_1 - s_{P_1 T_1}^*) \tag{11.22}$$

$$s_2 = s_0 + (s_{P_0 T_0}^* - s_0) + (s_{P_2 T_2}^* - s_{P_0 T_0}^*) + (s_2 - s_{P_2 T_2}^*) \tag{11.23}$$

These are entirely general expressions for the entropy at each state in terms of an arbitrary reference state value and a set of consistent calculations from that state to the actual desired state. One simplification of these equations would result from choosing the reference state to be a hypothetical ideal gas state at P_0 and T_0, thereby making the term

$$(s_{P_0 T_0}^* - s_0) = 0 \tag{11.24}$$

in each equation, which results in

$$s_0 = s_0^* \tag{11.25}$$

It should be apparent that this choice is a reasonable one, since whatever value is chosen for the correction term, Eq. 11.24, it will cancel out of the two equations when the change $s_2 - s_1$ is calculated, and the simplest value to choose is zero. In a similar manner, the simplest value to choose for the ideal gas reference value, Eq. 11.25, is zero, and we would commonly do that if there are no restrictions on choice, such as occur in the case of a chemical reaction.

Another point to be noted concerning reference states is related to the choice of P_0 and T_0, and for this purpose, let us substitute Eqs. 11.24 and 11.25 into Eqs. 11.22 and 11.23, and also assume constant specific heat, such that those equations can be written in the form

$$s_1 = s_0^* + C_{po} \ln\left(\frac{T_1}{T_0}\right) - R \ln\left(\frac{P_1}{P_0}\right) + (s_1 - s_{P_1 T_1}^*) \tag{11.26}$$

$$s_2 = s_0^* + C_{po} \ln\left(\frac{T_2}{T_0}\right) - R \ln\left(\frac{P_2}{P_0}\right) + (s_2 - s_{P_2 T_2}^*) \tag{11.27}$$

FIGURE 11.10

Example of mixing process.

Since the choice for P_0 and T_0 is arbitrary if there are no restrictions, such as would be the case with chemical reactions, it should be apparent from examining Eqs. 11.26 and 11.27 that the simplest choice would be for

$$P_0 = P_1 \text{ or } P_2, \qquad T_0 = T_1 \text{ or } T_2$$

It should be emphasized that inasmuch as the reference state was chosen as a hypothetical ideal gas at P_0, T_0, Eq. 11.24, it is immaterial how the real substance behaves at that pressure and temperature. As a result, there is no need to select a low value for the reference state pressure P_0.

Let us now extend these reference state developments to include real gas mixtures. Consider the mixing process shown in Fig. 11.10, with the states and amounts of each substance as given on the diagram. Proceeding with entropy expressions as was done above, we have

$$\bar{s}_1 = \bar{s}^*_{A_0} + \bar{C}_{po_A} \ln\left(\frac{T_1}{T_0}\right) - \bar{R} \ln\left(\frac{P_1}{P_0}\right) + (\bar{s}_1 - \bar{s}^*_{P_1 T_1})_A \qquad \textbf{(11.28)}$$

$$\bar{s}_2 = \bar{s}^*_{B_0} + \bar{C}_{po_B} \ln\left(\frac{T_2}{T_0}\right) - \bar{R} \ln\left(\frac{P_2}{P_0}\right) + (\bar{s}_2 - \bar{s}^*_{P_2 T_2})_B \qquad \textbf{(11.29)}$$

$$\bar{s}_3 = \bar{s}^*_{mix_0} + \bar{C}_{po_{mix}} \ln\left(\frac{T_3}{T_0}\right) - \bar{R} \ln\left(\frac{P_3}{P_0}\right) + (\bar{s}_3 - \bar{s}^*_{P_3 T_3})_{mix} \qquad \textbf{(11.30)}$$

in which

$$\bar{s}^*_{mix_0} = y_A \bar{s}^*_{A_0} + y_B \bar{s}^*_{B_0} - \bar{R}(y_A \ln y_A + y_B \ln y_B) \qquad \textbf{(11.31)}$$

$$\bar{C}_{po_{mix}} = y_A \bar{C}_{po_A} + y_B \bar{C}_{po_B} \qquad \textbf{(11.32)}$$

We note that when Eqs. 11.28–11.30 are substituted into the equation for the entropy change,

$$\dot{n}_3 \bar{s}_3 - \dot{n}_1 \bar{s}_1 - \dot{n}_2 \bar{s}_2$$

the arbitrary reference values $\bar{s}^*_{A_0}$, $\bar{s}^*_{B_0}$, P_0, and T_0 all cancel out of the result, which is, of course, necessary in view of their arbitrary nature. An ideal gas entropy of mixing expression, the final term in Eq. 11.31, remains in the result, establishing, in effect, the mixture reference value relative to its components. The remarks made earlier concerning the choices for reference state and the reference state entropies apply in this situation as well.

To summarize the development to this point, we find that a calculation of real mixture properties, as, for example, using Eq. 11.30, requires the establishment of a hypothetical ideal gas reference state, a consistent ideal gas calculation to the conditions of the real mixture, and finally a correction that accounts for the real behavior of the mixture at that state. We note that this last term is the only place that the real behavior is introduced, and this is therefore the term that must be calculated by the pseudo-pure substance model to be used.

In treating a real gas mixture as a pseudo-pure substance, there are two approaches that we will follow to represent the P-v-T behavior: use of the generalized charts, and use of an analytical equation of state. With the generalized charts, we need to have a model that provides a set of pseudocritical pressure and temperature in terms of the mixture component values. Many such models have been proposed and utilized over the years, but the simplest is that suggested by W. B. Kay in 1936, in which

$$(P_c)_{mix} = \sum_i y_i P_{ci}, \qquad (T_c)_{mix} = \sum_i y_i T_{ci} \qquad (11.33)$$

This is the only pseudocritical model that we will consider in this text. A number of others are more accurate in representing real gas mixture behavior, but are more complicated to use.

The other approach to be considered is that of using an analytical equation of state, in which the equation for the mixture must be developed from that for the components. In other words, for an equation in which the constants are known for each of the components, we must develop a set of empirical combining rules that will then give a set of constants for the mixture as though it were a pseudo-pure substance. This problem has been studied for many equations of state, using experimental data for the real gas mixtures, and various empirical rules have been proposed. For example, for both the van der Waals equation, Eq. 10.53, and the Redlich-Kwong equation, Eq. 10.59, the two pure substance constants a and b are commonly combined according to the relations

$$a_m = \left(\sum_i y_i a_i^{1/2}\right)^2, \qquad b_m = \sum_i y_i b_i \qquad (11.34)$$

The following example illustrates the use of these two approaches to treating real gas mixtures as pseudo-pure substances.

Example 11.8

A mixture of 59.39% CO_2 and 40.61% CH_4 (mole basis) is maintained at 310.94 K, 86.19 bar, at which condition the specific volume has been measured as 0.2205 m^3/kmol. Calculate the percent deviation if the specific volume had been calculated by (a) Kay's rule and (b) van der Waals' equation of state.

System: Gas mixture.

State: P, v, T known.

Model: (a) Kay's rule. (b) van der Waals' equation.

Solution

(a) For convenience, let

$$CO_2 = A, \qquad CH_4 = B$$

Then

$$T_{c_A} = 304.2 \text{ K}, \qquad P_{c_A} = 7.39 \text{ MPa}$$
$$T_{c_B} = 191.1 \text{ K}, \qquad P_{c_B} = 4.64 \text{ MPa}$$

For Kay's rule, Eq. 11.33,

$$T_{c_m} = \sum_i y_i T_{c_i} = y_A T_{c_A} + y_B T_{c_B}$$
$$= 0.5939(304.2) + 0.4061(191.1)$$
$$= 258.3 \text{ K}$$

$$P_{c_m} = \sum_i y_i P_{c_i} = y_A P_{c_A} + y_B P_{c_B}$$
$$= 0.5939(7.39) + 0.4061(4.64)$$
$$= 6.273 \text{ MPa}$$

Therefore, the pseudoreduced properties of the mixture are

$$T_{r_m} = \frac{T}{T_{c_m}} = \frac{310.94}{258.3} = 1.204$$

$$P_{r_m} = \frac{P}{P_{c_m}} = \frac{8.619}{6.273} = 1.374$$

From the generalized chart, Fig. A.5,

$$Z_m = 0.705$$

and

$$\bar{v} = \frac{Z_m \bar{R} T}{P} = \frac{0.705 \times 8.3144 \times 310.94}{8619}$$
$$= 0.2115 \text{ m}^3/\text{kmol}$$

The percent deviation from the experimental value is

$$\text{Percent deviation} = \left(\frac{0.2205 - 0.2115}{0.2205}\right) \times 100 = 4.08\%$$

(b) For van der Waals' equation, the pure substance constants are

$$a_A = \frac{27 \bar{R}^2 T_{c_A}^2}{64 P_{c_A}} = 365.185 \, \frac{\text{kPa m}^6}{\text{kmol}^2}$$

$$b_A = \frac{\bar{R} T_{c_A}}{8 P_{c_A}} = 0.042\,78 \text{ m}^3/\text{kmol}$$

and

$$a_B = \frac{27\bar{R}^2 T_{cB}^2}{64 P_{cB}} = 229.532 \frac{\text{kPa m}^6}{\text{kmol}^2}$$

$$b_B = \frac{\bar{R} T_{cB}}{8 P_{cB}} = 0.0428 \text{ m}^3/\text{kmol}$$

Therefore, for the mixture, from Eq. 11,34,

$$
\begin{aligned}
a_m &= (y_A \sqrt{a_A} + y_B \sqrt{a_B})^2 \\
&= (0.5939\sqrt{365.185} + 0.4061\sqrt{229.532})^2 \\
&= 306.315 \frac{\text{kPa m}^6}{\text{kmol}^2}
\end{aligned}
$$

$$
\begin{aligned}
b_m &= y_A b_A + y_B b_B \\
&= (0.5939 \times 0.042\,78 + 0.4061 \times 0.0428) \\
&= 0.042\,79 \text{ m}^3/\text{kmol}
\end{aligned}
$$

The equation of state for the mixture of this composition is

$$P = \frac{\bar{R} T}{\bar{v} - b_m} - \frac{a_m}{\bar{v}^2}$$

$$8619 = \frac{8.3144 \times 310.94}{\bar{v} - 0.042\,79} - \frac{306.315}{\bar{v}^2}$$

Solving for \bar{v} by trial and error,

$$\bar{v} = 0.2061 \text{ m}^3/\text{kmol}$$

$$\text{Percent deviation} = \left(\frac{0.2205 - 0.2061}{0.2205} \right) \times 100 = 6.5\%$$

As a point of interest from the ideal-gas law, $\bar{v} = 0.300$ m^3/kmol, which is a deviation of 36% from the measured value. Also, if we use the Redlich-Kwong equation of state, and follow the same procedure as for the van der Waals equation, the calculated specific volume of the mixture is 0.2127 m^3/kmol, which is in error by 3.5%.

We must be careful not to draw too general a conclusion from the results of this example. We have calculated percent deviation in v at only a single point for only one mixture. We do note, however, that the various methods used give quite different results. From a more general study of these models for a number of mixtures, we find that the results found here are fairly typical, at least qualitatively. Kay's rule is very useful because it is fairly accurate and yet relatively simple. The van der Waals equation is too simplified an expression to accurately represent P-v-T behavior except at moderate densities, but it is useful to demonstrate the procedures followed in utilizing more complex analytical equations of state. The Redlich-Kwong equation is considerably better, and is still relatively simple to use.

The more sophisticated generalized behavior models and empirical equations of state will represent mixture P-v-T behavior to within about 1% over a wide range of density, but they are of course more difficult to use than the methods considered in Example 11.8. The generalized models have the advantage of being easier to use, and they are suitable for hand computations. Calculations with the complex empirical equations of state become very involved, but have the advantage of expressing the P-v-T composition relations in analytical form, which is of great value when using a digital computer for such calculations.

The theoretical virial equation of state, Eq. 10.64, can be applied to mixtures, in which case the virial coefficients are dependent on composition as well as on temperature. For a mixture of A and B, the second virial coefficient becomes

$$B(T, y) = y_A{}^2 B_A(T) + 2y_A y_B B_{AB}(T) + y_B{}^2 B_B(T) \qquad (11.35)$$

in which the interaction coefficient $B_{AB}(T)$ results from forces between unlike molecules. For the Lennard–Jones (6-12) potential, this can be found from the parameters

$$\epsilon_{AB} = (\epsilon_A \epsilon_B)^{1/2}$$

$$b_{0AB} = \tfrac{1}{8}(b_{0A}^{1/3} + b_{0B}^{1/3})^3 \qquad (11.36)$$

and the values in Table A.12.

11.9 PARTIAL MOLAL PROPERTIES

In this section we begin to examine the second approach to the analysis of real mixtures that was mentioned in Section 11.7, that in which mixture extensive properties are treated as a summation of the contributions of the components, at the conditions at which the components exist in the mixture. This requires a detailed examination of the concept of partial molal properties, which is the subject of the following paragraphs. The term partial property was first used in connection with the expression for the internal energy of a mixture, Eq. 11.3. In the general case, any extensive property X is a function of the temperature and pressure of the mixture and the number of moles of each component. Thus, for a mixture of two components,

$$X = f(T, P, n_A, n_B)$$

Therefore,

$$dX_{T,P} = \left(\frac{\partial X}{\partial n_A}\right)_{T,P,n_B} dn_A + \left(\frac{\partial X}{\partial n_B}\right)_{T,P,n_A} dn_B \qquad (11.37)$$

Since at constant temperature and pressure an extensive property is directly proportional to the mass, Eq. 11.37 can be integrated to give

$$X_{T,P} = \bar{X}_A n_A + \bar{X}_B n_B \qquad (11.38)$$

where

$$\bar{X}_A = \left(\frac{\partial X}{\partial n_A}\right)_{T,P,n_B}, \qquad \bar{X}_B = \left(\frac{\partial X}{\partial n_B}\right)_{T,P,n_A} \qquad \textbf{(11.39)}$$

\bar{X} is defined as the partial molal property for a component in a mixture. It is particularly important to note that the partial molal property is defined under conditions of constant temperature and pressure. We also note that the general expression Eq. 11.38 is of the same form as Eq. 11.3.

The general extensive property X discussed above can be any of the properties V, H, U, S, A or G. As an example of the characteristics of partial properties, consider it to be the volume V. Then,

$$V = n_A \bar{V}_A + n_B \bar{V}_B \qquad \textbf{(11.40)}$$

where, by definition,

$$\bar{V}_A \equiv \left(\frac{\partial V}{\partial n_A}\right)_{T,P,n_B}, \qquad \bar{V}_B \equiv \left(\frac{\partial V}{\partial n_B}\right)_{T,P,n_A} \qquad \textbf{(11.41)}$$

For the special case of pure substance A,

$$\bar{V}_A = \left(\frac{\partial V}{\partial n_A}\right)_{T,P} = \bar{v}_A$$

and

$$V_A = \bar{v}_A n_A \qquad \textbf{(11.42)}$$

Thus the partial volume of A when no B is present reduces to the specific volume of pure A as would naturally be expected.

Let us now consider the general case of a real gas mixture, for which we can express the specific volume by dividing Eq. 11.40 by n, resulting in

$$\bar{v} = \frac{V}{n} = y_A \bar{V}_A + y_B \bar{V}_B$$

$$= \bar{V}_B + y_A(\bar{V}_A - \bar{V}_B) \qquad \textbf{(11.43)}$$

Suppose that the specific volume of this mixture at a given temperature and pressure varies with mixture composition in the manner shown in Fig. 11.11. From the relationships in Eqs. 11.41–11.43, we note that the specific volume varies from the value \bar{v}_B for pure B to the value \bar{v}_A for pure A, and that at any mixture composition, the partial molal volumes of A and B are equal to the respective intercepts of the tangent to the volume curve at that mixture composition point.

For the special case of an ideal gas mixture of A and B,

$$\bar{V}_A = \left(\frac{\partial V}{\partial n_A}\right)_{T,P,n_B} = \frac{\bar{R}T}{P} = \bar{v}$$

and

$$\bar{V}_B = \left(\frac{\partial V}{\partial n_B}\right)_{T,P,n_A} = \frac{\bar{R}T}{P} = \bar{v}$$

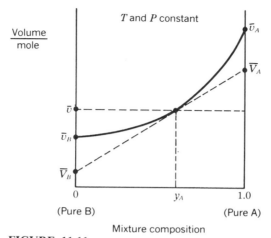

FIGURE 11.11

Specific volume–composition diagram for a mixture of two components.

Since \bar{v}, \bar{v}_A, \bar{v}_B are the same per mole,

$$\bar{V}_A = \bar{V}_B = \bar{v} = \bar{v}_A = \bar{v}_B = \frac{\bar{R}T}{P} \qquad \textbf{(11.44)}$$

Therefore, for a mixture of ideal gases we conclude from Eq. 11.40 that

$$V = n_A \bar{v}_A + n_B \bar{v}_B = V_A + V_B$$

which is Amagat's rule of additive volumes. We also note from Eq. 11.44 that a plot corresponding to that of Fig. 11.11 for an ideal gas mixture would be a horizontal straight line, where all the volumetric values in Eq. 11.44 are equal and are the same, regardless of the mixture composition.

The physical significance of the partial molal volume may be visualized by considering a reasonably large volume of a mixture of components A and B. Let a very small amount of A be added while the temperature, pressure and moles of B remain constant. The composition of the mixture will be essentially unchanged. The volume of the mixture will be increased by an amount equal to the specific volume of component A in the mixture (the partial molal volume of A) multiplied by the number of moles of component A added. Thus, in this case,

$$dV = \left(\frac{\partial V}{\partial n_A}\right)_{T,P,n_B} dn_A = \bar{V}_A \, dn_A$$

Similar analyses can of course be performed for the other extensive thermodynamic properties mentioned above. It can also be shown that the thermodynamic relations that apply to a mixture as a whole also hold for relations between partial

molal properties for a given component. For example, consider the relation

$$dG = dH - T\,dS - S\,dT$$

Let us again consider a mixture of components A and B, and let T, P, and n_B be held constant, and let n_A change by the amount dn_A. In this case, the preceding equation becomes

$$\left(\frac{\partial G}{\partial n_A}\right)_{T,P,n_B} = \left(\frac{\partial H}{\partial n_A}\right)_{T,P,n_B} - T\left(\frac{\partial S}{\partial n_A}\right)_{T,P,n_B}$$

or

$$\bar{G}_A = \bar{H}_A - T\bar{S}_A \tag{11.45}$$

Similarly

$$\bar{H}_A = \bar{U}_A + P\bar{V}_A \tag{11.46}$$

$$\bar{A}_A = \bar{U}_A - T\bar{S}_A \tag{11.47}$$

11.10 CHANGE IN PROPERTIES UPON MIXING

In this section, we will develop expressions for the changes in volume, enthalpy, and entropy upon mixing of two pure substances at constant temperature and pressure. We have previously seen that there is no volume change upon mixing of ideal gases at constant temperature and pressure. If two real gases each at T, P were mixed, however, the volume of the mixture would not necessarily equal the sum of the volumes of the constituents. The same is true for solutions involving liquids and solids. The change of volume on mixing, ΔV_{mix}, can be expressed as

$$\Delta V_{\text{mix}} = V_{\text{mixture}} - V_{\text{components}}$$

For a mixture of components A and B

$$\begin{aligned}
\Delta V_{\text{mix}} &= (\bar{V}_A n_A + \bar{V}_B n_B) - (\bar{v}_A n_A + \bar{v}_B n_B) \\
&= (\bar{V}_A - \bar{v}_A)n_A + (\bar{V}_B - \bar{v}_B)n_B
\end{aligned} \tag{11.48}$$

Similarly the enthalpy change of mixing is given by

$$\Delta H_{\text{mix}} = (\bar{H}_A - \bar{h}_A)n_A + (\bar{H}_B - \bar{h}_B)n_B \tag{11.49}$$

and the entropy change of mixing by

$$\Delta S_{\text{mix}} = (\bar{S}_A - \bar{s}_A)n_A + (\bar{S}_B - \bar{s}_B)n_B \tag{11.50}$$

For a mixture of ideal gases, the special case previously discussed, we concluded (Eq. 11.44) that

$$(\bar{V}_A - \bar{v}_A) = 0$$
$$(\bar{V}_B - \bar{v}_B) = 0$$

Therefore,

$$\Delta V_{\text{mix}} = 0 \tag{11.51}$$

Also, for a mixture of ideal gases

$$(\bar{H}_A - \bar{h}_A) = 0$$
$$(\bar{H}_B - \bar{h}_B) = 0$$

and therefore

$$\Delta H_{\text{mix}} = 0 \tag{11.52}$$

But, the entropy of mixing for ideal gases is given by the relation

$$(\bar{S}_A - \bar{s}_A) = -\bar{R} \ln y_A$$
$$(\bar{S}_B - \bar{s}_B) = -\bar{R} \ln y_B$$
$$\Delta S_{\text{mix}} = -\bar{R}(n_A \ln y_A + n_B \ln y_B) \tag{11.53}$$

in which \bar{S}_A is the partial entropy of A in the mixture at T, P, while \bar{s}_A is that for pure A at the same temperature and pressure, and similarly for component B. This is the expression that was derived in Example 11.2.

11.11 THE THERMODYNAMIC PROPERTY RELATION FOR VARIABLE COMPOSITION

In Section 10.2 we considered several forms of the property relation for systems of fixed mass, such as

$$dU = T\,dS - P\,dV$$

We note that in this equation temperature is the intensive property or potential function associated with entropy and pressure is the intensive property associated with volume. Now suppose we have a chemical reaction in which the amounts of components A and B change. How would we modify the property relation for this situation? Intuitively we might write the equation

$$dU = T\,dS - P\,dV + \mu_A\,dn_A + \mu_B\,dn_B \tag{11.54}$$

where μ_A is the intensive property or potential function associated with n_A, and similarly μ_B for n_B. This potential function is called the chemical potential.

To derive an expression for this chemical potential, we examine Eq. 11.54 and conclude that it is reasonable to write an expression for U in the form

$$U = f(S, V, n_A, n_B)$$

Therefore,

$$dU = \left(\frac{\partial U}{\partial S}\right)_{V,n_A,n_B} dS + \left(\frac{\partial U}{\partial V}\right)_{S,n_A,n_B} dV + \left(\frac{\partial U}{\partial n_A}\right)_{S,V,n_B} dn_A + \left(\frac{\partial U}{\partial n_B}\right)_{S,V,n_A} dn_B$$

Since the expressions

$$\left(\frac{\partial U}{\partial S}\right)_{V, n_A, n_B} \quad \text{and} \quad \left(\frac{\partial U}{\partial V}\right)_{S, n_A, n_B}$$

imply constant composition, it follows from Eq. 10.15 that

$$\left(\frac{\partial U}{\partial S}\right)_{V, n_A, n_B} = T \quad \text{and} \quad \left(\frac{\partial U}{\partial V}\right)_{S, n_A, n_B} = -P$$

Thus

$$dU = T\, dS - P\, dV + \left(\frac{\partial U}{\partial n_A}\right)_{S, V, n_B} dn_A + \left(\frac{\partial U}{\partial n_B}\right)_{S, V, n_A} dn_B \quad \textbf{(11.55)}$$

On comparing this equation with Eq. 11.54 it follows that the chemical potential can be defined by the relation

$$\mu_A = \left(\frac{\partial U}{\partial n_A}\right)_{S, V, n_B}, \quad \mu_B = \left(\frac{\partial U}{\partial n_B}\right)_{S, V, n_A} \quad \textbf{(11.56)}$$

We can also relate the chemical potential to the partial molal Gibbs function. To do so we proceed as follows.

$$G = U + PV - TS$$
$$dG = dU + P\, dV + V\, dP - T\, dS - S\, dT$$

Substituting Eq. 11.54 into this relation we have

$$dG = -S\, dT + V\, dP + \mu_A\, dn_A + \mu_B\, dn_B \quad \textbf{(11.57)}$$

This equation suggests that we write an expression for G in the following form.

$$G = f(T, P, n_A, n_B)$$

Proceeding as we did in the case of a similar expression for internal energy, Eq. 11.55, we have

$$dG = \left(\frac{\partial G}{\partial T}\right)_{P, n_A, n_B} dT + \left(\frac{\partial G}{\partial P}\right)_{T, n_A, n_B} dP + \left(\frac{\partial G}{\partial n_A}\right)_{T, P, n_B} dn_A + \left(\frac{\partial G}{\partial n_B}\right)_{T, P, n_A} dn_B$$

$$= -S\, dT + V\, dP + \left(\frac{\partial G}{\partial n_A}\right)_{T, P, n_B} dn_A + \left(\frac{\partial G}{\partial n_B}\right)_{T, P, n_A} dn_B$$

Comparing this with Eq. 11.57 it follows that

$$\mu_A = \left(\frac{\partial G}{\partial n_A}\right)_{T, P, n_B}, \quad \mu_B = \left(\frac{\partial G}{\partial n_B}\right)_{T, P, n_A}$$

Note that, since partial molal properties are defined at constant temperature and pressure, the quantities $(\partial G/\partial n_A)_{T, P, n_B}$ and $(\partial G/\partial n_B)_{T, P, n_A}$ are the partial molal Gibbs functions for the two components. That is, the chemical potential is equal to the partial molal Gibbs function.

$$\mu_A = \bar{G}_A = \left(\frac{\partial G}{\partial n_A}\right)_{T,P,n_B}, \qquad \mu_B = \bar{G}_B = \left(\frac{\partial G}{\partial n_B}\right)_{T,P,n_A} \qquad (11.58)$$

In terms of the enthalpy

$$H = U + PV$$

Equation 11.54 becomes

$$dH = T\,dS + V\,dP + \mu_A\,dn_A + \mu_B\,dn_B \qquad (11.59)$$

Following the same procedure as before, we find that

$$\mu_A = \left(\frac{\partial H}{\partial n_A}\right)_{S,P,n_B} \qquad (11.60)$$

In terms of the Helmholtz function

$$A = U - TS$$

the property relation is

$$dA = -S\,dT - P\,dV + \mu_A\,dn_A + \mu_B\,dn_B \qquad (11.61)$$

As before, we find an expression for the chemical potential

$$\mu_A = \left(\frac{\partial A}{\partial n_A}\right)_{T,V,n_B} \qquad (11.62)$$

Thus, we have found four different expressions for the chemical potential in terms of other properties, Eqs. 11.56, 11.58, 11.60 and 11.62. Of these, only the one, Eq. 11.58 satisfies the definition of a partial property. The partial molal Gibbs function is an extremely important property in the thermodynamic analysis of chemical reactions because at constant temperature and pressure (the conditions under which many chemical reactions occur) it is a measure of the chemical potential or the driving force tending to cause a chemical reaction to take place.

Finally, we also realize that the surface, magnetic, and other terms considered in Chapter 10 may be added to the property relation, Eq. 11.54, thereby giving a completely general form of the relation. This introduces a problem with identification of the chemical potential, and this matter is discussed in the following section.

11.12 A GENERAL DEFINITION OF GIBBS FUNCTION AND ENTHALPY

In the previous section we extended the thermodynamic property relation to systems of variable composition, and found that the chemical potential can be identified with the partial molal Gibbs function according to Eq. 11.58. Let us now consider the general situation, in which we have a substance that is not a simple compressible substance, as in Eq. 10.17, in addition to allowing variable

composition, as in Eq. 11.54. We can then write the general thermodynamic property relation

$$dU = T\,dS - P\,dV + \mathcal{T}\,dL + \mathcal{S}\,d\mathcal{A} + \mu_0\mathcal{H}\,d(V\mathcal{M}) + \mathcal{E}\,dZ$$
$$+ \sum_i \mu_i\,dn_i + \dots \quad \textbf{(11.63)}$$

The problem that arises in connection with this general relationship is that Eq. 11.58 is not valid if the Gibbs function is defined in the usual manner, according to

$$G = U + PV - TS \quad \textbf{(11.64)}$$

To demonstrate that this is in fact the case, it is sufficient to consider only one of the additional effects. Let us analyze a system of two components A and B in which surface effects are of importance. For this case Eq. 11.63 reduces to

$$dU = T\,dS - P\,dV + \mathcal{S}\,d\mathcal{A} + \mu_A\,dn_A + \mu_B\,dn_B \quad \textbf{(11.65)}$$

From Eqs. 11.64 and 11.65

$$dG = -S\,dT + V\,dP + \mathcal{S}\,d\mathcal{A} + \mu_A\,dn_A + \mu_B\,dn_B \quad \textbf{(11.66)}$$

which suggests a functional relationship of the form

$$G = f(T, P, \mathcal{A}, n_A, n_B)$$

Proceeding as in Section 11.11, we conclude that

$$\mu_A = \left(\frac{\partial G}{\partial n_A}\right)_{T,P,\mathcal{A},n_B} \quad \textbf{(11.67)}$$

There is nothing incorrect about this result, but we do realize that it is not consistent with the concept of a partial molal property since the extensive parameter \mathcal{A} is held constant. Instead, the corresponding intensive property \mathcal{S} should be held constant as are the intensive properties T and P if this quantity is to be a partial molal property.

We can explain why this difficulty occurs by analogy to the situation of boundary movement work when the mass is not constant. In that case we recall a boundary movement work (flow work) associated with pushing mass across a control surface, this being in addition to the ordinary boundary movement work. In the property relation, that effect $P\bar{V}_A$ appears as part of the chemical potential. In the present case, in addition to the ordinary surface extension work $\mathcal{S}\,d\mathcal{A}$, there is then an additional work associated with the creation of new surface, \mathcal{S} $\bar{\mathcal{A}}_A\,dn_A$, as the mass of component A changes. Thus, it is appropriate to treat the $\mathcal{S}\,\mathcal{A}$ term as analogous to the PV term. Let us therefore define a general Gibbs function for this situation as

$$G = U + PV - TS - \mathcal{S}\,\mathcal{A} \quad \textbf{(11.68)}$$

Then,

$$dG = dU + P\,dV + V\,dP - T\,dS - S\,dT - \mathcal{S}\,d\mathcal{A} - \mathcal{A}\,d\mathcal{S}$$

and from Eq. 11.66,

$$dG = -S\,dT + V\,dP - \mathscr{A}\,d\mathscr{S} + \mu_A\,dn_A + \mu_B\,dn_B \tag{11.69}$$

We note that Eq. 11.69 suggests the relation

$$G = f(T, P, \mathscr{S}, n_A, n_B)$$

which, proceeding as before, leads to

$$\mu_A = \left(\frac{\partial G}{\partial n_A}\right)_{T,P,\mathscr{S},n_B} = \bar{G}_A \tag{11.70}$$

and we conclude that if G is defined according to Eq. 11.68, then the chemical potential in Eqs. 11.65 and 11.69 is equal to the partial molal Gibbs function. It also follows for this case that

$$\mu_A = \bar{G}_A = \bar{U}_A + P\bar{V}_A - T\bar{S}_A - \mathscr{S}\,\bar{\mathscr{A}}_A \tag{11.71}$$

where the partial properties are all defined holding the intensive property \mathscr{S} constant.

Having defined a general Gibbs function by Eq. 11.68 for the case of surface tension and variable mass, it is necessary to define enthalpy in a similar manner by

$$H = G + TS = U + PV - \mathscr{S}\,\mathscr{A} \tag{11.72}$$

and the property relation Eq. 11.65 can then be written in terms of H in a consistent manner. It should also be noted that there is no effect on the Helmholtz function, as that property does not include the energy terms associated with mass transfer modes of work.

It is not difficult to extend our analysis to the general case associated with Eq. 11.63. That expression contains the various quasiequilibrium work modes represented in Eq. 4.13, which can be written in general variables as

$$\delta W = -\sum_k F_k\,dX_k \tag{11.73}$$

In this equation, F_k is the intensive property or driving force, and X_k is the corresponding extensive property affected by F_k. A general definition of the Gibbs function is then

$$G = U - TS - \sum_k F_k X_k \tag{11.74}$$

and the enthalpy is

$$H = G + TS = U - \sum_k F_k X_k \tag{11.75}$$

The general thermodynamic property relation, Eq. 11.63, is

$$dU = T\,dS + \sum_k F_k\,dX_k + \sum_i \mu_i\,dn_i \tag{11.76}$$

or, in terms of G, from Eq. 11.74

$$dG = -S\,dT - \sum_k X_k\,dF_k + \sum_i \mu_i\,dn_i \tag{11.77}$$

The chemical potential of component i in Eqs. 11.76 and 11.77 is

$$\mu_i = \left(\frac{\partial G}{\partial n_i}\right)_{T,F_k,n_{j \ne i}} \tag{11.78}$$

The property relation can similarly be expressed in terms of enthalpy, using its definition Eq. 11.75, or in terms of the Helmholtz function.

11.13 FUGACITY IN A MIXTURE AND ITS RELATION TO OTHER PROPERTIES

Let us define the fugacity of component A in a mixture by a similar procedure to that for a pure substance in Section 10.12. At constant temperature we define \bar{f}_A by

$$(d\bar{G}_A)_T = \bar{R}T\,d(\ln \bar{f}_A)_T \tag{11.79}$$

along with the requirement that

$$\lim_{p \to 0} \left(\frac{\bar{f}_A}{y_A P}\right) = 1 \tag{11.80}$$

so that as pressure approaches zero, the fugacity of component A approaches the ideal gas mixture partial pressure of component A. The fugacity of a component in a mixture as defined by these equations is not a true partial property according to the definition of Eq. 11.39. Nevertheless, we include the bar above the symbol as a reminder that the substance involved is a component of a mixture. Note that in the sense that f is essentially a pseudopressure, \bar{f} can be considered a pseudopartial pressure. The fugacity of component B in the mixture is defined by a pair of equations analogous to Eqs. 11.79 and 11.80.

To determine the relation of fugacity of a component to other properties, consider first the property relation written in terms of the Gibbs function. For a mixture of components A and B, this relation is given by Eq. 11.57.

At constant T and n_B this expression reduces to

$$dG_{T,n_B} = V\,dP_{T,n_B} + \bar{G}_A\,dn_{A\,T,n_B}$$

Taking a Maxwell cross-partial derivative we have

$$\left(\frac{\partial \bar{G}_A}{\partial P}\right)_{T,n_A,n_B} = \left(\frac{\partial V}{\partial n_A}\right)_{T,P,n_B} = \bar{V}_A \tag{11.81}$$

Therefore, from Eqs. 11.79 and 11.81 at constant temperature and composition,

$$(d\bar{G}_A)_{T,n_A,n_B} = \bar{R}T d(\ln \bar{f}_A)_{T,n_A,n_B} = \bar{V}_A\,dP_{T,n_A,n_B} \tag{11.82}$$

For pure substance A at the same constant temperature,

$$(d\bar{g}_A) = \bar{R}T\,d(\ln f_A)_T = \bar{v}_A\,dP_T \tag{11.83}$$

In order to derive an expression that permits us to evaluate the fugacity of component A in the mixture from measurable quantities, let us subtract Eq. 11.83 from Eq. 11.82, and integrate the resultant expression from P^* to P (where P^* is a very low pressure) at constant temperature and composition.

$$\int_{\ln \bar{f}^*_A/f^*_A}^{\ln \bar{f}_A/f_A} \bar{R}T\,d\left(\ln \frac{\bar{f}_A}{f_A}\right) = \int_{P^*\to 0}^{P} (\bar{V}_A - \bar{v}_A)\,dP \tag{11.84}$$

At this low pressure P^*, $\bar{f}^*_A = y_A P^*$ (Eq. 11.80) and $f^*_A = P^*$. Therefore as $P^* \to 0$,

$$\left(\ln \frac{\bar{f}^*_A}{f^*_A}\right) \to \ln \left(\frac{y_A P^*}{P^*}\right) = \ln y_A$$

Therefore we can write

$$\int_{\ln y_A}^{\ln \bar{f}_A/f_A} \bar{R}T\,d\left(\ln \frac{\bar{f}_A}{f_A}\right) = \int_{P^*\to 0}^{P} (\bar{V}_A - \bar{v}_A)\,dP$$

This integration results in the expression

$$\bar{R}T \ln \left(\frac{\bar{f}_A}{y_A f_A}\right) = \int_0^P (\bar{V}_A - \bar{v}_A)\,dP \tag{11.85}$$

in which \bar{f}_A is the fugacity of component A in the mixture of given composition at the given temperature and pressure P, whereas f_A is the fugacity of pure A at the same temperature and pressure. Equation 11.85 expresses the relation between \bar{f}_A and f_A in terms of the difference between \bar{V}_A and \bar{v}_A, a measurable quantity. It is also convenient to find a relation between \bar{f}_A and f_A in terms of the other quantities discussed in Sections 11.9 and 11.10, namely, $\bar{H}_A - \bar{h}_A$ and $\bar{S}_A - \bar{s}_A$, for such relations will permit evaluation of one or more of these quantities in terms of $\bar{V}_A - \bar{v}_A$.

First we shall determine the partial molal Gibbs function of component A in a mixture at T, P, with respect to a state at which the Gibbs function is known. Let us integrate Eq. 11.79 at constant T from P^* to P, where again P^* is sufficiently low to assume ideal gas behavior,

$$\int_{\bar{G}^*_A}^{\bar{G}_A} d\bar{G}_A = \int_{\bar{f}^*_A \to y_A P^*}^{\bar{f}_A} \bar{R}T\,d(\ln \bar{f}_A)_T \tag{11.86}$$

This integration yields the result

$$\bar{G}_A = \bar{G}^*_A + \bar{R}T \ln \frac{\bar{f}_A}{y_A P^*} = \bar{G}^*_A - \bar{R}T \ln y_A + \bar{R}T \ln \frac{\bar{f}_A}{P^*}$$

We have noted, Eq. 11.45, that

$$\bar{G}_A = \bar{H}_A - T\bar{S}_A$$

Therefore,

$$\bar{G}_A = \bar{H}_A^* - T\bar{S}_A^* - \bar{R}T \ln y_A + \bar{R}T \ln \frac{\bar{f}_A}{P^*} \qquad (11.87)$$

But, at low pressures,

$$\bar{H}_A^* = \bar{h}_A^*$$

Also, from Eq. 11.53,

$$\bar{S}_A^* + \bar{R} \ln y_A = \bar{s}_A^*$$

Therefore,

$$\bar{G}_A = \bar{h}_A^* - T\bar{s}_A^* + \bar{R}T \ln \frac{\bar{f}_A}{P^*}$$

$$\bar{G}_A = \bar{g}_A^* + \bar{R}T \ln \frac{\bar{f}_A}{P^*} \qquad (11.88)$$

To determine the change in \bar{G}_A with respect to temperature, we take another Maxwell cross-partial derivative of Eq. 11.57,

$$\left(\frac{\partial \bar{G}_A}{\partial T}\right)_{P,n_A,n_B} = -\left(\frac{\partial S}{\partial n_A}\right)_{T,P,n_B} = -\bar{S}_A \qquad (11.89)$$

or, at constant pressure and composition,

$$d\bar{G}_A = -\bar{S}_A \, dT \qquad (11.90)$$

But, from Eq. 11.45,

$$\bar{S}_A = \frac{\bar{H}_A - \bar{G}_A}{T}$$

Substituting,

$$(d\bar{G}_A)_{P,n_A,n_B} = -\frac{\bar{H}_A - \bar{G}_A}{T} \, dT_{P,n_A,n_B}$$

which may be rearranged to the form

$$\frac{T \, d\bar{G}_A - \bar{G}_A \, dT}{T^2} = -\frac{\bar{H}_A}{T^2} \, dT_{P,n_A,n_B} \qquad (11.91)$$

We note that Eq. 11.91 is, in effect,

$$d\left(\frac{\bar{G}_A}{T}\right)_{P,n_A,n_B} = -\frac{\bar{H}_A}{T^2} \, dT_{P,n_A,n_B} \qquad (11.92)$$

By the same procedure for pure A at constant pressure, we can write

$$d\left(\frac{\bar{g}_A}{T}\right)_P = -\frac{\bar{h}_A}{T^2}\,dT_P \tag{11.93}$$

Now, let us divide Eq. 11.88 by T and differentiate at constant pressure and composition.

$$d\left(\frac{\bar{G}_A}{T}\right)_{P,n_A,n_B} = d\left(\frac{\bar{g}_A^*}{T}\right)_{P,n_A,n_B} + \bar{R}\,d(\ln \bar{f}_A)_{P,n_A,n_B} \tag{11.94}$$

Substituting Eqs. 11.92 and 11.93, we see that at constant pressure and composition,

$$(d \ln \bar{f}_A)_{P,n_A,n_B} = \frac{\bar{h}_A^* - \bar{H}_A}{\bar{R}T^2}\,dT_{P,n_A,n_B} \tag{11.95}$$

Again, by the same procedure for pure A at constant pressure,

$$(d \ln f_A)_P = \frac{\bar{h}_A^* - \bar{h}_A}{\bar{R}T^2}\,dT_P \tag{11.96}$$

Combining Eqs. 11.95 and 11.96,

$$d\left(\ln \frac{\bar{f}_A}{f_A}\right)_{P,n_A,n_B} = \frac{\bar{h}_A - \bar{H}_A}{\bar{R}T^2}\,dT_{P,n_A,n_B} \tag{11.97}$$

which gives another relation between \bar{f}_A and f_A at a given T, P, in this case in terms of the difference between $(\bar{h}_A - \bar{H}_A)$.

Finally, let us determine the relation between \bar{f}_A and f_A in terms of the difference $(\bar{S}_A - \bar{s}_A)$. For pure component A,

$$\bar{g}_A = \bar{h}_A - T\bar{s}_A$$

Also,

$$\bar{G}_A = \bar{H}_A - T\bar{S}_A$$

Combining these equations,

$$(\bar{S}_A - \bar{s}_A) = \left(\frac{\bar{H}_A - \bar{h}_A}{T}\right) - \left(\frac{\bar{G}_A - \bar{g}_A}{T}\right) \tag{11.98}$$

Substituting Eqs. 10.92 and 11.88, we have

$$(\bar{S}_A - \bar{s}_A) = \left(\frac{\bar{H}_A - \bar{h}_A}{T}\right) - \bar{R}\ln\left(\frac{\bar{f}_A}{f_A}\right) \tag{11.99}$$

which can also be written in the form

$$(\bar{S}_A - \bar{s}_A) = \left(\frac{\bar{H}_A - \bar{h}_A}{T}\right) - \bar{R}\ln\left(\frac{\bar{f}_A}{y_A f_A}\right) - \bar{R}\ln y_A \tag{11.100}$$

so that the second term on the right side of the equation is similar to the first term of Eq. 11.85.

11.14 THE IDEAL SOLUTION

There are many mixtures and solutions for which the change of volume on mixing is negligibly small. These are referred to as ideal solutions. By this definition a mixture of ideal gases is an ideal solution, and this is quite acceptable terminology. However, ideal solutions also include certain solid solutions and liquid solutions as well as mixtures of nonideal gases. A number of significant simplications results from this assumption.

We have noted (Eq. 11.44) that for a mixture of ideal gases consisting of components A and B,

$$(\bar{V}_A - \bar{v}_A) = (\bar{V}_B - \bar{v}_B) = 0$$

Let us define an ideal solution as any solution or mixture for which

$$(\bar{V}_A - \bar{v}_A) = 0, \qquad (\bar{V}_B - \bar{v}_B) = 0 \tag{11.101}$$

at the pressure of the mixture and for all lower pressures as well. This definition therefore requires that

$$\Delta V_{mix} = 0$$

as in the case of mixtures of ideal gases.

Since, in accordance with Eq. 11.40, the volume of a mixture of components A and B is

$$V = \bar{V}_A n_A + \bar{V}_B n_B$$

the assumption of an ideal solution, Eq. 11.101, is equivalent to assuming Amagat's rule of additive volumes,

$$V = \bar{v}_A n_A + \bar{v}_B n_B$$

where \bar{v}_A and \bar{v}_B are the molal specific volumes of pure A and pure B each in the same phase as the mixture, and each at P and T, the pressure and temperature of the mixture. It should be noted that in the case of an ideal solution, \bar{v}_A and \bar{v}_B are the actual specific volumes of these components, and no assumption of ideal gas behavior is made.

The fugacity of a component in an ideal solution is readily determined by reference to Eq. 11.85.

$$\bar{R}T \ln\left(\frac{\bar{f}_A}{y_A f_A}\right) = \int_0^P (\bar{V}_A - \bar{v}_A)\, dP$$

From the definition of an ideal solution, $(\bar{V}_A - \bar{v}_A) = 0$, it follows that

$$\bar{f}_A = y_A f_A \tag{11.102}$$

Similarly, for component B,

$$\bar{f}_B = y_B f_B \tag{11.103}$$

For the general case of an ideal solution we can write

$$\bar{f}_i = y_i f_i \tag{11.104}$$

Eq. 11.104 constitutes the Lewis–Randall rule, which holds for an ideal solution. Note the similar appearance of these equations and those for the partial pressures of ideal gas mixture components as discussed in Section 11.1. This is consistent with the concept that fugacity is a pseudopressure.

The enthalpy of mixing for a mixture of components A and B is given by Eq. 11.49.

$$\Delta H_{\mathrm{mix}} = (\bar{H}_A - \bar{h}_A)n_A + (\bar{H}_B - \bar{h}_B)n_B$$

The fact that the enthalpy of mixing is zero for an ideal solution can be demonstrated by consideration of Eq. 11.97, which was written for constant composition.

$$d\left(\ln \frac{\bar{f}_A}{f_A}\right)_{P,n_A,n_B} = \left(\frac{\bar{h}_A - \bar{H}_A}{\bar{R}T^2}\right) dT_{P,n_A,n_B}$$

Substituting Eq. 11.85 it follows that for an ideal solution,

$$(\bar{h}_A - \bar{H}_A) = 0 \tag{11.105}$$

A similar expression can be written for component B, and therefore one concludes that the enthalpy of mixing for an ideal solution is equal to zero.

$$\Delta H_{\mathrm{mix}} = 0$$

We had reached the same conclusion for a mixture of ideal gases, but for an ideal solution \bar{h}_A and \bar{h}_B are actual enthalpies, and no assumption of ideal gas behavior is made.

The entropy of mixing for an ideal solution can be found by consideration of Eq. 11.100.

$$\bar{S}_A - \bar{s}_A = \left(\frac{\bar{H}_A - \bar{h}_A}{T}\right) - \bar{R} \ln \left(\frac{\bar{f}_A}{y_A f_A}\right) - \bar{R} \ln y_A$$

Since $\bar{H}_A - \bar{h}_A = 0$ and $\bar{f}_A = y_A f_A$, it follows that for an ideal solution

$$(\bar{S}_A - \bar{s}_A) = -\bar{R} \ln y_A$$
$$(\bar{S}_B - \bar{s}_B) = -\bar{R} \ln y_B \tag{11.106}$$

in which \bar{s}_A and \bar{s}_B are for pure A and B at T, P, the pressure and temperature of the mixture.

The entropy of mixing, Eq. 11.50, is

$$\Delta S_{\mathrm{mix}} = (\bar{S}_A - \bar{s}_A)n_A + (\bar{S}_B - \bar{s}_B)n_B$$

Therefore, for an ideal solution the entropy of mixing is given by the relation

$$\Delta S_{\mathrm{mix}} = -\bar{R}(n_A \ln y_A + n_B \ln y_B)$$

Note that the same expression for the entropy of mixing was obtained for a mixture of ideal gases, Eq. 11.14, although for an ideal solution \bar{s}_A and \bar{s}_B are actual entropies and not ideal gas entropies.

It also follows that since $\bar{V}_A - \bar{v}_A = 0$ and $\bar{V}_B - \bar{v}_B = 0$ for an ideal solution, the specific volume–composition diagram is a straight line for an ideal solution.

 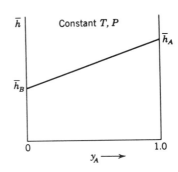

FIGURE 11.12

Typical specific volume–composition and enthalpy–composition diagrams for an ideal solution.

The same would be true for an enthalpy–composition diagram. These are shown in Fig. 11.12.

Example 11.9

A gas mixture consisting of 75% Ar and 25% C_2H_4 on a mole basis is stored at 25°C, 8.25 MPa in a cylinder having a volume of 0.5 m³. Determine the mass of gas in the tank assuming an ideal solution, and compare these results with those obtained from assuming a mixture of ideal gases.

Solution

For the ideal solution model, from Eqs. 11.40 and 11.101,

$$V = (n\bar{v})_{Ar} + (n\bar{v})_{C_2H_4}$$

or

$$\bar{v} = (y\bar{v})_{Ar} + (y\bar{v})_{C_2H_4}$$

If we had tables of thermodynamic properties that gave the specific volume of Ar and C_2H_4 at 25°C and 8.25 MPa, we would simply find the gas-phase molal specific volume \bar{v} of each component at this temperature and pressure, and proceed with the solution. In the absence of these data we can use the generalized charts, which will yield fairly accurate results. For the Ar at 25°C and 8.25 MPa,

$$T_r = \frac{298.2}{151.0} = 1.975, \qquad P_r = \frac{8.25}{4.86} = 1.70$$

For the C_2H_4 at 25°C and 8.25 MPa,

$$T_r = \frac{298.2}{282.4} = 1.06, \qquad P_r = \frac{8.25}{5.12} = 1.61$$

From the generalized compressibility chart, Fig. A.5,

$$Z_{Ar} = 0.97, \qquad Z_{C_2H_4} = 0.35$$

Therefore,

$$\bar{v}_{Ar} = \frac{Z\bar{R}T}{P} = \frac{0.97 \times 8.3144 \times 298.2}{8250} = 0.2915 \text{ m}^3/\text{kmol}$$

$$\bar{v}_{C_2H_4} = \frac{Z\bar{R}T}{P} = \frac{0.35 \times 8.3144 \times 298.2}{8250} = 0.1052 \text{ m}^3/\text{kmol}$$

$$\bar{v} = 0.75(0.2915) + 0.25(0.1052)$$
$$= 0.2449 \text{ m}^3/\text{kmol}$$

$$M = (yM)_{Ar} + (yM)_{C_2H_4} = 0.75(39.95) + 0.25(28.0) = 36.96$$

$$v = \frac{\bar{v}}{M} = \frac{0.2449}{36.96} = 0.006\ 626 \text{ m}^3/\text{kg}$$

$$m = \frac{V}{v} = \frac{0.5}{0.006\ 626} = 75.5 \text{ kg}$$

Assuming this to be a mixture of ideal gases we can use either the partial pressure or additive volume approach. Let us use the former.

$$P_{Ar} = y_{Ar}P = 0.75(8.25) = 6.188 \text{ MPa}$$

$$m_{Ar} = \frac{P_{Ar}V}{RT} = \frac{6188 \times 0.5}{0.208\ 13 \times 298.2} = 49.85 \text{ kg}$$

$$P_{C_2H_4} = y_{C_2H_4}P = 0.25(8.25) = 2.063 \text{ MPa}$$

$$m_{C_2H_4} = \frac{P_{C_2H_4}V}{RT} = \frac{2063 \times 0.5}{0.296\ 37 \times 298.2} = 11.67 \text{ kg}$$

$$m = 49.85 + 11.67 = 61.52 \text{ kg}$$

This compares with 75.5 kg for the assumption of an ideal solution.

We note two difficulties in particular from the results of this example. First, it is quite tedious to utilize the ideal solution model with the generalized charts, as it requires evaluating the entire problem separately for each of the components. Second, it required that we use the specific volume \bar{v}_A of pure gaseous A at the temperature and pressure of the gas mixture, and similarly for component B. However, there are certain pressures and temperatures at which the mixture would exist as a gas, but one of the components would, as a pure substance, be a solid or liquid. A familiar example is an air–water vapor mixture at ambient conditions. This poses the question of how to evaluate the properties of this component as a gas at the pressure and temperature of the mixture. There is no simple solution for this problem. In view of these difficulties, we may very well question the use of this model, even though the results are certainly more accurate than assuming ideal gas mixture behavior. The answer to this question is that there are other mixture models, described earlier in Section 11.8, that are much simpler to use with the generalized charts, and which are at least as accurate as the ideal solution for P-v-T, enthalpy, and entropy calculations. The ideal solution

does find important application, however, in connection with phase and chemical equilibria correlations, as will be noted in Chapter 13. It should also be pointed out that examination of the ideal solution is beneficial from an educational standpoint, as it is the general model of mixtures of which the ideal gas mixture is a special case. Thus, we are able to develop an intuitive feeling for the validity of assuming ideal gas when analyzing gas mixture behavior.

11.15 ACTIVITY

In our consideration of equilibrium in a following chapter we will find it convenient to make use of an additional property, the activity. It is appropriate, however, to introduce this property at this point.

We have noted that the partial molal Gibbs function for a component in a mixture can be expressed in terms of the low-pressure ideal gas state, as was done in Eq. 11.88.

$$\bar{G}_A = \bar{g}_A^* + \bar{R}T \ln (\bar{f}_A/P^*)$$

A more general approach would be to express the partial Gibbs functions in terms of a state at which we consider the mixture to behave as an ideal solution instead of an ideal gas mixture. The temperature of this state is the same as that of the mixture, and the pressure is the standard-state pressure P°, a reference pressure at the system temperature. Standard-state values were mentioned briefly in Chapter 7 in connection with entropy, and will be discussed more generally at a later point. It should be pointed out that this state may very well be a hypothetical one. That is, the mixture at T and P° may not actually behave as an ideal solution, but as long as we consistently calculate the mixture properties and develop the equations as though it did, the fact that it is a hypothetical reference state will introduce no difficulties. The significance of this statement will become evident as we proceed further.

The procedure for expressing \bar{G}_A is similar to that followed in the development of Eq. 11.88. In this case Eq. 11.79 is integrated at constant temperature and composition from the pressure P° to the mixture pressure P,

$$\int_{\bar{G}_A}^{\bar{G}_A} (d\bar{G}_A)_T = \int_{\bar{f}_A = y_A f_A^\circ}^{\bar{f}_A} \bar{R}T\, d(\ln \bar{f}_A)_T$$

in which f_A° is the fugacity of pure A at T, P°. Therefore,

$$\bar{G}_A = \bar{G}_A^\circ + \bar{R}T \ln \left(\frac{\bar{f}_A}{y_A f_A^\circ}\right)$$

$$= \bar{G}_A^\circ - \bar{R}T \ln y_A + \bar{R}T \ln \left(\frac{\bar{f}_A}{f_A^\circ}\right)$$

$$= \bar{H}_A^\circ - T\bar{S}_A^\circ - \bar{R}T \ln y_A + \bar{R}T \ln \left(\frac{\bar{f}_A}{f_A^\circ}\right) \qquad \textbf{(11.107)}$$

Since we assume an ideal solution at $P°$, from Eq. 11.105,

$$\bar{H}_A° = \bar{h}_A°$$ (11.108)

and from Eq. 11.106,

$$\bar{S}_A° + \bar{R} \ln y_A = \bar{s}_A°$$ (11.109)

$$\bar{G}_A = \bar{h}_A° - T\bar{s}_A° + \bar{R}T \ln \left(\frac{\bar{f}_A}{f_A°}\right)$$

Therefore,

$$\bar{G}_A = \bar{g}_A° + \bar{R}T \ln \left(\frac{\bar{f}_A}{f_A°}\right)$$ (11.110)

Thus we have an expression for the partial molal Gibbs function in terms of known values. It is from this important equation that the equilibrium constant is defined in Chapter 13. Consequently Eq. 11.110 is a very powerful relation. The values $f_A°$ and $\bar{g}_A°$ are for pure substance A at the temperature T and pressure $P°$, which is referred to as the standard state pressure. For gaseous mixtures, $P°$ is commonly taken as 0.1 MPa. For liquid and vapor phases in two-phase systems (discussed in Chapter 13), the standard state for each component is usually taken as the pure substance in that phase at the pressure of the mixture.

Examining Eq. 11.110, we find it convenient to define a quantity called activity. The activity a_A of component A in a mixture at T, P is defined as

$$a_A = \frac{\bar{f}_A}{f_A°}$$ (11.111)

The activity of component B is, of course, similarly defined. Equation 11.110 may now be written in the convenient form

$$\bar{G}_A = \bar{g}_A° + \bar{R}T \ln a_A$$ (11.112)

Several special cases for evaluating \bar{G}_A from this equation have already been considered. For example, in the case for which the mixture can be assumed an ideal gas mixture at the standard-state pressure, the equation reduces to Eq. 11.88. If the mixture can be assumed an ideal solution at pressure P as well as at $P°$, then from Eq. 11.102,

$$a_A = \frac{y_A f_A}{f_A°}$$ (11.113)

If conditions are such that an ideal gas mixture can be assumed at both $P°$ and P, then from Eq. 11.80,

$$a_A = \frac{y_A P}{P°}$$ (11.114)

PROBLEMS

11.1 At a certain point in a coal gasification process, the gas is sampled, and the sample stored in a 50-in^3 cylinder. An analysis of this mixture yields the following results:

Component	Percent by Volume
H_2	35
CO	50
CO_2	15

Determine:
(a) The mixture composition on a mass basis.
(b) The mass of sample in the cylinder at 1 atm, 70 F.
(c) The heat transfer required to heat the sample at constant volume from the initial state to 200 F.

11.2 The gas mixture of Problem 11.1 is compressed in a reversible adiabatic process from the initial state in the sample cylinder to a volume of 10 in.3 Determine the final temperature of the mixture and the work done during this process.

11.3 Two insulated tanks A and B are connected by a valve. Tank A has a volume of 2 m^3 and initially contains oxygen at 400 kPa, 10°C. Tank B has a volume of 10 m^3 and initially contains nitrogen at 200 kPa, 50°C. The valve is now opened and remains open until the resulting gas mixture comes to a uniform state. Determine:
(a) The final pressure and temperature.
(b) The entropy change for the process.

11.4 Consider the compression of a fuel–air mixture that takes place in the cylinder of a carbureted internal combustion engine. Assume that the fuel is ethane and that the air–fuel ratio on a mass basis is 15 to 1. The engine has a compression ratio of 9 to 1 and before the compression stroke begins the pressure in the cylinder is 88 kPa and the temperature is 35°C. Determine the pressure and temperature after compression and the work of compression per kilogram of mixture.

11.5 A spherical balloon that supports an interior pressure directly proportional to its diameter has a diameter of 3 ft and contains argon gas at 30 lbf/in.2, 100 F. The balloon is connected by a valve to a rigid 20-ft^3 tank containing carbon dioxide at 15 lbf/in.2, 200 F. The valve is now opened, and eventually the balloon and tank reach a uniform state, at which the pressure is 25 lbf/in.2 Considering the balloon and tank as a system, calculate the final temperature and the heat transfer for the process.

11.6 Hydrogen and nitrogen are mixed in an adiabatic, steady-flow process in the ratio 2 kmol of hydrogen per kmol of nitrogen. The hydrogen enters

at 500 kPa, 20°C, and the nitrogen enters at 500 kPa, 200°C. The pressure after mixing is 490 kPa. Determine the temperature of the mixture and the net entropy change per kilomole of mixture.

11.7 A large SSSF air separation plant facility takes in ambient air (79% N_2 and 21% O_2 by volume) at 14.7 lbf/in.2, 70 F, at the rate of 2 lb mol/s, and discharges a stream of pure O_2 gas at 30 lbf/in.2, 200 F, and a stream of pure N_2 gas at 14.7 lbf/in.2, 70 F. The plant operates on an electrical power input of 2000 kW. Calculate the net rate of entropy change for the process.

11.8 An insulated 5-ft^3 tank contains nitrogen gas at 30 lbf/in.2, 80 F. Carbon dioxide at 100 lbf/in.2, 200 F flows from a pipe into the tank until the pressure inside reaches 60 lbf/in.2, at which time the valve is closed. Calculate the final temperature inside the tank and the change in entropy for the process.

11.9 The only known sources of helium are the atmosphere (mole fraction approximately 5×10^{-6}) and natural gas. A large unit is being constructed to separate 100 m^3/s of natural gas, assumed to be 0.001 He mole fraction and the remainder CH_4, with the gas entering the unit at 150 kPa, 10°C. Pure helium exits at 100 kPa, 20°C, and pure methane exits at 140 kPa, 30°C. Any heat transfer is with the surroundings at 20°C. Is an electrical power input of 3000 kW sufficient to drive this unit?

11.10 An insulated, rigid, 0.2-m^3 vessel is divided into two equal parts A and B by an insulated partition that will support a pressure difference of up to 400 kPa before breaking. Side A contains methane and side B contains carbon dioxide, each initially at 1 MPa, 30°C. A valve on side B is now opened, and carbon dioxide flows out. It may be assumed that any carbon dioxide remaining inside B undergoes a reversible adiabatic expansion. Eventually, the partition breaks, and at this time the valve is closed. Calculate the net entropy change for the process that begins when the valve is closed.

11.11 A room of dimensions 12 ft × 20 ft × 8 ft contains an air–water vapor mixture at 70 F, 14.7 lbf/in.2, with a relative humidity of 40%. Calculate the following:
(a) The humidity ratio.
(b) The dew point temperature.
(c) The total mass of water vapor in the room.

11.12 The products of combustion of a hydrocarbon fuel with air are analyzed, with the results as follows:

Component	Percent by Volume
CO_2	10.4
CO	0.1
H_2O	11.8
O_2	3.4
N_2	74.3

This gas mixture passes through a SSSF heat exchanger at ambient pressure at the rate of 0.1 kg/s. What is the dew-point temperature? If the mixture is cooled 10°C below the dew-point temperature, how long will it take to collect 10 kg of liquid water?

11.13 One method of removing moisture from atmospheric air is to cool the air so that the moisture condenses or freezes out. Suppose that a certain experiment requires air with a humidity ratio of 0.0001. To what temperature must the air be cooled at a pressure of 100 kPa in order to achieve this humidity? To what temperature must it be cooled if the pressure is 10 MPa instead of 100 kPa?

11.14 Consider a rigid 10-ft^3 tank containing an air–water vapor mixture at 90 F, 14.7 lbf/in.2, with a relative humidity of 70%. This system is now cooled until the water just begins to condense. Determine the final temperature in the tank and the heat transfer for the process.

11.15 An air–water vapor mixture enters an air-conditioning unit at 110 kPa, 30°C, with a relative humidity of 50%. The mass flow rate of dry air entering the unit is 1 kg/s. The air–vapor mixture leaves the air-conditioning unit at 105 kPa, 10°C, with a relative humidity of 100%, and liquid condensate also leaves the unit at 10°C. Determine the rate of heat transfer for this process.

11.16 A saturated air–water vapor mixture at 70 F, 14.7 lbf/in.2, is contained in a closed 80-ft^3 tank in equilibrium with 1 lbm of liquid water. The tank is now heated to 175 F.
(a) Is there any liquid water at the final state?
(b) Calculate the heat transfer for the process.

11.17 An air–water vapor mixture enters a heater-humidifier unit at 40 F, 14.7 lbf/in.2, 50% relative humidity, at the volumetric flow rate of 5 ft^3/s. Liquid water at 50 F is sprayed into the mixture and heat is transferred such that the air–vapor mixture leaves the unit at 90 F, 14.7 lbf/in.2, 50% relative humidity. Determine the following:
(a) The mass flow rate of liquid water sprayed into the mixture.
(b) The rate of heat transfer to the unit.

11.18 A cylinder fitted with a piston restrained by a linear spring (force directly proportional to distance moved) contains a saturated air–water vapor mixture at 120 kPa, 0.1 m^3 volume, and at ambient temperature, 20°C. The piston area is 0.2 m^2 and the spring constant is 20 kN/m. This cylinder is attached by a valve to a line flowing dry air at 800 kPa, 40°C. The valve is now opened and air flows into the cylinder until the pressure reaches 200 kPa, at which point the temperature inside is 40°C. Determine the following:
(a) The relative humidity at the final state.
(b) The mass of air entering the cylinder.
(c) The work done during the process, considering the cylinder as a control volume.

11.19 Consider the system and process described in Problem 11.18. For this process:
(a) Calculate the heat transfer.
(b) Show that the process does not violate the second law.

11.20 Consider an air–water vapor mixture having a dry-bulb temperature of 90 F, a wet-bulb (adiabatic saturation) temperature of 70 F, and a pressure of 14.7 lbf/in.2 Determine the humidity ratio and relative humidity of this mixture, using the following:
(a) A first law analysis of the adiabatic saturation process.
(b) The psychrometric chart, Fig. A.4.

11.21 A combination air cooler and dehumidification unit receives outside ambient air at 100 F, 14.7 lbf/in.2, 90% relative humidity. This air–water vapor mixture is first cooled to a low temperature to condense the proper amount of water, after which the air–vapor mixture is heated, leaving the unit at 70 F, 14.7 lbf/in.2, 30% relative humidity. The volume flow rate of the air-vapor mixture at the exit conditions is 0.2 ft^3/s.
(a) Find the temperature to which the mixture is initially cooled, and the mass of water condensed per pound of dry air. Show the process that the water undergoes on a T-s diagram.
(b) Assuming that all the liquid water condensed leaves the unit at the minimum temperature, calculate the overall heat transfer rate.

11.22 Air at 40°C, 300 kPa, with a relative humidity of 35%, is to be expanded in a reversible adiabatic nozzle. To how low a pressure can the gas be expanded if no condensation is to take place? What is the exit velocity at this condition?

11.23 Air at 35°C, 100 kPa, 10% relative humidity, passes through a chamber into which cooling water at 15°C is sprayed in a steady-state, steady-flow, adiabatic process. The amount of cooling water added is such that when it evaporates into the air stream the exiting air–water vapor mixture will be 25°C, 100 kPa. Determine the exiting relative humidity and the irreversibility of the process per kilogram of dry air.

11.24 Ambient air at 14.7 lbf/in.2, 90 F, 40% relative humidity, is to be conditioned to 14.7 lbf/in.2, 70 F, 50% relative humidity in a SSSF process in the following manner. One stream is cooled at constant pressure to 40 F, and any liquid water condensate is trapped out. The air–vapor stream from this cooler is then mixed with an unconditioned stream to produce the desired state. Calculate the following:
(a) The ratio of the mass flow rates of the two streams.
(b) The required heat transfer in the cooler.

11.25 A rigid 10-ft^3 vessel initially contains an air–water vapor mixture at 20 lbf/in.2, 100 F, with a relative humidity of 10%. A supply line connected to this vessel by a valve carries steam at 100 lbf/in.2, 400 F. The valve is now opened and steam flows into the vessel until the relative humidity of the resultant air–water vapor mixture is 90%, at which time the valve

is closed. Heat is transferred from the vessel such that the temperature inside remains at 100 F during this process. Determine the heat transfer for the process, the mass of steam entering the vessel, and the final pressure inside the vessel.

11.26 A vertical cylinder is fitted with a piston held in place by a pin. The initial volume is 10 ft^3 and the cylinder contains an air–water vapor mixture at 14.7 lbf/in.2, 75 F, with a wet bulb temperature of 60 F. The pin is now removed, and at the same time a valve on the bottom of the cylinder is opened, allowing the mixture to flow out. A cylinder pressure of 20 lbf/in.2 is required to balance the piston. The valve is closed when the cylinder volume reaches 5 ft^3, at which point the temperature inside is 60 F.

(a) Is there any liquid water in the cylinder at the final state?

(b) Calculate the heat transfer to the cylinder during the process.

(c) Taking a control volume around the cylinder, calculate the entropy change of the control volume and that of the surroundings.

11.27 A problem of current interest is that of thermal pollution of river and lake water by using the water for power plant condenser cooling water and then returning it to the source at a higher temperature. The use of a cooling tower presents a solution to this particular problem. Consider the case shown in Fig. 11.13, in which 2.5 kg/s of water at 40°C (from the condenser) enters the top of the cooling tower, and the cooled water leaves the bottom at 20°C. The air–water vapor mixture enters the bottom of the cooling tower at 100 kPa, and has a dry-bulb temperature of 22°C and a wet-bulb temperature of 15°C. The air–water vapor mixture leaves the tower at 95 kPa, 30°C, with a relative humidity of 80%. Determine the kilograms of dry air per second that must be used and the fraction of the incoming water that evaporates. Assume the process to be adiabatic.

FIGURE 11.13
Sketch for Problem 11.27.

11.28 A steady supply of air is available from a line at 200 kPa, 30°C. This line is supplied by compressing ambient air (20°C, 100 kPa, 15°C wet-bulb

temperature), and no provision has been made to remove water vapor. The line is connected by a valve to a 500-L tank that initially contains water at 80°C, 99% quality. The valve is now opened until the pressure inside reaches 200 kPa, at which point the temperature inside is 40°C.
(a) How much liquid water is in the tank at the final state?
(b) Calculate the heat transfer from the tank.
(c) Calculate the net entropy change for the tank-filling process (not including the air line compressor).

11.29 Ambient air is at a condition of 14.7 lbf/in.², 95 F, 50% relative humidity. It is desired to produce a steady stream of air at 14.7 lbf/in.², 73 F, 70% relative humidity by first cooling one stream to an appropriate temperature to condense out the proper amount of water and then mixing this conditioned stream adiabatically with a second one at ambient conditions. What is the ratio of the two flow rates, and what is the temperature to which the first stream must be cooled?

11.30 One means for condensing steam during an emergency blowdown of a nuclear reactor is by use of a containment vessel as shown in Fig. 11.14. The vessel is insulated and has a total volume of 30 m³, with liquid water initially occupying 1 m³. The initial state inside the vessel is 35°C, 100 kPa. Now, during a short period of time, 55 kg of water enters the vessel at an average state of $P_i = 700$ kPa, $x_i = 0.50$. Find the temperature and pressure inside the vessel at the end of this period of time.

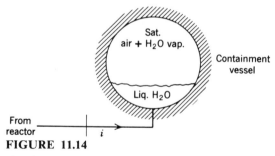

FIGURE 11.14
Sketch for Problem 11.30.

11.31 A gas mixture of 75% ethylene and 25% ethane on a mole basis enters a SSSF compressor at 100 F, 5 atm, at the rate of 0.3 ft³/s. The power input to the compressor is 10 hp and the gas mixture exits at 150 F, 20 atm. Assuming Kay's rule for the gas mixture, determine the following:
(a) Mass flow rate.
(b) Heat transfer rate from the compressor.

11.32 Consider the following reference state condition: the entropy of real saturated liquid methane at $-100°C$ is to be taken as 100 kJ/kmol K, and the entropy of hypothetical ideal gas ethane at $-100°C$, 1 MPa, is to be taken as 200 kJ/kmol K. Calculate the entropy per kmol of a real gas mixture of

50% methane, 50% ethane (mole basis) at 20°C, 4 MPa, in terms of the specified reference state values, and assuming Kay's rule for the real mixture behavior.

11.33 A cylinder fitted with a piston contains a gas mixture, 50% CO_2 and 50% C_2H_6 (mole basis) at 700 kPa, 35°C, at which point the cylinder volume is 5 L. This mixture is now compressed to 5.5 MPa in a reversible isothermal process. Calculate the heat transfer and work for the process, using the following model for the gas mixture:
(a) Ideal gas mixture.
(b) Kay's rule and the generalized charts.
(c) The van der Waals equation of state.

11.34 Consider a gas mixture of 50% nitrogen and 50% argon, on a mole basis, contained in a 1-ft³ volume at 320 R, 20 atm. Calculate the mass of this mixture, using the virial equation of state, and determine the percent error if ideal gas behavior is assumed.

11.35 Consider the gas mixture described in Problem 11.34. Calculate the mass of this mixture, assuming Kay's rule and the generalized charts, and also by using the Redlich–Kwong equation of state to represent the mixture behavior.

11.36 A gas mixture of ethane (component A) and propane (component B) is contained in the cylinder as shown in Fig. 11.15. The initial state is 30°C (room temperature), 70 kPa, 600 L, and a mole fraction of ethane of 0.20. The line contains pure ethane at 30°C, 7 MPa. The mass of the piston and weights is such that a cylinder pressure of 3.5 MPa would be required to float the piston. The valve is now cracked open and the pin is pulled from the piston. When the valve is closed, the temperature inside the cylinder is 65°C and the mole fraction of ethane is 0.60. Calculate the following:
(a) The final volume in the cylinder.
(b) The net entropy change for the process.

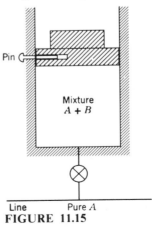

Line Pure A
FIGURE 11.15
Sketch for Problem 11.36.

11.37 Consider a gas mixture, 60% nitrogen and 40% oxygen on a mole basis, flowing through a long pipe. At the pipe inlet, the mixture is at 600 lbf/in.2, 300 R, with a velocity of 100 ft/s. If the pressure drop in the pipe is estimated to be 100 lbf/in.2, determine the exit temperature and velocity. Assume that the mixture behavior is represented by Kay's rule and the generalized charts.

11.38 It is frequently necessary to have available a gas mixture of a certain composition (for example, for use in calibrating gas analysis instruments). Such a mixture is prepared for storage by initially charging one of the components into a tank at some given pressure such that when the second component is then added until a final specified pressure is reached the mixture composition will be of the desired proportions. It is desired to prepare a gas mixture of 80% ethylene and 20% carbon dioxide (mole basis) at 10 MPa, 25°C, in an uninsulated, rigid, 50-L tank. The tank is initially to contain CO_2 at 25°C and some pressure P_1. The valve to a line flowing C_2H_4 at 25°C, 10 MPa, is now opened slightly, and remains open until the tank reaches 10 MPa, at which point the temperature can be assumed to be 25°C. Assume that the gas mixture so prepared can be represented by Kay's rule and the generalized charts.
 (a) Given the desired final state, what is the initial pressure of the carbon dioxide, P_1?
 (b) Determine the heat transfer and the net entropy change for the process of charging ethylene into the tank.

11.39 One technique for preparing a gas mixture of a desired composition was described in Problem 11.38. Another is the following. Two uninsulated 1-ft^3 tanks are located in a room at ambient temperature, 75 F. One tank contains ethylene gas at 1500 lbf/in.2, and the other contains carbon dioxide at pressure P_1. The valve connecting the tanks is now opened, and after a long time the two tanks contain a homogeneous gas mixture with a composition of 80% ethylene, 20% carbon dioxide (mol basis). Assume that the gas mixture can be represented by Kay's rule and the generalized charts.
 (a) What is the initial pressure of the carbon dioxide, P_1?
 (b) What is the final pressure of the gas mixture?
 (c) Determine the heat transfer and the net entropy change for this mixing process.

11.40 An insulated, rigid 5-ft^3 tank contains a gas mixture, 85% methane and 15% ethane on a mole basis, at 500 lbf/in.2, 100 F. A valve on the tank is now opened accidentally, and the pressure inside quickly drops to 300 lbf/in.2 before the valve is closed. Assume that the gas mixture is homogeneous. and that its behavior can be represented by Kay's rule and the generalized charts.
 (a) Calculate the mass that escaped from the tank.
 (b) Eventually, the temperature inside the tank returns to that of the surroundings, 100 F. What is the pressure at this time?

11.41 Repeat Problem 11.40 using the Redlich–Kwong equation of state to represent the real gas mixture behavior, instead of Kay's rule and the generalized charts.

11.42 The following data have been measured for carbon dioxide, methane, and their mixtures at 8.618 MPa, 37.8°C:

y_{CO_2}	y_{CH_4}	Z
1.0000	0.0000	0.3117
0.7961	0.2039	0.6262
0.5939	0.4061	0.7350
0.3944	0.6056	0.8084
0.1528	0.8472	0.8634
0.0000	1.0000	0.8892

(a) Determine the partial molal volumes of the components for a mixture of 50% CO_2, 50% CH_4 at this pressure and temperature.

(b) Evaluate the assumption of ideal solution for this system at the given temperature and pressure.

11.43 Repeat Problem 11.32, assuming that the real gas mixture behavior is represented by the ideal solution model (and the generalized charts) instead of by Kay's rule.

11.44 Repeat Problem 11.39, assuming ideal solution for the gas mixture, instead of Kay's rule.

11.45 Capsule A containing 1 L of saturated liquid butane at 110°C is located inside rigid tank B, which has a total volume (including A) of 51 L. Tank B initially contains ethane at 1.4 MPa, and is at room temperature, 25°C. The capsule now breaks and the two substances eventually form a homogeneous gas mixture at 25°C. Determine the final pressure in the tank and the irreversibility of the process.

11.46 A gas mixture, 80% ethylene and 20% propane on a mole basis, enters a SSSF device at 30°C, 3.5 MPa, at the rate of 4 L/s. Two separate streams leave the device; one is pure ethylene gas at 30°C, 2.8 MPa, and the other is pure liquid propane at 65°C, 3.5 MPa. Determine the minimum power input, in kW, required to operate this device.

11.47 Two heavily insulated rigid tanks, A and B, each of 1-ft^3 volume, are connected by a valve. Tank A initially contains methane at 80 F, 3500 lbf/in.2, and tank B initially contains carbon dioxide at 80 F, 200 lbf/in.2 The valve is now opened and methane flows quickly from A to B until the pressure inside A drops to 2500 lbf/in.2, at which time the valve is closed. Assuming that tank B now contains a homogeneous mixture and that tank A contains only methane, determine the final temperature in A, and the final temperature, pressure, and composition in B. Specify any assumptions and thermodynamic models used in the analysis and solution.

11.48 Boil-off vapor from a liquid propane storage tank at $0°C$ is throttled into an adiabatic mixing chamber and mixed with nitrogen in the ratio 4 kmol C_3H_8/1 kmol N_2, as shown in Fig. 11.16. The resulting mixture then enters a compressor, the exit state from which is 4.2 MPa, $60°C$. It is claimed that the compression process is adiabatic. Using the second law, show whether or not this is possible.

FIGURE 11.16
Sketch for Problem 11.48.

12

CHEMICAL REACTIONS

Many thermodynamic problems involve chemical reactions. Among the most familiar of these is the combustion of hydrocarbon fuels, for this process is utilized in most of our power-generating devices. However, we can all think of a host of other processes involving chemical reactions, including those that occur in the human body.

It is our purpose in this chapter to consider a first and second law analysis of systems undergoing a chemical reaction. In many respects, this chapter is simply an extension of our previous consideration of the first and second laws. However, a number of new terms are introduced, and it will also be necessary to introduce the third law of thermodynamics.

In this chapter the combustion process is considered in detail. There are two reasons for this emphasis. The first is that the combustion process is of great significance in many problems and devices with which the engineer is concerned. The second is that the combustion process provides an excellent vehicle for teaching the basic principles of the thermodynamics of chemical reactions. The student should keep both of these objectives in mind as the study of this chapter progresses.

Chemical equilibrium will be considered in Chapter 13 and, therefore, the matter of dissociation will be deferred until then.

12.1 FUELS

A thermodynamics textbook is not the place for a detailed treatment of fuels. However, some knowledge of them is a prerequisite to a consideration of combustion, and this section is therefore devoted to a brief discussion of some of the hydrocarbon fuels. Most fuels fall into one of three categories—coal, liquid hydrocarbons, or gaseous hydrocarbons.

Coal consists of the remains of vegetation deposits of past geologic ages, after subjection to biochemical actions, high pressure, temperature, and submersion. The characteristics of coal vary considerably with location, and even within a given mine there is some variation in composition.

The analysis of a sample of coal is given on one of two bases: the proximate analysis specifies, on a mass basis, the relative amounts of moisture, volatile matter, fixed carbon, and ash; the ultimate analysis specifies, on a mass basis, the relative amounts of carbon, sulfur, hydrogen, nitrogen, oxygen, and ash. The ultimate analysis may be given on an "as received" basis or on a dry basis. In the

TABLE 12.1

Characteristics of Some of the Hydrocarbon Families

Family	Formula	Structure	Saturated
Paraffin	C_nH_{2n+2}	Chain	Yes
Olefin	C_nH_{2n}	Chain	No
Diolefin	C_nH_{2n-2}	Chain	No
Naphthene	C_nH_{2n}	Ring	Yes
Aromatic			
Benzene	C_nH_{2n-6}	Ring	No
Naphthalene	C_nH_{2n-12}	Ring	No

latter case the ultimate analysis does not include the moisture as determined by the proximate analysis.

There are also a number of other properties of coal that are important in evaluating a coal for a given use. Some of these are the fusibility of the ash, the grindability or ease of pulverization, the weathering characteristics, and size.

Most liquid and gaseous hydrocarbon fuels are a mixture of many different hydrocarbons. For example, gasoline consists primarily of a mixture of about 40 hydrocarbons, with many others present in very small quantities. In discussing hydrocarbon fuels, therefore, brief consideration should be given to the most important families of hydrocarbons, which are summarized in Table 12.1.

Three terms should be defined. The first pertains to the structure of the molecule. The important types are the ring and chain structures; the difference between the two is illustrated in Fig. 12.1. The same figure illustrates the definition of saturated and unsaturated hydrocarbons. An unsaturated hydrocarbon has two or more adjacent carbon atoms joined by a double or triple bond, whereas in a saturated hydrocarbon all the carbon atoms are joined by a single bond. The third term to be defined is an isomer. Two hydrocarbons with the same number of carbon and hydrogen atoms and different structures are called isomers. Thus there are several different octanes (C_8H_{18}), each having 8 carbon atoms and 18 hydrogen atoms, but each has a different structure.

The various hydrocarbon families are identified by a common suffix. The compounds comprising the paraffin family all end in "-ane" (as propane and octane). Similarly, the compounds comprising the olefin family end in "-ylene" or "-ene"

Chain structure,
saturated

Chain structure,
unsaturated

Ring structure,
saturated

FIGURE 12.1

Molecular structure of some hydrocarbon fuels.

(as propene and octene), and the diolefin family ends in "-diene" (as butadiene). The naphthene family has the same chemical formula as the olefin family, but has a ring rather than chain structure. The hydrocarbons in the naphthene family are named by adding the prefix "cyclo-" (as cyclopentane).

The aromatic family includes the benzene series (C_nH_{2n-6}) and the naphthalene series (C_nH_{2n-12}). The benzene series has a ring structure and is unsaturated.

Alcohols are sometimes used as a fuel in internal combustion engines. The characteristic feature of the alcohol family is that one of the hydrogen atoms is replaced by an OH radical. Thus methyl alcohol, also called methanol, is CH_3OH.

Most liquid hydrocarbon fuels are mixtures of hydrocarbons that are derived from crude oil through distillation and cracking processes. Thus, from a given crude oil, a variety of different fuels can be produced, some of the common ones being gasoline, kerosene, diesel fuel, and fuel oil. Within each of these classifications there is a wide variety of grades, and each is made up of a large number of different hydrocarbons. The important distinction between these fuels is the distillation curve, Fig. 12.2. The distillation curve is obtained by slowly heating a sample of fuel so that it vaporizes. The vapor is then condensed and the amount measured. The more volatile hydrocarbons are vaporized first, and thus the temperature of the nonvaporized fraction increases during the process. The distillation curve, which is a plot of the temperature of the nonvaporized fraction versus the amount of vapor condensed, is an indication of the volatility of the fuel.

In dealing with combustion of liquid fuels it is convenient to express the composition in terms of a single hydrocarbon, even though it is a mixture of many hydrocarbons. Thus gasoline is usually considered to be octane, C_8H_{18}, and diesel fuel is considered to be dodecane, $C_{12}H_{26}$. The composition of a hydrocarbon fuel may also be given in terms of percentage of carbon and hydrogen.

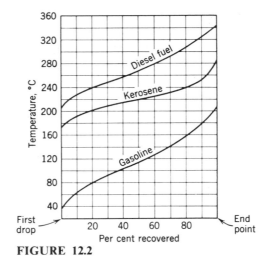

FIGURE 12.2

Typical distillation curves of some hydrocarbon fuels.

TABLE 12.2
Volumetric Analyses of Some Typical Gaseous Fuels

Constituent	Various Natural Gases				Producer Gas from Bituminous Coal	Carbureted Water Gas	Coke-Oven Gas
	A	B	C	D			
Methane	93.9	60.1	67.4	54.3	3.0	10.2	32.1
Ethane	3.6	14.8	16.8	16.3			
Propane	1.2	13.4	15.8	16.2			
Butanes plus[a]	1.3	4.2		7.4			
Ethene						6.1	3.5
Benzene						2.8	0.5
Hydrogen					14.0	40.5	46.5
Nitrogen		7.5		5.8	50.9	2.9	8.1
Oxygen					0.6	0.5	0.8
Carbon monoxide					27.0	34.0	6.3
Carbon dioxide					4.5	3.0	2.2

[a] This includes butane and all heavier hydrocarbons.

The two primary sources of gaseous hydrocarbon fuels are natural gas wells and certain chemical manufacturing processes. Table 12.2 gives the composition of a number of gaseous fuels. The major constituent of natural gas is methane, which distinguishes it from manufactured gas.

At the present time, there is a considerable effort being devoted to develop more economical processes for producing gaseous and also liquid hydrocarbon fuels from coal, and also from oil shale and tar sands deposits. Several alternative techniques have been demonstrated to be feasible, and these resources promise to provide an increasing proportion of our fuel supply in future years.

12.2 THE COMBUSTION PROCESS

The combustion process involves the oxidation of constituents in the fuel that are capable of being oxidized, and can therefore be represented by a chemical equation. During a combustion process the mass of each element remains the same. Thus writing chemical equations and solving problems involving quantities of the various constituents basically involves the conservation of mass of each element. A brief review of this subject, particularly as it applies to the combustion process, is presented in this chapter.

Consider first the reaction of carbon with oxygen.

$$\begin{array}{cc} \text{Reactants} & \text{Products} \\ C + O_2 & \rightarrow \quad CO_2 \end{array}$$

In dealing with English units, this equation states that 1 lb mol of carbon reacts with 1 lb mol of oxygen to form 1 lb mol of carbon dioxide. This also means that 12 lbm of carbon react with 32 lbm of oxygen to form 44 lbm of carbon dioxide. If SI units are being used, we would instead interpret the reaction equation as 1 kmol of carbon reacting with 1 kmol of oxygen to form 1 kmol of carbon dioxide, and also that 12 kg of carbon react with 32 kg of oxygen to form 44 kg of carbon dioxide. Throughout this chapter, we will mention one interpretation or the other when dealing with equations written on a "mole" basis, but should keep in mind that either system of units can be used.

In the reaction equation written above, it is noted that all the initial substances that undergo the combustion process are called the reactants, and the substances that result from the combustion process are called the products.

When a hydrocarbon fuel is burned, both the carbon and the hydrogen are oxidized. Consider the combustion of methane as an example.

$$CH_4 + 2O_2 \rightarrow CO_2 + 2H_2O \qquad \text{(12.1)}$$

In this case the products of combustion include both carbon dioxide and water. The water may be in the vapor, liquid, or solid phases, depending on the temperature and pressure of the products of combustion.

It should be pointed out that in the combustion process there are many intermediate products formed during the chemical reaction. In this book we are concerned with the initial and final products and not with the intermediate products, but this aspect is very important in a detailed consideration of combustion.

In most combustion processes the oxygen is supplied as air rather than as pure oxygen. The compostion of air on a molal basis is approximately 21% oxygen, 78% nitrogen, and 1% argon. We assume that the nitrogen and the argon do not undergo chemical reaction (except for dissociation, which will be considered in Chapter 13). They do leave at the same temperature as the other products, however, and therefore undergo a change of state if the products are at a temperature other than the original air temperature. It should be pointed out that at the high temperatures achieved in internal combustion engines, there is actually some reaction between the nitrogen and oxygen, and this gives rise to the air pollution problem associated with the oxides of nitrogen in the engine exhaust.

In combustion calculations involving air, the argon is usually neglected, and the air is considered to be composed of 21% oxygen and 79% nitrogen by volume. When this assumption is made, the nitrogen is sometimes referred to as atmospheric nitrogen. Atmospheric nitrogen has a molecular weight of 28.16 (which takes the argon into account) as compared to 28.013 for pure nitrogen. This distinction will not be made in this text, and we will consider the 79% nitrogen to be pure nitrogen.

The assumption that air is 21.0% oxygen and 79.0% nitrogen by volume leads to the conclusion that for each mole of oxygen, $79.0/21.0 = 3.76$ moles of nitrogen are involved. Therefore, when the oxygen for the combustion of methane is supplied as air, the reaction can be written

$$CH_4 + 2O_2 + 2(3.76)N_2 \rightarrow CO_2 + 2H_2O + 7.52N_2 \qquad \text{(12.2)}$$

The minimum amount of air that supplies sufficient oxygen for the complete combustion of all the carbon, hydrogen, and any other elements in the fuel that may oxidize is called the theoretical air. When complete combustion is achieved with theoretical air, the products contain no oxygen. In practice, it is found that complete combustion is not likely to be achieved unless the amount of air supplied is somewhat greater than the theoretical amount. The amount of air actually supplied is expressed in terms of percent theoretical air. Thus 150% theoretical air means that the air actually supplied is 1.5 times the theoretical air. The complete combustion of methane with 150% theoretical air is written

$$CH_4 + 2(1.5)O_2 + 2(3.76)(1.5)N_2 \rightarrow CO_2 + 2H_2O + O_2 + 11.28N_2 \quad (12.3)$$

The amount of air actually supplied may also be expressed in terms of percent excess air. The excess air is the amount of air supplied over and above the theoretical air. Thus 150% theoretical air is equivalent to 50% excess air. The terms theoretical air and excess air are both in current usage.

Two important parameters applied to combustion processes are the air–fuel ratio (designated AF) and its reciprocal, the fuel–air ratio (designated FA). The air–fuel ratio is usually expressed on a mass basis, but a mole basis is also used at times. The theoretical air–fuel ratio is the ratio of the mass (or moles) of theoretical air to the mass (or moles) of fuel.

When the amount of air supplied is less than the theoretical air required, the combustion is incomplete. If there is only a slight deficiency of air, the usual result is that some of the carbon unites with the oxygen to form carbon monoxide (CO) instead of carbon dioxide (CO_2). If the air supplied is considerably less than the theoretical air, there may also be some hydrocarbons in the products of combustion.

Even when some excess air is supplied there may be small amounts of carbon monoxide present, the exact amount depending on a number of factors including the mixing and turbulence during combustion. Thus the combustion of methane with 110% theoretical air might be as follows:

$$CH_4 + 2(1.1)O_2 + 2(1.1)3.76N_2 \rightarrow$$
$$+ 0.95CO_2 + 0.05CO + 2H_2O + 0.225O_2 + 8.27N_2 \quad (12.4)$$

The material covered so far in this section is illustrated by the following examples.

Example 12.1

Calculate the theoretical air–fuel ratio for the combustion of octane, C_8H_{18}.

Solution

The combustion equation is

$$C_8H_{18} + 12.5O_2 + 12.5(3.76)N_2 \rightarrow 8CO_2 + 9H_2O + 47.0N_2$$

The air–fuel ratio on a mole basis is

$$AF = \frac{12.5 + 47.0}{1} = 59.5 \text{ kmol air/kmol fuel (or lb mol air/lb mol fuel)}$$

The theoretical air–fuel ratio on a mass basis is found by introducing the molecular weight of the air and fuel.

$$AF = \frac{59.5(28.97)}{114.2} = 15.0 \text{ kg air/kg fuel (or lbm air/lbm fuel)}$$

Example 12.2

Determine the molal analysis of the products of combustion when octane, C_8H_{18}, is burned with 200 per cent theoretical air, and determine the dew point of the products if the pressure is 14.7 $lbf/in.^2$

Solution

The equation for the combustion of octane with 200% theoretical air is

$$C_8H_{18} + 12.5(2)O_2 + 12.5(2)(3.76)N_2 \rightarrow 8CO_2 + 9H_2O + 12.5O_2 + 94.0N_2$$

Total moles of product = $8 + 9 + 12.5 + 94.0 = 123.5$
Molal analysis of products:

$$
\begin{array}{lll}
CO_2 = 8/123.5 & = & 6.47\% \\
H_2O = 9/123.5 & = & 7.29 \\
O_2 = 12.5/123.5 & = & 10.12 \\
N_2 = 94/123.5 & = & \underline{76.12} \\
& & 100.00\%
\end{array}
$$

The partial pressure of the H_2O is $14.7(0.0729) = 1.072 \ lbf/in.^2$
The saturation temperature corresponding to this pressure is 104 F, which is also the dew-point temperature.

The water condensed from the products of combustion usually contains some dissolved gases and therefore may be quite corrosive. For this reason the products of combustion are often kept above the dew point until discharged to the atmosphere.

Example 12.3

Producer gas from bituminous coal (see Table 12.2) is burned with 20% excess air. Calculate the air–fuel ratio on a volumetric basis and on a mass basis.

Solution

To calculate the theoretical air requirement, let us write the combustion equation for the combustible substances in 1 kmol of fuel.

$$0.14H_2 + 0.070O_2 \rightarrow 0.14H_2O$$
$$0.27CO + 0.135O_2 \rightarrow 0.27CO_2$$
$$0.03CH_4 + 0.06O_2 \rightarrow 0.03CO_2 + 0.06H_2O$$

$$\underline{0.265} = \text{kmol oxygen required/kmol fuel}$$
$$\underline{-0.006} = \text{oxygen in fuel/kmol fuel}$$
$$0.259 = \text{kmol oxygen required from air/kmol fuel}$$

Therefore, the complete combustion equation for 1 kmol of fuel is

$$\overbrace{0.14H_2 + 0.27CO + 0.03CH_4 + 0.006O_2 + 0.509N_2 + 0.045CO_2}^{\text{fuel}}$$

$$\overbrace{+\ 0.259O_2 + 0.259(3.76)N_2}^{\text{air}} \rightarrow 0.20H_2O + 0.345CO_2 + 1.482N_2$$

$$\left(\frac{\text{kmol air}}{\text{kmol fuel}}\right)_{\text{theo}} = 0.259 \times \frac{1}{0.21} = 1.233$$

If the air and fuel are at the same pressure and temperature, this also represents the ratio of the volume of air to the volume of fuel.

$$\text{For 20\% excess air,} \quad \frac{\text{kmol air}}{\text{kmol fuel}} = 1.233 \times 1.200 = 1.48$$

The air–fuel ratio on a mass basis is

$$AF = \frac{1.48(28.97)}{0.14(2) + 0.27(28) + 0.03(16) + 0.006(32) + 0.509(28) + 0.045(44)}$$

$$= \frac{1.48(28.97)}{24.74} = 1.73 \text{ kg air/kg fuel (or lbm air/lbm fuel)}$$

An analysis of the products of combustion affords a very simple method for calculating the actual amount of air supplied in a combustion process. There are various experimental methods by which such an analysis can be made. Some yield results on a "dry" basis; that is, the fractional analysis of all the components, except for water vapor. Other experimental procedures give results that include the water vapor. In this presentation we are not concerned with the experimental devices and procedures, but rather with the use of such information in a thermodynamic analysis of the chemical reaction. The examples that follow illustrate how an analysis of the products can be used to determine the chemical reaction and the composition of the fuel.

In using the analysis of the products of combustion to obtain the actual fuel–air ratio, the basic principle is the conservation of the mass of each of the elements. Thus, in changing from reactants to products, we can make a carbon balance, hydrogen balance, oxygen balance, and nitrogen balance (plus any other elements that may be involved). Furthermore, we recognize that there is a definite ratio between the amounts of some of these elements. Thus the ratio between the nitrogen and oxygen supplied in the air is fixed, as well as the ratio between carbon and hydrogen if the composition of a hydrocarbon fuel is known.

Example 12.4

Methane (CH_4) is burned with atmospheric air. The analysis of the products on a dry basis is as follows:

CO_2	10.00%
O_2	2.37
CO	0.53
N_2	87.10
	100.00%

Calculate the air–fuel ratio and the percent theoretical air, and determine the combustion equation.

Solution

The solution involves writing the combustion equation for 100 kmol of dry products, introducing letter coefficients for the unknown quantities, and then solving for them.

From the analysis of the products, the following equation can be written, keeping in mind that this analysis is on a dry basis.

$$aCH_4 + bO_2 + cN_2 \rightarrow 10.0CO_2 + 0.53CO + 2.37O_2 + dH_2O + 87.1N_2$$

A balance for each of the elements involved will enable us to solve for all the unknown coefficients:

Nitrogen balance: $c = 87.1$

Since all the nitrogen comes from the air,

$$\frac{c}{b} = 3.76 \qquad b = \frac{87.1}{3.76} = 23.16$$

Carbon balance: $a = 10.00 + 0.53 = 10.53$
Hydrogen balance: $d = 2a = 21.06$
Oxygen balance: All the unknown coefficients have been solved for, and in this case the oxygen balance provides a check on the accuracy. Thus b can also be determined by an oxygen balance.

$$b = 10.00 + \frac{0.53}{2} + 2.37 + \frac{21.06}{2} = 23.16$$

Substituting these values for a, b, c, and d we have

$$10.53CH_4 + 23.16O_2 + 87.1N_2$$
$$\rightarrow 10.0CO_2 + 0.53CO + 2.37O_2 + 21.06H_2O + 87.1N_2$$

Dividing through by 10.53 yields the combustion equation per kmol of fuel.

$$CH_4 + 2.2O_2 + 8.27N_2 \rightarrow 0.95CO_2 + 0.05CO + 2H_2O + 0.225O_2 + 8.27N_2$$

The air–fuel ratio on a mole basis is

$$2.2 + 8.27 = 10.47 \text{ kmol air/kmol fuel}$$

The air–fuel ratio on a mass basis is found by introducing the molecular weights.

$$AF = \frac{10.47 \times 28.97}{16.0} = 18.97 \text{ kg air/kg fuel (or lbm air/lbm fuel)}$$

The theoretical air–fuel ratio is found by writing the combustion equation for theoretical air.

$$CH_4 + 2O_2 + 2(3.76)N_2 \rightarrow CO_2 + 2H_2O + 7.52N_2$$

$$AF_{theo} = \frac{(2 + 7.52)28.97}{16.0} = 17.23 \text{ kg air/kg fuel (or lbm air/lbm fuel)}$$

The percent theoretical air is $\frac{18.97}{17.23} = 110\%$

Example 12.5

Coal from Jenkin, Kentucky, has the following ultimate analysis on a dry basis, percent by mass:

Sulfur	0.6%
Hydrogen	5.7
Carbon	79.2
Oxygen	10.0
Nitrogen	1.5
Ash	3.0

This coal is to be burned with 30% excess air. Calculate the air–fuel ratio on a mass basis.

Solution

One approach to this problem is to write the combustion equation for each of the combustible elements per 100 kg of fuel. The molal composition per 100 kg of fuel is found first.

$$\text{kmol S/100 kg fuel} = \frac{0.6}{32} = 0.02$$

$$\text{kmol H}_2\text{/100 kg fuel} = \frac{5.7}{2} = 2.85$$

$$\text{kmol C/100 kg fuel} = \frac{79.2}{12} = 6.60$$

$$\text{kmol O}_2\text{/100 kg fuel} = \frac{10}{32} = 0.31$$

$$\text{kmol N}_2\text{/100 kg fuel} = \frac{1.5}{28} = 0.05$$

The combustion equations for the combustible elements are now written, which enables us to find the theoretical oxygen required.

$$0.02S + 0.02O_2 \rightarrow 0.02SO_2$$
$$2.85H_2 + 1.42O_2 \rightarrow 2.85H_2O$$
$$6.60C + 6.60O_2 \rightarrow 6.60CO_2$$

$$\overline{8.04} \text{ kmol } O_2 \text{ required}/100 \text{ kg fuel}$$
$$-\ 0.31 \text{ kmol } O_2 \text{ in fuel}/100 \text{ kg fuel}$$
$$\overline{7.73} \text{ kmol } O_2 \text{ from air}/100 \text{ kg fuel}$$

$$AF_{\text{theo}} = \frac{[7.73 + 7.73(3.76)]28.97}{100} = 10.63 \text{ kg air/kg fuel}$$

For 30% excess air the air–fuel ratio is

$$AF = 1.3 \times 10.63 = 13.82 \text{ kg air/kg fuel}$$

12.3 ENTHALPY OF FORMATION

In the first eleven chapters of this book the problems considered always involved a fixed chemical composition and never involved a change of composition due to chemical reaction. Therefore, in dealing with a thermodynamic property, we made use of tables of thermodynamic properties for the given substance, and in each of these tables the thermodynamic properties were given relative to some arbitrary base. In the steam tables, for example, the internal energy of saturated liquid at 0.01°C is assumed to be zero. This procedure is quite adequate when no change in composition is involved, because we are concerned with the changes in the properties of a given substance, and the properties at the condition of the reference state cancel out in the calculation. When dealing with the matter of reference states in Section 11.8, we noted that for a given substance (perhaps a component of a mixture), we are free to choose a reference state condition, for example, a hypothetical ideal gas, as long as we then carry out a consistent calculation from that state and condition to the real desired state. We also noted there that we are free to choose a reference state value, as long as there is no subsequent inconsistency in the calculation of the change in a property due to a chemical reaction, with a resulting change in the amount of a given substance. Now that we are to include the possibility of a chemical reaction, it will become necessary to choose these reference state values on a common and consistent basis. We will use as our reference state a temperature of 25°C, a pressure of 0.1 MPa, and a hypothetical ideal gas condition for those substances that are gases. In English units, the reference-state temperature is the same, 77 F, but the reference-state pressure is taken as 1 standard atmosphere, slightly different than that used with SI units. The discussion below will refer to the SI values, but the procedure is entirely equivalent when dealing with the English system.

Consider the simple SSSF combustion process shown in Fig. 12.3. This idealized reaction involves the combustion of solid carbon with gaseous (ideal gas) oxygen,

FIGURE 12.3

Example of combustion process.

each of which enters the control volume at the reference state, 25°C and 0.1 MPa. The carbon dioxide (ideal gas) formed by the reaction leaves the chamber at the reference state, 25°C and 0.1 MPa. If the heat transfer could be accurately measured, it would be found to be $-393\,522$ kJ/kmol of carbon dioxide formed. The chemical reaction can be written

$$C + O_2 \rightarrow CO_2$$

Applying the first law to this process we have

$$Q_{c.v.} + H_R = H_P \qquad (12.5)$$

where the subscripts R and P refer to the reactants and products, respectively. We will find it convenient to also write the first law for such a process in the form

$$Q_{c.v.} + \sum_R n_i \bar{h}_i = \sum_P n_e \bar{h}_e \qquad (12.6)$$

where the summations refer, respectively, to all the reactants or all the products.

Thus, a measurement of the heat transfer would give us the difference between the enthalpy of the products and the reactants, where each is at the reference state condition. However, suppose that we assign the value of zero to the enthalpy of all the elements at the reference state. In this case, the enthalpy of the reactants is zero, and

$$Q_{c.v.} = H_P = -393\,522 \text{ kJ/kmol}$$

The enthalpy of (hypothetical) ideal gas carbon dioxide at 25°C, 0.1 MPa pressure (with reference to this arbitrary base in which the enthalpy of the elements is chosen to be zero), is called the enthalpy of formation. We designate this with the symbol \bar{h}_f°. Thus, for carbon dioxide

$$\bar{h}_f^\circ = -393\,522 \text{ kJ/kmol}$$

The enthalpy of carbon dioxide in any other state, relative to this base in which the enthalpy of the elements is zero, would be found by adding the change of enthalpy between ideal gas at 25°C, 0.1 MPa, and the given state to the enthalpy of formation. That is, the enthalpy at any temperature and pressure, $\bar{h}_{T,P}$, is

$$\bar{h}_{T,P} = (\bar{h}_f^\circ)_{298,\,0.1\,\text{MPa}} + (\Delta\bar{h})_{298,\,0.1\,\text{MPa}\rightarrow T,P} \qquad (12.7)$$

where the term $(\Delta \bar{h})_{298,\,0.1\,MPa \to T,P}$ represents the difference in enthalpy between any given state and the enthalpy of ideal gas at 298.15 K, 0.1 MPa. For convenience we usually drop the subscripts in the examples that follow.

The procedure that we have demonstrated for carbon dioxide can be applied to any compound.

Table A.13 gives values of the enthalpy of formation for a number of substances in the units kJ/kmol.

Three further observations should be made in regard to enthalpy of formation.

1. We have demonstrated the concept of enthalpy of formation in terms of the measurement of the heat transfer in an idealized chemical reaction in which a compound is formed from the elements. Actually, the enthalpy of formation is usually found by the application of statistical thermodynamics, using observed spectroscopic data.

2. The justification of this procedure of arbitrarily assigning the value of zero to the enthalpy of the elements at 25°C, 0.1 MPa, rests on the fact that in the absence of nuclear reactions the mass of each element is conserved in a chemical reaction. No conflicts or ambiguities arise with this choice of reference state, and it proves to be very convenient in studying chemical reactions from a thermodynamic point of view.

3. In certain cases an element or compound can exist in more than one state at 25°C, 0.1 MPa. Carbon, for example, can be in the form of graphite or diamond. It is essential that the state to which a given value is related be clearly identified. Thus, in Table A.13, the enthalpy of formation of graphite is given the value of zero, and the enthalpy of each substance that contains carbon is given relative to this base. Another example is that oxygen may exist in the monatomic or diatomic form, or also as ozone, O_3. The value chosen as zero is for the form that is chemically stable at the reference state, which in the case of oxygen is the diatomic form. Then each of the others must have an enthalpy of formation consistent with the chemical reaction and heat transfer for the reaction that produces that form of oxygen.

It will be noted from Table A.13 that two values are given for the enthalpy of formation for water; one is for liquid water and the other for gaseous (hypothetical ideal gas) water, both at the reference state of 25°C, 0.1 MPa. It is convenient to use the hypothetical ideal gas reference in conjunction with the ideal gas table property changes given in Table A.11, and to use the real liquid reference in connection with real water property changes as given in the steam tables, Table A.1. The real liquid reference state properties are obtained from those at the hypothetical ideal gas reference by following a procedure of calculation as described in Section 11.8. The same procedure can be followed for other substances that have a saturation pressure less than 0.1 MPa at the reference temperature of 25°C.

Frequently students are bothered by the minus sign when the enthalpy of formation is negative. For example, the enthalpy of formation of carbon dioxide is negative. This is quite evident because the heat transfer was negative during the

steady-flow chemical reaction, and the enthalpy of the carbon dioxide must be less than the sum of enthalpy of the carbon and oxygen initially, both of which are assigned the value of zero. This is quite analogous to the situation we would have in the steam tables if we let the enthalpy of saturated vapor be zero at 0.1 MPa pressure, for in this case the enthalpy of the liquid would be negative, and we would simply use the negative value for the enthalpy of the liquid when solving problems.

12.4 FIRST LAW ANALYSIS OF REACTING SYSTEMS

The significance of the enthalpy of formation is that it is most convenient in performing a first law analysis of a reacting system, for the enthalpies of different substances can be added or subtracted, since they are all given relative to the same base.

In such problems we will write the first law for a steady-state, steady-flow process in the form

$$Q_{c.v.} + H_R = W_{c.v.} + H_P$$

or

$$Q_{c.v.} + \sum_R n_i \bar{h}_i = W_{c.v.} + \sum_P n_e \bar{h}_e$$

where R and P refer to the reactants and products, respectively. In each problem it is necessary to choose one parameter as the basis of the solution. Usually this is taken as 1 kmol of fuel.

Example 12.6

Consider the following reaction, which occurs in a steady-state, steady-flow process.

$$CH_4 + 2O_2 \rightarrow CO_2 + 2H_2O(l)$$

The reactants and products are each at a total pressure of 1 atm and 77 F. Determine the heat transfer per lb mol of fuel entering the combustion chamber.

Control volume: Combustion chamber.

Inlet state: P and T known; state fixed.

Exit state: P and T known; state fixed.

Process: SSSF.

Model: Three gases ideal gases; real liquid water.

Analysis:

First law:

$$Q_{c.v.} + \sum_R n_i \bar{h}_i = \sum_P n_e \bar{h}_e$$

Solution

Using values from Table A.13,

$$\sum_R n_i \bar{h}_i = (\bar{h}_f^\circ)_{CH_4} = -32\,211 \text{ Btu/lb mol fuel}$$

$$\sum_P n_e \bar{h}_e = (\bar{h}_f^\circ)_{CO_2} + 2(\bar{h}_f^\circ)_{H_2O(l)}$$

$$= -169\,297 + 2(-122\,971) = -415\,239 \text{ Btu/lb mol fuel}$$

$$Q_{c.v.} = -415\,239 - (-32\,211) = -383\,028 \text{ Btu/lb mol fuel}$$

In most cases, however, the substances that comprise the reactants and products in a chemical reaction are not at a temperature of 25°C and a pressure of 0.1 MPa (the state at which the enthalpy of formation is given). Therefore the change of enthalpy between 25°C and 0.1 MPa and the given state must be known. In the case of a solid or liquid, this change of enthalpy can usually be found from a table of thermodynamic properties or from specific heat data. In the case of gases, this change of enthalpy can usually be found by one of the following procedures.

1. Assume ideal gas behavior between 25°C, 0.1 MPa, and the given state. In this case, the enthalpy is a function of the temperature only, and can be found by use of an equation of C_{po} or from tabulated values of enthalpy as a function of temperature (which assumes ideal gas behavior). Table A.9 gives an equation for \bar{C}_{po} for a number of substances and Table A.11 gives values of $(\bar{h}^\circ - \bar{h}_{298}^\circ)$ (that is, the $\Delta\bar{h}$ of Eq. 12.7) in kJ/kmol, (\bar{h}_{298}° refers to 25°C or 298.15 K. For simplicity this is designated \bar{h}_{298}°.) The superscript $^\circ$ is used to designate that this is the enthalpy at 0.1 MPa pressure, based on ideal gas behavior, that is, the standard-state enthalpy.

2. If a table of thermodynamic properties is available, $\Delta\bar{h}$ can be found directly from these tables if a real substance behavior reference state is being used, such as that described above for liquid water. If a hypothetical ideal gas reference state is being used, then it is necessary to account for the real substance correction to properties at that state to gain entry to the tables.

3. If the deviation from ideal gas behavior is significant, but no tables of thermodynamic properties are available, the value for $\Delta\bar{h}$ can be found from the generalized charts and the values for \bar{C}_{po} or $\Delta\bar{h}$ at 0.1 MPa pressure as indicated above.

Thus, in general, for applying the first law to a steady-state, steady-flow process involving a chemical reaction and negligible changes in kinetic and potential energy, we can write

$$Q_{c.v.} + \sum_R n_i(\bar{h}_f^\circ + \Delta\bar{h})_i = W_{c.v.} + \sum_P n_e(\bar{h}_f^\circ + \Delta\bar{h})_e \tag{12.8}$$

Example 12.7

Calculate the enthalpy of water (on a kilomole basis) at 3.5 MPa, 300°C, relative to the 25°C and 0.1 MPa base, using the following procedures.

1. Assume the steam to be an ideal gas with the value of \bar{C}_{po} given in the Appendix, Table A.9.

2. Assume the steam to be an ideal gas with the value for $\Delta\bar{h}$ as given in the Appendix, Table A.11.

3. The steam tables.

4. The specific heat behavior given in **2** above and the generalized charts.

Solution

For each of these procedures, we can write

$$\bar{h}_{T,P} = (\bar{h}_f^{\circ} + \Delta\bar{h})$$

The only difference is in the procedure by which we calculate $\Delta\bar{h}$. From Table A.13 we note that

$$(\bar{h}_f^{\circ})_{H_2O(g)} = -241\ 827\ \text{kJ/kmol}$$

1. Using the specific heat equation for $H_2O(g)$ from Table A.9,

$$\bar{C}_{po} = 143.05 - 183.54\ \theta^{0.25} + 82.751\ \theta^{0.5} - 3.6989\ \theta\ \text{kJ/kmol K}$$

 where

$$\theta = T/100$$

 Therefore,

$$\Delta\bar{h} = \int_{298.15}^{573.15} \bar{C}_{po}(T)\,dT$$
$$= \int_{2.9815}^{5.7315} \bar{C}_{po}(\theta)100\,d\theta$$
$$= 9517\ \text{kJ/kmol}$$
$$\bar{h}_{T,P} = -241\ 827 + 9517 = -232\ 310\ \text{kJ/kmol}$$

2. Using Table A.11 for $H_2O(g)$,

$$\Delta\bar{h} = 9537\ \text{kJ/kmol}$$
$$\bar{h}_{T,P} = -241\ 827 + 9537 = -232\ 290\ \text{kJ/kmol}$$

3. Using the steam tables, either the liquid reference or the gaseous reference state may be used.
 For the liquid,

$$\Delta\bar{h} = 18.015(2977.5 - 104.9) = 51\ 750\ \text{kJ/kmol}$$
$$\bar{h}_{T,P} = -285\ 838 + 51\ 750 = -234\ 088\ \text{kJ/kmol}$$

For the gas,

$$\Delta \bar{h} = 18.015(2977.5 - 2547.2) = 7752 \text{ kJ/kmol}$$
$$\bar{h}_{T,P} = -241\,827 + 7752 = -234\,075 \text{ kJ/kmol}$$

The very small difference results from using the enthalpy of saturated vapor at 25°C (which is almost but not exactly an ideal gas) in calculating the Δh.

4. When using the generalized charts we use the notation introduced in Chapter 10.

$$\bar{h}_{T,P} = \bar{h}_f^\circ - (\bar{h}_2^* - \bar{h}_2) + (\bar{h}_2^* - \bar{h}_1^*) + (\bar{h}_1^* - \bar{h}_1)$$

where subscript 2 refers to the state at 3.5 MPa, 300°C, and state 1 refers to the state at 0.1 MPa, 25°C.

From part 2, $\bar{h}_2^* - \bar{h}_1^* = 9537 \text{ kJ/kmol}$.

$$\bar{h}_1^* - \bar{h}_1 = 0 \quad \text{(ideal gas reference)}$$
$$P_{r2} = \frac{3.5}{22.09} = 0.158, \qquad T_{r2} = \frac{573.2}{647.3} = 0.886$$

From the generalized enthalpy chart,

$$\frac{\bar{h}_2^* - \bar{h}_2}{T_c} = 2.3; \qquad \bar{h}_2^* - \bar{h}_2 = 2.3\,(647.3) = 1489 \text{ kJ/kmol}$$
$$\bar{h}_{T,P} = -241\,827 - 1489 + 9537 = -233\,779 \text{ kJ/kmol}$$

The particular approach that is used in a given problem will depend on the data available for the given substance.

Example 12.8

A small gas turbine uses $C_8H_{18}(l)$ for fuel, and 400 per cent theoretical air. The air and fuel enter at 77 F, and the products of combustion leave at 1100 F. The output of the engine and the fuel consumption are measured and it is found that the specific fuel consumption is one pound of fuel per horsepower-hour. Determine the heat transfer from the engine per pound mole of fuel. Assume complete combustion.

Control volume: Gas turbine engine.

Inlet states: T known for fuel and air.

Exit state: T known for combustion products.

Process: SSSF.

Model: All gases ideal gases, Table A.11; liquid octane, Table A.13.

Analysis: The combustion equation is

$$C_8H_{18}(l) + 4(12.5)O_2 + 4(12.5)(3.76)N_2 \rightarrow 8CO_2 + 9H_2O + 37.5O_2 + 188.0N_2$$

First law:

$$Q_{c.v.} + \sum_R n_i(\bar{h}_f^\circ + \Delta\bar{h})_i = W_{c.v.} + \sum_P n_e(\bar{h}_f^\circ + \Delta\bar{h})_e$$

Solution

Since the air is composed of elements and enters at 77 F, the enthalpy of the reactants is equal to that of the fuel.

$$\sum_R n_i[\bar{h}_f^\circ + \Delta\bar{h}]_i = (\bar{h}_f^\circ)_{C_8H_{18}(l)} = -107\,532 \text{ Btu/lb mol.}$$

Considering the products

$$\sum_P n_e(\bar{h}_f^\circ + \Delta\bar{h})_e = n_{CO_2}(\bar{h}_f^\circ + \Delta\bar{h})_{CO_2} + n_{H_2O}(\bar{h}_f^\circ + \Delta\bar{h})_{H_2O} + n_{O_2}(\Delta\bar{h})_{O_2} + n_{N_2}(\Delta\bar{h})_{N_2}$$

$$= 8(-169\,297 + 11\,315) + 9(-104\,036 + 8869)$$
$$+ 37.5(7792) + 188(7384)$$
$$= -439\,069 \text{ Btu/lb mol fuel.}$$

$$W_{c.v.} = 2544 \times 114.23 = 290\,601 \text{ Btu/lb mol fuel}$$

Therefore, from the first law,

$$Q_{c.v.} = -439\,969 + 290\,601 - (-107\,532)$$
$$= -41\,836 \text{ Btu/lb mol fuel.}$$

Example 12.9

A mixture of 1 kmol of gaseous ethene and 3 kmol of oxygen at 25°C reacts in a constant-volume bomb. Heat is transferred until the products are cooled to 600 K. Determine the amount of heat transfer from the system.

System: Constant-volume bomb.

Initial state: T known.

Final state: T known.

Process: Constant volume.

Model: Ideal gas mixtures, Tables A.11, A.13.

Analysis: The chemical reaction is

$$C_2H_4 + 3O_2 \rightarrow 2CO_2 + 2H_2O(g)$$

First law:

$$Q + U_R = U_P$$
$$Q + \sum_R n(\bar{h}_f^\circ + \Delta\bar{h} - \bar{R}T) = \sum_P n(\bar{h}_f^\circ + \Delta\bar{h} - \bar{R}T)$$

Solution

Using values from Tables A.11 and A.13,

$$\sum_{R} n(\bar{h}_f^\circ + \Delta\bar{h} - \bar{R}T) = (\bar{h}_f^\circ - \bar{R}T)_{C_2H_4} - n_{O_2}(\bar{R}T)_{O_2} = (\bar{h}_f^\circ)_{C_2H_4} - 4\bar{R}T$$
$$= 52\,283 - 4 \times 8.3144 \times 298.2 = 42\,366 \text{ kJ}$$

$$\sum_{P} n(\bar{h}_f^\circ + \Delta\bar{h} - \bar{R}T) = 2[(\bar{h}_f^\circ)_{CO_2} + \Delta\bar{h}_{CO_2}] + 2[(\bar{h}_f^\circ)_{H_2O(g)} + \Delta\bar{h}_{H_2O(g)}] - 4\bar{R}T$$
$$= 2(-393\,522 + 12\,916) + 2(-241\,827 + 10\,498)$$
$$-4 \times 8.3144 \times 600$$
$$= -1\,243\,824 \text{ kJ}$$

Therefore,

$$Q = -1\,243\,824 - 42\,366 = -1\,286\,190 \text{ kJ}$$

In the case of a real gas mixture, a pseudocritical method such as Kay's rule, Eq. 11.33, could be used to evaluate the nonideal gas contribution to enthalpy at the temperature and pressure of the mixture and this value added to the ideal gas mixture enthalpy at that temperature, as in the procedure developed in Section 11.8.

12.5 ADIABATIC FLAME TEMPERATURE

Consider a given combustion process that takes place adiabatically and with no work or changes in kinetic or potential energy involved. For such a process the temperature of the products is referred to as the adiabatic flame temperature. With the assumptions of no work and no changes in kinetic or potential energy, this is the maximum temperature that can be achieved for the given reactants because any heat transfer from the reacting substances and any incomplete combustion would tend to lower the temperature of the products.

For a given fuel and given pressure and temperature of the reactants, the maximum adiabatic flame temperature that can be achieved is with a stoichiometric mixture. The adiabatic flame temperature can be controlled by the amount of excess air that is used. This is important, for example, in gas turbines, where the maximum permissible temperature is determined by metallurgical considerations in the turbine, and close control of the temperature of the products is essential.

Example 12.10 shows how the adiabatic flame temperature may be found. The dissociation that takes place in the combustion products, which has a significant effect on the adiabatic flame temperature, will be considered in the next chapter.

Example 12.10

Liquid octane at 77 F is burned with 400 per cent theoretical air at 77 F in a steady-flow process. Determine the adiabatic flame temperature.

Control volume: Combustion chamber.

Inlet states: T known for fuel and air.

Process: SSSF.

Model: Gases ideal gases, Table A.11; liquid octane, Table A.13.

Analysis: The reaction is

$$C_8H_{18}(l) + 4(12.5)O_2 + 4(12.5)(3.76)N_2 \rightarrow$$
$$8CO_2 + 9H_2O(g) + 37.5O_2 + 188.0N_2$$

First law: Since the process is adiabatic,

$$H_R = H_P$$
$$\sum_R n_i(\bar{h}_f^\circ + \Delta\bar{h})_i = \sum_P n_e(\bar{h}_f^\circ + \Delta\bar{h})_e$$

where $\Delta\bar{h}_e$ refers to each constituent in the products at the adiabatic flame temperature.

Solution

From Tables A.11 and A.13,

$$H_R = \sum_R n_i[\bar{h}_f^\circ + \Delta\bar{h}]_i = (\bar{h}_f^\circ)_{C_8H_{18}(l)} = -107\,532 \text{ Btu/lb mol fuel}$$
$$H_P = \sum_P n_e[\bar{h}_f^\circ + \Delta\bar{h}]_e$$
$$= 8[-169\,297 + \Delta\bar{h}_{CO_2}] + 9[-104\,036 + \Delta\bar{h}_{H_2O}] + 37.5\Delta\bar{h}_{O_2} + 188.0\Delta\bar{h}_{N_2}$$

By trial-and-error solution, a temperature of the products is found that satisfies this equation. Assume

$$T_P = 1620 \text{ R}$$
$$H_P = \sum_P n_e[\bar{h}_f^\circ + \Delta\bar{h}]_e$$
$$= 8(-169\,297 + 12\,063) + 9(-104\,036 + 9432) + 37.5(8281) + 188.0(7839)$$
$$= -325\,039$$

Assume

$$T_P = 1800 \text{ R}$$
$$H_P = \sum_P n_e[\bar{h}_f^\circ + \Delta\bar{h}]_e$$
$$= 8(-169\,297 + 14\,371) + 9(-104\,036 + 11\,176) + 37.5(9769) + 188(9232)$$
$$= +26\,807$$

Since $H_P = H_R = -107\,532$, we find, by interpolation that the adiabatic flame temperature is 1731 R.

12.6 ENTHALPY AND INTERNAL ENERGY OF COMBUSTION; HEAT OF REACTION

The enthalpy of combustion, \bar{h}_{RP}, is defined as the difference between the enthalpy of the products and the enthalpy of the reactants when complete combustion occurs at a given temperature and pressure. That is,

$$\bar{h}_{RP} = H_P - H_R$$
$$\bar{h}_{RP} = \sum_P n_e(\bar{h}_f^{\circ} + \Delta\bar{h})_e - \sum_R n_i(\bar{h}_f^{\circ} + \Delta\bar{h})_i \tag{12.9}$$

The usual parameter for expressing the enthalpy of combustion is a unit mass of fuel, such as a kilogram (h_{RP}) or a kilomole (\bar{h}_{RP}) of fuel.

The tabulated values of the enthalpy of combustion of fuels are usually given for a temperature of 25°C and a pressure of 0.1 MPa. The enthalpy of combustion for a number of hydrocarbon fuels at this temperature and pressure, which we designate h_{RP_0}, is given in Table 12.3.

The internal energy of combustion is defined in a similar manner.

$$\bar{u}_{RP} = U_P - U_R$$
$$= \sum_P n_e(\bar{h}_f^{\circ} + \Delta\bar{h} - P\bar{v})_e - \sum_R n_i(\bar{h}_f^{\circ} + \Delta\bar{h} - P\bar{v})_i \tag{12.10}$$

When all the gaseous constituents can be considered as ideal gases, and the volume of the liquid and solid constituents is negligible compared to the value of the gaseous constituents, this relation for \bar{u}_{RP} reduces to

$$\bar{u}_{RP} = \bar{h}_{RP} - \bar{R}T(n_{\text{gaseous products}} - n_{\text{gaseous reactants}}) \tag{12.11}$$

Frequently the term "heating value" or "heat of reaction" is used. This represents the heat transferred from the chamber during combustion or reaction at constant temperature. In the case of a constant pressure or steady-flow process, we conclude from the first law of thermodynamics that it is equal to the negative of the enthalpy of combustion. For this reason this heat transfer is sometimes designated the constant-pressure heating value for combustion processes.

In the case of a constant-volume process, the heat transfer is equal to the negative of the internal energy of combustion. This is sometimes designated the constant-volume heating value in the case of combustion.

When the term heating value is used, the terms "higher" and "lower" heating value are used. The higher heating value is the heat transfer with liquid water in the products, and the lower heating value is the heat transfer with vapor water in the products.

Example 12.11

Calculate the enthalpy of combustion of propane at 25°C on both a kilomole and kilogram basis under the following conditions.

1. Liquid propane with liquid water in the products.
2. Liquid propane with gaseous water in the products.

3. Gaseous propane with liquid water in the products.
4. Gaseous propane with gaseous water in the products.

This example is designed to show how enthalpy of combustion can be determined from enthalpies of formation. Each of the four values corresponds to the appropriate value given in Table 12.3.

The enthalpy of evaporation of propane is 370 kJ/kg.

Analysis and Solution

The basic combustion equation is

$$C_3H_8 + 5O_2 \rightarrow 3CO_2 + 4H_2O$$

From Table A.13, $(\bar{h}_f^\circ)_{C_3H_8(g)} = -103\,847$ kJ/kmol. Therefore,

$$(\bar{h}_f^\circ)_{C_3H_8(l)} = -103\,847 - 44.097(370) = -120\,163 \text{ kJ/kmol}$$

1. Liquid propane–liquid water:

$$\bar{h}_{RP_0} = 3(\bar{h}_f^\circ)_{CO_2} + 4(\bar{h}_f^\circ)_{H_2O(l)} - (\bar{h}_f^\circ)_{C_3H_8(l)}$$
$$= 3(-393\,522) + 4(-285\,838) - (-120\,163)$$
$$= -2\,203\,755 \text{ kJ/kmol} = -\frac{2\,203\,755}{44.097} = -49\,975 \text{ kJ/kg}$$

The higher heating value of liquid propane is therefore 49 975 kJ/kg.

2. Liquid propane–gaseous water:

$$\bar{h}_{RP_0} = 3(\bar{h}_f^\circ)_{CO_2} + 4(\bar{h}_f^\circ)_{H_2O(g)} - (\bar{h}_f^\circ)_{C_3H_8(l)}$$
$$= 3(-393\,522) + 4(-241\,827) - (-120\,163)$$
$$= -2\,027\,711 \text{ kJ/kmol} = -\frac{2\,027\,711}{44.097} = -45\,983 \text{ kJ/kg}$$

The lower heating value of liquid propane is 45 983 kJ/kg.

3. Gaseous propane–liquid water:

$$\bar{h}_{RP_0} = 3(\bar{h}_f^\circ)_{CO_2} + 4(\bar{h}_f^\circ)_{H_2O(l)} - (\bar{h}_f^\circ)_{C_3H_8(g)}$$
$$= 3(-393\,522) + 4(-285\,838) - (-103\,847)$$
$$= -2\,220\,071 \text{ kJ/kmol} = -\frac{2\,220\,071}{44.097} = -50\,345 \text{ kJ/kg}$$

The higher heating value of gaseous propane is 50 345 kJ/kg.

4. Gaseous propane–gaseous water:

$$\bar{h}_{RP_0} = 3(\bar{h}_f^\circ)_{CO_2} + 4(\bar{h}_f^\circ)_{H_2O(g)} - (\bar{h}_f^\circ)_{C_3H_8(g)}$$
$$= 3(-393\,522) + 4(-241\,827) - (-103\,847)$$
$$= -2\,044\,027 \text{ kJ/kmol} = -\frac{2\,044\,027}{44.097} = -46\,353 \text{ kJ/kg}$$

The lower heating value of gaseous propane is 46 353 kJ/kg.

TABLE 12.3

Enthalpy of Combustion of Some Hydrocarbons at 25°C

Hydrocarbon	Formula	Liquid Water in Products (Negative of Higher Heating Value)		Vapor Water in Products (Negative of Lower Heating Value)	
		Liquid Hydro-carbon, kJ/kg fuel	Gaseous Hydro-carbon, kJ/kg fuel	Liquid Hydro-carbon, kJ/kg fuel	Gaseous Hydro-carbon, kJ/kg fuel
Paraffin Family					
Methane	CH_4		−55 496		−50 010
Ethane	C_2H_6		−51 875		−47 484
Propane	C_3H_8	−49 975	−50 345	−45 983	−46 353
Butane	C_4H_{10}	−49 130	−49 500	−45 344	−45 714
Pentane	C_5H_{12}	−48 643	−49 011	−44 983	−45 351
Hexane	C_6H_{14}	−48 308	−48 676	−44 733	−45 101
Heptane	C_7H_{16}	−48 071	−48 436	−44 557	−44 922
Octane	C_8H_{18}	−47 893	−48 256	−44 425	−44 788
Decane	$C_{10}H_{22}$	−47 641	−48 000	−44 239	−44 598
Dodecane	$C_{12}H_{26}$	−47 470	−47 828	−44 109	−44 467
Olefin Family					
Ethene	C_2H_4		−50 296		−47 158
Propene	C_3H_6		−48 917		−45 780
Butene	C_4H_8		−48 453		−45 316
Pentene	C_5H_{10}		−48 134		−44 996
Hexene	C_6H_{12}		−47 937		−44 800
Heptene	C_7H_{14}		−47 800		−44 662
Octene	C_8H_{16}		−47 693		−44 556
Nonene	C_9H_{18}		−47 612		−44 475
Decene	$C_{10}H_{20}$		−47 547		−44 410
Alkylbenzene Family					
Benzene	C_6H_6	−41 831	−42 266	−40 141	−40 576
Methylbenzene	C_7H_8	−42 437	−42 847	−40 527	−40 937
Ethylbenzene	C_8H_{10}	−42 997	−43 395	−40 924	−41 322
Propylbenzene	C_9H_{12}	−43 416	−43 800	−41 219	−41 603
Butylbenzene	$C_{10}H_{14}$	−43 748	−44 123	−41 453	−41 828

Example 12.12

Calculate the enthalpy of combustion of gaseous propane at 500 K. (At this temperature all the water formed during combustion will be in the vapor phase.) This example will demonstrate how the enthalpy of combustion of propane varies with temperature. The average constant-pressure specific heat of propane between 25°C and 500 K is 2.1 kJ/kg K.

Analysis: The combustion equation is

$$C_3H_8(g) + 5O_2 \rightarrow 3CO_2 + 4H_2O(g)$$

The enthalpy of combustion is, from Eq. 12.9,

$$(\bar{h}_{RP})_T = \sum_P n_e(\bar{h}_f^\circ + \Delta\bar{h})_e - \sum_R n_i(\bar{h}_f^\circ + \Delta\bar{h})_i$$

Solution

$$\bar{h}_{R500} = [\bar{h}_f^\circ + \bar{C}_{p\,av}(\Delta T)]_{C_3H_8(g)} + n_{O_2}(\Delta\bar{h})_{O_2}$$
$$= -103\,847 + 2.1 \times 44.097(500 - 298.2) + 5(6088)$$
$$= -54\,720 \text{ kJ}$$

$$\bar{h}_{P500} = n_{CO_2}(\bar{h}_f^\circ + \Delta\bar{h})_{CO_2} + n_{H_2O}(\bar{h}_f^\circ + \Delta\bar{h})_{H_2O}$$
$$= 3(-393\,522 + 8314) + 4(-241\,827 + 6920)$$
$$= -2\,095\,252 \text{ kJ}$$

$$\bar{h}_{RP500} = -2\,095\,252 - (-54\,720) = -2\,040\,532 \text{ kJ/kmol}$$

$$h_{RP500} = \frac{-2\,040\,532}{44.097} = -46\,274 \text{ kJ/kg}$$

This compares with a value of $-46\,353$ at 25°C.

This problem could also have been solved using the given value of the enthalpy of combustion at 25°C by noting that

$$(\bar{h}_{RP})_{500} = (H_P)_{500} - (H_R)_{500}$$
$$= n_{CO_2}(\bar{h}_f^\circ + \Delta\bar{h})_{CO_2} + n_{H_2O}(\bar{h}_f^\circ + \Delta\bar{h})_{H_2O}$$
$$\quad - [\bar{h}_f^\circ + \bar{C}_{p\,av}(\Delta T)]_{C_3H_8(g)} - n_{O_2}(\Delta\bar{h})_{O_2}$$
$$= \bar{h}_{RP_0} + n_{CO_2}\Delta\bar{h}_{CO_2} + n_{H_2O}\Delta\bar{h}_{H_2O}$$
$$\quad - \bar{C}_{p\,av}(\Delta T)_{C_3H_8(g)} - n_{O_2}\Delta\bar{h}_{O_2}$$

$$(\bar{h}_{RP})_{500} = -46\,353 \times 44.097 + 3(8314) + 4(6920)$$
$$\quad -2.1 \times 44.097(500 - 298.2) - 5(6088)$$
$$= -2\,040\,532 \text{ kJ/kmol}$$

$$(h_{RP})_{500} = \frac{-2\,040\,532}{44.097} = -46\,274 \text{ kJ/kg}$$

12.7 THE THIRD LAW OF THERMODYNAMICS AND ABSOLUTE ENTROPY

As we consider a second law analysis of chemical reactions, we face the same problem we did in regard to the first law: What base shall we use for the entropy of the various substances? This question leads directly to a consideration of the third law of thermodynamics.

The third law of thermodynamics was formulated during the early part of the twentieth century. The initial work was done primarily by W. H. Nernst (1864–1941) and Max Planck (1858–1947). The third law deals with the entropy of substances at the absolute zero of temperature, and in essence states that the entropy of a perfect crystal is zero at the absolute zero of temperature. From a statistical point of view this means that the crystal structure is such that it has the maximum degree of order. Further, since the temperature is absolute zero, the thermal energy is minimum. It also follows that a substance that does not have a perfect crystalline structure at absolute zero, but instead has a degree of randomness, such as a solid solution or a glassy solid, has a finite value of entropy at absolute zero. The experimental evidence on which the third law rests is primarily data on chemical reactions at low temperatures and measurements of heat capacity at temperatures approaching absolute zero. It should be noted that in contrast to the first and second laws, which lead, respectively, to the properties of internal energy and entropy, the third law deals only with the question of entropy at absolute zero. However, the implications of the third law are quite profound, particularly in regard to chemical equilibrium.

For our consideration at this point, the particular relevance of the third law is that it provides an absolute base from which the entropy of each substance can be measured. The entropy relative to this base is termed the absolute entropy. The increase in entropy between absolute zero and any given state can be found either from calorimetric data or by procedures based on statistical thermodynamics. The calorimetric method involves precise measurements of specific heat data over the temperature range, as well as the energy associated with phase transformations. These results are in agreement with the calculations based on statistical thermodynamics and observed molecular data.

Table A.13 gives the absolute entropy at 77 F and 1 atm or at 25°C and 0.1 MPa pressure for a number of substances. Table A.11 gives the absolute entropy for a number of gases at 1 atm or at 0.1 MPa pressure and various temperatures. In the case of gases, these are the hypothetical ideal gas values in all these tables. The pressure P° of 1 atm or of 0.1 MPa is termed the standard-state pressure, and the absolute entropy as given in these tables is designated \bar{s}°. The temperature is designated with a subscript such as \bar{s}°_{1000}, in which the temperature is in Rankine or in kelvin.

If the value of the absolute entropy is known at the standard-state pressure of 1 atm or of 0.1 MPa and a given temperature, it is a straightforward procedure to calculate the entropy change from this state (whether hypothetical ideal gas or real substance) to another desired state, following the procedure described in

Section 11.8. If the substance is one of those listed in Table A.11, then

$$\bar{s}_{T,P} = \bar{s}_T^\circ - \bar{R}\ln\left(\frac{P}{P^\circ}\right) + (\bar{s}_{T,P} - \bar{s}_{T,P}^*) \tag{12.12}$$

In this expression, the first term on the right side is the value from Table A.11, the second is the ideal gas term to account for a change in pressure from P° to P, and the third is the real substance behavior correction, as given in the generalized entropy chart, Fig. A.8. If the real substance behavior is to be evaluated from an equation of state or thermodynamic table of properties, then the ideal gas pressure change term should be made to a low pressure P^*, at which ideal gas behavior is a reasonable assumption but it is also listed in the tables. Then,

$$\bar{s}_{T,P} = \bar{s}_T^\circ - \bar{R}\ln\left(\frac{P^*}{P^\circ}\right) + (\bar{s}_{T,P} - \bar{s}_{T,P^*}^*) \tag{12.13}$$

If the substance is not one of those listed in Table A.11, and the absolute entropy is known only at one temperature T_0, as given in Table A.13, for example, then it will be necessary to calculate \bar{s}_T° from

$$\bar{s}_T^\circ = \bar{s}_{T_0}^\circ + \int_{T_0}^T \frac{\bar{C}_{po}}{T}\,dT \tag{12.14}$$

and then proceed with the calculation of Eq. 12.12 or 12.13.

It should be noted that if Eq. 12.12 is being used to calculate the absolute entropy of a substance in a region in which the ideal gas model is a valid representation of the behavior of that substance, then the last term on the right side of Eq. 12.12 simply drops out of the calculation.

In the case of calculating the absolute entropy of a mixture of ideal gases at T, P, the mixture entropy is given in terms of the component partial entropies as

$$\bar{s}_{\text{mix}}^* = \sum_i y_i \bar{S}_i^* \tag{12.15}$$

where

$$\bar{S}_i^* = \bar{s}_{i_T}^\circ - \bar{R}\ln\left(\frac{P}{P^\circ}\right) - \bar{R}\ln y_i$$

$$= \bar{s}_{i_{T^-}}^\circ - \bar{R}\ln\left(\frac{y_i P}{P^\circ}\right) \tag{12.16}$$

For a real gas mixture, a correction can be added to the ideal gas entropy calculated from Eqs. 12.15 and 12.16 by using a pseudocritical method such as Kay's rule, Eq. 11.33, which results in the expression

$$\bar{s}_{\text{mix}} = \bar{s}_{\text{mix}}^* + (\bar{s} - \bar{s}^*)_{T,P} \tag{12.17}$$

in which the second term on the right side is the correction term from the generalized entropy chart. An alternate approach for a real gas mixture would be to

assume ideal solution for the mixture, in which case

$$\bar{s}_{\text{mix}} = \sum_i y_i \bar{S}_i \qquad (12.18)$$

Each partial entropy \bar{S}_i in Eq. 12.18 is expressed as

$$\bar{S}_i = \bar{s}_i - \bar{R} \ln y_i \qquad (12.19)$$

and each \bar{s}_i in Eq. 12.19 is given by Eq. 12.12. It is seen that Kay's rule is considerably simpler, and is in general as good an approximation.

12.8 SECOND LAW ANALYSIS OF REACTING SYSTEMS

The concepts of reversible work, irreversibility, and availability were introduced in Chapter 8. These concepts involved both the first and second laws of thermodynamics. At the conclusion of Chapter 8 the Gibbs function was introduced and the statement was made that this property would be particularly relevant in dealing with chemical reactions. We proceed now to develop this matter further, and we will be particularly concerned with determining the maximum work (availability) that can be done through a combustion process and with examining the irreversibilities associated with such processes.

The reversible work, Eq. 8.9, for a steady-state, steady-flow process, in the absence of changes in kinetic and potential energy, is

$$W_{\text{rev}} = \sum m_i(h_i - T_0 s_i) - \sum m_e(h_e - T_0 s_e)$$

Applying this to an SSSF process involving a chemical reaction, and introducing the symbols we have been using in this chapter, we have

$$W_{\text{rev}} = \sum_R n_i(\bar{h}_f^\circ + \Delta\bar{h} - T_0\bar{s})_i - \sum_P n_e(\bar{h}_f^\circ + \Delta\bar{h} - T_0\bar{s})_e \qquad (12.20)$$

Similarly, the irreversibility for such a process becomes

$$I = \sum_P n_e T_0 \bar{s}_e - \sum_R n_i T_0 \bar{s}_i - Q_{\text{c.v.}} \qquad (12.21)$$

The availability, Ψ, in the absence of kinetic and potential energy changes, for an SSSF process was defined, Eq. 8.16, as

$$\Psi = (h - T_0 s) - (h_0 - T_0 s_0)$$

It was also indicated in Chapter 8 that when an SSSF chemical reaction takes place in such a manner that both the reactants and products are in temperature equilibrium with the surroundings, the Gibbs function $(g = h - Ts)$ becomes a significant variable, Eq. 8.28. For such a process, in the absence of changes in kinetic and potential energy, the reversible work is given by the relation

$$W_{\text{rev}} = \sum_R n_i \bar{g}_i - \sum_P n_e \bar{g}_e \qquad (12.22)$$

Since the Gibbs function is so relevant for processes involving chemical reactions, the Gibbs function of formation, \bar{g}_f°, has been defined by a procedure similar to that used in defining the enthalpy of formation. That is, the Gibbs function of each of the elements at 25°C and 0.1 MPa pressure (or 77 F, 1 atm) is assumed to be zero, and the Gibbs function of each substance is found relative to this base. Table A.13 lists the Gibbs function of formation of a number of substances at 25°C and 0.1 MPa (or 77 F, 1 atm) pressure. It is also evident that the Gibbs function can be found directly from data for \bar{h}_f° and \bar{s}° at the given temperature, as indicated in the following example.

Example 12.13

Determine the Gibbs function of formation of carbon dioxide in English units.

Analysis: Consider the reaction

$$C + O_2 \rightarrow CO_2$$

Assume that the carbon and oxygen are each initially at 77 F and 1 atm pressure, and that the CO_2 is finally at 77 F and 1 atm pressure.

The change in Gibbs function for this reaction is found first.

$$G_P - G_R = (H_P - H_R) - T_0(S_P - S_R)$$

$$\sum_P n_e(\bar{g}_f^\circ)_e - \sum_R n_i(\bar{g}_f^\circ)_i = \sum_P n_e(\bar{h}_f^\circ)_e - \sum_R n_i(\bar{h}_f^\circ)_i - T_0\left[\sum_P n_e(\bar{s}_{537}^\circ)_e - \sum_R n_i(\bar{s}_{537}^\circ)_i\right]$$

Solution

$$G_P - G_R = (\bar{h}_f^\circ)_{CO_2} - 536.7[(\bar{s}_{537}^\circ)_{CO_2} - (\bar{s}_{537}^\circ)_C - (\bar{s}_{537}^\circ)_{O_2}]$$
$$= -169\,297 - 536.7(51.072 - 1.359 - 49.004)$$
$$= -169\,297 - 380 = -169\,677 \text{ Btu/lb mol.}$$

Since the Gibbs function of the reactants, G_R, is zero (in accordance with the assumption that the Gibbs function of the elements is zero at 77 F and 1 atm pressure), it follows that

$$G_P = (\bar{g}_f^\circ)_{CO_2} = -169\,677 \text{ Btu/lb mol}$$

This is the value given in Table A.13.

Let us now consider the question of the maximum work that can be done during a chemical reaction. For example, consider 1 kmol of hydrocarbon fuel and the necessary air for complete combustion, each at 0.1 MPa pressure and 25°C, the pressure and temperature of the surroundings. What is the maximum work that can be done as this fuel reacts with the air? From our considerations in Chapter 8 we conclude that the maximum work would be done if this chemical

reaction took place reversibly and the products were finally in pressure and temperature equilibrium with the surroundings. From Eq. 12.22 we conclude this reversible work could be calculated from the relation

$$W_{rev} = \sum_R n_i \bar{g}_i - \sum_P n_e \bar{g}_e$$

However, since the final state is in equilibrium with the surroundings, we could consider this amount of work to be the availability of the fuel and air.

Example 12.14

Ethene (g) at 25°C and 0.1 MPa pressure is burned with 400% theoretical air at 25°C and 0.1 MPa pressure. Assume that this reaction takes place reversibly at 25°C and that the products leave at 25°C and 0.1 MPa pressure. To further simplify this problem assume that the oxygen and nitrogen are separated before the reaction takes place (each at 0.1 MPa, 25°C), that the constituents in the products are separated, and that each is at 25°C and 0.1 MPa. Thus the reaction takes place as shown in Fig. 12.4. For purposes of comparison between this and the two subsequent examples, we consider all of the water in the products to be a gas (a hypothetical situation in this example and Example 12.16).

Determine the reversible work for this process (that is, the work that would be done if this chemical action took place reversibly and isothermally).

Control volume: Combustion chamber.

Inlet states: P, T known for each gas.

Exit states: P, T known for each gas.

Model: All ideal gases, Tables A.11 and A.13.

Sketch: Fig. 12.4.

Analysis: The equation for this chemical reaction is

$$C_2H_4(g) + 3(4)O_2 + 3(4)(3.76)N_2 \rightarrow 2CO_2 + 2H_2O(g) + 9O_2 + 45.1N_2$$

The reversible work for this process is equal to the decrease in Gibbs function during this reaction, Eq. 12.22. The values for the Gibbs function can be taken directly from Table A.13, since each is at 25°C and 0.1 MPa pressure. From Eq. 12.22,

$$W_{rev} = \sum_R n_i \bar{g}_i - \sum_P n_e \bar{g}_e$$

FIGURE 12.4
Sketch for Example 12.14.

Solution

Since each of the reactants and each of the products are at 0.1 MPa pressure and 25°C, this reduces to

$$W_{rev} = (\bar{g}_f^\circ)_{C_2H_4} - 2(\bar{g}_f^\circ)_{CO_2} - 2(\bar{g}_f^\circ)_{H_2O(g)}$$
$$= 68\ 142 - 2(-394\ 407) - 2(-228\ 583)$$
$$= 1\ 314\ 122 \text{ kJ/kmol } C_2H_4$$
$$= \frac{1\ 314\ 122}{28.054} = 46\ 843 \text{ kJ/kg}$$

Therefore we might say that when 1 kg of ethene is at 25°C, 0.1 MPa pressure, the temperature and pressure of the surroundings, it has an availability of 46 843 kJ.

Thus it would seem logical to rate the efficiency of a device designed to do work by utilization of a combustion process, such as an internal combustion engine or a steam power plant, as the ratio of the actual work to the reversible work, or in the above example, the decrease in Gibbs function for the chemical reaction, instead of comparing the actual work to the heating value, as is commonly done. This is, in fact, the basic principle of the second law efficiency, which was introduced in connection with availability analysis in Chapter 8. As noted from Example 12.14, the difference between the decrease in Gibbs function and the heating value is small, which is typical for hydrocarbon fuels, such that the difference in the two types of efficiencies will not usually be large. We must always be careful, however, when discussing efficiencies to note the definition of the efficiency under consideration.

It is of particular interest to study the irreversibility that takes place during a combustion process. The following examples are presented to illustrate this matter, where we consider the same hydrocarbon fuel that was used in Example 12.14—ethene (g) at 25°C and 0.1 MPa. We determined the availability and found that it was 46 843 kJ/kg. Now let us burn this fuel with 400% theoretical air in a steady-state, steady-flow adiabatic process. We can determine the irreversibility of this process in two ways. The first is to calculate the increase in entropy for the process. Since the process is adiabatic, the increase in entropy is due entirely to the irreversibilities for the process and we can find the irreversibility from Eq. 12.21. We can also calculate the availability of the products of combustion at the adiabatic flame temperature, and note that they are less than the availability of the fuel and air before the combustion process, the difference being the irreversibility that occurs during the combustion process.

Example 12.15

Consider the same combustion process as in Example 12.14, but let it take place adiabatically, and assume that each constituent in the products is at 0.1 MPa pressure and at the adiabatic flame temperature. This combustion process is shown schematically in Fig. 12.5. The temperature of the surroundings is 25°C.

FIGURE 12.5
Sketch for Example 12.15.

For this combustion process determine the increase in entropy during combustion and the availability of the products of combustion.

Control volume: Combustion chamber.

Inlet states: P, T known for each gas.

Exit states: P known for each gas.

Sketch: Fig. 12.5.

Model: All ideal gases, Table A.11; Table A.13 for ethene.

Analysis: The combustion equation is

$$C_2H_4 + 12O_2 + 12(3.76)N_2 \rightarrow 2CO_2 + 2H_2O(g) + 9O_2 + 45.1N_2$$

The adiabatic flame temperature is determined first.
First law:

$$H_R = H_P$$
$$\sum_R n_i(\bar{h}_f^\circ)_i = \sum_P n_e(\bar{h}_f^\circ + \Delta\bar{h})_e$$

Solution

$$52\ 283 = 2(-393\ 522 + \Delta\bar{h}_{CO_2}) + 2(-241\ 827 + \Delta\bar{h}_{H_2O(g)})$$
$$+ 9\Delta\bar{h}_{O_2} + 45.1\Delta\bar{h}_{N_2}$$

By a trial-and-error solution we find the adiabatic flame temperature to be 1016 K.

We now proceed to find the change in entropy during this adiabatic combustion process.

$$S_R = \sum_R (n_i\bar{s}_i^\circ)_{298} = (\bar{s}_{C_2H_4}^\circ + 12\bar{s}_{O_2}^\circ + 45.1\bar{s}_{N_2}^\circ)_{298}$$

$$= 219.548 + 12(205.142) + 45.1(191.611)$$

$$= 11\ 322.908 \text{ kJ/kmol (fuel) K}$$

$$S_P = \sum_P (n_e\bar{s}_e^\circ)_{1016} = (2\bar{s}_{CO_2}^\circ + 2\bar{s}_{H_2O(g)}^\circ + 9\bar{s}_{O_2}^\circ + 45.1\bar{s}_{N_2}^\circ)_{1016}$$

$$= 2(270.162) + 2(233.344) + 9(244.120) + 45.1(228.670)$$

$$= 13\ 517.109 \text{ kJ/kmol (fuel) K}$$

$$S_P - S_R = 2194.201 \text{ kJ/kmol (fuel) K}$$

Since this is an adiabatic process, the increase in entropy indicates the irreversibility of the adiabatic combustion process. This irreversibility can be found from Eq. 12.21.

$$I = T_0 \left(\sum_P n_e \bar{s}_e - \sum_R n_i \bar{s}_i \right)$$

$$= 298.15 \times 2194.201 = 654\,201 \text{ kJ/kmol}$$

$$= \frac{654\,201}{28.054} = 23\,319 \text{ kJ/kg fuel}$$

Therefore, the availability after the combustion process is

$$\Psi_P = 46\,843 - 23\,319 = 23\,524 \text{ kJ/kg}$$

The availability of the products can also be found from the relation

$$\Psi_P = \sum_P \left[(\bar{h}_e - T_0 \bar{s}_e) - (\bar{h}_0 - T_0 \bar{s}_0) \right]$$

Since in this problem the products are separated and each is at 0.1 MPa pressure and the adiabatic flame temperature of 1016 K, this reduces to

$$\Psi_P = \sum_P n_e [(\bar{h}_e^\circ - \bar{h}_0^\circ) - T_0(\bar{s}_e^\circ - \bar{s}_0^\circ)]$$

$$= 2(34\,283) + 2(26\,648) + 9(23\,269) + 45.1(21\,988)$$
$$- 298.15[2(270.162 - 213.795) + 2(233.344 - 188.833)$$
$$+ 9(244.120 - 205.142) + 45.1(228.670 - 191.611)]$$
$$= 659\,942 \text{ kJ/kmol} = 23\,524 \text{ kJ/kg}$$

In other words, if every process after the adiabatic combustion process were reversible, the maximum amount of work that could be done is 23 524 kJ/kg fuel. This compares to a value of 46 843 kJ/kg for the reversible isothermal reaction. This means that if we had an engine that had the indicated adiabatic combustion process, and if all other processes were completely reversible, the efficiency would be about 50%.

In the two prior examples we made the assumption, for purposes of simplifying the calculation, that the constituents in the reactants and products were separated, and each was at 0.1 MPa pressure. This of course is not a realistic problem, and in the following example, Example 12.14 is repeated with the assumption that the reactants and products each consist of a mixture at 0.1 MPa pressure.

Example 12.16

Consider the same combustion process of Example 12.14, but assume that the reactants consist of a mixture at 0.1 MPa pressure and 25°C and that the products

FIGURE 12.6

Sketch for Example 12.16.

also consist of a mixture at 0.1 MPa and 25°C. Thus the combustion process is as shown in Fig. 12.6.

Determine the work that would be done if this combustion process took place reversibly and in pressure and temperature equilibrium with the surroundings.

Control volume: Combustion chamber.

Inlet state: P, T known.

Exit state: P, T known.

Sketch: Fig. 12.6.

Model: Reactants—ideal gas mixture, Table A.11. Products—ideal gas mixture, Table A.11.

Analysis: The combustion equation, as noted previously, is

$$C_2H_4(g) + 3(4)O_2 + 3(4)(3.76)N_2 \rightarrow 2CO_2 + 2H_2O(g) + 9O_2 + 45.1N_2$$

In this example the reactants are an ideal gas mixture at 0.1 MPa, 25°C, and the products are also an ideal gas mixture at 0.1 MPa, 25°C. Therefore, in both cases the mixture entropy is given by Eq. 12.15, in which the partial entropies are given by Eq. 12.16,

$$\bar{S}_i = \bar{s}^\circ_{i\,T} - \bar{R} \ln\left(\frac{y_i P}{P^\circ}\right)$$

In this particular case, the mixture pressure P and the standard-state pressure P° are the same. The values of standard-state entropy are taken from Tables A.11 and A.13.

Solution

For the reactants:

	n_i	y_i	$-\bar{R} \ln\left(\dfrac{y_i P}{P^\circ}\right)$	\bar{s}°_i	\bar{S}_i
C_2H_4	1	0.017 21	33.774	219.548	253.322
O_2	12	0.206 54	13.114	205.142	218.256
N_2	45.1	0.776 25	2.106	191.611	193.717

For the products:

	n_i	y_i	$-\bar{R} \ln \left(\dfrac{y_i P}{P^\circ}\right)$	\bar{s}_i°	\bar{S}_i
CO_2	2	0.034 42	28.011	213.795	241.806
H_2O	2	0.034 42	28.011	188.833	216.844
O_2	9	0.154 91	15.506	205.142	220.648
N_2	45.1	0.776 25	2.106	191.611	193.717

With the assumption of ideal gas behavior, the enthalpy of each constituent is equal to the enthalpy of formation at 25°C. The values of entropy are as calculated above. Therefore, from Eq. 12.20,

$$W_{rev} = \sum_R n_i (\bar{h}_f^\circ)_i - \sum_P n_e (\bar{h}_f^\circ)_e - T_0 \left(\sum_R n_i \bar{s}_i - \sum_P n_e \bar{s}_e \right)$$

$$= (\bar{h}_f^\circ)_{C_2H_4} - 2(\bar{h}_f^\circ)_{CO_2} - 2(\bar{h}_f^\circ)_{H_2O(g)}$$
$$\quad - 298.15(\bar{S}_{C_2H_4} + 12\bar{S}_{O_2} + 45.1\bar{S}_{N_2} - 2\bar{S}_{CO_2} - 2\bar{S}_{H_2O(g)} - 9\bar{S}_{O_2} - 45.1\bar{S}_{N_2})$$

$$= 52\,283 - 2(-393\,522) - 2(-241\,827)$$
$$\quad - 298.15[253.322 + 12(218.256) + 45.1(193.717)$$
$$\quad - 2(241.806) - 2(216.844) - 9(220.648) - 45.1(193.717)]$$

$$= 1\,332\,145 \text{ kJ/kmol}$$

$$= \frac{1\,332\,145}{28.054} = 47\,485 \text{ kJ/kg}$$

Note that this is essentially the same value that was obtained in Example 12.14, when the reactants and products were each separated and at 0.1 MPa pressure.

These examples raise the question of the possibility of a reversible chemical reaction. Some reactions can be made to approach reversibility by having them take place in an electrolytic cell, as described in Chapter 1. When a potential exactly equal to the electromotive force of the cell is applied, no reaction takes place. When the applied potential is increased slightly, the reaction proceeds in one direction, and if the applied potential is decreased slightly, the reaction proceeds in the opposite direction. The work involved is the electrical energy supplied or delivered.

Consider a reversible reaction occurring at constant temperature equal to that of its environment. The work output of the fuel cell would then be

$$W = -\left(\sum n_e \bar{g}_e - \sum n_i \bar{g}_i \right) = -\Delta G$$

where the ΔG is the change in Gibbs function for the overall chemical reaction. We also realize that the work is given in terms of the charged electrons flowing through an electrical potential \mathscr{E} as

$$W = \mathscr{E} n_e N_0 e$$

in which n_e is the number of kilomoles of electrons flowing through the external circuit, and

$$N_0 e = 6.022\,045 \times 10^{26} \text{ elec/kmol} \times 1.602\,189 \times 10^{-22} \text{ kJ/elec V}$$
$$= 96\,485 \text{ kJ/kmol V}$$

Thus, for a given reaction, the maximum (reversible reaction) electrical potential \mathscr{E}° of a fuel cell at a given temperature is

$$\mathscr{E}^\circ = \frac{-\Delta G}{96\,485 n_e} \tag{12.23}$$

Example 12.17

Calculate the reversible EMF at 25°C for the hydrogen–oxygen fuel cell described in Section 1.2.

Solution

The anode side reaction was stated to be

$$2H_2 \rightarrow 4H^+ + 4e^-$$

and that at the cathode is

$$4H^+ + 4e^- + O_2 \rightarrow 2H_2O$$

Therefore, the overall reaction is, in kilomoles,

$$2H_2 + O_2 \rightarrow 2H_2O$$

for which 4 kmol of electrons flow through the external circuit. Let us assume that each component is at its standard-state pressure of 0.1 MPa and that the water formed is liquid. Then

$$\Delta G_{25°C} = 2\bar{g}^\circ_{f_{H_2O}} - 2\bar{g}^\circ_{f_{H_2}} - \bar{g}^\circ_{f_{O_2}}$$
$$= 2(-237\,146) - 2(0) - 1(0)$$
$$= -474\,292 \text{ kJ}$$

Therefore, from Eq. 12.23,

$$\mathscr{E}^\circ = \frac{-(-474\,292)}{96\,485 \times 4} = 1.229 \text{ V}$$

Much effort is being directed toward the development of fuel cells in which carbon, hydrogen, or hydrocarbons will react with oxygen and produce electricity directly. If a fuel cell can be developed that utilizes a hydrocarbon fuel and has a sufficiently high efficiency and capacity (for a given volume or weight), drastic changes will be possible in our techniques for generating electricity on a commercial scale. At the present time, however, fuel cells are not competitive with conventional power plants for the production of electricity on a large scale.

12.9 EVALUATION OF ACTUAL COMBUSTION PROCESSES

In evaluating the performance of an actual combustion process a number of different parameters can be defined, depending on the nature of the process and the system considered. In the combustion chamber of a gas turbine, for example, the objective is to raise the temperature of the products to a given temperature (usually the maximum temperature the metals in the turbine can withstand). If we had a combustion process that was adiabatic and in which complete combustion was achieved, the temperature of the products would be the adiabatic flame temperature. Let us designate the fuel–air ratio needed to reach a given temperature under these conditions as the ideal fuel–air ratio. In the actual combustion chamber the combustion will be incomplete to some extent and there will be some heat transfer to the surroundings. Therefore, more fuel will be required to reach the given temperature, and this we designate as the actual fuel–air ratio. In this case, the combustion efficiency, η_{comb}, is defined as

$$\eta_{comb} = \frac{FA_{ideal}}{FA_{actual}} \qquad (12.24)$$

On the other hand, in the furnace of a steam generator (boiler), the purpose is to transfer the maximum possible amount of heat to the steam (water). In practice, the efficiency of a steam generator is defined as the ratio of the heat transferred to the steam to the higher heating value of the fuel. For coal this is the heating value as measured in a bomb calorimeter, which is the constant-volume heating value, and it corresponds to the internal energy of combustion. One observes a minor inconsistency, since the boiler involves a flow process, and the change in enthalpy is the significant factor. However, in most cases, the error thus introduced is less than the experimental error involved in measuring the heating value, and the efficiency of a steam generator is defined by the relation

$$\eta_{steam\ generator} = \frac{\text{heat transferred to steam/kg fuel}}{\text{higher heating value of the fuel}} \qquad (12.25)$$

In an internal combustion engine the purpose is to do work. The logical way to evaluate the performance of an internal combustion engine would be to compare the actual work done to the maximum work that would be done by a reversible change of state from the reactants to the products. This, as we noted previously, is called the second law efficiency.

However, in practice the efficiency of an internal combustion engine is defined as the ratio of the actual work to the negative of the enthalpy of combustion of the fuel (that is, the constant-pressure heating value). This is usually called the thermal efficiency, η_{th}:

$$\eta_{th} = \frac{w}{(-h_{RP_0})} = \frac{w}{\text{heating value}} \qquad (12.26)$$

The overall efficiency of a gas turbine or steam power plant is defined in the same way. It should be pointed out that in an internal combustion engine or fuel-burning steam power plant, the fact that the combustion is itself irreversible is a significant factor in the relatively low thermal efficiency of these devices.

One other factor should be pointed out regarding efficiency. We have noted that the enthalpy of combustion of a hydrocarbon fuel varies considerably with the phase of the water in the products (which leads to the concept of higher and lower heating values). Therefore, in considering the thermal efficiency of an engine, the heating value used to determine this efficiency must be borne in mind. Two engines made by different manufacturers may have identical performance, but if one manufacturer bases efficiency on the higher heating value and the other on the lower heating value, the latter will be able to claim a higher thermal efficiency. This claim is not significant, of course, as the performance is the same, and a consideration of the way in which the efficiency was defined would reveal this.

The whole matter of efficiencies of devices involving combustion processes is treated in detail in textbooks dealing with particular applications, and our discussion is intended only as an introduction to the subject. However, two examples will be cited to illustrate these remarks.

Example 12.18

The combustion chamber of a gas turbine uses a liquid hydrocarbon fuel that has an approximate composition of C_8H_{18}. On testing the following data are obtained.

$$T_{air} = 260 \text{ F}(720 \text{ R}) \qquad T_{products} = 1520 \text{ F}(1980 \text{ R})$$
$$V_{air} = 300 \text{ ft/s} \qquad V_{products} = 450 \text{ ft/s}$$
$$T_{fuel} = 120 \text{ F} \qquad FA_{actual} = 0.0211 \text{ lbm fuel/lbm air}$$

Calculate the combustion efficiency for this process.

Control volume: Combustion chamber.

Inlet states: T known for air and fuel.

Exit state: T known.

Model: Air and products—ideal gas, Table A.11. Fuel—Table A.13.

Analysis: For the ideal chemical reaction the heat transfer is zero. Therefore, writing the first law for a control volume that includes the combustion chamber, we have

$$H_R + KE_R = H_P + KE_P$$

$$H_R + KE_R = \sum_R n_i \left(\bar{h}_f^\circ + \Delta\bar{h} + \frac{MV^2}{2} \right)_i$$

$$= [\bar{h}_f^\circ + \bar{C}_p(120 - 77)]_{C_8H_{18}(l)} + n_{O_2} \left(\Delta\bar{h} + \frac{MV^2}{2} \right)_{O_2}$$

$$+ 3.76 n_{O_2} \left(\Delta\bar{h} + \frac{MV^2}{2} \right)_{N_2}$$

$$H_P + KE_P = \sum_P n_e \left(\bar{h}_f^\circ + \Delta\bar{h} + \frac{M\mathbf{V}^2}{2} \right)_e$$

$$= 8 \left(\bar{h}_f^\circ + \Delta\bar{h} + \frac{M\mathbf{V}^2}{2} \right)_{CO_2} + 9 \left(\bar{h}_f^\circ + \Delta\bar{h} + \frac{M\mathbf{V}^2}{2} \right)_{H_2O}$$

$$+ (n_{O_2} - 12.5) \left(\Delta\bar{h} + \frac{M\mathbf{V}^2}{2} \right)_{O_2} + 3.76 n_{O_2} \left(\Delta\bar{h} + \frac{M\mathbf{V}^2}{2} \right)_{N_2}$$

Solution

$$H_R + KE_R = -107\,532 + 0.5(114.23)(43) + n_{O_2} \left[1303 + \frac{32.0 \times (300)^2}{2 \times 32.17 \times 778} \right]$$

$$+ 3.76 n_{O_2} \left[1278 + \frac{28.02 \times (300)^2}{2 \times 32.17 \times 778} \right]$$

$$= (-105\,076 + 6355 n_{O_2}) \text{ Btu/lb mol fuel}$$

$$H_P + KE_P = 8 \left[-169\,297 + 16\,733 + \frac{44.01 \times (450)^2}{2 \times 32.17 \times 778} \right]$$

$$+ 9 \left[-104\,036 + 12\,978 + \frac{18.02 \times (450)^2}{2 \times 32.17 \times 778} \right]$$

$$+ (n_{O_2} - 12.5) \left[11\,279 + \frac{32.0 \times (450)^2}{2 \times 32.17 \times 778} \right]$$

$$+ 3.76 n_{O_2} \left[10\,651 + \frac{28.02 \times (450)^2}{2 \times 32.17 \times 778} \right]$$

$$= (-2\,174\,652 + 54\,243 n_{O_2}) \text{ Btu/lb mol fuel}$$

Therefore,

$$-105\,076 + 6355 n_{O_2} = -2\,174\,652 + 54\,243 n_{O_2}$$

$$n_{O_2} = \frac{2\,069\,576}{47\,888} = 43.2 \text{ lb mol } O_2/\text{lb mol fuel}$$

$$\text{lb mol air/lb mol fuel} = 4.76(43.2) = 205.8$$

$$FA_{\text{Ideal}} = \frac{114.23}{205.8 \times 28.97} = 0.0194$$

$$\eta_{\text{comb}} = \frac{0.0194}{0.0211} = 0.92 = 92\%$$

Example 12.19

In a certain steam power plant 325 000 kg of water per hour enter the boiler at a pressure of 12.5 MPa and a temperature of 200°C. Steam leaves the boiler at 9 MPa, 500°C. The power output of the turbine is 81 000 kW. Coal is used at the

rate of 26 700 kg/h, and has a higher heating value of 33 250 kJ/kg. Determine the efficiency of the steam generator and overall thermal efficiency of the plant.

In power plants the efficiency of both the boiler and the overall efficiency of the plant are based on the higher heating value of the fuel.

Solution

The efficiency of the boiler is defined by Eq. 12.25 as

$$\eta_{\text{steam generator}} = \frac{\text{heat transferred to } H_2O/\text{kg fuel}}{\text{higher heating value}}$$

Therefore,

$$\eta_{\text{steam generator}} = \frac{325\,000(3386.1 - 857.1)}{26\,700 \times 33\,250} \times 100 = 92.6\%$$

The thermal efficiency is defined by Eq. 12.26.

$$\eta_{\text{th}} = \frac{w}{\text{heating value}} = \frac{81\,000 \times 3600}{26\,700 \times 33\,250} \times 100 = 32.8\%$$

PROBLEMS

12.1 Liquid propane is burned with air, and a volumetric analysis of the products of combustion yields the following percent composition on a dry basis:

CO_2	8.6%
CO	0.6
O_2	7.2
N_2	83.6

Determine the percent of theoretical air used in this combustion process.

12.2 A sample of pine bark has the following ultimate analysis (on a dry basis) percent by mass:

H	5.6%
C	53.4
S	0.1
N	0.1
O	37.9
Ash	2.9

This bark is to be used as a fuel by burning it with 100% theoretical air in a furnace. Determine the air–fuel ratio on a mass basis.

12.3 An unknown hydrocarbon fuel is burned with air and a volumetric analysis of the combustion products yields the following percent composition

on a dry basis:

$$CO_2 \quad 10.5\%$$
$$O_2 \quad 5.3$$
$$N_2 \quad 84.2$$

Determine the composition of the fuel on a mass basis and the percent of theoretical air used in the combustion process.

12.4 The coal gasifier in an Integrated Gasification Combined Cycle power plant produces a gas mixture having the following volumetric percent composition:

$$CH_4 \quad 0.3\%$$
$$H_2 \quad 29.6$$
$$CO \quad 41.0$$
$$CO_2 \quad 10.0$$
$$N_2 \quad 0.8$$
$$H_2O \quad 17.0$$
$$H_2S \quad 1.1$$
$$NH_3 \quad 0.2$$

This gas is then cooled to 40°C, 3 MPa, and the H_2S and NH_3 are removed in water-scrubbers. Assuming that the resulting mixture, which is to be sent to the combustors, is saturated with water, determine the following:
(a) The mixture composition.
(b) The theoretical air–fuel ratio in the combustors.

12.5 Decane is burned with 110% theoretical air in a constant pressure process at 14.7 lbf/in.2 and the products are cooled to ambient temperature, 77 F.
(a) How many pounds of water are condensed per pound of fuel?
(b) Repeat part (a) assuming that the air used in the combustion has a relative humidity of 90%.

12.6 Many coals from the western United States have a high moisture content. Consider the following sample of Wyoming coal, for which the ultimate analysis, on an as received basis, is, by mass:

Moisture	28.9%
Ash	5.8
Sulfur	0.5
Nitrogen	0.7
Carbon	48.6
Hydrogen	3.5
Oxygen	12.0

This coal is to be burned in the steam generator of a large power plant with 120% theoretical air. Determine the air–fuel ratio on a mass basis.

12.7 The output gas mixture of a certain air-blown coal gasifier has the composition of producer gas, listed in Table 12.2. Consider the combustion of this gas with 120% theoretical air at 14.7 lbf/in.2 pressure.

(a) Determine the dew point of the products.

(b) How many pounds of water will be condensed per pound of fuel if the products are cooled 20 F below the dew-point temperature?

12.8 The hot exhaust gas from an internal combustion engine is analyzed and found to have the following percent composition on a volumetric basis:

CO_2 10%
CO 2
H_2O 13
O_2 3
N_2 72

This gas is to be fed to an exhaust gas reactor and mixed with a certain amount of air, as shown in Fig. 12.7, to eliminate the carbon monoxide. It is decided that a mole fraction of oxygen of 10% in the mixture at point 3 will ensure that no CO remains. What must the ratio of flows be entering the reactor?

FIGURE 12.7
Sketch for Problem 12.8.

12.9 A gas mixture of 50% ethane and 50% propane, by volume, enters a combustion chamber at 350 K, 10 MPa. Determine the enthalpy per kilomole of this mixture, relative to the thermochemical base of enthalpy.

12.10 A rigid chamber contains 1 lb mol of carbon monoxide and 2 lb mol of oxygen at 77 F, 2 atm. Combustion now occurs, with the resulting products consisting entirely of carbon dioxide and the excess oxygen at a temperature of 1700 F. Determine the following:

(a) The final pressure in the chamber.

(b) The heat transfer from the chamber during the process.

12.11 In a test of rocket propellant performance, liquid hydrazine (N_2H_4) at 100 kPa, 25°C, and oxygen gas at 100 kPa, 25°C, are fed to a combustion chamber in the ratio 0.5 kg O_2/kg N_2H_4. The heat transfer from the chamber to the surroundings is estimated to be 100 kJ/kg N_2H_4. Determine the temperature of the products exiting the chamber, assuming that only H_2O, H_2, and N_2 are present. The enthalpy of formation of liquid hydrazine is $+50\,417$ kJ/kmol.

12.12 Repeat Problem 12.11, assuming that saturated liquid oxygen at 90 K is used instead of 25°C oxygen gas in the combustion process. Use the generalized charts to determine the properties of liquid oxygen.

12.13 An automotive internal combustion engine burns liquid octane (C_8H_{18}) at the rate of 0.01 lbm/s, and uses 120% theoretical air. The air and fuel enter the engine at 77 F, and the combustion products leave the engine exhaust ports at 1160 F. It may be assumed that 90% of the carbon in the fuel burns to form CO_2 and the remaining 10% burns to form CO. It may also be assumed that the heat transfer from this engine is just equal to the work output of the engine. Determine the following:
(a) The power output of the engine.
(b) The composition of the combustion products and the dew point temperature.

12.14 In a new high-efficiency furnace, natural gas (assumed to be 90% methane and 10% ethane, by volume) and 110% theoretical air each enter at 25°C, 100 kPa, and the products exit the furnace at 50°C, 100 kPa.
(a) What is the heat transfer for this process?
(b) Repeat part (a) for an older-design furnace in which the combustion products exit at 250°C, 100 kPa.

12.15 One alternative to using petroleum or natural gas reserves as fuels is ethanol (C_2H_5OH), produced from grain by fermentation. Consider a combustion process in which liquid ethanol is burned with 120% theoretical air in a SSSF process. The reactants enter the combustion chamber at 77 F and the products exit at 140 F, 14.7 lbf/in.2 Calculate the heat transfer per lb mol of ethanol, using the enthalpy of formation of liquid ethanol of $-119\,441$ Btu/lb mol.

12.16 Hydrogen peroxide (H_2O_2) enters a gas generator at the rate of 0.1 kg/s and is decomposed to steam and oxygen. The resulting mixture is expanded through a turbine to atmospheric pressure as shown in Fig. 12.8. Determine the power output of the turbine and the heat transfer rate in the gas generator. The enthalpy of formation of liquid H_2O_2 is $-187\,583$ kJ/kmol.

FIGURE 12.8
Sketch for Problem 12.16.

12.17 Propane gas at 77 F is mixed with 200% theoretical air at 620 F and burned in an adiabatic SSSF combustor. What is the temperature of the products exiting the combustor?

12.18 Repeat Problem 12.17 assuming that the fuel is liquid propane instead of gaseous propane. Use the generalized charts to determine the enthalpy difference.

12.19 How much error is introduced if the adiabatic flame temperature of Problem 12.17 is calculated assuming that all substances have constant specific heat, with values taken from Table A.8?

12.20 Sulfur at 25°C is burned with 50% excess air, which enters the combustion chamber at 400 K. Assuming that all the sulfur is burned to sulfur dioxide (SO_2), calculate the adiabatic flame temperature. The enthalpy of formation of SO_2 is $-296\,831$ kJ/kmol, and the constant-pressure specific heat is given by the relation

$$\bar{C}_{po} = 32.238 + 0.0219T - 3.527 \times 10^{-6}T^2$$

where $\bar{C}_{po} =$ kJ/kmol K and $T =$ K.

12.21 Consider the gas mixture fed to the combustors in an Integrated Gasification Combined Cycle power plant, as described in Problem 12.4. If the adiabatic flame temperature is to be limited to 1500 K, what percent theoretical air should be used in the combustors?

12.22 Acetylene gas at 1 atm, 77 F is fed to the head of a cutting torch. Calculate the adiabatic flame temperature if the acetylene is burned with:
(a) 100% theoretical air at 77 F.
(b) 100% theoretical oxygen at 77 F.

12.23 A mixture of 80% ethane and 20% methane, mole basis, is throttled from 10 MPa, 65°C, to 100 kPa and fed to a combustion chamber where it undergoes complete combustion with air, which enters at 100 kPa, 600 K. The amount of air is such that the products of combustion exit at 100 kPa, 1200 K. Assuming the combustion process to be adiabatic and that all components behave as ideal gases except the fuel mixture, which behaves according to Kay's rule, determine the following:
(a) The percent theoretical air used in the process.
(b) The dew-point temperature of the products.

12.24 A study is to be made using liquid ammonia as the fuel in a gas turbine engine. Consider the compressor (air) and combustion chamber of this engine.
(a) Air enters the compressor at 100 kPa, 25°C, and is compressed to 1400 kPa. The isentropic efficiency of the compressor is 87%. Determine the exit temperature and the work input per kilomole.
(b) Two kilomoles of liquid ammonia at 25°C and x times theoretical air from the compressor enter the combustion chamber. What is x if the adiabatic flame temperature is to be fixed at 1400 K?

12.25 Gaseous propane at 77 F is reacted with moist air (air containing water vapor) at 260 F in a steady-state, steady-flow process. The combustion process is adiabatic and the exiting temperature is observed to be 1700 F. A sample of the products is tested and found to have a dew-point temperature of 160 F. Determine the percent theoretical air used and the relative humidity of this air. Assume complete combustion and also that the pressure is 1 atm throughout the process.

12.26 Consider natural gas B listed in Table 12.2. Calculate the enthalpy of combustion of this gas at 77 F, assuming that the products include:
(a) Vapor water.
(b) Liquid water.

12.27 Determine the higher heating value of the sample of Wyoming coal as specified in Problem 12.6.

12.28 Blast furnace gas in a steel mill is available at 500 F to be burned for the generation of steam. The composition of this gas is, on a volumetric basis:

H_2	2.4%
CH_4	0.1
CO	23.3
CO_2	14.4
N_2	56.4
H_2O	3.4

Determine the higher heating value (Btu/ft^3) of this gas at ambient temperature and pressure.

12.29 (a) Determine the enthalpy of formation of liquid benzene at 25°C.
(b) Liquid benzene at 25°C is burned with air at 500 K in a SSSF process. The products of combustion are cooled to 1400 K and have the following volumetric analysis on a dry basis:

CO_2	10.7%
CO	3.6
O_2	5.3
N_2	80.4

Calculate the heat transfer per kilomole of fuel during the combustion process.

12.30 The enthalpy of formation of magnesium oxide, MgO(s), is $-601\ 827$ kJ/ kmol at 25°C. The melting point of magnesium oxide is approximately 3000 K, and the increase in enthalpy between 298 K and 3000 K is 128 449 kJ/kmol. The enthalpy of sublimation at 3000 K is estimated at 418 000 kJ/kmol, and the specific heat of magnesium oxide vapor above 3000 K is estimated at 37.24 kJ/kmol K.
(a) Determine the enthalpy of combustion per kilogram of magnesium.
(b) Estimate the adiabatic flame temperature when magnesium is burned with theoretical oxygen.

12.31 One lb mol of carbon at 77 F, 2 atm, and 2 lb mol of oxygen at 260 F, 2 atm, enter a combustion chamber, and the products exit at 2060 F, 2 atm. Assuming complete combustion and a steady-state steady-flow process, determine the following:
(a) The heat transfer per lb mol of fuel.
(b) The net entropy change for the process, if the heat is transferred to the surroundings at 77 F.

12.32 A mixture of butane and 180% theoretical air enters a combustion chamber at 25°C, 300 kPa, and the products of combustion exit the chamber at 1200 K, 300 kPa. Assuming complete combustion, determine the heat transfer from the combustion chamber and the irreversibility for the process, both per kilomole of butane.

12.33 Consider the combustion of methanol (CH_3OH) with 25% excess air, after which the combustion products are passed through a heat exchanger and exit at 200 kPa, 40°C. Calculate the absolute entropy of the products exiting the heat exchanger per kilomole of methanol burned.

12.34 Calculate the irreversibility for the process described in Problem 12.10.

12.35 Consider the combustion of propane gas at 100 kPa, 25°C, with theoretical air at 100 kPa, 25°C, in a SSSF process. Assume that combustion is complete and that the products exit at 25°C. Determine the decrease in Gibbs function for the two following cases:
(a) The reactants and products are both separated into their various constituents and each constituent is at 100 kPa, 25°C. This is shown schematically in Fig. 12.9.
(b) The reactants and products each consist of a mixture at a total pressure of 100 kPa and a temperature of 25°C. This is shown schematically in Fig. 12.10.

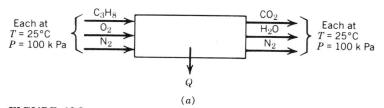

(a)

FIGURE 12.9

Sketch for Problem 12.35(a).

(b)

FIGURE 12.10

Sketch for Problem 12.35(b).

12.36 An inventor claims to have built a device that will take 10 lbm/h of water from the faucet at 50 F, 14.7 lbf/in.2 and produce separate streams of hydrogen and oxygen gas, each at 260 F, 25 lbf/in.2 He says that his device will operate in an 80 F room on 15-kW electrical power input. How do you evaluate this claim?

12.37 Two pound moles of ammonia are burned in a SSSF process with x lb mol of oxygen. The products, consisting of H_2O, N_2, and the excess O_2, exit at 400 F, 1000 lbf/in.2
(a) Calculate x if half the water in the products is condensed.
(b) Calculate the absolute entropy of the products at the exit conditions.

12.38 Consider one cylinder of a spark ignition, internal combustion engine. Before the compression stroke the cylinder is filled with a mixture of air and ethane. Assume that 150% theoretical air has been used, and that the pressure is 100 kPa and the temperature 25°C before compression. The compression ratio of the engine is 9 to 1.
(a) Determine the pressure and temperature after compression, assuming that the process is reversible and adiabatic. It may be assumed that the ethane behaves as an ideal gas.
(b) Assume that complete combustion is achieved while the piston remains at top dead center (that is, after the reversible adiabatic compression), and that the combustion process is adiabatic. Determine the temperature and pressure after combustion, and the increase in entropy during the combustion process.
(c) What is the irreversibility for this process?

12.39 Ammonia is to be produced by the process indicated in Fig. 12.11. A mixture of nitrogen and hydrogen at 25°C, 0.1 MPa, is compressed to the appropriate pressure, and enters the constant-pressure catalytic converter. Saturated liquid ammonia exits at 40°C. Calculate the heat transfer and the net entropy change for the overall process. The data for ideal gas ammonia at 25°C are:

$$\bar{h}_f^\circ = -46\ 191 \text{ kJ/kmol}$$
$$\bar{s}^\circ = 192.548 \text{ kJ/kmol K}$$
$$\bar{C}_{po} = 25.464 + 0.066\ 47T - 20.416 \times 10^{-6}T^2 \text{ kJ/kmol K}$$

FIGURE 12.11
Sketch for Problem 12.38.

12.40 Consider the combustion process described in Problem 12.23.
(a) Calculate the absolute entropy of the fuel mixture before it is throttled into the combustion chamber.
(b) Calculate the irreversibility for the overall process.

12.41 One lb mol of carbon (solid) and 1.5 lb mol of oxygen gas are contained in a rigid vessel at 77 F, 100 atm. The mixture is now ignited and the products, carbon dioxide plus the excess oxygen, are cooled to 120 F. Using Kay's rule to represent the behavior of the real gas mixture products, determine:
 (a) The final pressure.
 (b) The net entropy change, system plus surroundings, for the process.

12.42 Two kilomoles of ammonia at 25°C, 8 MPa, plus 200% theoretical oxygen at 25°C, 8 MPa, enter a SSSF combustion chamber. The products, a mixture of water, nitrogen, and oxygen, exit at 300°C, 8 MPa. Assuming that real gas mixture behavior is represented by Kay's rule and the generalized charts, calculate the following:
 (a) The heat transfer from the combustion chamber.
 (b) The net entropy change for the process.

12.43 Calculate the absolute entropy of saturated liquid carbon monoxide at 100 K.
 (a) Use the following information: Gas behavior—virial equation of state, second virial coefficient, Lennard–Jones (6–12) potential. Saturation—

$$\ln P_G = 13.82 - (753/T), \text{ kPa and K}$$

 (b) Compare the result of part (a) with the value found using the generalized charts.

12.44 A vertical cylinder fitted with a piston is divided into two parts, as shown in Fig. 12.12. Part A contains 1 kmol of saturated liquid butane at 320 K, and part B contains 100% theoretical oxygen at 275 K. External force on top of the piston maintains a pressure of 10 MPa on the oxygen. The partition separating A and B is now broken, and the two substances react to form carbon dioxide and water. During the reaction process, there is a heat transfer of 2.5×10^6 kJ out of the cylinder to the surroundings.
 (a) What is the final temperature? Initially, make a rough estimate of the temperature and use this value to decide on detailed models, assumptions, and approximations to be used in the final solution.
 (b) Calculate the absolute entropy of the cylinder contents at the final state.

FIGURE 12.12
Sketch for Problem 12.44.

12.45 In Example 12.17, a basic hydrogen–oxygen fuel cell reaction was analyzed at 25°C, 0.1 MPa. Repeat this calculation assuming that the fuel cell operates on air at 25°C, 0.1 MPa instead of on pure oxygen at this state.

12.46 Consider the hydrogen-air fuel cell shown in Fig. 12.13. The cell temperature is maintained at 540 F by rejecting heat to the surroundings. The following data have been taken:

$$\dot{W}_{elec.} = 3 \text{ kilowatts}$$

$T_1 = 77$ F	$P_1 = 100$ lbf/in.2	$\dot{V}_1 = 0.20$ ft^3/min
$T_2 = 77$ F	$P_2 = 100$ lbf/in.2	$\dot{V}_2 = 0.45$ ft^3/min
$T_3 = 540$ F	$P_3 = 90$ lbf/in.2	
$T_4 = 540$ F	$P_4 = 90$ lbf/in.2	$(y_{O_2})_4 = 0.05$

(a) Write the overall reaction equation, on a basis of lb mol/h for each substance.
(b) Calculate the heat transfer rate from the cell.
(c) Calculate the entropy flow at point 4.

FIGURE 12.13
Sketch for Problem 12.46.

12.47 Consider a methane–oxygen fuel cell in which the reaction at the anode is

$$CH_4 + 2H_2O \rightarrow CO_2 + 8e^- + 8H^+$$

The electrons produced by the reaction flow through the external load and the positive ions migrate through the electrolyte to the cathode, at which the reaction is

$$8e^- + 8H^+ + 2O_2 \rightarrow 4H_2O$$

(a) Calculate the reversible work and the reversible EMF for the methane–oxygen fuel cell operating at 25°C, 100 kPa.
(b) Repeat part (a), assuming that the fuel cell operates at 600 K instead of at room temperature.

12.48 A small air-cooled gasoline engine is tested, and the output is found to be 1.34 hp. The temperature of the products is measured and found to be 730 F. The products are analyzed with the following results, on a dry volumetric basis:

CO_2	11.4%
CO	2.9
O_2	1.6
N_2	84.1

The fuel may be considered to be liquid octane, and the fuel and air enter the engine at 77 F. The fuel flow rate to the engine is 1.20 lbm/h. Determine:

(a) The rate of heat transfer from the engine.

(b) The efficiency of the engine.

12.49 Consider the SSSF combustion of propane at 25°C with air at 400 K in which the products exit the combustion chamber at 1200 K. It may be assumed that the combustion efficiency is 90% and also that 95% of the carbon in the propane burns to form carbon dioxide, with the remaining 5% burning to form carbon monoxide. Determine the following:

(a) The ideal fuel–air ratio.

(b) The heat transfer from the combustion chamber.

12.50 Write a computer program to solve the following problem. It is desired to study the effect of percent theoretical air on the adiabatic flame temperature for a (variable) hydrocarbon fuel, assuming that the fuel and air enter the combustion chamber at 25°C, and also assuming complete combustion. Use equations from Table A.9 to represent the specific heat of the various products of combustion.

12.51 Write a computer program to solve the following problem. Repeat Problem 12.50, using polynomial equations for specific heat developed in the solution of Problem 5.85.

13
INTRODUCTION TO PHASE
AND CHEMICAL EQUILIBRIUM

In most of our considerations up to this point we have assumed that we are dealing either with systems that are in equilibrium or with those in which the deviation from equilibrium is infinitesimal, as in a quasiequilibrium or reversible process. For irreversible processes we made no attempt to describe the state of the system during the process but dealt only with the initial and final states of the system, at which time we considered the system to be in equilibrium.

In this chapter we examine the criteria for equilibrium and from them derive certain relations that will enable us, under certain conditions, to determine the properties of a system when it is in equilibrium. The specific systems we will consider are those involving equilibrium between phases and those involving chemical equilibrium in a single phase (homogeneous equilibrium) as well as certain related topics.

13.1 REQUIREMENTS FOR EQUILIBRIUM

As a general requirement for equilibrium we postulate that a system is in equilibrium when there is no possibility that it can do any work when it is isolated from its surroundings. In applying this criterion to a system it is helpful to divide the system into two or more subsystems, and consider the possibility of doing work by any conceivable interaction between these two subsystems. For example, in Fig. 13.1 a system has been divided into two systems and an engine, of any conceivable variety, placed between these subsystems. A system may be so defined as to include the immediate surroundings. In this case we can let the immediate surroundings be a subsystem and thus consider the general case of the equilibrium between a system and its surroundings.

The first requirement for equilibrium is that the two subsystems have the same temperature, for otherwise we could operate a heat engine between the two systems and do work. Thus we conclude that one requirement for equilibrium is that a system must be at a uniform temperature to be in equilibrium. It is also evident that there must be no unbalanced mechanical forces between the two systems, or else one could operate a turbine or piston engine between the two systems and do work.

However, we would like to establish general criteria for equilibrium that would apply to all simple compressible systems, including those that undergo chemical

517

FIGURE 13.1

Two subsystems that communicate through an engine.

reactions. We will find that the Gibbs function is a particularly significant property in defining the criteria for equilibrium.

Let us first consider a qualitative example to illustrate this point. Consider a natural gas well that is 1 km deep, and let us assume that the temperature of the gas is constant throughout the gas well. Suppose we have analyzed the composition of the gas at the top of the well, and we would like to know the composition of the gas at the bottom of the well. Furthermore, let us assume that equilibrium conditions prevail in the well. If this is true we would expect that an engine such as that shown in Fig. 13.2 (which operates on the basis of the pressure and composition change with elevation and does not involve combustion) would not be capable of doing any work.

If we consider a steady-state, steady-flow process for a control volume around this engine, we could apply Eq. 8.28:

$$\dot{W}_{rev} = \dot{m}_i\left(g_i + \frac{\mathbf{V}_i^2}{2} + gZ_i\right) - \dot{m}_e\left(g_e + \frac{\mathbf{V}_e^2}{2} + gZ_e\right)$$

FIGURE 13.2

Illustration showing the relation between reversible work and the criteria for equilibrium.

However,

$$\dot{W}_{rev} = 0, \qquad \dot{m}_i = \dot{m}_e \qquad \text{and} \qquad \frac{\mathbf{V}_i^2}{2} = \frac{\mathbf{V}_e^2}{2}$$

Then we can write

$$g_i + gZ_i = g_e + gZ_e$$

and the requirement for equilibrium in the well between two levels that are a distance dZ apart would be

$$dg_T + g\,dZ_T = 0$$

In contrast to a deep gas well, most of the system that we consider are of such size that ΔZ is negligibly small, and therefore we consider the pressure in the system to be uniform.

This leads to the general statement of equilibrium that applies to simple compressible systems that may undergo a change in chemical composition, namely, that at equilibrium

$$dG_{T,P} = 0 \qquad\qquad (13.1)$$

In the case of a chemical reaction, it is helpful to think of the equilibrium state as the state in which the Gibbs function is a minimum. For example, consider a system consisting initially of n_A moles of substance A and n_B moles of substance B, which react in accordance with the relation

$$v_A A + v_B B \rightleftharpoons v_C C + v_D D$$

Let the reaction take place at constant pressure and temperature. If we plot G for this system as a function of n_A, the number of moles of A present, we would have a curve as shown in Fig. 13.3. At the minimum point on the curve, $dG_{T,P} = 0$, and this will be the equilibrium composition for this system at the given temperature and pressure. The subject of chemical equilibrium will be developed further in Section 13.6.

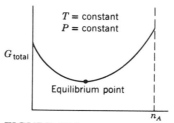

FIGURE 13.3

Illustration of the requirement for chemical equilibrium.

13.2 EQUILIBRIUM BETWEEN TWO PHASES OF A PURE SUBSTANCE

As another example of this requirement for equilibrium, let us consider the equilibrium between two phases of a pure substance. Consider a system consisting of two phases of a pure substance at equilibrium. We know that under these conditions the two phases are at the same pressure and temperature. Consider the change of state associated with a transfer of dn moles from phase 1 to phase 2 while the temperature and pressure remain constant. That is,

$$dn^1 = -dn^2$$

The Gibbs function of this system is given by

$$G = f(T, P, n^1, n^2)$$

where n^1 and n^2 designate the number of moles in each phase. Therefore,

$$dG = \left(\frac{\partial G}{\partial T}\right)_{P,n^1,n^2} dT + \left(\frac{\partial G}{\partial P}\right)_{T,n^1,n^2} dP + \left(\frac{\partial G}{\partial n^1}\right)_{T,P,n^2} dn^1 + \left(\frac{\partial G}{\partial n^2}\right)_{T,P,n^1} dn^2$$

By definition,

$$\left(\frac{\partial G}{\partial n^1}\right)_{T,P,n^2} = \bar{g}^1 \qquad \left(\frac{\partial G}{\partial n^2}\right)_{T,P,n^1} = \bar{g}^2$$

Therefore, at constant temperature and pressure,

$$dG = \bar{g}^1 \, dn^1 + \bar{g}^2 \, dn^2 = dn^1(\bar{g}^1 - \bar{g}^2)$$

Now at equilibrium (Eq. 13.1)

$$dG_{T,P} = 0$$

Therefore, at equilibrium, we have

$$\bar{g}^1 = \bar{g}^2 \qquad\qquad (13.2)$$

That is, under equilibrium conditions, the Gibbs function of each phase of a pure substance is equal. Let us check this by determining the Gibbs function of saturated liquid (water) and saturated vapor (steam) at 14.7 lbf/in.[2] From the steam tables:

For the liquid:

$$g_f = h_f - Ts_f = 180.16 - 671.7 \times 0.3123 = -29.5 \text{ Btu/lbm}$$

For the vapor:

$$g_g = h_g - Ts_g = 1150.5 - 671.7 \times 1.7567 = -29.5 \text{ Btu/lbm}$$

Equation 13.2 can also be derived by applying the relation

$$T \, ds = dh - v \, dP$$

to the change of phase that takes place at constant pressure and temperature. For this process this relation can be integrated as follows:

$$\int_f^g T\,ds = \int_f^g dh$$
$$T(s_g - s_f) = (h_g - h_f)$$
$$h_f - Ts_f = h_g - Ts_g$$
$$g_f = g_g$$

The Clapeyron equation, which was derived in Section 10.3, can be derived by an alternate method by considering the fact that the Gibbs functions of two phases in equilibrium are equal. In Chapter 10 we considered the relation (Eq. 10.9) for a simple compressible substance:

$$dg = v\,dP - s\,dT$$

Consider a system that consists of a saturated liquid and a saturated vapor in equilibrium, and let this system undergo a change of pressure dP. The corresponding change in temperature, as determined from the vapor-pressure curve, is dT. Both phases will undergo the change in Gibbs function, dg, but since the phases always have the same value of the Gibbs function when they are in equilibrium, it follows that

$$dg_f = dg_g$$

But, from Eq. 10.9,

$$dg = v\,dP - s\,dT$$

it follows that

$$dg_f = v_f\,dP - s_f\,dT$$
$$dg_g = v_g\,dP - s_g\,dT$$

Since

$$dg_f = dg_g$$

it follows that

$$v_f\,dP - s_f\,dT = v_g\,dP - s_g\,dT$$
$$dP(v_g - v_f) = dT(s_g - s_f) \tag{13.3}$$
$$\frac{dP}{dT} = \frac{s_{fg}}{v_{fg}} = \frac{h_{fg}}{Tv_{fg}}$$

In summary, when different phases of a pure substance are in equilibrium, each phase has the same value of the Gibbs function per unit mass. This fact is relevant to different solid phases of a pure substance and is important in metallurgical applications of thermodynamics. Example 13.1 illustrates this principle.

Example 13.1

What pressure is required to make diamonds from graphite at a temperature of 25°C? The following data are given for a temperature of 25°C and a pressure of 0.1 MPa.

	Graphite	**Diamond**
\bar{g}	0	2867.8 kJ/kmol
v	0.000 444 m³/kg	0.000 284 m³/kg
β_T	0.304×10^{-6} 1/MPa	0.016×10^{-6} 1/MPa

Analysis and Solution

The basic principle in the solution is that graphite and diamond can exist in equilibrium when they have the same value of the Gibbs function. At 0.1 MPa pressure the Gibbs function of the diamond is greater than that of the graphite. However, the rate of increase in Gibbs function with pressure is greater for the graphite than for the diamond and, therefore, at some pressure they can exist in equilibrium, and our problem is to find this pressure.

We have already considered the relation

$$dg = v\, dP - s\, dT$$

Since we are considering a process that takes place at constant temperature, this reduces to

$$dg_T = v\, dP_T \tag{a}$$

Now at any pressure P and the given temperature, the specific volume can be found from the following relation, which utilizes the isothermal compressibility factor.

$$v = v° + \int_{P=0.1}^{P} \left(\frac{\partial v}{\partial P}\right)_T dP = v° + \int_{P=0.1}^{P} \frac{v}{v}\left(\frac{\partial v}{\partial P}\right)_T dP$$

$$= v° - \int_{P=0.1}^{P} v\beta_T\, dP \tag{b}$$

The superscript ° will be used in this example to indicate the properties at a pressure of 0.1 MPa and a temperature of 25°C.

The specific volume changes only slightly with pressure, so that $v \approx v°$. Also, we assume that β_T is constant and that we are considering a very high pressure. With these assumptions this equation can be integrated to give

$$v = v° - v°\beta_T P = v°(1 - \beta_T P) \tag{c}$$

We can now substitute this into Eq. a to give the relation

$$dg_T = [v°(1 - \beta_T P)]\, dP_T$$

$$g - g° = v°(P - P°) - v°\beta_T \frac{(P^2 - P°^2)}{2} \tag{d}$$

If we assume that $P° \ll P$ this reduces to

$$g - g° = v°\left(P - \frac{\beta_T P^2}{2}\right)$$ (e)

For the graphite, $g° = 0$ and we can write

$$g_G = v_G°\left[P - (\beta_T)_G \frac{P^2}{2}\right]$$

For the diamond, $g°$ has a definite value and we have

$$g_D = g_D° + v_D°\left[P - (\beta_T)_D \frac{P^2}{2}\right]$$

But, at equilibrium the Gibbs function of the graphite and diamond are equal:

$$g_G = g_D$$

Therefore,

$$v_G°\left[P - (\beta_T)_G \frac{P^2}{2}\right] = g_D° + v_D°\left[P - (\beta_T)_D \frac{P^2}{2}\right]$$

$$(v_G° - v_D°)P - [v_G°(\beta_T)_G - v_D°(\beta_T)_D]\frac{P^2}{2} = g_D°$$

$$(0.000\,444 - 0.000\,284)P$$

$$- (0.000\,444 \times 0.304 \times 10^{-6} - 0.000\,284 \times 0.016 \times 10^{-6})P^2/2 = \frac{2867.8}{12.011 \times 1000}$$

Solving this for P we find

$$P = 1493 \text{ MPa}$$

That is, at 1493 MPa, 25°C, graphite and diamond can exist in equilibrium, and the possibility exists for conversion from graphite to diamonds.

The preceding example could also have been evaluated in terms of the fugacities instead of the Gibbs functions for the two forms of carbon. Similarly, the equilibrium requirement for liquid and vapor water discussed earlier could be expressed in terms of the fugacities of the two phases. That is,

$$f^L = f^V = f^{sat}$$

where f^{sat} is the fugacity of the substance determined at the given temperature and its saturation pressure P^{sat}.

The fugacity of a pure compressed liquid or solid phase at pressures moderately greater than the saturation pressure can be calculated with considerable accuracy as follows. For a pure substance at constant temperature,

$$dg_T = RT(d \ln f)_T = v\, dP_T$$

Integrating at constant temperature between the saturation state and the pressure P, we have

$$\int_{f^{\text{sat}}}^{f} RT(d \ln f)_T = \int_{P^{\text{sat}}}^{P} v \, dP_T$$

$$RT \ln \frac{f}{f^{\text{sat}}} = \int_{P^{\text{sat}}}^{P} v \, dP_T$$

If we assume that v is a constant, which often holds with considerable accuracy for liquids and solids, we have

$$RT \ln \frac{f}{f^{\text{sat}}} \approx v(P - P^{\text{sat}}) \tag{13.4}$$

Equation 13.4 is used to determine the fugacity coefficients of compressed liquid presented in the generalized chart given in Appendix Fig. A.7.

Since v is small for the liquid and solid phases, the quantity $v(P - P^{\text{sat}})$ is small for moderate changes of pressure. Therefore, for liquids and solids at moderate pressures, we conclude that

$$f^L \approx f^{\text{sat}} \quad \text{and} \quad f^S \approx f^{\text{sat}} \tag{13.5}$$

13.3 EQUILIBRIUM OF A MULTICOMPONENT, MULTIPHASE SYSTEM

As an introduction to a consideration of equilibrium in multicomponent, multiphase systems, let us consider a system consisting of two components and two phases. To show the general characteristics of such a system, consider a mixture of oxygen and nitrogen at a pressure of 0.1 MPa and in the range of temperature where both the liquid and vapor phases are present.

Figure 13.4 shows the composition of the liquid and vapor phases that are in equilibrium as a function of temperature at a pressure of 0.1 MPa. The upper line, marked "vapor line," gives the composition of the vapor phase, and the lower line gives the composition of the liquid phase. If we have pure nitrogen, the boiling point is 77.3 K. If we have pure oxygen, the boiling point is 90.2 K. If we have a mixture of nitrogen and oxygen with both phases present at a temperature of 84 K, the vapor will have a composition of 64% nitrogen and 36% oxygen. The liquid will have a composition of 30% nitrogen and 70% oxygen.

Consider a mixture consisting of 21% oxygen and 79% nitrogen (approximately the composition of air) and at a pressure of 0.1 MPa and an initial temperature of 74 K. At this state the mixture will be in the liquid phase. Let this liquid be slowly heated while the pressure remains constant. By referring to Fig. 13.4 we note that when the temperature reaches 78.8 K, the first bubble will form. This vapor will have a composition of approximately 6% oxygen and 94% nitrogen. As more heat is added, the temperature increases and the mole fraction of oxygen in the liquid increases. When the last drop of liquid remains, the vapor will have a

FIGURE 13.4

Equilibrium diagram for liquid–vapor phases of the nitro-
gen–oxygen system at a pressure of 0.1 MPa.

composition of 21% oxygen and 79% nitrogen, the temperature will be 81.9 K, and
the liquid composition will be 54% oxygen and 46% nitrogen.

To understand the phase behavior of a system such as that described here, it
may be helpful to briefly discuss the more general phase diagram. We recall that
the phase behavior for a pure substance is dictated by the saturation line on $P-T$
coordinates, such as in Fig. 3.2. For a two-component system A and B, overall
composition is now an additional independent property, and phase behavior is
represented in terms of surfaces on a pressure–composition–temperature diagram,
as shown in Fig. 13.5a. This diagram includes three constant-temperature planes
to indicate that there is a lower vapor surface and an upper liquid surface that
together form an equilibrium envelope between the two pure substance saturation
lines. Only within this envelope is it possible to have two phases present in equi-
librium. Beneath the envelope the two-component system is all superheated vapor
and above the envelope the system is all compressed liquid.

We realize now that a constant-pressure plane (0.1 MPa) of the three-
dimensional phase diagram, as viewed from above, results in the temperature–
composition diagram, Fig. 13.4, discussed in some detail previously. The other
two projections of such a diagram are as shown in Figs. 13.5b and 13.5c.

Let us now consider the situation in a two-component system as shown in
Fig. 13.5 when the temperature is increased to a value greater than the critical

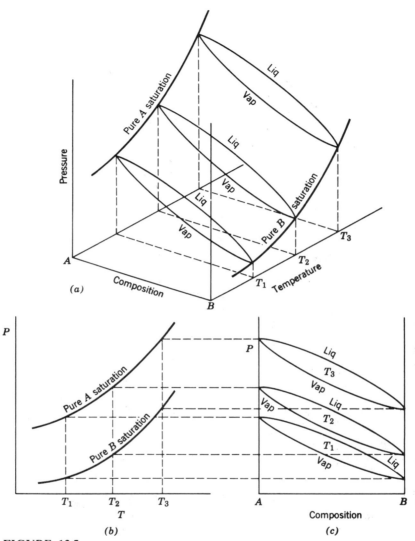

FIGURE 13.5

Phase diagram for a two-component system.

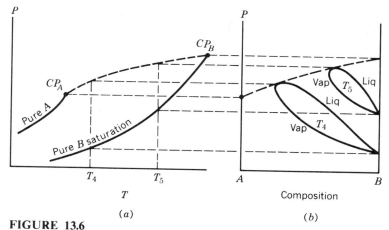

FIGURE 13.6

Two-component system in the critical region.

temperature of one of the components. The pressure–temperature and pressure–composition projections now appear as shown in Figs. 13.6a and 13.6b for the two temperatures T_4 and T_5. We note that as the temperature is increased toward the critical temperature of component B, there is an increasing range of composition for which the two-phase envelope is not intersected as the pressure is increased at a constant temperature.

In studying the characteristics of a two-component system, we must first ask the question, "What is the requirement for phase equilibrium in this system?" We can proceed in a manner analogous to that followed in Section 13.2 for a pure substance, with superscripts 1 and 2 denoting the two phases, and in this case subscripts A and B referring to the two components. It follows from applying Eq. 11.57 to each phase that

$$dG^1 = -S^1\,dT + V^1\,dP + \mu_A{}^1 dn_A{}^1 + \mu_B{}^1\,dn_B{}^1$$
$$dG^2 = -S^2\,dT + V^2\,dP + \mu_A{}^2 dn_A{}^2 + \mu_B{}^2\,dn_B{}^2 \qquad \textbf{(13.6)}$$

Let us consider this two-phase, two-component mixture that is in equilibrium as a system, and let each phase be considered as a subsystem. One possible change of state that might occur is for a very small amount of component A to be transferred from phase 1 to phase 2 while the moles of B in each phase and the temperature and pressure remain constant. This is shown schematically in Fig. 13.7.

For this change of state

$$dn_A{}^2 = -dn_A{}^1 \qquad \textbf{(13.7)}$$

Since this system is in equilibrium, $dG = 0$. Therefore,

$$dG = dG^1 + dG^2 = 0 \qquad \textbf{(13.8)}$$

T and P
remain constant

FIGURE 13.7

An equilibrium mixture involving two components and two phases.

Since T, P, $n_B{}^1$, and $n_B{}^2$ are all constant, it follows from Eqs. 13.6 and 13.8 that

$$dG = \mu_A{}^1 \, dn_A{}^1 + \mu_A{}^2 \, dn_A{}^2$$
$$= \mu_A{}^1 \, dn_A{}^1 - \mu_A{}^2 \, dn_A{}^1 = dn_A{}^1(\mu_A{}^1 - \mu_A{}^2)$$
$$= 0$$

Therefore, at equilibrium

$$\mu_A{}^1 = \mu_A{}^2 \tag{13.9}$$

Thus the requirement for equilibrium is that the chemical potential of each component is the same in all phases. If the chemical potential is not the same in all phases, there will be a tendency for mass to pass from one phase to the other. When the chemical potential of a component is the same in both phases, there is no tendency for a net transfer of mass from one phase to the other.

This requirement for equilibrium is readily extended to multicomponent, multiphase systems, for the chemical potential of each component must be the same in all phases.

$$\mu_A{}^1 = \mu_A{}^2 = \mu_A{}^3 = \cdots \text{ for all phases}$$
$$\mu_B{}^1 = \mu_B{}^2 = \mu_B{}^3 = \cdots \text{ for all phases}$$

$$\tag{13.10}$$

for all components

Let us now consider alternate means of expressing the requirement for phase equilibrium at a given temperature and pressure. We found in Section 11.11, Eq. 11.58, that the chemical potential of component A is identical to the partial Gibbs

function \bar{G}_A of that component in the mixture. Thus, the requirement for equilibrium, Eq. 13.9, can also be expressed as

$$\bar{G}_A{}^1 = \bar{G}_A{}^2 \tag{13.11}$$

That is, at equilibrium the partial Gibbs function of a given component is the same in each phase. In this form the result is analogous to that for pure substance phase equilibrium, Eq. 13.2. It also follows from the definition of fugacity of a component in a mixture, Eqs. 11.79 and 11.80,

$$(d\bar{G}_A) = \bar{R}T d(\ln \bar{f}_A)_T$$
$$\lim_{P \to 0} (\bar{f}_A / y_A P) = 1$$

that the requirement for phase equilibrium at a given temperature and pressure, Eq. 13.9, can be expressed in terms of fugacity as

$$\bar{f}_A{}^1 = \bar{f}_A{}^2 \tag{13.12}$$

That is, at equilibrium the fugacity of a given component is the same in each phase. This is perhaps the most useful formulation of the requirement for equilibrium between phases at a given temperature and pressure, as we have previous experience in evaluating fugacities. A more general concept, however, is that of activity, which was defined and discussed in Section 11.15. Thus, the definition of activity, Eq. 11.111, could be substituted into Eq. 13.12 for the fugacity in each phase.

We find that the requirement for equilibrium can be expressed in numerous ways, in terms of different thermodynamic variables. The question of interest to us at this point is whether we can use such expressions to predict the equilibrium compositions of the different phases from the known properties of the pure substances that make up the system. To be able to do this, we need to utilize a model for each phase. In this section we will consider two models, namely, the Ideal Solution model and the Raoult's Rule–Ideal Gas model. For each model, we will restrict our discussion to a two-phase (liquid and vapor), two-component system as shown in Fig. 13.7. The results of this analysis are then readily extended to multiphase or multicomponent systems. In the analysis, we will distinguish liquid phase mole fractions from those for the vapor phase by using the symbol x for the liquid phase.

Ideal Solution

A frequently used mixture model is the Ideal Solution, in which the fugacity of component A in a mixture is expressed as the product of the mole fraction of A and the fugacity of pure A in the same phase as the mixture and at the pressure and temperature of the mixture (and similarly for component B). These specified conditions for pure A may very well result in a hypothetical state. The Ideal Solution has been discussed in detail in Section 11.14, but for our purposes here we need only be familiar with the fugacity equation described in Eq. 11.102.

Let us assume that both the liquid and vapor phases of a two-component system behave according to the Ideal Solution model. The requirement for equilibrium for component A is, from Eqs. 13.12 and 11.102,

$$x_A f_A{}^L = y_A f_A{}^V \tag{13.13}$$

Similarly, for component B,

$$x_B f_B{}^L = y_B f_B{}^V \tag{13.14}$$

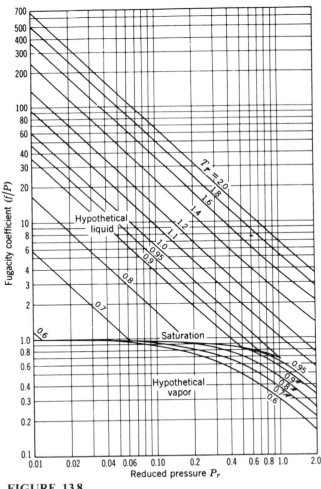

FIGURE 13.8

Hypothetical liquid and vapor fugacities.

We also realize that for the liquid phase

$$x_A + x_B = 1 \qquad (13.15)$$

and for the vapor phase

$$y_A + y_B = 1 \qquad (13.16)$$

Now, at a given temperature and pressure, the four pure substance fugacities in Eqs. 13.13 and 13.14 are fixed values. Therefore, Eqs. 13.13 through 13.16 comprise a set of four equations in four unknowns, which can be solved for the equilibrium composition of the liquid and vapor phases.

The problem that arises in evaluating the Ideal Solution model is that connected with hypothetical states. This is most easily recognized by referring to the phase diagram, Fig. 13.5b. For an equilibrium two-phase mixture of A and B, the system pressure P is less than $P_A{}^{\text{sat}}$ and greater than $P_B{}^{\text{sat}}$ at the given temperature T. We realize from examining Fig. 13.5b that at the conditions of P and T, pure A is a superheated vapor and pure B is a compressed liquid. As a result, the fugacities $f_A{}^V$ and $f_B{}^L$ are readily found by conventional methods, as for example from Fig. A.7. However, the fugacities $f_A{}^L$ and $f_B{}^V$ cannot be found, as pure liquid A and pure vapor B do not exist at the conditions of P and T. These are the hypothetical states. To determine the fugacities for these hypothetical states, vapor-phase fugacities have been extrapolated into the liquid region and liquid-phase fugacities extrapolated into the vapor region in a manner so as to give reasonable generalized correlations from Eqs. 13.13 and 13.14 with experimental phase equilibrium data. These extrapolated hypothetical values are presented in Fig. 13.8.

The following example should help to clarify the determination of fugacities in the Ideal Solution model.

Example 13.2

Calculate the phase compositions in the liquid-vapor system methane-carbon monoxide at -150 F, 600 lbf/in.2, assuming Ideal Solution in both phases.

Analysis and Solution

Let CO be component A and CH_4 be component B. Using critical point properties from Table A.7.

$$T_{rA} = \frac{310}{240} = 1.293, \qquad P_{rA} = \frac{600}{507} = 1.182$$

$$T_{rB} = \frac{310}{343.9} = 0.901, \qquad P_{rB} = \frac{600}{673} = 0.891$$

From Fig. A.7, it is seen that pure A is a gas and pure B is a liquid at the given state. Therefore, from Fig. A.7,

$$f_A{}^V = 0.85 \times 600 = 510$$
$$f_B{}^L = 0.45 \times 600 = 270$$

Pure liquid A and pure gaseous B are the hypothetical states. From Fig. 13.8,

$$f_A{}^L = 1.6 \times 600 = 960$$
$$f_B{}^V = 0.6 \times 600 = 360$$

Substituting these values into the equilibrium equations, Eqs. 13.13 and 13.14, and solving,

$$x_A = \frac{510}{960}\, y_A = 0.531 y_A$$

$$x_B = \frac{360}{270}\, y_B = 1.333 y_B$$

$$x_A + x_B = 0.531 y_A + 1.333(1 - y_A) = 1$$
$$y_A = 0.415, \qquad x_A = 0.221$$

When the temperature is greater than the critical temperature of one of the components, the situation shown in Fig. 13.6, the hypothetical liquid-phase fugacity can become very large, as can be seen from the values for T_r greater than unity in Fig. 13.8. The result of a very large value for $f_A{}^L$ is to drive the liquid-phase mole fraction x_A to a very small value. In an extreme case, the liquid phase becomes essentially pure B, and Eqs. 13.13 and 13.15 drop out of the equilibrium set, leaving Eqs. 13.14 and 13.16 to specify the two gas-phase mole fractions at a given temperature and pressure.

Raoult's Rule–Ideal Gas

This model is a special case of the Ideal Solution, and while not as accurate, it is nevertheless quite reasonable for the behavior of many systems at low pressure. For the liquid phase, we make two simplifying assumptions:

1. The fugacity of pure liquid A at T and P of the system is equal to the fugacity of saturated A (liquid or vapor) at the same T, and its corresponding saturation pressure $P_A{}^{\text{sat}}$, or

$$f_A{}^L = f_A{}^{\text{sat}} \tag{13.17}$$

This is equivalent to assuming that the $\int v\, dP$ correction of Eq. 13.4 is negligibly small.

2. Pure saturated vapor A at T and $P_A{}^{\text{sat}}$ behaves as an ideal gas, or

$$f_A{}^{\text{sat}} = P_A{}^{\text{sat}} \tag{13.18}$$

Combining Eqs. 13.17 and 13.18, we obtain, for the two components A and B,

$$f_A{}^L = P_A{}^{\text{sat}}, \qquad f_B{}^L = P_B{}^{\text{sat}} \tag{13.19}$$

which, when combined with the Ideal Solution model, Eq. 11.102, is termed Raoult's Rule.

For the vapor phase, we assume that pure gas A and pure gas B behave as ideal gases at T and P, so that

$$f_A^V = P, \qquad f_B^V = P \qquad (13.20)$$

Now, substituting Eqs. 13.19 and 13.20 for each component into the result for the Ideal Solution model, Eqs. 13.13 and 13.14, we obtain the result for the Raoult's Rule–Ideal Gas model:

$$x_A P_A^{sat} = y_A P \qquad (13.21)$$
$$x_B P_B^{sat} = y_B P \qquad (13.22)$$

which together with

$$x_A + x_B = 1 \qquad (13.23)$$
$$y_A + y_B = 1 \qquad (13.24)$$

forms a set of four equations in four unknowns at a given T and P (P_A^{sat} and P_B^{sat} depend only on T). This set of equations can then be solved to determine the equilibrium compositions in each phase.

The three additional assumptions made in this model should clearly indicate its limitations as compared with the Ideal Solution, which is the more accurate, especially at higher pressures.

Example 13.3

Air (assumed to be 21% O_2, 79% N_2) is cooled to 80 K, 0.1 MPa pressure. Calculate the composition of the liquid and vapor phases at this condition, assuming the Raoult's Rule–Ideal Gas model, and compare the results with Fig. 13.4. At 80 K,

$$P_{N_2}^{sat} = 0.1370 \text{ MPa}, \qquad P_{O_2}^{sat} = 0.030\,06 \text{ MPa}$$

Analysis and Solution

For convenience let $N_2 = A$ and $O_2 = B$ in the calculations. Using Eqs. 13.21 and 13.22,

$$x_A P_A^{sat} = y_A P$$
$$x_B P_B^{sat} = y_B P$$

Therefore, at 80 K, 0.1 MPa pressure,

$$x_A(0.137) = y_A(0.1)$$
$$x_B(0.030\,06) = y_B(0.1)$$

But,

$$x_A + x_B = 1$$
$$y_A + y_B = 1$$

Substituting for y_A, y_B in the last equation,

$$\frac{0.137}{0.1} x_A + \frac{0.030\,06}{0.1} x_B = 1.37x_A + 0.3006(1 - x_A) = 1$$

$$x_A = 0.654, \qquad y_A = 0.896$$

From Fig. 13.4, which is based on experimental data, we find that at 80 K, 0.1 MPa,

$$x_A = 0.66, \qquad y_A = 0.89$$

Thus we conclude that for this system at this temperature and pressure, Raoult's Rule gives quite accurate results.

13.4 THE GIBBS PHASE RULE (WITHOUT CHEMICAL REACTION)

The Gibbs phase rule, which was derived by Professor J. Willard Gibbs of Yale University in 1875, ranks among the truly significant contributions to physical science. In this section we consider the Gibbs phase rule for a system that does not involve a chemical reaction. For such a system the Gibbs phase rule is

$$\mathscr{P} + \mathscr{V} = \mathscr{C} + 2 \tag{13.25}$$

where \mathscr{P} is the number of phases present, \mathscr{V} is the variance, and \mathscr{C} the number of components present. The term variance designates the number of intensive properties that must be specified to completely fix the state of the system. For example, in Example 13.3 we considered a two-phase mixture of oxygen and nitrogen. For this system the number of phases \mathscr{P} equals 2, the number of components \mathscr{C} is 2, and therefore the variance \mathscr{V} is 2. This is evident from Fig. 13.4, for this diagram is for a fixed pressure (0.1 MPa), and the fixing of one additional intensive property, such as temperature, mole fraction of a given component in the liquid phase, or mole fraction of a given component in the vapor phase, will completely determine all other intensive properties and thus fix the state of the system (although not the relative amounts of the two phases.)

Let us consider further the application of the Gibbs phase rule to a pure substance. In this case $\mathscr{C} = 1$. When we have one phase present, such as superheated vapor, $\mathscr{P} = 1$, and we conclude that $\mathscr{V} = \mathscr{C} + 2 - 1 = 2$. That is, two intensive properties must be specified to fix the state. We are already familiar with the superheated vapor tables for a number of substances and recognize that these tables are presented with pressure and temperature as the two independent properties. Such a system is referred to as a bivariant system.

Suppose we have two phases of a pure substance in equilibrium, such as saturated liquid and saturated vapor. In this case $\mathscr{C} = 1$, $\mathscr{P} = 2$, and $\mathscr{V} = 1 + 2 - 2 = 1$. That is, one intensive property fixes the value of all other intensive properties and thus determines the state of each phase. Again, from our familiarity with tables

of thermodynamic properties, we recall that either pressure or temperature is used as the independent intensive property for tabulating liquid–vapor equilibrium data for a pure substance. Such a system is known as monovariant.

Consider also the triple point of a pure substance. In this case $\mathscr{C} = 1$ and $\mathscr{P} = 3$. Therefore $\mathscr{V} = 1 + 2 - 3 = 0$. That is, at the triple point all intensive properties are fixed, and if any intensive property is varied, we are no longer at the triple point. This is known as an invariant system.

It might be well at this point to again emphasize that a pure substance can have several phases in the solid state. For example, Fig. 3.6 shows the various phases of water. Note that there are several states at which three phases exist in equilibrium, and each of these is a triple point.

The application of the phase rule to a two-component, two-phase system has already been made. As a final application let us consider a two-component, three-phase system. For such a system $\mathscr{V} = 2 + 2 - 3 = 1$. That is, it is a monovariant system, and specifying one property, such as pressure or temperature, fixes the state of the system.

The validity of the Gibbs phase rule can be outlined by considering a system consisting of \mathscr{C} components and \mathscr{P} phases in equilibrium at a given temperature and pressure. Assuming that each component exists in each phase, the state of the system could be completely specified if the concentration of each component in each phase and the temperature and pressure were specified. This would be a total of $\mathscr{C}\mathscr{P} + 2$ intensive properties.

We know, however, that these are not all independent intensive properties. If we determine the number of equations we have between these intensive properties, we can subtract this from the $\mathscr{C}\mathscr{P} + 2$ intensive properties and find the number of independent intensive properties or, as it has been defined, the variance. The fact that at equilibrium the chemical potential of each component is the same in all phases gives rise to a total of $\mathscr{C}(\mathscr{P} - 1)$ equations between the intensive properties of the mixture.

Furthermore, the fact that the sum of the mole fractions equals unity in each of the \mathscr{P} phases gives \mathscr{P} additional equations. Therefore the variance is

$$\mathscr{V} = \mathscr{C}\mathscr{P} + 2 - \mathscr{P} - \mathscr{C}(\mathscr{P} - 1) = \mathscr{C} + 2 - \mathscr{P}$$

13.5 METASTABLE EQUILIBRIUM

Although the limited scope of this book precludes an extensive treatment of metastable equilibrium, a brief introduction to the subject is presented in this section. Let us first consider an example of metastable equilibrium.

Consider a slightly superheated vapor, such as steam, expanding in a convergent-divergent nozzle, as shown in Fig. 13.9. Assuming the process is reversible and adiabatic, the steam will follow path 1-a on the T-s diagram, and at point a we would expect condensation to occur. However, if point a is reached in the divergent section of the nozzle, it is observed that no condensation occurs until point b is reached, and at this point the condensation occurs very abruptly in what is referred

FIGURE 13.9

Illustration of the phenomenon of supersaturation in a nozzle.

to as a condensation shock. Between points a and b the steam exists as a vapor, but the temperature is below the saturation temperature for the given pressure. This is known as a metastable state. The possibility of a metastable state exists with any phase transformation. The dotted lines on the equilibrium diagram shown in Fig. 13.10 represent possible metastable states for solid–liquid–vapor equilibrium.

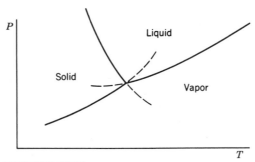

FIGURE 13.10

Metastable states for solid–liquid–vapor equilibrium.

FIGURE 13.11

Schematic diagram illustrating a metastable state.

The nature of a metastable state is often pictured schematically by the kind of diagram shown in Fig. 13.11. The ball is in a stable position (the "metastable state") for small displacements, but with a large displacement it moves to a new equilibrium position. The steam expanding in the nozzle is in a metastable state between a and b. This means that droplets smaller than a certain critical size will re-evaporate, and only when droplets of larger than this critical size have formed (this corresponds to moving the ball out of the depression) will the new equilibrium state appear.

13.6 CHEMICAL EQUILIBRIUM

We now turn our attention to chemical equilibrium and consider first a chemical reaction involving only one phase. This is referred to as a homogeneous chemical reaction. It may be helpful to visualize this as a gaseous phase, but the basic considerations apply to any phase.

Consider a vessel, Fig. 13.12, that contains four compounds, A, B, C, and D, which are in chemical equilibrium at a given pressure and temperature. For example, these might consist of CO_2, H_2, CO, and H_2O in equilibrium. Let the number

Components
A, B, C, D
in chemical
equilibrium

FIGURE 13.12

Schematic diagram for consideration of chemical equilibrium.

of moles of each component be designated n_A, n_B, n_C, and n_D. Further, let the chemical reaction that takes place between these four constituents be

$$v_A A + v_B B \rightleftharpoons v_C C + v_D D \tag{13.26}$$

where the v's are the stoichiometric coefficients. It should be emphasized that there is a very definite relation between the v's (the stoichiometric coefficients), whereas the n's (the number of moles present) for any constituent can be varied simply by varying the amount of that component in the reaction vessel.

Let us now consider how the requirement for equilibrium, namely, that $dG_{T,P} = 0$ at equilibrium, applies to a homogeneous chemical reaction. When we considered phase equilibrium (Section 13.3) we proceeded by assuming that the two phases were in equilibrium at a given temperature and pressure, and then let a small quantity of one component be transferred from one phase to the other. In a similar manner, let us assume that the four components are in chemical equilibrium and then assume that from this equilibrium state, while the temperature and pressure remain constant, the reaction proceeds an infinitesimal amount toward the right as Eq. 13.26 is written. This results in a decrease in the moles of A and B and an increase in the moles of C and D. Let us designate the degree of reaction by ϵ, and define the degree of reaction by the relations

$$\begin{aligned} dn_A &= -v_A \, d\epsilon \\ dn_B &= -v_B \, d\epsilon \\ dn_C &= +v_C \, d\epsilon \\ dn_D &= +v_D \, d\epsilon \end{aligned} \tag{13.27}$$

That is, the change in the number of moles of any component during a chemical reaction is given by the product of the stoichiometric coefficients (the v's) and the degree of reaction.

Let us evaluate the change in the Gibbs function associated with this chemical reaction that proceeds to the right in the amount $d\epsilon$. In doing so we use, as would be expected, the Gibbs function of each component in the mixture—the partial molal Gibbs function (or its equivalent, the chemical potential):

$$dG_{T,P} = \bar{G}_C \, dn_C + \bar{G}_D \, dn_D + \bar{G}_A \, dn_A + \bar{G}_B \, dn_B$$

Substituting Eq. 13.27, we have

$$dG_{T,P} = (v_C \bar{G}_C + v_D \bar{G}_D - v_A \bar{G}_A - v_B \bar{G}_B) \, d\epsilon \tag{13.28}$$

We have already considered (Section 11.15) how to evaluate the partial molal Gibbs function of a given component in terms of the Gibbs function of the pure component at the standard state and the activity of that component. For the component i, we can write (Eq. 11.112)

$$\bar{G}_i = \bar{g}_i^\circ + \bar{R}T \ln a_i$$

Substituting this relation into Eq. 13.28 we have

$$\begin{aligned} dG_{T,P} = [&v_C(\bar{g}_C^\circ + \bar{R}T \ln a_C) + v_D(\bar{g}_D^\circ + \bar{R}T \ln a_D) \\ &- v_A(\bar{g}_A^\circ + \bar{R}T \ln a_A) - v_B(\bar{g}_B^\circ + \bar{R}T \ln a_B)] \, d\epsilon \end{aligned} \tag{13.29}$$

Let us define ΔG° as follows:

$$\Delta G^\circ = \nu_C \bar{g}_C^\circ + \nu_D \bar{g}_D^\circ - \nu_A \bar{g}_A^\circ - \nu_B \bar{g}_B^\circ \qquad (13.30)$$

That is, ΔG° is the change in the Gibbs function that would occur if the chemical reaction given by Eq. 13.26 (which involves the stoichiometric amounts of each component) proceeded completely from left to right, with the reactants A and B initially separated and at temperature T and the standard state pressure and the products C and D finally separated and at temperature T and the standard state pressure. Note also that ΔG° for a given reaction is a function of only the temperature. This will be most important to bear in mind as we proceed with our developments of homogeneous chemical equilibrium. Let us now digress from our development to consider an example involving the calculation of ΔG°.

Example 13.4

Determine the value of ΔG° for the reaction $2H_2O \rightleftharpoons 2H_2 + O_2$ at 298 K and at 2000 K, with the water in the gaseous phase.

Solution

We take 0.1 MPa as our standard-state pressure, and recall that we assume further that \bar{g} for all the elements is zero at 0.1 MPa pressure and 298 K. Therefore, at 298 K,

$$(\bar{g}_f^\circ)_{H_2} = 0, \qquad (\bar{g}_f^\circ)_{O_2} = 0$$

From Table A.13, at 298 K,

$$(\bar{g}_f^\circ)_{H_2O} = -228\,583 \text{ kJ/kmol}$$
$$\Delta G^\circ = 2(\bar{g}_f^\circ)_{H_2} + (\bar{g}_f^\circ)_{O_2} - 2(\bar{g}_f^\circ)_{H_2O}$$
$$= 0 + 0 - 2(-228\,583) = 457\,166 \text{ kJ}$$

At 2000 K, from Table A.11,

$$\bar{g}_{H_2}^\circ = \bar{g}_{2000}^\circ - \bar{g}_{298}^\circ = (\bar{h}_{2000}^\circ - \bar{h}_{298}^\circ) - (2000\bar{s}_{2000}^\circ - 298.15\bar{s}_{298}^\circ)$$
$$= 52\,932 - (2000 \times 188.406 - 298.15 \times 130.684)$$
$$= -284\,917 \text{ kJ/kmol}$$

Similarly, for the oxygen at 2000 K,

$$\bar{g}_{O_2}^\circ = (\bar{h}_{2000}^\circ - \bar{h}_{298}^\circ) - (2000\bar{s}_{2000}^\circ - 298.15\bar{s}_{298}^\circ)$$
$$= 59\,199 - (2000 \times 268.764 - 298.15 \times 205.142)$$
$$= -417\,166 \text{ kJ/kmol}$$

For the water,

$$\bar{g}_{H_2O}^\circ = \bar{g}_f^\circ + \bar{g}_{2000}^\circ - \bar{g}_{298}^\circ$$
$$= -228\,583 + 72\,689 - (2000 \times 264.681 - 298.15 \times 188.833)$$
$$= -628\,955 \text{ kJ/kmol}$$

Therefore, at 2000 K,

$$\Delta G^\circ = 2(-284\,917) + (-417\,166) - 2(-628\,955) = 270\,910 \text{ kJ}$$

The value for ΔG° at 2000 K can alternately be evaluated by using the fact that at constant T,

$$\Delta G^\circ = \Delta H^\circ - T\,\Delta S^\circ$$

At 2000 K,

$$\Delta H^\circ = 2(\bar{h}^\circ_{2000} - \bar{h}^\circ_{298})_{H_2} + (\bar{h}^\circ_{2000} - \bar{h}^\circ_{298})_{O_2} - 2(\bar{h}^\circ_f + \bar{h}^\circ_{2000} - \bar{h}^\circ_{298})_{H_2O}$$
$$= 2(52\,932) + (59\,199) - 2(-241\,827 + 72\,689)$$
$$= 503\,340 \text{ kJ}$$
$$\Delta S^\circ = 2(\bar{s}^\circ_{2000})_{H_2} + (\bar{s}^\circ_{2000})_{O_2} - 2(\bar{s}^\circ_{2000})_{H_2O}$$
$$= 2(188.406) + (268.764) - 2(264.681)$$
$$= 116.214 \text{ kJ/K}$$

Therefore,

$$\Delta G^\circ = 503\,340 - 2000 \times 116.214 = 270\,910 \text{ kJ}$$

Note that while the two methods are identical in result, the second requires fewer calculations, and is therefore simpler to use.

Returning now to our development, substituting Eq. 13.30 into Eq. 13.29 and rearranging we can write

$$dG_{T,P} = \left\{ \Delta G^\circ + \bar{R}T \ln\left[\frac{a_C{}^{v_C} a_D{}^{v_D}}{a_A{}^{v_A} a_B{}^{v_B}}\right] \right\} d\epsilon \qquad \textbf{(13.31)}$$

At equilibrium $dG_{T,P} = 0$. Therefore, since $d\epsilon$ is arbitrary,

$$\ln\left[\frac{a_C{}^{v_C} a_D{}^{v_D}}{a_A{}^{v_A} a_B{}^{v_B}}\right] = -\frac{\Delta G^\circ}{\bar{R}T} \qquad \textbf{(13.32)}$$

For convenience, we define the equilibrium constant K as

$$\ln K = -\frac{\Delta G^\circ}{\bar{R}T} \qquad \textbf{(13.33)}$$

which we note must be a function of temperature only for a given reaction, since ΔG° is given by Eq. 13.30 in terms of the properties of the pure substances at a given temperature and the standard-state pressure.

Combining Eqs. 13.32 and 13.33, we have

$$K = \frac{a_C{}^{v_C} a_D{}^{v_D}}{a_A{}^{v_A} a_B{}^{v_B}} \qquad \textbf{(13.34)}$$

which is the chemical equilibrium equation corresponding to the reaction equation, Eq. 13.26.

Example 13.5

Determine the equilibrium constant K, expressed as $\ln K$, for the reaction $2H_2O \rightleftharpoons 2H_2 + O_2$ at 298 K and at 2000 K.

Solution

We have already found, in Example 13.4, $\Delta G°$ for this reaction at these two temperatures. Therefore, at 298 K,

$$(\ln K)_{298} = -\frac{\Delta G°_{298}}{\bar{R}T} = \frac{-457\ 166}{8.3144 \times 298.15} = -184.42$$

At 2000 K we have

$$(\ln K)_{2000} = \frac{-\Delta G°_{2000}}{\bar{R}T} = \frac{-270\ 910}{8.3144 \times 2000} = -16.292$$

Table A.14 gives the values of the equilibrium constant for a number of reactions. Note again that for each reaction the value of the equilibrium constant is determined from the properties of each of the pure constituents at the standard-state pressure and is a function of temperature only.

For other reaction equations, the chemical equilibrium constant can be calculated as in Example 13.5 or can be determined analytically in the following manner. Consider the general reaction Eq. 13.26. The standard-state Gibbs function for each constituent can be expressed by writing Eq. 11.93 at the pressure $P°$. For component A

$$d\left(\frac{\bar{g}_A°}{T}\right)_{P°} = -\frac{\bar{h}_A°}{T^2}\ dT_{P°} \tag{13.35}$$

Therefore, for the reaction given by Eq. 13.26,

$$d\left(\frac{\Delta G°}{T}\right)_{P°} = -\frac{\Delta H°}{T^2}\ dT_{P°} \tag{13.36}$$

where $\Delta G°$ is defined by Eq. 13.30 and $\Delta H°$ similarly by

$$\Delta H° = v_C\bar{h}_C° + v_D\bar{h}_D° - v_A\bar{h}_A° - v_B\bar{h}_B° \tag{13.37}$$

Now, substituting the definition of the equilibrium constant Eq. 13.33 into Eq. 13.36,

$$d\ln K = \frac{\Delta H°}{\bar{R}T^2}\ dT_{P°} \tag{13.38}$$

which is termed the van't Hoff equation. In integrating this equation, we must be careful to note that $\Delta H°$ as defined by Eq. 13.37 is a function of temperature.

In applying the concept of the equilibrium constant to the determination of the equilibrium composition for a chemical reaction, we must be able to determine

the activity of the various constituents in the mixture. The most general model that we will utilize in this text is the Ideal Solution. For this model, the activity of each component is, from Eq. 11.113,

$$a_i = \frac{y_i f_i}{f_i^\circ}$$

If we further assume that all of the standard-state values are for gases, then the standard-state fugacity f_i° is equal to the standard-state pressure P°. The chemical equilibrium equation, Eq. 13.34, then can be rewritten for an ideal solution as

$$K = \frac{a_C^{\nu_C} a_D^{\nu_D}}{a_A^{\nu_A} a_B^{\nu_B}} = \frac{y_C^{\nu_C} y_D^{\nu_D}}{y_A^{\nu_A} y_B^{\nu_B}} \left(\frac{P}{P^\circ}\right)^{\nu_C + \nu_D - \nu_A - \nu_B} \left[\frac{(f/P)_C^{\nu_C} (f/P)_D^{\nu_D}}{(f/P)_A^{\nu_A} (f/P)_B^{\nu_B}}\right] \quad (13.39)$$

We have written the equation in this form because it demonstrates quite clearly the influence of various factors on the equilibrium composition (the y's). That is, we know that temperature and pressure both influence the equilibrium composition. From Eq. 13.39 it can be seen that the influence of temperature enters through the value of K (which is a function of temperature only) and the influence of pressure through the term $(P/P^\circ)^{\nu_C + \nu_D - \nu_A - \nu_B}$ (both temperature and pressure influence the various fugacity coefficients f/P).

Let us now consider a special case of the ideal solution, the ideal gas model, which will be found to be appropriate for most of our examples and applications. If each component behaves as an ideal gas, then each fugacity coefficient f/P in Eq. 13.39 is unity, and the chemical equilibrium equation reduces to the form

$$K = \frac{y_C^{\nu_C} y_D^{\nu_D}}{y_A^{\nu_A} y_B^{\nu_B}} \left(\frac{P}{P^\circ}\right)^{\nu_C + \nu_D - \nu_A - \nu_B} \quad (13.40)$$

We now consider a number of examples that illustrate the procedure for determining the equilibrium composition for a homogeneous reaction and the influence of certain variables on the equilibrium composition.

Example 13.6

One kilomole of carbon at $25°C$ and 0.1 MPa pressure reacts with 1 kmol of oxygen at $25°C$ and 0.1 MPa pressure to form an equilibrium mixture of CO_2, CO, and O_2 at 3000 K, 0.1 MPa pressure, in a steady-flow process. Determine the equilibrium composition and the heat transfer for this process.

Control volume: Combustion chamber.

Inlet states: P, T known for carbon and for oxygen.

Exit state: P, T known.

Process: SSSF.

Sketch: Fig. 13.13.

Model: Table A.13 for carbon; ideal gases, Tables A.11 and A.13.

FIGURE 13.13
Sketch for Example 13.6.

Analysis and Solution

It is convenient to view the overall process as though it occurred in two separate steps, a combustion process followed by a heating and dissociation of the combustion product carbon dioxide, as indicated in Fig. 13.13. This two-step process is represented as follows:

$$\text{Combustion:} \qquad \qquad C + O_2 \rightarrow CO_2$$
$$\text{Dissociation reaction:} \qquad 2CO_2 \rightleftharpoons 2CO + O_2$$

That is, the energy released by the combustion of C and O_2 heats the CO_2 formed to high temperature, resulting in the dissociation of part of the CO_2 to CO and O_2. Thus the overall reaction can be written

$$C + O_2 \rightarrow aCO_2 + bCO + dO_2$$

where the unknown coefficients a, b, and d must be found by solution of the equilibrium equation associated with the dissociation reaction. Once this is accomplished, we can write the first law for a control volume around the combustion chamber to calculate the heat transfer.

From the combustion equation we find that the initial composition for the dissociation reaction is 1 kmol CO_2. Therefore, letting $2z$ be the number of kilomoles of CO_2 dissociated, we find

	$2CO_2$	\rightleftharpoons	$2CO$	$+$	O_2
Initial:	1		0		0
Change:	$-2z$		$+2z$		$+z$
At equilibrium:	$(1-2z)$		$2z$		z

Therefore, the overall reaction is

$$C + O_2 \rightarrow (1-2z)CO_2 + 2zCO + zO_2$$

and the total number of kilomoles at equilibrium is

$$n = (1 - 2z) + 2z + z = 1 + z$$

The equilibrium mole fractions are

$$y_{CO_2} = \frac{1 - 2z}{1 + z}, \qquad y_{CO} = \frac{2z}{1 + z}, \qquad y_{O_2} = \frac{z}{1 + z}$$

From Table A.14 we find that the value of the equilibrium constant at 3000 K for the dissociation reaction considered here is

$$\ln K = -2.222, \qquad K = 0.1084$$

Substituting these quantities along with $P = 0.1$ MPa into Eq. 13.40, we have the equilibrium equation:

$$K = 0.1084 = \frac{y_{CO}^2 y_{O_2}}{y_{CO_2}^2} \left(\frac{P}{P^\circ}\right)^{2+1-2}$$

$$= \frac{\left(\dfrac{2z}{1+z}\right)^2 \left(\dfrac{z}{1+z}\right)}{\left(\dfrac{1-2z}{1+z}\right)^2}(1)$$

or, in more convenient form,

$$\frac{K}{(P/P^\circ)} = \frac{0.1084}{1} = \left(\frac{2z}{1-2z}\right)^2 \left(\frac{z}{1+z}\right)$$

To obtain the physically meaningful root of this mathematical relation, we note that the number of moles of each component must be greater than zero. Thus the root of interest to us must lie in the range

$$0 \leqslant z \leqslant 0.5$$

Solving the equilibrium equation by trial and error, we find

$$z = 0.2186$$

Therefore, the overall process is

$$C + O_2 \rightarrow 0.5628 CO_2 + 0.4372 CO + 0.2186 O_2$$

where the equilibrium mole fractions are

$$y_{CO_2} = \frac{0.5628}{1.2186} = 0.4618$$

$$y_{CO} = \frac{0.4372}{1.2186} = 0.3588$$

$$y_{O_2} = \frac{0.2186}{1.2186} = 0.1794$$

The heat transfer from the combustion chamber to the surroundings can be calculated using the enthalpies of formation and Table A.11. For this process,

$$H_R = (\bar{h}_f^\circ)_C + (\bar{h}_f^\circ)_{O_2} = 0 + 0 = 0$$

The equilibrium products leave the chamber at 3000 K. Therefore,

$$H_P = n_{CO_2}(\bar{h}_f^\circ + \bar{h}_{3000}^\circ - \bar{h}_{298}^\circ)_{CO_2} + n_{CO}(\bar{h}_f^\circ + \bar{h}_{3000}^\circ - \bar{h}_{298}^\circ)_{CO} + n_{O_2}(\bar{h}_f^\circ + \bar{h}_{3000}^\circ - \bar{h}_{298}^\circ)_{O_2}$$
$$= 0.5628(-393\ 522 + 152\ 862) + 0.4372(-110\ 529 + 93\ 542) + 0.2186(98\ 098)$$
$$= -121\ 426\ kJ$$

Substituting into the first law,

$$Q_{c.v.} = H_P - H_R$$
$$= -121\ 426\ kJ/kmol\ C\ burned$$

Example 13.7

One lb mol of carbon at 77 F reacts with 2 lb mol of oxygen at 77 F to form an equilibrium mixture of CO_2, CO, and O_2 at 5400 R, 1 atm pressure. Determine the equilibrium composition.

Control volume: Combustion chamber.

Inlet states: T known for carbon and for oxygen.

Exit state: P, T known.

Process: SSSF.

Model: Ideal gas mixture at equilibrium.

Analysis and Solution

The overall process can be imagined to occur in two steps as in the previous example. The combustion process is

$$C + 2O_2 \rightarrow CO_2 + O_2$$

and the subsequent dissociation reaction is

	$2CO_2$	$\rightleftharpoons 2CO$	$+ O_2$
Initial:	1	0	1
Change:	$-2z$	$+2z$	$+z$
At equilibrium:	$(1-2z)$	$2z$	$(1+z)$

We find that in this case the overall process is

$$C + 2O_2 \rightarrow (1-2z)CO_2 + 2zCO + (1+z)O_2$$

and the total number of moles at equilibrium is

$$n = (1-2z) + 2z + (1+z) = 2+z$$

The mole fractions are

$$y_{CO_2} = \frac{1 - 2z}{2 + z}, \qquad y_{CO} = \frac{2z}{2 + z}, \qquad y_{O_2} = \frac{1 + z}{2 + z}$$

The equilibrium constant for the reaction $2CO_2 \rightleftharpoons 2CO + O_2$ at 5400 R (3000 K) was found in Example 13.6 to be 0.1084. Therefore, the equilibrium equation is, substituting these quantities and $P° = 0.1$ MPa, $P = 1$ atm $= 0.1013$ MPa,

$$K = 0.1084 = \frac{y_{CO}^2 y_{O_2}}{y_{CO_2}^2} \left(\frac{P}{P°} \right)^{2 + 1 - 2}$$

$$= \frac{\left(\dfrac{2z}{2 + z} \right)^2 \left(\dfrac{1 + z}{2 + z} \right)}{\left(\dfrac{1 - 2z}{2 + z} \right)^2} (1.013)$$

or

$$\frac{K}{(P/P°)} = \frac{0.1084}{1.013} = \left(\frac{2z}{1 - 2z} \right)^2 \left(\frac{1 + z}{2 + z} \right)$$

We note that for the number of moles of each component to be greater than zero,

$$0 \leqslant z \leqslant 0.5$$

Solving the equilibrium equation for z, we find

$$z = 0.1545$$

so that the overall process is

$$C + 2O_2 \rightarrow 0.691CO_2 + 0.309CO + 1.1545O_2$$

The mole fractions of the components in the equilibrium mixture are

$$y_{CO_2} = \frac{0.691}{2.1545} = 0.3207$$

$$y_{CO} = \frac{0.309}{2.1545} = 0.1434$$

$$y_{O_2} = \frac{1.1545}{2.1545} = 0.5359$$

The heat transferred from the chamber in this process could be found by the same procedure followed in Example 13.6, considering the overall process as expressed above.

13.7 SIMULTANEOUS REACTIONS

In developing the equilibrium equation and equilibrium constant expressions of Section 13.6, it was assumed that there was only a single chemical reaction equation relating the substances present in the system. To demonstrate the more general situation in which there is more than one chemical reaction, we will now analyze a system involving two simultaneous reactions by a procedure analogous to that followed in Section 13.6. These results are then readily extended to systems involving several simultaneous reactions.

Consider a mixture of substances A, B, C, D, L, M, and N as indicated in Fig. 13.14. These substances are assumed to exist at a condition of chemical equilibrium at temperature T and pressure P, and are related by the two independent reactions

$$(1) \quad v_{A_1}A + v_B B \rightleftharpoons v_C C + v_D D \qquad (13.41)$$

$$(2) \quad v_{A_2} A + v_L L \rightleftharpoons v_M M + v_N N \qquad (13.42)$$

Components
A, B, C, D, L, M, N
in chemical
equilibrium

FIGURE 13.14

**Sketch demonstrating
simultaneous reactions.**

We have considered the situation where one of the components (substance A) is involved in each of the reactions in order to demonstrate the effect of this condition on the resulting equations. As in the previous section, the changes in amounts of the components are related by the various stoichiometric coefficients (which are not the same as the number of moles of each substance present in the vessel). We also realize that the coefficients v_{A_1} and v_{A_2} are not necessarily the same. That is, substance A does not in general take part in each of the reactions to the same extent.

Development of the requirement for equilibrium is completely analogous to that of Section 13.6. We consider that each reaction proceeds an infinitesimal amount toward the right side. This results in a decrease in the number of moles of A, B, and L, and an increase in the moles of C, D, M, and N. Letting the degrees of reaction be ϵ_1 and ϵ_2 for reactions 1 and 2, respectively, the changes in the

number of moles are, for infinitesimal shifts from the equilibrium composition,

$$dn_A = -v_{A_1} d\epsilon_1 - v_{A_2} d\epsilon_2$$
$$dn_B = -v_B d\epsilon_1$$
$$dn_L = -v_L d\epsilon_2$$
$$dn_C = +v_C d\epsilon_1 \qquad \text{(13.43)}$$
$$dn_D = +v_D d\epsilon_1$$
$$dn_M = +v_M d\epsilon_2$$
$$dn_N = +v_N d\epsilon_2$$

The change in Gibbs function for the mixture in the vessel at constant temperature and pressure is

$$dG_{T,P} = \bar{G}_A \, dn_A + \bar{G}_B \, dn_B + \bar{G}_C \, dn_C + \bar{G}_D \, dn_D + \bar{G}_L \, dn_L + \bar{G}_M \, dn_M + \bar{G}_N \, dn_N$$

Substituting the expressions of Eq. 13.43 and collecting terms,

$$dG_{T,P} = (v_C \bar{G}_C + v_D \bar{G}_D - v_{A_1} \bar{G}_A - v_B \bar{G}_B) \, d\epsilon_1$$
$$+ (v_M \bar{G}_M + v_N \bar{G}_N - v_{A_2} \bar{G}_A - v_L \bar{G}_L) \, d\epsilon_2 \qquad \text{(13.44)}$$

It is convenient to again express each of the partial molal Gibbs functions in terms of the activity as

$$\bar{G}_i = \bar{g}_i^\circ + \bar{R}T \ln a_i$$

Equation 13.44 written in this form becomes

$$dG_{T,P} = \left[\Delta G_1^\circ + \bar{R}T \ln \frac{a_C^{v_C} a_D^{v_D}}{a_A^{v_{A_1}} a_B^{v_B}} \right] d\epsilon_1$$
$$+ \left[\Delta G_2^\circ + \bar{R}T \ln \frac{a_M^{v_M} a_N^{v_N}}{a_A^{v_{A_2}} a_L^{v_L}} \right] d\epsilon_2 \qquad \text{(13.45)}$$

In this equation the standard-state change in Gibbs function for each reaction is defined as

$$\Delta G_1^\circ = v_C \bar{g}_C^\circ + v_D \bar{g}_D^\circ - v_{A_1} \bar{g}_A^\circ - v_B \bar{g}_B^\circ \qquad \text{(13.46)}$$
$$\Delta G_2^\circ = v_M \bar{g}_M^\circ + v_N \bar{g}_N^\circ - v_{A_2} \bar{g}_A^\circ - v_L \bar{g}_L^\circ \qquad \text{(13.47)}$$

Equation 13.45 expresses the change in Gibbs function of the system at constant T, P, for infinitesimal degrees of reaction of both reactions 1 and 2, Eqs. 13.41 and 13.42. The requirement for equilibrium is that $dG_{T,P} = 0$. Therefore, since reactions 1 and 2 are independent, $d\epsilon_1$ and $d\epsilon_2$ can be independently varied. It follows that at equilibrium each of the bracketed terms of Eq. 13.45 must be zero. Defining equilibrium constants for the two reactions by

$$\ln K_1 = -\frac{\Delta G_1^\circ}{\bar{R}T} \qquad \text{(13.48)}$$

and

$$\ln K_2 = -\frac{\Delta G_2^\circ}{\bar{R}T} \qquad \text{(13.49)}$$

We find that at equilibrium

$$K_1 = \frac{a_C{}^{\nu_C} a_D{}^{\nu_D}}{a_A{}^{\nu_{A_1}} a_B{}^{\nu_B}}$$

(13.50)

and

$$K_2 = \frac{a_M{}^{\nu_M} a_N{}^{\nu_N}}{a_A{}^{\nu_{A_2}} a_L{}^{\nu_L}}$$

(13.51)

can be expressed in terms of the mole fractions for the appropriate model of the mixture, after which these expressions must be solved simultaneously for the equilibrium composition of the mixture. The following example is presented to demonstrate and clarify this procedure.

Example 13.8

One kilomole of water vapor is heated to 3000 K, 0.1 MPa pressure. Determine the equilibrium composition, assuming that H_2O, H_2, O_2, and OH are present.

Control volume: Heat exchanger.
Exit state: P, T known.
Model: Ideal gas mixture at equilibrium.

Analysis and Solution

There are two independent reactions relating the four components of the mixture at equilibrium. These can be written as

(1) $2H_2O \rightleftharpoons 2H_2 + O_2$
(2) $2H_2O \rightleftharpoons H_2 + 2OH$

Let $2a$ be the number of kilomoles of water dissociating according to reaction (1) during the heating, and $2b$ the number of kilomoles of water dissociating according to reaction (2). Since the initial composition is 1 kmol water, the changes according to the two reactions are

Change: (1) $2H_2O \rightleftharpoons 2H_2 + O_2$
 $\;-2a \quad\; +2a \quad +a$

Change: (2) $2H_2O \rightleftharpoons H_2 + 2OH$
 $\;-2b \quad\; +b \quad +2b$

Therefore, the number of kilomoles of each component at equilibrium is its initial number plus the change, so that at equilibrium

$$n_{H_2O} = 1 - 2a - 2b$$
$$n_{H_2} = 2a + b$$
$$n_{O_2} = a$$
$$\underline{n_{OH} = 2b}$$
$$n = 1 + a + b$$

The overall chemical reaction that occurs during the heating process can be written

$$H_2O \rightarrow (1 - 2a - 2b)H_2O + (2a + b)H_2 + aO_2 + 2bOH$$

The right-hand side of this expression is the equilibrium composition of the system. Since the number of kilomoles of each substance must necessarily be greater than zero, we find that the possible values of a and b are restricted to

$$a \geqslant 0$$
$$b \geqslant 0$$
$$(a + b) \leqslant 0.5$$

The two equilibrium equations are, assuming the mixture to behave as an ideal gas,

$$K_1 = \frac{y_{H_2}^2 y_{O_2}}{y_{H_2O}^2} \left(\frac{P}{P^\circ}\right)^{2+1-2}$$

$$K_2 = \frac{y_{H_2} y_{OH}^2}{y_{H_2O}^2} \left(\frac{P}{P^\circ}\right)^{1+2-2}$$

Since the mole fraction of each component is the ratio of the number of kilomoles of the component to the total number of kilomoles of the mixture, these equations can be written in the form

$$K_1 = \frac{\left(\dfrac{2a + b}{1 + a + b}\right)^2 \left(\dfrac{a}{1 + a + b}\right)}{\left(\dfrac{1 - 2a - 2b}{1 + a + b}\right)^2} \left(\frac{P}{P^\circ}\right)$$

$$= \left(\frac{2a + b}{1 - 2a - 2b}\right)^2 \left(\frac{a}{1 + a + b}\right)\left(\frac{P}{P^\circ}\right)$$

and

$$K_2 = \frac{\left(\dfrac{2a + b}{1 + a + b}\right)\left(\dfrac{2b}{1 + a + b}\right)^2}{\left(\dfrac{1 - 2a - 2b}{1 + a + b}\right)^2} \left(\frac{P}{P^\circ}\right)$$

$$= \left(\frac{2a + b}{1 + a + b}\right)\left(\frac{2b}{1 - 2a - 2b}\right)^2 \left(\frac{P}{P^\circ}\right)$$

giving two equations in the two unknowns a and b, since $P = 0.1$ MPa and the values of K_1, K_2 are known. From Table A.14 at 3000 K, we find

$$K_1 = 0.002\,087, \qquad K_2 = 0.002\,812$$

Therefore, the equations can be solved simultaneously for a and b. The values satisfying the equations are

$$a = 0.0540, \qquad b = 0.0543$$

Substituting these values into the expressions for the number of kilomoles of each component and of the mixture, we find the equilibrium mole fractions to be

$$y_{H_2O} = 0.7069$$
$$y_{H_2} = 0.1464$$
$$y_{O_2} = 0.0487$$
$$y_{OH} = 0.0980$$

The methods used in this section can readily be extended to equilibrium systems involving more than two independent reactions. In each case, the number of simultaneous equilibrium equations is equal to the number of independent reactions. Solution of a large set of nonlinear simultaneous equations naturally becomes quite difficult, however, and is not easily accomplished by hand calculations. Instead, solution of these problems is normally made using iterative procedures on a digital computer.

13.8 IONIZATION

In the final section of this chapter, we consider the equilibrium of systems involving ionized gases, or plasmas, a field that has found increasing application and study in recent years. In previous sections we have discussed chemical equilibrium, with a particular emphasis on molecular dissociation, as for example the reaction

$$N_2 \rightleftharpoons 2N$$

which occurs to an appreciable extent for most molecules only at high temperature (of the order of magnitude 3000 to 10 000 K). At still higher temperatures, such as those found in electric arcs, the gas becomes ionized. That is, some of the atoms lose an electron, according to the reaction

$$N \rightleftharpoons N^+ + e^-$$

where N^+ denotes a singly ionized nitrogen atom (one that has lost one electron and consequently has a positive charge) and e^- represents the free electron. As temperature is increased still further, many of the ionized atoms lose another electron, according to the reaction

$$N^+ \rightleftharpoons N^{++} + e^-$$

and thus become doubly ionized. As the temperature is increased further, the process continues until a temperature is reached at which all the electrons have been stripped from the nucleus.

Ionization generally is appreciable only at high temperature. However, dissociation and ionization both tend to occur to greater extents at low pressure, and consequently dissociation and ionization may be appreciable in such environments as the upper atmosphere, even at moderate temperature. Other effects such as radiation will also cause ionization, but these effects are not considered here.

The problems involved with analyzing the composition in a plasma become much more difficult than those for an ordinary chemical reaction, since in an electric field the free electrons in the mixture do not exchange energy with the positive ions and neutral atoms at the same rate as they do with the field. Consequently, in a plasma in an electric field, the electron gas is not at exactly the same temperature as the heavy particles. However, for moderate fields, it is a reasonable approximation to assume a condition of thermal equilibrium in the plasma, at least for preliminary calculations. Under this condition, we can treat the ionization equilibrium in exactly the same manner as an ordinary chemical equilibrium analysis.

At these extremely high temperatures, we may assume that the plasma behaves as an ideal gas mixture of neutral atoms, positive ions, and electron gas. Thus, for the ionization of some atomic species A,

$$A \rightleftharpoons A^+ + e^- \qquad (13.52)$$

we may write the ionization-equilibrium equation in the form

$$K = \frac{y_{A^+} y_{e^-}}{y_A} \left(\frac{P}{P^\circ}\right)^{1+1-1} \qquad (13.53)$$

The ionization-equilibrium constant K is defined in the ordinary manner

$$\ln K = -\frac{\Delta G^\circ}{\bar{R} T} \qquad (13.54)$$

and is a function of temperature only. The standard-state Gibbs function change for reaction 13.52 is found from

$$\Delta G^\circ = \bar{g}^\circ_{A^+} + \bar{g}^\circ_{e^-} - \bar{g}^\circ_A \qquad (13.55)$$

The standard-state Gibbs function for each component at the given plasma temperature can be calculated using the procedures of statistical thermodynamics, so that ionization-equilibrium constants can be tabulated as functions of temperature.

Solution of the ionization-equilibrium equation, Eq. 13.53, is then accomplished in the same manner as for an ordinary chemical-reaction equilibrium.

Example 13.9

Calculate the equilibrium composition if argon gas is heated in an arc to 10 000 K, 1 kPa, assuming the plasma to consist of Ar, Ar$^+$, e^-. The ionization-equilibrium constant for the reaction

$$Ar \rightleftharpoons Ar^+ + e^-$$

at this temperature is 0.000 42.

Control volume: Heating arc.

Exit state: P, T known.

Model: Ideal gas mixture at equilibrium.

Analysis and Solution

Consider an initial composition of 1 kmol neutral argon, and let z be the number of kilomoles ionized during the heating process. Therefore,

$$\text{Ar} \rightleftharpoons \text{Ar}^+ + e^-$$

	Ar	Ar^+	e^-
Initial:	1	0	0
Change:	$-z$	$+z$	$+z$
Equilibrium:	$(1-z)$	z	z

and

$$n = (1-z) + z + z = 1 + z$$

Since the number of kilomoles of each component must be positive, the variable z is restricted to the range

$$0 \leqslant z \leqslant 1$$

The equilibrium mole fractions are

$$y_{\text{Ar}} = \frac{n_{\text{Ar}}}{n} = \frac{1-z}{1+z}$$

$$y_{\text{Ar}^+} = \frac{n_{\text{Ar}^+}}{n} = \frac{z}{1+z}$$

$$y_{e^-} = \frac{n_{e^-}}{n} = \frac{z}{1+z}$$

The equilibrium equation is

$$K = \frac{y_{\text{Ar}^+} y_{e^-}}{y_{\text{Ar}}} \left(\frac{P}{P^\circ}\right)^{1+1-1} = \frac{\left(\dfrac{z}{1+z}\right)\left(\dfrac{z}{1+z}\right)}{\left(\dfrac{1-z}{1+z}\right)} \left(\frac{P}{P^\circ}\right)$$

so that, at 10 000 K, 1 kPa,

$$0.000\,42 = \left(\frac{z^2}{1-z^2}\right)(0.01)$$

Solving,

$$z = 0.2008$$

and the composition is found to be

$$y_{\text{Ar}} = 0.6656$$
$$y_{\text{Ar}^+} = 0.1672$$
$$y_{e^-} = 0.1672$$

Simultaneous reactions, such as simultaneous molecular dissociation and ionization reactions or multiple ionization reactions, can be analyzed in the same manner as the ordinary simultaneous chemical reactions of Section 13.7. In doing

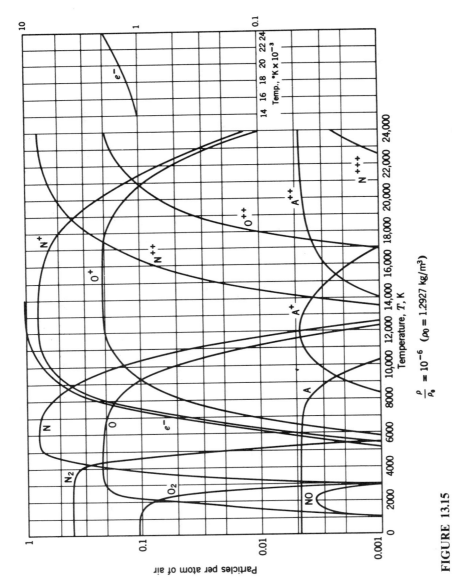

FIGURE 13.15

Equilibrium composition of air [W. E. Moeckel and K. C. Weston, NACA TN 4265 (1958)].

so, we again make the assumption of thermal equilibrium in the plasma, which, as mentioned before, is, in many cases, a reasonable approximation. Figure 13.15 shows the equilibrium composition of air at high temperature and low density, and indicates the overlapping regions of the various dissociation and ionization processes.

PROBLEMS

13.1 Consider a 1-mile deep gas well containing pure ethylene gas at a uniform temperature of 60 F. The pressure is measured at the top of the well and found to be 50 atm. What is the pressure at the bottom of the well?

13.2 Consider a 2-km-deep gas well containing a gas mixture of methane and ethane at a uniform temperature of 30°C. The pressure at the top of the well is 14 MPa and the composition on a mole basis is 90% methane, 10% ethane. Determine the pressure and composition at the bottom of the well, assuming the following:
(a) Ideal gas mixture.
(b) Ideal solution.

13.3 Determine the phase compositions in the liquid–vapor equilibrium system nitrogen–methane at 115 K, 500 kPa, assuming the following:
(a) Raoult's Rule–ideal gas.
(b) Ideal solution in both phases.

13.4 Ethylene gas is burned with air and the products exit at 120°C, 3.5 MPa. Assuming complete combustion, what is the minimum percent theoretical air if there is to be no condensation of the products?

13.5 As the first step in an air liquefaction plant, ambient air at 77 F, 14.7 lbf/in.2, 50% relative humidity, is compressed to a pressure of 100 atm. What is the dew-point temperature of the mixture exiting the compressor?

13.6 Consider the liquid–vapor equilibrium system nitrogen–methane at a temperature of 150 K. Plot to scale the liquid and vapor phase-boundary lines on a pressure vs. composition diagram.

13.7 A mixture of 3 lb mol hydrogen and 1 lb mol carbon monoxide is contained in a rigid vessel at 77 F and an unknown pressure, P_1. A chemical reaction now occurs, and the products consist solely of methane and water, at 400 F and 40 atm pressure.
(a) Determine the initial pressure P_1.
(b) Calculate $S_2 - S_1$ for the contents of the vessel.

13.8 Consider a SSSF process concerning the production of a synthetic fuel, methyl alcohol (CH_3OH), from waste material that has initially been partially oxidized to produce a gas mixture of one-third hydrogen and two-thirds carbon monoxide, by mole. This mixture enters the reactor at 600 K,

6 MPa, at the rate of 3 kmol/s, and steam at 300°C, 6 MPa, enters the reactor separately at the rate of 1 kmol/s. Exiting the reactor is a mixture of methyl alcohol and carbon dioxide at 450 K, 6 MPa. What is the rate of entropy change for this process? For ideal gas CH_3OH at 25°C:

$$\bar{h}_f^\circ = -201\,167 \text{ kJ/kmol}$$
$$\bar{s}^\circ = 239.70 \text{ kJ/kmol K}$$
$$\bar{C}_{po} = 45.02 \text{ kJ/kmol K}$$

13.9 Liquid hydrazine, N_2H_4, at 25°C and 130% theoretical air at 25°C, 6 MPa, enter a SSSF combustion chamber. The products, consisting of H_2O, N_2, and O_2, exit at 450 K, 6 MPa. Calculate the heat transfer from the combustion chamber per kilomole of hydrazine. The enthalpy of formation of liquid hydrazine is $+50\,417$ kJ/kmol.

13.10 Hydrogen gas enters a steady-state steady-flow combustion chamber at 77 F, 50 atm. A separate line supplies 80% theoretical air, which is throttled from 77 F, 150 atm into the chamber. The products, consisting of H_2O, H_2 and N_2, exit from the chamber at 350 F, 50 atm. What is the irreversibility of this process?

13.11 At a certain point in a synthetic fuel production process, the product, consisting of equal amounts of ethanol and carbon dioxide, is stored in a 1-m^3 tank at 400 K, 5 MPa, before being processed further. Determine the following:
(a) The absolute entropy per kilomole of product at this state.
(b) The total amount of product in the tank.
For ideal gas ethanol:

$$\bar{h}_f^\circ = -235\,308 \text{ kJ/kmol} \qquad \bar{s}_{25°C}^\circ = 282.111 \text{ kJ/kmol K}$$
$$\bar{C}_{po} = 115 \text{ kJ/kmol K}$$

13.12 Consider a hydrogen–oxygen fuel cell in which the ideal gases hydrogen and oxygen each enter at temperature T and 100 kPa, and ideal gas water exits at temperature T and 100 kPa. Plot to scale the maximum EMF of the fuel cell vs. the temperature T over the temperature range 25°C to 600 K.

13.13 Plot to scale the values of ln K vs. $1/T$ for the reaction $2CO_2 \rightleftharpoons 2CO + O_2$. Write an equation for the ln K as a function of temperature.

13.14 Calculate the equilibrium constant for the reaction $H_2 \rightleftharpoons 2H$ at temperatures of 298 K, 3000 R, and 6000 K.

13.15 Oxygen gas is heated from room temperature to 4000 K, 500 kPa, at which state the diatomic species has partially dissociated to the monatomic form. Determine the equilibrium composition at this state.

13.16 Air (assumed to be 79% nitrogen and 21% oxygen) is heated in a SSSF process at a constant pressure of 14.7 lbf/in.2, and some NO is formed. At what temperature will the mole fraction of NO be 0.001?

13.17 A mixture of 1 kmol carbon dioxide, 2 kmol carbon monoxide, and 2 kmol oxygen at 25°C and 150 kPa pressure is heated in a constant-pressure SSSF process to 3000 K. Assuming that only these same substances are present in the exiting chemical equilibrium mixture, determine the composition of that mixture.

13.18 Repeat Problem 13.17 for the case where the initial mixture also includes 2 kmol of nitrogen, which does not dissociate during the process.

13.19 A gas mixture of 1.5 lb mol CO and 1 lb mol H_2 enters a catalytic reactor along with 1 lb mol of H_2O. Exiting the reactor is a chemical equilibrium gas mixture of CO, H_2, H_2O and CO_2 at 900 R, 5 atm. Determine the equilibrium composition.

13.20 Catalytic gas generators are frequently used to decompose a liquid, providing a desired gas mixture (spacecraft control systems, fuel cell gas supply, and so forth). Consider pure liquid hydrazine, N_2H_4, fed to a gas generator, from which exits a chemical equilibrium gas mixture of N_2, H_2, and NH_3 at 100°C, 350 kPa. Calculate the percent composition of the equilibrium mixture. For the reaction $N_2 + 3H_2 \rightleftharpoons 2NH_3$ at 100°C, ln K = 6.982.

13.21 A gas mixture of 1 lb mol carbon monoxide, 1 lb mol nitrogen and 1 lb mol oxygen at 77 F, 20 lbf/in.2, is heated in a constant-pressure SSSF process. The exit mixture can be assumed to be in chemical equilibrium, with CO_2, CO, O_2 and N_2 present, and the mole fraction of CO_2 at this point is 0.176. Calculate the heat transfer for the process.

13.22 One approach to using a hydrocarbon fuel in a fuel cell is to "reform" the hydrocarbon to obtain hydrogen, which is then fed to the fuel cell. As a part of the analysis of such a procedure, consider the reforming reaction.

$$CH_4 + H_2O \rightleftharpoons 3H_2 + CO$$

(a) Determine the equilibrium constant for this reaction at a temperature of 800 K.

(b) One kilomole methane and 4 kilomoles water are fed to a catalytic reformer. A chemical equilibrium mixture of CH_4, H_2O, H_2, and CO exits at 800 K, 100 kPa. Determine the equilibrium composition of this mixture.

13.23 In a test of a gas turbine combustor, saturated liquid methane at 100 kPa is to be burned with excess air to hold the adiabatic flame temperature to 1600 K. It is assumed that the products consist of a chemical equilibrium mixture of CO_2, H_2O, N_2, O_2, and NO. Determine the percent excess air used in the combustion, and the percentage of NO in the products.

13.24 One lb mol of solid carbon and b lb mol of oxygen gas are contained in a closed rigid tank at 77 F, 25 lbf/in.2 The carbon is now ignited and the final state is a chemical equilibrium gas mixture of CO_2, O_2 and CO at 5300 F, at which state the mole fraction of CO_2 is 0.20. Calculate b, the initial number of lb mol of O_2, and the final pressure in the tank, P_2.

13.25 Methane at 25°C, 100 kPa, is burned with 200% theoretical oxygen at 400 K, 100 kPa, in an adiabatic SSSF process, and the products of combustion exit at 100 kPa. Assume that the only significant dissociation reaction in the products is that of carbon dioxide going to carbon monoxide and oxygen. Determine the equilibrium composition of the products and also their temperature at the combustor exit.

13.26 Calculate the irreversibility for the adiabatic combustion process described in Problem 13.25.

13.27 Acetylene gas at 77 F is burned with 140% theoretical air, which enters the burner at 77 F, 14.7 lbf/in.2, 80% relative humidity. The combustion products form a chemical equilibrium mixture of CO_2, H_2O, NO, O_2 and N_2 at 3500 F, 14.7 lbf/in.2 This mixture is then cooled to 1340 F very rapidly, such that the composition does not change. Determine the following:
(a) The mole fraction of NO in the products.
(b) The heat transfer for the overfall process.

13.28 A step in the production of a synthetic liquid fuel from organic waste matter is the following conversion process: 1 kmol of ethylene gas (converted from the waste) at 25°C, 5 MPa, and 2 kmol of steam at 300°C, 5 MPa, enter a catalytic reactor. An ideal gas, chemical equilibrium mixture of ethanol, ethylene, and water leaves the reactor at 700 K, 5 MPa. (Thermochemical properties for ideal gas ethanol have been given in Problem 13.11.)
(a) Determine the mixture composition.
(b) Calculate the heat transfer for the reactor.
(c) Would it be more desirable to operate this system with different inlet proportions?

13.29 A mixture of 1 lb mol of CO and 100% theoretical air enters a SSSF combustor at 720 R, 1.5 atm, and a chemical equilibrium mixture of CO_2, CO, O_2 and N_2 exits at 1.5 atm and the adiabatic flame temperature. This mixture now enters a heat exchanger, from which a chemical equilibrium mixture exits at 1800 R, 1.5 atm. Calculate the following:
(a) The irreversibility for the combustor process.
(b) The heat transfer from the products in the heat exchanger.
(c) The irreversibility for the heat exchanger process, assuming that the heat transfer is with the surroundings at 77 F.

13.30 An important step in the manufacture of chemical fertilizer is the production of ammonia, according to the reaction

$$N_2 + 3H_2 \rightleftharpoons 2NH_3$$

(a) Using the data given below, calculate the equilibrium constant for this reaction at 150°C.
(b) For an initial composition of 25% nitrogen, 75% hydrogen, on a mole basis, calculate the equilibrium composition at 150°C, 5 MPa.

For NH$_3$:

$$(\bar{h}_f^\circ)_{298} = -46\ 191 \text{ kJ/kmol} \qquad (\bar{g}_f^\circ)_{298} = -16\cdot622 \text{ kJ/kmol}$$
$$\bar{C}_{po} = 25.464 + 0.0664\ T - 20.416 \times 10^{-6}\ T^2 \text{ kJ/kmol K}$$

13.31 Repeat Problem 13.30 for a system pressure of 15 MPa, assuming the equilibrium mixture to behave as an ideal solution.

13.32 Consider the following simplification of the production of a synthetic fuel (methyl alcohol) from coal: a gas mixture of 50% carbon monoxide, 50% hydrogen leaves a coal gasifier at 400 K, 1 MPa and enters a catalytic converter. A chemical equilibrium gas mixture of methanol, carbon monoxide, and hydrogen leaves the converter at the same temperature and pressure. (Thermochemical properties for ideal gas methanol have been given in Problem 13.8.)

(a) Calculate the equilibrium constant for an appropriate reaction equation, and determine the equilibrium composition of the mixture leaving the converter.

(b) Would it be more desirable to operate the converter at ambient pressure?

(c) Would it be more desirable to operate the converter at ambient temperature?

13.33 Consider a coal gasifier proposed for supplying a gaseous fuel to a gas turbine power plant. Fifty lbm per second of dry coal (represented thermochemically as 48 lbm C plus 2 lbm H) enter the gasifier, along with 4.76 lb mol/s of air and 2 lb mol/s of steam. The output stream from this unit is a chemical equilibrium gas mixture containing H$_2$, CO, N$_2$, CH$_4$ and CO$_2$ at 1160 F, 10 atm.

(a) Set up the reaction and equilibrium equation(s) for this system and calculate the appropriate equilibrium constant(s).

(b) Determine the composition of the gas mixture leaving the gasifier.

13.34 Ethane is burned with 150% theoretical air in a gas turbine combustor and the products exiting consist of a chemical equilibrium mixture of CO$_2$, H$_2$O, O$_2$, N$_2$ and NO at 2800 F, 150 lbf/in.2

(a) Determine the mole fraction of NO in the products.

(b) Is it reasonable to ignore CO in the products?

13.35 A mixture of 1 kmol water and 1 kmol oxygen at 400 K is heated to 3000 K, 200 kPa, in a SSSF process. Determine the equilibrium composition at the outlet of the heat exchanger, assuming that the mixture consists of H$_2$O, H$_2$, O$_2$, and OH.

13.36 One lb mol of air (assumed to be 78% nitrogen, 21% oxygen and 1% argon) at room temperature is heated to 7200 R, 30 lbf/in.2 Find the equilibrium composition at this state, assuming that only N$_2$, O$_2$, NO, O and Ar are present.

13.37 Butane is burned with 200% theoretical air, and the products of combustion, an equilibrium mixture containing only CO_2, O_2, H_2O, N_2, NO and NO_2, exit from the combustion chamber at 2520 R, 20 atm. Determine the equilibrium composition at this state.

13.38 Consider a coal gasifier in which 0.5 kmol of oxygen and x kmol of water enter for each kilomole of coal (assumed to be carbon). Leaving the gasifier is a homogeneous chemical equilibrium ideal gas mixture at T and P. This mixture contains CH_4, H_2, H_2O, CO, and CO_2.
 (a) Determine the number of independent reaction equations for this system.
 (b) Set up the equilibrium equations and specify the procedure followed to solve for the composition.
 (c) Comment on any restrictions on x, and on whether it is desirable to have x as small or as large as possible.

13.39 Acetylene gas and x times theoretical air $(x \geqslant 1)$ at room temperature and 500 kPa are burned at constant pressure in an adiabatic SSSF process. The flame temperature is 2600 K and the combustion products are assumed to consist of N_2, O_2, CO_2, H_2O, CO, and NO. Determine the value of x.

13.40 One kilomole of water vapor at 100 kPa, 400 K, is heated to 3000 K in a constant-pressure SSSF processs. Determine the final composition, assuming that H_2O, H_2, H, O_2, and OH are present at equilibrium.

13.41 Methane is burned with theoretical oxygen is a SSSF process, and the products exit the combustion chamber at 5300 F, 100 lbf/in.² Calculate the equilibrium composition at this state, assuming that only CO_2, CO, H_2O, H_2, O_2 and OH are present.

13.42 Operation of an MHD converter requires an electrically conducting gas. It is proposed to utilize helium gas "seeded" with 1.0 mole percent cesium, as shown in Fig. 13.16. The cesium is partly ionized $(Cs \rightleftharpoons Cs^+ + e^-)$ by heating the mixture to 2500 K, 1 MPa, in a nuclear reactor, in order to provide free electrons. No helium is ionized in this process, so that the mixture entering the converter consists of He, Cs, Cs^+, and e^-. To analyze the converter process, it is necessary to know precisely the mole fraction of electrons in the mixture. Determine this fraction. At 2500 K, $\ln K = -13.4$ for the cesium ionization reaction listed above.

FIGURE 13.16
Sketch for Problem 13.42.

13.43 One kilomole of argon gas at room temperature is heated to 20 000 K, 100 kPa. Assume that the plasma at this condition consists of an equilibrium mixture of Ar, Ar^+, Ar^{++}, and e^- according to the simultaneous reactions:

$$(1) \ Ar \rightleftharpoons Ar^+ + e^-$$
$$(2) \ Ar^+ \rightleftharpoons Ar^{++} + e^-$$

The ionization-equilibrium constants for these reactions at 20 000 K have been calculated from spectroscopic data as: $\ln K_1 = 3.11$, $\ln K_2 = -4.92$. Determine the equilibrium composition of the plasma.

13.44 Plot to scale the equilibrium composition of nitrogen at 10 kPa pressure over the temperature range 5000 K to 15 000 K, assuming that N_2, N, N^+, and e^- are present. For the ionization reaction $N \rightleftharpoons N^+ + e^-$, the ionization-equilibrium constant K has been calculated from spectroscopic data as:

T (K)	K
10 000	6.26×10^{-4}
12 000	1.51×10^{-2}
14 000	0.151
16 000	0.92

13.45 Write a computer program to solve the following problem. One kilomole of carbon at 25°C is burned with b kilomoles of oxygen in a constant-pressure adiabatic process. The products consist of an equilibrium mixture of CO_2, CO, and O_2. It is desired to determine the flame temperature for various combinations of b and the pressure P, assuming variable specific heat for the components; equations from Table A.9.

13.46 Write a computer program to solve the following problem. One kilomole of water is heated to temperature T and pressure P, at which exists a chemical equilibrium mixture of H_2O, H_2, and O_2. It is desired to find the equilibrium composition at this T, P by direct determination of the minimum value of the system Gibbs function in terms of the amount of water that has dissociated (see Fig. 13.3), and to compare this result with that found using the analytical procedure developed in Section 13.6.

13.47 Write a computer program to solve the following problem. Repeat Problem 13.46 for the case where the equilibrium mixture also includes OH, in addition to the other components given in that problem.

13.48 Write a computer program to solve the following problem. It is desired to study the influence of percent theoretical air on the adiabatic flame temperature of methane (with particular interest in the concentration of NO in the products), assuming that the products consist of a chemical equilibrium mixture of CO_2, H_2O, O_2, N_2, and NO only. Assume variable specific heat for the gases, using polynomial equations of the type developed in the solution of Problem 5.85.

14
FLOW THROUGH NOZZLES AND BLADE PASSAGES

This chapter deals with the thermodynamic aspects of one-dimensional flow through nozzles and blade passages. In addition, the momentum equation for the control volume is developed and applied to these same problems. The sonic velocity is defined in terms of thermodynamic properties, and the importance of the Mach number as a variable in compressible flow is noted.

14.1 STAGNATION PROPERTIES

In dealing with problems involving flow, many discussions and equations can be simplified by introducing the concept of the isentropic stagnation state and the properties associated with it. The isentropic stagnation state is the state a flowing fluid would attain if it underwent a reversible adiabatic deceleration to zero velocity. This state is designated in this chapter with the subscript 0. From the first law for a steady-state, steady-flow process we conclude that

$$h + \frac{\mathbf{V}^2}{2} = h_0 \qquad (14.1)$$

The actual and the isentropic stagnation states for a typical gas or vapor are shown on the h-s diagram of Fig. 14.1. Sometimes it is advantageous to make a distinction between the actual and the isentropic stagnation states. The actual stagnation state is the state achieved after an actual deceleration to zero velocity (as at the nose of a body placed in a fluid stream), and there may be irreversibilities associated with the deceleration process. Therefore, the term stagnation property is sometimes reserved for the properties associated with the actual state, and the term total property is used for the isentropic stagnation state.

It is evident from Fig. 14.1 that the enthalpy is the same for both the actual and isentropic stagnation states (assuming that the actual process is adiabatic). Therefore, for an ideal gas, the actual stagnation temperature is the same as the isentropic stagnation temperature. However, the actual stagnation pressure may be less than the isentropic stagnation pressure and for this reason the term total pressure (meaning isentropic stagnation pressure) has particular meaning compared to the actual stagnation pressure.

562

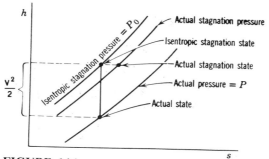

FIGURE 14.1

Enthalpy–entropy diagram illustrating the definition of stagnation state.

Example 14.1

Air flows in a duct at a pressure of 20 lbf/in.2 with a velocity of 600 ft/s. The temperature of the air is 80 F. Determine the isentropic stagnation pressure and temperature.

Analysis and Solution

If we assume that the air is an ideal gas with constant specific heat as given in Table A.8, the calculation is as follows. From Eq. 14.1

$$\frac{V^2}{2} = h_0 - h = C_{po}(T_0 - T)$$

$$\frac{(600)^2}{64.34 \times 778} = 0.240(T_0 - T)$$

$$T_0 = 570\ R = 110\ F$$

The stagnation pressure can be found from the relation

$$\frac{T_0}{T} = \left(\frac{P_0}{P}\right)^{(k-1)/k}$$

$$\frac{570}{540} = \left(\frac{P_0}{P}\right)^{0.286}$$

$$\frac{P_0}{P} = 1.210$$

$$P_0 = 20(1.210) = 24.2\ lbf/in.^2$$

The Air Tables, Table A.10, which are abridged from Keenan and Kayes' *Gas Tables*, could also have been used, and then the variation of specific heat with temperature would have been taken into account. Since the actual and stagnation states have the same entropy, we proceed as follows: Using Table 1 of the *Gas*

Tables (Appendix Table A.10),

$$T = 540 \text{ R} \qquad h = 129.06 \qquad P_r = 1.3860$$

$$h_0 = h + \frac{\mathbf{V}^2}{2} = 129.06 + \frac{(600)^2}{64.34 \times 778} = 136.26$$

$$T_0 = 570 \text{ R} \qquad P_{r0} = 1.6748$$

$$P_0 = P \times \frac{P_{r0}}{P_r} = 20.0 \times \frac{1.6748}{1.3860} = 24.2 \text{ lbf/in.}^2$$

14.2 THE MOMENTUM EQUATION FOR THE CONTROL VOLUME

Before proceeding it will be advantageous to develop the momentum equation for the control volume. Newton's second law states that the sum of the external forces acting on a body in a given direction is proportional to the rate of change of momentum in the given direction. Writing this in equation form for the x-direction we have

$$\sum F_x \propto \frac{d(m\mathbf{V}_x)}{dt}$$

For the system of units used in this book, this proportionality can be written directly as an equality.

$$\sum F_x = \frac{d(m\mathbf{V}_x)}{dt} \tag{14.2}$$

Equation 14.2 has been written for a body of fixed mass, or in thermodynamic parlance, for a system. We now proceed to write the momentum equation for a control volume, and follow a procedure similar to that used in writing the continuity equation and the first and second laws of thermodynamics for a control volume.

Consider the system and control volume shown in Fig. 14.2. Let the control volume be fixed relative to its coordinate frame. During the time interval δt, the mass δm_i enters the control volume with a velocity $(\mathbf{V}_r)_i$ and velocity components $(\mathbf{V}_x)_i$, $(\mathbf{V}_y)_i$, and $(\mathbf{V}_z)_i$. During this same time interval the mass δm_e leaves the control volume, with velocity $(\mathbf{V}_r)_e$ and velocity components $(\mathbf{V}_x)_e$, $(\mathbf{V}_y)_e$, and $(\mathbf{V}_z)_e$.

If we write the x-momentum equation for the system during this time interval we have

$$\left(\sum F_x\right)_{\text{av}} = \frac{\Delta(m\mathbf{V}_x)}{\delta t} = \frac{(m\mathbf{V}_x)_2 - (m\mathbf{V}_x)_1}{\delta t} \tag{14.3}$$

Let $(m\mathbf{V}_x)_t = x$-momentum in the control volume at time t.

$(m\mathbf{V}_x)_{t+\delta t} = x$-momentum in the control volume at time $t + \delta t$.

FIGURE 14.2

Schematic diagram for the development of the momentum equation for a control volume.

Then

$(m\mathbf{V}_x)_1 = (m\mathbf{V}_x)_t + (\mathbf{V}_x)_i\,\delta m_i = x\text{-momentum of the system at time } t.$

$(m\mathbf{V}_x)_2 = (m\mathbf{V}_x)_{t+\delta t} + (\mathbf{V}_x)_e\,\delta m_e = x\text{-momentum of the system at time } t + \delta t.$

It follows that

$$(m\mathbf{V}_x)_2 - (m\mathbf{V}_x)_1 = \left[(m\mathbf{V}_x)_{t+\delta t} - (m\mathbf{V}_x)_t\right] + \left[(\mathbf{V}_x)_e\,\delta m_e - (\mathbf{V}_x)_i\,\delta m_i\right] \quad \textbf{(14.4)}$$

The first bracketed term on the right side of Eq. 14.4 represents the change of x-momentum within the control volume during the time interval δt, and the second bracketed term in that equation represents the x-directional momentum flow across the control surface during δt. Now, dividing Eq. 14.4 by δt and substituting into Eq. 14.3,

$$\left(\sum F_x\right)_{\text{av}} = \frac{(m\mathbf{V}_x)_{t+\delta t} - (m\mathbf{V}_x)_t}{\delta t} + \frac{(\mathbf{V}_x)_e\,\delta m_e - (\mathbf{V}_x)_i\,\delta m_i}{\delta t} \quad \textbf{(14.5)}$$

We now establish the limit for each of the terms in this expression as $\delta t \to 0$.

$$\lim_{\delta t \to 0}\left(\sum F_x\right)_{\text{av}} = \sum F_x$$

$$\lim_{\delta t \to 0}\left[\frac{(m\mathbf{V}_x)_{t+\delta t} - (m\mathbf{V}_x)_t}{\delta t}\right] = \frac{d(m\mathbf{V}_x)_{\text{c.v.}}}{dt}$$

$$\lim_{\delta t \to 0}\left[\frac{(\mathbf{V}_x)_e\,\delta m_e - (\mathbf{V}_x)_i\,\delta m_i}{\delta t}\right] = \sum \dot{m}_e(\mathbf{V}_e)_x - \sum \dot{m}_i(\mathbf{V}_i)_x \quad \textbf{(14.6)}$$

Thus, as $\delta t \to 0$, we have a rate form of the momentum equation for the control volume.

$$\sum F_x = \frac{d(m\mathbf{V}_x)_{\text{c.v.}}}{dt} + \sum \dot{m}_e(\mathbf{V}_e)_x - \sum \dot{m}_i(\mathbf{V}_i)_x \quad \textbf{(14.7)}$$

Similar equations can also be written for the y- and z-directions.

$$\sum F_y = \frac{d(m\mathbf{V}_y)_{\text{c.v.}}}{dt} + \sum \dot{m}_e(\mathbf{V}_e)_y - \sum \dot{m}_i(\mathbf{V}_i)_y \tag{14.8}$$

$$\sum F_z = \frac{d(m\mathbf{V}_z)_{\text{c.v.}}}{dt} + \sum \dot{m}_e(\mathbf{V}_e)_z - \sum \dot{m}_i(\mathbf{V}_i)_z \tag{14.9}$$

In this chapter we will be concerned primarily with steady-state, steady-flow processes in which there is a single flow with uniform properties into the control volume, and a single flow with uniform properties out of the control volume. The SSSF assumption means that the rate of momentum change for the control volume terms in Eqs. 14.7, 14.8, and 14.9 are equal to zero. That is,

$$\frac{d(m\mathbf{V}_x)_{\text{c.v.}}}{dt} = 0 \qquad \frac{d(m\mathbf{V}_y)_{\text{c.v.}}}{dt} = 0 \qquad \frac{d(m\mathbf{V}_z)_{\text{c.v.}}}{dt} = 0 \tag{14.10}$$

Therefore for the SSSF process the momentum equation for the control volume, assuming uniform properties at each state, reduces to the form

$$\sum F_x = \sum \dot{m}_e(\mathbf{V}_e)_x - \sum \dot{m}_i(\mathbf{V}_i)_x \tag{14.11}$$
$$\sum F_y = \sum \dot{m}_e(\mathbf{V}_e)_y - \sum \dot{m}_i(\mathbf{V}_i)_y \tag{14.12}$$
$$\sum F_z = \sum \dot{m}_e(\mathbf{V}_e)_z - \sum \dot{m}_i(\mathbf{V}_i)_z \tag{14.13}$$

Further, for the special case in which there is a single flow into and out of the control volume, these equations reduce to

$$\sum F_x = \dot{m}[(\mathbf{V}_e)_x - (\mathbf{V}_i)_x] \tag{14.14}$$
$$\sum F_y = \dot{m}[(\mathbf{V}_e)_y - (\mathbf{V}_i)_y] \tag{14.15}$$
$$\sum F_z = \dot{m}[(\mathbf{V}_e)_z - (\mathbf{V}_i)_z] \tag{14.16}$$

Example 14.2

On a level floor a man is pushing a wheelbarrow (Fig. 14.3) into which sand is falling at the rate of 1 kg/s. The man is walking at the rate of 1 m/s and the sand has a velocity of 10 m/s as it falls into the wheelbarrow. Determine the force the man must exert on the wheelbarrow and the force the floor exerts on the wheelbarrow due to the falling sand.

Analysis and Solution

Consider a control surface around the wheelbarrow. Consider first the x-direction. From Eq. 14.7

$$\sum F_x = \frac{d(m\mathbf{V}_x)_{\text{c.v.}}}{dt} + \sum \dot{m}_e(\mathbf{V}_e)_x - \sum \dot{m}_i(\mathbf{V}_i)_x$$

Let us analyze this problem from the point of view of an observer riding on the wheelbarrow. For this observer, \mathbf{V}_x of the material in the wheelbarrow is zero

FIGURE 14.3
Sketch for Example 14.2.

and therefore,

$$\frac{d(m\mathbf{V}_x)_{\text{c.v.}}}{dt} = 0$$

However, for this observer the sand crossing the control surface has an x-component velocity of -1 m/s, and \dot{m}, the mass flow out of the control volume, is -1 kg/s. Therefore,

$$F_x = (1 \text{ kg/s}) \times (1 \text{ m/s}) = 1\text{N}$$

If one considers this from the point of view of an observer who is stationary on the earth's surface we conclude that \mathbf{V}_x of the falling sand is zero and therefore

$$\sum \dot{m}_e(\mathbf{V}_e)_x - \sum \dot{m}_i(\mathbf{V}_i)_x = 0$$

However, for this observer there is a change of momentum within the control volume, namely,

$$\sum F_x = \frac{d(m\mathbf{V}_x)_{\text{c.v.}}}{dt} = (1 \text{ m/s}) \times (1 \text{ kg/s}) = 1\text{N}$$

Next consider the vertical (y) direction.

$$\sum F_y = \frac{d(m\mathbf{V}_y)_{\text{c.v.}}}{dt} + \sum \dot{m}_e(\mathbf{V}_e)_y - \sum \dot{m}_i(\mathbf{V}_i)_y$$

For both the stationary and moving observer the first term drops out because \mathbf{V}_y of the mass within the control volume is zero. However, for the mass crossing the control surface, $\mathbf{V}_y = 10$ m/s and

$$\dot{m} = -1 \text{ kg/s}$$

Therefore,

$$F_y = (10 \text{ m/s}) \times (-1 \text{ kg/s}) = -10\text{N}$$

The minus sign indicates that the force is in the opposite direction to \mathbf{V}_y.

14.3 FORCES ACTING ON A CONTROL SURFACE

In the last section we considered the momentum equation for the control volume. We now wish to evaluate the net force on a control surface that causes this change in momentum. Let us do this by considering the system shown in Fig. 14.4, which involves a pipe bend. The control surface is designated by the dotted lines, and is so chosen that at the point where the fluid crosses the system boundary the flow is perpendicular to the control surface. The shear forces at the section where the fluid crosses the boundary of the system are assumed to be negligible. Figure 14.4a shows the velocities and Fig. 14.4b shows the forces involved. The force R is the result of all external forces on the system, except for the pressure of the surroundings. The pressure of the surroundings, P_0, acts on the entire boundary except at \mathscr{A}_i and \mathscr{A}_e, where the fluid crosses the control surface, P_i and P_e represent the absolute pressures at these points.

The net forces acting on the system in the x- and y-directions, F_x and F_y, are the sum of the pressure forces and the external force R in their respective directions. The influence of the pressure of the surroundings, P_0, is most easily taken into account by noting that it acts over the entire system boundary except at \mathscr{A}_i and \mathscr{A}_e. Therefore, we can write

$$\sum F_x = (P_i \mathscr{A}_i)_x - (P_0 \mathscr{A}_i)_x + (P_e \mathscr{A}_e)_x - (P_0 \mathscr{A}_e)_x + R_x$$
$$\sum F_y = (P_i \mathscr{A}_i)_y - (P_0 \mathscr{A}_i)_y + (P_e \mathscr{A}_e)_y - (P_0 \mathscr{A}_e)_y + R_y$$

This equation may be simplified by combining the pressure terms.

$$\sum F_x = [(P_i - P_0) \mathscr{A}_i]_x + [(P_e - P_0) \mathscr{A}_e]_x + R_x$$
$$\sum F_y = [(P_i - P_0) \mathscr{A}_i]_y + [(P_e - P_0) \mathscr{A}_e]_y + R_y \qquad \textbf{(14.17)}$$

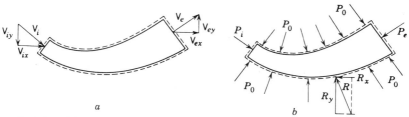

a b

FIGURE 14.4
Forces acting on a control surface.

The proper sign for each pressure and force must of course be used in all calculations.

Equations 14.11, 14.12, and 14.17 may be combined to give

$$\sum F_x = \sum \dot{m}_e(\mathbf{V}_e)_x - \sum \dot{m}_i(\mathbf{V}_i)_x$$
$$= \sum [(P_i - P_0)\mathscr{A}_i]_x + \sum [(P_e - P_0)\mathscr{A}_e]_x + R_x$$
$$\sum F_y = \sum \dot{m}_e(\mathbf{V}_e)_y - \sum \dot{m}_i(\mathbf{V}_i)_y$$
$$= \sum [(P_i - P_0)\mathscr{A}_i]_y + \sum [(P_e - P_0)\mathscr{A}_e]_y + R_y \qquad \textbf{(14.18)}$$

If there is a single flow across the control surface, Eqs. 14.14, 14.15, and 14.17 can be combined to give

$$\sum F_x = \dot{m}(\mathbf{V}_e - \mathbf{V}_i)_x = [(P_i - P_0)\mathscr{A}_i]_x + [(P_e - P_0)\mathscr{A}_e]_x + R_x$$
$$\sum F_y = \dot{m}(\mathbf{V}_e - \mathbf{V}_i)_y = [(P_i - P_0)\mathscr{A}_i]_y + [(P_e - P_0)\mathscr{A}_e]_y + R_y \qquad \textbf{(14.19)}$$

A similar equation could be written for the z-direction. These equations are very useful in analyzing the forces involved in a control volume analysis.

Example 14.3

A jet engine is being tested on a test stand (Fig. 14.5). The inlet area to the compressor is 1.8 ft^2, and air enters the compressor at 14.0 lbf/in.2, 300 ft/s. The pressure of the atmosphere is 14.7 lbf/in.2 The exit area of the engine is 1 ft^2, and the products of combustion leave the exit plane at a pressure of 18 lbf/in.2 and a velocity of 1300 ft/s. The air-fuel ratio is 50 lbm air/lbm fuel, and the fuel enters with a low velocity. The rate of air flow entering the engine is 30 lbm/s. Determine the thrust on the engine.

Analysis and Solution

In the solution that follows it is assumed that forces and velocities to the right are positive.

$\mathscr{A}_i = 1.8$ ft^2
$V_i = 300$ ft/s
$P_i = 14.0$ lbf/in.2
$\dot{m}_i = 30$ lbm/s

Fuel 0.6 lbm/s

$\mathscr{A}_e = 1$ ft^2
$V_e = 1300$ ft/s
$P_e = 18$ lbf/in.2
$\dot{m}_e = 30.6$ lbm/s

R_x

FIGURE 14.5
Sketch for Example 14.3.

Using Eq. 14.19

$$R_x + [(P_i - P_0)\mathscr{A}_i]_x + [(P_e - P_0)\mathscr{A}_e]_x = (\dot{m}_e\mathbf{V}_e - \dot{m}_i\mathbf{V}_i)_x$$

$$R_x + [(14.0 - 14.7)144 \text{ lbf/ft}^2 \times 1.8 \text{ ft}^2] - [(18.0 - 14.7)\,144 \text{ lbf/ft}^2 \times 1.0 \text{ ft}^2]$$

$$= \frac{(30.6 \times 1300 - 30 \times 300) \text{ lbm ft/s}^2}{32.17 \text{ lbm ft/lbf s}^2}$$

$$R_x = 957 + 180 + 475 = 1612 \text{ lbf}$$

(Note that the momentum of the fuel entering has been neglected.)

14.4 ADIABATIC, ONE-DIMENSIONAL, STEADY-STATE STEADY FLOW OF AN INCOMPRESSIBLE FLUID THROUGH A NOZZLE

A nozzle is a device in which the kinetic energy of a fluid is increased in an adiabatic process. This increase involves a decrease in pressure and is accomplished by the proper change in flow area. A diffuser is a device that has the opposite function, namely, to increase the pressure by decelerating the fluid. In this section we discuss both nozzles and diffusers, but to minimize words we shall use only the term nozzle.

Consider the nozzle shown in Fig. 14.6, and assume an adiabatic, one-dimensional, steady-state steady-flow process of an incompressible fluid. From the continuity equation we conclude that

$$\dot{m}_e = \dot{m}_i = \rho\mathscr{A}_i\mathbf{V}_i = \rho\mathscr{A}_e\mathbf{V}_e$$

or

$$\frac{\mathscr{A}_i}{\mathscr{A}_e} = \frac{\mathbf{V}_e}{\mathbf{V}_i} \qquad (14.20)$$

The first law for this process is

$$h_e - h_i + \frac{\mathbf{V}_e{}^2 - \mathbf{V}_i{}^2}{2} + (Z_e - Z_i)g = 0 \qquad (14.21)$$

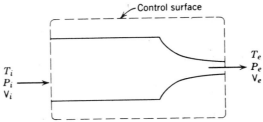

FIGURE 14.6
Schematic sketch of a nozzle.

From the second law we conclude that $s_e \geq s_i$, where the equality holds for a reversible process. Therefore, from the relation

$$T\,ds = dh - v\,dP$$

we conclude that for the reversible process

$$h_e - h_i = \int_i^e v\,dP \tag{14.22}$$

If we assume that the fluid is incompressible, Eq. 14.22 can be integrated to give

$$h_e - h_i = v(P_e - P_i) \tag{14.23}$$

Substituting this in Eq. 14.21 we have

$$v(P_e - P_i) + \frac{\mathbf{V}_e{}^2 - \mathbf{V}_i{}^2}{2} + (Z_e - Z_i)g = 0 \tag{14.24}$$

This is of course the Bernoulli equation, which was derived in Section 7.14, Eq. 7.62, and for the reversible, adiabatic, one-dimensional, steady-state steady flow of an incompressible fluid through a nozzle the Bernoulli equation represents a combined statement of the first and second laws of thermodynamics.

Example 14.4

Water enters the diffuser in a pump casing with a velocity of 30 m/s, a pressure of 350 kPa, and a temperature of 25°C. It leaves the diffuser with a velocity of 7 m/s and a pressure of 600 kPa. Determine the exit pressure for a reversible diffuser with these inlet conditions and exit velocity. Determine the increase in enthalpy, internal energy, and entropy for the actual diffuser.

Analysis and Solution

Consider first a control surface around a reversible diffuser with the given inlet conditions and exit velocity. Equation 14.24, the Bernoulli equation, is a statement of the first and second laws of thermodynamics for this process. Since there is no change in elevation this equation reduces to

$$v[(P_e)_s - P_i] + \frac{\mathbf{V}_e{}^2 - \mathbf{V}_i{}^2}{2} = 0$$

where $(P_e)_s$ represents the exit pressure for the reversible diffuser. From the steam tables, $v = 0.001\,003$ m³/kg.

$$P_{es} - P_i = \frac{(30)^2 - (7)^2}{0.001\,003 \times 2 \times 1000} = 424\,\text{kPa}$$

$$P_{es} = 774\,\text{kPa}$$

Next consider a control surface around the actual diffuser. The change in enthalpy can be found from the first law for this process, Eq. 14.21.

$$h_e - h_i = \frac{\mathbf{V}_i^2 - \mathbf{V}_e^2}{2} = \frac{(30)^2 - (7)^2}{2 \times 1000} = 0.4255 \text{ kJ/kg}$$

The change in internal energy can be found from the definition of enthalpy, $h_e - h_i = (u_e - u_i) + (P_e v_e - P_i v_i)$.

Thus, for an incompressible fluid

$$
\begin{aligned}
u_e - u_i &= h_e - h_i - v(P_e - P_i) \\
&= 0.4255 - 0.001\,003(600 - 350) \\
&= 0.174\,75 \text{ kJ/kg}
\end{aligned}
$$

The change of entropy can be approximated from the familiar relation

$$T\,ds = du + P\,dv$$

by assuming that the temperature is constant (which is approximately true in this case) and noting that for an incompressible fluid $dv = 0$. With these assumptions

$$s_e - s_i = \frac{u_e - u_i}{T} = \frac{0.174\,75}{298.2} = 0.000\,586 \text{ kJ/kg K}$$

Since this is an irreversible adiabatic process, the entropy will increase, as the above calculation indicates.

14.5 VELOCITY OF SOUND IN AN IDEAL GAS

When a pressure disturbance occurs in a compressible fluid, the disturbance travels with a velocity that depends on the state of the fluid. A sound wave is a very small pressure disturbance; the velocity of sound, also called the sonic velocity, is an important parameter in compressible-fluid flow. We proceed now to determine an expression for the sonic velocity of an ideal gas in terms of the properties of the gas.

Let a disturbance be set up by the movement of the piston at the end of the tube, Fig. 14.7a. A wave travels down the tube with a velocity c, which is the sonic velocity. Assume that after the wave has passed the properties of the gas have changed an infinitesimal amount and that the gas is moving with the velocity $d\mathbf{V}$ toward the wave front.

In Fig. 14.7b this process is shown from the point of view of an observer who travels with the wave front. Consider the control surface shown in Fig. 14.7b. From the first law for this steady-state steady-flow process we can write

$$h + \frac{c^2}{2} = (h + dh) + \frac{(c - d\mathbf{V})^2}{2}$$

$$dh - c\,d\mathbf{V} = 0 \tag{14.25}$$

FIGURE 14.7

**Diagram illustrating sonic velocity. (a) Stationary observer.
(b) Observer traveling with wave front.**

From the continuity equation we can write

$$\rho \mathscr{A} c = (\rho + d\rho)\mathscr{A}(c - d\mathbf{V})$$
$$c\,d\rho - \rho\,d\mathbf{V} = 0 \tag{14.26}$$

Consider also the relation between properties

$$T\,ds = dh - \frac{dP}{\rho}$$

If the process is isentropic, $ds = 0$, and this equation can be combined with Eq. 14.25 to give the relation

$$\frac{dP}{\rho} - c\,d\mathbf{V} = 0 \tag{14.27}$$

This can be combined with Eq. 14.26 to give the relation

$$\frac{dP}{d\rho} = c^2$$

Since we have assumed the process to be isentropic this is better written as a partial derivative.

$$\left(\frac{\partial P}{\partial \rho}\right)_s = c^2 \tag{14.28}$$

An alternate derivation is to introduce the momentum equation. For the control volume of Fig. 14.7b the momentum equation is

$$P\mathscr{A} - (P + dP)\mathscr{A} = \dot{m}(c - d\mathbf{V} - c) = \rho \mathscr{A} c(c - d\mathbf{V} - c)$$
$$d\mathbf{V} = \rho c\,dP \tag{14.29}$$

On combining this with Eq. 14.26 we obtain Eq. 14.28.

$$\left(\frac{\partial P}{\partial \rho}\right)_s = c^2$$

It will be of particular advantage to solve Eq. 14.28 for the velocity of sound in an ideal gas.

When an ideal gas undergoes an isentropic change of state, we found in Chapter 7 that, for this process, assuming constant specific heat.

$$\frac{dP}{P} - k\frac{d\rho}{\rho} = 0$$

or

$$\left(\frac{\partial P}{\partial \rho}\right)_s = \frac{kP}{\rho}$$

Substituting this equation in Eq. 14.28 we have an equation for the velocity of sound in an ideal gas,

$$c^2 = \frac{kP}{\rho} \tag{14.30}$$

Since for an ideal gas

$$\frac{P}{\rho} = RT$$

this equation may also be written

$$c^2 = kRT \tag{14.31}$$

Example 14.5

Determine the velocity of sound in air at 80 F and at 1000 F.

Analysis and Solution

Using Eq. 14.31

$$c = \sqrt{kRT}$$
$$= \sqrt{1.4 \times 32.17 \text{ lbm ft/lbf } s^2 \times 53.34 \text{ ft lbf/lbm R} \times 540 \text{ R}}$$
$$= 1138 \text{ ft/s}$$

Similarly, at 1000 F, using $k = 1.4$,

$$c = \sqrt{1.4 \times 32.17 \times 53.34 \times 1460}$$
$$= 1862 \text{ ft/s}$$

Note the significant increase in sonic velocity as the temperature increases.

The Mach number, M, is defined as the ratio of the actual velocity **V** to the sonic velocity c.

$$M = \frac{\mathbf{V}}{c} \tag{14.32}$$

When $M > 1$ the flow is supersonic; when $M < 1$ the flow is subsonic; and when $M = 1$ the flow is sonic. The importance of the Mach number as a parameter in fluid-flow problems will be evident in the paragraphs that follow.

14.6 REVERSIBLE, ADIABATIC, ONE-DIMENSIONAL STEADY FLOW OF AN IDEAL GAS THROUGH A NOZZLE

A nozzle or diffuser with both a converging and diverging section is shown in Fig. 14.8. The minimum cross-sectional area is called the throat.

Our first consideration concerns the conditions that determine whether a nozzle or diffuser should be converging or diverging, and the conditions that prevail at the throat. For the control volume shown the following relations can be written. First law:

$$dh + \mathbf{V}\, d\mathbf{V} = 0 \tag{14.33}$$

Property relation:

$$T\, ds = dh - \frac{dP}{\rho} = 0 \tag{14.34}$$

Continuity equation:

$$\rho \mathscr{A} \mathbf{V} = \dot{m} = \text{constant}$$

$$\frac{d\rho}{\rho} + \frac{d\mathscr{A}}{\mathscr{A}} + \frac{d\mathbf{V}}{\mathbf{V}} = 0 \tag{14.35}$$

Combining Eqs. 14.33 and 14.34 we have

$$dh = \frac{dP}{\rho} = -\mathbf{V}\, d\mathbf{V}$$

$$d\mathbf{V} = -\frac{1}{\rho \mathbf{V}}\, dP$$

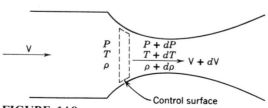

FIGURE 14.8

One-dimensional, reversible, adiabatic steady flow through a nozzle.

Substituting this in Eq. 14.35

$$\frac{d\mathscr{A}}{\mathscr{A}} = \left(-\frac{d\rho}{\rho} - \frac{d\mathbf{V}}{\mathbf{V}}\right) = -\frac{d\rho}{\rho}\left(\frac{dP}{dP}\right) + \frac{1}{\rho\mathbf{V}^2}\,dP$$

$$= \frac{-dP}{\rho}\left(\frac{d\rho}{dP} - \frac{1}{\mathbf{V}^2}\right) = \frac{dP}{\rho}\left(-\frac{1}{(dP/d\rho)} + \frac{1}{\mathbf{V}^2}\right)$$

Since the flow is isentropic

$$\frac{dP}{d\rho} = c^2 = \frac{\mathbf{V}^2}{M^2}$$

and therefore,

$$\frac{d\mathscr{A}}{\mathscr{A}} = \frac{dP}{\rho\mathbf{V}^2}(1 - M^2) \qquad\qquad \textbf{(14.36)}$$

This is a very significant equation, for from it we can draw the following conclusions about the proper shape for nozzles and diffusers:

For a nozzle, $dP < 0$. Therefore,
 for a subsonic nozzle, $M < 1$, $d\mathscr{A} < 0$, and the nozzle is converging;
 for a supersonic nozzle, $M > 1$, $d\mathscr{A} > 0$, and the nozzle is diverging.

For a diffuser, $dP > 0$. Therefore,
 for a subsonic diffuser, $M < 1$, $d\mathscr{A} > 0$, and the diffuser is diverging;
 for a supersonic diffuser, $M > 1$, $d\mathscr{A} < 0$, and the diffuser is converging.

When $M = 1$, $d\mathscr{A} = 0$, which means that sonic velocity can be achieved only at the throat of a nozzle or diffuser. These conclusions are summarized in Fig. 14.9.

We will now develop a number of relations between the actual properties, stagnation properties, and Mach number. These relations are very useful in dealing with isentropic flow of an ideal gas in a nozzle.

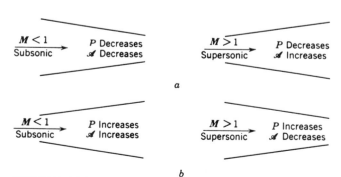

FIGURE 14.9
Required area changes for (a) nozzles and (b) diffusers.

Equation 14.1 gives the relation between enthalpy, stagnation enthalpy, and kinetic energy.

$$h + \frac{\mathbf{V}^2}{2} = h_0$$

For an ideal gas with constant specific heat Eq. 14.1 can be written

$$\mathbf{V}^2 = 2C_{po}(T_0 - T) = 2\frac{kRT}{k - 1}\left(\frac{T_0}{T} - 1\right)$$

Since

$$c^2 = kRT$$

$$\mathbf{V}^2 = \frac{2c^2}{k - 1}\left(\frac{T_0}{T} - 1\right)$$

$$\frac{\mathbf{V}^2}{c^2} = M^2 = \frac{2}{k - 1}\left(\frac{T_0}{T} - 1\right)$$

$$\frac{T_0}{T} = 1 + \frac{(k - 1)}{2}M^2 \tag{14.37}$$

For an isentropic process,

$$\left(\frac{T_0}{T}\right)^{k/(k-1)} = \frac{P_0}{P} \qquad \left(\frac{T_0}{T}\right)^{1/(k-1)} = \frac{\rho_0}{\rho}$$

Therefore,

$$\frac{P_0}{P} = \left[1 + \frac{(k - 1)}{2}M^2\right]^{k/(k-1)} \tag{14.38}$$

$$\frac{\rho_0}{\rho} = \left[1 + \frac{(k - 1)}{2}M^2\right]^{1/(k-1)} \tag{14.39}$$

Values of P/P_0, ρ/ρ_0, and T/T_0 are given as a function of M in Tables 30 to 35 of the *Gas Tables*, each table being for a given value of k. Table A.16 of the Appendix has been abstracted from Table 30 of the *Gas Tables*, and applies to an ideal gas with $k = 1.4$.

The conditions at the throat of the nozzle can be found by noting that $M = 1$ at the throat. The properties at the throat are denoted by an asterisk (*). Therefore,

$$\frac{T^*}{T_0} = \frac{2}{k + 1} \tag{14.40}$$

$$\frac{P^*}{P_0} = \left(\frac{2}{k + 1}\right)^{k/(k-1)} \tag{14.41}$$

$$\frac{\rho^*}{\rho_0} = \left(\frac{2}{k + 1}\right)^{1/(k-1)} \tag{14.42}$$

TABLE 14.1
Critical Pressure, Density, and Temperature Ratios for Isentropic Flow of an Ideal Gas

	$k = 1.1$	$k = 1.2$	$k = 1.3$	$k = 1.4$	$k = 1.67$
P^*/P_0	0.5847	0.5644	0.5457	0.5283	0.4867
ρ^*/ρ_0	0.6139	0.6209	0.6276	0.6340	0.6497
T^*/T_0	0.9524	0.9091	0.8696	0.8333	0.7491

These properties at the throat of a nozzle when $M = 1$ are frequently referred to as critical pressure, critical temperature, and critical density and the ratios given by Eqs. 14.40, 14.41, and 14.42 are referred to as the critical-temperature ratio, critical-pressure ratio, and critical-density ratio. Table 14.1 gives these ratios for various values of k.

14.7 MASS RATE OF FLOW OF AN IDEAL GAS THROUGH AN ISENTROPIC NOZZLE

We now turn our attention to a consideration of the mass rate of flow per unit area, \dot{m}/\mathscr{A}, in a nozzle. From the continuity equation we proceed as follows:

$$
\begin{aligned}
\frac{\dot{m}}{\mathscr{A}} = \rho \mathbf{V} &= \frac{P\mathbf{V}}{RT} \sqrt{\frac{kT_0}{kT_0}} \\
&= \frac{P\mathbf{V}}{\sqrt{kRT}} \sqrt{\frac{k}{R}} \sqrt{\frac{T_0}{T}} \sqrt{\frac{1}{T_0}} \\
&= \frac{PM}{\sqrt{T_0}} \sqrt{\frac{k}{R}} \sqrt{1 + \frac{k-1}{2} M^2}
\end{aligned}
\qquad (14.43)
$$

By substituting Eq. 14.38 into Eq. 14.42 the flow per unit area can be expressed in terms of stagnation pressure, stagnation temperature, Mach number, and gas properties.

$$
\frac{\dot{m}}{\mathscr{A}} = \frac{P_0}{\sqrt{T_0}} \sqrt{\frac{k}{R}} \times \frac{M}{\left(1 + \dfrac{k-1}{2} M^2\right)^{(k+1)/2(k-1)}}
\qquad (14.44)
$$

At the throat, $M = 1$, and therefore the flow per unit area at the throat, \dot{m}/\mathscr{A}^*, can be found by setting $M = 1$ in Eq. 14.44.

$$
\frac{\dot{m}}{\mathscr{A}^*} = \frac{P_0}{\sqrt{T_0}} \sqrt{\frac{k}{R}} \times \frac{1}{\left(\dfrac{k+1}{2}\right)^{(k+1)/2(k-1)}}
\qquad (14.45)
$$

The area ratio $\mathscr{A}/\mathscr{A}^*$ can be obtained by dividing Eq. 14.45 by Eq. 14.44.

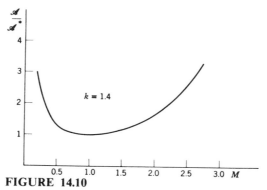

FIGURE 14.10

Area ratio as a function of Mach number for a reversible, adiabatic nozzle.

$$\frac{\mathscr{A}}{\mathscr{A}^*} = \frac{1}{M}\left[\left(\frac{2}{k+1}\right)\left(1 + \frac{k-1}{2}M^2\right)\right]^{(k+1)/2(k-1)} \tag{14.46}$$

The area ratio $\mathscr{A}/\mathscr{A}^*$ is the ratio of the area at the point where the Mach number is M to the throat area, and values of $\mathscr{A}/\mathscr{A}^*$ as a function of Mach number are given in Tables 30 through 35 of the *Gas Tables* and in Table A.16 in the Appendix. Figure 14.10 shows a plot of $\mathscr{A}/\mathscr{A}^*$ vs. M, which is in accordance with our previous conclusion that a subsonic nozzle is converging and a supersonic nozzle is diverging.

The final point to be made regarding the isentropic flow of an ideal gas through a nozzle involves the effect of varying the back pressure (the pressure outside the nozzle exit) on the mass rate of flow.

Consider first a convergent nozzle as shown in Fig. 14.11, which also shows the pressure ratio P/P_0 along the length of the nozzle. The conditions upstream

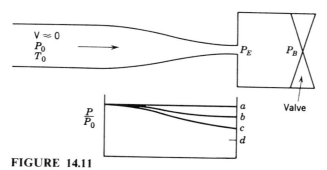

FIGURE 14.11

Pressure ratio as a function of back pressure for a convergent nozzle.

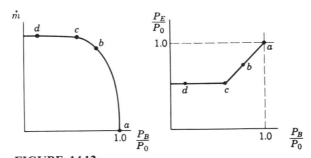

FIGURE 14.12

Mass rate of flow and exit pressure as a function of back pressure for a convergent nozzle.

are the stagnation conditions, which are assumed to be constant. The pressure at the exit plane of the nozzle is designated P_E, and the back pressure P_B. Let us consider how the mass rate of flow \dot{m} and the exit plane pressure P_E/P_0 vary as the back pressure P_B is decreased. These quantities are plotted in Fig. 14.12.

When $P_B/P_0 = 1$ there is of course no flow, and $P_E/P_0 = 1$ as designated by point a. Next let the back pressure P_B be lowered to that designated by point b, so that P_B/P_0 is greater than the critical-pressure ratio. The mass rate of flow has a certain value and $P_E = P_B$. The exit Mach number is less than 1. Next let the back pressure be lowered to the critical pressure, designated by point c. The Mach number at the exit is now unity, and P_E is equal to P_B. When P_B is decreased below the critical pressure, designated by point d, there is no further increase in the mass rate of flow, and P_E remains constant at a value equal to the critical pressure, and the exit Mach number is unity. The drop in pressure from P_E to P_B takes place outside the nozzle exit. Under these conditions the nozzle is said to be choked, which means that for given stagnation conditions the nozzle is passing the maximum possible mass flow.

Consider next a convergent–divergent nozzle in a similar arrangement, Fig. 14.13. Point a designates the condition when $P_B = P_0$ and there is no flow. When P_B is decreased to the pressure indicated by point b, so that P_B/P_0 is less than 1 but considerably greater than the critical-pressure ratio, the velocity increases in the convergent section, but $M < 1$ at the throat. Therefore, the diverging section acts as a subsonic diffuser in which the pressure increases and velocity decreases. Point c designates the back pressure at which $M = 1$ at the throat, but the diverging section acts as a subsonic diffuser (with $M = 1$ at the inlet) in which the pressure increases and velocity decreases. Point d designates one other back pressure that permits isentropic flow, and in this case the diverging section acts as a supersonic nozzle, with a decrease in pressure and an increase in velocity. Between the back pressures designated by points c and d, an isentropic solution is not possible, and shock waves will be present. This matter is discussed in the section that follows. When the back pressure is decreased below that designated by point d, the exit-

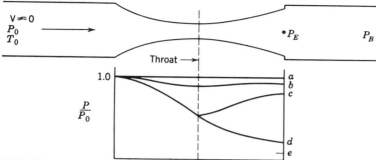

FIGURE 14.13

Nozzle pressure ratio as a function of back pressure for a reversible, convergent–divergent nozzle.

plane pressure P_E remains constant, and the drop in pressure from P_E to P_B takes place outside the nozzle. This is designated by point e.

Example 14.6

A convergent nozzle has an exit area of 500 mm². Air enters the nozzle with a stagnation pressure of 1000 kPa and a stagnation temperature of 360 K. Determine the mass rate of flow for back pressures of 800 kPa, 528 kPa, and 300 kPa, assuming isentropic flow.

Analysis and Solution

For air $k = 1.4$ and Table 30 of the *Gas Tables* (or Table A.16 in the Appendix) may be used. The critical-pressure ratio, P^*/P_0, is 0.528. Therefore, for a back pressure of 528 kPa, $M = 1$ at the nozzle exit and the nozzle is choked. Decreasing the back pressure below 528 kPa will not increase the flow.

For a back pressure of 528 kPa,

$$\frac{T^*}{T} = 0.8333 \qquad T^* = 300 \text{ K}$$

At the exit

$$\mathbf{V} = c = \sqrt{kRT}$$
$$= \sqrt{1.4 \times 0.287 \times 300 \times 1000} = 347.2 \text{ m/s}$$
$$\rho^* = \frac{P^*}{RT^*} = \frac{528}{0.287 \times 300} = 6.1324 \text{ kg/m}^3$$
$$\dot{m} = \rho \mathscr{A} \mathbf{V}$$

Applying this relation to the throat section.

$$\dot{m} = 6.1324 \times 500 \times 10^{-6} \times 347.2 = 1.0646 \text{ kg/s}$$

For a back pressure of 800 kPa, $P_E/P_0 = 0.8$ (subscript E designates the properties in the exit plane). From Table A.16.

$$M_E = 0.573 \qquad T_E/T_0 = 0.9381$$
$$T_E = 337.7 \text{ K}$$
$$c_E = \sqrt{kRT_E} = \sqrt{1.4 \times 0.287 \times 337.7 \times 1000} = 368.4 \text{ m/s}$$
$$\mathbf{V}_E = M_E c_E = 211.1 \text{ m/s}$$
$$\rho_E = \frac{P_E}{RT_E} = \frac{800}{0.287 \times 337.7} = 8.2542 \text{ kg/m}^3$$
$$\dot{m} = \rho \mathscr{A} \mathbf{V}$$

Applying this relation to the exit section,

$$\dot{m} = 8.2542 \times 500 \times 10^{-6} \times 211.1 = 0.8712 \text{ kg/s}$$

For a back pressure less than the critical pressure, which in this case is 528 kPa, the nozzle is choked and the mass rate of flow is the same as that for the critical pressure. Therefore, for an exhaust pressure of 300 kPa, the mass rate of flow is 1.0646 kg/s.

Example 14.7

A converging–diverging nozzle has an exit area to throat area ratio of 2. Air enters this nozzle with a stagnation pressure of 1000 kPa and a stagnation temperature of 360 K. The throat area is 500 mm². Determine the mass rate of flow, exit pressure, exit temperature, exit Mach number, and exit velocity for the following conditions:

(a) Sonic velocity at the throat, diverging section acting as a nozzle. (Corresponds to point d in Fig. 14.13.)

(b) Sonic velocity at the throat, diverging section acting as a diffuser. (Corresponds to point c in Fig. 14.13.)

Analysis and Solution

(a) In Table A.16 of the Appendix we find that there are two Mach numbers listed for $\mathscr{A}/\mathscr{A}^* = 2$. One of these is greater than unity and one is less than unity. When the diverging section acts as a supersonic nozzle we use the value for $M > 1$. The following are from Table A.16.

$$\frac{\mathscr{A}_E}{\mathscr{A}^*} = 2.0 \qquad M_E = 2.197 \qquad \frac{P_E}{P_0} = 0.0939 \qquad \frac{T_E}{T_0} = 0.5089$$

Therefore,

$$P_E = 0.0939(1000) = 93.9 \text{ kPa}$$
$$T_E = 0.5089(360) = 183.2 \text{ K}$$
$$c_E = \sqrt{kRT_E} = \sqrt{1.4 \times 0.287 \times 183.2 \times 1000} = 271.3 \text{ m/s}$$
$$\mathbf{V}_E = M_E c_E = 2.197(271.3) = 596.1 \text{ m/s}$$

The mass rate of flow can be determined by considering either the throat section or the exit section. However, in general it is preferable to determine the mass rate of flow from conditions at the throat. Since in this case $M = 1$ at the throat, the calculation is identical to the calculation for the flow in the convergent nozzle of Example 14.6 when it is choked.

(b) The following are from Table A.16.

$$\frac{\mathcal{A}_E}{\mathcal{A}^*} = 2.0 \qquad M = 0.308 \qquad \frac{P_E}{P_0} = 0.936 \qquad \frac{T_E}{T_0} = 0.9812$$

$$P_E = 0.936(1000) = 936 \text{ kPa}$$

$$T_E = 0.9812(360) = 353.3 \text{ K}$$

$$c_E = \sqrt{kRT_E} = \sqrt{1.4 \times 0.287 \times 353.3 \times 1000} = 376.8 \text{ m/s}$$

$$\mathbf{V}_E = M_E c_E = 0.308(376.8) = 116 \text{ m/s}$$

Since $M = 1$ at the throat, the mass rate of flow is the same as in (a), which is also equal to the flow in the convergent nozzle of Example 14.6 when it is choked.

In the example above a solution assuming isentropic flow is not possible if the back pressure is between 936 kPa and 93.9 kPa. If the back pressure is in this range there will be either a normal shock in the nozzle or oblique shock waves outside the nozzle. The matter of normal shock waves is considered in the following section.

14.8 NORMAL SHOCK IN AN IDEAL GAS FLOWING THROUGH A NOZZLE

A shock wave involves an extremely rapid and abrupt change of state. In a normal shock this change of state takes place across a plane normal to the direction of the flow. Figure 14.14 shows a control surface that includes such a normal shock. We can now determine the relations that govern the flow. Assuming steady-state, steady-flow we can write the following relations, where subscripts x and y denote the conditions upstream and downstream of the shock, respectively. Note that no heat and work cross the control surface.

FIGURE 14.14

One-dimensional normal shock.

First law:

$$h_x + \frac{\mathbf{V}_x^{\,2}}{2} = h_y + \frac{\mathbf{V}_y^{\,2}}{2} = h_{0x} = h_{0y} \tag{14.47}$$

Continuity equation:

$$\frac{\dot{m}}{\mathscr{A}} = \rho_x \mathbf{V}_x = \rho_y \mathbf{V}_y \tag{14.48}$$

Momentum equation:

$$\mathscr{A}(P_x - P_y) = \dot{m}(\mathbf{V}_y - \mathbf{V}_x) \tag{14.49}$$

Second law: Since the process is adiabatic

$$s_y - s_x \geqslant 0 \tag{14.50}$$

The energy and continuity equations can be combined to give an equation that when plotted on the h-s diagram is called the Fanno line. Similarly, the momentum and continuity equations can be combined to give an equation the plot of which on the h-s diagram is known as the Rayleigh line. Both of these lines are shown on the h-s diagram of Fig. 14.15. It can be shown that the point of maximum entropy on each line, points a and b, corresponds to $M = 1$. The lower part of each line corresponds to supersonic velocities, and the upper part to subsonic velocities.

The two points where all three equations are satisfied are points x and y, x being in the supersonic region and y in the subsonic region. Since the second law

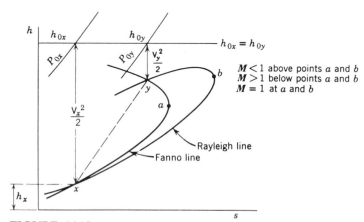

FIGURE 14.15

End states for a one-dimensional normal shock on an enthalpy–entropy diagram.

requires that $s_y - s_x \geq 0$ in an adiabatic process, we conclude that the normal shock can proceed only from x to y. This means that the velocity changes from supersonic ($M > 1$) before the shock to subsonic ($M < 1$) after the shock.

The equations governing normal shock waves will now be developed. If we assume constant specific heats we conclude from Eq. 14.47, the energy equation, that

$$T_{0x} = T_{0y} \tag{14.51}$$

That is, there is no change in stagnation temperature across a normal shock. Introducing Eq. 14.37

$$\frac{T_{0x}}{T_x} = 1 + \frac{k-1}{2} M_x{}^2 \qquad \frac{T_{0y}}{T_y} = 1 + \frac{k-1}{2} M_y{}^2$$

and substituting into Eq. 14.51 we have

$$\frac{T_y}{T_x} = \frac{1 + \dfrac{k-1}{2} M_x{}^2}{1 + \dfrac{k-1}{2} M_y{}^2} \tag{14.52}$$

The equation of state, the definition of Mach number, and the relation $c = \sqrt{kRT}$ can be introduced into the continuity equation as follows:

$$\rho_x \mathbf{V}_x = \rho_y \mathbf{V}_y$$

But

$$\rho_x = \frac{P_x}{RT_x} \qquad \rho_y = \frac{P_y}{RT_y}$$

$$\frac{T_y}{T_x} = \frac{P_y \mathbf{V}_y}{P_x \mathbf{V}_x} = \frac{P_y M_y c_y}{P_x M_x c_x} = \frac{P_y M_y \sqrt{T_y}}{P_x M_x \sqrt{T_x}}$$

$$= \left(\frac{P_y}{P_x}\right)^2 \left(\frac{M_y}{M_x}\right)^2 \tag{14.53}$$

Combining Eqs. 14.52 and 14.53, which involves combining the energy equation and the continuity equation, gives the equation of the Fanno line.

$$\frac{P_y}{P_x} = \frac{M_x \sqrt{1 + \dfrac{k-1}{2} M_x{}^2}}{M_y \sqrt{1 + \dfrac{k-1}{2} M_y{}^2}} \tag{14.54}$$

The momentum and continuity equations can be combined as follows to give the equation of the Rayleigh line.

$$P_x - P_y = \frac{\dot{m}}{\mathscr{A}}(\mathbf{V}_y - \mathbf{V}_x) = \rho_y \mathbf{V}_y^{\,2} - \rho_x \mathbf{V}_x^{\,2}$$

$$P_x + \rho_x \mathbf{V}_x^{\,2} = P_y + \rho_y \mathbf{V}_y^{\,2}$$

$$P_x + \rho_x M_x^{\,2} c_x^{\,2} = P_y + \rho_y M_y^{\,2} c_y^{\,2}$$

$$P_x + \frac{P_x M_x^{\,2}}{RT_x}(kRT_x) = P_y + \frac{P_y M_y^{\,2}}{RT_y}(kRT_y)$$

$$P_x(1 + kM_x^{\,2}) = P_y(1 + kM_y^{\,2})$$

$$\frac{P_y}{P_x} = \frac{1 + kM_x^{\,2}}{1 + kM_y^{\,2}} \tag{14.55}$$

Equations 14.54 and 14.55 can be combined to give the following equation relating M_x and M_y.

$$M_y^{\,2} = \frac{M_x^{\,2} + \dfrac{2}{k-1}}{\dfrac{2k}{k-1}M_x^{\,2} - 1} \tag{14.56}$$

Tables 48 through 53 of the *Gas Tables* give the normal shock functions, which include M_y as a function of M_x. Table A.17 of the Appendix has been abstracted from Table 48 of the *Gas Tables*, and applies to an ideal gas with $k = 1.4$. Note that M_x is always supersonic and M_y is always subsonic, which agrees with the previous statement that in a normal shock the velocity changes from supersonic to subsonic. These tables also give the pressure, density, temperature, and stagnation pressure ratios across a normal shock as a function of M_x. These are found from Eqs. 14.52 and 14.53 and the equation of state. Note that there is always a drop in stagnation pressure across a normal shock and an increase in the static pressure.

Example 14.8

Consider the convergent–divergent nozzle of Example 14.7 in which the diverging section acts as a supersonic nozzle (Fig. 14.16). Assume that a normal shock stands

FIGURE 14.16
Sketch for Example 14.8.

in the exit plane of the nozzle. Determine the static pressure and temperature and the stagnation pressure just downstream of the normal shock.

Sketch: Fig. 14.16.

Analysis and Solution

From Table 48 of the *Gas Tables* (Table A.17 of the Appendix)

$$M_x = 2.197 \quad M_y = 0.547 \quad \frac{P_y}{P_x} = 5.46 \quad \frac{T_y}{T_x} = 1.854 \quad \frac{P_{0y}}{P_{0x}} = 0.630$$

$$P_y = 5.46 \times P_x = 5.46(93.9) = 512.7 \text{ kPa}$$

$$T_y = 1.854 \times T_x = 1.854(183.2) = 339.7 \text{ K}$$

$$P_{0y} = 0.631 \times P_{0x} = 0.630(1000) = 630 \text{ kPa}$$

In the light of this example we can conclude the discussion concerning the flow through a convergent–divergent nozzle. Figure 14.13 is repeated here as Fig. 14.17 for convenience, except that points f, g, and h have been added. Consider point d. We have already noted that with this back pressure the exit plane pressure P_E is just equal to the back pressure P_B, and isentropic flow is maintained in the nozzle. Let the back pressure be raised to that designated by point f. The exit-plane pressure P_E is not influenced by this increase in back pressure, and the increase in pressure from P_E to P_B takes place outside the nozzle. Let the back pressure be raised to that designated by point g, which is just sufficient to cause a normal shock to stand in the exit plane of the nozzle. The exit-plane pressure P_E (downstream of the shock) is equal to the back pressure P_B, and $M < 1$ leaving the nozzle. This is the case in Example 14.8. Now let the back pressure be raised to that corresponding to point h. As the back pressure is raised from g to h the normal shock moves into the nozzle as indicated. Since $M < 1$ downstream of the normal shock, the diverging part of the nozzle that is downstream of the shock

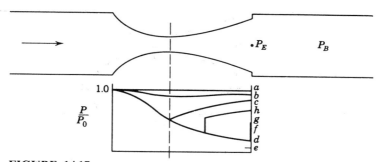

FIGURE 14.17

Nozzle pressure ratio as a function of back pressure for a convergent–divergent nozzle.

acts as a subsonic diffuser. As the back pressure is increased from h to c the shock moves further upstream and disappears at the nozzle throat where the back pressure corresponds to c. This is reasonable since there are no supersonic velocities involved when the back pressure corresponds to c, and hence no shock waves are possible.

Example 14.9

Consider the convergent–divergent nozzle of Examples 14.7 and 14.8. Assume that there is a normal shock wave standing at the point where $M = 1.5$. Determine the exit-plane pressure, temperature, and Mach number. Assume isentropic flow except for the normal shock (Fig. 14.18).

Sketch: Fig. 14.18.

Analysis and Solution

The properties at point x can be determined from Table A.16, because the flow is isentropic to point x.

$$M_x = 1.5 \qquad \frac{P_x}{P_{0x}} = 0.2724 \qquad \frac{T_x}{T_{0x}} = 0.6897 \qquad \frac{\mathscr{A}_x}{\mathscr{A}_x^*} = 1.1762$$

Therefore,

$$P_x = 0.2724(1000) = 272.4 \text{ kPa}$$
$$T_x = 0.6897(360) = 248.3 \text{ K}$$

The properties at point y can be determined from the normal shock functions, Table A.17.

$$M_y = 0.7011 \qquad \frac{P_y}{P_x} = 2.4583 \qquad \frac{T_y}{T_x} = 1.320 \qquad \frac{P_{0y}}{P_{0x}} = 0.9298$$
$$P_y = 2.4583 P_x = 2.4583(272.4) = 669.6 \text{ kPa}$$
$$T_y = 1.320 T_x = 1.320(248.3) = 327.8 \text{ K}$$
$$P_{0y} = 0.9298 P_{0x} = 0.9298(1000) = 929.8 \text{ kPa}$$

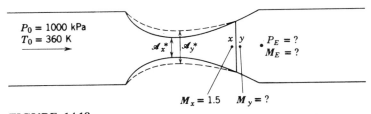

FIGURE 14.18
Sketch for Example 14.9.

Since there is no change in stagnation temperature across a normal shock,

$$T_{0x} = T_{0y} = 360 \text{ K}$$

From y to E the diverging section acts as a subsonic diffuser. In solving this problem it is convenient to think of the flow at y as having come from an isentropic nozzle having a throat area \mathscr{A}_y^*. Such a hypothetical nozzle is shown by the dotted line. From the table of isentropic flow functions, Table A.16, we find the following for $M_y = 0.7011$.

$$M_y = 0.7011 \quad \frac{\mathscr{A}_y}{\mathscr{A}_y^*} = 1.0938 \quad \frac{P_y}{P_{0y}} = 0.7202 \quad \frac{T_y}{T_{0y}} = 0.9105$$

From the statement of the problem

$$\frac{\mathscr{A}_E}{\mathscr{A}_x^*} = 2.0$$

Also, since the flow from y to E is isentropic,

$$\frac{\mathscr{A}_E}{\mathscr{A}_E^*} = \frac{\mathscr{A}_E}{\mathscr{A}_y^*} = \frac{\mathscr{A}_E}{\mathscr{A}_x^*} \times \frac{\mathscr{A}_x^*}{\mathscr{A}_x} \times \frac{\mathscr{A}_x}{\mathscr{A}_y} \times \frac{\mathscr{A}_y}{\mathscr{A}_y^*}$$

$$= \frac{\mathscr{A}_E}{\mathscr{A}_y^*} = 2.0 \times \frac{1}{1.1762} \times 1 \times 1.0938 = 1.860$$

From the table of isentropic flow functions for $\mathscr{A}/\mathscr{A}^* = 1.860$ and $M < 1$

$$M_E = 0.339 \quad \frac{P_E}{P_{0E}} = 0.9222 \quad \frac{T_E}{T_{0E}} = 0.9771$$

$$\frac{P_E}{P_{0E}} = \frac{P_E}{P_{0y}} = 0.9222$$

$$P_E = 0.9222(P_{0y}) = 0.9222(929.8) = 857.5 \text{ kPa}$$

$$T_E = 0.9771(T_{0E}) = 0.9771(360) = 351.7 \text{ K}$$

In conclusion it should be pointed out that in considering the normal shock we have ignored the effect of viscosity and thermal conductivity, which are certain to be present. The actual shock wave will occur over some finite thickness. However, the development as given here gives a very good qualitative picture of normal shocks, and also provides a basis for fairly accurate quantitative results.

14.9 FLOW OF A VAPOR THROUGH A NOZZLE

In this section we consider a vapor to be a substance that is in the gaseous phase but with limited superheat. Therefore, the vapor will probably deviate significantly from the ideal gas relations, and the possibility of condensation must be considered. The most familiar example is the flow of steam through the nozzles of a steam turbine.

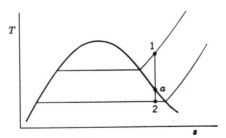

FIGURE 14.19
Reversible, adiabatic expansion of a vapor.

The principles that have been developed for the isentropic flow of an ideal gas apply also to the isentropic flow of a vapor. However, because the vapor deviates from the ideal gas relationships, the appropriate tables of thermodynamic properties must be used. Further, the possibility of condensation must be borne in mind, as indicated in Fig. 14.19. If steam expands isentropically from state 1 to state 2, and if equilibrium is maintained throughout the nozzle, condensation would begin at point a, and at pressures below this a mixture of liquid droplets and vapor would be present. In an actual nozzle the formation of the liquid tends to be delayed due to an effect known as supersaturation. This will be considered in a later paragraph.

Let us consider first the isentropic flow of steam in a nozzle without condensation. The value of the specific-heat ratio k for steam varies, but $k = 1.3$ is a good approximation over a considerable range. Therefore, the critical pressure ratio, P^*/P_0, can be found by the relation

$$\frac{P^*}{P_0} = \left(\frac{2}{k+1}\right)^{k/(k-1)} = 0.545 \qquad \textbf{(14.57)}$$

This is also the value given in Table 14.1. Knowing, therefore, the critical-pressure ratio, the throat area for a given flow can be calculated. The exit area of the nozzle can be calculated in a similar manner. The example below illustrates this procedure.

Example 14.10

Steam at a stagnation pressure of 100 lbf/in.² and a stagnation temperature of 600 F expands in a nozzle to 20 lbf/in.² Determine the throat area and exit area required for a flow of 20 000 lbm/h, assuming reversible adiabatic flow.

Analysis and Solution

First consider the properties and flow at the throat.

$$\frac{P^*}{P_0} = 0.545 \qquad\qquad P^* = 54.5 \text{ lbf/in.}^2$$

$$s^* = s_0 = 1.7582 \text{ Btu/lbm R} \qquad h_0 = 1329.3 \text{ Btu/lbm}$$

$$T^* = 460 \text{ F} \qquad\qquad h^* = 1264.0 \text{ Btu/lbm}$$

$$h_0 = h^* + \frac{\mathbf{V}^{*2}}{2} \qquad\qquad \mathbf{V}^* = \sqrt{2}\sqrt{h_0 - h^*}$$

$$\mathbf{V}^* = 223.7\sqrt{1329.3 - 1264.0} = 223.7\sqrt{65.3} = 1800 \text{ ft/s}$$

$$v^* = 9.961 \text{ (from the steam tables)}$$

$$\dot{m}v^* = \mathscr{A}^*\mathbf{V}^*$$

$$\mathscr{A}^* = \frac{20\,000 \times 9.961}{3600 \times 1800} = 0.0307 \text{ ft}^2 = 4.42 \text{ in.}^2$$

At the nozzle exit

$$P_E = 20 \text{ lbf/in.}^2 \qquad s_E = s_0 = 1.7582 \text{ Btu/lbm R}$$

Therefore,

$$h_E = 1174.9 \text{ Btu/lbm} \qquad v_E = 21.28 \text{ ft}^3/\text{lbm}$$

$$\mathbf{V}_E = \sqrt{2}\sqrt{h_0 - h_E} = 223.7\sqrt{1329.3 - 1174.9} = 223.7\sqrt{154.4}$$

$$= 2780 \text{ ft/s.}$$

$$\dot{m}v_E = \mathscr{A}_E\mathbf{V}_E$$

$$\mathscr{A}_E = \frac{20\,000 \times 21.28}{3600 \times 2780} = 0.0426 \text{ ft}^2 = 6.13 \text{ in.}^2$$

For steam initially saturated the critical-pressure ratio is usually taken as

$$\frac{P^*}{P_0} = 0.577 \tag{14.58}$$

If the flow with initially saturated steam is assumed to be isentropic there will be a certain amount of liquid entrained with the vapor. Calculation of throat and exit areas under most conditions is similar to the example above.

When the steam flowing through a nozzle is initially superheated, there will be a certain point in the nozzle where the steam becomes saturated. However, as noted in Section 13.5 in connection with a discussion of metastable equilibrium, if the point at which condensation would occur under equilibrium conditions (point a in Fig. 13.9) occurs in the divergent section of the nozzle, a condition of metastable equilibrium exists. That is, the formation of droplets is delayed, and the vapor temperature is less than the saturation temperature for the given pressure. This is frequently referred to as supersaturation.

This phenomenon of supersaturation is observed only in the diverging portions of the nozzle. In a nozzle in which supersaturation occurs the rate of mass flow might be slightly greater than that obtained in an isentropic flow without supersaturation.

Supersaturation may also be present in a supersonic wind tunnel. An abrupt condensation of the moisture in the air would disturb the flow pattern, and for this reason most of the moisture is removed from the air to be used in the tunnel.

A similar phenomenon may occur in the flow of a saturated liquid through a nozzle or valve. When a saturated liquid flows through a nozzle or valve some of it will vaporize as the pressure is reduced. In practice the formation of the vapor is delayed, because a metastable state exists.

14.10 NOZZLE AND DIFFUSER COEFFICIENTS

Up to this point we have considered only isentropic flow and normal shocks. As was pointed out in Chapter 7, isentropic flow through a nozzle provides a standard to which the performance of an actual nozzle can be compared. For nozzles, the three important parameters by which actual flow can be compared to the ideal flow are nozzle efficiency, velocity coefficient, and discharge coefficient. These are defined as follows:

The nozzle efficiency η_N is defined as

$$\eta_N = \frac{\text{Actual kinetic energy at nozzle exit}}{\text{Kinetic energy at nozzle exit with isentropic flow to same exit pressure}}$$

$$(14.59)$$

The efficiency can be defined in terms of properties. On the h-s diagram of Fig. 14.20 state $0i$ represents the stagnation state of the fluid entering the nozzle; state e represents the actual state at the nozzle exit; and state s represents the state that would have been achieved at the nozzle exit if the flow had been reversible and adiabatic to the same exit pressure. Therefore, in terms of these states the nozzle efficiency is

$$\eta_N = \frac{h_{0i} - h_e}{h_{0i} - h_s}$$

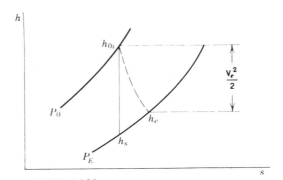

FIGURE 14.20

Enthalpy–entropy diagram showing the effects of irreversibility in a nozzle.

Nozzle efficiencies vary in general from 90 to 99 percent. Large nozzles usually have higher efficiencies than small nozzles, and nozzles with straight axes have higher efficiencies than nozzles with curved axes. The irreversibilities, which cause the departure from isentropic flow, are primarily due to frictional effects, and are confined largely to the boundary layer. The rate of change of cross-sectional area along the nozzle axis (that is, the nozzle contour) is an important parameter in the design of an efficient nozzle, particularly in the divergent section. Detailed consideration of this matter is beyond the scope of this text, and the reader is referred to standard references on the subject.

The velocity coefficient C_V is defined as

$$C_V = \frac{\text{Actual velocity at nozzle exit}}{\text{Velocity at nozzle exit with isentropic flow and same exit pressure}} \qquad (14.60)$$

It follows that the velocity coefficient is equal to the square root of the nozzle efficiency

$$C_V = \sqrt{\eta_N} \qquad (14.61)$$

The coefficient of discharge C_D is defined by the relation

$$C_D = \frac{\text{Actual mass rate of flow}}{\text{Mass rate of flow with isentropic flow}}$$

In determining the mass rate of flow with isentropic conditions, the actual back pressure is used if the nozzle is not choked. If the nozzle is choked, the isentropic mass rate of flow is based on isentropic flow and sonic velocity at the minimum section (that is, sonic velocity at the exit of a convergent nozzle and at the throat of a convergent–divergent nozzle).

The performance of a diffuser is usually given in terms of diffuser efficiency, which is best defined with the aid of an h-s diagram. On the h-s diagram of Fig. 14.21 states 1 and 01 are the actual and stagnation states of the fluid entering the diffuser. States 2 and 02 are the actual and stagnation states of the fluid leaving the diffuser. State 3 is not attained in the diffuser, but it is the state that has the same entropy as the initial state and the pressure of the isentropic stagnation state leaving the diffuser. The efficiency of the diffuser η_D is defined as

$$\eta_D = \frac{\Delta h_s}{\mathbf{V}_1{}^2/2} = \frac{h_3 - h_1}{h_{01} - h_1} = \frac{h_3 - h_1}{h_{02} - h_1} \qquad (14.62)$$

If we assume an ideal gas with constant specific heat this reduces to

$$\eta_D = \frac{T_3 - T_1}{T_{02} - T_1} = \frac{\dfrac{(T_3 - T_1)}{T_1} T_1}{\dfrac{\mathbf{V}_1{}^2}{2 C_{po}}}$$

$$C_{po} = \frac{kR}{k-1} \qquad T_1 = \frac{c_1{}^2}{kR} \qquad \mathbf{V}_1{}^2 = M_1{}^2 c_1{}^2 \qquad \frac{T_3}{T_1} = \left(\frac{P_{02}}{P_1}\right)^{(k-1)/k}$$

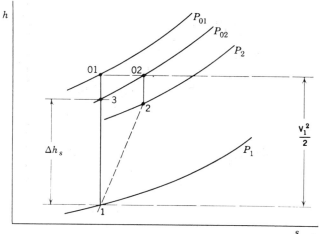

FIGURE 14.21

Enthalpy–entropy diagram showing the definition of diffuser efficiency.

Therefore,

$$\eta_D = \frac{\left(\dfrac{P_{02}}{P_1}\right)^{(k-1)/k} - 1}{\dfrac{k-1}{2} M_1^2}$$

$$\left(\frac{P_{02}}{P_1}\right)^{(k-1)/k} = \left(\frac{P_{01}}{P_1}\right)^{(k-1)/k} \times \left(\frac{P_{02}}{P_{01}}\right)^{(k-1)/k}$$

$$= \left(1 + \frac{k-1}{2} M_1^2\right)\left(\frac{P_{02}}{P_{01}}\right)^{(k-1)/k}$$

$$\eta_D = \frac{\left(1 + \dfrac{k-1}{2} M_1^2\right)\left(\dfrac{P_{02}}{P_{01}}\right)^{(k-1)/k} - 1}{\dfrac{k-1}{2} M_1^2} \qquad (14.63)$$

14.11 NOZZLES AND ORIFICES AS FLOW-MEASURING DEVICES

The mass rate of flow of a fluid flowing in a pipe is frequently determined by measuring the pressure drop across a nozzle or orifice in the line, as shown in Fig. 14.22. The ideal process for such a nozzle or orifice is assumed to be isentropic

FIGURE 14.22

Nozzles and orifices as flow-measuring devices.

flow through a nozzle that has the measured pressure drop from inlet to exit and a minimum cross-sectional area equal to the minimum area of the nozzle or orifice. The actual flow is related to the ideal flow by the coefficient of discharge, which is defined by Eq. 14.62.

The pressure difference measured across an orifice depends upon the location of the pressure taps as indicated in Fig. 14.22. Since the ideal flow is based on the measured pressure difference, it follows that the coefficient of discharge depends on the locations of the pressure taps. Also, the coefficient of discharge for a sharp-edged orifice is considerably less than that for a well-rounded nozzle, primarily due to a contraction of the stream, known as the vena contracta, as it flows through a sharpedged orifice.

There are two approaches to determining the discharge coefficient of a nozzle or orifice. One is to follow a standard design procedure, such as the ones established by the American Society of Mechanical Engineers,[1] and use the coefficient of discharge given for a particular design. A more accurate method is to calibrate a given nozzle or orifice, and determine the discharge coefficient for a given installation by accurately measuring the actual mass rate of flow. The procedure to be followed will depend on the accuracy desired and other factors involved (such as time, expense, availability of calibration facilities) in a given situation.

For incompressible fluids flowing through an orifice the ideal flow for a given pressure drop can be found by the procedure outlined in Section 14.4. Actually, it is advantageous to combine Eqs. 14.24 and 14.20 to give the following relation, which is valid for reversible flow.

$$v(P_2 - P_1) + \frac{\mathbf{V}_2^2 - \mathbf{V}_1^2}{2} = v(P_2 - P_1) + \frac{\mathbf{V}_2^2 - (\mathscr{A}_2/\mathscr{A}_1)^2\mathbf{V}_2^2}{2} = 0 \quad \textbf{(14.64)}$$

[1] *Fluid Meters, Their Theory and Application*, ASME, 1959. *Flow Measurement*, ASME, 1959.

or

$$v(P_2 - P_1) + \frac{\mathbf{V}_2^{\,2}}{2}\left[1 - \left(\frac{\mathscr{A}_2}{\mathscr{A}_1}\right)^2\right] = 0$$

$$\mathbf{V}_2 = \sqrt{\frac{2v(P_1 - P_2)}{[1 - (\mathscr{A}_2/\mathscr{A}_1)^2]}} \tag{14.65}$$

For an ideal gas it is frequently advantageous to use the following simplified procedure when the pressure drop across an orifice or nozzle is small. Consider the nozzle shown in Fig. 14.23. From the first law we conclude that

$$h_i + \frac{\mathbf{V}_i^{\,2}}{2} = h_e + \frac{\mathbf{V}_e^{\,2}}{2}$$

Assuming constant specific heat, this reduces to

$$\frac{\mathbf{V}_e^{\,2} - \mathbf{V}_i^{\,2}}{2} = h_i - h_e = C_{po}(T_i - T_e)$$

Let ΔP and ΔT be the decrease in pressure and temperature across the nozzle. Since we are considering reversible adiabatic flow we note that

$$\frac{T_e}{T_i} = \left(\frac{P_e}{P_i}\right)^{(k-1)/k}$$

or

$$\frac{T_i - \Delta T}{T_i} = \left(\frac{P_i - \Delta P}{P_i}\right)^{(k-1)/k}$$

$$1 - \frac{\Delta T}{T_i} = \left(1 - \frac{\Delta P}{P_i}\right)^{(k-1)/k}$$

Using the binomial expansion on the right side of the equation we have

$$1 - \frac{\Delta T}{T_i} = 1 - \frac{k-1}{k}\frac{\Delta P}{P_i} - \frac{k-1}{2k^2}\frac{\Delta P^2}{P_i^2}\cdots$$

If $\Delta P/P_i$ is small this reduces to

$$\frac{\Delta T}{T_i} = \frac{k-1}{k}\frac{\Delta P}{P_i}$$

FIGURE 14.23
Analysis of a nozzle as a flow-measuring device.

Substituting this into the first-law equation we have

$$\frac{\mathbf{V}_e^2 - \mathbf{V}_i^2}{2} = C_{po} \frac{k-1}{k} \Delta P \frac{T_i}{P_i}$$

But for an ideal gas

$$C_{po} = \frac{kR}{k-1} \qquad \text{and} \qquad v_i = R \frac{T_i}{P_i}$$

Therefore,

$$\frac{\mathbf{V}_e^2 - \mathbf{V}_i^2}{2} = v_i \Delta P$$

which is the same as Eq. 14.64, which was developed for incompressible flow. Therefore, when the pressure drop across a nozzle or orifice is small, the flow can be calculated with high accuracy by assuming incompressible flow.

The Pitot tube, Fig. 14.24, is an important instrument for measuring the velocity of a fluid. In calculating the flow with a Pitot tube it is assumed that the fluid is decelerated isentropically in front of the Pitot tube, and therefore the stagnation pressure of the free stream can be measured.

Applying the first law to this process we have

$$h + \frac{\mathbf{V}^2}{2} = h_0$$

If we assume incompressible flow for this isentropic process, the first law reduces to (because $T\,ds = dh - v\,dp$)

$$\frac{\mathbf{V}^2}{2} = h_0 - h = v(P_0 - P)$$

FIGURE 14.24

Schematic arrangement of a Pitot tube.

or

$$\mathbf{V} = \sqrt{2v(P_0 - P)} \tag{14.66}$$

If we consider the compressible flow of an ideal gas with constant specific heat, the velocity can be found from the relations

$$\frac{\mathbf{V}^2}{2} = h_0 - h = C_{po}(T_0 - T) = C_{po}T\left(\frac{T_0}{T} - 1\right)$$

$$= C_{po}T\left[\left(\frac{P_0}{P}\right)^{(k-1)/k} - 1\right] \tag{14.67}$$

It is of interest to know the error introduced by assuming incompressible flow when using the Pitot tube to measure the velocity of an ideal gas. To do so we introduce Eq. 14.38 and rearrange it as follows:

$$\frac{P_0}{P} = \left(1 + \frac{k-1}{2}M^2\right)^{k/(k-1)} = \left[1 + \left(\frac{k-1}{2}\right)\left(\frac{\mathbf{V}^2}{c^2}\right)\right]^{k/(k-1)} \tag{14.68}$$

But

$$\frac{\mathbf{V}^2}{2} + C_{po}T = C_{po}T_0$$

$$\frac{\mathbf{V}^2}{2} + \frac{kRc^2}{(k-1)kR} = \frac{kRc_0^2}{(k-1)kR}$$

$$1 + \frac{2c^2}{(k-1)\mathbf{V}^2} = \frac{2c_0^2}{(k-1)\mathbf{V}^2}, \quad \text{where} \quad c_0 = \sqrt{kRT_0}$$

$$\frac{c^2}{\mathbf{V}^2} = \frac{k-1}{2}\left[\left(\frac{2}{k-1}\right)\left(\frac{c_0^2}{\mathbf{V}^2}\right) - 1\right] = \frac{c_0^2}{\mathbf{V}^2} - \frac{k-1}{2}$$

or

$$\frac{c^2}{\mathbf{V}^2} = \frac{c_0^2}{\mathbf{V}^2} - \frac{k-1}{2} \tag{14.69}$$

Substituting this into Eq. 14.68 and rearranging

$$\frac{P}{P_0} = \left[1 - \frac{k-1}{2}\left(\frac{\mathbf{V}}{c_0}\right)^2\right]^{k/(k-1)} \tag{14.70}$$

Expanding this equation by the binomial theorem, and including terms through $(\mathbf{V}/c_0)^4$, we have

$$\frac{P}{P_0} = 1 - \frac{k}{2}\left(\frac{\mathbf{V}}{c_0}\right)^2 + \frac{k}{8}\left(\frac{\mathbf{V}}{c_0}\right)^4$$

On rearranging this we have

$$\frac{P_0 - P}{\rho_0 \mathbf{V}^2/2} = 1 - \frac{1}{4}\left(\frac{\mathbf{V}}{c_0}\right)^2 \tag{14.71}$$

TABLE 14.2

V/c_0	Approximate Room-Temperature Velocity, m/s	Error in Pressure for a Given Velocity, %	Error in Velocity for a Given Pressure, %
0.0	0	0	0
0.1	35	0.25	−0.13
0.2	70	1.0	−0.5
0.3	105	2.25	−1.2
0.4	140	4.0	−2.1
0.5	175	6.25	−3.3

For incompressible flow the corresponding equation is

$$\frac{P_0 - P}{\rho_0 V^2/2} = 1$$

Therefore, the second term on the right side of Eq. 14.71 represents the error involved if incompressible flow is assumed. The error in pressure for a given velocity and the error in velocity for a given pressure that would result from assuming incompressible flow are given in Table 14.2

14.12 FLOW THROUGH BLADE PASSAGES

We will now apply the principles we have been considering in this chapter to the flow of fluids through blade passages. We will limit our consideration to uniform, parallel, one-dimensional flow.

There are a number of different devices that involve flow through blade passages. Some of the familiar ones are steam and gas turbines, axial-flow compressors, and certain types of blowers, pumps, and torque converters. Some of these devices, such as the steam turbine, are made up of a number of stages. One stage consists of a nozzle (or blade passages that act as a nozzle) and the moving row of blades that follows. Sometimes the blades are also called buckets, but in this text the term blade will be used.

In the analysis of such problems it is necessary to discuss both absolute and relative velocities. By absolute velocity we mean the velocity a stationary observer would detect. The relative velocity is the velocity an observer traveling with the blade would detect, which is designated with a subscript R. The velocity of the blade is designated V_B.

It is advantageous to show these velocities on a vector diagram. Figure 14.25 shows typical vector diagrams for both a turbine and a compressor. In each case the first sketch shows the velocity vectors in relation to the blades, and the second sketch shows how these vector diagrams are usually drawn. In these diagrams V_1 represents the velocity of the fluid entering the blade passage, and α designates

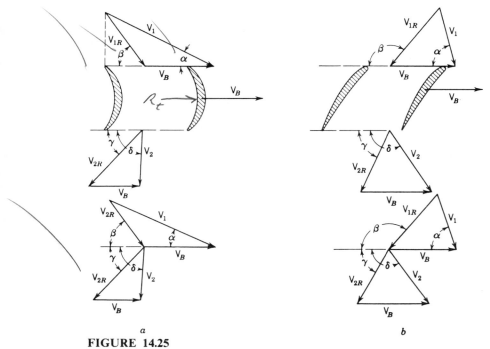

a *b*

FIGURE 14.25

Velocity vector diagrams for (*a*) a turbine and (*b*) a compressor.

the angle at which it enters. \mathbf{V}_{1R} represents the relative velocity of the fluid entering the blade passage and β the angle at which it enters. Similarly, \mathbf{V}_2 and \mathbf{V}_{2R} represent the velocity and relative velocity of the fluid leaving at the angles δ and γ, respectively.

In analyzing the flow through a blade passage it is convenient to consider the control surface shown in Fig. 14.26, which is shown from both a stationary and a moving observer's point of view.

Let us begin the analysis by writing the first law for both cases, assuming steady, adiabatic flow through the blade passage. The stationary observer is aware that work is being done on the blade, and the steady-flow equation for the first law is

$$h_1 + \frac{\mathbf{V}_1{}^2}{2} = h_2 + \frac{\mathbf{V}_2{}^2}{2} + w \tag{14.72}$$

The moving observer is not aware of any work being done on the blade, and in this case the first law is

$$h_1 + \frac{\mathbf{V}_{1R}{}^2}{2} = h_2 + \frac{\mathbf{V}_{2R}{}^2}{2} \tag{14.73}$$

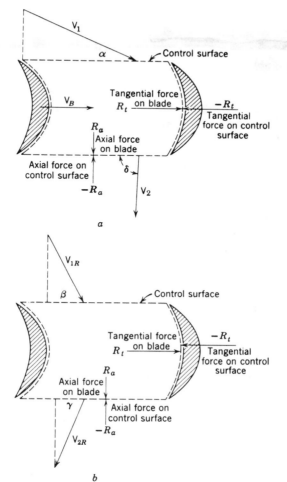

FIGURE 14.26

Analysis of forces on a turbine blade. (*a*) Stationary observer. (*b*) Observer moving with the blade.

By combining Eqs. 14.72 and 14.73, we can derive the relation

$$w = \frac{(\mathbf{V}_1{}^2 - \mathbf{V}_{1R}{}^2) - (\mathbf{V}_2{}^2 - \mathbf{V}_{2R}{}^2)}{2} \tag{14.74}$$

Applying the second law of thermodynamics to this process we conclude, since the process is adiabatic, that

$$s_2 \geqslant s_1$$

We will next consider the momentum equation, and evaluate both the tangential force R_t and the axial force R_a on the blade. For our sign convention we assume that tangential velocities and forces in the direction of the movement of the blade are positive and axial velocities and forces in the direction of the axial component of the velocity are positive.

Let us first apply the momentum equation in the tangential direction to the control volume as viewed by the stationary observer. In doing so we make the reasonable assumption that the pressure forces in the tangential direction balance. Then, using Eq. 14.19

$$R_t = \dot{m}(-\mathbf{V}_2 \cos \delta - \mathbf{V}_1 \cos \alpha)$$

or

$$R_t = \dot{m}(\mathbf{V}_1 \cos \alpha + \mathbf{V}_2 \cos \delta) \tag{14.75}$$

Similarly, for the control volume as viewed by the moving observer we have

$$-R_t = \dot{m}(-\mathbf{V}_{2R} \cos \gamma - \mathbf{V}_{1R} \cos \beta)$$

or

$$R_t = \dot{m}(\mathbf{V}_{1R} \cos \beta + \mathbf{V}_{2R} \cos \gamma) \tag{14.76}$$

Equation 14.76 could have been developed from Eq. 14.75 and from the geometry of the vector diagram by noting that

$$\mathbf{V}_1 \cos \alpha - \mathbf{V}_B = \mathbf{V}_{1R} \cos \beta$$

and

$$\mathbf{V}_2 \cos \delta + \mathbf{V}_B = \mathbf{V}_{2R} \cos \gamma$$

Therefore,

$$\mathbf{V}_1 \cos \alpha + \mathbf{V}_2 \cos \delta = \mathbf{V}_{1R} \cos \beta + \mathbf{V}_{2R} \cos \gamma \tag{14.77}$$

The work done by the control volume on the blade in time dt as the blade moves a distance dx is

$$\delta W = R_t \, dx$$

The rate at which work is done, or the power, is

$$\frac{\delta W}{dt} = R_t \frac{dx}{dt} = R_t \mathbf{V}_B$$

Substituting for R_t from Eqs. 14.75 and 14.76

$$\frac{\delta W}{dt} = \dot{W}_{c.v.} = \dot{m}\mathbf{V}_B(\mathbf{V}_1 \cos \alpha + \mathbf{V}_2 \cos \delta)$$

$$\frac{\delta W}{\delta t} = \dot{W}_{c.v.} = \dot{m}\mathbf{V}_B(\mathbf{V}_{1R} \cos \beta + \mathbf{V}_{2R} \cos \gamma) \tag{14.78}$$

The work w per unit mass of fluid flowing is therefore

$$\dot{W}_{c.v.}/\dot{m} = w = \mathbf{V}_B(\mathbf{V}_1 \cos \alpha + \mathbf{V}_2 \cos \delta) \tag{14.79}$$

or

$$\dot{W}_{c.v.}/\dot{m} = w = \mathbf{V}_B(\mathbf{V}_{1R} \cos \beta + \mathbf{V}_{2R} \cos \gamma) \tag{14.80}$$

The axial thrust on the blade can be found by applying the momentum equation in the axial direction. We must recognize that the pressure at the inlet section may be different from the pressure at the exit section. Consider first the control volume as viewed by the stationary observer. Again we use Eq. 14.19.

$$R_a = \dot{m}(\mathbf{V}_2 \sin \delta - \mathbf{V}_1 \sin \alpha) - (P_1 \mathscr{A}_1 - P_2 \mathscr{A}_2) \tag{14.81}$$

As viewed by the moving observer we have

$$R_a = \dot{m}(\mathbf{V}_{2R} \sin \gamma - \mathbf{V}_{1R} \sin \beta) - (P_1 \mathscr{A}_1 - P_2 \mathscr{A}_2) \tag{14.82}$$

Since there is no motion in the axial direction there is no work involved. However, adequate thrust bearings must be provided to handle the axial force. It is important to note in Eqs. 14.81 and 14.82 that if the pressure changes across the moving blade, the axial forces will be significantly influenced. This will be discussed more fully in the next section when the difference between an impulse and a reaction stage is considered.

Example 14.11

Steam enters the blade passage of one stage of a steam turbine with a velocity of 550 m/s and at an angle (α) of 20°. The steam leaves the blade (as seen by the moving observer) at an angle (γ) of 50°. There is no pressure change across the blade and there are no irreversibilities (which means that there is no change in the magnitude of the relative velocity during the flow through the blade passage). Determine the work per kilogram of steam and the axial thrust. The blade speed is 250 m/s.

Analysis and Solution

We first draw the vector diagram as shown in Fig. 14.27.

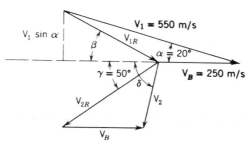

FIGURE 14.27
Sketch for Example 14.11.

FIGURE 14.28
Sketch for Example 14.11.

$$\mathbf{V}_{1R} \sin \beta = \mathbf{V}_1 \sin \alpha = 550 \sin 20° = 550(0.342) = 188.1 \text{ m/s}$$
$$\mathbf{V}_{1R} \cos \beta + \mathbf{V}_B = \mathbf{V}_1 \cos \alpha = 550 \cos 20° = 550(0.9397) = 516.8 \text{ m/s}$$
$$\mathbf{V}_{1R} \cos \beta = 516.8 - 250 = 266.8 \text{ m/s}$$
$$\tan \beta = \frac{\mathbf{V}_{1R} \sin \beta}{\mathbf{V}_{1R} \cos \beta} = \frac{188.1}{266.8} = 0.7050$$
$$\beta = 35.185°$$
$$\mathbf{V}_{1R} \sin \beta = 188.1$$
$$\mathbf{V}_{1R} = \frac{188.1}{\sin 35.185°} = \frac{188.1}{0.5762} = 326.4 \text{ m/s}$$
$$\mathbf{V}_{2R} = \mathbf{V}_{1R} = 326.4 \text{ m/s}$$
$$\mathbf{V}_2 \sin \delta = \mathbf{V}_{2R} \sin \gamma = 326.4 \sin 50° = 326.4(0.7660) = 250.1 \text{ m/s}$$
$$\mathbf{V}_2 \cos \delta + \mathbf{V}_B = \mathbf{V}_{2R} \cos \gamma = 326.4 \cos 50° = 326.4(0.6428) = 209.8 \text{ m/s}$$
$$\mathbf{V}_2 \cos \delta = 209.8 - 250 = -40.2 \text{ m/s}$$

This implies that we drew our vector diagram slightly wrong and it should have been as shown in Fig. 14.28

$$\cot \delta = \frac{\mathbf{V}_2 \cos \delta}{\mathbf{V}_2 \sin \delta} = \frac{-40.2}{250.1} = -0.1607$$
$$\delta = 99.13°$$
$$\mathbf{V}_2 = \sqrt{(\mathbf{V}_2 \sin \delta)^2 + (\mathbf{V}_2 \cos \delta)^2}$$
$$= \sqrt{(250.1)^2 + (40.2)^2} = 253.3 \text{ m/s}$$

14.13 IMPULSE AND REACTION STAGES FOR TURBINES

Two terms used in connection with turbines are impulse stage and reaction stage. A turbine stage is defined as the fixed nozzle or blade and the moving blades that follow it. Figure 14.29a shows an impulse stage. In an impulse stage the entire

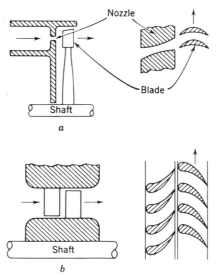

FIGURE 14.29

Schematic arrangement for (*a*) an impulse stage and (*b*) a reaction stage.

pressure drop takes place in a stationary nozzle, and the pressure remains constant as the fluid flows through the blade passage. There is a decrease in the kinetic energy of the fluid as it flows through the blade passage, and the enthalpy will increase due to irreversibilities associated with the fluid flow.

In the pure reaction stage the entire pressure drop occurs as the fluid flows through the moving blades. Thus, the moving blade acts as a nozzle, and the blade passage must have the proper contour for a nozzle (converging if the exit pressure is greater than the critical pressure and converging-diverging if the exit pressure is less than the critical pressure). In the pure reaction stage the only purpose of the stationary blade is to direct the fluid into the moving blade at the proper angle and velocity.

The pure reaction stage is used relatively infrequently. Rather, most turbines that are classified as reaction turbines have a pressure and enthalpy drop in both the fixed and moving blades. The degree of reaction is defined as the fraction of the enthalpy drop that occurs in the moving blades. Thus, in the 50 percent reaction stage, which is very commonly used, half of the enthalpy drop across the stage occurs in the fixed blade and the other half in the moving blade.

A few comparisons between the impulse and reaction blade may be noted here. Since there is a pressure drop across both the fixed and moving blades in the reaction stage, there is a tendency for leakage to occur across the tip of the blade. Therefore, very close clearances must be maintained at the blade tips. Further, the pressure drop across the moving blade of the reaction stage gives rise to axial forces that must be balanced in order to prevent axial motion in reaction turbines.

In an impulse stage it is possible to utilize only part of the periphery for admission of steam. This is usually termed partial admission. In fact, in turbines that utilize an impulse stage for the first stage the power output can be controlled by opening and closing nozzles. The main advantage of the reaction stage is that lower fluid velocities are utilized and higher efficiencies can be attained at lower velocities. This will be discussed more fully in the following sections.

14.14 SOME FURTHER CONSIDERATIONS OF IMPULSE STAGES

Figure 14.30 shows a typical velocity diagram for an impulse stage. The blade velocity coefficient k_B is defined as the ratio of the relative velocity leaving the blade to the relative velocity entering the blade. That is,

$$k_B = \frac{\mathbf{V}_{2R}}{\mathbf{V}_{1R}} \qquad (14.83)$$

In the ideal impulse stage $\mathbf{V}_{2R} = \mathbf{V}_{1R}$ and $k_B = 1$, but in the actual impulse stage \mathbf{V}_{2R} will be less than \mathbf{V}_{1R} due to irreversibilities in the fluid as it flows through the blade passage.

The blade efficiency η_B is defined as the ratio of the actual work per kilogram of fluid flowing to the kinetic energy of the fluid entering the blade passage.

$$\eta_B = \frac{w}{\mathbf{V}_1{}^2/2} \qquad (14.84)$$

Thus, a blade efficiency of 100 percent means that the work is exactly equal to the kinetic energy of the fluid entering the blade, and the kinetic energy of the fluid leaving the blade is zero.

Quite obviously, the fluid must have some axial velocity if it is to flow out the blade passage. However, it is of interest to consider a turbine having a zero axial

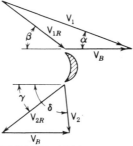

FIGURE 14.30

A velocity vector diagram for an impulse blade.

velocity and to determine the blade-speed ratio that will give an efficiency of 100 percent. The blade-speed ratio is the ratio of the blade speed V_B to the velocity of the fluid leaving the nozzle V_1 and the ratio is designated V_B/V_1.

Figure 14.31a shows a reversible impulse stage that has a very small entrance and blade exit angle, and a blade-speed ratio of 0.5. It is evident that as the angles α and γ (the inlet and blade exit angles) approach zero the exit velocity V_2 will also approach zero. Such a stage would have an efficiency of 100 percent. Figures 14.31b and c show reversible impulse stages with essentially zero angles, but with blade-speed ratios considerably less and considerably greater than 0.5. In both of these the exit velocity is large, and therefore the blade efficiency is considerably less than 100 percent. We conclude, therefore, that for a reversible, zero-angle impulse turbine the optimum blade-speed ratio is 0.5. The curve marked impulse stage on Fig. 14.34 shows the variation of blade efficiency with blade-speed ratio for a reversible stage.

An expression for the blade efficiency as a function of the angles α and γ (the inlet angle and blade exit angles), the blade-speed ratio, and the blade velocity coefficient can be derived.

From Eq. 14.79

$$w = V_B(V_1 \cos \alpha + V_2 \cos \delta)$$
$$V_{2R} = k_B V_{1R}$$

From the vector diagram we note that

$$V_2 \cos \delta = V_{2R} \cos \gamma - V_B = k_B V_{1R} \cos \gamma - V_B$$
$$V_{1R} = \frac{V_1 \cos \alpha - V_B}{\cos \beta}$$

Therefore,

$$w = V_B \left(V_1 \cos \alpha + k_B \cos \gamma \frac{V_1 \cos \alpha - V_B}{\cos \beta} - V_B \right)$$
$$= V_B \left[(V_1 \cos \alpha - V_B) \left(1 + \frac{k_B \cos \gamma}{\cos \beta} \right) \right]$$
$$\eta_B = \frac{w}{V_1^2/2} = \frac{2V_B}{V_1^2}(V_1 \cos \alpha - V_B)\left(1 + k_B \frac{\cos \gamma}{\cos \beta} \right)$$
$$= \frac{2V_B^2}{V_1^2}\left(\frac{V_1}{V_B} \cos \alpha - 1 \right)\left(1 + k_B \frac{\cos \gamma}{\cos \beta} \right)$$

a b c

FIGURE 14.31

A velocity vector diagram illustrating optimum blade-speed ratio.

But

$$\cos \beta = \frac{\mathbf{V}_1 \cos \alpha - \mathbf{V}_B}{\sqrt{(\mathbf{V}_1 \cos \alpha - \mathbf{V}_B)^2 + (\mathbf{V}_1 \sin \alpha)^2}}$$

$$= \frac{\cos \alpha - \mathbf{V}_B/\mathbf{V}_1}{\sqrt{(\cos \alpha - \mathbf{V}_B/\mathbf{V}_1)^2 + \sin^2 \alpha}}$$

$$\eta_B = 2 \frac{\mathbf{V}_B}{\mathbf{V}_1} \left(\cos \alpha - \frac{\mathbf{V}_B}{\mathbf{V}_1} \right) \left[1 + \frac{k_B \cos \gamma}{\dfrac{\cos \alpha - \mathbf{V}_B/\mathbf{V}_1}{\sqrt{(\cos \alpha - \mathbf{V}_B/\mathbf{V}_1)^2 + \sin^2 \alpha}}} \right] \qquad (14.85)$$

This relation gives the blade efficiency as a function of the blade-speed ratio $\mathbf{V}_B/\mathbf{V}_1$, the angles α and γ, and the blade velocity coefficient k_B. From this relation it can be shown that for the usual values of angles and blade velocity coefficient, the optimum efficiency is obtained with blade velocity coefficient of approximately 0.5, which was the value obtained for the reversible, zero-angle impulse stage.

This leads to consideration of the velocity-compounded turbine. Suppose one was attempting to build a turbine having one impulse stage, and that it was to receive steam at 700 kPa, 250°C, and exhaust at a pressure of 15 kPa. The velocity of the steam leaving the nozzle would be about 1100 m/s. To have maximum

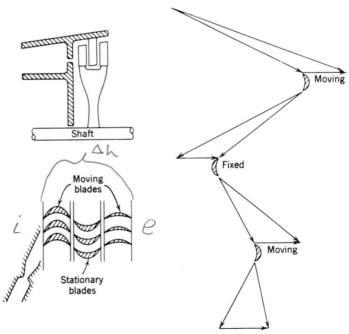

FIGURE 14.32
A velocity-compounded impulse stage.

efficiency a blade-speed ratio of 0.5 must be used, and this would require a blade speed of 550 m/s. Blade speeds of this magnitude result in very high stresses due to centrifugal force, and further, the irreversibilities associated with fluid flow increase as the fluid velocity increases.

For these reasons velocity compounding is used, and a schematic diagram and a vector diagram for a velocity-compounded stage are shown in Fig. 14.32. The stationary blade serves to change the direction of the fluid so that it will enter the second row of moving blades properly. It can be shown that for a reversible, zero-angle turbine having two moving rows the blade-speed ratio for optimum efficiency is 1/4. Sometimes velocity-compounded turbines are built with as many as three moving rows.

Velocity-compounded turbines have a lower efficiency than comparable turbines that have a number of impulse or reaction stages. However, they are relatively inexpensive and are therefore frequently used in small sizes where initial cost is more important than efficiency. A velocity-compounded stage is also frequently used as the first stage in large steam turbines in order to have a large pressure and temperature drop before the steam enters the moving blades.

14.15 SOME FURTHER CONSIDERATIONS OF REACTION STAGES

First a remark should be made regarding the blade efficiency of a reaction stage. Both the fixed blades and moving blades act as a nozzle. Therefore, the efficiency of the moving blades can be defined in the same way as the efficiency of a nozzle if the relative velocities are used.

The second matter to consider is the blade-speed ratio for maximum efficiency of a reaction stage. Let us consider a reversible reaction stage in which the isentropic enthalpy drops across the fixed and moving blades are equal (that is, a 50 percent reaction stage). For a zero-angle turbine it is evident from Fig. 14.33 that for zero kinetic energy leaving the blade, a blade-speed ratio of 1.0 is required. Thus, maximum efficiency in a reaction stage is obtained for a blade-speed ratio of 1.0.

This leads us to a comparison of the impulse stage with the reaction stage regarding the blade-speed ratio for maximum efficiency.

If we compare reversible impulse and reaction stages for a given velocity from the stationary nozzle or blade, it is evident that for maximum efficiency the blade

FIGURE 14.33

A velocity vector diagram for a reaction stage.

velocity flow the reaction stage would be twice the blade velocity of the impulse stage. However, this is not a good comparison because the reaction stage would have a greater enthalpy drop (due to the enthalpy drop in the moving row) than the impulse stage.

A better comparison can be made by considering equal enthalpy drops across the stage. Let this enthalpy drop be denoted Δh_s, and let the velocity \mathbf{V}_0 be defined by the relation

$$\mathbf{V}_0 = \sqrt{2\Delta h_s}$$

That is, \mathbf{V}_0 would be the velocity leaving the nozzle of the reversible impulse stage that has this enthalpy drop. That is, for the impulse stage

$$\mathbf{V}_1 = \mathbf{V}_0$$

For the reversible 50 percent reaction stage the velocity leaving the fixed blades would be

$$\mathbf{V}_1 = \sqrt{\tfrac{1}{2} \times 2\Delta h_s} = \frac{\mathbf{V}_0}{\sqrt{2}}$$

For maximum efficiency in the impulse stage, $\mathbf{V}_B/\mathbf{V}_1 = 0.5$. Therefore, for the impulse stage

$$\frac{\mathbf{V}_B}{\mathbf{V}_1} = \frac{\mathbf{V}_B}{\mathbf{V}_0} = 0.5$$

For maximum efficiency in the reaction stage, $\mathbf{V}_B/\mathbf{V}_1 = 1.0$. Therefore, for the reaction stage

$$\frac{\mathbf{V}_B}{\mathbf{V}_1} = \frac{\mathbf{V}_B\sqrt{2}}{\mathbf{V}_0} \qquad \frac{\mathbf{V}_B}{\mathbf{V}_0} = \frac{1}{\sqrt{2}}$$

Thus, for a given enthalpy drop per stage, the reaction stage requires a higher blade speed for maximum efficiency than the impulse stage. Conversely, for a given blade speed the reaction turbine requires more stages than an impulse stage, and

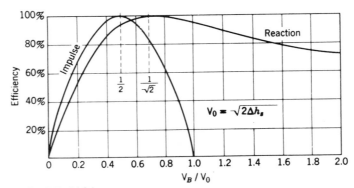

FIGURE 14.34

Efficiency of an ideal impulse and reaction stage as a function of $\mathbf{V}_B/\mathbf{V}_0$.

in this case the fluid velocity is lower in the reaction stage than in the impulse stage.

Figure 14.34 shows the blade efficiency as a function of V_B/V_0 for an impulse stage and reaction stage.

PROBLEMS

14.1 Air leaves the compressor of a jet engine at a temperature of 300 F, a pressure of 20 lbf/in.2, and a velocity of 400 ft/s. Determine the isentropic stagnation temperature and pressure.

14.2 The products of combustion of a jet engine leave the engine with a velocity relative to the plane of 400 m/s, a temperature of 480°C, and a pressure of 85 kPa. Assuming that $k = 1.34$, $C_p = 1.13$ kJ/kg K for the products, determine the stagnation pressure and temperature of the products (relative to the airplane).

14.3 Steam leaves a nozzle with a velocity of 800 ft/s. The stagnation pressure is 100 lbf/in.2, and the stagnation temperature is 500 F. What is the static pressure and temperature?

14.4 Water is pumped from a lake and discharged through a nozzle as shown in Fig. 14.35. At the pump discharge the pressure is 700 kPa, and the temperature is 25°C. The nozzle is located 10 m above the pump. Assume 100 kPa atmospheric pressure and reversible flow throughout. Determine the velocity of the water leaving the nozzle.

10 m

$P = 700$ kPa
$T = 25°C$

FIGURE 14.35
Sketch for Problem 14.4.

14.5 In a water turbine that utilizes nozzles, the water enters the nozzle at 50 lbf/in.2, 80 F and leaves at standard atmospheric pressure. If the flow through the nozzle is reversible and adiabatic, determine the velocity and kinetic energy per lbm of water leaving the nozzle.

14.6 Air is expanded in a nozzle from 2 MPa, 200°C, to 0.2 MPa. The mass rate of flow through the nozzle is 5 kg/s. Assume the flow to be reversible and adiabatic.

(a) Determine specific volume, velocity, Mach number, and cross-sectional area for each 0.2 MPa decrease of pressure, and plot these as a function of pressure.

(b) Determine the throat and exit areas for the nozzle.

14.7 Consider the nozzle of Problem 14.6 and determine what back pressure will cause a normal shock to stand in the exit plane of the nozzle. What is the mass rate of flow under these conditions?

14.8 At what Mach number will the normal shock occur in the nozzle of Problem 14.6 if the back pressure is 1.4 MPa?

14.9 Consider the nozzle of Problem 14.6. What back pressure will be required to cause subsonic flow throughout the entire nozzle with $M = 1$ at the throat?

14.10 Determine the mass rate of flow through the nozzle of Problem 14.6 for a back pressure of 1.9 MPa.

14.11 A nozzle is designed assuming reversible adiabatic flow with an exit Mach number of 2.6 while flowing air having a stagnation pressure and temperature of 300 lbf/in.2 and 300 F, respectively. The mass rate of flow is 10 lbm/s, and k may be assumed to be 1.40 and constant.

(a) Determine the exit pressure, temperature, and area, and the throat area.

(b) Suppose that the back pressure at the nozzle exit is raised to 200 lbf/in.2, and that the flow remains isentropic except for a normal shock wave. Determine the exit Mach number and temperature, and the mass flow through the nozzle.

14.12 A jet plane travels through the air with a speed of 600 miles/h at an altitude of 20 000 ft, where the pressure is 5.75 lbf/in.2 and the temperature is −12°C. Consider the diffuser of the engine. The air leaves the diffuser with a velocity of 300 ft/s. Determine the pressure and temperature leaving the diffuser, and the ratio of inlet to exit area of the diffuser, assuming the flow to be reversible and adiabatic.

14.13 The products of combustion enter the nozzle of a jet engine at a total pressure of 18 lbf/in.2, and a total temperature of 1200 F. The atmospheric pressure is 6.75 lbf/in.2 The nozzle is convergent, and the rate of flow is 50 lbm/s. Assume the flow to be reversible and adiabatic. Determine the exit area of the nozzle.

14.14 Air is expanded in a nozzle from 700 kPa, 200°C, to 150 kPa in a nozzle having an efficiency of 90%. The mass rate of flow is 4 kg/s. Determine

the exit area of the nozzle, the exit velocity, and the increase of entropy per kilogram of fluid. Compare these results with those of the reversible adiabatic nozzle of Problem 14.6.

14.15 Repeat Problem 14.12 assuming a diffuser efficiency of 80%.

14.16 Consider the diffuser of a supersonic aircraft flying at $M = 1.4$ at such an altitude that the temperature is $-20°C$, and the atmosphere pressure is 50 kPa. Consider two possible ways in which this diffuser might operate, and for each case calculate the throat area required for a flow of 50 kg/s.

 (a) The diffuser operates as a reversible adiabatic diffuser with subsonic exit velocity.

 (b) A normal shock stands at the entrance to the diffuser. Except for the normal shock the flow is reversible and adiabatic, and the exit velocity is subsonic. This is shown in Fig. 14.36. Assume a convergent–divergent diffuser with $M = 1$ at the throat.

FIGURE 14.36
Sketch for Problem 14.16.

14.17 Air enters a diffuser with a velocity of 600 ft/s, a static pressure of 10 lbf/in.², and a temperature of 20 F. The velocity leaving the diffuser is 200 ft/s and the static pressure at the diffuser exit is 11.7 lbf/in.² Determine the static temperature at the diffuser exit and the diffuser efficiency. Compare the stagnation pressures at the diffuser inlet and exit.

14.18 Steam at a pressure of 600 kPa and a temperature of 300°C expands to a pressure of 150 kPa in a nozzle having an efficiency of 90%. The mass rate of flow is 10 kg/s. Determine the nozzle exit area.

14.19 Steam at an initial pressure of 100 lbf/in.² and a temperature of 600 F flows through a convergent-divergent nozzle having a throat area of 0.5 in.² The pressure at the exit plane is 20 lbf/in.² and the exit velocity is 2600 ft/s. The flow from the nozzle entrance to the throat is reversible and adiabatic. Determine the exit area of the nozzle, the over-all nozzle efficiency, and the increase in entropy per lbm of fluid.

14.20 A sharp-edged orifice is used to measure the flow of air in a pipe. The pipe diameter is 100 mm and the diameter of the orifice is 25 mm. Upstream of the orifice, the absolute pressure is 150 kPa and the temperature is 35°C. The pressure drop across the orifice is 12 kPa, and the coefficient of discharge is 0.62. Determine the mass rate of flow in the pipeline.

14.21 A steam turbine utilizes convergent nozzles. An estimate of the rate of steam flow is to be made from the pressure drop across the nozzles of one stage. The inlet conditions to these nozzles are 600 kPa, 250°C. The exit pressure is 400 kPa. The coefficient of discharge is estimated to be 0.94.

The total exit area of the nozzles in this stage is 0.005 m². Determine the rate of flow under these conditions.

14.22 The coefficient of discharge of a sharp-edged orifice is determined at one set of conditions by use of an accurately calibrated gasometer. The orifice has a diameter of 0.75 in. and the pipe diameter is 2.0 in. The absolute upstream pressure is 30 lbf/in.² and the pressure drop across the orifice is 3.2 in. of mercury. The temperature of the air entering the orifice is 80 F. The mass rate of flow as measured by the gasometer is 5.30 lbm/min. What is the coefficient of discharge of the orifice under these conditions?

14.23 Consider the flow of air through an impulse blade passage. The air enters the blade passage at an angle of 18° with a velocity of 460 m/s, a pressure of 110 kPa, and a temperature of 90°C. The blade velocity is 250 m/s and the air leaves the blade passage at an angle of 45° relative to the blade. The mass rate of flow is 10 kg/s, and it is assumed that the flow is reversible and adiabatic.
(a) Draw the velocity diagram to scale.
(b) What is the power output of the turbine?
(c) If the blade wheel to which the blades are attached has a diameter of 0.3 m, what is the length of each blade?

14.24 Steam enters an impulse turbine in which all processes are assumed to be reversible and adiabatic. The inlet pressure is 100 lbf/in.² and the inlet temperature is 800 F. The exhaust pressure is 16 lbf/in.² The steam leaves the nozzle and enters the turbine at an angle of 20°. The blade-speed ratio is 0.5 and the blade exit angle is 50°. Determine the blade efficiency of this turbine.

14.25 Consider an impulse turbine having the same inlet conditions as Problem 14.24. The nozzle efficiency is 92% and the blade velocity coefficient is 0.96. The blade-speed ratio is 0.5. Determine the work per lbm of steam and draw the velocity diagram to scale.

14.26 Consider a two stage, reversible, velocity-compounded impulse turbine that has the same inlet conditions as Problem 14.24. Assume a blade-velocity coefficient of 0.25, and an inlet angle of 20°. Both the moving and fixed blades have equal inlet and exit angles (i.e., the angle at which the steam enters relative to the blade is equal to the relative angle at which it leaves). Determine the blade efficiency of the turbine and draw the velocity diagram to scale.

14.27 Consider a single-stage reaction turbine having equal enthalpy drops across the fixed and moving blades. Consider the same inlet conditions and exit pressure as Problem 14.24. All processes are reversible and adiabatic. The blade inlet angle is 20° and the blade-speed ratio is 0.9.
(a) What is the pressure at the exit of the fixed blades?

(b) Draw the velocity vector diagram to scale.

(c) What is the net work per lbm of steam flowing through the turbine?

14.28 A single-stage air turbine having 50% reaction operates with inlet pressure and temperature of 350 kPa, 1000 K, and exhaust pressure of 100 kPa. The fixed-blade exit angle is 18°, and exhaust from the turbine is in the axial direction. The blade-speed ratio is 0.9, and all processes are reversible and adiabatic. Determine:

(a) All velocities and angles, and draw the velocity vector diagram.

(b) The work per kilogram of air flowing through the turbine.

APPENDIX
TABLES, FIGURES, AND CHARTS

TABLES

FIGURES AND CHARTS

S.H. Vap

ex, A 400 8.353 D
 B 415 x
 G 440 8.775 E

TABLE A.1E Thermodynamic Properties of Steam [a] **(English Units)**

TABLE A.1.1E Saturated Steam: Temperature Table (English Units)

Temp. F. T	Press. lbf/ sq. in. P	Specific Volume ft³/lbm		Internal Energy Btu/lbm			Enthalpy Btu/lbm			Entropy Btu/lbm R		
		Sat. Liquid v_f	Sat. Vapor v_g	Sat. Liquid u_f	Evap. u_{fg}	Sat. Vapor u_g	Sat. Liquid h_f	Evap. h_{fg}	Sat. Vapor h_g	Sat. Liquid s_f	Evap. s_{fg}	Sat. Vapor s_g
32.018	0.088 66	0.016 022	3302	0.00	1021.2	1021.2	0.01	1075.4	1075.4	0.000 00	2.1869	2.1869
35	0.099 92	0.016 021	2948	2.99	1019.2	1022.2	3.00	1073.7	1076.7	0.006 07	2.1704	2.1764
40	0.121 66	0.016 020	2445	8.02	1015.8	1023.9	8.02	1070.9	1078.9	0.016 17	2.1430	2.1592
45	0.147 48	0.016 021	2037	13.04	1012.5	1025.5	13.04	1068.1	1081.1	0.026 18	2.1162	2.1423
50	0.178 03	0.016 024	1704.2	18.06	1009.1	1027.2	18.06	1065.2	1083.3	0.036 07	2.0899	2.1259
60	0.2563	0.016 035	1206.9	28.08	1002.4	1030.4	28.08	1059.6	1087.7	0.055 55	2.0388	2.0943
70	0.3632	0.016 051	867.7	38.09	995.6	1033.7	38.09	1054.0	1092.0	0.074 63	1.9896	2.0642
80	0.5073	0.016 073	632.8	48.08	988.9	1037.0	48.09	1048.3	1096.4	0.093 32	1.9423	2.0356
90	0.6988	0.016 099	467.7	58.07	982.2	1040.2	58.07	1042.7	1100.7	0.111 65	1.8966	2.0083
100	0.9503	0.016 130	350.0	68.04	975.4	1043.5	68.05	1037.0	1105.0	0.129 63	1.8526	1.9822
110	1.2763	0.016 166	265.1	78.02	968.7	1046.7	78.02	1031.3	1109.3	0.147 30	1.8101	1.9574
120	1.6945	0.016 205	203.0	87.99	961.9	1049.9	88.00	1025.5	1113.5	0.164 65	1.7690	1.9336
130	2.225	0.016 247	157.17	97.97	955.1	1053.0	97.98	1019.8	1117.8	0.181 72	1.7292	1.9109
140	2.892	0.016 293	122.88	107.95	948.2	1056.2	107.96	1014.0	1121.9	0.198 51	1.6907	1.8892
150	3.722	0.016 343	96.99	117.95	941.3	1059.3	117.96	1008.1	1126.1	0.215 03	1.6533	1.8684
160	4.745	0.016 395	77.23	127.94	934.4	1062.3	127.96	1002.2	1130.1	0.231 30	1.6171	1.8484
170	5.996	0.016 450	62.02	137.95	927.4	1065.4	137.97	996.2	1134.2	0.247 32	1.5819	1.8293
180	7.515	0.016 509	50.20	147.97	920.4	1068.3	147.99	990.2	1138.2	0.263 11	1.5478	1.8109

[a] Abridged from Joseph H. Keenan, Frederick G. Keyes, Philip G. Hill, and Joan G. Moore, *Steam Tables* (New York: John Wiley & Sons, Inc., 1969).

TABLE 1.1E (Continued) Saturated Steam: Temperature Table (English Units)

Temp. F. T	Press. lbf/sq. in. P	Specific Volume ft³/lbm		Internal Energy Btu/lbm			Enthalpy Btu/lbm			Entropy Btu/lbm R		
		Sat. Liquid v_f	Sat. Vapor v_g	Sat. Liquid u_f	Evap. u_{fg}	Sat. Vapor u_g	Sat. Liquid h_f	Evap. h_{fg}	Sat. Vapor h_g	Sat. Liquid s_f	Evap. s_{fg}	Sat. Vapor s_g
190	9.343	0.016 570	40.95	158.00	913.3	1071.3	158.03	984.1	1142.1	0.278 66	1.5146	1.7932
200	11.529	0.016 634	33.63	168.04	906.2	1074.2	168.07	977.9	1145.9	0.294 00	1.4822	1.7762
210	14.125	0.016 702	27.82	178.10	898.9	1077.0	178.14	971.6	1149.7	0.309 13	1.4508	1.7599
212	14.698	0.016 716	26.80	180.11	897.5	1077.6	180.16	970.3	1150.5	0.312 13	1.4446	1.7567
220	17.188	0.016 772	23.15	188.17	891.7	1079.8	188.22	965.3	1153.5	0.324 06	1.4201	1.7441
230	20.78	0.016 845	19.386	198.26	884.3	1082.6	198.32	958.8	1157.1	0.338 80	1.3901	1.7289
240	24.97	0.016 922	16.327	208.36	876.9	1085.3	208.44	952.3	1160.7	0.353 35	1.3609	1.7143
250	29.82	0.017 001	13.826	218.49	869.4	1087.9	218.59	945.6	1164.2	0.367 72	1.3324	1.7001
260	35.42	0.017 084	11.768	228.64	861.8	1090.5	228.76	938.8	1167.6	0.381 93	1.3044	1.6864
270	41.85	0.017 170	10.066	238.82	854.1	1093.0	238.95	932.0	1170.9	0.395 97	1.2771	1.6731
280	49.18	0.017 259	8.650	249.02	846.3	1095.4	249.18	924.9	1174.1	0.409 86	1.2504	1.6602
290	57.53	0.017 352	7.467	259.25	838.5	1097.7	259.44	917.8	1177.2	0.423 60	1.2241	1.6477
300	66.98	0.017 448	6.472	269.52	830.5	1100.0	269.73	910.4	1180.2	0.437 20	1.1984	1.6356
310	77.64	0.017 548	5.632	279.81	822.3	1102.1	280.06	903.0	1183.0	0.450 67	1.1731	1.6238
320	89.60	0.017 652	4.919	290.14	814.1	1104.2	290.43	895.3	1185.8	0.464 00	1.1483	1.6123
330	103.00	0.017 760	4.312	300.51	805.7	1106.2	300.84	887.5	1188.4	0.477 22	1.1238	1.6010
340	117.93	0.017 872	3.792	310.91	797.1	1108.0	311.30	879.5	1190.8	0.490 31	1.0997	1.5901
350	134.53	0.017 988	3.346	321.35	788.4	1109.8	321.80	871.3	1193.1	0.503 29	1.0760	1.5793
360	152.92	0.018 108	2.961	331.84	779.6	1111.4	332.35	862.9	1195.2	0.516 17	1.0526	1.5688
370	173.23	0.018 233	2.628	342.37	770.6	1112.9	342.96	854.2	1197.2	0.528 94	1.0295	1.5585
380	195.60	0.018 363	2.339	352.95	761.4	1114.3	353.62	845.4	1199.0	0.541 63	1.0067	1.5483
390	220.2	0.018 498	2.087	363.58	752.0	1115.6	364.34	836.2	1200.6	0.554 22	0.9841	1.5383
400	247.1	0.018 638	1.8661	374.27	742.4	1116.6	375.12	826.8	1202.0	0.566 72	0.9617	1.5284
410	276.5	0.018 784	1.6726	385.01	732.6	1117.6	385.97	817.2	1203.1	0.579 16	0.9395	1.5187

420	308.5	0.018 936	1.5024	395.81	722.5	1118.3	396.89	807.2	1204.1	0.591 52	0.9175	1.5091
430	343.3	0.019 094	1.3521	406.68	712.2	1118.9	407.89	796.9	1204.8	0.603 81	0.8957	1.4995
440	381.2	0.019 260	1.2192	417.62	701.7	1119.3	418.98	786.3	1205.3	0.616 05	0.8740	1.4900
450	422.1	0.019 433	1.1011	428.6	690.9	1119.5	430.2	775.4	1205.6	0.6282	0.8523	1.4806
460	466.3	0.019 614	0.9961	439.7	679.8	1119.6	441.4	764.1	1205.5	0.6404	0.8308	1.4712
470	514.1	0.019 803	0.9025	450.9	668.4	1119.4	452.8	752.4	1205.2	0.6525	0.8093	1.4618
480	565.5	0.020 002	0.8187	462.2	656.7	1118.9	464.3	740.3	1204.6	0.6646	0.7878	1.4524
490	620.7	0.020 211	0.7436	473.6	644.7	1118.3	475.9	727.8	1203.7	0.6767	0.7663	1.4430
500	680.0	0.020 43	0.6761	485.1	632.3	1117.4	487.7	714.8	1202.5	0.6888	0.7448	1.4335
520	811.4	0.020 91	0.5605	508.5	606.2	1114.8	511.7	687.3	1198.9	0.7130	0.7015	1.4145
540	961.5	0.021 45	0.4658	532.6	578.4	1111.0	536.4	657.5	1193.8	0.7374	0.6576	1.3950
560	1131.8	0.022 07	0.3877	557.4	548.4	1105.8	562.0	625.0	1187.0	0.7620	0.6129	1.3749
580	1324.3	0.022 78	0.3225	583.1	515.9	1098.9	588.6	589.3	1178.0	0.7872	0.5668	1.3540
600	1541.0	0.023 63	0.2677	609.9	480.1	1090.0	616.7	549.7	1166.4	0.8130	0.5187	1.3317
620	1784.4	0.024 65	0.2209	638.3	440.2	1078.5	646.4	505.0	1151.4	0.8398	0.4677	1.3075
640	2057.1	0.025 93	0.1805	668.7	394.5	1063.2	678.6	453.4	1131.9	0.8681	0.4122	1.2803
660	2362	0.027 67	0.144 59	702.3	340.0	1042.3	714.4	391.1	1105.5	0.8990	0.3493	1.2483
680	2705	0.030 32	0.111 27	741.7	269.3	1011.0	756.9	309.8	1066.7	0.9350	0.2718	1.2068
700	3090	0.036 66	0.074 38	801.7	145.9	947.7	822.7	167.5	990.2	0.9902	0.1444	1.1346
705.44	3204	0.050 53	0.050 53	872.6	0	872.6	902.5	0	902.5	1.0580	0	1.0580

TABLE A.1.2E Saturated Steam: Pressure Table (English Units)

Press. Lbf/ Sq. In. P	Temp. Fahr. T	Specific Volume ft³/lbm		Internal Energy Btu/lbm			Enthalpy Btu/lbm			Entropy Btu/lbm R		
		Sat Liquid v_f	Sat. Vapor v_g	Sat. Liquid u_f	Evap. u_{fg}	Sat. Vapor u_g	Sat. Liquid h_f	Evap. h_{fg}	Sat. Vapor h_g	Sat. Liquid s_f	Evap. s_{fg}	Sat. Vapor s_g
1.0	101.70	0.016 136	333.6	69.74	974.3	1044.0	69.74	1036.0	1105.8	0.132 66	1.8453	1.9779
2.0	126.04	0.016 230	173.75	94.02	957.8	1051.8	94.02	1022.1	1116.1	0.174 99	1.7448	1.9198
3.0	141.43	0.016 300	118.72	109.38	947.2	1056.6	109.39	1013.1	1122.5	0.200 89	1.6852	1.8861
4.0	152.93	0.016 358	90.64	120.88	939.3	1060.2	120.89	1006.4	1127.3	0.219 83	1.6426	1.8624
5.0	162.21	0.016 407	73.53	130.15	932.9	1063.0	130.17	1000.9	1131.0	0.234 86	1.6093	1.8441
6.0	170.03	0.016 451	61.98	137.98	927.4	1065.4	138.00	996.2	1134.2	0.247 36	1.5819	1.8292
8.0	182.84	0.016 526	47.35	150.81	918.4	1069.2	150.84	988.4	1139.3	0.267 54	1.5383	1.8058
10	193.19	0.016 590	38.42	161.20	911.0	1072.2	161.23	982.1	1143.3	0.283 58	1.5041	1.7877
14.696	211.99	0.016 715	26.80	180.10	897.5	1077.6	180.15	970.4	1150.5	0.312 12	1.4446	1.7567
15	213.03	0.016 723	26.29	181.14	896.8	1077.9	181.19	969.7	1150.9	0.313 67	1.4414	1.7551
20	227.96	0.016 830	20.09	196.19	885.8	1082.0	196.26	960.1	1156.4	0.335 80	1.3962	1.7320
25	240.08	0.016 922	16.306	208.44	876.9	1085.3	208.52	952.2	1160.7	0.353 45	1.3607	1.7142
30	250.34	0.017 004	13.748	218.84	869.2	1088.0	218.93	945.4	1164.3	0.368 21	1.3314	1.6996
35	259.30	0.017 073	11.900	227.93	862.4	1090.3	228.04	939.3	1167.4	0.380 93	1.3064	1.6873
40	267.26	0.017 146	10.501	236.03	856.2	1092.3	236.16	933.8	1170.0	0.392 14	1.2845	1.6767
45	274.46	0.017 209	9.403	243.37	850.7	1094.0	243.51	928.8	1172.3	0.402 18	1.2651	1.6673
50	281.03	0.017 269	8.518	250.08	845.5	1095.6	250.24	924.2	1174.4	0.411 29	1.2476	1.6589
55	287.10	0.017 325	7.789	256.28	840.8	1097.0	256.46	919.9	1176.3	0.419 63	1.2317	1.6513
60	292.73	0.017 378	7.177	262.06	836.3	1098.3	262.25	915.8	1178.0	0.427 33	1.2170	1.6444
65	298.00	0.017 429	6.657	267.46	832.1	1099.5	267.67	911.9	1179.6	0.434 50	1.2035	1.6380
70	302.96	0.017 478	6.209	272.56	828.1	1100.6	272.79	908.3	1181.0	0.441 20	1.1909	1.6321
75	307.63	0.017 524	5.818	277.37	824.3	1101.6	277.61	904.8	1182.4	0.447 49	1.1790	1.6265
80	312.07	0.017 570	5.474	281.95	820.6	1102.6	282.21	901.4	1183.6	0.453 44	1.1679	1.6214
85	316.29	0.017 613	5.170	286.30	817.1	1103.5	286.58	898.2	1184.8	0.459 07	1.1574	1.6165

90	320.31	0.017 655	4.898	290.46	813.8	1104.3	290.76	895.1	1185.9	0.464 42	1.1475	1.6119
95	324.16	0.017 696	4.654	294.45	810.6	1105.0	294.76	892.1	1186.9	0.469 52	1.1380	1.6076
100	327.86	0.017 736	4.434	298.28	807.5	1105.8	298.61	889.2	1187.8	0.474 39	1.1290	1.6034
110	334.82	0.017 813	4.051	305.52	801.6	1107.1	305.88	883.7	1189.6	0.483 55	1.1122	1.5957
120	341.30	0.017 886	3.730	312.27	796.0	1108.3	312.67	878.5	1191.1	0.492 01	1.0966	1.5886
130	347.37	0.017 957	3.457	318.61	790.7	1109.4	319.04	873.5	1192.5	0.499 89	1.0822	1.5821
140	353.08	0.018 024	3.221	324.58	785.7	1110.3	325.05	868.7	1193.8	0.507 27	1.0688	1.5761
150	358.48	0.018 089	3.016	330.24	781.0	1111.2	330.75	864.2	1194.9	0.514 22	1.0562	1.5704
160	363.60	0.018 152	2.836	335.63	776.4	1112.0	336.16	859.8	1196.0	0.520 78	1.0443	1.5651
170	368.47	0.018 214	2.676	340.76	772.0	1112.7	341.33	855.6	1196.9	0.527 00	1.0330	1.5600
180	373.13	0.018 273	2.533	345.68	767.7	1113.4	346.29	851.5	1197.8	0.532 92	1.0223	1.5553
190	377.59	0.018 331	2.405	350.39	763.6	1114.0	351.04	847.5	1198.6	0.538 57	1.0122	1.5507
200	381.86	0.018 387	2.289	354.9	759.6	1114.6	355.6	843.7	1199.3	0.5440	1.0025	1.5464
250	401.04	0.018 653	1.8448	375.4	741.4	1116.7	376.2	825.8	1202.1	0.5680	0.9594	1.5274
300	417.43	0.018 896	1.5442	393.0	725.1	1118.2	394.1	809.8	1203.9	0.5883	0.9232	1.5115
350	431.82	0.019 124	1.3267	408.7	710.3	1119.0	409.9	795.0	1204.9	0.6060	0.8917	1.4978
400	444.70	0.019 340	1.1620	422.8	696.7	1119.5	424.2	781.2	1205.5	0.6218	0.8638	1.4856
450	456.39	0.019 547	1.0326	435.7	683.9	1119.6	437.4	768.2	1205.6	0.6360	0.8385	1.4746
500	467.13	0.019 748	0.9283	447.7	671.7	1119.4	449.5	755.8	1205.3	0.6490	0.8154	1.4645
550	477.07	0.019 943	0.8423	458.9	660.2	1119.1	460.9	743.9	1204.8	0.6611	0.7941	1.4551
600	486.33	0.020 13	0.7702	469.4	649.1	1118.6	471.7	732.4	1204.1	0.6723	0.7742	1.4464
700	503.23	0.020 51	0.6558	488.9	628.2	1117.0	491.5	710.5	1202.0	0.6927	0.7378	1.4305
800	518.36	0.020 87	0.5691	506.6	608.4	1115.0	509.7	689.6	1199.3	0.7110	0.7050	1.4160
900	532.12	0.021 23	0.5009	523.0	589.6	1112.6	526.6	669.5	1196.0	0.7277	0.6750	1.4027
1000	544.75	0.021 59	0.4459	538.4	571.5	1109.9	542.4	650.0	1192.4	0.7432	0.6471	1.3903
1200	567.37	0.022 32	0.3623	566.7	536.8	1103.5	571.7	612.3	1183.9	0.7712	0.5961	1.3673
1400	587.25	0.023 07	0.3016	592.7	503.3	1096.0	598.6	575.5	1174.1	0.7964	0.5497	1.3461
1600	605.06	0.023 86	0.2552	616.9	470.5	1087.4	624.0	538.9	1162.9	0.8196	0.5062	1.3258
1800	621.21	0.024 72	0.2183	640.0	437.6	1077.7	648.3	502.1	1150.4	0.8414	0.4645	1.3060
2000	636.00	0.025 65	0.18813	662.4	404.2	1066.6	671.9	464.4	1136.3	0.8623	0.4238	1.2861
2500	668.31	0.028 60	0.13059	717.7	313.4	1031.0	730.9	360.5	1091.4	0.9131	0.3196	1.2327
3000	695.52	0.034 31	0.08404	783.4	185.4	968.8	802.5	213.0	1015.5	0.9732	0.1843	1.1575
3203.6	705.44	0.050 53	0.05053	872.6	0	872.6	902.5	0	902.5	1.0580	0	1.0580

TABLE A.1.3E Superheated Vapor (English Units)

F	P = 1.0(101.70)				P = 5.0(162.21)				P = 10.0(193.19)			
	v	u	h	s	v	u	h	s	v	u	h	s
Sat	333.6	1044.0	1105.8	1.9779	73.53	1063.0	1131.0	1.8441	38.42	1072.2	1143.3	1.7877
200	392.5	1077.5	1150.1	2.0508	78.15	1076.3	1148.6	1.8715	38.85	1074.7	1146.6	1.7927
240	416.4	1091.2	1168.3	2.0775	83.00	1090.3	1167.1	1.8987	41.32	1089.0	1165.5	1.8205
280	440.3	1105.0	1186.5	2.1028	87.83	1104.3	1185.5	1.9244	43.77	1103.3	1184.3	1.8467
320	464.2	1118.9	1204.8	2.1269	92.64	1118.3	1204.0	1.9487	46.20	1117.6	1203.1	1.8714
360	488.1	1132.9	1223.2	2.1500	97.45	1132.4	1222.6	1.9719	48.62	1131.8	1221.8	1.8948
400	511.9	1147.0	1241.8	2.1720	102.24	1146.6	1241.2	1.9941	51.03	1146.1	1240.5	1.9171
440	535.8	1161.2	1260.4	2.1932	107.03	1160.9	1259.9	2.0154	53.44	1160.5	1259.3	1.9385
500	571.5	1182.8	1288.5	2.2235	114.20	1182.5	1288.2	2.0458	57.04	1182.2	1287.7	1.9690
600	631.1	1219.3	1336.1	2.2706	126.15	1219.1	1335.8	2.0930	63.03	1218.9	1335.5	2.0164
700	690.7	1256.7	1384.5	2.3142	138.08	1256.5	1384.3	2.1367	69.01	1256.3	1384.0	2.0601
800	750.3	1294.9	1433.7	2.3550	150.01	1294.7	1433.5	2.1775	74.98	1294.6	1433.3	2.1009
1000	869.5	1373.9	1534.8	2.4294	173.86	1373.9	1534.7	2.2520	86.91	1373.8	1534.6	2.1755
1200	988.6	1456.7	1639.6	2.4967	197.70	1456.6	1639.5	2.3192	98.84	1456.5	1639.4	2.2428
1400	1107.7	1543.1	1748.1	2.5584	221.54	1543.1	1748.1	2.3810	110.76	1543.0	1748.0	2.3045

F	P = 14.696(211.99)				P = 20(227.96)				P = 40(267.26)			
	v	u	h	s	v	u	h	s	v	u	h	s
Sat	26.80	1077.6	1150.5	1.7567	20.09	1082.0	1156.4	1.7320	10.501	1092.3	1170.0	1.6767
240	28.00	1087.9	1164.0	1.7764	20.47	1086.5	1162.3	1.7405				
280	29.69	1102.4	1183.1	1.8030	21.73	1101.4	1181.8	1.7676	10.711	1097.3	1176.6	1.6857
320	31.36	1116.8	1202.1	1.8280	22.98	1116.0	1201.0	1.7930	11.360	1112.8	1196.9	1.7124
360	33.02	1131.2	1221.0	1.8516	24.21	1130.6	1220.1	1.8168	11.996	1128.0	1216.8	1.7373
400	34.67	1145.6	1239.9	1.8741	25.43	1145.1	1239.2	1.8395	12.623	1143.0	1236.4	1.7606
440	36.31	1160.1	1258.8	1.8956	26.64	1159.6	1258.2	1.8611	13.243	1157.8	1255.8	1.7828
500	38.77	1181.8	1287.3	1.9263	28.46	1181.5	1286.8	1.8919	14.164	1180.1	1284.9	1.8140

T	$P = 60(292.73)$				$P = 80(312.07)$				$P = 100(327.86)$			
600	42.86	1218.6	1335.2	1.9737	31.47	1218.4	1334.8	1.9395	15.685	1217.3	1333.4	1.8621
700	46.93	1256.1	1383.8	2.0175	34.47	1255.9	1383.5	1.9834	17.196	1255.1	1382.4	1.9063
800	51.00	1294.4	1433.1	2.0584	37.46	1294.3	1432.9	2.0243	18.701	1293.7	1432.1	1.9474
1000	59.13	1373.7	1534.5	2.1330	43.44	1373.5	1534.3	2.0989	21.70	1373.1	1533.8	2.0223
1200	67.25	1456.5	1639.3	2.2003	49.41	1456.4	1639.2	2.1663	24.69	1456.1	1638.9	2.0897
1400	75.36	1543.0	1747.9	2.2621	55.37	1542.9	1747.9	2.2281	27.68	1542.7	1747.6	2.1515
1600	83.47	1633.2	1860.2	2.3194	61.33	1633.2	1860.1	2.2854	30.66	1633.0	1859.9	2.2089

T	$P = 120(341.30)$				$P = 140(353.08)$				$P = 160(363.60)$			
Sat	7.177	1098.3	1178.0	1.6444	5.474	1102.6	1183.6	1.6214	4.434	1105.8	1187.8	1.6034
320	7.485	1109.5	1192.6	1.6634	5.544	1106.0	1188.0	1.6271				
360	7.924	1125.3	1213.3	1.6893	5.886	1122.5	1209.7	1.6541	4.662	1119.7	1205.9	1.6259
400	8.353	1140.8	1233.5	1.7134	6.217	1138.5	1230.6	1.6790	4.934	1136.2	1227.5	1.6517
440	8.775	1156.0	1253.4	1.7360	6.541	1154.2	1251.0	1.7022	5.199	1152.3	1248.5	1.6755
500	9.399	1178.6	1283.0	1.7678	7.017	1177.2	1281.1	1.7346	5.587	1175.7	1279.1	1.7085
600	10.425	1216.3	1332.1	1.8165	7.794	1215.3	1330.7	1.7838	6.216	1214.2	1329.3	1.7582
700	11.440	1254.4	1381.4	1.8609	8.561	1253.6	1380.3	1.8285	6.834	1252.8	1379.2	1.8033
800	12.448	1293.0	1431.2	1.9022	9.321	1292.4	1430.4	1.8700	7.445	1291.8	1429.6	1.8449
1000	14.454	1372.7	1533.2	1.9773	10.831	1372.3	1532.6	1.9453	8.657	1371.9	1532.1	1.9204
1200	16.452	1455.8	1638.5	2.0448	12.333	1455.5	1638.1	2.0130	9.861	1455.2	1637.7	1.9882
1400	18.445	1542.5	1747.3	2.1067	13.830	1542.3	1747.0	2.0749	11.060	1542.0	1746.7	2.0502
1600	20.44	1632.8	1859.7	2.1641	15.324	1632.6	1859.5	2.1323	12.257	1632.4	1859.3	2.1076
1800	22.43	1726.7	1975.7	2.2179	16.818	1726.5	1975.5	2.1861	13.452	1726.4	1975.3	2.1614
2000	24.41	1824.0	2095.1	2.2685	18.310	1823.9	2094.9	2.2367	14.647	1823.7	2094.8	2.2121

T												
Sat	3.730	1108.3	1191.1	1.5886	3.221	1110.3	1193.8	1.5761	2.836	1112.0	1196.0	1.5651
360	3.844	1116.7	1202.0	1.6021	3.259	1113.5	1198.0	1.5812				
400	4.079	1133.8	1224.4	1.6288	3.466	1131.4	1221.2	1.6088	3.007	1128.8	1217.8	1.5911
450	4.360	1154.3	1251.2	1.6590	3.713	1152.4	1248.6	1.6399	3.228	1150.5	1246.1	1.6230

(Continued)

TABLE A.1.3E (*Continued*) Superheated Vapor (English Units)

F	v	u	h	s	v	u	h	s	v	u	h	s
	P = 120(341.30)				**P = 140(353.08)**				**P = 160(363.60)**			
500	4.633	1174.2	1277.1	1.6868	3.952	1172.7	1275.1	1.6682	3.440	1171.2	1273.0	1.6518
550	4.900	1193.8	1302.6	1.7127	4.184	1192.6	1300.9	1.6944	3.646	1191.3	1299.2	1.6784
600	5.164	1213.2	1327.8	1.7371	4.412	1212.1	1326.4	1.7191	3.848	1211.1	1325.0	1.7034
700	5.682	1252.0	1378.2	1.7825	4.860	1251.2	1377.1	1.7648	4.243	1250.4	1376.0	1.7494
800	6.195	1291.2	1428.7	1.8243	5.301	1290.5	1427.9	1.8068	4.631	1289.9	1427.0	1.7916
1000	7.208	1371.5	1531.5	1.9000	6.173	1371.0	1531.0	1.8827	5.397	1370.6	1530.4	1.8677
1200	8.213	1454.9	1637.3	1.9679	7.036	1454.6	1636.9	1.9507	6.154	1454.3	1636.5	1.9358
1400	9.214	1541.8	1746.4	2.0300	7.895	1541.6	1746.1	2.0129	6.906	1541.4	1745.9	1.9980
1600	10.212	1632.3	1859.0	2.0875	8.752	1632.1	1858.8	2.0704	7.656	1631.9	1858.6	2.0556
1800	11.209	1726.2	1975.1	2.1413	9.607	1726.1	1975.0	2.1242	8.405	1725.9	1974.8	2.1094
2000	12.205	1823.6	2094.6	2.1919	10.461	1823.5	2094.5	2.1749	9.153	1823.3	2094.3	2.1601

F	v	u	h	s	v	u	h	s	v	u	h	s
	P = 180(373.13)				**P = 200(381.86)**				**P = 225(391.87)**			
Sat	2.533	1113.4	1197.8	1.5553	2.289	1114.6	1199.3	1.5464	2.043	1115.8	1200.8	1.5365
400	2.648	1126.2	1214.4	1.5749	2.361	1123.5	1210.8	1.5600	2.073	1119.9	1206.2	1.5427
450	2.850	1148.5	1243.4	1.6078	2.548	1146.4	1240.7	1.5938	2.245	1143.8	1237.3	1.5779
500	3.042	1169.6	1270.9	1.6372	2.724	1168.0	1268.8	1.6239	2.405	1165.9	1266.1	1.6087
550	3.228	1190.0	1297.5	1.6642	2.893	1188.7	1295.7	1.6512	2.558	1187.0	1293.5	1.6366
600	3.409	1210.0	1323.5	1.6893	3.058	1208.9	1322.1	1.6767	2.707	1207.5	1320.2	1.6624
700	3.763	1249.6	1374.9	1.7357	3.379	1248.8	1373.8	1.7234	2.995	1247.7	1372.4	1.7095
800	4.110	1289.3	1426.2	1.7781	3.693	1288.6	1425.3	1.7660	3.276	1287.8	1424.2	1.7523
900	4.453	1329.4	1477.7	1.8175	4.003	1328.9	1477.1	1.8055	3.553	1328.3	1476.2	1.7920
1000	4.793	1370.2	1529.8	1.8545	4.310	1369.8	1529.3	1.8425	3.827	1369.3	1528.6	1.8292
1200	5.467	1454.0	1636.1	1.9227	4.918	1453.7	1635.7	1.9109	4.369	1453.4	1635.3	1.8977
1400	6.137	1541.2	1745.6	1.9849	5.521	1540.9	1745.3	1.9732	4.906	1540.7	1744.9	1.9600

(continued — rows 1600, 1800, 2000 of three preceding pressure blocks)

T	v	u	h	s		v	u	h	s		v	u	h	s
1600	6.804	1631.7	1858.4	2.0425		6.123	1631.6	1858.2	2.0308		5.441	1631.3	1857.9	2.0177
1800	7.470	1725.8	1974.6	2.0964		6.722	1725.6	1974.4	2.0847		5.975	1725.4	1974.2	2.0716
2000	8.135	1823.2	2094.2	2.1470		7.321	1823.0	2094.0	2.1354		6.507	1822.9	2093.8	2.1223

P = 250(401.04) | P = 275(409.52) | P = 300(417.43)

T	v	u	h	s		v	u	h	s		v	u	h	s
Sat	1.8448	1116.7	1202.1	1.5274		1.6813	1117.5	1203.1	1.5192		1.5442	1118.2	1203.9	1.5115
450	2.002	1141.1	1233.7	1.5632		1.8026	1138.3	1230.0	1.5495		1.6361	1135.4	1226.2	1.5365
500	2.150	1163.8	1263.3	1.5948		1.9407	1161.7	1260.4	1.5820		1.7662	1159.5	1257.5	1.5701
550	2.290	1185.3	1291.3	1.6233		2.071	1183.6	1289.0	1.6110		1.8878	1181.9	1286.7	1.5997
600	2.426	1206.1	1318.3	1.6494		2.196	1204.7	1316.4	1.6376		2.004	1203.2	1314.5	1.6266
650	2.558	1226.5	1344.9	1.6739		2.317	1225.3	1343.2	1.6623		2.117	1224.1	1341.6	1.6516
700	2.688	1246.7	1371.1	1.6970		2.436	1245.7	1369.7	1.6856		2.227	1244.6	1368.3	1.6751
800	2.943	1287.0	1423.2	1.7401		2.670	1286.2	1422.1	1.7289		2.442	1285.4	1421.0	1.7187
900	3.193	1327.6	1475.3	1.7799		2.898	1327.0	1474.5	1.7689		2.653	1326.3	1473.6	1.7589
1000	3.440	1368.7	1527.9	1.8172		3.124	1368.2	1527.2	1.8064		2.860	1367.7	1526.5	1.7964
1200	3.929	1453.0	1634.8	1.8858		3.570	1452.6	1634.3	1.8751		3.270	1452.2	1633.8	1.8653
1400	4.414	1540.4	1744.6	1.9483		4.011	1540.1	1744.2	1.9376		3.675	1539.8	1743.8	1.9279
1600	4.896	1631.1	1857.6	2.0060		4.450	1630.9	1857.3	1.9954		4.078	1630.7	1857.0	1.9857
1800	5.376	1725.2	1974.0	2.0599		4.887	1725.0	1973.7	2.0493		4.479	1724.9	1973.5	2.0396
2000	5.856	1822.7	2093.6	2.1106		5.323	1822.5	2093.4	2.1000		4.879	1822.3	2093.2	2.0904

P = 350(431.82) | P = 400(444.70) | P = 450(456.39)

T	v	u	h	s		v	u	h	s		v	u	h	s
Sat	1.3267	1119.0	1204.9	1.4978		1.1620	1119.5	1205.5	1.4856		1.0326	1119.6	1205.6	1.4746
450	1.3733	1129.2	1218.2	1.5125		1.1745	1122.6	1209.6	1.4901					
500	1.4913	1154.9	1251.5	1.5482		1.2843	1150.1	1245.2	1.5282		1.1226	1145.1	1238.5	1.5097
550	1.5998	1178.3	1281.9	1.5790		1.3833	1174.6	1277.0	1.5605		1.2146	1170.7	1271.9	1.5436
600	1.7025	1200.3	1310.6	1.6068		1.4760	1197.3	1306.6	1.5892		1.2996	1194.3	1302.5	1.5732
650	1.8013	1221.6	1338.3	1.6323		1.5645	1219.1	1334.9	1.6153		1.3803	1216.6	1331.5	1.6000
700	1.8975	1242.5	1365.4	1.6562		1.6503	1240.4	1362.5	1.6397		1.4580	1238.2	1359.6	1.6248
800	2.085	1283.8	1418.8	1.7004		1.8163	1282.1	1416.6	1.6844		1.6077	1280.5	1414.4	1.6701

TABLE A.1.3E (Continued) Superheated Vapor (English Units)

F	P = 350(431.82)				P = 400(444.70)				P = 450(456.39)			
	v	u	h	s	v	u	h	s	v	u	h	s
900	2.267	1325.0	1471.8	1.7409	1.9776	1323.7	1470.1	1.7252	1.7524	1322.4	1468.3	1.7113
1000	2.446	1366.6	1525.0	1.7787	2.136	1365.5	1523.6	1.7632	1.8941	1364.4	1522.2	1.7495
1200	2.799	1451.5	1632.8	1.8478	2.446	1450.7	1631.8	1.8327	2.172	1450.0	1630.8	1.8192
1400	3.148	1539.3	1743.1	1.9106	2.752	1538.7	1742.4	1.8956	2.444	1538.1	1741.7	1.8823
1600	3.494	1630.2	1856.5	1.9685	3.055	1629.8	1855.9	1.9535	2.715	1629.3	1855.4	1.9403
1800	3.838	1724.5	1973.1	2.0225	3.357	1724.1	1972.6	2.0076	2.983	1723.7	1972.1	1.9944
2000	4.182	1822.0	2092.8	2.0733	3.658	1821.6	2092.4	2.0584	3.251	1821.3	2092.0	2.0453

F	P = 500(467.13)				P = 600(486.33)				P = 700(503.23)			
	v	u	h	s	v	u	h	s	v	u	h	s
Sat	0.9283	1119.4	1205.3	1.4645	0.7702	1118.6	1204.1	1.4464	0.6558	1117.0	1202.0	1.4305
500	0.9924	1139.7	1231.5	1.4923	0.7947	1128.0	1216.2	1.4592				
550	1.0792	1166.7	1266.6	1.5279	0.8749	1158.2	1255.4	1.4990	0.7275	1149.0	1243.2	1.4723
600	1.1583	1191.1	1298.3	1.5585	0.9456	1184.5	1289.5	1.5320	0.7929	1177.5	1280.2	1.5081
650	1.2327	1214.0	1328.0	1.5860	1.0109	1208.6	1320.9	1.5609	0.8520	1203.1	1313.4	1.5387
700	1.3040	1236.0	1356.7	1.6112	1.0727	1231.5	1350.6	1.5872	0.9073	1226.9	1344.4	1.5661
800	1.4407	1278.8	1412.1	1.6571	1.1900	1275.4	1407.6	1.6343	1.0109	1272.0	1402.9	1.6145
900	1.5723	1321.0	1466.5	1.6987	1.3021	1318.4	1462.9	1.6766	1.1089	1315.6	1459.3	1.6576
1000	1.7008	1363.3	1520.7	1.7371	1.4108	1361.2	1517.8	1.7155	1.2036	1358.9	1514.9	1.6970
1100	1.8271	1406.0	1575.1	1.7731	1.5173	1404.2	1572.7	1.7519	1.2960	1402.4	1570.2	1.7337
1200	1.9518	1449.2	1629.8	1.8072	1.6222	1447.7	1627.8	1.7861	1.3868	1446.2	1625.8	1.7682
1400	2.198	1537.6	1741.0	1.8704	1.8289	1536.5	1739.5	1.8497	1.5652	1535.3	1738.1	1.8321
1600	2.442	1628.9	1854.8	1.9285	2.033	1628.0	1853.7	1.9080	1.7409	1627.1	1852.6	1.8906
1800	2.684	1723.3	1971.7	1.9827	2.236	1722.6	1970.8	1.9622	1.9152	1721.8	1969.9	1.9449
2000	2.926	1820.9	2091.6	2.0335	2.438	1820.2	2090.8	2.0131	2.0887	1819.5	2090.1	1.9958

T	P = 800(518.36)				P = 1000(544.75)				P = 1250(572.56)			
Sat	0.5691	1115.0	1199.3	1.4160	0.4459	1109.9	1192.4	1.3903	0.3454	1101.7	1181.6	1.3619
550	0.6154	1138.8	1229.9	1.4469	0.4534	1114.8	1198.7	1.3966				
600	0.6776	1170.1	1270.4	1.4861	0.5140	1153.7	1248.8	1.4450	0.3786	1129.0	1216.6	1.3954
650	0.7324	1197.2	1305.6	1.5186	0.5637	1184.7	1289.1	1.4822	0.4267	1167.2	1266.0	1.4410
700	0.7829	1222.1	1338.0	1.5471	0.6080	1212.0	1324.6	1.5135	0.4670	1198.4	1306.4	1.4767
750	0.8306	1245.7	1368.6	1.5730	0.6490	1237.2	1357.3	1.5412	0.5030	1226.1	1342.4	1.5070
800	0.8764	1268.5	1398.2	1.5969	0.6878	1261.2	1388.5	1.5664	0.5364	1251.8	1375.8	1.5341
900	0.9640	1312.9	1455.6	1.6408	0.7610	1307.3	1448.1	1.6120	0.5984	1300.0	1438.4	1.5820
1000	1.0482	1356.7	1511.9	1.6807	0.8305	1352.2	1505.9	1.6530	0.6563	1346.4	1498.2	1.6244
1100	1.1300	1400.5	1567.8	1.7178	0.8976	1396.8	1562.9	1.6908	0.7116	1392.0	1556.6	1.6631
1200	1.2102	1444.6	1623.8	1.7526	0.9630	1441.5	1619.7	1.7261	0.7652	1437.5	1614.5	1.6991
1400	1.3674	1534.2	1736.6	1.8167	1.0905	1531.9	1733.7	1.7909	0.8689	1529.0	1730.0	1.7648
1600	1.5218	1626.2	1851.5	1.8754	1.2152	1624.4	1849.3	1.8499	0.9699	1622.2	1846.5	1.8243
1800	1.6749	1721.0	1969.0	1.9298	1.3384	1719.5	1967.2	1.9046	1.0693	1717.6	1965.0	1.8791
2000	1.8271	1818.8	2089.3	1.9808	1.4608	1817.4	2087.7	1.9557	1.1678	1815.7	2085.8	1.9304

T	P = 1500(596.39)				P = 1750(617.31)				P = 2000(636.00)			
Sat	0.2769	1091.8	1168.7	1.3359	0.2268	1080.2	1153.7	1.3109	0.18813	1066.6	1136.3	1.2861
600	0.2816	1096.6	1174.8	1.3416								
650	0.3329	1147.0	1239.4	1.4012	0.2627	1122.5	1207.6	1.3603	0.2057	1091.1	1167.2	1.3141
700	0.3716	1183.4	1286.6	1.4429	0.3022	1166.7	1264.6	1.4106	0.2487	1147.7	1239.8	1.3782
750	0.4049	1214.1	1326.5	1.4767	0.3341	1201.3	1309.5	1.4485	0.2803	1187.3	1291.1	1.4216
800	0.4350	1241.8	1362.5	1.5058	0.3622	1231.3	1348.6	1.4802	0.3071	1220.1	1333.8	1.4562
850	0.4631	1267.7	1396.2	1.5320	0.3878	1258.8	1384.4	1.5081	0.3312	1249.5	1372.0	1.4860
900	0.4897	1292.5	1428.5	1.5562	0.4119	1284.8	1418.2	1.5334	0.3534	1276.8	1407.6	1.5126
1000	0.5400	1340.4	1490.3	1.6001	0.4569	1334.3	1482.3	1.5789	0.3945	1328.1	1474.1	1.5598
1100	0.5876	1387.2	1550.3	1.6399	0.4990	1382.2	1543.8	1.6197	0.4325	1377.2	1537.2	1.6017
1200	0.6334	1433.5	1609.3	1.6765	0.5392	1429.4	1604.0	1.6571	0.4685	1425.2	1598.6	1.6398
1400	0.7213	1526.1	1726.3	1.7431	0.6158	1523.1	1722.6	1.7245	0.5368	1520.2	1718.8	1.7082
1600	0.8064	1619.9	1843.7	1.8031	0.6896	1617.6	1841.0	1.7850	0.6020	1615.4	1838.2	1.7692

(Continued)

TABLE A.1.3E *(Continued)* Superheated Vapor (English Units)

F	v	u	h	s	v	u	h	s	v	u	h	s
	P = 1500(596.39)				P = 1750(617.31)				P = 2000(636.00)			
1800	0.8899	1715.7	1962.7	1.8582	0.7617	1713.9	1960.5	1.8404	0.6656	1712.0	1958.3	1.8249
2000	0.9725	1814.0	2083.9	1.9096	0.8330	1812.3	2082.0	1.8919	0.7284	1810.6	2080.2	1.8765

F	v	u	h	s	v	u	h	s	v	u	h	s
	P = 2500(668.31)				P = 3000(695.52)				P = 3500			
Sat	0.130 59	1031.0	1091.4	1.2327	0.084 04	968.8	1015.5	1.1575				
650									0.024 91	663.5	679.7	0.8630
700	0.168 39	1098.7	1176.6	1.3073	0.097 71	1003.9	1058.1	1.1944	0.030 58	759.5	779.3	0.9506
750	0.2030	1155.2	1249.1	1.3686	0.148 31	1114.7	1197.1	1.3122	0.104 60	1058.4	1126.1	1.2440
800	0.2291	1195.7	1301.7	1.4112	0.175 72	1167.6	1265.2	1.3675	0.136 26	1134.7	1223.0	1.3226
850	0.2513	1229.5	1345.8	1.4456	0.197 31	1207.7	1317.2	1.4080	0.158 18	1183.4	1285.9	1.3716
900	0.2712	1259.9	1385.4	1.4752	0.2160	1241.8	1361.7	1.4414	0.176 25	1222.4	1336.5	1.4096
950	0.2896	1288.2	1422.2	1.5018	0.2328	1272.7	1402.0	1.4705	0.192 14	1256.4	1380.8	1.4416
1000	0.3069	1315.2	1457.2	1.5262	0.2485	1301.7	1439.6	1.4967	0.2066	1287.6	1421.4	1.4699
1100	0.3393	1366.8	1523.8	1.5704	0.2772	1356.2	1510.1	1.5434	0.2328	1345.2	1496.0	1.5193
1200	0.3696	1416.7	1587.7	1.6101	0.3036	1408.0	1576.6	1.5848	0.2566	1399.2	1565.3	1.5624
1400	0.4261	1514.2	1711.3	1.6804	0.3524	1508.1	1703.7	1.6571	0.2997	1501.9	1696.1	1.6368
1600	0.4795	1610.2	1832.6	1.7424	0.3978	1606.3	1827.1	1.7201	0.3395	1601.7	1821.6	1.7010
1800	0.5312	1708.2	1954.0	1.7986	0.4416	1704.5	1949.6	1.7769	0.3776	1700.8	1945.4	1.7583
2000	0.5820	1807.2	2076.4	1.8506	0.4844	1803.9	2072.8	1.8291	0.4147	1800.6	2069.2	1.8108

F	v	u	h	s	v	u	h	s	v	u	h	s
	P = 4000				P = 5000				P = 6000			
650	0.024 47	657.7	675.8	0.8574	0.023 77	648.0	670.0	0.8482	0.023 22	640.0	665.8	0.8405
700	0.028 67	742.1	763.4	0.9345	0.026 76	721.8	746.6	0.9156	0.025 63	708.1	736.5	0.9028

750	0.063 31	960.7	1007.5	1.1395	0.033 64	821.4	852.6	1.0049	0.029 78	788.6	821.7	0.9746
800	0.105 22	1095.0	1172.9	1.2740	0.059 32	987.2	1042.1	1.1583	0.039 42	896.9	940.7	1.0708
850	0.128 33	1156.5	1251.5	1.3352	0.085 56	1092.7	1171.9	1.2596	0.058 18	1018.8	1083.4	1.1820
900	0.146 22	1201.5	1309.7	1.3789	0.103 85	1155.1	1251.1	1.3190	0.075 88	1102.9	1187.2	1.2599
950	0.161 51	1239.2	1358.8	1.4144	0.118 53	1202.2	1311.9	1.3629	0.090 08	1162.0	1262.0	1.3140
1000	0.175 20	1272.9	1402.6	1.4449	0.131 20	1242.0	1363.4	1.3988	0.102 07	1209.1	1322.4	1.3561
1100	0.199 54	1333.9	1481.6	1.4973	0.153 02	1310.6	1452.2	1.4577	0.122 18	1286.4	1422.1	1.4222
1200	0.2213	1390.1	1553.9	1.5423	0.171 99	1371.6	1530.8	1.5066	0.139 27	1352.7	1507.3	1.4752
1300	0.2414	1443.7	1622.4	1.5823	0.189 18	1428.6	1603.7	1.5493	0.154 53	1413.3	1584.9	1.5206
1400	0.2603	1495.7	1688.4	1.6188	0.205 17	1483.2	1673.0	1.5876	0.168 54	1470.5	1657.6	1.5608
1600	0.2959	1597.1	1816.1	1.6841	0.2348	1587.9	1805.2	1.6551	0.194 20	1578.7	1794.3	1.6307
1800	0.3296	1697.1	1941.1	1.7420	0.2626	1689.8	1932.7	1.7142	0.218 01	1682.4	1924.5	1.6910
2000	0.3625	1797.3	2065.6	1.7948	0.2895	1790.8	2058.6	1.7676	0.240 87	1784.3	2051.7	1.7450

TABLE A.1.4E Compressed Liquid (English Units)

T		P = 500(467.13)				P = 1000(544.75)				P = 1500(596.39)		
	v	u	h	s	v	u	h	s	v	u	h	s
Sat	0.019 748	447.70	449.53	0.649 04	0.021 591	538.39	542.38	0.743 20	0.023 461	604.97	611.48	0.808 24
32	0.015 994	0.00	1.49	0.000 00	0.015 967	0.03	2.99	0.000 05	0.015 939	0.05	4.47	0.000 07
50	0.015 998	18.02	19.50	0.035 99	0.015 972	17.99	20.94	0.035 92	0.015 946	17.95	22.38	0.035 84
100	0.016 106	67.87	69.36	0.129 32	0.016 082	67.70	70.68	0.129 01	0.016 058	67.53	71.99	0.128 70
150	0.016 318	117.66	119.17	0.214 57	0.016 293	117.38	120.40	0.214 10	0.016 268	117.10	121.62	0.213 64
200	0.016 608	167.65	169.19	0.293 41	0.016 580	167.26	170.32	0.292 81	0.016 554	166.87	171.46	0.292 21
250	0.016 972	217.99	219.56	0.367 02	0.016 941	217.47	220.61	0.366 28	0.016 910	216.96	221.65	0.365 54
300	0.017 416	268.92	270.53	0.436 41	0.017 379	268.24	271.46	0.435 52	0.017 343	267.58	272.39	0.434 63
350	0.017 954	320.71	322.37	0.502 49	0.017 909	319.83	323.15	0.501 40	0.017 865	318.98	323.94	0.500 34
400	0.018 608	373.68	375.40	0.566 04	0.018 550	372.55	375.98	0.564 72	0.018 493	371.45	376.59	0.563 43
450	0.019 420	428.40	430.19	0.627 98	0.019 340	426.89	430.47	0.626 32	0.019 264	425.44	430.79	0.624 70
500					0.020 36	483.8	487.5	0.6874	0.020 024	481.8	487.4	0.6853
550									0.021 58	542.1	548.1	0.7469

T		P = 2000(636.00)				P = 3000(695.52)				P = 5000		
	v	u	h	s	v	u	h	s	v	u	h	s
Sat	0.025 649	662.40	671.89	0.862 27	0.034 310	783.45	802.50	0.973 20				
32	0.015 912	0.06	5.95	0.000 08	0.015 859	0.09	8.90	0.000 09	0.015 755	0.11	14.70	−0.000 01
50	0.015 920	17.91	23.81	0.035 75	0.015 870	17.84	26.65	0.035 55	0.015 773	17.67	32.26	0.035 08
100	0.016 034	67.37	73.30	0.128 39	0.015 987	67.04	75.91	0.127 77	0.015 897	66.40	81.11	0.126 51
200	0.016 527	166.49	172.60	0.291 62	0.016 476	165.74	174.89	0.290 46	0.016 376	164.32	179.47	0.288 18
300	0.017 308	266.93	273.33	0.433 76	0.017 240	265.66	275.23	0.432 05	0.017 110	263.25	279.08	0.428 75
400	0.018 439	370.38	377.21	0.562 16	0.018 334	368.32	378.50	0.559 70	0.018 141	364.47	381.25	0.555 06
450	0.019 191	424.04	431.14	0.623 13	0.019 053	421.36	431.93	0.620 11	0.018 803	416.44	433.84	0.614 51
500	0.020 14	479.8	487.3	0.6832	0.019 944	476.2	487.3	0.679 4	0.019 603	469.8	487.9	0.6724
560	0.021 72	551.8	559.8	0.7565	0.021 382	546.2	558.0	0.750 8	0.020 835	536.7	556.0	0.7411

600	0.023 30	605.4	614.0	0.8086	0.022 74	597.0	609.6	0.021 91	584.0	604.2	0.7876
640	0.024 75	654.3	668.0	0.8545				0.023 34	634.6	656.2	0.8357
680	0.028 79	728.4	744.3	0.9226				0.025 35	690.6	714.1	0.8873
700								0.026 76	721.8	746.6	0.9156
				0.8004							

TABLE A.1.5E Saturated Solid-Vapor (English Units)

Temp. Fahr. T	Press. lbf/in² P	Specific Volume		Internal Energy			Enthalpy			Entropy		
		Sat. Solid v_i	Sat. Vapor $v_g \times 10^{-3}$	Sat. Solid u_i	Subl. u_{ig}	Sat. Vapor u_g	Sat. Solid h_i	Subl. h_{ig}	Sat. Vapor h_g	Sat. Solid s_i	Subl. s_{ig}	Sat. Vapor s_g
32.018	0.0887	0.01747	3.302	-143.34	1164.6	1021.2	-143.34	1218.7	1075.4	-0.292	2.479	2.187
32	0.0886	0.01747	3.305	-143.35	1164.6	1021.2	-143.35	1218.7	1075.4	-0.292	2.479	2.187
30	0.0808	0.01747	3.607	-144.35	1164.9	1020.5	-144.35	1218.9	1074.5	-0.294	2.489	2.195
25	0.0641	0.01746	4.506	-146.84	1165.7	1018.9	-146.84	1219.1	1072.3	-0.299	2.515	2.216
20	0.0505	0.01745	5.655	-149.31	1166.5	1017.2	-149.31	1219.4	1070.1	-0.304	2.542	2.238
15	0.0396	0.01745	7.13	-151.75	1167.3	1015.5	-151.75	1219.7	1067.9	-0.309	2.569	2.260
10	0.0309	0.01744	9.04	-154.17	1168.1	1013.9	-154.17	1219.9	1065.7	-0.314	2.597	2.283
5	0.0240	0.01743	11.52	-156.56	1168.8	1012.2	-156.56	1220.1	1063.5	-0.320	2.626	2.306
0	0.0185	0.01743	14.77	-158.93	1169.5	1010.6	-158.93	1220.2	1061.2	-0.325	2.655	2.330
-5	0.0142	0.01742	19.03	-161.27	1170.2	1008.9	-161.27	1220.3	1059.0	-0.330	2.684	2.354
-10	0.0109	0.01741	24.66	-163.59	1170.9	1007.3	-163.59	1220.4	1056.8	-0.335	2.714	2.379
-15	0.0082	0.01740	32.2	-165.89	1171.5	1005.6	-165.89	1220.5	1054.6	-0.340	2.745	2.405
-20	0.0062	0.01740	42.2	-168.16	1172.1	1003.9	-168.16	1220.6	1052.4	-0.345	2.776	2.431
-25	0.0046	0.01739	55.7	-170.40	1172.7	1002.3	-170.40	1220.6	1050.2	-0.351	2.808	2.457
-30	0.0035	0.01738	74.1	-172.63	1173.2	1000.6	-172.63	1220.6	1048.0	-0.356	2.841	2.485
-35	0.0026	0.01737	99.2	-174.82	1173.8	998.9	-174.82	1220.6	1045.8	-0.361	2.874	2.513
-40	0.0019	0.01737	133.8	-177.00	1174.3	997.3	-177.00	1220.6	1043.6	-0.366	2.908	2.542

TABLE A.1SI Thermodynamic Properties of Steam[a] (SI Units)

TABLE A.1.1SI Saturated Steam: Temperature Table (SI Units)

Temp. °C T	Press. kPa P	Specific Volume m³/kg		Internal Energy kJ/kg			Enthalpy kJ/kg			Entropy kJ/kg K		
		Sat. Liquid v_f	Sat. Liquid v_g	Sat. Vapor u_f	Evap. u_{fg}	Sat. Vapor u_g	Sat. Liquid h_f	Evap. h_{fg}	Sat. Vapor h_g	Sat. Liquid s_f	Evap. s_{fg}	Sat. Vapor s_g
0.01	0.6113	0.001 000	206.14	.00	2375.3	2375.3	.01	2501.3	2501.4	.0000	9.1562	9.1562
5	0.8721	0.001 000	147.12	20.97	2361.3	2382.3	20.98	2489.6	2510.6	.0761	8.9496	9.0257
10	1.2276	0.001 000	106.38	42.00	2347.2	2389.2	42.01	2477.7	2519.8	.1510	8.7498	8.9008
15	1.7051	0.001 001	77.93	62.99	2333.1	2396.1	62.99	2465.9	2528.9	.2245	8.5569	8.7814
20	2.339	0.001 002	57.79	83.95	2319.0	2402.9	83.96	2454.1	2538.1	.2966	8.3706	8.6672
25	3.169	0.001 003	43.36	104.88	2304.9	2409.8	104.89	2442.3	2547.2	.3674	8.1905	8.5580
30	4.246	0.001 004	32.89	125.78	2290.8	2416.6	125.79	2430.5	2556.3	.4369	8.0164	8.4533
35	5.628	0.001 006	25.22	146.67	2276.7	2423.4	146.68	2418.6	2565.3	.5053	7.8478	8.3531
40	7.384	0.001 008	19.52	167.56	2262.6	2430.1	167.57	2406.7	2574.3	.5725	7.6845	8.2570
45	9.593	0.001 010	15.26	188.44	2248.4	2436.8	188.45	2394.8	2583.2	.6387	7.5261	8.1648
50	12.349	0.001 012	12.03	209.32	2234.2	2443.5	209.33	2382.7	2592.1	.7038	7.3725	8.0763
55	15.758	0.001 015	9.568	230.21	2219.9	2450.1	230.23	2370.7	2600.9	.7679	7.2234	7.9913
60	19.940	0.001 017	7.671	251.11	2205.5	2456.6	251.13	2358.5	2609.6	.8312	7.0784	7.9096
65	25.03	0.001 020	6.197	272.02	2191.1	2463.1	272.06	2346.2	2618.3	.8935	6.9375	7.8310
70	31.19	0.001 023	5.042	292.95	2176.6	2469.6	292.98	2333.8	2626.8	.9549	6.8004	7.7553
75	38.58	0.001 026	4.131	313.90	2162.0	2475.9	313.93	2321.4	2635.3	1.0155	6.6669	7.6824
80	47.39	0.001 029	3.407	334.86	2147.4	2482.2	334.91	2308.8	2643.7	1.0753	6.5369	7.6122
85	57.83	0.001 033	2.828	355.84	2132.6	2488.4	355.90	2296.0	2651.9	1.1343	6.4102	7.5445
90	70.14	0.001 036	2.361	376.85	2117.7	2494.5	376.92	2283.2	2660.1	1.1925	6.2866	7.4791
95	84.55	0.001 040	1.982	397.88	2102.7	2500.6	397.96	2270.2	2668.1	1.2500	6.1659	7.4159

[a] Adapted from Joseph H. Keenan, Frederick G. Keyes, Philip G. Hill, and Joan G. Moore, *Steam Tables* (New York: John Wiley & Sons, Inc., 1978).

TABLE A.1.1SI (*Continued*) Saturated Steam: Temperature Table (SI Units)

Temp. °C T	Press. MPa P	Specific Volume m³/kg		Internal Energy kJ/kg			Enthalpy kJ/kg			Entropy kJ/kg K		
		Sat. Liquid v_f	Sat. Vapor v_g	Sat. Liquid u_f	Evap. u_{fg}	Sat. Vapor u_g	Sat. Liquid h_f	Evap. h_{fg}	Sat. Vapor h_g	Sat. Liquid s_f	Evap. s_{fg}	Sat. Vapor s_g
100	0.101 35	0.001 044	1.6729	418.94	2087.6	2506.5	419.04	2257.0	2676.1	1.3069	6.0480	7.3549
105	0.120 82	0.001 048	1.4194	440.02	2072.3	2512.4	440.15	2243.7	2683.8	1.3630	5.9328	7.2958
110	0.143 27	0.001 052	1.2102	461.14	2057.0	2518.1	461.30	2230.2	2691.5	1.4185	5.8202	7.2387
115	0.169 06	0.001 056	1.0366	482.30	2041.4	2523.7	482.48	2216.5	2699.0	1.4734	5.7100	7.1833
120	0.198 53	0.001 060	0.8919	503.50	2025.8	2529.3	503.71	2202.6	2706.3	1.5276	5.6020	7.1296
125	0.2321	0.001 065	0.7706	524.74	2009.9	2534.6	524.99	2188.5	2713.5	1.5813	5.4962	7.0775
130	0.2701	0.001 070	0.6685	546.02	1993.9	2539.9	546.31	2174.2	2720.5	1.6344	5.3925	7.0269
135	0.3130	0.001 075	0.5822	567.35	1977.7	2545.0	567.69	2159.6	2727.3	1.6870	5.2907	6.9777
140	0.3613	0.001 080	0.5089	588.74	1961.3	2550.0	589.13	2144.7	2733.9	1.7391	5.1908	6.9299
145	0.4154	0.001 085	0.4463	610.18	1944.7	2554.9	610.63	2129.6	2740.3	1.7907	5.0926	6.8833
150	0.4758	0.001 091	0.3928	631.68	1927.9	2559.5	632.20	2114.3	2746.5	1.8418	4.9960	6.8379
155	0.5431	0.001 096	0.3468	653.24	1910.8	2564.1	653.84	2098.6	2752.4	1.8925	4.9010	6.7935
160	0.6178	0.001 102	0.3071	674.87	1893.5	2568.4	675.55	2082.6	2758.1	1.9427	4.8075	6.7502
165	0.7005	0.001 108	0.2727	696.56	1876.0	2572.5	697.34	2066.2	2763.5	1.9925	4.7153	6.7078
170	0.7917	0.001 114	0.2428	718.33	1858.1	2576.5	719.21	2049.5	2768.7	2.0419	4.6244	6.6663
175	0.8920	0.001 121	0.2168	740.17	1840.0	2580.2	741.17	2032.4	2773.6	2.0909	4.5347	6.6256
180	1.0021	0.001 127	0.194 05	762.09	1821.6	2583.7	763.22	2015.0	2778.2	2.1396	4.4461	6.5857
185	1.1227	0.001 134	0.174 09	784.10	1802.9	2587.0	785.37	1997.1	2782.4	2.1879	4.3586	6.5465
190	1.2544	0.001 141	0.156 54	806.19	1783.8	2590.0	807.62	1978.8	2786.4	2.2359	4.2720	6.5079
195	1.3978	0.001 149	0.141 05	828.37	1764.4	2592.8	829.98	1960.0	2790.0	2.2835	4.1863	6.4698
200	1.5538	0.001 157	0.127 36	850.65	1744.7	2595.3	852.45	1940.7	2793.2	2.3309	4.1014	6.4323
205	1.7230	0.001 164	0.115 21	873.04	1724.5	2597.5	875.04	1921.0	2796.0	2.3780	4.0172	6.3952
210	1.9062	0.001 173	0.104 41	895.53	1703.9	2599.5	897.76	1900.7	2798.5	2.4248	3.9337	6.3585

215	2.104	0.001 181	0.094 79	918.14	1682.9	2601.1	920.62	1879.9	2800.5	2.4714	3.8507	6.3221
220	2.318	0.001 190	0.086 19	940.87	1661.5	2602.4	943.62	1858.5	2802.1	2.5178	3.7683	6.2861
225	2.548	0.001 199	0.078 49	963.73	1639.6	2603.3	966.78	1836.5	2803.3	2.5639	3.6863	6.2503
230	2.795	0.001 209	0.071 58	986.74	1617.2	2603.9	990.12	1813.8	2804.0	2.6099	3.6047	6.2146
235	3.060	0.001 219	0.065 37	1009.89	1594.2	2604.1	1013.62	1790.5	2804.2	2.6558	3.5233	6.1791
240	3.344	0.001 229	0.059 76	1033.21	1570.8	2604.0	1037.32	1766.5	2803.8	2.7015	3.4422	6.1437
245	3.648	0.001 240	0.054 71	1056.71	1546.7	2603.4	1061.23	1741.7	2803.0	2.7472	3.3612	6.1083
250	3.973	0.001 251	0.050 13	1080.39	1522.0	2602.4	1085.36	1716.2	2801.5	2.7927	3.2802	6.0730
255	4.319	0.001 263	0.045 98	1104.28	1496.7	2600.9	1109.73	1689.8	2799.5	2.8383	3.1992	6.0375
260	4.688	0.001 276	0.042 21	1128.39	1470.6	2599.0	1134.37	1662.5	2796.9	2.8838	3.1181	6.0019
265	5.081	0.001 289	0.038 77	1152.74	1443.9	2596.6	1159.28	1634.4	2793.6	2.9294	3.0368	5.9662
270	5.499	0.001 302	0.035 64	1177.36	1416.3	2593.7	1184.51	1605.2	2789.7	2.9751	2.9551	5.9301
275	5.942	0.001 317	0.032 79	1202.25	1387.9	2590.2	1210.07	1574.9	2785.0	3.0208	2.8730	5.8938
280	6.412	0.001 332	0.030 17	1227.46	1358.7	2586.1	1235.99	1543.6	2779.6	3.0668	2.7903	5.8571
285	6.909	0.001 348	0.027 77	1253.00	1328.4	2581.4	1262.31	1511.0	2773.3	3.1130	2.7070	5.8199
290	7.436	0.001 366	0.025 57	1278.92	1297.1	2576.0	1289.07	1477.1	2766.2	3.1594	2.6227	5.7821
295	7.993	0.001 384	0.023 54	1305.2	1264.7	2569.9	1316.3	1441.8	2758.1	3.2062	2.5375	5.7437
300	8.581	0.001 404	0.021 67	1332.0	1231.0	2563.0	1344.0	1404.9	2749.0	3.2534	2.4511	5.7045
305	9.202	0.001 425	0.019 948	1359.3	1195.9	2555.2	1372.4	1366.4	2738.7	3.3010	2.3633	5.6643
310	9.856	0.001 447	0.018 350	1387.1	1159.4	2546.4	1401.3	1326.0	2727.3	3.3493	2.2737	5.6230
315	10.547	0.001 472	0.016 867	1415.5	1121.1	2536.6	1431.0	1283.5	2714.5	3.3982	2.1821	5.5804
320	11.274	0.001 499	0.015 488	1444.6	1080.9	2525.5	1461.5	1238.6	2700.1	3.4480	2.0882	5.5362
330	12.845	0.001 561	0.012 996	1505.3	993.7	2498.9	1525.3	1140.6	2665.9	3.5507	1.8909	5.4417
340	14.586	0.001 638	0.010 797	1570.3	894.3	2464.6	1594.2	1027.9	2622.0	3.6594	1.6763	5.3357
350	16.513	0.001 740	0.008 813	1641.9	776.6	2418.4	1670.6	893.4	2563.9	3.7777	1.4335	5.2112
360	18.651	0.001 893	0.006 945	1725.2	626.3	2351.5	1760.5	720.5	2481.0	3.9147	1.1379	5.0526
370	21.03	0.002 213	0.004 925	1844.0	384.5	2228.5	1890.5	441.6	2332.1	4.1106	.6865	4.7971
374.14	22.09	0.003 155	0.003 155	2029.6	0	2029.6	2099.3	0	2099.3	4.4298	0	4.4298

TABLE A.1.2SI Saturated Steam: Pressure Table (SI Units)

Press. kPa P	Temp. °C T	Specific Volume		Internal Energy			Enthalpy			Entropy		
		Sat. Liquid v_f	Sat. Vapor v_g	Sat. Liquid u_f	Evap. u_{fg}	Sat. Vapor u_g	Sat. Liquid h_f	Evap. h_{fg}	Sat. Vapor h_g	Sat. Liquid s_f	Evap. s_{fg}	Sat. Vapor s_g
0.6113	0.01	0.001 000	206.14	.00	2375.3	2375.3	.01	2501.3	2501.4	.0000	9.1562	9.1562
1.0	6.98	0.001 000	129.21	29.30	2355.7	2385.0	29.30	2484.9	2514.2	.1059	8.8697	8.9756
1.5	13.03	0.001 001	87.98	54.71	2338.6	2393.3	54.71	2470.6	2525.3	.1957	8.6322	8.8279
2.0	17.50	0.001 001	67.00	73.48	2326.0	2399.5	73.48	2460.0	2533.5	.2607	8.4629	8.7237
2.5	21.08	0.001 002	54.25	88.48	2315.9	2404.4	88.49	2451.6	2540.0	.3120	8.3311	8.6432
3.0	24.08	0.001 003	45.67	101.04	2307.5	2408.5	101.05	2444.5	2545.5	.3545	8.2231	8.5776
4.0	28.96	0.001 004	34.80	121.45	2293.7	2415.2	121.46	2432.9	2554.4	.4226	8.0520	8.4746
5.0	32.88	0.001 005	28.19	137.81	2282.7	2420.5	137.82	2423.7	2561.5	.4764	7.9187	8.3951
7.5	40.29	0.001 008	19.24	168.78	2261.7	2430.5	168.79	2406.0	2574.8	.5764	7.6750	8.2515
10	45.81	0.001 010	14.67	191.82	2246.1	2437.9	191.83	2392.8	2584.7	.6493	7.5009	8.1502
15	53.97	0.001 014	10.02	225.92	2222.8	2448.7	225.94	2373.1	2599.1	.7549	7.2536	8.0085
20	60.06	0.001 017	7.649	251.38	2205.4	2456.7	251.40	2358.3	2609.7	.8320	7.0766	7.9085
25	64.97	0.001 020	6.204	271.90	2191.2	2463.1	271.93	2346.3	2618.2	.8931	6.9383	7.8314
30	69.10	0.001 022	5.229	289.20	2179.2	2468.4	289.23	2336.1	2625.3	.9439	6.8247	7.7686
40	75.87	0.001 027	3.993	317.53	2159.5	2477.0	317.58	2319.2	2636.8	1.0259	6.6441	7.6700
50	81.33	0.001 030	3.240	340.44	2143.4	2483.9	340.49	2305.4	2645.9	1.0910	6.5029	7.5939
75	91.78	0.001 037	2.217	384.31	2112.4	2496.7	384.39	2278.6	2663.0	1.2130	6.2434	7.4564
MPa												
0.100	99.63	0.001 043	1.6940	417.36	2088.7	2506.1	417.46	2258.0	2675.5	1.3026	6.0568	7.3594
0.125	105.99	0.001 048	1.3749	444.19	2069.3	2513.5	444.32	2241.0	2685.4	1.3740	5.9104	7.2844
0.150	111.37	0.001 053	1.1593	466.94	2052.7	2519.7	467.11	2226.5	2693.6	1.4336	5.7897	7.2233
0.175	116.06	0.001 057	1.0036	486.80	2038.1	2524.9	486.99	2213.6	2700.6	1.4849	5.6868	7.1717
0.200	120.23	0.001 061	0.8857	504.49	2025.0	2529.5	504.70	2201.9	2706.7	1.5301	5.5970	7.1271

0.225	124.00	0.001 064	0.7933	520.47	2013.1	2533.6	520.72	2191.3	2712.1	1.5706	5.5173	7.0878
0.250	127.44	0.001 067	0.7187	535.10	2002.1	2537.2	535.37	2181.5	2716.9	1.6072	5.4455	7.0527
0.275	130.60	0.001 070	0.6573	548.59	1991.9	2540.5	548.89	2172.4	2721.3	1.6408	5.3801	7.0209
0.300	133.55	0.001 073	0.6058	561.15	1982.4	2543.6	561.47	2163.8	2725.3	1.6718	5.3201	6.9919
0.325	136.30	0.001 076	0.5620	572.90	1973.5	2546.4	573.25	2155.8	2729.0	1.7006	5.2646	6.9652
0.350	138.88	0.001 079	0.5243	583.95	1965.0	2548.9	584.33	2148.1	2732.4	1.7275	5.2130	6.9405
0.375	141.32	0.001 081	0.4914	594.40	1956.9	2551.3	594.81	2140.8	2735.6	1.7528	5.1647	6.9175
0.40	143.63	0.001 084	0.4625	604.31	1949.3	2553.6	604.74	2133.8	2738.6	1.7766	5.1193	6.8959
0.45	147.93	0.001 088	0.4140	622.77	1934.9	2557.6	623.25	2120.7	2743.9	1.8207	5.0359	6.8565
0.50	151.86	0.001 093	0.3749	639.68	1921.6	2561.2	640.23	2108.5	2748.7	1.8607	4.9606	6.8213
0.55	155.48	0.001 097	0.3427	655.32	1909.2	2564.5	655.93	2097.0	2753.0	1.8973	4.8920	6.7893
0.60	158.85	0.001 101	0.3157	669.90	1897.5	2567.4	670.56	2086.3	2756.8	1.9312	4.8288	6.7600
0.65	162.01	0.001 104	0.2927	683.56	1886.5	2570.1	684.28	2076.0	2760.3	1.9627	4.7703	6.7331
0.70	164.97	0.001 108	0.2729	696.44	1876.1	2572.5	697.22	2066.3	2763.5	1.9922	4.7158	6.7080
0.75	167.78	0.001 112	0.2556	708.64	1866.1	2574.7	709.47	2057.0	2766.4	2.0200	4.6647	6.6847
0.80	170.43	0.001 115	0.2404	720.22	1856.6	2576.8	721.11	2048.0	2769.1	2.0462	4.6166	6.6628
0.85	172.96	0.001 118	0.2270	731.27	1847.4	2578.7	732.22	2039.4	2771.6	2.0710	4.5711	6.6421
0.90	175.38	0.001 121	0.2150	741.83	1838.6	2580.5	742.83	2031.1	2773.9	2.0946	4.5280	6.6226
0.95	177.69	0.001 124	0.2042	751.95	1830.2	2582.1	753.02	2023.1	2776.1	2.1172	4.4869	6.6041
1.00	179.91	0.001 127	0.194 44	761.68	1822.0	2583.6	762.81	2015.3	2778.1	2.1387	4.4478	6.5865
1.10	184.09	0.001 133	0.177 53	780.09	1806.3	2586.4	781.34	2000.4	2781.7	2.1792	4.3744	6.5536
1.20	187.99	0.001 139	0.163 33	797.29	1791.5	2588.8	798.65	1986.2	2784.8	2.2166	4.3067	6.5233
1.30	191.64	0.001 144	0.151 25	813.44	1777.5	2591.0	814.93	1972.7	2787.6	2.2515	4.2438	6.4953
1.40	195.07	0.001 149	0.140 84	828.70	1764.1	2592.8	830.30	1959.7	2790.0	2.2842	4.1850	6.4693
1.50	198.32	0.001 154	0.131 77	843.16	1751.3	2594.5	844.89	1947.3	2792.2	2.3150	4.1298	6.4448
1.75	205.76	0.001 166	0.113 49	876.46	1721.4	2597.8	878.50	1917.9	2796.4	2.3851	4.0044	6.3896
2.00	212.42	0.001 177	0.099 63	906.44	1693.8	2600.3	908.79	1890.7	2799.5	2.4474	3.8935	6.3409
2.25	218.45	0.001 187	0.088 75	933.83	1668.2	2602.0	936.49	1865.2	2801.7	2.5035	3.7937	6.2972
2.5	223.99	0.001 197	0.079 98	959.11	1644.0	2603.1	962.11	1841.0	2803.1	2.5547	3.7028	6.2575
3.0	233.90	0.001 217	0.066 68	1004.78	1599.3	2604.1	1008.42	1795.7	2804.2	2.6457	3.5412	6.1869

(Continued)

TABLE A.1.2SI (*Continued*) Saturated Steam: Pressure Table (SI Units)

Press. MPa P	Temp. °C T	Specific Volume		Internal Energy			Enthalpy			Entropy		
		Sat. Liquid v_f	Sat. Vapor v_g	Sat. Liquid u_f	Evap. u_{fg}	Sat. Vapor u_g	Sat. Liquid h_f	Evap. h_{fg}	Sat. Vapor h_g	Sat. Liquid s_f	Evap. s_{fg}	Sat. Vapor s_g
3.5	242.60	0.001 235	0.057 07	1045.43	1558.3	2603.7	1049.75	1753.7	2803.4	2.7253	3.4000	6.1253
4	250.40	0.001 252	0.049 78	1082.31	1520.0	2602.3	1087.31	1714.1	2801.4	2.7964	3.2737	6.0701
5	263.99	0.001 286	0.039 44	1147.81	1449.3	2597.1	1154.23	1640.1	2794.3	2.9202	3.0532	5.9734
6	275.64	0.001 319	0.032 44	1205.44	1384.3	2589.7	1213.35	1571.0	2784.3	3.0267	2.8625	5.8892
7	285.88	0.001 351	0.027 37	1257.55	1323.0	2580.5	1267.00	1505.1	2772.1	3.1211	2.6922	5.8133
8	295.06	0.001 384	0.023 52	1305.57	1264.2	2569.8	1316.64	1441.3	2758.0	3.2068	2.5364	5.7432
9	303.40	0.001 418	0.020 48	1350.51	1207.3	2557.8	1363.26	1378.9	2742.1	3.2858	2.3915	5.6772
10	311.06	0.001 452	0.018 026	1393.04	1151.4	2544.4	1407.56	1317.1	2724.7	3.3596	2.2544	5.6141
11	318.15	0.001 489	0.015 987	1433.7	1096.0	2529.8	1450.1	1255.5	2705.6	3.4295	2.1233	5.5527
12	324.75	0.001 527	0.014 263	1473.0	1040.7	2513.7	1491.3	1193.6	2684.9	3.4962	1.9962	5.4924
13	330.93	0.001 567	0.012 780	1511.1	985.0	2496.1	1531.5	1130.7	2662.2	3.5606	1.8718	5.4323
14	336.75	0.001 611	0.011 485	1548.6	928.2	2476.8	1571.1	1066.5	2637.6	3.6232	1.7485	5.3717
15	342.24	0.001 658	0.010 337	1585.6	869.8	2455.5	1610.5	1000.0	2610.5	3.6848	1.6249	5.3098
16	347.44	0.001 711	0.009 306	1622.7	809.0	2431.7	1650.1	930.6	2580.6	3.7461	1.4994	5.2455
17	352.37	0.001 770	0.008 364	1660.2	744.8	2405.0	1690.3	856.9	2547.2	3.8079	1.3698	5.1777
18	357.06	0.001 840	0.007 489	1698.9	675.4	2374.3	1732.0	777.1	2509.1	3.8715	1.2329	5.1044
19	361.54	0.001 924	0.006 657	1739.9	598.1	2338.1	1776.5	688.0	2464.5	3.9388	1.0839	5.0228
20	365.81	0.002 036	0.005 834	1785.6	507.5	2293.0	1826.3	583.4	2409.7	4.0139	.9130	4.9269
21	369.89	0.002 207	0.004 952	1842.1	388.5	2230.6	1888.4	446.2	2334.6	4.1075	.6938	4.8013
22	373.80	0.002 742	0.003 568	1961.9	125.2	2087.1	2022.2	143.4	2165.6	4.3110	.2216	4.5327
22.09	374.14	0.003 155	0.003 155	2029.6	0	2029.6	2099.3	0	2099.3	4.4298	0	4.4298

TABLE A.1.3SI Superheated Vapor (SI Units)

T	P = .010 MPa (45.81)				P = .050 MPa (81.33)				P = .10 MPa (99.63)			
	v	u	h	s	v	u	h	s	v	u	h	s
Sat.	14.674	2437.9	2584.7	8.1502	3.240	2483.9	2645.9	7.5939	1.6940	2506.1	2675.5	7.3594
50	14.869	2443.9	2592.6	8.1749								
100	17.196	2515.5	2687.5	8.4479	3.418	2511.6	2682.5	7.6947	1.6958	2506.7	2676.2	7.3614
150	19.512	2587.9	2783.0	8.6882	3.889	2585.6	2780.1	7.9401	1.9364	2582.8	2776.4	7.6134
200	21.825	2661.3	2879.5	8.9038	4.356	2659.9	2877.7	8.1580	2.172	2658.1	2875.3	7.8343
250	24.136	2736.0	2977.3	9.1002	4.820	2735.0	2976.0	8.3556	2.406	2733.7	2974.3	8.0333
300	26.445	2812.1	3076.5	9.2813	5.284	2811.3	3075.5	8.5373	2.639	2810.4	3074.3	8.2158
400	31.063	2968.9	3279.6	9.6077	6.209	2968.5	3278.9	8.8642	3.103	2967.9	3278.2	8.5435
500	35.679	3132.3	3489.1	9.8978	7.134	3132.0	3488.7	9.1546	3.565	3131.6	3488.1	8.8342
600	40.295	3302.5	3705.4	10.1608	8.057	3302.2	3705.1	9.4178	4.028	3301.9	3704.7	9.0976
700	44.911	3479.6	3928.7	10.4028	8.981	3479.4	3928.5	9.6599	4.490	3479.2	3928.2	9.3398
800	49.526	3663.8	4159.0	10.6281	9.904	3663.6	4158.9	9.8852	4.952	3663.5	4158.6	9.5652
900	54.141	3855.0	4396.4	10.8396	10.828	3854.9	4396.3	10.0967	5.414	3854.8	4396.1	9.7767
1000	58.757	4053.0	4640.6	11.0393	11.751	4052.9	4640.5	10.2964	5.875	4052.8	4640.3	9.9764
1100	63.372	4257.5	4891.2	11.2287	12.674	4257.4	4891.1	10.4859	6.337	4257.3	4891.0	10.1659
1200	67.987	4467.9	5147.8	11.4091	13.597	4467.8	5147.7	10.6662	6.799	4467.7	5147.6	10.3463
1300	72.602	4683.7	5409.7	11.5811	14.521	4683.6	5409.6	10.8382	7.260	4683.5	5409.5	10.5183

T	P = .20 MPa (120.23)				P = .30 MPa (133.55)				P = .40 MPa (143.63)			
	v	u	h	s	v	u	h	s	v	u	h	s
Sat.	.8857	2529.5	2706.7	7.1272	.6058	2543.6	2725.3	6.9919	.4625	2553.6	2738.6	6.8959
150	.9596	2576.9	2768.8	7.2795	.6339	2570.8	2761.0	7.0778	.4708	2564.5	2752.8	6.9299
200	1.0803	2654.4	2870.5	7.5066	.7163	2650.7	2865.6	7.3115	.5342	2646.8	2860.5	7.1706
250	1.1988	2731.2	2971.0	7.7086	.7964	2728.7	2967.6	7.5166	.5951	2726.1	2964.2	7.3789
300	1.3162	2808.6	3071.8	7.8926	.8753	2806.7	3069.3	7.7022	.6548	2804.8	3066.8	7.5662
400	1.5493	2966.7	3276.6	8.2218	1.0315	2965.6	3275.0	8.0330	.7726	2964.4	3273.4	7.8985

TABLE A.1.3SI (*Continued*) Superheated Vapor (SI Units)

T	P = .20 MPa (120.23)				P = .30 MPa (133.55)				P = .40 MPa (143.63)			
	v	u	h	s	v	u	h	s	v	u	h	s
500	1.7814	3130.8	3487.1	8.5133	1.1867	3130.0	3486.0	8.3251	.8893	3129.2	3484.9	8.1913
600	2.013	3301.4	3704.0	8.7770	1.3414	3300.8	3703.2	8.5892	1.0055	3300.2	3702.4	8.4558
700	2.244	3478.8	3927.6	9.0194	1.4957	3478.4	3927.1	8.8319	1.1215	3477.9	3926.5	8.6987
800	2.475	3663.1	4158.2	9.2449	1.6499	3662.9	4157.8	9.0576	1.2372	3662.4	4157.3	8.9244
900	2.706	3854.5	4395.8	9.4566	1.8041	3854.2	4395.4	9.2692	1.3529	3853.9	4395.1	9.1362
1000	2.937	4052.5	4640.0	9.6563	1.9581	4052.3	4639.7	9.4690	1.4685	4052.0	4639.4	9.3360
1100	3.168	4257.0	4890.7	9.8458	2.1121	4256.8	4890.4	9.6585	1.5840	4256.5	4890.2	9.5256
1200	3.399	4467.5	5147.3	10.0262	2.2661	4467.2	5147.1	9.8389	1.6996	4467.0	5146.8	9.7060
1300	3.630	4683.2	5409.3	10.1982	2.4201	4683.0	5409.0	10.0110	1.8151	4682.8	5408.8	9.8780

T	P = .50 MPa (151.86)				P = .60 MPa (158.85)				P = .80 MPa (170.43)			
	v	u	h	s	v	u	h	s	v	u	h	s
Sat.	.3749	2561.2	2748.7	6.8213	.3157	2567.4	2756.8	6.7600	.2404	2576.8	2769.1	6.6628
200	.4249	2642.9	2855.4	7.0592	.3520	2638.9	2850.1	6.9665	.2608	2630.6	2839.3	6.8158
250	.4744	2723.5	2960.7	7.2709	.3938	2720.9	2957.2	7.1816	.2931	2715.5	2950.0	7.0384
300	.5226	2802.9	3064.2	7.4599	.4344	2801.0	3061.6	7.3724	.3241	2797.2	3056.5	7.2328
350	.5701	2882.6	3167.7	7.6329	.4742	2881.2	3165.7	7.5464	.3544	2878.2	3161.7	7.4089
400	.6173	2963.2	3271.9	7.7938	.5137	2962.1	3270.3	7.7079	.3843	2959.7	3267.1	7.5716
500	.7109	3128.4	3483.9	8.0873	.5920	3127.6	3482.8	8.0021	.4433	3126.0	3480.6	7.8673
600	.8041	3299.6	3701.7	8.3522	.6697	3299.1	3700.9	8.2674	.5018	3297.9	3699.4	8.1333
700	.8969	3477.5	3925.9	8.5952	.7472	3477.0	3925.3	8.5107	.5601	3476.2	3924.2	8.3770
800	.9896	3662.1	4156.9	8.8211	.8245	3661.8	4156.5	8.7367	.6181	3661.1	4155.6	8.6033
900	1.0822	3853.6	4394.7	9.0329	.9017	3853.4	4394.4	8.9486	.6761	3852.8	4393.7	8.8153
1000	1.1747	4051.8	4639.1	9.2328	.9788	4051.5	4638.8	9.1485	.7340	4051.0	4638.2	9.0153

(continued from previous page)

T	v	u	h	s		v	u	h	s		v	u	h	s
1100	1.2672	4256.3	4889.9	9.4224		1.0559	4256.1	4889.6	9.3381		.7919	4255.6	4889.1	9.2050
1200	1.3596	4466.8	5146.6	9.6029		1.1330	4466.5	5146.3	9.5185		.8497	4466.1	5145.9	9.3855
1300	1.4521	4682.5	5408.6	9.7749		1.2101	4682.3	5408.3	9.6906		.9076	4681.8	5407.9	9.5575

P = 1.00 MPa (179.91)

T	v	u	h	s
Sat.	.194 44	2583.6	2778.1	6.5865
200	.2060	2621.9	2827.9	6.6940
250	.2327	2709.9	2942.6	6.9247
300	.2579	2793.2	3051.2	7.1229
350	.2825	2875.2	3157.7	7.3011
400	.3066	2957.3	3263.9	7.4651
500	.3541	3124.4	3478.5	7.7622
600	.4011	3296.8	3697.9	8.0290
700	.4478	3475.3	3923.1	8.2731
800	.4943	3660.4	4154.7	8.4996
900	.5407	3852.2	4392.9	8.7118
1000	.5871	4050.5	4637.6	8.9119
1100	.6335	4255.1	4888.6	9.1017
1200	.6798	4465.6	5145.4	9.2822
1300	.7261	4681.3	5407.4	9.4543

P = 1.20 MPa (187.99)

T	v	u	h	s
Sat.	.163 33	2588.8	2784.8	6.5233
200	.169 30	2612.8	2815.9	6.5898
250	.192 34	2704.2	2935.0	6.8294
300	.2138	2789.2	3045.8	7.0317
350	.2345	2872.2	3153.6	7.2121
400	.2548	2954.9	3260.7	7.3774
500	.2946	3122.8	3476.3	7.6759
600	.3339	3295.6	3696.3	7.9435
700	.3729	3474.4	3922.0	8.1881
800	.4118	3659.7	4153.8	8.4148
900	.4505	3851.6	4392.2	8.6272
1000	.4892	4050.0	4637.0	8.8274
1100	.5278	4254.6	4888.0	9.0172
1200	.5665	4465.1	5144.9	9.1977
1300	.6051	4680.9	5407.0	9.3698

P = 1.40 MPa (195.07)

T	v	u	h	s
Sat.	.140 84	2592.8	2790.0	6.4693
200	.143 02	2603.1	2803.3	6.4975
250	.163 50	2698.3	2927.2	6.7467
300	.182 28	2785.2	3040.4	6.9534
350	.2003	2869.2	3149.5	7.1360
400	.2178	2952.5	3257.5	7.3026
500	.2521	3121.1	3474.1	7.6027
600	.2860	3294.4	3694.8	7.8710
700	.3195	3473.6	3920.8	8.1160
800	.3528	3659.0	4153.0	8.3431
900	.3861	3851.1	4391.5	8.5556
1000	.4192	4049.5	4636.4	8.7559
1100	.4524	4254.1	4887.5	8.9457
1200	.4855	4464.7	5144.4	9.1262
1300	.5186	4680.4	5406.5	9.2984

P = 1.60 MPa (201.41)

T	v	u	h	s
Sat.	.123 80	2596.0	2794.0	6.4218
225	.132 87	2644.7	2857.3	6.5518
250	.141 84	2692.3	2919.2	6.6732
300	.158 62	2781.1	3034.8	6.8844
350	.174 56	2866.1	3145.4	7.0694
400	.190 05	2950.1	3254.2	7.2374
500	.2203	3119.5	3472.0	7.5390
600	.2500	3293.3	3693.2	7.8080
700	.2794	3472.7	3919.7	8.0535

P = 1.80 MPa (207.15)

T	v	u	h	s
Sat.	.110 42	2598.4	2797.1	6.3794
225	.116 73	2636.6	2846.7	6.4808
250	.124 97	2686.0	2911.0	6.6066
300	.140 21	2776.9	3029.2	6.8226
350	.154 57	2863.0	3141.2	7.0100
400	.168 47	2947.7	3250.9	7.1794
500	.195 50	3117.9	3469.8	7.4825
600	.2220	3292.1	3691.7	7.7523
700	.2482	3471.8	3918.5	7.9983

P = 2.00 MPa (212.42)

T	v	u	h	s
Sat.	.099 63	2600.3	2799.5	6.3409
225	.103 77	2628.3	2835.8	6.4147
250	.111 44	2679.6	2902.5	6.5453
300	.125 47	2772.6	3023.5	6.7664
350	.138 57	2859.8	3137.0	6.9563
400	.151 20	2945.2	3247.6	7.1271
500	.175 68	3116.2	3467.6	7.4317
600	.199 60	3290.9	3690.1	7.7024
700	.2232	3470.9	3917.4	7.9487

(Continued)

TABLE A.1.3SI (Continued) Superheated Vapor (SI Units)

T	P = 1.60 MPa (201.41)				P = 1.80 MPa (207.15)				P = 2.00 MPa (212.42)			
	v	u	h	s	v	u	h	s	v	u	h	s
800	.3086	3658.3	4152.1	8.2808	.2742	3657.6	4151.2	8.2258	.2467	3657.0	4150.3	8.1765
900	.3377	3850.5	4390.8	8.4935	.3001	3849.9	4390.1	8.4386	.2700	3849.3	4389.4	8.3895
1000	.3668	4049.0	4635.8	8.6938	.3260	4048.5	4635.2	8.6391	.2933	4048.0	4634.6	8.5901
1100	.3958	4253.7	4887.0	8.8837	.3518	4253.2	4886.4	8.8290	.3166	4252.7	4885.9	8.7800
1200	.4248	4464.2	5143.9	9.0643	.3776	4463.7	5143.4	9.0096	.3398	4463.3	5142.9	8.9607
1300	.4538	4679.9	5406.0	9.2364	.4034	4679.5	5405.6	9.1818	.3631	4679.0	5405.1	9.1329

T	P = 2.50 MPa (223.99)				P = 3.00 MPa (233.90)				P = 3.50 MPa (242.60)			
	v	u	h	s	v	u	h	s	v	u	h	s
Sat.	.079 98	2603.1	2803.1	6.2575	.066 68	2604.1	2804.2	6.1869	.057 07	2603.7	2803.4	6.1253
225	.080 27	2605.6	2806.3	6.2639								
250	.087 00	2662.6	2880.1	6.4085	.070 58	2644.0	2855.8	6.2872	.058 72	2623.7	2829.2	6.1749
300	.098 90	2761.6	3008.8	6.6438	.081 14	2750.1	2993.5	6.5390	.068 42	2738.0	2977.5	6.4461
350	.109 76	2851.9	3126.3	6.8403	.090 53	2843.7	3115.3	6.7428	.076 78	2835.3	3104.0	6.6579
400	.120 10	2939.1	3239.3	7.0148	.099 36	2932.8	3230.9	6.9212	.084 53	2926.4	3222.3	6.8405
450	.130 14	3025.5	3350.8	7.1746	.107 87	3020.4	3344.0	7.0834	.091 96	3015.3	3337.2	7.0052
500	.139 98	3112.1	3462.1	7.3234	.116 19	3108.0	3456.5	7.2338	.099 18	3103.0	3450.9	7.1572
600	.159 30	3288.0	3686.3	7.5960	.132 43	3285.0	3682.3	7.5085	.113 24	3282.1	3678.4	7.4339
700	.178 32	3468.7	3914.5	7.8435	.148 38	3466.5	3911.7	7.7571	.126 99	3464.3	3908.8	7.6837
800	.197 16	3655.3	4148.2	8.0720	.164 14	3653.5	4145.9	7.9862	.140 56	3651.8	4143.7	7.9134
900	.215 90	3847.9	4387.6	8.2853	.179 80	3846.5	4385.9	8.1999	.154 02	3845.0	4384.1	8.1276
1000	.2346	4046.7	4633.1	8.4861	.195 41	4045.4	4631.6	8.4009	.167 43	4044.1	4630.1	8.3288
1100	.2532	4251.5	4884.6	8.6762	.210 98	4250.3	4883.3	8.5912	.180 80	4249.2	4881.9	8.5192
1200	.2718	4462.1	5141.7	8.8569	.226 52	4460.9	5140.5	8.7720	.194 15	4459.8	5139.3	8.7000
1300	.2905	4677.8	5404.0	9.0291	.242 06	4676.6	5402.8	8.9442	.207 49	4675.5	5401.7	8.8723

P = 4.0 MPa (250.40)

Sat.	.049 78	2602.3	2801.4	6.0701
275	.054 57	2667.9	2886.2	6.2285
300	.058 84	2725.3	2960.7	6.3615
350	.066 45	2826.7	3092.5	6.5821
400	.073 41	2919.9	3213.6	6.7690
450	.080 02	3010.2	3330.3	6.9363
500	.086 43	3099.5	3445.3	7.0901
600	.098 85	3279.1	3674.4	7.3688
700	.110 95	3462.1	3905.9	7.6198
800	.122 87	3650.0	4141.5	7.8502
900	.134 69	3843.6	4382.3	8.0647
1000	.146 45	4042.9	4628.7	8.2662
1100	.158 17	4248.0	4880.6	8.4567
1200	.169 87	4458.6	5138.1	8.6376
1300	.181 56	4674.3	5400.5	8.8100

P = 4.5 MPa (257.49)

Sat.	.044 06	2600.1	2798.3	6.0198
275	.047 30	2650.3	2863.2	6.1401
300	.051 35	2712.0	2943.1	6.2828
350	.058 40	2817.8	3080.6	6.5131
400	.064 75	2913.3	3204.7	6.7047
450	.070 74	3005.0	3323.3	6.8746
500	.076 51	3095.3	3439.6	7.0301
600	.087 65	3276.0	3670.5	7.3110
700	.098 47	3459.9	3903.0	7.5631
800	.109 11	3648.3	4139.3	7.7942
900	.119 65	3842.2	4380.6	8.0091
1000	.130 13	4041.6	4627.2	8.2108
1100	.140 56	4246.8	4879.3	8.4015
1200	.150 98	4457.5	5136.9	8.5825
1300	.161 39	4673.1	5399.4	8.7549

P = 5.0 MPa (263.99)

Sat.	.039 44	2597.1	2794.3	5.9734
275	.041 41	2631.3	2838.3	6.0544
300	.045 32	2698.0	2924.5	6.2084
350	.051 94	2808.7	3068.4	6.4493
400	.057 81	2906.6	3195.7	6.6459
450	.063 30	2999.7	3316.2	6.8186
500	.068 57	3091.0	3433.8	6.9759
600	.078 69	3273.0	3666.5	7.2589
700	.088 49	3457.6	3900.1	7.5122
800	.098 11	3646.6	4137.1	7.7440
900	.107 62	3840.7	4378.8	7.9593
1000	.117 07	4040.4	4625.7	8.1612
1100	.126 48	4245.6	4878.0	8.3520
1200	.135 87	4456.3	5135.7	8.5331
1300	.145 26	4672.0	5398.2	8.7055

P = 6.0 MPa (275.64)

Sat.	.032 44	2589.7	2784.3	5.8892
300	.036 16	2667.2	2884.2	6.0674
350	.042 23	2789.6	3043.0	6.3335
400	.047 39	2892.9	3177.2	6.5408
450	.052 14	2988.9	3301.8	6.7193
500	.056 65	3082.2	3422.2	6.8803
550	.061 01	3174.6	3540.6	7.0288
600	.065 25	3266.9	3658.4	7.1677
700	.073 52	3453.1	3894.2	7.4234
800	.081 60	3643.1	4132.7	7.6566
900	.089 58	3837.8	4375.3	7.8727
1000	.097 49	4037.8	4622.7	8.0751
1100	.105 36	4243.3	4875.4	8.2661

P = 7.0 MPa (285.88)

Sat.	.027 37	2580.5	2772.1	5.8133
300	.029 47	2632.2	2838.4	5.9305
350	.035 24	2769.4	3016.0	6.2283
400	.039 93	2878.6	3158.1	6.4478
450	.044 16	2978.0	3287.1	6.6327
500	.048 14	3073.4	3410.3	6.7975
550	.051 95	3167.2	3530.9	6.9486
600	.055 65	3260.7	3650.3	7.0894
700	.062 83	3448.5	3888.3	7.3476
800	.069 81	3639.5	4128.2	7.5822
900	.076 69	3835.0	4371.8	7.7991
1000	.083 50	4035.3	4619.8	8.0020
1100	.090 27	4240.9	4872.8	8.1933

P = 8.0 MPa (295.06)

Sat.	.023 52	2569.8	2758.0	5.7432
300	.024 26	2590.9	2785.0	5.7906
350	.029 95	2747.7	2987.3	6.1301
400	.034 32	2863.8	3138.3	6.3634
450	.038 17	2966.7	3272.0	6.5551
500	.041 75	3064.3	3398.3	6.7240
550	.045 16	3159.8	3521.0	6.8778
600	.048 45	3254.4	3642.0	7.0206
700	.054 81	3443.9	3882.4	7.2812
800	.060 97	3636.0	4123.8	7.5173
900	.067 02	3832.1	4368.3	7.7351
1000	.073 01	4032.8	4616.9	7.9384
1100	.078 96	4238.6	4870.3	8.1300

(Continued)

TABLE A.1.3SI *(Continued)* Superheated Vapor (SI Units)

T	P = 6.0 MPa (275.64) v	u	h	s	P = 7.0 MPa (285.88) v	u	h	s	P = 8.0 MPa (295.06) v	u	h	s
1200	.113 21	4454.0	5133.3	8.4474	.097 03	4451.7	5130.9	8.3747	.084 89	4449.5	5128.5	8.3115
1300	.121 06	4669.6	5396.0	8.6199	.103 77	4667.3	5393.7	8.5473	.090 80	4665.0	5391.5	8.4842

T	P = 9.0 MPa (303.40) v	u	h	s	P = 10.0 MPa (311.06) v	u	h	s	P = 12.5 MPa (327.89) v	u	h	s
Sat.	.020 48	2557.8	2742.1	5.6772	.018 026	2544.4	2724.7	5.6141	.013 495	2505.1	2673.8	5.4624
325	.023 27	2646.6	2856.0	5.8712	.019 861	2610.4	2809.1	5.7568				
350	.025 80	2724.4	2956.6	6.0361	.022 42	2699.2	2923.4	5.9443	.016 126	2624.6	2826.2	5.7118
400	.029 93	2848.4	3117.8	6.2854	.026 41	2832.4	3096.5	6.2120	.020 00	2789.3	3039.3	6.0417
450	.033 50	2955.2	3256.6	6.4844	.029 75	2943.4	3240.9	6.4190	.022 99	2912.5	3199.8	6.2719
500	.036 77	3055.2	3386.1	6.6576	.032 79	3045.8	3373.7	6.5966	.025 60	3021.7	3341.8	6.4618
550	.039 87	3152.2	3511.0	6.8142	.035 64	3144.6	3500.9	6.7561	.028 01	3125.0	3475.2	6.6290
600	.042 85	3248.1	3633.7	6.9589	.038 37	3241.7	3625.3	6.9029	.030 29	3225.4	3604.0	6.7810
650	.045 74	3343.6	3755.3	7.0943	.041 01	3338.2	3748.2	7.0398	.032 48	3324.4	3730.4	6.9218
700	.048 57	3439.3	3876.5	7.2221	.043 58	3434.7	3870.5	7.1687	.034 60	3422.9	3855.3	7.0536
800	.054 09	3632.5	4119.3	7.4596	.048 59	3628.9	4114.8	7.4077	.038 69	3620.0	4103.6	7.2965
900	.059 50	3829.2	4364.8	7.6783	.053 49	3826.3	4361.2	7.6272	.042 67	3819.1	4352.5	7.5182
1000	.064 85	4030.3	4614.0	7.8821	.058 32	4027.8	4611.0	7.8315	.046 58	4021.6	4603.8	7.7237
1100	.070 16	4236.3	4867.7	8.0740	.063 12	4234.0	4865.1	8.0237	.050 45	4228.2	4858.8	7.9165
1200	.075 44	4447.2	5126.2	8.2556	.067 89	4444.9	5123.8	8.2055	.054 30	4439.3	5118.0	8.0987
1300	.080 72	4662.7	5389.2	8.4284	.072 65	4460.5	5387.0	8.3783	.058 13	4654.8	5381.4	8.2717

T	P = 15.0 MPa (342.24) v	u	h	s	P = 17.5 MPa (354.75) v	u	h	s	P = 20.0 MPa (365.81) v	u	h	s
Sat.	.010 337	2455.5	2610.5	5.3098	.007 920	2390.2	2528.8	5.1419	.005 834	2293.0	2409.7	4.9269
350	.011 470	2520.4	2692.4	5.4421								

P = 15.0 MPa

T (°C)	v	u	h	s
400	.015 649	2740.7	2975.5	5.8811
450	.018 445	2879.5	3156.2	6.1404
500	.020 80	2996.6	3308.6	6.3443
550	.022 93	3104.7	3448.6	6.5199
600	.024 91	3208.6	3582.3	6.6776
650	.026 80	3310.3	3712.3	6.8224
700	.028 61	3410.9	3840.1	6.9572
800	.032 10	3610.9	4092.4	7.2040
900	.035 46	3811.9	4343.8	7.4279
1000	.038 75	4015.4	4596.6	7.6348
1100	.042 00	4222.6	4852.6	7.8283
1200	.045 23	4433.8	5112.3	8.0108
1300	.048 45	4649.1	5376.0	8.1840

P = 17.5 MPa

T (°C)	v	u	h	s
400	.012 447	2685.0	2902.9	5.7213
450	.015 174	2844.2	3109.7	6.0184
500	.017 358	2970.3	3274.1	6.2383
550	.019 288	3083.9	3421.4	6.4230
600	.021 06	3191.5	3560.1	6.5866
650	.022 74	3296.0	3693.9	6.7357
700	.024 34	3398.7	3824.6	6.8736
800	.027 38	3601.8	4081.1	7.1244
900	.030 31	3804.7	4335.1	7.3507
1000	.033 16	4009.3	4589.5	7.5589
1100	.035 97	4216.9	4846.4	7.7531
1200	.038 76	4428.3	5106.6	7.9360
1300	.041 54	4643.5	5370.5	8.1093

P = 20.0 MPa

T (°C)	v	u	h	s
400	.009 942	2619.3	2818.1	5.5540
450	.012 695	2806.2	3060.1	5.9017
500	.014 768	2942.9	3238.2	6.1401
550	.016 555	3062.4	3393.5	6.3348
600	.018 178	3174.0	3537.6	6.5048
650	.019 693	3281.4	3675.3	6.6582
700	.021 13	3386.4	3809.0	6.7993
800	.023 85	3592.7	4069.7	7.0544
900	.026 45	3797.5	4326.4	7.2830
1000	.028 97	4003.1	4582.5	7.4925
1100	.031 45	4211.3	4840.2	7.6874
1200	.033 91	4422.8	5101.0	7.8707
1300	.036 36	4638.0	5365.1	8.0442

P = 25.0 MPa

T (°C)	v	u	h	s
375	.001 973 1	1798.7	1848.0	4.0320
400	.006 004	2430.1	2580.2	5.1418
425	.007 881	2609.2	2806.3	5.4723
450	.009 162	2720.7	2949.7	5.6744
500	.011 123	2884.3	3162.4	5.9592
550	.012 724	3017.5	3335.6	6.1765
600	.014 137	3137.9	3491.4	6.3602
650	.015 433	3251.6	3637.4	6.5229
700	.016 646	3361.3	3777.5	6.6707
800	.018 912	3574.3	4047.1	6.9345
900	.021 045	3783.0	4309.1	7.1680
1000	.023 10	3990.9	4568.5	7.3802
1100	.025 12	4200.2	4828.2	7.5765

P = 30.0 MPa

T (°C)	v	u	h	s
375	.001 789 2	1737.8	1791.5	3.9305
400	.002 790	2067.4	2151.1	4.4728
425	.005 303	2455.1	2614.2	5.1504
450	.006 735	2619.3	2821.4	5.4424
500	.008 678	2820.7	3081.1	5.7905
550	.010 168	2970.3	3275.4	6.0342
600	.011 446	3100.5	3443.9	6.2331
650	.012 596	3221.0	3598.9	6.4058
700	.013 661	3335.8	3745.6	6.5606
800	.015 623	3555.5	4024.2	6.8332
900	.017 448	3768.5	4291.9	7.0718
1000	.019 196	3978.8	4554.7	7.2867
1100	.020 903	4189.2	4816.3	7.4845

P = 35.0 MPa

T (°C)	v	u	h	s
375	.001 700 3	1702.9	1762.4	3.8722
400	.002 100	1914.1	1987.6	4.2126
425	.003 428	2253.4	2373.4	4.7747
450	.004 961	2498.7	2672.4	5.1962
500	.006 927	2751.9	2994.4	5.6282
550	.008 345	2921.0	3213.0	5.9026
600	.009 527	3062.0	3395.5	6.1179
650	.010 575	3189.8	3559.9	6.3010
700	.011 533	3309.8	3713.5	6.4631
800	.013 278	3536.7	4001.5	6.7450
900	.014 883	3754.0	4274.9	6.9886
1000	.016 410	3966.7	4541.1	7.2064
1100	.017 895	4178.3	4804.6	7.4057

(Continued)

TABLE A.1.3SI *(Continued)* Superheated Vapor (SI Units)

T	v	u	h	s	v	u	h	s	v	u	h	s
	P = 25.0 MPa				P = 30.0 MPa				P = 35.0 MPa			
1200	.027 11	4412.0	5089.9	7.7605	.022 589	4401.3	5079.0	7.6692	.019 360	4390.7	5068.3	7.5910
1300	.029 10	4626.9	5354.4	7.9342	.024 266	4616.0	5344.0	7.8432	.020 815	4605.1	5333.6	7.7653
	P = 40.0 MPa				P = 50.0 MPa				P = 60.0 MPa			
375	.001 640 7	1677.1	1742.8	3.8290	.001 559 4	1638.6	1716.6	3.7639	.001 502 8	1609.4	1699.5	3.7141
400	.001 907 7	1854.6	1930.9	4.1135	.001 730 9	1788.1	1874.6	4.0031	.001 633 5	1745.4	1843.4	3.9318
425	.002 532	2096.9	2198.1	4.5029	.002 007	1959.7	2060.0	4.2734	.001 816 5	1892.7	2001.7	4.1626
450	.003 693	2365.1	2512.8	4.9459	.002 486	2159.6	2284.0	4.5884	.002 085	2053.9	2179.0	4.4121
500	.005 622	2678.4	2903.3	5.4700	.003 892	2525.5	2720.1	5.1726	.002 956	2390.6	2567.9	4.9321
550	.006 984	2869.7	3149.1	5.7785	.005 118	2763.6	3019.5	5.5485	.003 956	2658.8	2896.2	5.3441
600	.008 094	3022.6	3346.4	6.0114	.006 112	2942.0	3247.6	5.8178	.004 834	2861.1	3151.2	5.6452
650	.009 063	3158.0	3520.6	6.2054	.006 966	3093.5	3441.8	6.0342	.005 595	3028.8	3364.5	5.8829
700	.009 941	3283.6	3681.2	6.3750	.007 727	3230.5	3616.8	6.2189	.006 272	3177.2	3553.5	6.0824
800	.011 523	3517.8	3978.7	6.6662	.009 076	3479.8	3933.6	6.5290	.007 459	3441.5	3889.1	6.4109
900	.012 962	3739.4	4257.9	6.9150	.010 283	3710.3	4224.4	6.7882	.008 508	3681.0	4191.5	6.6805
1000	.014 324	3954.6	4527.6	7.1356	.011 411	3930.5	4501.1	7.0146	.009 480	3906.4	4475.2	6.9127
1100	.015 642	4167.4	4793.1	7.3364	.012 496	4145.7	4770.5	7.2184	.010 409	4124.1	4748.6	7.1195
1200	.016 940	4380.1	5057.7	7.5224	.013 561	4359.1	5037.2	7.4058	.011 317	4338.2	5017.2	7.3083
1300	.018 229	4594.3	5323.5	7.6969	.014 616	4572.8	5303.6	7.5808	.012 215	4551.4	5284.3	7.4837

TABLE A.1.4SI Compressed Liquid (SI Units)

T	P = 5 MPa (263.99)				P = 10 MPa (311.06)				P = 15 MPa (342.24)			
	v	u	h	s	v	u	h	s	v	u	h	s
Sat.	.001 285 9	1147.8	1154.2	2.9202	.001 452 4	1393.0	1407.6	3.3596	.001 658 1	1585.6	1610.5	3.6848
0	.000 997 7	.04	5.04	.0001	.000 995 2	.09	10.04	.0002	.000 992 8	.15	15.05	.0004
20	.000 999 5	83.65	88.65	.2956	.000 997 2	83.36	93.33	.2945	.000 995 0	83.06	97.99	.2934
40	.001 005 6	166.95	171.97	.5705	.001 003 4	166.35	176.38	.5686	.001 001 3	165.76	180.78	.5666
60	.001 014 9	250.23	255.30	.8285	.001 012 7	249.36	259.49	.8258	.001 010 5	248.51	263.67	.8232
80	.001 026 8	333.72	338.85	1.0720	.001 024 5	332.59	342.83	1.0688	.001 022 2	331.48	346.81	1.0656
100	.001 041 0	417.52	422.72	1.3030	.001 038 5	416.12	426.50	1.2992	.001 036 1	414.74	430.28	1.2955
120	.001 057 6	501.80	507.09	1.5233	.001 054 9	500.08	510.64	1.5189	.001 052 2	498.40	514.19	1.5145
140	.001 076 8	586.76	592.15	1.7343	.001 073 7	584.68	595.42	1.7292	.001 070 7	582.66	598.72	1.7242
160	.001 098 8	672.62	678.12	1.9375	.001 095 3	670.13	681.08	1.9317	.001 091 8	667.71	684.09	1.9260
180	.001 124 0	759.63	765.25	2.1341	.001 119 9	756.65	767.84	2.1275	.001 115 9	753.76	770.50	2.1210
200	.001 153 0	848.1	853.9	2.3255	.001 148 0	844.5	856.0	2.3178	.001 143 3	841.0	858.2	2.3104
220	.001 186 6	938.4	944.4	2.5128	.001 180 5	934.1	945.9	2.5039	.001 174 8	929.9	947.5	2.4953
240	.001 226 4	1031.4	1037.5	2.6979	.001 218 7	1026.0	1038.1	2.6872	.001 211 4	1020.8	1039.0	2.6771
260	.001 274 9	1127.9	1134.3	2.8830	.001 264 5	1121.1	1133.7	2.8699	.001 255 0	1114.6	1133.4	2.8576
280					.001 321 6	1220.9	1234.1	3.0548	.001 308 4	1212.5	1232.1	3.0393
300					.001 397 2	1328.4	1342.3	3.2469	.001 377 0	1316.6	1337.3	3.2260
320									.001 472 4	1431.1	1453.2	3.4247
340									.001 631 1	1567.5	1591.9	3.6546

TABLE A.1.4SI(*Continued*) Compressed Liquid (SI Units)

T	P = 20 MPa (365.81) v	u	h	s	P = 30 MPa v	u	h	s	P = 50 MPa v	u	h	s
Sat.	.002 036	1785.6	1826.3	4.0139								
0	.000 990 4	.19	20.01	.0004	.000 985 6	.25	29.82	.0001	.000 976 6	.20	49.03	−.0014
20	.000 992 8	82.77	102.62	.2923	.000 988 6	82.17	111.84	.2899	.000 980 4	81.00	130.02	.2848
40	.000 999 2	165.17	185.16	.5646	.000 995 1	164.04	193.89	.5607	.000 987 2	161.86	211.21	.5527
60	.001 008 4	247.68	267.85	.8206	.001 004 2	246.06	276.19	.8154	.000 996 2	242.98	292.79	.8052
80	.001 019 9	330.40	350.80	1.0624	.001 015 6	328.30	358.77	1.0561	.001 007 3	324.34	374.70	1.0440
100	.001 033 7	413.39	434.06	1.2917	.001 029 0	410.78	441.66	1.2844	.001 020 1	405.88	456.89	1.2703
120	.001 049 6	496.76	517.76	1.5102	.001 044 5	493.59	524.93	1.5018	.001 034 8	487.65	539.39	1.4857
140	.001 067 8	580.69	602.04	1.7193	.001 062 1	576.88	608.75	1.7098	.001 051 5	569.77	622.35	1.6915
160	.001 088 5	665.35	687.12	1.9204	.001 082 1	660.82	693.28	1.9096	.001 070 3	652.41	705.92	1.8891
180	.001 112 0	750.95	773.20	2.1147	.001 104 7	745.59	778.73	2.1024	.001 091 2	735.69	790.25	2.0794
200	.001 138 8	837.7	860.5	2.3031	.001 130 2	831.4	865.3	2.2893	.001 114 6	819.7	875.5	2.2634
220	.001 169 3	925.9	949.3	2.4870	.001 159 0	918.3	953.1	2.4711	.001 140 8	904.7	961.7	2.4419
240	.001 204 6	1016.0	1040.0	2.6674	.001 192 0	1006.9	1042.6	2.6490	.001 170 2	990.7	1049.2	2.6158
260	.001 246 2	1108.6	1133.5	2.8459	.001 230 3	1097.4	1134.3	2.8243	.001 203 4	1078.1	1138.2	2.7860
280	.001 296 5	1204.7	1230.6	3.0248	.001 275 5	1190.7	1229.0	2.9986	.001 241 5	1167.2	1229.3	2.9537
300	.001 359 6	1306.1	1333.3	3.2071	.001 330 4	1287.9	1327.8	3.1741	.001 286 0	1258.7	1323.0	3.1200
320	.001 443 7	1415.7	1444.6	3.3979	.001 399 7	1390.7	1432.7	3.3539	.001 338 8	1353.3	1420.2	3.2868
340	.001 568 4	1539.7	1571.0	3.6075	.001 492 0	1501.7	1546.5	3.5426	.001 403 2	1452.0	1522.1	3.4557
360	.001 822 6	1702.8	1739.3	3.8772	.001 626 5	1626.6	1675.4	3.7494	.001 483 8	1556.0	1630.2	3.6291
380					.001 869 1	1781.4	1837.5	4.0012	.001 588 4	1667.2	1746.6	3.8101

TABLE A.1.5SI Saturated Solid-Vapor (SI Units)

Temp. °C T	Press. kPa P	Specific Volume		Internal Energy			Enthalpy			Entropy		
		Sat. Solid $v_i \times 10^3$	Sat. Vapor v	Sat. Solid u_i	Subl. u_{ig}	Sat. Vapor u_g	Sat. Solid h_i	Subl. h_{ig}	Sat. Vapor h_g	Sat. Solid s_i	Subl. s_{ig}	Sat. Vapor s_g
.01	.6113	1.0908	206.1	−333.40	2708.7	2375.3	−333.40	2834.8	2501.4	−1.221	10.378	9.156
0	.6108	1.0908	206.3	−333.43	2708.8	2375.3	−333.43	2834.8	2501.3	−1.221	10.378	9.157
−2	.5176	1.0904	241.7	−337.62	2710.2	2372.6	−337.62	2835.3	2497.7	−1.237	10.456	9.219
−4	.4375	1.0901	283.8	−341.78	2711.6	2369.8	−341.78	2835.7	2494.0	−1.253	10.536	9.283
−6	.3689	1.0898	334.2	−345.91	2712.9	2367.0	−345.91	2836.2	2490.3	−1.268	10.616	9.348
−8	.3102	1.0894	394.4	−350.02	2714.2	2364.2	−350.02	2836.6	2486.6	−1.284	10.698	9.414
−10	.2602	1.0891	466.7	−354.09	2715.5	2361.4	−354.09	2837.0	2482.9	−1.299	10.781	9.481
−12	.2176	1.0888	553.7	−358.14	2716.8	2358.7	−358.14	2837.3	2479.2	−1.315	10.865	9.550
−14	.1815	1.0884	658.8	−362.15	2718.0	2355.9	−362.15	2837.6	2475.5	−1.331	10.950	9.619
−16	.1510	1.0881	786.0	−366.14	2719.2	2353.1	−366.14	2837.9	2471.8	−1.346	11.036	9.690
−18	.1252	1.0878	940.5	−370.10	2720.4	2350.3	−370.10	2838.2	2468.1	−1.362	11.123	9.762
−20	.1035	1.0874	1128.6	−374.03	2721.6	2347.5	−374.03	2838.4	2464.3	−1.377	11.212	9.835
−22	.0853	1.0871	1358.4	−377.93	2722.7	2344.7	−377.93	2838.6	2460.6	−1.393	11.302	9.909
−24	.0701	1.0868	1640.1	−381.80	2723.7	2342.0	−381.80	2838.7	2456.9	−1.408	11.394	9.985
−26	.0574	1.0864	1986.4	−385.64	2724.8	2339.2	−385.64	2838.9	2453.2	−1.424	11.486	10.062
−28	.0469	1.0861	2413.7	−389.45	2725.8	2336.4	−389.45	2839.0	2449.5	−1.439	11.580	10.141
−30	.0381	1.0858	2943	−393.23	2726.8	2333.6	−393.23	2839.0	2445.8	−1.455	11.676	10.221
−32	.0309	1.0854	3600	−396.98	2727.8	2330.8	−396.98	2839.1	2442.1	−1.471	11.773	10.303
−34	.0250	1.0851	4419	−400.71	2728.7	2328.0	−400.71	2839.1	2438.4	−1.486	11.872	10.386
−36	.0201	1.0848	5444	−404.40	2729.6	2325.2	−404.40	2839.1	2434.7	−1.501	11.972	10.470
−38	.0161	1.0844	6731	−408.06	2730.5	2322.4	−408.06	2839.0	2430.9	−1.517	12.073	10.556
−40	.0129	1.0841	8354	−411.70	2731.3	2319.6	−411.70	2838.9	2427.2	−1.532	12.176	10.644

TABLE A.2E Thermodynamic Properties of Ammonia[a] (English Units)

TABLE A.2.1E Saturated Ammonia (English Units)

Temp. F	Abs. Press. lbf/in.² P	Specific Volume ft³/lbm			Enthalpy Btu/lbm			Entropy Btu/lbm R		
		Sat. Liquid v_f	Evap. v_{fg}	Sat. Vapor v_g	Sat. Liquid h_f	Evap. h_{fg}	Sat. Vapor h_g	Sat. Liquid s_f	Evap. s_{fg}	Sat. Vapor s_g
−60	5.55	0.022 80	44.707	44.73	−21.2	610.8	589.6	−0.0517	1.5286	1.4769
−55	6.54	0.022 90	38.357	38.38	−15.9	607.5	591.6	−0.0386	1.5017	1.4631
−50	7.67	0.023 00	33.057	33.08	−10.6	604.3	593.7	−0.0256	1.4753	1.4497
−45	8.95	0.023 10	28.597	28.62	−5.3	600.9	595.6	−0.0127	1.4495	1.4368
−40	10.41	0.023 22	24.837	24.86	0	597.6	597.6	0.000	1.4242	1.4242
−35	12.05	0.023 33	21.657	21.68	5.3	594.2	599.5	0.0126	1.3994	1.4120
−30	13.90	0.023 50	18.947	18.97	10.7	590.7	601.4	0.0250	1.3751	1.4001
−25	15.98	0.023 60	16.636	16.66	16.0	587.2	603.2	0.0374	1.3512	1.3886
−20	18.30	0.023 70	14.656	14.68	21.4	583.6	605.0	0.0497	1.3277	1.3774
−15	20.88	0.023 81	12.946	12.97	26.7	580.0	606.7	0.0618	1.3044	1.3664
−10	23.74	0.023 93	11.476	11.50	32.1	576.4	608.5	0.0738	1.2820	1.3558
−5	26.92	0.024 06	10.206	10.23	37.5	572.6	610.1	0.0857	1.2597	1.3454
0	30.42	0.024 19	9.092	9.116	42.9	568.9	611.8	0.0975	1.2377	1.3352
5	34.27	0.024 32	8.1257	8.150	48.3	565.0	613.3	0.1092	1.2161	1.3253
10	38.51	0.024 46	7.2795	7.304	53.8	561.1	614.9	0.1208	1.1949	1.3157
15	43.14	0.024 60	6.5374	6.562	59.2	557.1	616.3	0.1323	1.1739	1.3062
20	48.21	0.024 74	5.8853	5.910	64.7	553.1	617.8	0.1437	1.1532	1.2969
25	53.73	0.024 88	5.3091	5.334	70.2	548.9	619.1	0.1551	1.1328	1.2879
30	59.74	0.025 03	4.8000	4.825	75.7	544.8	620.5	0.1663	1.1127	1.2790
35	66.26	0.025 18	4.3478	4.373	81.2	540.5	621.7	0.1775	1.0929	1.2704

40	73.32	0.025 33	3.9457	3.971	86.8	536.2	623.0	0.1885	1.0733	1.2618
45	80.96	0.025 48	3.5885	3.614	92.3	531.8	624.1	0.1996	1.0539	1.2535
50	89.19	0.025 64	3.2684	3.294	97.9	527.3	625.2	0.2105	1.0348	1.2453
55	98.06	0.025 81	2.9822	3.008	103.5	522.8	626.3	0.2214	1.0159	1.2373
60	107.6	0.025 97	2.7250	2.751	109.2	518.1	627.3	0.2322	0.9972	1.2294
65	117.8	0.026 14	2.4939	2.520	114.8	513.4	628.2	0.2430	0.9786	1.2216
70	128.8	0.026 32	2.2857	2.312	120.5	508.6	629.1	0.2537	0.9603	1.2140
75	140.5	0.026 50	2.0985	2.125	126.2	503.7	629.9	0.2643	0.9422	1.2065
80	153.0	0.026 68	1.9283	1.955	132.0	498.7	630.7	0.2749	0.9242	1.1991
85	166.4	0.026 87	1.7741	1.801	137.8	493.6	631.4	0.2854	0.9064	1.1918
90	180.6	0.027 07	1.6339	1.661	143.5	488.5	632.0	0.2958	0.8888	1.1846
95	195.8	0.027 27	1.5067	1.534	149.4	483.2	632.6	0.3062	0.8713	1.1775
100	211.9	0.027 47	1.3915	1.419	155.2	477.8	633.0	0.3166	0.8539	1.1705
105	228.9	0.027 69	1.2853	1.313	161.1	472.3	633.4	0.3269	0.8366	1.1635
110	247.0	0.027 90	1.1891	1.217	167.0	466.7	633.7	0.3372	0.8194	1.1566
115	266.2	0.028 13	1.0999	1.128	173.0	460.9	633.9	0.3474	0.8023	1.1497
120	286.4	0.028 36	1.0186	1.047	179.0	455.0	634.0	0.3576	0.7851	1.1427
125	307.8	0.028 60	0.9444	0.973	185.1	448.9	634.0	0.3679	0.7679	1.1358

[a] Reprinted by permission from National Bureau of Standards Circular No. 142, *Tables of Thermodynamic Properties of Ammonia.*

TABLE A.22.E Superheated Ammonia (English Units)

Abs. Press. lbf/in² (Sat. Temp.)		Temperature, F											
		0	20	40	60	80	100	120	140	160	180	200	220
10 (−41.34)	v	28.58	29.90	31.20	32.49	33.78	35.07	36.35	37.62	38.90	40.17	41.45	
	h	618.9	629.1	639.3	649.5	659.7	670.0	680.3	690.6	701.1	711.6	722.2	
	s	1.477	1.499	1.520	1.540	1.559	1.578	1.596	1.614	1.631	1.647	1.664	
15 (−27.29)	v	18.92	19.82	20.70	21.58	22.44	23.31	24.17	25.03	25.88	26.74	27.59	
	h	617.2	627.8	638.2	648.5	658.9	669.2	679.6	690.0	700.5	711.1	721.7	
	s	1.427	1.450	1.471	1.491	1.511	1.529	1.548	1.566	1.583	1.599	1.616	
20 (−16.64)	v	14.09	14.78	15.45	16.12	16.78	17.43	18.08	18.73	19.37	20.02	20.66	21.3
	h	615.5	626.4	637.0	647.5	658.0	668.5	678.9	689.4	700.0	710.6	721.2	732.0
	s	1.391	1.414	1.436	1.456	1.476	1.495	1.513	1.531	1.549	1.565	1.582	1.598
25 (−7.96)	v	11.19	11.75	12.30	12.84	13.37	13.90	14.43	14.95	15.47	15.99	16.50	17.02
	h	613.8	625.0	635.8	646.5	657.1	667.7	678.2	688.8	699.4	710.1	720.8	731.6
	s	1.362	1.386	1.408	1.429	1.449	1.468	1.486	1.504	1.522	1.539	1.555	1.571
30 (−.57)	v	9.25	9.731	10.20	10.65	11.10	11.55	11.99	12.43	12.87	13.30	13.73	14.16
	h	611.9	623.5	634.6	645.5	656.2	666.9	677.5	688.2	698.8	709.6	720.3	731.1
	s	1.337	1.362	1.385	1.406	1.426	1.446	1.464	1.482	1.500	1.517	1.533	1.550

35 (5.89)	v	—	8.287	8.695	9.093	9.484	9.869	10.25	10.63	11.00	11.38	11.75	12.12
	h	—	622.0	633.4	644.4	655.3	666.1	676.8	687.6	698.3	709.1	719.9	730.7
	s	—	1.341	1.365	1.386	1.407	1.427	1.445	1.464	1.481	1.498	1.515	1.531
40 (11.66)	v	—	7.203	7.568	7.922	8.268	8.609	8.945	9.278	9.609	9.938	10.27	10.59
	h	—	620.4	632.1	643.4	654.4	665.3	676.1	686.9	697.7	708.5	719.4	730.3
	s	—	1.323	1.347	1.369	1.390	1.410	1.429	1.447	1.465	1.482	1.499	1.515
45 (16.87)	v	—	6.363	6.694	7.014	7.326	7.632	7.934	8.232	8.528	8.822	9.115	9.406
	h	—	618.8	630.8	642.3	653.5	664.6	675.5	686.3	697.2	708.0	718.9	729.9
	s	—	1.307	1.331	1.354	1.375	1.395	1.414	1.433	1.450	1.468	1.485	1.501
50 (21.67)	v	—	—	5.988	6.280	6.564	6.843	7.117	7.387	7.655	7.921	8.185	8.448
	h	—	—	629.5	641.2	652.6	663.7	674.7	685.7	696.6	707.5	718.5	729.4
	s	—	—	1.317	1.340	1.361	1.382	1.401	1.420	1.437	1.455	1.472	1.488
60 (30.21)	v	—	—	4.933	5.184	5.428	5.665	5.897	6.126	6.352	6.576	6.798	7.019
	h	—	—	626.8	639.0	650.7	662.1	673.3	684.4	695.5	706.5	717.5	728.6
	s	—	—	1.291	1.315	1.337	1.358	1.378	1.397	1.415	1.432	1.449	1.466

(Continued)

TABLE A.2.2E (*Continued*) **Superheated Ammonia (English Units)**

Abs. Press. lbf/in.² (Sat. Temp.)		Temperature, F											
		60	80	100	120	140	160	180	200	240	280	320	360
70 (37.7)	v	4.401	4.615	4.822	5.025	5.224	5.420	5.615	5.807	6.187	6.563	—	—
	h	636.6	648.7	660.4	671.8	683.1	694.3	705.5	716.6	738.9	761.4	—	—
	s	1.294	1.317	1.338	1.358	1.377	1.395	1.413	1.430	1.463	1.494	—	—
80 (44.4)	v	3.812	4.005	4.190	4.371	4.548	4.722	4.893	5.063	5.398	5.73	—	—
	h	634.3	646.7	658.7	670.4	681.8	693.2	704.4	715.6	738.1	760.7	—	—
	s	1.275	1.298	1.320	1.340	1.360	1.378	1.396	1.414	1.447	1.478	—	—
90 (50.47)	v	3.353	3.529	3.698	3.862	4.021	4.178	4.332	4.484	4.785	5.081	—	—
	h	631.8	644.7	657.0	668.9	680.5	692.0	703.4	714.7	737.3	760.0	—	—
	s	1.257	1.281	1.304	1.325	1.344	1.363	1.381	1.400	1.432	1.464	—	—
100 (56.05)	v	2.985	3.149	3.304	3.454	3.600	3.743	3.883	4.021	4.294	4.562	—	—
	h	629.3	642.6	655.2	667.3	679.2	690.8	702.3	713.7	736.5	759.4	—	—
	s	1.241	1.266	1.289	1.310	1.331	1.349	1.368	1.385	1.419	1.451	—	—
140 (74.79)	v	—	2.166	2.288	2.404	2.515	2.622	2.727	2.830	3.030	3.227	3.420	—
	h	—	633.8	647.8	661.1	673.7	686.0	698.0	709.9	733.3	756.7	780.0	—
	s	—	1.214	1.240	1.263	1.284	1.305	1.324	1.342	1.376	1.409	1.440	—

180	v	—	—	1.720	1.818	1.910	1.999	2.084	2.167	2.328	2.484	2.637	—
(89.78)	h	—	—	639.9	654.4	668.0	681.0	693.6	705.9	730.1	753.9	777.7	—
	s	—	—	1.199	1.225	1.248	1.269	1.289	1.308	1.344	1.377	1.408	—
220	v	—	—	—	1.443	1.525	1.601	1.675	1.745	1.881	2.012	2.140	2.265
(102.42)	h	—	—	—	647.3	662.0	675.8	689.1	701.9	726.8	751.1	775.3	779.5
	s	—	—	—	1.192	1.217	1.239	1.260	1.280	1.317	1.351	1.383	1.413
240	v	—	—	—	1.302	1.380	1.452	1.521	1.587	1.741	1.835	1.954	2.069
(108.09)	h	—	—	—	643.5	658.8	673.1	686.7	699.8	725.1	749.8	774.1	798.4
	s	—	—	—	1.176	1.203	1.226	1.248	1.268	1.305	1.339	1.371	1.402
260	v	—	—	—	1.182	1.257	1.326	1.391	1.453	1.572	1.686	1.796	1.904
(113.42)	h	—	—	—	639.5	655.6	670.4	684.4	697.7	723.4	748.4	772.9	797.4
	s	—	—	—	1.162	1.189	1.213	1.235	1.256	1.294	1.329	1.361	1.391
280	v	—	—	—	1.078	1.151	1.217	1.279	1.339	1.451	1.558	1.661	1.762
(118.45)	h	—	—	—	635.4	652.2	667.6	681.9	695.6	721.8	747.0	771.7	796.3
	s	—	—	—	1.147	1.176	1.201	1.224	1.245	1.283	1.318	1.351	1.382

TABLE A.2SI Thermodynamic Properties of Ammoniaa (SI Units)

TABLE A.2.1SI Saturated Ammonia (SI Units)

Temp. °C	Abs. Press. kPa P	Specific Volume m³/kg Sat. Liquid v_f	Evap. v_{fg}	Sat. Vapor v_g	Enthalpy kJ/kg Sat. Liquid h_f	Evap. h_{fg}	Sat. Vapor h_g	Entropy kJ/kg K Sat. Liquid s_f	Evap. s_{fg}	Sat. Vapor s_g
−50	40.88	0.001 424	2.6239	2.6254	−44.3	1416.7	1372.4	−0.1942	6.3502	6.1561
−48	45.96	0.001 429	2.3518	2.3533	−35.5	1411.3	1375.8	−0.1547	6.2696	6.1149
−46	51.55	0.001 434	2.1126	2.1140	−26.6	1405.8	1379.2	−0.1156	6.1902	6.0746
−44	57.69	0.001 439	1.9018	1.9032	−17.8	1400.3	1382.5	−0.0768	6.1120	6.0352
−42	64.42	0.001 444	1.7155	1.7170	−8.9	1394.7	1385.8	−0.0382	6.0349	5.9967
−40	71.77	0.001 449	1.5506	1.5521	0.0	1389.0	1389.0	0.0000	5.9589	5.9589
−38	79.80	0.001 454	1.4043	1.4058	8.9	1383.3	1392.2	0.0380	5.8840	5.9220
−36	88.54	0.001 460	1.2742	1.2757	17.8	1377.6	1395.4	0.0757	5.8101	5.8858
−34	98.05	0.001 465	1.1582	1.1597	26.8	1371.8	1398.5	0.1132	5.7372	5.8504
−32	108.37	0.001 470	1.0547	1.0562	35.7	1365.9	1401.6	0.1504	5.6652	5.8156
−30	119.55	0.001 476	0.9621	0.9635	44.7	1360.0	1404.6	0.1873	5.5942	5.7815
−28	131.64	0.001 481	0.8790	0.8805	53.6	1354.0	1407.6	0.2240	5.5241	5.7481
−26	144.70	0.001 487	0.8044	0.8059	62.6	1347.9	1410.5	0.2605	5.4548	5.7153
−24	158.78	0.001 492	0.7373	0.7388	71.6	1341.8	1413.4	0.2967	5.3864	5.6831
−22	173.93	0.001 498	0.6768	0.6783	80.7	1335.6	1416.2	0.3327	5.3188	5.6515
−20	190.22	0.001 504	0.6222	0.6237	89.7	1329.3	1419.0	0.3684	5.2520	5.6205
−18	207.71	0.001 510	0.5728	0.5743	98.8	1322.9	1421.7	0.4040	5.1860	5.5900
−16	226.45	0.001 515	0.5280	0.5296	107.8	1316.5	1424.4	0.4393	5.1207	5.5600
−14	246.51	0.001 521	0.4874	0.4889	116.9	1310.0	1427.0	0.4744	5.0561	5.5305
−12	267.95	0.001 528	0.4505	0.4520	126.0	1303.5	1429.5	0.5093	4.9922	5.5015

-10	290.85	0.001 534	0.4169	0.4185	135.2	1296.8	1432.0	0.5440	4.9290	5.4730
-8	315.25	0.001 540	0.3863	0.3878	144.3	1290.1	1434.4	0.5785	4.8664	5.4449
-6	341.25	0.001 546	0.3583	0.3599	153.5	1283.3	1436.8	0.6128	4.8045	5.4173
-4	368.90	0.001 553	0.3328	0.3343	162.7	1276.4	1439.1	0.6469	4.7432	5.3901
-2	398.27	0.001 559	0.3094	0.3109	171.9	1269.4	1441.3	0.6808	4.6825	5.3633
0	429.44	0.001 566	0.2879	0.2895	181.1	1262.4	1443.5	0.7145	4.6223	5.3369
2	462.49	0.001 573	0.2683	0.2698	190.4	1255.2	1445.6	0.7481	4.5627	5.3108
4	497.49	0.001 580	0.2502	0.2517	199.6	1248.0	1447.6	0.7815	4.5037	5.2852
6	534.51	0.001 587	0.2335	0.2351	208.9	1240.6	1449.6	0.8148	4.4451	5.2599
8	573.64	0.001 594	0.2182	0.2198	218.3	1233.2	1451.5	0.8479	4.3871	5.2350
10	614.95	0.001 601	0.2040	0.2056	227.6	1225.7	1453.3	0.8808	4.3295	5.2104
12	658.52	0.001 608	0.1910	0.1926	237.0	1218.1	1455.1	0.9136	4.2725	5.1861
14	704.44	0.001 616	0.1789	0.1805	246.4	1210.4	1456.8	0.9463	4.2159	5.1621
16	752.79	0.001 623	0.1677	0.1693	255.9	1202.6	1458.5	0.9788	4.1597	5.1385
18	803.66	0.001 631	0.1574	0.1590	265.4	1194.7	1460.0	1.0112	4.1039	5.1151
20	857.12	0.001 639	0.1477	0.1494	274.9	1186.7	1461.5	1.0434	4.0486	5.0920
22	913.27	0.001 647	0.1388	0.1405	284.4	1178.5	1462.9	1.0755	3.9937	5.0692
24	972.19	0.001 655	0.1305	0.1322	294.0	1170.3	1464.3	1.1075	3.9392	5.0467
26	1033.97	0.001 663	0.1228	0.1245	303.6	1162.0	1465.6	1.1394	3.8850	5.0244
28	1098.71	0.001 671	0.1156	0.1173	313.2	1153.6	1466.8	1.1711	3.8312	5.0023
30	1166.49	0.001 680	0.1089	0.1106	322.9	1145.0	1467.9	1.2028	3.7777	4.9805
32	1237.41	0.001 689	0.1027	0.1044	332.6	1136.4	1469.0	1.2343	3.7246	4.9589
34	1311.55	0.001 698	0.0969	0.0986	342.3	1127.6	1469.9	1.2656	3.6718	4.9374
36	1389.03	0.001 707	0.0914	0.0931	352.1	1118.7	1470.8	1.2969	3.6192	4.9161
38	1469.92	0.001 716	0.0863	0.0880	361.9	1109.7	1471.5	1.3281	3.5669	4.8950
40	1554.33	0.001 726	0.0815	0.0833	371.7	1100.5	1472.2	1.3591	3.5148	4.8740
42	1642.35	0.001 735	0.0771	0.0788	381.6	1091.2	1472.8	1.3901	3.4630	4.8530
44	1734.09	0.001 745	0.0728	0.0746	391.5	1081.7	1473.2	1.4209	3.4112	4.8322
46	1829.65	0.001 756	0.0689	0.0707	401.5	1072.0	1473.5	1.4518	3.3595	4.8113
48	1929.13	0.001 766	0.0652	0.0669	411.5	1062.2	1473.7	1.4826	3.3079	4.7905
50	2032.62	0.001 777	0.0617	0.0635	421.7	1052.0	1473.7	1.5135	3.2561	4.7696

a Adapted from National Bureau of Standards Circular No. 142, *Tables of Thermodynamic Properties of Ammonia.*

TABLE A.2.2SI Superheated Ammonia (SI Units)

Abs. Press. kPa (Sat. Temp.) °C		−20	−10	0	10	20	30	40	50	60	70	80	100
50 (−46.54)	v	2.4474	2.5481	2.6482	2.7479	2.8473	2.9464	3.0453	3.1441	3.2427	3.3413	3.4397	
	h	1435.8	1457.0	1478.1	1499.2	1520.4	1541.7	1563.0	1584.5	1606.1	1627.8	1649.7	
	s	6.3256	6.4077	6.4865	6.5625	6.6360	6.7073	6.7766	6.8441	6.9099	6.9743	7.0372	
75 (−39.18)	v	1.6233	1.6915	1.7591	1.8263	1.8932	1.9597	2.0261	2.0923	2.1584	2.2244	2.2903	
	h	1433.0	1454.7	1476.1	1497.5	1518.9	1540.3	1561.8	1583.4	1605.1	1626.9	1648.9	
	s	6.1190	6.2028	6.2828	6.3597	6.4339	6.5058	6.5756	6.6434	6.7096	6.7742	6.8373	
100 (−33.61)	v	1.2110	1.2631	1.3145	1.3654	1.4160	1.4664	1.5165	1.5664	1.6163	1.6659	1.7155	1.8145
	h	1430.1	1452.2	1474.1	1495.7	1517.3	1538.9	1560.5	1582.2	1604.1	1626.0	1648.0	1692.6
	s	5.9695	6.0552	6.1366	6.2144	6.2894	6.3618	6.4321	6.5003	6.5668	6.6316	6.6950	6.8177
125 (−29.08)	v	0.9635	1.0059	1.0476	1.0889	1.1297	1.1703	1.2107	1.2509	1.2909	1.3309	1.3707	1.4501
	h	1427.2	1449.8	1472.0	1493.9	1515.7	1537.5	1559.3	1581.1	1603.0	1625.0	1647.2	1691.8
	s	5.8512	5.9389	6.0217	6.1006	6.1763	6.2494	6.3201	6.3887	6.4555	6.5206	6.5842	6.7072
150 (−25.23)	v	0.7984	0.8344	0.8697	0.9045	0.9388	0.9729	1.0068	1.0405	1.0740	1.1074	1.1408	1.2072
	h	1424.1	1447.3	1469.8	1492.1	1514.1	1536.1	1558.0	1580.0	1602.0	1624.1	1646.3	1691.1
	s	5.7526	5.8424	5.9266	6.0066	6.0831	6.1568	6.2280	6.2970	6.3641	6.4295	6.4933	6.6167

Temperature, °C

200	v	0.6199	0.6471	0.6738	0.7001	0.7261	0.7519	0.7774	0.8029	0.8282	0.8533	0.9035
(−18.86)	h	1442.0	1465.5	1488.4	1510.9	1533.2	1555.5	1577.7	1599.9	1622.2	1644.6	1689.6
	s	5.6863	5.7737	5.8559	5.9342	6.0091	6.0813	6.1512	6.2189	6.2849	6.3491	6.4732
250	v	0.4910	0.5135	0.5354	0.5568	0.5780	0.5989	0.6196	0.6401	0.6605	0.6809	0.7212
(−13.67)	h	1436.6	1461.0	1484.5	1507.6	1530.3	1552.9	1575.4	1597.8	1620.3	1642.8	1688.2
	s	5.5609	5.6517	5.7365	5.8165	5.8928	5.9661	6.0368	6.1052	6.1717	6.2365	6.3613
300	v		0.4243	0.4430	0.4613	0.4792	0.4968	0.5143	0.5316	0.5488	0.5658	0.5997
(−9.23)	h		1456.3	1480.6	1504.2	1527.4	1550.3	1573.0	1595.7	1618.4	1641.1	1686.7
	s		5.5493	5.6366	5.7186	5.7963	5.8707	5.9423	6.0114	6.0785	6.1437	6.2693
350	v		0.3605	0.3770	0.3929	0.4086	0.4239	0.4391	0.4541	0.4689	0.4837	0.5129
(−5.35)	h		1451.5	1476.5	1500.7	1524.4	1547.6	1570.7	1593.6	1616.5	1639.3	1685.2
	s		5.4600	5.5502	5.6342	5.7135	5.7890	5.8615	5.9314	5.9990	6.0647	6.1910
400	v		0.3125	0.3274	0.3417	0.3556	0.3692	0.3826	0.3959	0.4090	0.4220	0.4478
(−1.89)	h		1446.5	1472.4	1497.2	1521.3	1544.9	1568.3	1591.5	1614.5	1637.6	1683.7
	s		5.3803	5.4735	5.5597	5.6405	5.7173	5.7907	5.8613	5.9296	5.9957	6.1228
450	v		0.2752	0.2887	0.3017	0.3143	0.3266	0.3387	0.3506	0.3624	0.3740	0.3971
(1.26)	h		1441.3	1468.1	1493.6	1518.2	1542.2	1565.9	1589.3	1612.6	1635.8	1682.2
	s		5.3078	5.4042	5.4926	5.5752	5.6532	5.7275	5.7989	5.8678	5.9345	6.0623

(Continued)

TABLE A.2.2SI (*Continued*) Superheated Ammonia (SI Units)

Abs. Press. kPa (Sat. Temp.) °C		20	30	40	50	60	70	80	100	120	140	160	180
							Temperature, °C						
500 (4.14)	v	0.2698	0.2813	0.2926	0.3036	0.3144	0.3251	0.3357	0.3565	0.3771	0.3975		
	h	1489.9	1515.0	1539.5	1563.4	1587.1	1610.6	1634.0	1680.7	1727.5	1774.7		
	s	5.4314	5.5157	5.5950	5.6704	5.7425	5.8120	5.8793	6.0079	6.1301	6.2472		
600 (9.29)	v	0.2217	0.2317	0.2414	0.2508	0.2600	0.2691	0.2781	0.2957	0.3130	0.3302		
	h	1482.4	1508.6	1533.8	1558.5	1582.7	1606.6	1630.4	1677.7	1724.9	1772.4		
	s	5.3222	5.4102	5.4923	5.5697	5.6436	5.7144	5.7826	5.9129	6.0363	6.1541		
700 (13.81)	v	0.1874	0.1963	0.2048	0.2131	0.2212	0.2291	0.2369	0.2522	0.2672	0.2821		
	h	1474.5	1501.9	1528.1	1553.4	1578.2	1602.6	1626.8	1674.6	1722.4	1770.2		
	s	5.2259	5.3179	5.4029	5.4826	5.5582	5.6303	5.6997	5.8316	5.9562	6.0749		
800 (17.86)	v	0.1615	0.1696	0.1773	0.1848	0.1920	0.1991	0.2060	0.2196	0.2329	0.2459	0.2589	
	h	1466.3	1495.0	1522.2	1548.3	1573.7	1598.6	1623.1	1671.6	1719.8	1768.0	1816.4	
	s	5.1387	5.2351	5.3232	5.4053	5.4827	5.5562	5.6268	5.7603	5.8861	6.0057	6.1202	
900 (21.54)	v		0.1488	0.1559	0.1627	0.1693	0.1757	0.1820	0.1942	0.2061	0.2178	0.2294	
	h		1488.0	1516.2	1543.0	1569.1	1594.4	1619.4	1668.5	1717.1	1765.7	1814.4	
	s		5.1593	5.2508	5.3354	5.4147	5.4897	5.5614	5.6968	5.8237	5.9442	6.0594	

1000 (24.91)	v	0.1321	0.1388	0.1450	0.1511	0.1570	0.1627	0.1739	0.1847	0.1954	0.2058	0.2162
	h	1480.6	1510.0	1537.7	1564.4	1590.3	1615.6	1665.4	1714.5	1763.4	1812.4	1861.7
	s	5.0889	5.1840	5.2713	5.3525	5.4292	5.5021	5.6392	5.7674	5.8888	6.0047	6.1159
1200 (30.96)	v		0.1129	0.1185	0.1238	0.1289	0.1338	0.1434	0.1526	0.1616	0.1705	0.1792
	h		1497.1	1526.6	1554.7	1581.7	1608.0	1659.2	1709.2	1758.9	1808.5	1858.2
	s		5.0629	5.1560	5.2416	5.3215	5.3970	5.5379	5.6687	5.7919	5.9091	6.0214
1400 (36.28)	v		0.0944	0.0995	0.1042	0.1088	0.1132	0.1216	0.1297	0.1376	0.1452	0.1528
	h		1483.4	1515.1	1544.7	1573.0	1600.2	1652.8	1703.9	1754.3	1804.5	1854.7
	s		4.9534	5.0530	5.1434	5.2270	5.3053	5.4501	5.5836	5.7087	5.8273	5.9406
1600 (41.05)	v			0.0851	0.0895	0.0937	0.0977	0.1053	0.1125	0.1195	0.1263	0.1330
	h			1502.9	1534.4	1564.0	1592.3	1646.4	1698.5	1749.7	1800.5	1851.2
	s			4.9584	5.0543	5.1419	5.2232	5.3722	5.5084	5.6355	5.7555	5.8699
1800 (45.39)	v			0.0739	0.0781	0.0820	0.0856	0.0926	0.0992	0.1055	0.1116	0.1177
	h			1490.0	1523.5	1554.6	1584.1	1639.8	1693.1	1745.1	1796.5	1847.7
	s			4.8693	4.9715	5.0635	5.1482	5.3018	5.4409	5.5699	5.6914	5.8069
2000 (49.38)	v			0.0648	0.0688	0.0725	0.0760	0.0824	0.0885	0.0943	0.0999	0.1054
	h			1476.1	1512.0	1544.9	1575.6	1633.2	1687.6	1740.4	1792.4	1844.1
	s			4.7834	4.8930	4.9902	5.0786	5.2371	5.3793	5.5104	5.6333	5.7499

TABLE A.3E Thermodynamic Properties of Freon-12 (Dichlorodifluoromethane)[a]

TABLE A.3.1E Saturated Freon-12 (English Units)

Temp. F T	Abs. Press. lbf/in.² P	Specific Volume ft³/lbm			Enthalpy Btu/lbm			Entropy Btu/lbm R		
		Sat. Liquid v_f	Evap. v_{fg}	Sat. Vapor v_g	Sat. Liquid h_f	Evap. h_{fg}	Sat. Vapor h_g	Sat. Liquid s_f	Evap. s_{fg}	Sat. Vapor s_g
−130	0.412 24	0.009 736	70.7203	70.730	−18.609	81.577	62.968	−0.049 83	0.247 43	0.197 60
−120	0.641 90	0.009 816	46.7312	46.741	−16.565	80.617	64.052	−0.043 72	0.237 31	0.193 59
−110	0.970 34	0.009 899	31.7671	31.777	−14.518	79.663	65.145	−0.037 79	0.227 80	0.190 02
−100	1.4280	0.009 985	21.1541	22.164	−12.466	78.714	66.248	−0.032 00	0.218 83	0.186 83
−90	2.0509	0.010 073	15.8109	15.821	−10.409	77.764	67.355	−0.026 37	0.210 34	0.183 98
−80	2.8807	0.010 164	11.5228	11.533	−8.3451	76.812	68.467	−0.020 86	0.202 29	0.181 43
−70	3.9651	0.010 259	8.5584	8.5687	−6.2730	75.853	69.580	−0.015 48	0.194 64	0.179 16
−60	5.3575	0.010 357	6.4670	6.4774	−4.1919	74.885	70.693	−0.010 21	0.187 16	0.177 14
−50	7.1168	0.010 459	4.9637	4.9742	−2.1011	73.906	71.805	−0.005 06	0.180 38	0.175 33
−40	9.3076	0.010 564	3.8644	3.8750	0	72.913	72.913	0	0.173 73	0.173 73
−30	11.999	0.010 674	3.0478	3.0585	2.1120	71.903	74.015	0.004 96	0.167 33	0.172 29
−20	15.267	0.010 788	2.4321	2.4429	4.2357	70.874	75.110	0.009 83	0.161 19	0.171 02
−10	19.189	0.010 906	1.9628	1.9727	6.3716	69.824	76.196	0.014 62	0.155 27	0.169 89
0	23.849	0.011 030	1.5979	1.6089	8.5207	68.750	77.271	0.019 32	0.149 56	0.168 88
10	29.335	0.011 160	1.3129	1.3241	10.684	67.651	78.335	0.023 95	0.144 03	0.167 98
20	35.736	0.011 296	1.0875	1.0988	12.863	66.522	79.385	0.028 52	0.138 67	0.167 19
30	43.148	0.011 438	0.907 36	0.918 80	15.058	65.361	80.419	0.033 01	0.133 47	0.166 48
40	51.667	0.011 588	0.761 98	0.773 57	17.273	64.163	81.436	0.037 45	0.128 41	0.165 86
50	61.394	0.011 746	0.643 62	0.655 37	19.507	62.926	82.433	0.041 84	0.234 46	0.165 30
60	72.433	0.011 913	0.546 48	0.558 39	21.766	61.643	83.409	0.046 18	0.118 61	0.164 79

70	84.888	0.012 089	0.466 09	0.478 18	24.050	60.309	84.359	0.050 48	0.113 86	0.164 34
80	98.870	0.012 277	0.399 07	0.411 35	26.365	58.917	85.282	0.054 75	0.109 17	0.163 92
90	114.49	0.012 478	0.342 81	0.355 29	28.713	57.461	86.174	0.059 00	0.104 53	0.163 53
100	131.86	0.012 693	0.295 25	0.307 94	31.100	55.929	87.029	0.063 23	0.099 92	0.163 15
110	151.11	0.012 924	0.255 77	0.267 69	33.531	54.313	87.844	0.067 45	0.095 34	0.162 79
120	172.35	0.013 174	0.220 19	0.233 26	36.013	52.597	88.610	0.071 68	0.090 73	0.162 41
130	195.71	0.013 447	0.190 19	0.203 64	38.553	50.768	89.321	0.075 83	0.086 09	0.162 02
140	221.32	0.013 746	0.164 24	0.177 99	41.162	48.805	89.967	0.080 21	0.081 38	0.161 59
150	249.31	0.014 078	0.141 56	0.155 64	43.850	46.684	90.534	0.084 53	0.076 57	0.161 10
160	279.82	0.014 449	0.121 59	0.136 04	46.633	44.373	91.006	0.088 93	0.072 60	0.160 53
170	313.00	0.014 871	0.103 86	0.118 73	49.529	41.830	91.359	0.093 42	0.066 43	0.159 85
180	349.00	0.015 360	0.087 94	0.103 30	52.562	38.999	91.561	0.098 04	0.060 96	0.159 00
190	387.98	0.015 942	0.073 476	0.089 418	55.769	35.792	91.561	0.102 84	0.055 11	0.157 93
200	430.09	0.016 659	0.060 069	0.076 728	59.203	32.075	91.278	0.107 89	0.048 62	0.156 51
210	475.52	0.017 601	0.047 242	0.064 843	62.959	27.599	90.558	0.113 32	0.039 21	0.154 53
220	524.43	0.018 986	0.035 154	0.053 140	67.246	21.790	89.036	0.119 43	0.032 06	0.151 49
230	577.03	0.021 854	0.017 581	0.039 435	72.893	12.229	85.122	0.127 39	0.017 73	0.145 12
233.6 (critical)	596.9	0.028 70	0	0.028 70	78.86	0	78.86	0.135 90	0	0.135 90

TABLE A.3.2E Superheated Freon-12 (English Units)

Temp. F	5 lbf/in.² v	5 lbf/in.² h	5 lbf/in.² s	10 lbf/in.² v	10 lbf/in.² h	10 lbf/in.² s	15 lbf/in.² v	15 lbf/in.² h	15 lbf/in.² s
0	8.0611	78.582	0.196 63	3.9809	78.246	0.184 71	2.6201	77.902	0.177 51
20	8.4265	81.309	0.202 44	4.1691	81.014	0.190 61	2.7494	80.712	0.183 49
40	8.7903	84.090	0.208 12	4.3556	83.828	0.196 35	2.8770	83.561	0.189 31
60	9.1528	86.922	0.213 67	4.5408	86.689	0.201 97	3.0031	86.451	0.194 98
80	9.5142	89.806	0.219 12	4.7248	89.596	0.207 46	3.1281	89.383	0.200 51
100	9.8747	92.738	0.224 45	4.9079	92.548	0.212 83	3.2521	92.357	0.205 93
120	10.234	95.717	0.229 68	5.0903	95.546	0.218 09	3.3754	95.373	0.211 22
140	10.594	98.743	0.234 81	5.2720	98.586	0.223 25	3.4981	98.429	0.216 40
160	10.952	101.812	0.239 85	5.4533	101.669	0.228 30	3.6202	101.525	0.221 48
180	11.311	104.925	0.244 79	5.6341	104.793	0.233 26	3.7419	104.661	0.226 46
200	11.668	108.079	0.249 64	5.8145	107.957	0.238 13	3.8632	107.835	0.231 35
220	12.026	111.272	0.254 41	5.9946	111.159	0.242 91	3.9841	111.046	0.236 14

Temp. F	20 lbf/in.² v	20 lbf/in.² h	20 lbf/in.² s	25 lbf/in.² v	25 lbf/in.² h	25 lbf/in.² s	30 lbf/in.² v	30 lbf/in.² h	30 lbf/in.² s
20	2.0391	80.403	0.178 29	1.6125	80.088	0.174 14	1.3278	79.765	0.170 65
40	2.1373	83.289	0.184 19	1.6932	83.012	0.180 12	1.3969	82.730	0.176 71
60	2.2340	86.210	0.189 92	1.7723	85.965	0.185 91	1.4644	85.716	0.182 57
80	2.3295	89.168	0.195 50	1.8502	88.950	0.191 55	1.5306	88.729	0.188 26
100	2.4241	92.164	0.200 95	1.9271	91.968	0.197 04	1.5957	91.770	0.193 79
120	2.5179	95.198	0.206 28	2.0032	95.021	0.202 40	1.6600	94.843	0.199 18
140	2.6110	98.270	0.211 49	2.0786	98.110	0.207 63	1.7237	97.948	0.204 45
160	2.7036	101.380	0.216 59	2.1535	101.234	0.212 76	1.7868	101.086	0.209 60
180	2.7957	104.528	0.221 59	2.2279	104.393	0.217 78	1.8494	104.258	0.214 63
200	2.8874	107.712	0.226 49	2.3019	107.588	0.222 69	1.9116	107.464	0.219 57
220	2.9789	110.932	0.231 30	2.3756	110.817	0.227 52	1.9735	110.702	0.224 40
240	3.0700	114.186	0.236 02	2.4491	114.080	0.232 25	2.0351	113.973	0.229 15

	35 lbf/in.²			40 lbf/in.²			50 lbf/in.²		
40	1.1850	82.442	0.173 75	1.0258	82.148	0.171 12	0.802 48	81.540	0.166 55
60	1.2442	85.463	0.179 68	1.0789	85.206	0.177 12	0.847 13	84.676	0.172 71
80	1.3021	88.504	0.185 42	1.1306	88.277	0.182 92	0.890 25	87.811	0.178 62
100	1.3589	91.570	0.191 00	1.1812	91.367	0.188 54	0.932 16	90.953	0.184 34
120	1.4148	94.663	0.196 43	1.2309	94.480	0.194 01	0.973 13	94.110	0.189 88
140	1.4701	97.785	0.201 72	1.2798	97.620	0.199 33	1.0133	97.286	0.195 27
160	1.5248	100.938	0.206 89	1.3282	100.788	0.204 53	1.0529	100.485	0.200 51
180	1.5789	104.122	0.211 95	1.3761	103.985	0.209 61	1.0920	103.708	0.205 63
200	1.6327	107.338	0.216 90	1.4236	107.212	0.214 57	1.1307	106.958	0.210 64
220	1.6862	110.586	0.221 75	1.4707	110.469	0.219 44	1.1690	110.235	0.215 53
240	1.7394	113.865	0.226 51	1.5176	113.757	0.224 20	1.2070	113.539	0.220 32
260	1.7923	117.175	0.231 17	1.5642	117.074	0.228 88	1.2447	116.871	0.225 02

	60 lbf/in.²			70 lbf/in.²			80 lbf/in.²		
60	0.692 10	84.126	0.168 92	0.580 88	83.552	0.165 56	—	—	—
80	0.729 64	87.330	0.174 97	0.614 58	86.832	0.171 75	0.527 95	86.316	0.168 85
100	0.765 88	90.528	0.180 79	0.646 85	90.091	0.177 68	0.557 34	89.640	0.174 89
120	0.801 10	93.731	0.186 41	0.678 03	93.343	0.183 39	0.585 56	92.945	0.180 70
140	0.835 51	96.945	0.191 86	0.708 36	96.597	0.188 91	0.612 86	96.242	0.186 29
160	0.869 28	100.776	0.197 16	0.738 00	99.862	0.194 27	0.639 43	99.542	0.191 70
180	0.902 52	103.427	0.202 33	0.767 08	103.141	0.199 48	0.665 43	102.851	0.196 96
200	0.935 31	106.700	0.207 36	0.795 71	106.439	0.204 55	0.690 95	106.174	0.202 07
220	0.967 75	109.997	0.212 29	0.823 97	109.756	0.209 51	0.716 09	109.513	0.207 06
240	0.999 88	113.319	0.217 10	0.851 91	113.096	0.214 35	0.740 90	112.872	0.211 93
260	1.031 80	116.666	0.221 82	0.879 59	116.459	0.219 09	0.765 44	116.251	0.216 69
280	1.063 40	120.039	0.226 44	0.907 05	119.846	0.223 73	0.789 75	119.652	0.221 35

(Continued)

TABLE A.3.2E (*Continued*) **Superheated Freon-12 (English Units)**

Temp. F	90 lbf/in²			100 lbf/in²			125 lbf/in²		
	v	h	s	v	h	s	v	h	s
100	0.487 49	89.175	0.172 34	0.431 38	88.694	0.169 96	0.329 43	87.407	0.164 55
120	0.513 46	92.536	0.178 24	0.455 62	92.116	0.175 97	0.350 86	91.008	0.170 87
140	0.538 45	95.879	0.183 91	0.478 81	95.507	0.181 72	0.370 98	94.537	0.176 86
160	0.562 68	99.216	0.189 38	0.501 18	98.884	0.187 26	0.390 15	98.023	0.182 58
180	0.586 29	102.557	0.194 69	0.522 91	102.257	0.192 62	0.408 57	101.484	0.188 07
200	0.609 41	105.905	0.199 84	0.544 13	105.633	0.197 82	0.426 42	104.934	0.193 38
220	0.632 13	109.267	0.204 86	0.564 92	109.018	0.202 87	0.443 80	108.380	0.198 53
240	0.654 51	112.644	0.209 76	0.585 38	112.415	0.207 80	0.460 81	111.829	0.203 53
260	0.676 62	116.040	0.214 55	0.605 54	115.828	0.212 61	0.477 50	115.287	0.208 40
280	0.698 49	119.456	0.219 23	0.625 46	119.258	0.217 31	0.493 94	118.756	0.213 16
300	0.720 16	122.892	0.223 81	0.645 18	122.707	0.221 91	0.510 16	122.238	0.217 80
320	0.741 66	126.349	0.228 30	0.664 72	126.176	0.226 41	0.526 19	125.737	0.222 35

Temp. F	150 lbf/in²			175 lbf/in²			200 lbf/in²		
	v	h	s	v	h	s	v	h	s
120	0.280 07	89.800	0.166 29	—	—	—	—	—	—
140	0.298 45	93.498	0.172 56	0.245 95	92.373	0.168 59	0.205 79	91.137	0.164 80
160	0.315 66	97.112	0.178 49	0.261 98	96.142	0.174 78	0.221 21	95.100	0.171 30
180	0.332 00	100.675	0.184 15	0.276 97	99.823	0.180 62	0.235 35	98.921	0.177 37
200	0.347 69	104.206	0.189 58	0.291 20	103.447	0.186 20	0.248 60	102.652	0.183 11
220	0.362 85	107.720	0.194 83	0.304 85	107.036	0.191 56	0.261 17	106.325	0.188 60
240	0.377 61	111.226	0.199 92	0.318 04	110.605	0.196 74	0.273 23	109.962	0.193 87
260	0.392 03	114.732	0.204 85	0.330 87	114.162	0.201 75	0.284 89	113.576	0.198 96
280	0.406 17	118.242	0.209 67	0.343 39	117.717	0.206 62	0.296 23	117.178	0.203 90
300	0.420 08	121.761	0.214 36	0.355 67	121.273	0.211 37	0.307 30	120.775	0.208 70
320	0.433 79	125.290	0.218 94	0.367 73	124.835	0.215 99	0.318 15	124.373	0.213 37
340	0.447 33	128.833	0.223 43	0.379 63	128.407	0.220 52	0.328 81	127.974	0.217 93

Temp	250 lbf/in.²			300 lbf/in.²			400 lbf/in.²		
160	0.162 49	92.717	0.164 62	—	—	—	—	—	—
180	0.176 05	96.925	0.171 30	0.134 82	94.556	0.165 37	—	—	—
200	0.188 24	100.930	0.177 47	0.146 97	98.975	0.172 17	0.091 005	93.718	0.160 92
220	0.199 52	104.809	0.183 26	0.157 74	103.136	0.178 38	0.103 16	99.046	0.168 88
240	0.210 14	108.607	0.188 77	0.167 61	107.140	0.184 19	0.113 00	103.735	0.175 68
260	0.220 27	112.351	0.194 04	0.176 85	111.043	0.189 69	0.121 63	108.105	0.181 83
280	0.230 01	116.060	0.199 13	0.185 62	114.879	0.194 95	0.129 49	112.286	0.187 56
300	0.239 44	119.747	0.204 05	0.194 02	118.670	0.200 00	0.136 80	116.343	0.192 98
320	0.248 62	123.420	0.208 82	0.202 14	122.430	0.204 89	0.143 72	120.318	0.198 14
340	0.257 59	127.088	0.213 46	0.210 02	126.171	0.209 63	0.150 32	124.235	0.203 10
360	0.266 39	130.754	0.217 99	0.217 70	129.900	0.214 23	0.156 68	128.112	0.207 89
380	0.275 04	134.423	0.222 41	0.225 22	133.624	0.218 72	0.162 85	131.961	0.212 58

Temp	500 lbf/in.²			600 lbf/in.²		
220	0.064 207	92.397	0.156 83	—	—	—
240	0.077 620	99.218	0.166 72	0.047 488	91.024	0.153 35
260	0.087 054	104.526	0.174 21	0.061 922	99.741	0.165 66
280	0.094 923	109.277	0.180 72	0.070 859	105.637	0.173 74
300	0.101 90	113.729	0.186 66	0.078 059	110.729	0.180 53
320	0.108 29	117.997	0.192 21	0.084 333	115.420	0.186 63
340	0.114 26	122.143	0.197 46	0.090 017	119.871	0.192 27
360	0.119 92	126.205	0.202 47	0.095 289	124.167	0.197 57
380	0.125 33	130.207	0.207 30	0.100 25	128.355	0.202 62
400	0.130 54	134.166	0.211 96	0.104 98	132.466	0.207 46
420	0.135 59	138.096	0.216 48	0.109 52	136.523	0.212 13
440	0.140 51	142.004	0.220 87	0.113 91	140.539	0.216 64

TABLE A.3SI Thermodynamic Properties of Freon-12 (Dichlorodifluoromethane)[a]

TABLE A.3.1SI Saturated Freon-12 (SI Units)

Temp. °C	Abs. Press. MPa P	Specific Volume m³/kg Sat. Liquid v_f	Evap. v_{fg}	Sat. Vapor v_g	Enthalpy kJ/kg Sat. Liquid h_f	Evap. h_{fg}	Sat. Vapor h_g	Entropy kJ/kg K Sat. Liquid s_f	Evap. s_{fg}	Sat. Vapor s_g
−90	0.0028	0.000 608	4.414 937	4.415 545	−43.243	189.618	146.375	−0.2084	1.0352	0.8268
−85	0.0042	0.000 612	3.036 704	3.037 316	−38.968	187.608	148.640	−0.1854	0.9970	0.8116
−80	0.0062	0.000 617	2.137 728	2.138 345	−34.688	185.612	150.924	−0.1630	0.9609	0.7979
−75	0.0088	0.000 622	1.537 030	1.537 651	−30.401	183.625	153.224	−0.1411	0.9266	0.7855
−70	0.0123	0.000 627	1.126 654	1.127 280	−26.103	181.640	155.536	−0.1197	0.8940	0.7744
−65	0.0168	0.000 632	0.840 534	0.841 166	−21.793	179.651	157.857	−0.0987	0.8630	0.7643
−60	0.0226	0.000 637	0.637 274	0.637 910	−17.469	177.653	160.184	−0.0782	0.8334	0.7552
−55	0.0300	0.000 642	0.490 358	0.491 000	−13.129	175.641	162.512	−0.0581	0.8051	0.7470
−50	0.0391	0.000 648	0.382 457	0.383 105	−8.772	173.611	164.840	−0.0384	0.7779	0.7396
−45	0.0504	0.000 654	0.302 029	0.302 682	−4.396	171.558	167.163	−0.0190	0.7519	0.7329
−40	0.0642	0.000 659	0.241 251	0.241 910	−0.000	169.479	169.479	−0.0000	0.7269	0.7269
−35	0.0807	0.000 666	0.194 732	0.195 398	4.416	167.368	171.784	0.0187	0.7027	0.7214
−30	0.1004	0.000 672	0.158 703	0.159 375	8.854	165.222	174.076	0.0371	0.6795	0.7165
−25	0.1237	0.000 679	0.130 487	0.131 166	13.315	163.037	176.352	0.0552	0.6570	0.7121
−20	0.1509	0.000 685	0.108 162	0.108 847	17.800	160.810	178.610	0.0730	0.6352	0.7082
−15	0.1826	0.000 693	0.090 326	0.091 018	22.312	158.534	180.846	0.0906	0.6141	0.7046
−10	0.2191	0.000 700	0.075 946	0.076 646	26.851	156.207	183.058	0.1079	0.5936	0.7014
−5	0.2610	0.000 708	0.064 255	0.064 963	31.420	153.823	185.243	0.1250	0.5736	0.6986

0	0.3086	0.000 716	0.054 673	0.055 389	36.022	151.376	187.397	0.1418	0.5542	0.6960
5	0.3626	0.000 724	0.046 761	0.047 485	40.659	148.859	189.518	0.1585	0.5351	0.6937
10	0.4233	0.000 733	0.040 180	0.040 914	45.337	146.265	191.602	0.1750	0.5165	0.6916
15	0.4914	0.000 743	0.034 671	0.035 413	50.058	143.586	193.644	0.1914	0.4983	0.6897
20	0.5673	0.000 752	0.030 028	0.030 780	54.828	140.812	195.641	0.2076	0.4803	0.6879
25	0.6516	0.000 763	0.026 091	0.026 854	59.653	137.933	197.586	0.2237	0.4626	0.6863
30	0.7449	0.000 774	0.022 734	0.023 508	64.539	134.936	199.475	0.2397	0.4451	0.6848
35	0.8477	0.000 786	0.019 855	0.020 641	69.494	131.805	201.299	0.2557	0.4277	0.6834
40	0.9607	0.000 798	0.017 373	0.018 171	74.527	128.525	203.051	0.2716	0.4104	0.6820
45	1.0843	0.000 811	0.015 220	0.016 032	79.647	125.074	204.722	0.2875	0.3931	0.6806
50	1.2193	0.000 826	0.013 344	0.014 170	84.868	121.430	206.298	0.3034	0.3758	0.6792
55	1.3663	0.000 841	0.011 701	0.012 542	90.201	117.565	207.766	0.3194	0.3582	0.6777
60	1.5259	0.000 858	0.010 253	0.011 111	95.665	113.443	209.109	0.3355	0.3405	0.6760
65	1.6988	0.000 877	0.008 971	0.009 847	101.279	109.024	210.303	0.3518	0.3224	0.6742
70	1.8858	0.000 897	0.007 828	0.008 725	107.067	104.255	211.321	0.3683	0.3038	0.6721
75	2.0874	0.000 920	0.006 802	0.007 723	113.058	99.068	212.126	0.3851	0.2845	0.6697
80	2.3046	0.000 946	0.005 875	0.006 821	119.291	93.373	212.665	0.4023	0.2644	0.6667
85	2.5380	0.000 976	0.005 029	0.006 005	125.818	87.047	212.865	0.4201	0.2430	0.6631
90	2.7885	0.001 012	0.004 246	0.005 258	132.708	79.907	212.614	0.4385	0.2200	0.6585
95	3.0569	0.001 056	0.003 508	0.004 563	140.068	71.658	211.726	0.4579	0.1946	0.6526
100	3.3440	0.001 113	0.002 790	0.003 903	148.076	61.768	209.843	0.4788	0.1655	0.6444
105	3.6509	0.001 197	0.002 045	0.003 242	157.085	49.014	206.099	0.5023	0.1296	0.6319
110	3.9784	0.001 364	0.001 098	0.002 462	168.059	28.425	196.484	0.5322	0.0742	0.6064
112	4.1155	0.001 792	0.000 005	0.001 797	174.920	0.151	175.071	0.5651	0.0004	0.5655

ᵃ Copyright 1955 and 1956, E. I. du Pont de Nemours & Company, Inc. Reprinted by permission. Adapted from English units.

TABLE A.3.2SI Superheated Freon-12 (SI Units)

Temp. °C	0.05 MPa			0.10 MPa			0.15 MPa		
	v m³/kg	h kJ/kg	s kJ/kg K	v m³/kg	h kJ/kg	s kJ/kg K	v m³/kg	h kJ/kg	s kJ/kg K
−20.0	0.341 857	181.042	0.7912	0.167 701	179.861	0.7401			
−10.0	0.356 227	186.757	0.8133	0.175 222	185.707	0.7628	0.114 716	184.619	0.7318
0.0	0.370 508	192.567	0.8350	0.182 647	191.628	0.7849	0.119 866	190.660	0.7543
10.0	0.384 716	198.471	0.8562	0.189 994	197.628	0.8064	0.124 932	196.762	0.7763
20.0	0.398 863	204.469	0.8770	0.197 277	203.707	0.8275	0.129 930	202.927	0.7977
30.0	0.412 959	210.557	0.8974	0.204 506	209.866	0.8482	0.134 873	209.160	0.8186
40.0	0.427 012	216.733	0.9175	0.211 691	216.104	0.8684	0.139 768	215.463	0.8390
50.0	0.441 030	222.997	0.9372	0.218 839	222.421	0.8883	0.144 625	221.835	0.8591
60.0	0.455 017	229.344	0.9565	0.225 955	228.815	0.9078	0.149 450	228.277	0.8787
70.0	0.468 978	235.774	0.9755	0.233 044	235.285	0.9269	0.154 247	234.789	0.8980
80.0	0.482 917	242.282	0.9942	0.240 111	241.829	0.9457	0.159 020	241.371	0.9169
90.0	0.496 838	248.868	1.0126	0.247 159	248.446	0.9642	0.163 774	248.020	0.9354

Temp. °C	0.20 MPa			0.25 MPa			0.30 MPa		
	v m³/kg	h kJ/kg	s kJ/kg K	v m³/kg	h kJ/kg	s kJ/kg K	v m³/kg	h kJ/kg	s kJ/kg K
0.0	0.088 608	189.669	0.7320	0.069 752	188.644	0.7139	0.057 150	187.583	0.6984
10.0	0.092 550	195.878	0.7543	0.073 024	194.969	0.7366	0.059 984	194.034	0.7216
20.0	0.096 418	202.135	0.7760	0.076 218	201.322	0.7587	0.062 734	200.490	0.7440
30.0	0.100 228	208.446	0.7972	0.079 350	207.715	0.7801	0.065 418	206.969	0.7658
40.0	0.103 989	214.814	0.8178	0.082 431	214.153	0.8010	0.068 049	213.480	0.7869
50.0	0.107 710	221.243	0.8381	0.085 470	220.642	0.8214	0.070 635	220.030	0.8075
60.0	0.111 397	227.735	0.8578	0.088 474	227.185	0.8413	0.073 185	226.627	0.8276
70.0	0.115 055	234.291	0.8772	0.091 449	233.785	0.8608	0.075 705	233.273	0.8473
80.0	0.118 690	240.910	0.8962	0.094 398	240.443	0.8800	0.078 200	239.971	0.8665
90.0	0.122 304	247.593	0.9149	0.097 327	247.160	0.8987	0.080 673	246.723	0.8853
100.0	0.125 901	254.339	0.9332	0.100 238	253.936	0.9171	0.083 127	253.530	0.9038
110.0	0.129 483	261.147	0.9512	0.103 134	260.770	0.9352	0.085 566	260.391	0.9220

	0.40 MPa			0.50 MPa			0.60 MPa		
20.0	0.045 836	198.762	0.7199	0.035 646	196.935	0.6999			
30.0	0.047 971	205.428	0.7423	0.037 464	203.814	0.7230	0.030 422	202.116	0.7063
40.0	0.050 046	212.095	0.7639	0.039 214	210.656	0.7452	0.031 966	209.154	0.7291
50.0	0.052 072	218.779	0.7849	0.040 911	217.484	0.7667	0.033 450	216.141	0.7511
60.0	0.054 059	225.488	0.8054	0.042 565	224.315	0.7875	0.034 887	223.104	0.7723
70.0	0.056 014	232.230	0.8253	0.044 184	231.161	0.8077	0.036 285	230.062	0.7929
80.0	0.057 941	239.012	0.8448	0.045 774	238.031	0.8275	0.037 653	237.027	0.8129
90.0	0.059 846	245.837	0.8638	0.047 340	244.932	0.8467	0.038 995	244.009	0.8324
100.0	0.061 731	252.707	0.8825	0.048 886	251.869	0.8656	0.040 316	251.016	0.8514
110.0	0.063 600	259.624	0.9008	0.050 415	258.845	0.8840	0.041 619	258.053	0.8700
120.0	0.065 455	266.590	0.9187	0.051 929	265.862	0.9021	0.042 907	265.124	0.8882
130.0	0.067 298	273.605	0.9364	0.053 430	272.923	0.9198	0.044 181	272.231	0.9061

	0.70 MPa			0.80 MPa			0.90 MPa		
40.0	0.026 761	207.580	0.7148	0.022 830	205.924	0.7016	0.019 744	204.170	0.6982
50.0	0.028 100	214.745	0.7373	0.024 068	213.290	0.7248	0.020 912	211.765	0.7131
60.0	0.029 387	221.854	0.7590	0.025 247	220.558	0.7469	0.022 012	219.212	0.7358
70.0	0.030 632	228.931	0.7799	0.026 380	227.766	0.7682	0.023 062	226.564	0.7575
80.0	0.031 843	235.997	0.8002	0.027 477	234.941	0.7888	0.024 072	233.856	0.7785
90.0	0.033 027	243.066	0.8199	0.028 545	242.101	0.8088	0.025 051	241.113	0.7987
100.0	0.034 189	250.146	0.8392	0.029 588	249.260	0.8283	0.026 005	248.355	0.8184
110.0	0.035 332	257.247	0.8579	0.030 612	256.428	0.8472	0.026 937	255.593	0.8376
120.0	0.036 458	264.374	0.8763	0.031 619	263.613	0.8657	0.027 851	262.839	0.8562
130.0	0.037 572	271.531	0.8943	0.032 612	270.820	0.8838	0.028 751	270.100	0.8745
140.0	0.038 673	278.720	0.9119	0.033 592	278.055	0.9016	0.029 639	277.381	0.8923
150.0	0.039 764	285.946	0.9292	0.034 563	285.320	0.9189	0.030 515	284.687	0.9098

TABLE A.3.2SI (*Continued*) Superheated Freon-12 (SI Units)

Temp. °C	v m³/kg	h kJ/kg	s kJ/kg K	v m³/kg	h kJ/kg	s kJ/kg K	v m³/kg	h kJ/kg	s kJ/kg K
	1.00 MPa			1.20 MPa			1.40 MPa		
50.0	0.018 366	210.162	0.7021	0.014 483	206.661	0.6812			
60.0	0.019 410	217.810	0.7254	0.015 463	214.805	0.7060	0.012 579	211.457	0.6876
70.0	0.020 397	225.319	0.7476	0.016 368	222.687	0.7293	0.013 448	219.822	0.7123
80.0	0.021 341	232.739	0.7689	0.017 221	230.398	0.7514	0.014 247	227.891	0.7355
90.0	0.022 251	240.101	0.7895	0.018 032	237.995	0.7727	0.014 997	235.766	0.7575
100.0	0.023 133	247.430	0.8094	0.018 812	245.518	0.7931	0.015 710	243.512	0.7785
110.0	0.023 993	254.743	0.8287	0.019 567	252.993	0.8129	0.016 393	251.170	0.7988
120.0	0.024 835	262.053	0.8475	0.020 301	260.441	0.8320	0.017 053	258.770	0.8183
130.0	0.025 661	269.369	0.8659	0.021 018	267.875	0.8507	0.017 695	266.334	0.8373
140.0	0.026 474	276.699	0.8839	0.021 721	275.307	0.8689	0.018 321	273.877	0.8558
150.0	0.027 275	284.047	0.9015	0.022 412	282.745	0.8867	0.018 934	281.411	0.8738
160.0	0.028 068	291.419	0.9187	0.023 093	290.195	0.9041	0.019 535	288.946	0.8914

Temp. °C	v m³/kg	h kJ/kg	s kJ/kg K	v m³/kg	h kJ/kg	s kJ/kg K	v m³/kg	h kJ/kg	s kJ/kg K
	1.60 MPa			1.80 MPa			2.00 MPa		
70.0	0.011 208	216.650	0.6959	0.009 406	213.049	0.6794	0.008 704	218.859	0.6909
80.0	0.011 984	225.177	0.7204	0.010 187	222.498	0.7057	0.009 406	228.056	0.7166
90.0	0.012 698	233.390	0.7433	0.010 884	230.835	0.7298	0.010 035	236.760	0.7402
100.0	0.013 366	241.397	0.7651	0.011 526	239.155	0.7524	0.010 615	245.154	0.7624
110.0	0.014 000	249.264	0.7859	0.012 126	247.264	0.7739	0.011 159	253.341	0.7835
120.0	0.014 608	257.035	0.8059	0.012 697	255.228	0.7944	0.011 676	261.384	0.8037
130.0	0.015 195	264.742	0.8253	0.013 244	263.094	0.8141	0.012 172	269.327	0.8232
140.0	0.015 765	272.406	0.8440	0.013 772	270.891	0.8332	0.012 651	277.201	0.8420
150.0	0.016 320	280.044	0.8623	0.014 284	278.642	0.8518	0.013 116	285.027	0.8603
160.0	0.016 864	287.669	0.8801	0.014 784	286.364	0.8698	0.013 570	292.822	0.8781
170.0	0.017 398	295.290	0.8975	0.015 272	294.069	0.8874	0.014 013	300.598	0.8955
180.0	0.017 923	302.914	0.9145	0.015 752	301.767	0.9046			

Temp	2.50 MPa			3.00 MPa			3.50 MPa		
90.0	0.006 595	219.562	0.6823						
100.0	0.007 264	229.852	0.7103	0.005 231	220.529	0.6770			
110.0	0.007 837	239.271	0.7352	0.005 886	232.068	0.7075	0.004 324	222.121	0.6750
120.0	0.008 351	248.192	0.7582	0.006 419	242.208	0.7336	0.004 959	234.875	0.7078
130.0	0.008 827	256.794	0.7798	0.006 887	251.632	0.7573	0.005 456	245.661	0.7349
140.0	0.009 273	265.180	0.8003	0.007 313	260.620	0.7793	0.005 884	255.524	0.7591
150.0	0.009 697	273.414	0.8200	0.007 709	269.319	0.8001	0.006 270	264.846	0.7814
160.0	0.010 104	281.540	0.8390	0.008 083	277.817	0.8200	0.006 626	273.817	0.8023
170.0	0.010 497	289.589	0.8574	0.008 439	286.171	0.8391	0.006 961	282.545	0.8222
180.0	0.010 879	297.583	0.8752	0.008 782	294.422	0.8575	0.007 279	291.100	0.8413
190.0	0.011 250	305.540	0.8926	0.009 114	302.597	0.8753	0.007 584	299.528	0.8597
200.0	0.011 614	313.472	0.9095	0.009 436	310.718	0.8927	0.007 878	307.864	0.8775

Temp	4.00 MPa		
120.0	0.003 736	224.863	0.6771
130.0	0.004 325	238.443	0.7111
140.0	0.004 781	249.703	0.7386
150.0	0.005 172	259.904	0.7630
160.0	0.005 522	269.492	0.7854
170.0	0.005 845	278.684	0.8063
180.0	0.006 147	287.602	0.8262
190.0	0.006 434	296.326	0.8453
200.0	0.006 708	304.906	0.8636
210.0	0.006 972	313.380	0.8813
220.0	0.007 228	321.774	0.8985
230.0	0.007 477	330.108	0.9152

TABLE A.4SI Thermodynamic Properties of Nitrogen[a]

TABLE A.4.1SI Saturated Nitrogen (SI Units)

Temp. K	Press. MPa P	Specific Volume m³/kg			Enthalpy kJ/kg			Entropy kJ/kg K		
		Sat. Liquid v_f	Evap. v_{fg}	Sat. Vapor v_g	Sat. Liquid h_f	Evap. h_{fg}	Sat. Vapor h_g	Sat. Liquid s_f	Evap. s_{fg}	Sat. Vapor s_g
63.143	0.01253	0.001 152	1.480 060	1.481 212	−150.348	215.188	64.840	2.4310	3.4076	5.8386
65	0.01742	0.001 162	1.093 173	1.094 335	−146.691	213.291	66.600	2.4845	3.2849	5.7694
70	0.03858	0.001 189	0.525 785	0.526 974	−136.569	207.727	71.158	2.6345	2.9703	5.6048
75	0.07612	0.001 221	0.280 970	0.282 191	−126.287	201.662	75.375	2.7755	2.6915	5.4670
77.347	0.101325	0.001 237	0.215 504	0.216 741	−121.433	198.645	77.212	2.8390	2.5706	5.4096
80	0.1370	0.001 256	0.162 794	0.164 050	−115.926	195.089	79.163	2.9083	2.4409	5.3492
85	0.2291	0.001 296	0.100 434	0.101 730	−105.461	187.892	82.431	3.0339	2.2122	5.2461
90	0.3608	0.001 340	0.064 950	0.066 290	−94.817	179.894	85.077	3.1535	2.0001	5.1536
95	0.5411	0.001 392	0.043 504	0.044 896	−83.895	170.877	86.982	3.2688	1.7995	5.0683
100	0.7790	0.001 452	0.029 861	0.031 313	−72.571	160.562	87.991	3.3816	1.6060	4.9876
105	1.0843	0.001 524	0.020 745	0.022 269	−60.691	148.573	87.882	3.4930	1.4150	4.9080
110	1.4673	0.001 613	0.014 402	0.016 015	−48.027	134.319	86.292	3.6054	1.2209	4.8263
115	1.9395	0.001 797	0.009 696	0.011 493	−34.157	116.701	82.544	3.7214	1.0145	4.7359
120	2.5135	0.001 904	0.006 130	0.008 034	−18.017	93.092	75.075	3.8450	0.7803	4.6253
125	3.2079	0.002 323	0.002 568	0.004 891	+6.202	50.114	56.316	4.0356	0.3989	4.4345
126.1	3.4000	0.003 184	0.000 000	0.003 184	+30.791	0.000	30.791	4.2269	0.0000	4.2269

[a] Adapted from "Thermodynamic Properties of Nitrogen Including Liquid and Vapor Phases from 63 K to 2000 K with Pressures to 10,000 Bar," R. T. Jacobsen and R. R. Stewart, *Jour. of Phys. and Chem. Ref. Data*, **2**: 757–922 (1973).

TABLE A.4.2SI Superheated Nitrogen (SI Units)

Temp. K	v m³/kg	h kJ/kg	s kJ/kg K	v m³/kg	h kJ/kg	s kJ/kg K	v m³/kg	h kJ/kg	s kJ/kg K
	0.1 MPa			0.2 MPa			0.5 MPa		
100	0.290 978	101.965	5.6944	0.142 475	100.209	5.4767	0.055 520	94.345	5.1706
125	0.367 217	128.505	5.9313	0.181 711	127.371	5.7194	0.073 422	123.824	5.4343
150	0.442 619	154.779	6.1228	0.220 014	153.962	5.9132	0.090 150	151.470	5.6361
175	0.517 576	180.935	6.2841	0.257 890	180.314	6.0760	0.106 394	178.434	5.8025
200	0.592 288	207.029	6.4234	0.295 531	206.537	6.2160	0.122 394	205.063	5.9447
225	0.666 552	233.085	6.5460	0.332 841	232.690	6.3388	0.138 173	231.459	6.0690
250	0.741 375	259.122	6.6561	0.370 418	258.796	6.4491	0.154 006	257.828	6.1801
275	0.815 563	285.144	6.7550	0.407 619	284.876	6.5485	0.169 642	284.076	6.2800
300	0.890 205	311.158	6.8457	0.445 047	310.937	6.6393	0.185 346	310.273	6.3715

Temp. K	v m³/kg	h kJ/kg	s kJ/kg K	v m³/kg	h kJ/kg	s kJ/kg K	v m³/kg	h kJ/kg	s kJ/kg K
	1.0 MPa			2.0 MPa			4.0 MPa		
125	0.033 065	117.422	5.1872	0.014 021	101.489	4.8878			
150	0.041 884	147.176	5.4042	0.019 546	137.916	5.1547	0.008 234	115.716	4.8384
175	0.050 125	175.255	5.5779	0.024 155	168.709	5.3449	0.011 186	154.851	5.0804
200	0.058 096	202.596	5.7237	0.028 436	197.609	5.4992	0.013 648	187.521	5.2553
225	0.065 875	229.526	5.8502	0.035 697	225.578	5.6309	0.015 894	217.757	5.3976
250	0.073 634	256.220	5.9632	0.036 557	253.032	5.7469	0.018 060	246.793	5.5202
275	0.081 260	282.720	6.0639	0.040 485	280.132	5.8501	0.020 133	275.056	5.6277
300	0.088 899	309.173	6.1563	0.044 398	307.014	5.9436	0.022 178	302.848	5.7248

	6.0 MPa			8.0 MPa			10.0 MPa		
150	0.004 413	87.090	4.5667	0.002 917	61.903	4.3518	0.002 388	48.687	4.2287
175	0.006 913	140.183	4.8966	0.004 863	125.536	4.7470	0.003 750	112.489	4.6239
200	0.008 772	177.447	5.0961	0.006 390	167.680	4.9726	0.005 016	158.578	4.8709
225	0.010 396	210.139	5.2410	0.007 691	202.867	5.1384	0.006 104	196.079	5.0474
250	0.011 934	240.806	5.3796	0.008 903	235.141	5.2750	0.007 112	229.861	5.1900
275	0.013 383	270.222	5.4917	0.010 034	265.676	5.3910	0.008 046	261.450	5.3103
300	0.014 800	298.907	5.5916	0.011 133	295.219	5.4942	0.008 950	291.800	5.4163

	15.0 MPa			20.0 MPa		
150	0.001 956	36.922	4.0798	0.001 781	33.637	3.9956
175	0.002 603	92.284	4.4213	0.002 186	83.453	4.3029
200	0.003 369	140.886	4.6813	0.002 685	130.291	4.5535
225	0.004 106	182.034	4.8752	0.003 208	172.307	4.7511
250	0.004 808	218.710	5.0303	0.003 728	210.456	4.9127
275	0.005 461	252.465	5.1845	0.004 223	245.640	5.0467
300	0.006 091	284.523	5.2707	0.004 704	278.942	5.1629

TABLE A.5SI Thermodynamic Properties of Saturated and Superheated Methane[a] (SI Units)

Temp. K	\bar{v} m³/kmol	\bar{h} kJ/kmol	\bar{s} kJ/kmol K	\bar{v} m³/kmol	\bar{h} kJ/kmol	\bar{s} kJ/kmol K	\bar{v} m³/kmol	\bar{h} kJ/kmol	\bar{s} kJ/kmol K
	0.05 MPa			0.10 MPa			0.50 MPa		
f 103.70	0.036 97	4 148.8	75.064						
g 103.70	16.895	12 563.9	156.261						
f 114.47				0.037 94	4 575.7	79.060			
g 114.47				8.953	12 781.6	152.697			
f 135.32							0.041 53	5 961.6	90.212
g 135.32							2.007	13 319.4	144.585
150	24.763	14 130.3	168.736	12.290	14 101.2	162.865	2.301	13 867.8	148.498
175	28.964	14 970.3	173.908	14.414	14 946.5	168.073	2.769	14 780.3	154.072
200	33.151	15 811.1	178.392	16.523	15 790.5	172.578	3.218	15 659.3	158.768
225	37.330	16 656.2	182.368	18.624	16 637.8	176.567	3.657	16 530.1	162.871
250	41.501	17 509.9	185.960	20.719	17 493.1	180.168	4.090	17 402.2	166.546
275	45.675	18 377.2	189.262	22.811	18 361.4	183.475	4.520	18 283.4	169.905
300	49.843	19 253.0	192.340	24.900	19 247.9	186.558	4.946	19 179.9	173.024
350	58.174	21 108.2	198.018	29.073	21 094.0	192.241	5.793	21 040.5	178.757
400	66.500	23 074.8	203.260	33.241	23 061.0	197.488	6.636	23 017.7	184.033

	1.0 MPa			1.5 MPa			2.0 MPa		
f 149.11	0.044 57	6 830.9	96.174						
g 149.11	1.021	13 493.9	140.859						
f 158.46				0.047 18	7 465.9	100.150			
g 158.46				0.675	13 525.2	138.401			
f 165.84							0.049 79	8 000.1	103.298
g 165.84							0.489 36	13 485.5	136.374
175	1.307	14 553.7	147.426	0.818	14 293.5	143.021	0.560 35	13 938.2	139.327
200	1.553	15 489.6	152.428	1.001	15 307.1	148.442	0.714 98	15 110.8	145.336
225	1.786	16 395.5	156.697	1.167	16 255.1	152.910	0.848 13	16 108.7	150.041
250	2.012	17 232.0	160.475	1.325	17 178.8	156.434	0.971 92	17 062.7	154.062
275	2.234	18 191.0	163.902	1.478	18 097.2	160.304	1.090 42	18 001.9	157.643
300	2.452	19 101.3	167.070	1.629	19 022.2	163.523	1.205 56	18 942.3	160.916
350	2.884	20 982.0	172.864	1.923	20 923.9	169.383	1.429 52	20 865.6	166.842
400	3.310	22 973.0	178.178	2.182	22 929.2	174.735	1.648 29	22 885.5	172.233

TABLE A.5SI (*Continued*) Saturated and Superheated Methane[a] (SI Units)

Temp. K	3.0 MPa			4.0 MPa			5.0 MPa		
	\bar{v} m³/kmol	\bar{h} kJ/kmol	\bar{s} kJ/kmol K	\bar{v} m³/kmol	\bar{h} kJ/kmol	\bar{s} kJ/kmol K	\bar{v} m³/kmol	\bar{h} kJ/kmol	\bar{s} kJ/kmol K
f 177.25	0.055 86	8 930.4	108.408						
g 177.25	0.296 41	13.238.8	132.719						
f 186.09				0.065 58	9 883.5	113.317			
g 186.09				0.185 41	12 691.7	128.411			
200	0.430 43	14 664.3	140.333	0.232 02	14 107.0	135.797	0.182 63	13 288.3	130.563
225	0.533 97	15 796.9	145.677	0.375 96	15 457.7	142.181	0.280 82	15 084.0	139.084
250	0.624 74	16 822.0	149.999	0.451 17	16 571.3	146.878	0.347 79	16 310.6	144.261
275	0.709 26	17 807.2	153.756	0.518 94	17 608.5	150.833	0.405 82	17 401.7	148.442
300	0.790 05	18 780.5	157.144	0.582 62	18 617.2	154.345	0.459 32	18 453.4	152.086
350	0.944 93	20 748.6	163.209	0.702 91	20 631.9	160.555	0.558 96	20 516.2	158.445
400	1.094 45	22 798.3	168.680	0.817 75	22 711.7	166.106	0.653 03	22 626.4	164.078
450	1.240 73	24 965.9	173.772	0.934 33	24 900.6	171.249	0.743 87	24 836.4	169.270

	6.0 MPa			8.0 MPa			10.0 MPa		
200	0.098 54	11 775.7	122.296	0.065 92	10 596.5	115.648	0.060 16	10 339.5	113.737
225	0.215 54	14 669.9	136.143	0.135 02	13 719.8	130.400	0.085 35	12 847.8	125.525
250	0.277 85	16 040.2	141.931	0.191 91	15 471.5	137.811	0.142 47	14 896.1	134.189
275	0.329 46	17 202.4	146.365	0.235 69	16 787.8	142.836	0.180 64	16 374.3	139.834
300	0.376 11	18 289.5	150.150	0.273 85	17 963.1	146.929	0.213 42	17 641.7	144.249
350	0.461 71	20 401.8	156.663	0.341 97	20 177.8	153.760	0.270 80	19 961.1	151.406
400	0.541 67	22 542.3	162.377	0.404 34	22 379.2	159.638	0.322 48	22 222.9	157.446
450	0.618 41	24 773.4	167.619	0.463 51	24 651.7	164.978	0.371 02	24 536.0	162.881
500	0.693 20	27 113.1	172.558	0.520 79	27 022.3	169.982	0.417 70	26 936.6	167.949

[a] Adapted from "The Thermophysical Properties of Methane, from 90 to 500 K at Pressures to 700 Bar," R. D. Goodwin, National Bureau of Standards Technical Note 653 (1974).

TABLE A.6E Critical Constants[a] (English Units)

Substance	Formula	Molecular Weight	Temp. R	Pressure lbf/in.2	Volume, ft^3/lb-mole
Ammonia	NH_3	17.03	729.8	1636	1.16
Argon	Ar	39.948	272	705	1.20
Bromine	Br_2	159.808	1052	1500	2.17
Carbon dioxide	CO_2	44.01	547.5	1071	1.51
Carbon monoxide	CO	28.011	240	507	1.49
Chlorine	Cl_2	70.906	751	1120	1.99
Deuterium (Normal)	D_2	4.00	69.1	241	. . .
Helium	He	4.003	9.5	33.2	0.926
Helium3	He	3.00	6.01	16.9	. . .
Hydrogen (Normal)	H_2	2.016	59.9	188.1	1.04
Krypton	Kr	83.80	376.9	798	1.48
Neon	Ne	20.183	80.1	395	0.668
Nitrogen	N_2	28.013	227.1	492	1.44
Nitrous oxide	N_2O	44.013	557.4	1054	1.54
Oxygen	O_2	31.999	278.6	736	1.25
Sulfur dioxide	SO_2	64.063	775.2	1143	1.95
Water	H_2O	18.015	1165.3	3204	0.90
Xenon	Xe	131.30	521.55	852	1.90
Benzene	C_6H_6	78.115	1012	714	4.17
n-Butane	C_4H_{10}	58.124	765.2	551	4.08
Carbon tetrachloride	CCl_4	153.82	1001.5	661	4.42
Chloroform	$CHCl_3$	119.38	965.8	794	3.85
Dichlorodifluoromethane	CCl_2F_2	120.91	692.4	582	3.49
Dichlorofluoromethane	$CHCl_2F$	102.92	813.0	749	3.16
Ethane	C_2H_6	30.020	549.8	708	2.37
Ethyl alcohol	C_2H_5OH	46.07	929.0	926	2.68
Ethylene	C_2H_4	28.054	508.3	742	1.99
n-Hexane	C_6H_{14}	86.178	914.2	439	5.89
Methane	CH_4	16.043	343.9	673	1.59
Methyl alcohol	CH_3OH	32.042	923.7	1154	1.89
Methyl chloride	CH_3Cl	50.488	749.3	968	2.29
Propane	C_3H_8	44.097	665.9	617	3.20
Propene	C_3H_6	42.081	656.9	670	2.90
Propyne	C_3H_4	40.065	722	776	. . .
Trichlorofluoromethane	CCl_3F	137.37	848.1	635	3.97

[a] K. A. Kobe and R. E. Lynn, Jr. *Chem. Rev.*, **52**: 117–236 (1953).

TABLE A.6SI Critical Constants[a] (SI Units)

Substance	Formula	Molecular Weight	Temp. K	Pressure MPa	Volume m³/kmol
Ammonia	NH_3	17.03	405.5	11.28	.0724
Argon	Ar	39.948	151	4.86	.0749
Bromine	Br_2	159.808	584	10.34	.1355
Carbon Dioxide	CO_2	44.01	304.2	7.39	.0943
Carbon Monoxide	CO	28.011	133	3.50	.0930
Chlorine	Cl_2	70.906	417	7.71	.1242
Deuterium (normal)	D_2	4.00	38.4	1.66	—
Helium	He	4.003	5.3	0.23	.0578
Helium[3]	He	3.00	3.3	0.12	—
Hydrogen (normal)	H_2	2.016	33.3	1.30	.0649
Krypton	Kr	83.80	209.4	5.50	.0924
Neon	Ne	20.183	44.5	2.73	.0417
Nitrogen	N_2	28.013	126.2	3.39	.0899
Nitrous Oxide	N_2O	44.013	309.7	7.27	.0961
Oxygen	O_2	31.999	154.8	5.08	.0780
Sulfur Dioxide	SO_2	64.063	430.7	7.88	.1217
Water	H_2O	18.015	647.3	22.09	.0568
Xenon	Xe	131.30	289.8	5.88	.1186
Benzene	C_6H_6	78.115	562	4.92	.2603
n-Butane	C_4H_{10}	58.124	425.2	3.80	.2547
Carbon Tetrachloride	CCl_4	153.82	556.4	4.56	.2759
Chloroform	$CHCl_3$	119.38	536.6	5.47	.2403
Dichlorodifluoromethane	CCl_2F_2	120.91	384.7	4.01	.2179
Dichlorofluoromethane	$CHCl_2F$	102.92	451.7	5.17	.1973
Ethane	C_2H_6	30.070	305.5	4.88	.1480
Ethyl Alcohol	C_2H_5OH	46.07	516	6.38	.1673
Ethylene	C_2H_4	28.054	282.4	5.12	.1242
n-Hexane	C_6H_{14}	86.178	507.9	3.03	.3677
Methane	CH_4	16.043	191.1	4.64	.0993
Methyl Alcohol	CH_3OH	32.042	513.2	7.95	.1180
Methyl Chloride	CH_3Cl	50.488	416.3	6.68	.1430
Propane	C_3H_8	44.097	370	4.26	.1998
Propene	C_3H_6	42.081	365	4.62	.1810
Propyne	C_3H_4	40.065	401	5.35	—
Trichlorofluoromethane	CCl_3F	137.37	471.2	4.38	.2478

[a] K. A. Kobe and R. E. Lynn, Jr., *Chem. Rev.*, **52**: 117–236 (1953).

TABLE A.7E Properties of Various Solids and Liquids[a] (English Units)

Solid	C_p Btu/lbm R	ρ lbm/ft^3	Liquid	C_p Btu/lbm R	ρ lbm/ft^3
Aluminum	0.215	170	Ammonia	1.146	38
Copper	0.092	555	Ethanol	0.587	49
Granite	0.243	170	Freon-12	0.233	82
Graphite	0.170	155	Mercury	0.033	847
Iron	0.107	490	Methanol	0.609	49
Lead	0.030	705	Oil (light)	0.430	57
Rubber (soft)	0.439	70	Water	1.000	62
Silver	0.056	655			
Tin	0.052	360			
Wood (most)	0.420	22–45			

[a] Values at 80 F.

TABLE A.7SI Properties of Various Solids and Liquids[a] (SI Units)

Solid	C_p kJ/kg K	ρ kg/m^3	Liquid	C_p kJ/kg K	ρ kg/m^3
Aluminum	0.900	2700	Ammonia	4.800	602
Copper	0.386	8900	Ethanol	2.456	783
Granite	1.017	2700	Freon-12	0.977	1310
Graphite	0.711	2500	Mercury	0.139	13 560
Iron	0.450	7840	Methanol	2.550	787
Lead	0.128	11 310	Oil (light)	1.800	910
Rubber (soft)	1.840	1100	Water	4.184	997
Silver	0.235	10 470			
Tin	0.217	5730			
Wood (most)	1.760	350–700			

[a] Values at 25°C.

TABLE A.8E Properties of Various Ideal Gasesa (English Units)

Gas	Chemical Formula	Molecular Weight	R ft-lbf/ lbm R	C_{po} Btu/ lbm R	C_{vo} Btu/ lbm R	k
Air	—	28.97	53.34	0.240	0.171	1.400
Argon	Ar	39.948	38.68	0.1253	0.0756	1.667
Butane	C_4H_{10}	58.124	26.58	0.415	0.381	1.09
Carbon Dioxide	CO_2	44.01	35.10	0.203	0.158	1.285
Monoxide	CO	28.01	55.16	0.249	0.178	1.399
Ethane	C_2H_6	30.07	51.38	0.427	0.361	1.183
Ethylene	C_2H_4	28.054	55.07	0.411	0.340	1.208
Helium	He	4.003	386.0	1.25	0.753	1.667
Hydrogen	H_2	2.016	766.4	3.43	2.44	1.404
Methane	CH_4	16.04	96.35	0.532	0.403	1.32
Neon	Ne	20.183	76.55	0.246	0.1477	1.667
Nitrogen	N_2	28.013	55.15	0.248	0.177	1.400
Octane	C_8H_{18}	114.23	13.53	0.409	0.392	1.044
Oxygen	O_2	31.999	48.28	0.219	0.157	1.395
Propane	C_3H_8	44.097	35.04	0.407	0.362	1.124
Steam	H_2O	18.015	85.76	0.445	0.335	1.329

a (C_{po}, C_{vo}, and k are at 80 F.)

TABLE A.8SI Properties of Various Ideal Gasesa (SI Units)

Gas	Chemical Formula	Molecular Weight	R $\dfrac{kJ}{kg\,K}$	C_{po} $\dfrac{kJ}{kg\,K}$	C_{vo} $\dfrac{kJ}{kg\,K}$	k
Air	—	28.97	0.287 00	1.0035	0.7165	1.400
Argon	Ar	39.948	0.208 13	0.5203	0.3122	1.667
Butane	C_4H_{10}	58.124	0.143 04	1.7164	1.5734	1.091
Carbon Dioxide	CO_2	44.01	0.188 92	0.8418	0.6529	1.289
Carbon Monoxide	CO	28.01	0.296 83	1.0413	0.7445	1.400
Ethane	C_2H_6	30.07	0.276 50	1.7662	1.4897	1.186
Ethylene	C_2H_4	28.054	0.296 37	1.5482	1.2518	1.237
Helium	He	4.003	2.077 03	5.1926	3.1156	1.667
Hydrogen	H_2	2.016	4.124 18	14.2091	10.0849	1.409
Methane	CH_4	16.04	0.518 35	2.2537	1.7354	1.299
Neon	Ne	20.183	0.411 95	1.0299	0.6179	1.667
Nitrogen	N_2	28.013	0.296 80	1.0416	0.7448	1.400
Octane	C_8H_{18}	114.23	0.072 79	1.7113	1.6385	1.044
Oxygen	O_2	31.999	0.259 83	0.9216	0.6618	1.393
Propane	C_3H_8	44.097	0.188 55	1.6794	1.4909	1.126
Steam	H_2O	18.015	0.461 52	1.8723	1.4108	1.327

a (C_{po}, C_{vo}, and k are at 300 K.)

TABLE A.9E Constant Pressure Specific Heats of Various Ideal Gases[a] (English Units)

$$\bar{C}_{po} = \frac{Btu}{lb\ mole\ R} \qquad \theta = \frac{T(Rankine)}{180}$$

Gas		Range R	Max Error %
N_2	$\bar{C}_{po} = 9.3355 - 122.56\theta^{-1.5} + 256.38\theta^{-2} - 196.08\theta^{-3}$	540–6300	0.43
O_2	$\bar{C}_{po} = 8.9465 + 4.8044 \times 10^{-3}\theta^{1.5} - 42.679\theta^{-1.5} + 56.615\theta^{-2}$	540–6300	0.30
H_2	$\bar{C}_{po} = 13.505 - 167.96\theta^{-0.75} + 278.44\theta^{-1} - 134.01\theta^{-1.5}$	540–6300	0.60
CO	$\bar{C}_{po} = 16.526 - 0.16841\theta^{0.75} - 47.985\theta^{-0.5} + 42.246\theta^{-0.75}$	540–6300	0.42
OH	$\bar{C}_{po} = 19.490 - 14.185\theta^{0.25} + 4.1418\theta^{0.75} - 1.0196\theta$	540–6300	0.43
NO	$\bar{C}_{po} = 14.169 - 0.40861\theta^{0.5} - 16877\theta^{-0.5} + 17.899\theta^{-1.5}$	540–6300	0.34
H_2O	$\bar{C} = 34.190 - 43.868\theta^{0.25} + 19.778\theta^{0.5} - 0.88407\theta$	540–6300	0.43
CO_2	$\bar{C}_{po} = -0.89286 + 7.2967\theta^{0.5} - 0.98074\theta + 5.7835 \times 10^{-3}\theta^2$	540–6300	0.19
NO_2	$\bar{C}_{po} = 11.005 + 51.650\theta^{-0.5} - 86.916\theta^{-0.75} + 55.580\theta^{-2}$	540–6300	0.26
CH_4	$\bar{C}_{po} = -160.82 + 105.10\theta^{0.25} - 5.9452\theta^{0.75} + 77.408\theta^{-0.5}$	540–3600	0.15
C_2H_4	$\bar{C}_{po} = -22.800 + 29.433\theta^{0.5} - 8.5185\theta^{0.75} + 43.683\theta^{-3}$	540–3600	0.07
C_2H_6	$\bar{C}_{po} = 1.648 + 4.124\theta - 0.153\theta^2 + 1.74 \times 10^{-3}\theta^3$	540–2700	0.83
C_3H_8	$\bar{C}_{po} = -0.966 + 7.279\theta - 0.3755\theta^2 + 7.58 \times 10^{-3}\theta^3$	540–2700	0.40
C_4H_{10}	$\bar{C}_{po} = 0.945 + 8.873\theta - 0.438\theta^2 + 8.36 \times 10^{-3}\theta^3$	540–2700	0.54

TABLE A.9SI Constant-Pressure Specific Heats of Various Ideal Gases[a] (SI Units)

$$C_{po} = kJ/kmol\ K \qquad \theta = T(Kelvin)/100$$

Gas		Range K	Max. Error %
N_2	$\bar{C}_{po} = 39.060 - 512.79\theta^{-1.5} + 1072.7\theta^{-2} - 820.40\theta^{-3}$	300–3500	0.43
O_2	$\bar{C}_{po} = 37.432 + 0.020\ 102\theta^{1.5} - 178.57\theta^{-1.5} + 236.88\theta^{-2}$	300–3500	0.30
H_2	$\bar{C}_{po} = 56.505 - 702.74\theta^{-0.75} + 1165.0\theta^{-1} - 560.70\theta^{-1.5}$	300–3500	0.60
CO	$\bar{C}_{po} = 69.145 - 0.704\ 63\theta^{0.75} - 200.77\theta^{-0.5} + 176.76\theta^{-0.75}$	300–3500	0.42
OH	$\bar{C}_{po} = 81.546 - 59.350\theta^{0.25} + 17.329\theta^{0.75} - 4.2660\theta$	300–3500	0.43
NO	$\bar{C}_{po} = 59.283 - 1.7096\theta^{0.5} - 70.613\theta^{-0.5} + 74.889\theta^{-1.5}$	300–3500	0.34
H_2O	$\bar{C}_{po} = 143.05 - 183.54\theta^{0.25} + 82.751\theta^{0.5} - 3.6989\theta$	300–3500	0.43
CO_2	$\bar{C}_{po} = -3.7357 + 30.529\theta^{0.5} - 4.1034\theta + 0.024\ 198\theta^2$	300–3500	0.19
NO_2	$\bar{C}_{po} = 46.045 + 216.10\theta^{-0.5} - 363.66\theta^{-0.75} + 232.550\theta^{-2}$	300–3500	0.26
CH_4	$\bar{C}_{po} = -672.87 + 439.74\theta^{0.25} - 24.875\theta^{0.75} + 323.88\theta^{-0.5}$	300–2000	0.15
C_2H_4	$\bar{C}_{po} = -95.395 + 123.15\theta^{0.5} - 35.641\theta^{0.75} + 182.77\theta^{-3}$	300–2000	0.07
C_2H_6	$\bar{C}_{po} = 6.895 + 17.26\theta - 0.6402\theta^2 + 0.007\ 280\theta^3$	300–1500	0.83
C_3H_8	$\bar{C}_{po} = -4.042 + 30.46\theta - 1.571\theta^2 + 0.031\ 71\theta^3$	300–1500	0.40
C_4H_{10}	$\bar{C}_{po} = 3.954 + 37.12\theta - 1.833\theta^2 + 0.034\ 98\theta^3$	300–1500	0.54

[a] From T. C. Scott and R. E. Sonntag, Univ. of Michigan, unpublished (1971), except C_2H_6, C_3H_8, C_4H_{10} from K. A. Kobe, Petroleum Refiner 28 No. 2, 113 (1949).

TABLE A.10E Ideal Gas Integrals for Air[a] (English Units)

T, R	h, Btu/lbm	P_r	u, Btu/lbm	v_r	$s°$ Btu/lbm R
200	47.67	0.04320	33.96	1714.9	0.363 03
220	52.46	0.06026	37.38	1352.5	0.385 84
240	57.25	0.08165	40.80	1088.8	0.406 66
260	62.03	0.10797	44.21	892.0	0.425 82
280	66.82	0.13986	47.63	741.6	0.443 56
300	71.61	0.17795	51.04	624.5	0.460 07
320	76.40	0.22290	54.46	531.8	0.475 50
340	81.18	0.27545	57.87	457.2	0.490 02
360	85.97	0.3363	61.29	396.6	0.503 69
380	90.75	0.4061	64.70	346.6	0.516 63
400	95.53	0.4858	68.11	305.0	0.528 90
420	100.32	0.5760	71.52	270.1	0.540 58
440	105.11	0.6776	74.93	240.6	0.551 72
460	109.90	0.7913	78.36	215.33	0.562 35
480	114.69	0.9182	81.77	193.65	0.572 55
500	119.48	1.0590	85.20	174.90	0.582 33
520	124.27	1.2147	88.62	158.58	0.591 73
540	129.06	1.3860	92.04	144.32	0.600 78
560	133.86	1.5742	95.47	131.78	0.609 50
580	138.66	1.7800	98.90	120.70	0.617 93
600	143.47	2.005	102.34	110.88	0.626 07
620	148.28	2.249	105.78	102.12	0.633 95
640	153.09	2.514	109.21	94.30	0.641 59
660	157.92	2.801	112.67	87.27	0.649 02
680	162.73	3.111	116.12	80.96	0.656 21
700	167.56	3.446	119.58	75.25	0.663 21
720	172.39	3.806	123.04	70.07	0.670 02
740	177.23	4.193	126.51	65.38	0.676 65
760	182.08	4.607	129.99	61.10	0.683 12
780	186.94	5.051	133.47	57.20	0.689 42
800	191.81	5.526	136.97	53.63	0.695 58
820	196.69	6.033	140.47	50.35	0.701 60
840	201.56	6.573	143.98	47.34	0.707 47
860	206.46	7.149	147.50	44.57	0.713 23
880	211.35	7.761	151.02	42.01	0.718 86
900	216.26	8.411	154.57	39.64	0.724 38
920	221.18	9.102	158.12	37.44	0.729 79
940	226.11	9.834	161.68	35.41	0.735 09
960	231.06	10.610	165.26	33.52	0.740 30
980	236.02	11.430	168.83	31.76	0.745 40
1000	240.98	12.298	172.43	30.12	0.750 42
1020	245.97	13.215	176.04	28.59	0.755 36
1040	250.95	14.182	179.66	27.17	0.760 19
1060	255.96	15.203	183.29	25.82	0.764 96
1080	260.97	16.278	186.93	24.58	0.769 64

TABLE A.10E (*Continued*) **Ideal Gas Integrals for Aira (English Units)**

T, R	h, Btu/lbm	P_r	u, Btu/lbm	v_r	$s°$ Btu/lbm R
1100	265.99	17.413	190.58	23.40	0.774 26
1120	271.03	18.604	194.25	22.30	0.778 80
1140	276.08	19.858	197.94	21.27	0.783 26
1160	281.14	21.18	201.63	20.293	0.787 67
1180	286.21	22.56	205.33	19.377	0.792 01
1200	291.30	24.01	209.05	18.514	0.796 28
1220	296.41	25.53	212.78	17.700	0.800 50
1240	301.52	27.13	216.53	16.932	0.804 66
1260	306.65	28.80	220.28	16.205	0.808 76
1280	311.79	30.55	224.05	15.518	0.812 80
1300	316.94	32.39	227.83	14.868	0.816 80
1320	322.11	34.31	231.63	14.253	0.820 75
1340	327.29	36.31	235.43	13.670	0.824 64
1360	332.48	38.41	239.25	13.118	0.828 48
1380	337.68	40.59	243.08	12.593	0.832 29
1400	342.90	42.88	246.93	12.095	0.836 04
1420	348.14	45.26	250.79	11.622	0.839 75
1440	353.37	47.75	254.66	11.172	0.843 41
1460	358.63	50.34	258.54	10.743	0.847 04
1480	363.89	53.04	262.44	10.336	0.850 62
1500	369.17	55.86	266.34	9.948	0.854 16
1520	374.47	58.78	270.26	9.578	0.857 67
1540	379.77	61.83	274.20	9.226	0.861 13
1560	385.08	65.00	278.13	8.890	0.864 56
1580	390.40	68.30	282.09	8.569	0.867 94
1600	395.74	71.73	286.06	8.263	0.871 30
1620	401.09	75.29	290.04	7.971	0.874 62
1640	406.45	78.99	294.03	7.691	0.877 91
1660	411.82	82.83	298.02	7.424	0.881 16
1680	417.20	86.82	302.04	7.168	0.884 39
1700	422.59	90.95	306.06	6.924	0.887 58
1720	428.00	95.24	310.09	6.690	0.890 74
1740	433.41	99.69	314.13	6.465	0.893 87
1760	438.83	104.30	318.18	6.251	0.896 97
1780	444.26	109.08	322.24	6.045	0.900 03
1800	449.71	114.03	326.32	5.847	0.903 08
1820	455.17	119.16	330.40	5.658	0.906 09
1840	460.63	124.47	334.50	5.476	0.909 08
1860	466.12	129.95	338.61	5.302	0.912 03
1880	471.60	135.64	342.73	5.134	0.914 97
1900	477.09	141.51	346.85	4.974	0.917 88
1920	482.60	147.59	350.98	4.819	0.920 76
1940	488.12	153.87	355.12	4.670	0.923 62
1960	493.64	160.37	359.28	4.527	0.926 45
1980	499.17	167.07	363.43	4.390	0.929 26

TABLE A.10E (*Continued*) **Ideal Gas Integrals for Air**[a] **(English Units)**

T, R	h, Btu/lbm	P_r	u, Btu/lbm	v_r	$s°$ Btu/lbm R
2000	504.71	174.00	367.61	4.258	0.932 05
2020	510.26	181.16	371.79	4.130	0.934 81
2040	515.82	188.54	375.98	4.008	0.937 56
2060	521.39	196.16	380.18	3.890	0.940 26
2080	526.97	204.02	384.39	3.777	0.942 96
2100	532.55	212.1	388.60	3.667	0.945 64
2120	538.15	220.5	392.83	3.561	0.948 29
2140	543.74	229.1	397.05	3.460	0.950 92
2160	549.35	238.0	401.29	3.362	0.953 52
2180	554.97	247.2	405.53	3.267	0.956 11
2200	560.59	256.6	409.78	3.176	0.958 68
2220	566.23	266.3	414.05	3.088	0.961 23
2240	571.86	276.3	418.31	3.003	0.963 76
2260	577.51	286.6	422.59	2.921	0.966 26
2280	583.16	297.2	426.87	2.841	0.968 76
2300	588.82	308.1	431.16	2.765	0.971 23
2320	594.49	319.4	435.46	2.691	0.973 69
2340	600.16	330.9	439.76	2.619	0.976 11
2360	605.84	342.8	444.07	2.550	0.978 53
2380	611.53	355.0	448.38	2.483	0.980 92
2400	617.22	367.6	452.70	2.419	0.983 31
2420	622.92	380.5	457.02	2.356	0.985 67
2440	628.62	393.7	461.36	2.296	0.988 02
2460	634.34	407.3	465.70	2.237	0.990 35
2480	640.05	421.3	470.05	2.180	0.992 66
2500	645.78	435.7	474.40	2.125	0.994 97
2550	660.12	473.3	485.31	1.9956	1.000 64
2600	674.49	513.5	496.26	1.8756	1.006 23
2650	688.90	556.3	507.25	1.7646	1.011 72
2700	703.35	601.9	518.26	1.6617	1.017 12
2750	717.83	650.4	529.31	1.5662	1.022 44
2800	732.33	702.0	540.40	1.4775	1.027 67
2850	746.88	756.7	551.52	1.3951	1.032 82
2900	761.45	814.8	562.66	1.3184	1.037 88
2950	776.05	876.4	573.84	1.2469	1.042 88
3000	790.68	941.4	585.04	1.1803	1.047 79
3050	805.34	1010.5	596.28	1.1181	1.052 64
3100	820.03	1083.4	607.53	1.0600	1.057 41
3150	834.75	1160.5	618.82	1.0056	1.062 12
3200	849.48	1241.7	630.12	.9546	1.066 76

[a] Abridged from Table 1 in *Gas Tables*, by Joseph H. Keenan and Joseph Kaye. Copyright 1948, by Joseph H. Keenan and Joseph Kaye. Published by John Wiley & Sons, Inc., New York.

TABLE A.10SI Ideal Gas Integrals for Air[a] (SI Units)

T, K	h kJ/kg	P_r	u, kJ/kg	v_r	$s°$ kJ/kg K
100	99.76	0.029 90	71.06	2230	1.4143
110	109.77	0.041 71	78.20	1758.4	1.5098
120	119.79	0.056 52	85.34	1415.7	1.5971
130	129.81	0.074 74	92.51	1159.8	1.6773
140	139.84	0.096 81	99.67	964.2	1.7515
150	149.86	0.123 18	106.81	812.0	1.8206
160	159.87	0.154 31	113.95	691.4	1.8853
170	169.89	0.190 68	121.11	594.5	1.9461
180	179.92	0.232 79	128.28	515.6	2.0033
190	189.94	0.281 14	135.40	450.6	2.0575
200	199.96	0.3363	142.56	396.6	2.1088
210	209.97	0.3987	149.70	351.2	2.1577
220	219.99	0.4690	156.84	312.8	2.2043
230	230.01	0.5477	163.98	280.0	2.2489
240	240.03	0.6355	171.15	251.8	2.2915
250	250.05	0.7329	178.29	227.45	2.3325
260	260.09	0.8405	185.45	206.26	2.3717
270	270.12	0.9590	192.59	187.74	2.4096
280	280.14	1.0889	199.78	171.45	2.4461
290	290.17	1.2311	206.92	157.07	2.4813
300	300.19	1.3860	214.09	144.32	2.5153
310	310.24	1.5546	221.27	132.96	2.5483
320	320.29	1.7375	228.45	122.81	2.5802
330	330.34	1.9352	235.65	113.70	2.6111
340	340.43	2.149	242.86	105.51	2.6412
350	350.48	2.379	250.05	98.11	2.6704
360	360.58	2.626	257.23	91.40	2.6987
370	370.67	2.892	264.47	85.31	2.7264
380	380.77	3.176	271.72	79.77	2.7534
390	390.88	3.481	278.96	74.71	2.7796
400	400.98	3.806	286.19	70.07	2.8052
410	411.12	4.153	293.45	65.83	2.8302
420	421.26	4.522	300.73	61.93	2.8547
430	431.43	4.915	308.03	58.34	2.8786
440	441.61	5.332	315.34	55.02	2.9020
450	451.83	5.775	322.66	51.96	2.9249
460	462.01	6.245	329.99	49.11	2.9473
470	472.25	6.742	337.34	46.48	2.9693
480	482.48	7.268	344.74	44.04	2.9909
490	492.74	7.824	352.11	41.76	3.0120
500	503.02	8.411	359.53	39.64	3.0328
510	513.32	9.031	366.97	37.65	3.0532
520	523.63	9.684	374.39	35.80	3.0733
530	533.98	10.372	381.88	34.07	3.0930
540	544.35	11.097	389.40	32.45	3.1124

TABLE A.10SI (*Continued*) Ideal Gas Integrals for Air[a] (SI Units)

T, K	h, kJ/kg	P_r	u, kJ/kg	v_r	$s°$ kJ/kg K
550	554.75	11.858	396.89	30.92	3.1314
560	565.17	12.659	404.44	29.50	3.1502
570	575.57	13.500	411.98	28.15	3.1686
580	586.04	14.382	419.56	26.89	3.1868
590	596.53	15.309	427.17	25.70	3.2047
600	607.02	16.278	434.80	24.58	3.2223
610	617.53	17.297	442.43	23.51	3.2397
620	628.07	18.360	450.13	22.52	3.2569
630	638.65	19.475	457.83	21.57	3.2738
640	649.21	20.64	465.55	20.674	3.2905
650	659.84	21.86	473.32	19.828	3.3069
660	670.47	23.13	481.06	19.026	3.3232
670	681.15	24.46	488.88	18.266	3.3392
680	691.82	25.85	496.65	17.543	3.3551
690	702.52	27.29	504.51	16.857	3.3707
700	713.27	28.80	512.37	16.205	3.3861
710	724.01	30.38	520.26	15.585	3.4014
720	734.20	31.92	527.72	15.027	3.4156
730	745.62	33.72	536.12	14.434	3.4314
740	756.44	35.50	544.05	13.900	3.4461
750	767.30	37.35	552.05	13.391	3.4607
760	778.21	39.27	560.08	12.905	3.4751
770	789.10	41.27	568.10	12.440	3.4894
780	800.03	43.35	576.15	11.998	3.5035
790	810.98	45.51	584.22	11.575	3.5174
800	821.94	47.75	592.34	11.172	3.5312
810	832.96	50.08	600.46	10.785	3.5449
820	843.97	52.49	608.62	10.416	3.5584
830	855.01	55.00	616.79	10.062	3.5718
840	866.09	57.60	624.97	9.724	3.5850
850	877.16	60.29	633.21	9.400	3.5981
860	888.28	63.09	641.44	9.090	3.6111
870	899.42	65.98	649.70	8.792	3.6240
880	910.56	68.98	658.00	8.507	3.6367
890	921.75	72.08	666.31	8.233	3.6493
900	932.94	75.29	674.63	7.971	3.6619
910	944.15	78.61	682.98	7.718	3.6743
920	955.38	82.05	691.33	7.476	3.6865
930	966.64	85.60	699.73	7.244	3.6987
940	977.92	89.28	708.13	7.020	3.7108
950	989.22	93.08	716.57	6.805	3.7227
960	1000.53	97.00	725.01	6.599	3.7346
970	1011.88	101.06	733.48	6.400	3.7463
980	1023.25	105.24	741.99	6.209	3.7580
990	1034.63	109.57	750.48	6.025	3.7695

TABLE A.10SI (*Continued*) Ideal Gas Integrals for Air[a] (SI Units)

T, K	kJ/kg	P_r	u, kJ/kg	v_r	$s°$ kJ/kg K
1000	1046.03	114.03	759.02	5.847	3.7810
1020	1068.89	123.12	775.67	5.521	3.8030
1040	1091.85	133.34	793.35	5.201	3.8259
1060	1114.85	143.91	810.61	4.911	3.8478
1080	1137.93	155.15	827.94	4.641	3.8694
1100	1161.07	167.07	845.34	4.390	3.8906
1120	1184.28	179.71	862.85	4.156	3.9116
1140	1207.54	193.07	880.37	3.937	3.9322
1160	1230.90	207.24	897.98	3.732	3.9525
1180	1254.34	222.2	915.68	3.541	3.9725
1200	1277.79	238.0	933.40	3.362	3.9922
1220	1301.33	254.7	951.19	3.194	4.0117
1240	1324.89	272.3	969.01	3.037	4.0308
1260	1348.55	290.8	986.92	2.889	4.0497
1280	1372.25	310.4	1004.88	2.750	4.0684
1300	1395.97	330.9	1022.88	2.619	4.0868
1320	1419.77	352.5	1040.93	2.497	4.1049
1340	1443.61	375.3	1059.03	2.381	4.1229
1360	1467.50	399.1	1077.17	2.272	4.1406
1380	1491.43	424.2	1095.36	2.169	4.1580
1400	1515.41	450.5	1113.62	2.072	4.1753
1420	1539.44	478.0	1131.90	1.9808	4.1923
1440	1563.49	506.9	1150.23	1.8942	4.2092
1460	1587.61	537.1	1168.61	1.8124	4.2258
1480	1611.80	568.8	1187.03	1.7350	4.2422
1500	1635.99	601.9	1205.47	1.6617	4.2585
1520	1660.23	636.5	1223.96	1.5922	4.2746
1540	1684.49	672.8	1242.53	1.5264	4.2904
1560	1708.82	710.5	1261.11	1.4639	4.3061
1580	1733.17	750.0	1279.72	1.4047	4.3216
1600	1757.55	791.2	1298.35	1.3485	4.3369
1620	1782.02	834.1	1317.07	1.2950	4.3521
1640	1806.44	878.9	1335.78	1.2442	4.3671
1660	1830.96	925.6	1354.55	1.1959	4.3819
1680	1855.47	974.2	1373.34	1.1499	4.3967
1700	1880.06	1024.8	1392.16	1.1061	4.4112
1720	1904.65	1077.4	1411.01	1.0645	4.4256
1740	1929.30	1132.2	1429.91	1.0248	4.4398
1760	1953.96	1189.3	1448.86	0.9869	4.4539
1780	1978.64	1248.4	1467.77	0.9507	4.4679
1800	2003.34	1310.0	1486.76	0.9162	4.4817

[a] Adapted from Table 1 in *Gas Tables*, by Joseph H. Keenan and Joseph Kaye. Copyright 1948, by Joseph H. Keenan and Joseph Kaye. Published by John Wiley & Sons, Inc., New York.

TABLE A.11E[a] **Enthalpy of Formation at 77 F, Ideal-Gas Enthalpy, and Absolute Entropy at One Atmosphere Pressure**

Temp. R	Nitrogen, Diatomic (N_2) (9/30/65) $(h_f^\circ)_{537} = 0$ Btu/lb mole $M = 28.013$		Nitrogen, Monatomic (N) (3/31/61) $(\bar{h}^\circ_f)_{537} = 203\ 337$ Btu/lb mole $M = 14.007$	
	$(\bar{h}^\circ - \bar{h}^\circ_{537})$ Btu/lb mole	\bar{s}° Btu/lb mole R	$(\bar{h}^\circ - \bar{h}^\circ_{537})$ Btu/lb mole	\bar{s}° Btu/lb mole R
0	− 3 730	0	− 2 666	0
180	− 2 483	38.170	− 1 771	31.187
360	− 1 229	42.992	− 878	34.631
537	0	45.770	0	36.614
540	23	45.813	16	36.645
720	1 278	47.818	911	38.074
900	2 543	49.386	1 805	39.183
1 080	3 825	50.685	2 700	40.089
1 260	5 135	51.806	3 593	40.855
1 440	6 473	52.798	4 487	41.518
1 620	7 839	53.692	5 382	42.103
1 800	9 232	54.507	6 277	42.627
1 980	10 651	55.258	7 171	43.100
2 160	12 092	55.955	8 066	43.532
2 340	13 552	56.604	8 959	43.930
2 520	15 030	57.212	9 853	44.298
2 700	16 522	57.784	10 748	44.641
2 880	18 027	58.324	11 642	44.962
3 060	19 544	58.835	12 537	45.263
3 240	21 073	59.320	13 430	45.547
3 420	22 608	59.782	14 324	45.815
3 600	24 152	60.222	15 219	46.070
3 780	25 704	60.642	16 114	46.313
3 960	27 263	61.045	17 008	46.544
4 140	28 827	61.431	17 903	46.765
4 320	30 395	61.802	18 799	46.977
4 500	31 970	62.159	19 694	47.180

TABLE A.11Ea (*Continued*) **Enthalpy of Formation at 77 F, Ideal-Gas Enthalpy, and Absolute Entropy at One Atmosphere Pressure**

Temp. R	Nitrogen, Diatomic (N$_2$) (9/30/65) $(h_f^\circ)_{537} = 0$ Btu/lb mole $M = 28.013$		Nitrogen, Monatomic (N) (3/31/61) $(\bar{h}^\circ{}_f)_{537} = 203\,337$ Btu/lb mole $M = 14.007$	
	$(\bar{h}^\circ - \bar{h}^\circ_{537})$ Btu/lb mole	\bar{s}° Btu/lb mole R	$(\bar{h}^\circ - \bar{h}^\circ_{537})$ Btu/lb mole	\bar{s}° Btu/lb mole R
4 680	33 548	62.503	20 590	47.375
4 860	35 131	62.835	21 488	47.563
5 040	36 716	63.155	22 387	47.745
5 220	38 304	63.465	23 285	47.920
5 400	39 897	63.765	24 187	48.090
5 760	43 090	64.337	25 994	48.414
6 120	46 294	64.877	27 812	48.720
6 480	49 509	65.387	29 644	49.011
6 840	52 731	65.871	31 491	49.288
7 200	55 960	66.331	33 356	49.554
7 560	59 198	66.770	35 244	49.810
7 920	62 442	67.189	37 157	50.057
8 280	65 693	67.591	39 098	50.297
8 640	68 951	67.976	41 069	50.530
9 000	72 214	68.346	43 070	50.757
9 360	75 483	68.702	45 106	50.978
9 720	78 759	69.045	47 178	51.195
10 180	82 042	69.377	49 284	51.408
10 540	85 331	69.698	51 426	51.617
10 800	88 627	70.008	53 604	51.822

a The thermochemical data in Table A.11 are calculated from the JANAF Thermochemical Tables, Thermal Research Laboratory, The Dow Chemical Company, Midland, Michigan. The date each table was issued is indicated.

TABLE A.11E[a] (*Continued*) Enthalpy of Formation at 77 F, Ideal-Gas Enthalpy, and Absolute Entropy at One Atmosphere Pressure

Temp. R	Oxygen, Diatomic (O_2) (9/30/65) $(\bar{h}_f^\circ)_{537} = 0$ Btu/lb mole $M = 31.999$		Oxygen, Monatomic (O) (6/30/62) $(\bar{h}_f^\circ)_{537} = 107\ 206$ Btu/lb mole $M = 16.00$	
	$(\bar{h}^\circ - \bar{h}_{537}^\circ)$ Btu/lb mole	\bar{s}° Btu/lb mole R	$(\bar{h}^\circ - \bar{h}_{537}^\circ)$ Btu/lb mole	\bar{s}° Btu/lb mole R
0	− 3 735	0	− 2 894	0
180	− 2 486	41.395	− 1 944	32.466
360	− 1 233	46.218	− 941	36.340
537	0	49.004	0	38.468
540	23	49.047	18	38.501
720	1 303	51.091	950	39.991
900	2 619	52.722	1 868	41 131
1 080	3 978	54.098	2 779	42.054
1 260	5 378	55.297	3 686	42.831
1 440	6 815	56.361	4 590	43.501
1 620	8 280	57.320	5 494	44.092
1 800	9 769	58.192	6 394	44.619
1 980	11 279	58.991	7 292	45.095
2 160	12 805	59.729	8 192	45.529
2 340	14 348	60.415	9 088	45.928
2 520	15 903	61.055	9 986	46.298
2 700	17 471	61.656	10 883	46.642
2 880	19 049	62.222	11 779	46.963
3 060	20 637	62.757	12 676	47.265
3 240	22 237	63.265	13 572	47.550
3 420	23 848	63.749	14 468	47.819
3 600	25 468	64.210	15 365	48.074
3 780	27 097	64.652	16 261	48.317
3 960	28 739	65.076	17 158	48.549
4 140	30 388	65.483	18 052	48.770
4 320	32 047	65.876	18 949	48.982
4 500	33 718	66.254	19 847	49.185

TABLE A.11E[a] (*Continued*) **Enthalpy of Formation at 77 F, Ideal-Gas Enthalpy, and Absolute Entropy at One Atmosphere Pressure**

	Oxygen, Diatomic (O_2) (9/30/65)		Oxygen, Monatomic (O) (6/30/62)	
	$(\bar{h}_f^\circ)_{537} = 0$ Btu/lb mole $M = 31.999$		$(\bar{h}_f^\circ)_{537} = 107\,206$ Btu/lb mole $M = 16.00$	
Temp. R	$(\bar{h}^\circ - \bar{h}_{537}^\circ)$ Btu/lb mole	\bar{s}° Btu/lb mole R	$(\bar{h}^\circ - \bar{h}_{537}^\circ)$ Btu/lb mole	\bar{s}° Btu/lb mole R
4 680	35 395	66.620	20 743	49.381
4 860	37 084	66.974	21 641	49.569
5 040	38 781	67.317	22 540	49.751
5 220	40 487	67.650	23 440	49.926
5 400	42 203	67.973	24 340	50.096
5 760	45 657	68.592	26 143	50 419
6 120	49 144	69.179	27 952	50.724
6 480	52 657	69.737	29 767	51.012
6 840	56 198	70 269	31 588	51.285
7 200	59 762	70.776	33 417	51.546
7 560	63 347	71.262	35 255	51.795
7 920	66 953	71.728	37 100	52.033
8 280	70 574	72.176	38 954	52.262
8 640	74 212	72.606	40 817	52.482
9 000	77 863	73.019	42 687	52.695
9 360	81 526	73.418	44 568	52.899
9 720	85 198	73.803	46 456	53.097
10 180	88 879	74.175	48 353	53.289
10 540	92 567	74.535	50 258	53.475
10 800	96 262	74.883	52 171	53.655

TABLE A.11E[a] (*Continued*) **Enthalpy of Formation at 77F, Ideal-Gas Enthalpy, and Absolute Entropy at One Atmosphere Pressure**

Temp. R	Carbon Dioxide (CO_2) (9/30/65) $(\bar{h}_f^\circ)_{537} = -169\,297$ Btu/lb mole $M = 44.01$		Carbon Monoxide (CO) (9/30/65) $(\bar{h}_f^\circ)_{537} = -47\,551$ Btu/lb mole $M = 28.01$	
	$(\bar{h}^\circ - \bar{h}_{537}^\circ)$ Btu/lb mole	\bar{s}° Btu/lb mole R	$(\bar{h}^\circ - \bar{h}_{537}^\circ)$ Btu/lb mole	\bar{s}° Btu/lb mole R
0	−4 028	0	−3 730	0
180	−2 777	42.758	−2 483	39.613
360	−1 469	47.769	−1 229	44.435
537	0	51.072	0	47.214
540	29	51.127	23	47.257
720	1 724	53.830	1 280	49.265
900	3 577	56.122	2 551	50.841
1 080	5 557	58.126	3 847	52.152
1 260	7 641	59.910	5 171	53.287
1 440	9 815	61.522	6 529	54.293
1 620	12 064	62.992	7 915	55.200
1 800	14 371	64.344	9 329	56.028
1 980	16 733	65.594	10 769	56.790
2 160	19 138	66.756	12 229	57.496
2 340	21 578	67.841	13 709	58.154
2 520	24 052	68.859	15 203	58.769
2 700	26 550	69.817	16 713	59.348
2 880	29 074	70.722	18 234	59.893
3 060	31 617	71.578	19 764	60.409
3 240	34 177	72.391	21 305	60.898
3 420	36 752	73.165	22 855	61.363
3 600	39 343	73.903	24 410	61.807
3 780	41 945	74.608	25 974	62.230
3 960	44 559	75.284	27 542	62.635
4 140	47 182	75.931	29 115	63.024
4 320	49 813	76.554	30 694	63.397
4 500	52 454	77.153	32 276	63.756

TABLE A.11E[a] (*Continued*) **Enthalpy of Formation at 77 F, Ideal-Gas Enthalpy, and Absolute Entropy at One Atmosphere Pressure**

Temp. R	Carbon Dioxide (CO_2) (9/30/65) $(\bar{h}^\circ_f)_{537} = -169\,297$ Btu/lb mole $M = 44.01$		Carbon Monoxide (CO) (9/30/65) $(\bar{h}^\circ_f)_{537} = -47\,551$ Btu/lb mole $M = 28.01$	
	$(\bar{h}^\circ - \bar{h}^\circ_{537})$ Btu/lb mole	\bar{s}° Btu/lb mole R	$(\bar{h}^\circ - \bar{h}^\circ_{537})$ Btu/lb mole	\bar{s}° Btu/lb mole R
4 680	55 103	77.730	33 863	64.102
4 860	57 758	78.286	35 453	64.435
5 040	60 421	78.824	37 048	64.757
5 220	63 088	79.344	38 644	65.069
5 400	65 763	79.848	40 243	65.370
5 760	71 127	80.810	43 450	65.945
6 120	76 513	81.717	46 669	66.487
6 480	81 914	82.574	49 894	66.999
6 840	87 332	83.388	53 129	67.485
7 200	92 768	84.162	56 369	67.946
7 560	98 219	84.901	59 618	68.387
7 920	103 682	85.607	62 874	68.807
8 280	109 159	86.284	66 134	69.210
8 640	114 651	86.933	69 403	69.596
9 000	120 155	87.557	72 675	69.967
9 360	125 674	88.158	75 953	70.325
9 720	131 207	88.738	79 238	70.669
10 180	136 757	89.299	82 528	71.001
10 540	142 322	89.841	85 822	71.322
10 800	147 902	90.367	89 123	71.633

TABLE A.11E[a] (*Continued*) Enthalpy of Formation at 77 F, Ideal-Gas Enthalpy, and Absolute Entropy at One Atmosphere Pressure

Temp. R	Water (H_2O)(3/31/61) $(\bar{h}_f^\circ)_{537} = -104\ 036$ Btu/lb mole $M = 18.015$		Hydroxyl (OH)(3/31/66) $(\bar{h}_f^\circ)_{537} = 16\ 978$ Btu/lb mole $M = 17.007$	
	$(\bar{h}^\circ - \bar{h}_{537}^\circ)$ Btu/lb mole	\bar{s}° Btu/lb mole R	$(\bar{h}^\circ - \bar{h}_{537}^\circ)$ Btu/lb mole	\bar{s}° Btu/lb mole R
0	−4 261	0	−3 946	0
180	−2 846	36.396	−2 641	35.726
360	−1 411	41.916	−1 280	40.985
537	0	45.106	0	43.880
540	27	45.155	23	43.925
720	1 485	47.484	1 305	45.974
900	2 977	49.334	2 578	47.551
1 080	4 516	50.891	3 847	48.837
1 260	6 102	52.249	5 121	49.927
1 440	7 740	53.464	6 401	50.877
1 620	9 432	54.570	7 695	51.724
1 800	11 176	55.592	9 005	52.491
1 980	12 978	56.545	10 336	53.195
2 160	14 832	57.441	11 684	53.847
2 340	16 736	58.288	13 054	54.455
2 520	18 691	59.092	14 441	55.027
2 700	20 691	59.859	15 849	55.566
2 880	22 734	60.591	17 273	56.077
3 060	24 817	61.293	18 715	56.563
3 240	26 935	61.965	20 173	57.025
3 420	29 088	62.612	21 643	57.467
3 600	31 271	63.234	23 128	57.891
3 780	33 484	63.834	24 626	58.296
3 960	35 723	64.412	26 136	58.686
4 140	37 985	64.971	27 655	59.062
4 320	40 270	65.511	29 185	59.424
4 500	42 575	66.034	30 724	59.773

TABLE A.11Ea (*Continued*) Enthalpy of Formation at 77 F, Ideal-Gas Enthalpy, and Absolute Entropy at One Atmosphere Pressure

Temp. R	Water (H_2O)(3/31/61) $(\bar{h}_f^\circ)_{537} = -104\,036$ Btu/lb mole $M = 18.015$		Hydroxyl (OH)(3/31/66) $(\bar{h}_f^\circ)_{537} = 16\,978$ Btu/lb mole $M = 17.007$	
	$(\bar{h}^\circ - \bar{h}_{537}^\circ)$ Btu/lb mole	\bar{s}° Btu/lb mole R	$(\bar{h}^\circ - \bar{h}_{537}^\circ)$ Btu/lb mole	\bar{s}° Btu/lb mole R
4 680	44 901	66.541	32 272	60.110
4 860	47 243	67.032	33 829	60.436
5 040	49 601	67.508	35 392	60.752
5 220	51 975	67.971	36 963	61.058
5 400	54 362	68.421	38 540	61.355
5 760	59 177	69.284	41 713	61.924
6 120	64 039	70.102	44 908	62.462
6 480	68 940	70.881	48 123	62.973
6 840	73 877	71.622	51 358	63.458
7 200	78 849	72.331	54 608	63.922
7 560	83 849	73.008	57 875	64.364
7 920	88 875	73.658	61 157	64.788
8 280	93 926	74.281	64 453	65.195
8 640	99 000	74.881	67 759	65.586
9 000	104 092	75.459	71 080	65.963
9 360	109 204	76.016	74 412	66.326
9 720	114 336	76.553	77 755	66.676
10 180	119 486	77.074	81 108	67.015
10 540	124 652	77.577	84 472	67.343
10 800	129 836	78.065	87 845	67.661

TABLE A.11E^a (*Continued*) **Enthalpy of Formation at 77 F, Ideal–Gas Enthalpy, and Absolute Entropy at One Atmosphere Pressure**

Temp. R	Hydrogen, Diatomic (H_2) (3/31/61) $(\bar{h}_f^\circ)_{537} = 0$ Btu/lb mole $M = 2.016$		Hydrogen, Monatomic (H) (9/30/65) $(\bar{h}_f^\circ)_{537} = 93\ 780$ Btu/lb mole $M = 1.008$	
	$(\bar{h}^\circ - \bar{h}_{537}^\circ)$ Btu/lb mole	\bar{s}° Btu/lb mole R	$(\bar{h}^\circ - \bar{h}_{537}^\circ)$ Btu/lb mole	\bar{s}° Btu/lb mole R
0	−3 643	0	−2 666	0
180	−2 277	24.387	−1 771	21.965
360	−1 192	28.520	−878	25.408
537	0	31.208	0	27.392
540	23	31.251	16	27.423
720	1 273	33.247	911	28.852
900	2 531	34.806	1 805	29.961
1 080	3 791	36.082	2 700	30.867
1 260	5 054	37.165	3 593	31.632
1 440	6 325	38.107	4 487	32.296
1 620	7 607	38.946	5 382	32.881
1 800	8 899	39.702	6 277	33.404
1 980	10 206	40.394	7 171	33.878
2 160	11 527	41.033	8 066	34.310
2 340	12 866	41.628	8 959	34.708
2 520	14 224	42.187	9 853	35.076
2 700	15 602	42.716	10 748	35.419
2 880	17 003	43.217	11 642	35.739
3 060	18 419	43.695	12 537	36.041
3 240	19 854	44.150	13 430	36.325
3 420	21 305	44.586	14 324	36.593
3 600	22 772	45.004	15 219	36.848
3 780	24 255	45.406	16 114	37.090
3 960	25 753	45 793	17 008	37.322
4 140	27 263	46.166	17 901	37.542
4 320	28 787	46.527	18 796	37.754
4 500	30 326	46.875	19 690	37.957

TABLE A.11E[a] (*Continued*) Enthalpy of Formation at 77 F, Ideal-Gas Enthalpy, and Absolute Entropy at One Atmosphere Pressure

| | Hydrogen, Diatomic (H_2) (3/31/61) | | Hydrogen, Monatomic (H) (9/30/65) | |
| | $(\bar{h}_f^\circ)_{537} = 0$ Btu/lb mole $M = 2.016$ | | $(\bar{h}_f^\circ)_{537} = 93\,780$ Btu/lb mole $M = 1.008$ | |
Temp. R	$(\bar{h}^\circ - \bar{h}_{537}^\circ)$ Btu/lb mole	\bar{s}° Btu/lb mole R	$(\bar{h}^\circ - \bar{h}_{537}^\circ)$ Btu/lb mole	\bar{s}° Btu/lb mole R
4 680	31 874	47.213	20 585	38.152
4 860	33 435	47.540	21 479	38.339
5 040	35 006	47.857	22 374	38.520
5 220	36 587	48.166	23 267	38.694
5 400	38 178	48.465	24 161	38.862
5 760	41 386	49.040	25 951	39.183
6 120	44 629	49.586	27 738	39.484
6 480	47 909	50.107	29 527	39.768
6 840	51 223	50.605	31 316	40.037
7 200	54 571	51.082	33 104	40.292
7 560	57 949	51.540	34 893	40.534
7 920	61 358	51.980	36 682	40.765
8 280	64 798	52.405	38 470	40.986
8 640	68 267	52.815	40 259	41.198
9 000	71 762	53.211	42 046	41.400
9 360	75 285	53.595	43 835	41.595
9 720	78 835	53.967	45 625	41.783
10 180	82 409	54.328	47 412	41.963
10 540	86 009	54.679	49 201	42.138
10 800	89 633	55.020	50 990	42.306

TABLE A.11E^a (*Continued*) Enthalpy of Formation at 77 F, Ideal-Gas Enthalpy, and Absolute Entropy at One Atmosphere Pressure

Temp. R	Nitric Oxide (NO) (6/30/63) $(\bar{h}_f^\circ)_{537} = 38\,974$ Btu/lb mole $M = 30.006$		Nitrogen Dioxide (NO$_2$) (6/30/63) $(\bar{h}_f^\circ)_{537} = 14\,508$ Btu/lb mole $M = 46.005$	
	$(\bar{h}^\circ - \bar{h}_{537}^\circ)$ Btu/lb mole	\bar{s}° Btu/lb mole R	$(\bar{h}^\circ - \bar{h}_{537}^\circ)$ Btu/lb mole	\bar{s}° Btu/lb mole R
0	−3 955	0	−4 387	0
180	−2 612	42.286	−2 956	48.356
360	−1 269	47.477	−1 507	53.925
537	0	50.347	0	57.324
540	23	50.392	29	57.379
720	1 309	52.444	1 699	60.041
900	2 606	54.053	3 506	62.279
1 080	3 935	55.397	5 438	64.234
1 260	5 296	56.562	7 472	65.975
1 440	6 689	57.596	9 589	67.544
1 620	8 113	58.528	11 770	68.971
1 800	9 563	59.377	14 004	70.278
1 980	11 036	60.157	16 277	71.482
2 160	12 528	60.878	18 583	72.597
2 340	14 036	61.548	20 916	73.634
2 520	15 559	62.175	23 272	74.604
2 700	17 093	62.763	25 645	75.513
2 880	18 637	63.317	28 031	76.369
3 060	20 191	63.840	30 433	77.178
3 240	21 751	64.335	32 845	77.943
3 420	23 319	64.806	35 264	78.670
3 600	24 892	65.255	37 692	79.362
3 780	26 471	65.683	40 127	80.022
3 960	28 057	66.092	42 568	80.653
4 140	29 644	66.484	45 014	81.257
4 320	31 237	66.861	47 466	81.837
4 500	32 834	67.223	49 921	82.394

TABLE A.11E[a] (*Continued*) **Enthalpy of Formation at 77 F, Ideal-Gas Enthalpy, and Absolute Entropy at One Atmosphere Pressure**

Temp. R	Nitric Oxide (NO) (6/30/63)		Nitrogen Dioxide (NO_2) (6/30/63)	
	$(\bar{h}_f^\circ)_{537} = 38\,974$ Btu/lb mole $M = 30.006$		$(\bar{h}_f^\circ)_{537} = 14\,508$ Btu/lb mole $M = 46.005$	
	$(\bar{h}^\circ - \bar{h}_{537}^\circ)$ Btu/lb mole	\bar{s}° Btu/lb mole R	$(\bar{h}^\circ - \bar{h}_{537}^\circ)$ Btu/lb mole	\bar{s}° Btu/lb mole R
4 680	34 432	67.571	52 380	82.930
4 860	36 036	67.907	54 842	83.446
5 040	37 640	68.232	57 307	83.944
5 220	39 249	68.545	59 774	84.425
5 400	40 860	68.849	62 244	84.890
5 760	44 087	69.427	67 190	85.777
6 120	47 324	69.973	72 144	86.611
6 480	50 569	70.488	77 103	87.398
6 840	53 820	70.976	82 067	88.144
7 200	57 078	71.440	87 035	88.852
7 560	60 341	71.882	92 007	89.525
7 920	63 612	72.305	96 982	90.168
8 280	66 886	72.709	101 961	90.783
8 640	70 168	73.097	106 942	91.372
9 000	73 453	73.470	111 924	91.937
9 360	76 741	73.828	116 910	92.480
9 720	80 037	74.173	121 896	93.003
10 180	83 335	74.507	126 884	93.507
10 540	86 639	74.829	131 873	93.993
10 800	89 946	75.140	136 863	94.463

TABLE A.11SI[a] Enthalpy of Formation at 25°C, Ideal-Gas Enthalpy, and Absolute Entropy at 0.1 MPa (1 bar) Pressure

Temp. K	Nitrogen, Diatomic (N_2) (9/30/65) $(\bar{h}_f^\circ)_{298} = 0$ kJ/kmol $M = 28.013$		Nitrogen, Monatomic (N) (3/31/61) $(\bar{h}_f^\circ)_{298} = 472\,646$ kJ/kmol $M = 14.007$	
	$(\bar{h}^\circ - \bar{h}_{298}^\circ)$ kJ/kmol	\bar{s}° kJ/kmol K	$(\bar{h}^\circ - \bar{h}_{298}^\circ)$ kJ/kmol	\bar{s}° kJ/kmol K
0	−8 669	0	−6 197	0
100	−5 770	159.813	−4 117	130.596
200	−2 858	179.988	−2 042	145.006
298	0	191.611	0	153.302
300	54	191.791	38	153.432
400	2 971	200.180	2 117	159.411
500	5 912	206.740	4 197	164.051
600	8 891	212.175	6 276	167.842
700	11 937	216.866	8 351	171.047
800	15 046	221.016	10 431	173.821
900	18 221	224.757	12 510	176.268
1000	21 460	228.167	14 590	178.461
1100	24 757	231.309	16 669	180.440
1200	28 108	234.225	18 749	182.247
1300	31 501	236.941	20 824	183.803
1400	34 936	239.484	22 903	185.452
1500	38 405	241.878	24 983	186.887
1600	41 903	244.137	27 062	188.230
1700	45 430	246.275	29 142	189.490
1800	48 982	248.304	31 217	190.678
1900	52 551	250.237	33 296	191.799
2000	56 141	252.078	35 376	192.866
2100	59 748	253.836	37 455	193.883
2200	63 371	255.522	39 535	194.850
2300	67 007	257.137	41 614	195.774
2400	70 651	258.689	43 698	196.661
2500	74 312	260.183	45 777	197.511

TABLE A.11SI[a] (*Continued*) **Enthalpy of Formation at 25°C, Ideal-Gas Enthalpy, and Absolute Entropy at 0.1 MPa (1 bar) Pressure**

Temp. K	Nitrogen, Diatomic (N_2) (9/30/65)		Nitrogen, Monatomic (N) (3/31/61)	
	$(\bar{h}_f^\circ)_{298} = 0$ kJ/kmol $M = 28.013$		$(\bar{h}_f^\circ)_{298} = 472\,646$ kJ/kmol $M = 14.007$	
	$(\bar{h}^\circ - \bar{h}_{298}^\circ)$ kJ/kmol	\bar{s}° kJ/kmol K	$(\bar{h}^\circ - \bar{h}_{298}^\circ)$ kJ/kmol	\bar{s}° kJ/kmol K
2600	77 973	261.622	47 861	198.326
2700	81 659	263.011	49 949	199.113
2800	85 345	264.350	52 036	199.875
2900	89 036	265.647	54 124	200.607
3000	92 738	266.902	56 220	201.318
3200	100 161	269.295	60 421	202.674
3400	107 608	271.555	64 647	203.954
3600	115 081	273.689	68 906	205.171
3800	122 570	275.714	73 199	206.330
4000	130 076	277.638	77 534	207.443
4200	137 603	279.475	81 923	208.514
4400	145 143	281.228	86 370	209.548
4600	152 699	282.910	90 881	210.552
4800	160 272	284.521	95 462	211.527
5000	167 858	286.069	100 115	212.477
5200	175 456	287.559	104 847	213.401
5400	183 071	288.994	109 663	214.309
5600	190 703	290.383	114 558	215.201
5800	198 347	291.726	119 537	216.075
6000	206 008	293.023	124 600	216.933

[a] The thermochemical data in Table A.11 are calculated from the JANAF Thermochemical Tables, Thermal Research Laboratory, The Dow Chemical Company, Midland, Michigan. The date each table was issued is indicated.

TABLE A.11SI[a] (*Continued*) Enthalpy of Formation at 25°C, Ideal-Gas Enthalpy, and Absolute Entropy at 0.1 MPa (1 bar) Pressure

Temp. K	Oxygen, Diatomic (O_2) (9/30/65) $(\bar{h}_f^\circ)_{298} = 0$ kJ/kmol $M = 31.999$		Oxygen, Monatomic (O) (6/30/62) $(\bar{h}_f^\circ)_{298} = 249\,195$ kJ/kmol $M = 16.00$	
	$(\bar{h}^\circ - \bar{h}_{298}^\circ)$ kJ/kmol	\bar{s}° kJ/kmol K	$(\bar{h}^\circ - \bar{h}_{298}^\circ)$ kJ/kmol	\bar{s}° kJ/kmol K
0	$-8\,682$	0	$-6\,728$	0
100	$-5\,778$	173.306	$-4\,519$	135.947
200	$-2\,866$	193.486	$-2\,188$	152.156
298	0	205.142	0	161.060
300	54	205.322	42	161.198
400	3 029	213.874	2 209	167.432
500	6 088	220.698	4 343	172.202
600	9 247	226.455	6 460	176.063
700	12 502	231.272	8 569	179.314
800	15 841	235.924	10 669	182.118
900	19 246	239.936	12 770	184.590
1000	22 707	243.585	14 862	186.795
1100	26 217	246.928	16 949	188.787
1200	29 765	250.016	19 041	190.603
1300	33 351	252.886	21 125	192.272
1400	36 966	255.564	23 213	193.820
1500	40 610	258.078	25 296	195.260
1600	44 279	260.446	27 380	196.603
1700	47 970	262.685	29 464	197.866
1800	51 689	264.810	31 547	199.059
1900	55 434	266.835	33 631	200.184
2000	59 199	268.764	35 715	201.251
2100	62 986	270.613	37 798	202.268
2200	66 802	272.387	39 882	203.238
2300	70 634	274.090	41 961	204.163
2400	74 492	275.735	44 045	205.050
2500	78 375	277.316	46 133	205.899

TABLE A.11SI[a] (*Continued*) **Enthalpy of Formation at 25°C, Ideal-Gas Enthalpy, and Absolute Entropy at 0.1 MPa (1 bar) Pressure**

Temp. K	Oxygen, Diatomic (O_2) (9/30/65) $(\bar{h}_f^\circ)_{298} = 0$ kJ/kmol M = 31.999		Oxygen, Monatomic (O) (6/30/62) $(\bar{h}_f^\circ)_{298} = 249\,195$ kJ/kmol M = 16.00	
	$(\bar{h}^\circ - \bar{h}_{298}^\circ)$ kJ/kmol	\bar{s}° kJ/kmol K	$(\bar{h}^\circ - \bar{h}_{298}^\circ)$ kJ/kmol	\bar{s}° kJ/kmol K
2600	82 274	278.848	48 216	206.720
2700	86 199	280.329	50 304	207.506
2800	90 144	281.764	52 392	208.268
2900	94 111	283.157	54 484	209.000
3000	98 098	284.508	56 576	209.711
3200	106 127	287.098	60 768	211.063
3400	114 232	289.554	64 973	212.339
3600	122 399	291.889	69 191	213.544
3800	130 629	294.115	73 425	214.686
4000	138 913	296.236	77 676	215.778
4200	147 248	298.270	81 948	216.820
4400	155 628	300.219	86 236	217.816
4600	164 046	302.094	90 546	218.774
4800	172 502	303.893	94 876	219.694
5000	180 987	305.621	99 224	220.585
5200	189 502	307.290	103 596	221.439
5400	198 037	308.901	107 985	222.267
5600	206 593	310.458	112 395	223.071
5800	215 166	311.964	116 821	223.849
6000	223 756	313.420	121 269	224.602

TABLE A.11SI[a] *(Continued)* **Enthalpy of Formation at 25°C, Ideal-Gas Enthalpy, and Absolute Entropy at 0.1 MPa (1 bar) Pressure**

Temp. K	Carbon Dioxide (CO_2) (9/30/65)		Carbon Monoxide (CO) (9/30/65)	
	$(\bar{h}_f^\circ)_{298} = -393\,522$ kJ/kmol $M = 44.01$		$(\bar{h}_f^\circ)_{298} = -110\,529$ kJ/kmol $M = 28.01$	
	$(\bar{h}^\circ - \bar{h}_{298}^\circ)$ kJ/kmol	\bar{s}° kJ/kmol K	$(\bar{h}^\circ - \bar{h}_{298}^\circ)$ kJ/kmol	\bar{s}° kJ/kmol K
0	$-9\,364$	0	$-8\,669$	0
100	$-6\,456$	179.109	$-5\,770$	165.850
200	$-3\,414$	199.975	$-2\,858$	186.025
298	0	213.795	0	197.653
300	67	214.025	54	197.833
400	4\,008	225.334	2\,975	206.234
500	8\,314	234.924	5\,929	212.828
600	12\,916	243.309	8\,941	218.313
700	17\,761	250.773	12\,021	223.062
800	22\,815	257.517	15\,175	227.271
900	28\,041	263.668	18\,397	231.066
1000	33\,405	269.325	21\,686	234.531
1100	38\,894	274.555	25\,033	237.719
1200	44\,484	279.417	28\,426	240.673
1300	50\,158	283.956	31\,865	243.426
1400	55\,907	288.216	35\,338	245.999
1500	61\,714	292.224	38\,848	248.421
1600	67\,580	296.010	42\,384	250.702
1700	73\,492	299.592	45\,940	252.861
1800	79\,442	302.993	49\,522	254.907
1900	85\,429	306.232	53\,124	256.852
2000	91\,450	309.320	56\,739	258.710
2100	97\,500	312.269	60\,375	260.480
2200	103\,575	315.098	64\,019	262.174
2300	109\,671	317.805	67\,676	263.802
2400	115\,788	320.411	71\,346	265.362
2500	121\,926	322.918	75\,023	266.865

TABLE A.11SI[a] (*Continued*) **Enthalpy of Formation at 25°C, Ideal-Gas Enthalpy, and Absolute Entropy at 0.1 MPa (1 bar) Pressure**

Temp. K	Carbon Dioxide (CO_2) (9/30/65) $(\bar{h}_f^\circ)_{298} = -393\,522$ kJ/kmol $M = 44.01$		Carbon Monoxide (CO) (9/30/65) $(\bar{h}_f^\circ)_{298} = -110\,529$ kJ/kmol $M = 28.01$	
	$(\bar{h}^\circ - \bar{h}_{298}^\circ)$ kJ/kmol	\bar{s}° kJ/kmol K	$(\bar{h}^\circ - \bar{h}_{298}^\circ)$ kJ/kmol	\bar{s}° kJ/kmol K
2600	128 085	325.332	78 714	268.312
2700	134 256	327.658	82 408	269.705
2800	140 444	329.909	86 115	271.053
2900	146 645	332.085	89 826	272.358
3000	152 862	334.193	93 542	273.618
3200	165 331	338.218	100 998	276.023
3400	177 849	342.013	108 479	278.291
3600	190 405	345.599	115 976	280.433
3800	202 999	349.005	123 495	282.467
4000	215 635	352.243	131 026	284.396
4200	228 304	355.335	138 578	286.241
4400	241 003	358.289	146 147	287.998
4600	253 734	361.122	153 724	289.684
4800	266 500	363.837	161 322	291.299
5000	279 295	366.448	168 929	292.851
5200	292 123	368.963	176 548	294.349
5400	304 984	371.389	184 184	295.789
5600	317 884	373.736	191 832	297.178
5800	330 821	376.004	199 489	298.521
6000	343 791	378.205	207 162	299.822

TABLE A.11SI[a] (*Continued*) Enthalpy of Formation at 25°C, Ideal-Gas Enthalpy, and Absolute Entropy at 0.1 MPa (1 bar) Pressure

Temp. K	Water (H_2O) (3/31/61) $(\bar{h}_f^\circ)_{298} = -241\,827$ kJ/kmol $M = 18.015$		Hydroxyl (OH) (3/31/66) $(\bar{h}_f^\circ)_{298} = 39\,463$ kJ/kmol $M = 17.007$	
	$(\bar{h}^\circ - \bar{h}_{298}^\circ)$ kJ/kmol	\bar{s}° kJ/kmol K	$(\bar{h}^\circ - \bar{h}_{298}^\circ)$ kJ/kmol	\bar{s}° kJ/kmol K
0	−9 904	0	−9 171	0
100	−6 615	152.390	−6 138	149.587
200	−3 280	175.486	−2 975	171.591
298	0	188.833	0	183.703
300	63	189.038	54	183.892
400	3 452	198.783	3 033	192.465
500	6 920	206.523	5 991	199.063
600	10 498	213.037	8 941	204.443
700	14 184	218.719	11 903	209.004
800	17 991	223.803	14 878	212.979
900	21 924	228.430	17 887	216.523
1000	25 978	232.706	20 933	219.732
1100	30 167	236.694	24 025	222.677
1200	34 476	240.443	27 158	225.405
1300	38 903	243.986	30 342	227.949
1400	43 447	247.350	33 568	230.342
1500	48 095	250.560	36 840	232.598
1600	52 844	253.622	40 150	234.736
1700	57 685	256.559	43 501	236.769
1800	62 609	259.371	46 890	238.702
1900	67 613	262.078	50 308	240.551
2000	72 689	264.681	53 760	242.325
2100	77 831	267.191	57 241	244.020
2200	83 036	269.609	60 752	245.652
2300	88 295	271.948	64 283	247.225
2400	93 604	274.207	67 839	248.739
2500	98 964	276.396	71 417	250.200

TABLE A.11SI[a] (*Continued*) Enthalpy of Formation at 25°C, Ideal-Gas Enthalpy, and Absolute Entropy at 0.1 MPa (1 bar) Pressure

Temp. K	Water (H_2O) (3/31/61) $(\bar{h}^\circ_f)_{298} = -241\,827$ kJ/kmol $M = 18.015$		Hydroxyl (OH) (3/31/66) $(\bar{h}^\circ_f)_{298} = 39\,463$ kJ/kmol $M = 17.007$	
	$(\bar{h}^\circ - \bar{h}^\circ_{298})$ kJ/kmol	\bar{s}° kJ/kmol K	$(\bar{h}^\circ - \bar{h}^\circ_{298})$ kJ/kmol	\bar{s}° kJ/kmol K
2600	104 370	278.517	75 015	251.610
2700	109 813	280.571	78 634	252.974
2800	115 294	282.563	82 266	254.296
2900	120 813	284.500	85 918	255.576
3000	126 361	286.383	89 584	256.819
3200	137 553	289.994	96 960	259.199
3400	148 854	293.416	104 387	261.450
3600	160 247	296.676	111 859	263.588
3800	171 724	299.776	119 378	265.618
4000	183 280	302.742	126 934	267.559
4200	194 903	305.575	134 528	269.408
4400	206 585	308.295	142 156	271.182
4600	218 325	310.901	149 816	272.885
4800	230 120	313.412	157 502	274.521
5000	241 957	315.830	165 222	276.099
5200	253 839	318.160	172 967	277.617
5400	265 768	320.407	180 736	279.082
5600	277 738	322.587	188 531	280.500
5800	289 746	324.692	196 351	281.873
6000	301 796	326.733	204 192	283.203

TABLE A.11SI^a (*Continued*) Enthalpy of Formation at 25°C, Ideal-Gas Enthalpy, and Absolute Entropy at 0.1 MPa (1 bar) Pressure

Temp. K	Hydrogen, Diatomic (H_2) (3/31/61) $(\bar{h}_f^\circ)_{298} = 0$ kJ/kmol $M = 2.016$		Hydrogen, Monatomic (H) (9/30/65) $(\bar{h}_f^\circ)_{298} = 217\,986$ kJ/kmol $M = 1.008$	
	$(\bar{h}^\circ - \bar{h}_{298}^\circ)$ kJ/kmol	\bar{s}° kJ/kmol K	$(\bar{h}^\circ - \bar{h}_{298}^\circ)$ kJ/kmol	\bar{s}° kJ/kmol K
0	−8 468	0	−6 197	0
100	−5 293	102.145	−4 117	92.011
200	−2 770	119.437	−2 042	106.417
298	0	130.684	0	114.718
300	54	130.864	38	114.847
400	2 958	139.215	2 117	120.826
500	5 883	145.738	4 197	125.466
600	8 812	151.077	6 276	129.257
700	11 749	155.608	8 351	132.458
800	14 703	159.549	10 431	135.236
900	17 682	163.060	12 510	137.684
1000	20 686	166.223	14 590	139.872
1100	23 723	169.118	16 669	141.855
1200	26 794	171.792	18 749	143.662
1300	29 907	174.281	20 824	145.328
1400	33 062	176.620	22 903	146.867
1500	36 267	178.833	24 983	148.303
1600	39 522	180.929	27 062	149.641
1700	42 815	182.929	29 142	150.905
1800	46 150	184.833	31 217	152.093
1900	49 522	186.657	33 296	153.215
2000	52 932	188.406	35 376	154.281
2100	56 379	190.088	37 455	155.294
2200	59 860	191.707	39 535	156.265
2300	63 371	193.268	41 610	157.185
2400	66 915	194.778	43 689	158.072
2500	70 492	196.234	45 769	158.922

TABLE A.11SI[a] (*Continued*) **Enthalpy of Formation at 25°C, Ideal-Gas Enthalpy, and Absolute Entropy at 0.1 MPa (1 bar) Pressure**

Temp. K	Hydrogen, Diatomic (H$_2$) (3/31/61) $(\bar{h}_f^\circ)_{298} = 0$ kJ/kmol $M = 2.016$		Hydrogen, Monatomic (H) (9/30/65) $(\bar{h}_f^\circ)_{298} = 217\,986$ kJ/kmol $M = 1.008$	
	$(\bar{h}^\circ - \bar{h}_{298}^\circ)$ kJ/kmol	\bar{s}° kJ/kmol K	$(\bar{h}^\circ - \bar{h}_{298}^\circ)$ kJ/kmol	\bar{s}° kJ/kmol K
2600	74 090	197.649	47 848	159.737
2700	77 718	199.017	49 928	160.520
2800	81 370	200.343	52 007	161.277
2900	85 044	201.636	54 082	162.005
3000	88 743	202.887	56 162	162.708
3200	96 199	205.293	60 321	164.051
3400	103 738	207.577	64 475	165.311
3600	111 361	209.757	68 634	166.499
3800	119 064	211.841	72 793	167.624
4000	126 846	213.837	76 948	168.691
4200	134 700	215.753	81 107	169.704
4400	142 624	217.594	85 266	170.670
4600	150 620	219.372	89 420	171.595
4800	158 682	221.087	93 579	172.482
5000	166 808	222.744	97 734	173.327
5200	174 996	224.351	101 893	174.143
5400	183 247	225.907	106 052	174.930
5600	191 556	227.418	110 207	175.683
5800	199 924	228.886	114 365	176.415
6000	208 346	230.313	118 524	177.118

TABLE A.11SI[a] (*Continued*) **Enthalpy of Formation at 25°C, Ideal-Gas Enthalpy, and Absolute Entropy at 0.1 MPa (1 bar) Pressure**

Temp. K	Nitric Oxide (NO) (6/30/63) $(\bar{h}_f^\circ)_{298} = 90\,592$ kJ/kmol $M = 30.006$		Nitrogen Dioxide (NO$_2$) (6/30/63) $(\bar{h}_f^\circ)_{298} = 33\,723$ kJ/kmol $M = 46.005$	
	$(\bar{h}^\circ - \bar{h}_{298}^\circ)$ kJ/kmol	\bar{s}° kJ/kmol K	$(\bar{h}^\circ - \bar{h}_{298}^\circ)$ kJ/kmol	\bar{s}° kJ/kmol K
0	−9 192	0	−10 196	0
100	−6 071	177.034	−6 870	202.431
200	−2 950	198.753	−3 502	225.732
298	0	210.761	0	239.953
300	54	210.950	67	240.183
400	3 042	219.535	3 950	251.321
500	6 058	226.267	8 150	260.685
600	9 146	231.890	12 640	268.865
700	12 309	236.765	17 368	276.149
800	15 548	241.091	22 288	282.714
900	18 857	244.991	27 359	288.684
1000	22 230	248.543	32 552	294.153
1100	25 652	251.806	37 836	299.190
1200	29 121	254.823	43 196	303.855
1300	32 627	257.626	48 618	308.194
1400	36 166	260.250	54 095	312.253
1500	39 731	262.710	59 609	316.056
1600	43 321	265.028	65 157	319.637
1700	46 932	267.216	70 739	323.022
1800	50 559	269.287	76 345	326.223
1900	54 204	271.258	81 969	329.265
2000	57 861	273.136	87 613	332.160
2100	61 530	274.927	93 274	334.921
2200	65 216	276.638	98 947	337.562
2300	68 906	278.279	104 633	340.089
2400	72 609	279.856	110 332	342.515
2500	76 320	281.370	116 039	344.846

TABLE A.11SI[a] (*Continued*) **Enthalpy of Formation at 25°C, Ideal-Gas Enthalpy, and Absolute Entropy at 0.1 MPa (1 bar) Pressure**

Temp. K	Nitric Oxide (NO) (6/30/63) $(\bar{h}_f^\circ)_{298} = 90\,592$ kJ/kmol $M = 30.006$		Nitrogen Dioxide (NO$_2$) (6/30/63) $(\bar{h}_f^\circ)_{298} = 33\,723$ kJ/kmol $M = 46.005$	
	$(\bar{h}^\circ - \bar{h}_{298}^\circ)$ kJ/kmol	\bar{s}° kJ/kmol K	$(\bar{h}^\circ - \bar{h}_{298}^\circ)$ kJ/kmol	\bar{s}° kJ/kmol K
2600	80 036	282.827	121 754	347.089
2700	83 764	284.232	127 478	349.248
2800	87 492	285.592	133 206	351.331
2900	91 232	286.902	138 942	353.344
3000	94 977	288.174	144 683	355.289
3200	102 479	290.592	156 180	359.000
3400	110 002	292.876	167 695	362.490
3600	117 545	295.031	179 222	365.783
3800	125 102	297.073	190 761	368.904
4000	132 675	299.014	202 309	371.866
4200	140 260	300.864	213 865	374.682
4400	147 863	302.634	225 430	377.372
4600	155 473	304.324	237 003	379.946
4800	163 101	305.947	248 580	382.410
5000	170 736	307.508	260 161	384.774
5200	178 381	309.006	271 751	387.046
5400	186 042	310.449	283 340	389.234
5600	193 707	311.847	294 934	391.343
5800	201 388	313.194	306 532	393.376
6000	209 074	314.495	318 130	395.343

TABLE A.12 The Reduced Second Virial Co-efficient $B^*(T^*)$ and Force Constants for the Lennard–Jones (6–12) Potential

TABLE A.12.1 The Reduced Second Virial Coefficient

T^*	$B^*(T^*)$	T^*	$B^*(T^*)$
0.3	− 27.881	2.6	− 0.266 13
0.4	− 13.799	2.8	− 0.184 51
0.5	− 8.7202	3.0	− 0.115 23
0.6	− 6.1980	3.2	− 0.055 79
0.7	− 4.7100	3.4	− 0.004 28
0.8	− 3.7342	3.6	+ 0.040 72
0.9	− 3.0471	3.8	0.080 33
1.0	− 2.5381	4.0	0.115 42
1.1	− 2.1464	4.2	0.146 68
1.2	− 1.8359	4.4	0.174 69
1.3	− 1.5841	4.6	0.199 90
1.4	− 1.3758	4.8	0.222 68
1.5	− 1.2009	5.0	0.243 34
1.6	− 1.0519	6.0	0.322 90
1.7	− 0.923 62	7.0	0.376 09
1.8	− 0.812 03	8.0	0.413 43
1.9	− 0.714 15	9.0	0.440 60
2.0	− 0.627 63	10.0	0.460 88
2.2	− 0.481 71	20.0	0.525 37
2.4	− 0.363 58	30.0	0.526 93

TABLE A.12.2[a] Force Constants from Experimental Virial Coefficient Data

Substance	ϵ/k, K	b_0, m^3/kmol
Ne	35.8	0.0262
Ar	119.0	0.0502
Kr	173.0	0.0583
Xe	225.3	0.0854
N_2	95.05	0.0635
O_2	117.5	0.0578
CO	100.2	0.0675
NO	131.0	0.0402
CO_2	186.0	0.118
N_2O	193.0	0.118
CH_4	148.1	0.0698
CF_4	152	0.131

[a] From J. O. Hirschfelder, C. F. Curtiss, and R. B. Bird, *Molecular Theory of Gases and Liquids,* John Wiley and Sons, New York, Second Printing, Corrected, with Notes Added, March 1964.

TABLE A.13E Enthalpy of Formation, Gibbs Function of Formation and Absolute Entropy of Various Substances at 77 F, 1 Atmosphere Pressure

Substance	Formula	M	State	\bar{h}_f° Btu/lb mole	\bar{g}_f° Btu/lb mole	\bar{s}° Btu/lb mole R
Carbon monoxide[a]	CO	28.011	gas	−47 551	−59 009	47.214
Carbon dioxide[a]	CO_2	44.011	gas	−169 297	−169 677	51.072
Water[a,b]	H_2O	18.015	gas	−104 036	−98 345	45.106
Water[b]	H_2O	18.015	liq.	−122 971	−102 042	16.716
Methane[a]	CH_4	16.043	gas	−32 211	−21 861	44.490
Acetylene[a]	C_2H_2	26.038	gas	+97 542	+89 987	48.004
Ethene[a]	C_2H_4	28.054	gas	+22 493	+29 306	52.447
Ethane[e]	C_2H_6	30.070	gas	−36 425	−14 148	54.85
Propane[e]	C_3H_8	44.097	gas	−44 676	−10 105	64.51
Butane[e]	C_4H_{10}	58.124	gas	−54 270	−7 380	74.12
Octane[e]	C_8H_{18}	114.23	gas	−89 680	+7 110	111.55
Octane[e]	C_8H_{18}	114.23	liq.	−107 532	+2 844	86.23
Carbon[a]	C	12.011	solid	0	0	1.359

[a] From JANAF *Thermochemical Data*, The Dow Chemical Company, Thermal Laboratory, Midland, Mich.
[b] From *Circular 500*, National Bureau of Standards.
[e] From F. D. Rossini et al, *API Research Project 44.*.

TABLE A.13SI Enthalpy of Formation, Gibbs Function of Formation and Absolute Entropy of Various Substances at 25°C, 0.1 MPa Pressure (SI Units)

Substance	Formula	M	State	\bar{h}_f° kJ/kmol	\bar{g}_f° kJ/kmol	\bar{s}° kJ/kmol K
Carbon monoxide[a]	CO	28.011	gas	−110 529	−137 182	197.653
Carbon dioxide[a]	CO_2	44.011	gas	−393 522	−394 407	213.795
Water[a,b]	H_2O	18.015	gas	−241 827	−228 583	188.833
Water[b]	H_2O	18.015	liq.	−285 838	−237 146	69.940
Methane[a]	CH_4	16.043	gas	−74 873	−50 783	186.256
Acetylene[a]	C_2H_2	26.038	gas	+226 731	+209 169	200.958
Ethene[c]	C_2H_4	28.054	gas	+52 283	+68 142	219.548
Ethane[c]	C_2H_6	30.070	gas	−84 667	−32 842	229.602
Propane[c]	C_3H_8	44.097	gas	−103 847	−23 414	270.019
Butane[c]	C_4H_{10}	58.124	gas	−126 148	−17 044	310.227
Octane[c]	C_8H_{18}	114.23	gas	−208 447	+16 599	466.835
Octane[c]	C_8H_{18}	114.23	liq.	−249 952	+6713	360.896
Carbon[a] (graphite)	C	12.011	solid	0	0	5.686

[a] From JANAF *Thermochemical Data*, The Dow Chemical Company, Thermal Laboratory, Midland, Mich.
[b] From *Circular 500*, The National Bureau of Standards.
[c] From F. D. Rossini et al., *API Research Project 44*.

TABLE A.14 Logarithms to the Base e of the Equilibrium Constant K

For the reaction $v_A A + v_B B \rightleftharpoons v_C C + v_D D$ the equilibrium constant K is defined as

$$K = \frac{a_C^{v_C} a_D^{v_D}}{a_A^{v_A} a_B^{v_B}}, \quad P^0 = 0.1 \text{ MPa}$$

Based on thermodynamic data given in the JANAF Thermochemical Tables, Thermal Research Laboratory, The Dow Chemical Company, Midland, Michigan.

Temp. K	$H_2 \rightleftharpoons 2H$	$O_2 \rightleftharpoons 2O$	$N_2 \rightleftharpoons 2N$	$2H_2O \rightleftharpoons 2H_2 + O_2$	$2H_2O \rightleftharpoons H_2 + 2OH$	$2CO_2 \rightleftharpoons 2CO + O_2$	$N_2 + O_2 \rightleftharpoons 2NO$
298	-164.005	-186.975	-367.480	-184.416	-212.416	-207.524	-70.104
500	-92.827	-105.630	-213.372	-105.382	-120.562	-115.232	-40.590
1000	-39.803	-45.150	-99.127	-46.326	-52.068	-47.058	-18.776
1200	-30.874	-35.005	-80.011	-36.364	-40.566	-35.742	-15.138
1400	-24.463	-27.742	-66.329	-29.218	-32.198	-27.684	-12.540
1600	-19.637	-22.285	-56.055	-23.842	-26.132	-21.660	-10.588
1800	-15.866	-18.030	-48.051	-19.652	-21.314	-16.994	-9.072
2000	-12.840	-14.622	-41.645	-16.290	-17.456	-13.270	-7.862
2200	-10.353	-11.827	-36.391	-13.536	-14.296	-10.240	-6.866
2400	-8.276	-9.497	-32.011	-11.238	-11.664	-7.720	-6.038
2600	-6.517	-7.521	-28.304	-9.296	-9.438	-5.602	-5.342
2800	-5.002	-5.826	-25.117	-7.624	-7.526	-3.788	-4.742
3000	-3.685	-4.357	-22.359	-6.172	-5.874	-2.222	-4.228
3200	-2.534	-3.072	-19.937	-4.902	-4.424	-0.858	-3.776
3400	-1.516	-1.935	-17.800	-3.782	-3.152	0.338	-3.380
3600	-0.609	-0.926	-15.898	-2.784	-2.176	1.402	-3.026
3800	0.202	-0.019	-14.199	-1.890	-1.002	2.352	-2.712
4000	0.934	0.796	-12.660	-1.084	-0.088	3.198	-2.432
4500	2.486	2.513	-9.414	0.624	1.840	4.980	-1.842
5000	3.725	3.895	-6.807	1.992	3.378	6.394	-1.372
5500	4.743	5.023	-4.666	3.120	4.636	7.542	-0.994
6000	5.590	5.963	-2.865	4.064	5.686	8.490	-0.682

TABLE A.15 Atomic Weights, and Melting and Boiling Points of the Elements[a] Based on the Assigned Relative Atomic Mass of $^{12}C = 12$ (The following values apply to elements as they exist in materials of terrestrial origin and to certain artificial elements.)

Name	Symbol	Atomic Number	Atomic Weight	Melting Point °C	Boiling Point °C
Actinium	Ac	89	227.028	1050	3200 ± 300
Aluminum	Al	13	26.98154	660.37	2467
Americium	Am	95	(243)	994 ± 4	2607
Antimony	Sb	51	121.75	630.74	1750
Argon	Ar	18	39.948	−189.2	−185.7
Arsenic (gray)	As	33	74.9216	817 (28 atm)	613 (sub.)
Astatine	At	85	(210)	302	337
Barium	Ba	56	137.33	725	1640
Berkelium	Bk	97	(247)	—	—
Beryllium	Be	4	9.01218	1278 ± 5	2970 (5 mm)
Bismuth	Bi	83	208.9804	271.3	1560 ± 5
Boron	B	5	10.81	2300	2550 (sub.)
Bromine	Br	35	79.904	−7.2	58.78
Cadmium	Cd	48	112.41	320.9	765
Calcium	Ca	20	40.08	839 ± 2	1484
Californium	Cf	98	(251)	—	—
Carbon	C	6	12.011	~3550	4827
Cerium	Ce	58	140.12	798 ± 3	3257
Cesium	Cs	55	132.9054	28.40 ± 0.01	669.3
Chlorine	Cl	17	35.453	−100.98	−34.6
Chromium	Cr	24	51.996	1857 ± 20	2672
Cobalt	Co	27	58.9332	1495	2870
Copper	Cu	29	63.546	1083.4 ± 0.2	2567
Curium	Cm	96	(247)	1340 ± 40	—

Element	Symbol	Atomic number	Atomic weight		
Dysprosium	Dy	66	162.50	1409	2335
Einsteinium	Es	99	(254)	—	—
Erbium	Er	68	167.26	1522	2510
Europium	Eu	63	151.96	822 ± 5	1597
Fermium	Fm	100	(257)	—	—
Fluorine	F	9	18.998 403	−219.62	−188.14
Francium	Fr	87	(223)	(27)	(677)
Gadolinium	Gd	64	157.25	1311 ± 1	3233
Gallium	Ga	31	69.72	29.78	2403
Germanium	Ge	32	72.59	937.4	2830
Gold	Au	79	196.9665	1064.43	3080
Hafnium	Hf	72	178.49	2227 ± 20	4602
Helium	He	2	4.002 60	−272.2 (26 atm)	−268.934
Holmium	Ho	67	164.9304	1470	2720
Hydrogen	H	1	1.0079	−259.14	−252.87
Indium	In	49	114.82	156.61	2080
Iodine	I	53	126.9045	113.5	184.35
Iridium	Ir	77	192.22	2410	4130
Iron	Fe	26	55.847	1535	2750
Krypton	Kr	36	83.80	−156.6	−152.30 ± 0.10
Lanthanum	La	57	138.9055	920 ± 5	3454
Lawrencium	Lr	103	(260)	—	—
Lead	Pb	82	207.2	327.502	1740
Lithium	Li	3	6.941	180.54	1342
Lutetium	Lu	71	174.967 ± 0.003	1656 ± 5	3315
Magnesium	Mg	12	24.305	648.8 ± 0.5	1090
Manganese	Mn	25	54.9380	1244 ± 3	1962
Mendelevium	Md	101	(257)	—	—
Mercury	Hg	80	200.59	−38.87	356.58
Molybdenum	Mo	42	95.94	2617	4612

TABLE A.15 (*Continued*) Atomic Weights, and Melting and Boiling Points of the Elements[a] Based on the Assigned Relative Mass of $^{12}C = 12$ (The following values apply to elements as they exist in materials of terrestrial origin and to certain artificial elements.)

Name	Symbol	Atomic Number	Atomic Weight	Melting Point °C	Boiling Point °C
Neodymium	Nd	60	144.24	1010	3127
Neon	Ne	10	20.179	−248.67	−246.048
Neptunium	Np	93	237.0482	640 ± 1	3902
Nickel	Ni	28	58.70	1453	2732
Niobium (Columbium)	Nb	41	92.9064	2468 ± 10	4742
Nitrogen	N	7	14.0067	−209.86	−195.8
Nobelium	No	102	(259)	—	
Osmium	Os	76	190.2	3045 ± 30	5027 ± 100
Oxygen	O	8	15.9994	−218.4	−182.962
Palladium	Pd	46	106.4	1554	3140
Phosphorus	P	15	30.973 76	44.1 (white)	280 (white)
Platinum	Pt	78	195.09	1772	3827 ± 100
Plutonium	Pu	94	(244)	641	3232
Polonium	Po	84	(209)	254	962
Potassium	K	19	39.0983	63.25	759.9
Praseodymium	Pr	59	140.9077	931 ± 4	3212
Promethium	Pm	61	(145)	~1080	2460 (?)
Protactinium	Pa	91	231.0359	<1600	—
Radium	Ra	88	226.0254	700	1140
Radon	Rn	86	(222)	−71	−61.8
Rhenium	Re	75	186.2	3180	5627 (est.)
Rhodium	Rh	45	102.9055	1966 ± 3	3727 ± 100
Rubidium	Rb	37	85.4678	38.89	686
Ruthenium	Ru	44	101.07	2310	3900

Samarium	Sm	62	150.4	1072 ± 5	1778
Scandium	Sc	21	44.9559	1539	2832
Selenium	Se	34	78.96	217	684.9 ± 1.0
Silicon	Si	14	28.0855	1410	2355
Silver	Ag	47	107.868	961.93	2212
Sodium	Na	11	22.98977	97.81 ± 0.03	882.9
Strontium	Sr	38	87.62	769	1384
Sulfur	S	16	32.06	112.8	444.674
Tantalum	Ta	73	180.9479	2996	5425 ± 100
Technetium	Tc	43	(97)	2172	4877
Tellurium	Te	52	127.60	449.5 ± 0.3	989.8 ± 3.8
Terbium	Tb	65	158.9254	1360 ± 4	3041
Thallium	Tl	81	204.37	303.5	1457 ± 10
Thorium	Th	90	232.0381	1750	~4790
Thulium	Tm	69	168.9342	1545 ± 15	1727
Tin	Sn	50	118.69	231.9681	2270
Titanium	Ti	22	47.90	1660 ± 10	3287
Tungsten	W	74	183.85	3410 ± 20	5660
Uranium	U	92	238.029	1132.3 ± 0.8	3818
Vanadium	V	23	50.9415	1890 ± 10	3380
Wolfram (see Tungsten)					
Xenon	Xe	54	131.30	−111.9	−107.1 ± 3
Ytterbium	Yb	70	173.04	824 ± 5	1193
Yttrium	Y	39	88.9059	1523 ± 8	3337
Zinc	Zn	30	65.38	419.58	907
Zirconium	Zr	40	91.22	1852 ± 2	4377

[a] Handbook of Chemistry and Physics, 62nd Ed. 1981–82. CRC Press, 1981.

TABLE A.16 One-Dimensional Isentropic Compressible-Flow Functions for an Ideal Gas with Constant Specific Heat and Molecular Weight and $k = 1.4$[a]

M	M^*	$\dfrac{\mathscr{A}}{\mathscr{A}^*}$	$\dfrac{P}{P_0}$	$\dfrac{\rho}{\rho_0}$	$\dfrac{T}{T_0}$
0	0	∞	1.000 00	1.000 00	1.000 00
0.10	0.109 43	5.8218	0.993 03	0.995 02	0.998 00
0.20	0.218 22	2.9635	0.972 50	0.980 27	0.992 06
0.30	0.325 72	2.0351	0.939 47	0.956 38	0.982 32
0.40	0.431 33	1.5901	0.895 62	0.924 28	0.968 99
0.50	0.534 52	1.3398	0.843 02	0.885 17	0.952 38
0.60	0.634 80	1.1882	0.784 00	0.840 45	0.932 84
0.70	0.731 79	1.094 37	0.720 92	0.791 58	0.910 75
0.80	0.825 14	1.038 23	0.656 02	0.740 00	0.886 52
0.90	0.914 60	1.008 86	0.591 26	0.687 04	0.860 58
1.00	1.000 00	1.000 00	0.528 28	0.633 94	0.833 33
1.10	1.081 24	1.007 93	0.468 35	0.581 69	0.805 15
1.20	1.1583	1.030 44	0.412 38	0.531 14	0.776 40
1.30	1.2311	1.066 31	0.360 92	0.482 91	0.747 38
1.40	1.2999	1.1149	0.314 24	0.437 42	0.718 39
1.50	1.3646	1.1762	0.272 40	0.394 98	0.689 65
1.60	1.4254	1.2502	0.235 27	0.355 73	0.661 38
1.70	1.4825	1.3376	0.202 59	0.319 69	0.633 72
1.80	1.5360	1.4390	0.174 04	0.286 82	0.606 80
1.90	1.5861	1.5552	0.149 24	0.256 99	0.580 72
2.00	1.6330	1.6875	0.127 80	0.230 05	0.555 56
2.10	1.6769	1.8369	0.109 35	0.205 80	0.531 35
2.20	1.7179	2.0050	0.093 52	0.184 05	0.508 13
2.30	1.7563	2.1931	0.079 97	0.164 58	0.485 91
2.40	1.7922	2.4031	0.068 40	0.147 20	0.464 68
2.50	1.8258	2.6367	0.058 53	0.131 69	0.444 44
2.60	1.8572	2.8960	0.050 12	0.117 87	0.425 17
2.70	1.8865	3.1830	0.042 95	0.105 57	0.406 84
2.80	1.9140	3.5001	0.036 85	0.094 62	0.389 41
2.90	1.9398	3.8498	0.031 65	0.084 89	0.372 86
3.00	1.9640	4.2346	0.027 22	0.076 23	0.357 14
3.50	2.0642	6.7896	0.013 11	0.045 23	0.289 86
4.00	2.1381	10.719	0.006 58	0.027 66	0.238 10
4.50	2.1936	16.562	0.003 46	0.017 45	0.198 02
5.00	2.2361	25.000	$189(10)^{-5}$	0.011 34	0.166 67
6.00	2.2953	53.180	$633(10)^{-6}$	0.005 19	0.121 95
7.00	2.3333	104.143	$242(10)^{-6}$	0.002 61	0.092 59
8.00	2.3591	190.109	$102(10)^{-6}$	0.001 41	0.072 46
9.00	2.3772	327.189	$474(10)^{-7}$	0.000 815	0.058 14
10.00	2.3904	535.938	$236(10)^{-7}$	0.000 495	0.047 62
∞	2.4495	∞	0	0	0

[a] Abridged from Table 30 in Gas Tables, by Joseph H. Keenan and Joseph Kaye. Copyright 1948, by Joseph H. Keenan and Joseph Kaye. Published by John Wiley & Sons, Inc., New York.

TABLE A.17 One-Dimensional Normal-Shock Functions for an Ideal Gas with Constant Specific Heat and Molecular Weight and $k = 1.4$[a]

M_x	M_y	$\dfrac{P_y}{P_x}$	$\dfrac{\rho_y}{\rho_x}$	$\dfrac{T_y}{T_x}$	$\dfrac{P_{oy}}{P_{ox}}$	$\dfrac{P_{oy}}{P_x}$
1.00	1.000 00	1.0000	1.0000	1.0000	1.000 00	1.8929
1.10	0.911 77	1.2450	1.1691	1.0649	0.998 92	2.1328
1.20	0.842 17	1.5133	1.3416	1.1280	0.992 80	2.4075
1.30	0.785 96	1.8050	1.5157	1.1909	0.979 35	2.7135
1.40	0.739 71	2.1200	1.6896	1.2547	0.958 19	3.0493
1.50	0.701 09	2.4583	1.8621	1.3202	0.929 78	3.4133
1.60	0.668 44	2.8201	2.0317	1.3880	0.895 20	3.8049
1.70	0.640 55	3.2050	2.1977	1.4583	0.855 73	4.2238
1.80	0.616 50	3.6133	2.3592	1.5316	0.812 68	4.6695
1.90	0.595 62	4.0450	2.5157	1.6079	0.767 35	5.1417
2.00	0.577 35	4.5000	2.6666	1.6875	0.720 88	5.6405
2.10	0.561 28	4.9784	2.8119	1.7704	0.674 22	6.1655
2.20	0.547 06	5.4800	2.9512	1.8569	0.628 12	6.7163
2.30	0.534 41	6.0050	3.0846	1.9468	0.583 31	7.2937
2.40	0.523 12	6.5533	3.2119	2.0403	0.540 15	7.8969
2.50	0.512 99	7.1250	3.3333	2.1375	0.499 02	8.5262
2.60	0.503 87	7.7200	3.4489	2.2383	0.460 12	9.1813
2.70	0.495 63	8.3383	3.5590	2.3429	0.423 59	9.8625
2.80	0.488 17	8.9800	3.6635	2.4512	0.389 46	10.569
2.90	0.481 38	9.6450	3.7629	2.5632	0.357 73	11.302
3.00	0.475 19	10.333	3.8571	2.6790	0.328 34	12.061
4.00	0.434 96	18.500	4.5714	4.0469	0.138 76	21.068
5.00	0.415 23	29.000	5.0000	5.8000	0.061 72	32.654
10.00	0.387 57	116.50	5.7143	20.388	0.003 04	129.217
∞	0.377 96	∞	6.000	∞	0	∞

[a] Abridged from Table 48 in *Gas Tables*, by Joseph H. Keenan and Joseph Kaye. Copyright 1948, by Joseph H. Keenan and Joseph Kaye. Published by John Wiley & Sons, Inc., New York.

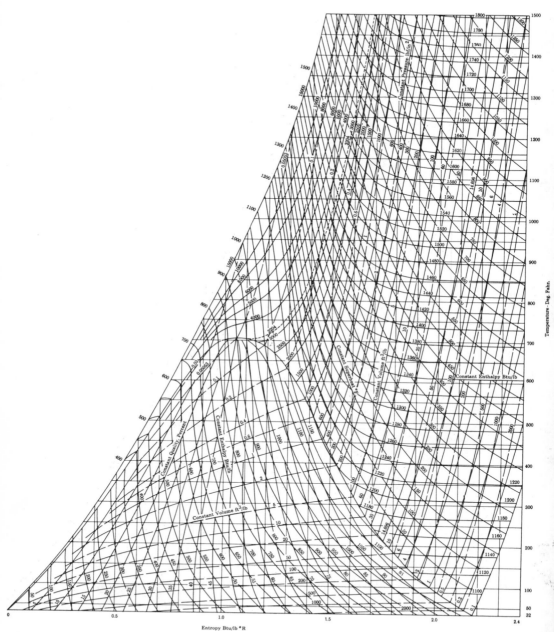

FIGURE A.1

Temperature–entropy chart for steam.

Entropy Btu/lb °R

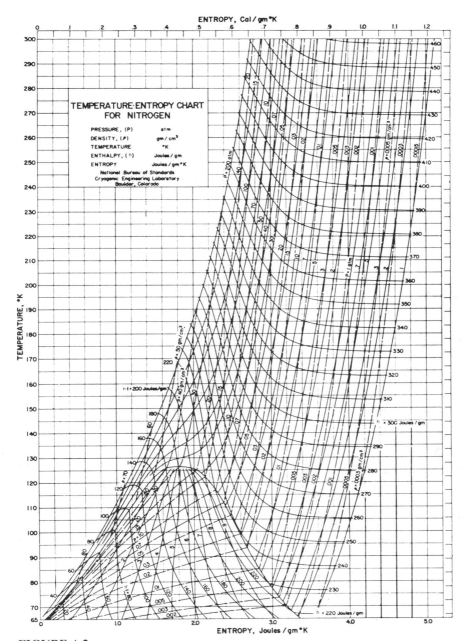

FIGURE A.2

Temperature–entropy diagram for nitrogen. (Reprinted by permission from ASHRAE Handbook of Fundamentals 1981.)

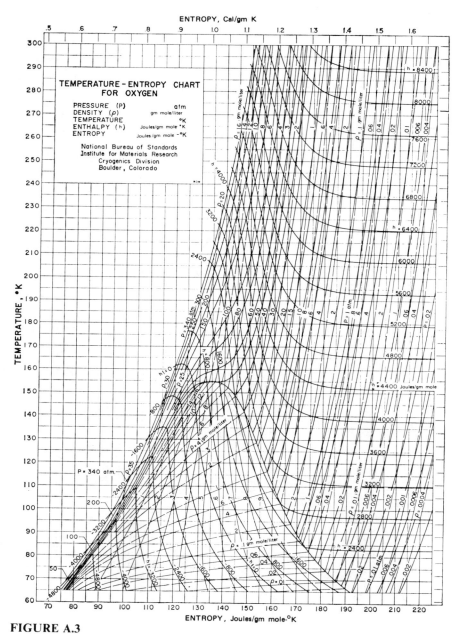

FIGURE A.3

Temperature–entropy diagram for oxygen. (Reprinted by permission from ASHRAE Handbook of Fundamentals 1981.)

FIGURE A.4

Psychrometric chart.

FIGURE A.5

Generalized compressibility chart.

FIGURE A.6

Generalized enthalpy correction chart (based on Figure A.5).

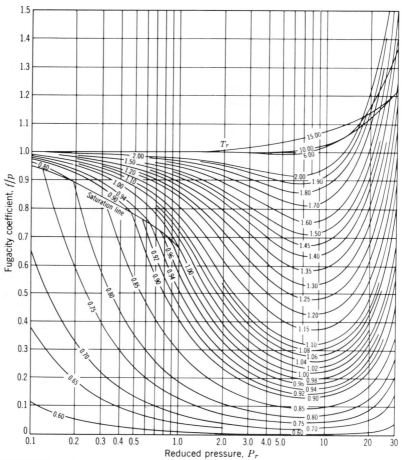

FIGURE A.7

Generalized fugacity coefficient chart (based on Figure A.5).

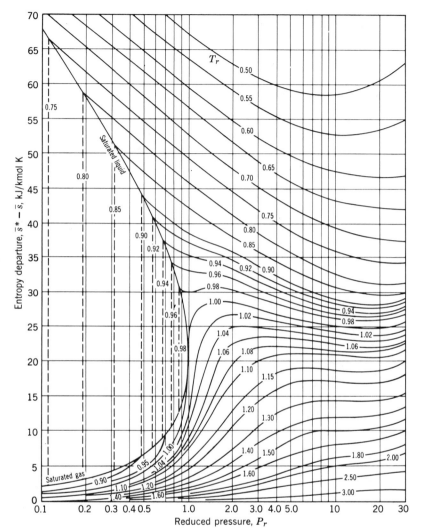

FIGURE A.8

Generalized entropy correction chart (based on Figure A.5).

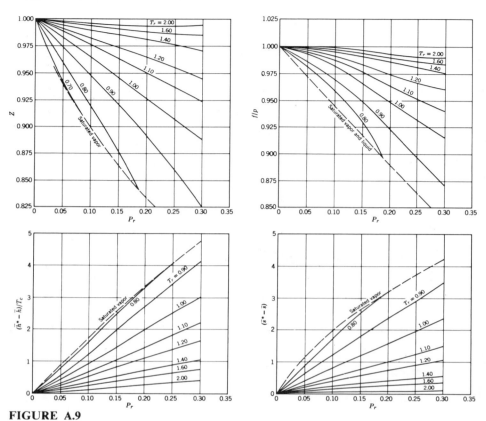

FIGURE A.9

Generalized charts at low pressure.

SOME SELECTED REFERENCES

H. B. Callen, *Thermodynamics*, John Wiley & Sons, New York, 1960.

G. N. Hatsopoulos and J. H. Keenan, *Principles of General Thermodynamics*, John Wiley & Sons, New York, 1965, 1981.

O. A. Hougen, K. M. Watson, and R. A. Ragatz, *Chemical Process Principles, Part Two: Thermodynamics*, Second Edition, John Wiley & Sons, New York, 1959.

J. H. Keenan, *Thermodynamics* M.I.T. Press, Cambridge, MA, 1970.

J. Kestin, *A Course in Thermodynamics*, Hemisphere Publishing Corp., Washington, D.C., 1966, 1979.

E. F. Obert, *Concepts of Thermodynamics*, McGraw-Hill Book Co., New York, 1960.

H. Reiss, *Methods of Thermodynamics*, Blaisdell Publishing Co., Waltham, MA, 1965.

W. C. Reynolds and H. C. Perkins, *Engineering Thermodynamics*, Second Edition, McGraw-Hill Book Co., New York, 1977.

A. H. Shapiro, *The Dynamics and Thermodynamics of Compressible Fluid Flow*, The Ronald Press Co., New York, 1953.

R. E. Sonntag and G. J. Van Wylen, *Introduction to Thermodynamics, Classical and Statistical*, Second Edition, John Wiley & Sons, New York, 1982.

K. Wark, *Thermodynamics*, Fourth Edition, McGraw-Hill Book Co., New York, 1983.

M. W. Zemansky, M. M. Abbott, and H. C. Van Ness, *Basic Engineering Thermodynamics*, McGraw-Hill Book Co., New York, 1966, 1975.

ANSWERS TO SELECTED PROBLEMS

2.3	490.33 N	**5.24**	-767.5 Btu
2.6	10.03 lbm	**5.27**	(a) 1250 Btu/lbm
2.9	27.2 lbf		(b) 248 Btu/lbm
2.12	0.578 lbf/in.2; 1.18 in.		(c) 203 Btu/lbm
2.15	69.5 lbf/in.2	**5.30**	(a) 24.7 lbf/in.2; -130.5 Btu
			(b) 30.5%
3.3	(a) 170 F	**5.33**	-40.6 kJ; -15.2 kJ
	(b) 25.2 lbf/in.2	**5.36**	845.3 kJ; 3761.4 kJ
3.6	22.34 m^3	**5.39**	0.0505 kg/s
3.9	(a) 7.63 kg	**5.42**	464.7 Btu/lbm
	(b) 10%	**5.45**	22.39 lbm/s
3.18	6.8 lbf/in.2	**5.48**	2.108×10^7 Btu/h
3.21	80 144.4 kg; 0.000 223	**5.51**	(a) 33 000 hp
3.24	103.44 ft^3/min; 1.2%, 99%		(b) -1.916×10^8 Btu/h;
3.27	41.3 lbf/in.2		4.75×10^7 Btu/h;
3.33	6.82%		2.291×10^8 Btu/h
			(c) 9.58×10^6 lbm/h
4.3	(b) 4680 ft lbf; 53.8%	**5.54**	(a) 108.2 kW
4.6	-449.4 kJ		(b) -41.4 kW
4.9	(a) 46.8 lbf/in.2	**5.57**	0.959
	(c) -15.1 Btu	**5.60**	(a) $-.549$ kW
4.12	(a) -444 ft lbf		(b) -8.654 kW
	(b) 233 F		(c) 5.397 kW
4.15	(a) 30 000 ft lbf	**5.63**	3.718 lbm
	(b) 7%	**5.66**	520 kPa
4.18	(a) 183 kPa	**5.69**	27.24 kg
	(b) -364.8 kJ	**5.72**	(a) 71.21 lbm
4.21	615 F; 221 800 ft lbf		(b) 4500 Btu
4.24	-44.9 Btu	**5.75**	6747 kJ
4.27	-8.92×10^{-6} J	**5.78**	(a) 39.5 lbf/in.2
			(b) 6.4 Btu
5.3	-192.5 kJ		(c) 8.9 Btu
5.6	-1974.3 Btu	**5.81**	224 039 kJ; 191 506 kJ
5.9	9.803 ft^3		
5.12	(a) 0.2446	**6.6**	0.587
	(b) 610.7 Btu	**6.9**	1.471 kW
5.15	81.7 kJ	**6.18**	300 J; 3.33×10^{-8}
5.18	403.7°C	**6.21**	(a) 0.91 kW
5.21	376.7 kJ		(b) 45.2°C

7.3 (b) 0.7794; 0.2760
(c) 351.2 kJ
7.6 − 245 ft lbf
7.9 70 m/s
7.12 705 Btu; 450 F
7.15 2.685 Btu/R
7.24 0.462 Btu/R
7.27 (a) 624.3 Btu
(b) 0.650 Btu/R
7.33 (a) 563.3 K; − 52.9 kJ
(b) 597.5 K; − 51.2 kJ
(c) 559.2 K; − 52.1 kJ
7.36 0.365 kJ/K
7.39 (a) − 46.4 Btu
(b) − 50.9 Btu
(c) − 49.2 Btu
7.42 (b) 228.7 K; − 48.8 kJ
(c) 76.9 kJ
7.45 6371 lbm/h
7.51 (a) 16.35 MW
(b) 4232 kW
7.54 (a) 385.5 Btu/lbm
(b) 56.6 lbf/in.2
7.60 (a) − 77.0 Btu/lbm; − 77.0 Btu/lbm;
− 0.1425 Btu/lbm R
(b) − 95.0 Btu/lbm; − 28.4 Btu/lbm;
− 0.0428 Btu/lbm R
(c) − 105.0 Btu/lbm; 0; 0
7.63 0.4178 kJ/K s
7.69 (a) 8910 kJ
(b) 7.538 kJ/K
7.72 (a) 14.026 MW
(b) 3157 kW
7.75 660 kPa
7.78 112 lbf/in.2; − 86.7 F
7.81 (a) 0.895
(b) 278.1 F
7.84 71 lbf/in.2; 599 R
7.87 83 lbf/in.2; 121.6 F; 15.5 Btu;
0.017 Btu/R
7.90 (a) 13.96 lbm
(b) 17 255 Btu
(c) − 0.64 Btu/R; 7.57 Btu/R
7.96 (a) 41.5 lbf/in.2
(b) 240 F; 493.5 F

8.3 12.8 Btu/lbm
8.6 − 61.7 kJ/kg; 0.3 kJ/kg
8.9 20.4 kJ/kg

8.12 255.9 kJ
8.15 2.694 Btu/lbm
8.18 (a) 670.3 kJ/kg
(b) 370.4 kJ/kg
(c) − 4.8 kJ/kg, 375.2 kJ/kg;
− 178.7 kJ/kg, 549.1 kJ/kg;
− 300.0 kJ/kg, 670.4 kJ/kg
8.21 257.5 kJ/kg
8.24 88.2%
8.27 60.21 kW
8.30 394 407 kJ

9.6 (a) 33 400 kW; − 484 kW
(b) 142 747 kg/s; 147 324 kg/s
(c) 0.027
9.9 0.35; 0
9.12 (a) 0.438
(b) 0.473
(c) 0.488
9.18 6.145 × 10^6 Btu/h; 2.791 × 10^6
Btu/h
9.24 5.195; 5.38
9.30 (a) 0.881
(b) 1.132
9.33 (a) 8.7 kJ/kg
(b) 5.325 kW
(c) 2.219
9.36 0.97 Btu
9.39 11.057 MPa; 3602 K; 0.498;
1865 kPa
9.42 4680 R
9.45 200 kJ/kg; 1000 kJ/kg; 0.80
9.48 (a) 3657 hp; 8499 hp; 0.509
(b) 32.7 lbf/in.2
9.51 (a) 848 K
(b) 0.273
9.57 206.7 Btu/lbm; 529.3 Btu/lbm; 0.609
9.60 592 kPa; 969 m/s
9.63 0.573

10.3 1.58%
10.21 (a) 455 K
(b) 118.8 kJ
10.36 3193 Btu
10.39 1600 lbf/in.2
10.42 1620 kPa
10.45 17.0 Btu/lbm; 14.6 Btu/lbm
10.48 − 30.0 Btu/lbm; − 87.7 Btu/lbm

10.51	**(a)** $-207\,380$ kJ
	(b) 217.8 kJ/kg
10.54	136 Btu
10.57	13.7 hp
11.3	**(a)** 233 kPa; 310.5 K
	(b) 6.238 kJ/K
11.6	353.9 K; 6.228 kJ/kmol K
11.12	49°C; 55 min.
11.15	-32.61 kW
11.18	**(a)** 0.13
	(b) 0.4359 kg
	(c) 25.6 kJ
11.21	**(a)** 37.1 F; 0.033 75 lbm
	(b) -2370 Btu/h
11.24	**(a)** 0.605
	(b) -19.6 Btu/lbm air
11.27	4.584 kg/s; 0.0278
11.30	53.9°C; 116 kPa
11.33	**(a)** -7.21 kJ; -7.21 kJ
	(b) -11.66 kJ; -7.67 kJ
	(c) -9.93 kJ; -7.81 kJ
11.36	**(a)** 19 L
	(b) 6.4439 kJ/K
11.39	**(a)** 728 lbf/in.2
	(b) 1010 lbf/in.2
	(c) 182 Btu; 1.57 Btu/R
11.42	**(a)** 0.1740 m^3/kmol; 0.2880 m^3/kmol
11.45	1.584 MPa; 23.9 kJ
12.3	82.7% C, 17.3% H; 131%
12.6	6.296 kg/kg coal
12.9	$-102\,448$ kJ/kmol
12.12	2795 K
12.15	$-522\,293$ Btu
12.18	3111 R
12.21	239%
12.24	**(a)** 684.5 K; 11 230 kJ/kmol
	(b) 2.794
12.27	20 986 kJ/kg

12.30	**(a)** $-24\,746$ kJ/kg
	(b) 4487 K
12.33	1838.75 kJ/K
12.39	$-172\,477$ kJ; 194.46 kJ/K
12.42	**(a)** $-547\,786$ kJ
	(b) 2141.6 kJ/K
12.45	1.219 V
12.48	**(a)** $-13\,923$ Btu/h
	(b) 13.8%
13.3	**(a)** Liq. 21.3% N_2; Vap. 82.8% N_2
	(b) Liq. 28.3% N_2; Vap. 83.5% N_2
13.9	$-511\,688$ kJ
13.15	52% O_2; 48% O
13.18	29.5% CO_2; 16.4% CO; 23.5% O_2; 30.6% N_2
13.21	75 386 Btu
13.24	4.19 lb mol; 273 lbf/in.2
13.27	**(a)** 0.006 45
	(b) $-359\,760$ Btu
13.30	**(a)** 9.874
	(b) 86.95% NH_3; 3.26% N_2; 9.79% H_2
13.33	**(b)** 17.6% CO; 5.3% H_2; 15.6% CH_4; 15.6% CO_2; 45.9% N_2
13.36	66.57% N_2; 4.73% O_2; 0.89% Ar; 5.27% NO; 22.54% O
13.39	1.07
13.42	3.885×10^{-5}
14.3	88 lbf/in.2; 466 F
14.6	**(b)** 1345 mm^2; 2598 mm^2
14.9	1859 kPa
14.12	10.34 lbf/in.2; 531 R; 0.463
14.15	9.52 lbf/in.2; 531 R; 0.427
14.18	0.019 35 m^2
14.21	3.735 kg/s
14.24	80%
14.27	**(a)** 43 lbf/in.2
	(c) 173.7 Btu/lbm

INDEX

745

$$A \quad \begin{bmatrix} 1 \\ x \\ C \end{bmatrix} \quad \begin{matrix} 10 \\ 11 \\ 15 \end{matrix} \begin{matrix} D \\ E \\ F \end{matrix}$$

$$\frac{1-x}{10-11} = \frac{1-6}{10-15}$$

$$1-x = \frac{(1-6)(10-11)}{10-15}$$

$$x = 1 - \frac{(1-6)(10-11)}{10-15} = A - (((A-C) \cdot (D-E)) \div (D-F))$$